Computational Fluid Dynamics

Computational Fluid Dynamics: Numerical Methods and Applications offers a comprehensive exploration of computational fluid dynamics (CFD) methods, focusing on numerical techniques for solving fluid flow problems through both finite difference and finite-volume methods. It explores numerical schemes, advection-diffusion equations, shock-capturing methods, and turbulence modeling.

Emphasizing stability and accuracy preservation as part of modern CFD practice, the book addresses both laminar and turbulent regimes. It spans the full spectrum of CFD from the fundamentals of discretized conservation laws to modern turbulence and LES modeling. The book also discusses advanced techniques such as high-order discretization, time integration, pressure-velocity coupling, and high-fidelity simulation.

Senior undergraduate aerospace and mechanical engineering students taking a CFD course will find this book particularly valuable.

Instructors will be able to utilize a Solutions Manual, Figure Slides, and Python codes for their course.

Dr. Naseem Uddin, CEng MIMechE, MIEAust CPENG, earned his PhD in aerospace engineering from Universität Stuttgart, Germany in 2008. He is a Senior Assistant Professor in the Mechanical Engineering Programme, Universiti Teknologi Brunei, Brunei (UTB), Brunei Darussalam. Previously, he worked as a full professor at NED University of Engineering and Technology, Pakistan. Dr. Uddin is a registered chartered engineer with Engineers Australia and the Institution of Mechanical Engineers, UK. He is also listed in the National Engineering Register (NER) of Australia and recognized as a professional engineer by the board of professional engineers of Queensland, Australia and by the Pakistan Engineering Council. He is a member of the American Society of Mechanical Engineers (ASME). Dr. Uddin has authored several textbooks, *Fluid Mechanics: A Problem-Solving Approach*; *Heat Transfer: A Systematic Learning Approach*; and *Thermodynamics: Fundamentals and Applications*, all published by CRC Press.

Computational Fluid Dynamics

Numerical Methods and Applications

Naseem Uddin

CRC Press
Taylor & Francis Group
Boca Raton London New York

CRC Press is an imprint of the
Taylor & Francis Group, an **informa** business

First edition published 2026
by CRC Press
2385 NW Executive Center Drive, Suite 320, Boca Raton FL 33431

and by CRC Press
4 Park Square, Milton Park, Abingdon, Oxon, OX14 4RN

CRC Press is an imprint of Taylor & Francis Group, LLC

ISBN: 978-1-041-12136-7 (hbk)
ISBN: 978-1-041-12138-1 (pbk)
ISBN: 978-1-003-66317-1 (ebk)
ISBN: 978-1-003-71688-4 (eBook+)

DOI: 10.1201/9781003663171

Typeset in CMR10 font
by KnowledgeWorks Global Ltd.

Publisher's note: This book has been prepared from camera-ready copy provided by the authors.

Access the Support Material: Routledge.com/9781041121367

Dedication

This book is dedicated to my wife and children

Contents

Nomenclature

A	Surface, Helmholtz free energy, Area, Amplification factor
A^+, A^-	Flux–split matrices
A_{ex}	Exact amplitude
A_{mod}	Modified amplitude factor per step
A^n_{mod}	Total amplitude after n steps using modified amplitude
a	Stretching parameter; sonic speed $a^2 = \gamma p/\rho$
$a^{(ex)}_{ij}$	Nonlinear anisotropy correction tensor
b_{ij}	Reynolds-stress anisotropy tensor
b	Constant in scalar-flux function
c	Phase speed; local sonic speed
c_g	Group velocity $d\omega/dk$
C_μ	Eddy viscosity model constant
c_p, c_v	Specific heats
D, d	Diameter
D_{eff}	Effective total dissipation: $D_{\text{eff}} = \nu k^2 + \xi k^4$
D_{ij}	Diffusive transport tensor
$\mathscr{D}, D_m$	Diffusion coefficient
$\mathtt{D}$	Discriminant of PDE
E	Total energy per unit volume
$E(f)$	One-dimensional turbulence energy spectrum
$E(k)$	Three-dimensional turbulence energy spectrum
e	Internal energy per unit mass
er_ω	Phase error of numerical scheme
$\mathbf{F}(\mathbf{U})$	Flux vector $[\rho u, \rho u^2 + p, u(E+p)]^T$
$G(x,x')$	Filter kernel in LES
h	Uniform distance between two adjacent grid points, timestep
I	Turbulent intensity
II_S	Second invariant of strain-rate tensor
II_Ω	Second invariant of rotation-rate tensor
i	Index in x direction, iota
j	Index in y direction
k	Turbulent kinetic energy, wavenumber (=1/length), heat capacity ratio, index in z direction, vector component index

M	Mach number
m	Mass
n_i	Outward unit normal
$\mathscr{O}(.)$, $O(.)$, $\mathrm{O}(.)$	Order of accuracy or truncation error
P	Characteristic polynomial or amplification factor for exact solution
P_{ij}	Production tensor
p	Pressure, dimensionless scaled eigenvalue = $\lambda\Delta t$
p'	First pressure correction
p''	Second pressure correction
Q	Conservative variable vector
q	Scalar conserved variable (e.g., ρ, u, p) in WENO reconstruction
R	Eigenvalue matrix
S_{ij}	Mean strain-rate tensor
S^{ij}	Non-dimensionalized strain-rate tensor
$S_{ik}S_{kj}$	Quadratic strain-rate product
$S_{ik}W_{kj}$	Strain and rotation-rate product
S_{kk}	Trace of strain-rate tensor
S	Shock propagation speed
s_i	stability limit on imaginary axis in Argand diagram
s_r	stability limit on real axis Argand diagram
$\mathbf{U}$	Vector of conserved variables $[\rho, \rho u, E]^T$
u, v, w	Velocity components in x, y, z directions
$u'v'$	Reynolds shear stress
$\langle u_i u_j \rangle$	Reynolds stress tensor
u'^2, v'^2, w'^2	Normal Reynolds stresses
u_τ	Frictional velocity
V	Vector of intermediate velocity
V	Control volume or thermodynamic volume
v_p	Phase speed ω/k
W_{ij}	Mean rotation-rate tensor
W^{ij}	Non-dimensionalized rotation-rate tensor
$W_{ik}W_{kj}$	Quadratic rotation-rate product
$\mathtt{W}$	Artificial compressibility parameter
w^L, w^R	Left and right states at a cell interface
x, y, z	Cartesian coordinates

Greek Symbols

α	Complex constant, $\alpha = \alpha_r + i\,\alpha_i$, eigenvalue of the system or Jacobian matrix
α_r	Real part of α

α_i	Imaginary part of α
β_1, β_2	Model coefficients
ΔA	Amplitude error
$\Delta\Theta$	Phase error
Δx	Grid spacing in x direction
Δy	Grid spacing in y direction
Δz	Grid spacing in z direction
Δt	Time step size
δ_{ij}	Kronecker delta
$\delta(k)$	dissipation rate of spectral mode
η	Kolmogorov length scale
ℓ	Integral length scale
ε	Hyperviscosity coefficient (numerical stabilization term), Volume-averaged turbulence kinetic energy dissipation rate
Γ	Conservation quantity
γ	Interpolation factor, heat capacities ratio
κ	Stiffness ratio, parameter controlling the bias or shape of the limiter function
λ	Thermal conductivity, roots, eigenvalues
μ	Absolute viscosity
μ_t	Turbulent (eddy) viscosity
ν	Kinematic viscosity
$\omega(k)$	dispersion function (temporal frequency) of spectral mode
π	≈ 3.14159
ψ	Stream function, conserved quantity in FV
ϕ	Arbitrary conserved variable in FD, initial profile, potential function
ρ	Density
σ	Diffusion number or mesh Fourier number, face area, cross-sectional area
τ_{ij}	Reynolds stress tensor
θ	Phase angle or phase shift
Θ_{ex}	Exact phase angle
Θ_{num}	Numerical phase angle
Θ_{mod}	Modified phase angle (numerical phase per step)
ξ, η	Characteristic (Riemann) coordinates $x \pm ct$
ξ	Coefficient from truncation error, $\xi = \Delta x^2/12$

Subscripts

$(\cdot)_B$	Bottom Node value
$(\cdot)_E$	East Node value

$(.)_e$	Eastern face value of a CV
$(\cdot)_k$	Quantity indexed by WENO substencil
$(.)_N$	North Node value
$(.)_n$	Northern face value of a CV
$(.)_S$	South Node value
$(.)_s$	Southern face value of a CV
$(.)_T$	Top Node value
$(.)_W$	West Node value
$(.)_w$	Western face value of a CV
$(.)_{sgs}$	Subgrid-scale
$(.)^L$	Left of discontinuity
$(\cdot)_{\text{mod, max}}$	maximum value of the modified quantity
$(.)^R$	Right of discontinuity
$(.)^{PR}$	Riemann Problem

Acronyms

AB	Adams-Bashforth method
ABM	Adams-Bashforth-Moulton method
ACM	Artificial compressibility method
ADI	Alternating Direction Implicit Method
ALE	Arbitrary Lagrangian-Eulerian method
AM	Adams-Moulton method
AMG	Algebraic Multigrid
AP_{mod}	Per-period amplification factor
BDF2	Second-order Backward Differentiation
BiCGSTAB	Bi-Conjugate Gradient Stabilized
CDS	Central Difference Scheme
CFD	Computational Fluid Dynamics
CFL	Courant-Friedrichs-Lewy number = $c\Delta t/\Delta x$
CIR	Courant-Isaacson-Rees scheme
CG	Conjugate Gradient
CN	Crank-Nicolson scheme
CSC	Compressed Sparse Column
CSR	Compressed Sparse Row
DNS	Direct Numerical Simulation
DRP	Dispersion Relation Preserving scheme
EARSM	Explicit Algebraic Reynolds Stress Model
EE	Explicit Euler Method
FD	Finite-difference scheme
FDO2	Finite Difference, Second Order
FDO4	Finite Difference, Fourth Order
FDO6	Finite Difference, Sixth Order
FDM	Finite-difference method

FTFS	Forward-in-Time Forward-in-Space
FFT	Fast Fourier Transform
FOU	First order upwind
FS	Fractional Step method
FV	Finite-volume
FVM	Finite-volume method
GMRES	Generalized Minimal Residual
GPPW	Grid Points Per Wavelength
HLL	Harten-Lax-van Leer
HLLC	Harten-Lax-van Leer Contact
HRSC	High-Resolution Shock-Capturing
ICE	Implicit Continuous Eulerian method
IE	Implicit Euler Method
ILU	Incomplete LU Factorization
IBM	Immersed Boundary Method
IMEX	Implicit-Explicit
KFD	Compact finite-difference scheme
KFD-O4	Compact (Keller) Finite Difference, Fourth Order
KFD-O6	Compact (Keller) Finite Difference, Sixth Order
LAF	Local Amplification Factor
LES	Large Eddy Simulation
LGMRES	Loose GMRES
LRR	Launder–Reece–Rodi Model
LF	Leapfrog scheme
LU	Lower-Upper Factorization
MAC	Marker and Cell method
MINRES	Minimum Residual
MUSCL	Monotone Upstream-centered Schemes for Conservation Laws
MUSTA	Multi-Stage Time-stepping Algorithm
NRBC	Non-reflective boundary conditions
ODE	Ordinary Differential Equation
OHOC	Optimized High-Order Compact scheme
PDE	Partial Differential Equation
PIMPLE	Pressure-Implicit with Multiple Correction Steps
PISO	Pressure-Implicit with Splitting of Operators
PIV	Particle Image Velocimetry
PPM	Piecewise Parabolic Method
QUICKEST	QUICK with Estimated Streaming Terms
QUICK	Quadratic Upstream Interpolation for Convective Kinematics
RK	Runge-Kutta method
RK2-O2	Runge-Kutta method Second Order

RK4-O4	Runge-Kutta method Fourth Order
RSM	Reynolds Stress Model
SA	Spalart–Allmaras
SDIRK	Singly Diagonally Implicit Runge–Kutta
SIMPLE	Semi-Implicit Method for Pressure-Linked Equations
SMAC	Simplified Marker and Cell method
SOR	Successive Over-Relaxation method
SSP	Strong Stability Preserving
SST	Shear Stress Transport
TDMA	Tri-Diagonal Matrix Algorithm
VNR	Von Neumann-Richtmyer
VOF	Volume of Fluid method
WENO	Weighted Essentially Non-Oscillatory
WNF–LSS	Wall-Normal-Free Linear Stress-Strain

Special Symbols

$\widehat{(.)}$	Resolved component in LES
$\widetilde{(.)}$	Unresolved component in LES
$(.)^{sgs}$	Subgrid component in LES
i	Imaginary unit, $i^2 = -1$.
$\overline{(\cdot)}$	Time-averaged quantity

Dimensinless Quantities

k^* or k^*_{mod}	modified wavenumber = $k \cdot \Delta x$
k^{*2}_{mod}	Modified wavenumber for second derivative
k^{*4}_{mod}	Modified wavenumber for fourth derivative
p	Dimensionless parameter for ODE $dy/dt = \alpha y$
R_Δ	Cell or mesh Reynolds number (Numerical Peclet number)
Re	Flow Reynolds number
Pe	Peclet number in FV also called R_Δ
y^+	$(= u_\tau\ \Delta\ \text{y}/\nu)$ Non-dimensional distance from wall
$(\cdot)^+$	Non-dimensional wall unit quantity

Author

Dr. Naseem Uddin earned his PhD in Aerospace Engineering from the University of Stuttgart, Germany. He is a Fellow of Engineers Australia and is recognized as a Chartered Engineer by both the Institution of Mechanical Engineers, UK, and Engineers Australia. He is listed on the National Engineering Register of Australia and is accredited as a professional engineer by the Board of Professional Engineers of Queensland and the Pakistan Engineering Council. He is also a member of the American Society of Mechanical Engineers (ASME, USA). Dr. Uddin currently serves as a Senior Assistant Professor in the Mechanical Engineering Programme at Universiti Teknologi Brunei, where he teaches Fluid Mechanics at both undergraduate and postgraduate levels. Prior to joining UTB, he held the position of Full Professor at the NED University of Engineering and Technology, Pakistan. His research interests primarily focus on heat transfer and computational turbulence. He is the author of three textbooks published by CRC Press: Fluid Mechanics: A Problem-Solving Approach, Heat Transfer: A Systematic Learning Approach, and Thermodynamics: Fundamentals and Applications.

Prologue

Computational Fluid Dynamics (CFD) uses numerical analysis and algorithms to solve and analyze problems involving fluid flows. With the rapid growth of computing technology, CFD has become essential in engineering and the sciences, offering predictive insights for applications ranging from aircraft aerodynamics and weather forecasting to industrial piping and biomedical flows. By simulating fluid behavior, CFD reduces the reliance on physical experiments, lowering costs and accelerating innovation across fields such as aerospace, automotive, energy, and medicine.

This book is intended for advanced undergraduate and graduate students in mechanical, aerospace, and chemical engineering, as well as practicing engineers and researchers who seek an application-oriented foundation in computational fluid dynamics. The material balances mathematical analysis, algorithm development, and hands-on problem-solving, providing a clear and practical approach to the core methods and concepts in modern CFD.

The text introduces fundamental numerical approaches through both finite-difference and finite-volume perspectives, beginning with their discretization philosophies and conservation properties. It then develops time-integration methods, evaluates stability using phase and amplitude analysis, and addresses numerical dispersion in the advection equation. The treatment of the diffusion equation follows, highlighting the implementation of boundary conditions. Building on these foundations, the book explores coupled transport phenomena, solution strategies for the advection-diffusion system, and the Burgers equation. Each derivation in finite-difference form is paired with a control-volume interpretation to prepare readers for extensions to the compressible and incompressible Navier-Stokes equations.

Instead of devoting separate discussions to different systems of equation solvers such as Jacobi, Gauss-Seidel, SOR, or conjugate gradient methods, this book adopts a practical approach by demonstrating the use of Python and established Python libraries to solve linear systems efficiently. By relying on well-tested, readily available solvers, readers can focus on applying CFD concepts and analyzing results without being delayed by the detailed development of numerical solution algorithms. It should be noted that methods for solving systems of equations are part of general numerical analysis and are not unique to CFD; those seeking deeper theoretical understanding or algorithmic details are encouraged to consult specialized texts in numerical linear algebra.

The use of Python programming and libraries throughout enables readers to translate numerical analysis into interactive exploration, facilitating both learning and practical application. The discussion of compressible flow focuses on the numerical treatment of hyperbolic conservation laws, starting with one-dimensional gas dynamics and advancing to finite-volume schemes for resolving discontinuities, such as the Godunov and Roe methods. Techniques including total variation diminishing limiters and essentially non-oscillatory reconstructions are presented to address post-shock oscillations. Classical test problems, including the Sod shock tube, are used to verify accuracy and robustness.

Several advanced topics in computational fluid dynamics are not covered in this book. Subjects such as machine learning-assisted CFD, reduced-order modeling, adaptive mesh refinement, and high-performance computing are not addressed, as the primary focus is on establishing core numerical methods and the theoretical foundation necessary for fluid flow simulation. Likewise, multiphase flows, reactive flows, fluid-structure interaction, and uncertainty quantification are omitted to maintain a manageable scope and allow deeper coverage of essential discretization techniques, solution algorithms, and turbulence modeling. Readers seeking expertise in these specialized or rapidly evolving areas are encouraged to consult dedicated texts and recent research literature. The intention is to provide a solid grounding in the principles and standard tools of CFD, which will support further study and adaptation as computational methods and application domains continue to advance.

This book focuses on structured grids for both finite difference and finite volume methods, as well as for compact finite difference schemes. This choice is made primarily for pedagogical reasons: structured grids enable a clearer introduction to discretization techniques, allow for more accessible mathematical analysis, and simplify the implementation of high-order and compact schemes. By building concepts on structured grids, readers can more easily grasp the underlying principles of numerical methods and boundary condition treatment before approaching the added complexity of unstructured or adaptive grids. Those seeking further study in unstructured grid techniques or advanced grid generation are encouraged to consult specialized resources after establishing a strong foundation with the material presented here.

This book is structured to support two university-level courses in computational fluid dynamics. Chapters 1 to 15 focus on finite difference and compact finite difference methods, forming a suitable foundation for a course on computational incompressible fluid dynamics. These chapters cover essential numerical methods, and the solution of the incompressible Navier-Stokes equations. Chapters 16 to 19 shift the focus to hyperbolic finite difference and finite volume schemes, emphasizing their application in computational gas dynamics, including shock-capturing methods and compressible flow simulations. Chapters 20 and 21, which address turbulence modeling and high fidelity simulations, can be included with either course depending on curriculum needs and emphasis. This organization allows instructors flexibility in tailoring the material for a variety of academic and research objectives.

The code examples discussed in this book are available on GitHub at `https://github.com/Odin2111/CFD_FD`. Additionally, instructors can access supplementary code resources for all end-of-chapter problems through the publisher's website.

I wish to extend my sincere gratitude to all my students whose dedication, rigorous inquiry, and collaborative spirit enriched this work. I am also grateful to the worthy reviewers and colleagues who provided valuable feedback and constructive criticism during the preparation of the manuscript. Finally, I wish to acknowledge the support of friends and family, whose encouragement and understanding have been essential throughout this process.

Uddin
2025.

1 Introduction to Computational Fluid Dynamics

In this chapter, we will discuss the foundational concepts of Computational Fluid Dynamics (CFD) and its application in solving fluid flow problems. After finishing this chapter, you should be able to:

- Understand the fundamental principles of CFD and its importance in modern engineering.
- Learn about the Finite Difference Method (FDM) and its application in solving fluid dynamics problems.
- Develop an understanding of advection, diffusion, and their mathematical representation in CFD.
- Be familiar with the advection-diffusion equation and its significance in fluid transport.
- Understand the differences between continuum and discontinuum models in fluid flow analysis.

In Computational Fluid Dynamics (CFD), the continuity, momentum (Navier-Stokes), and energy equations are solved simultaneously to predict fluid behavior. These equations are strongly coupled and inherently nonlinear, meaning that variations in one field directly impact the others. For instance, density ρ from the continuity equation directly affects momentum distribution, while velocity and pressure obtained from the Navier-Stokes equations influence the energy (temperature) distribution. Similarly, temperature or enthalpy calculated from the energy equation can alter fluid density and pressure, especially in compressible flow scenarios.

In general compressible flows, the primary unknown variables are velocity components, pressure, and temperature (or enthalpy). The governing set the continuity equation, the three momentum equations, and the energy equation, collectively provide five equations to determine these five unknown variables. Due to their complexity and interdependence, analytical solutions are limited to simplified cases. Consequently, CFD relies on numerical methods, such as finite volume or finite difference approaches, to discretize and iteratively solve these partial differential equations (PDEs) computationally.

Together, these conservation equations constitute the fundamental framework of CFD modeling, capturing the essential physical interactions within fluid flow. The accuracy and reliability of CFD simulations depend critically on accurately representing these equations and their interactions, thereby ensuring proper conservation of mass, momentum, and energy within the numerical solution.

DOI: 10.1201/9781003663171-1

1.1 THE FLOW AND ENERGY EQUATIONS

The dynamics of fluid flow and the accompanying energy exchange are governed by fundamental physical principles, mathematically expressed through conservation laws. Collectively known as the flow and energy equations, these include the continuity equation, the Navier-Stokes (momentum) equations, and the energy equation. Each describes distinct yet interconnected aspects of fluid motion, providing the foundational equations utilized in computational fluid dynamics (CFD) to simulate complex flow phenomena.

1.1.1 CONTINUITY EQUATION

The continuity equation represents the principle of mass conservation, which states that mass cannot be created or destroyed within any given control volume. It ensures that the net mass flux into and out of any fluid region equals the rate of mass accumulation within that region. In this section, we derive the continuity equation for compressible flow.

We develop the differential formulation for the continuity equation. Figure 1.1 shows the three dimensional control volume. For mass conservation we use the mass balance equations as:

$$[mass\ flow\ rate]_{out} - [mass\ flow\ rate]_{in} = mass\ inside\ CV$$

The control volume has dimensions Δx, Δy and Δz. The velocity and density associated at exit of control volume are indicated by symbol (e). The inlet velocity and density associated with x, y and z directions are (ρ u), (ρ v) and (ρ w). In order to estimate the out going mass flow rate we use the Taylor's series:

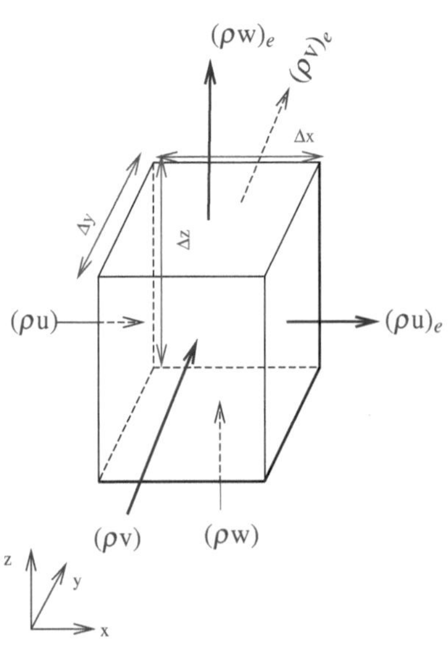

Figure 1.1: **Three-dimensional control volume for the derivation of the continuity equation.**

$$(\rho u)_e = (\rho u) + \frac{\partial(\rho u)}{\partial x}\frac{(\Delta x)^1}{1!} + \cancel{\frac{\partial^2(\rho u)_1}{\partial x^2}\frac{(\Delta x)^2}{2!} + \cdots}$$

$$(\rho v)_e = (\rho v) + \frac{\partial(\rho v)}{\partial y}\frac{(\Delta y)^1}{1!} + \cancel{\frac{\partial^2(\rho v)_1}{\partial y^2}\frac{(\Delta y)^2}{2!} + \cdots}$$

$$(\rho w)_e = (\rho w) + \frac{\partial(\rho w)}{\partial z}\frac{(\Delta z)^1}{1!} + \cancel{\frac{\partial^2(\rho w)_1}{\partial z^2}\frac{(\Delta z)^2}{2!} + \cdots}$$

We ignore the higher order terms and the above expression will be reduced to form

$$(\rho u)_e = (\rho u) + \frac{\partial(\rho u)}{\partial x}\Delta x$$

$$(\rho v)_e = (\rho v) + \frac{\partial(\rho v)}{\partial y}\Delta y$$

$$(\rho w)_e = (\rho w) + \frac{\partial(\rho w)}{\partial z}\Delta z$$

The net mass flow rate in x direction is

$$\Delta(\dot{m}_x) = \Delta y\Delta z\left[(\rho u)_e - (\rho u)\right] = \Delta y\Delta z\left[(\rho u) + \tfrac{\partial(\rho u)}{\partial x}\Delta x - (\rho u)\right] = \Delta y\Delta z\left[\tfrac{\partial(\rho u)}{\partial x}\Delta x\right]$$

Similarly, the net mass flow rate in y direction is

$$\Delta(\dot{m}_y) = \Delta x\Delta z\left[(\rho v)_e - (\rho v)\right] = \Delta x\Delta z\left[(\rho v) + \tfrac{\partial(\rho v)}{\partial y}\tfrac{(\Delta y)^1}{1!} - (\rho v)\right] = \Delta x\Delta z\left[\tfrac{\partial(\rho v)}{\partial y}\Delta y\right]$$

and, the net mass flow rate in z direction is

$$\Delta(\dot{m}_z) = \Delta x\Delta y\left[(\rho w)_e - (\rho w)\right] = \Delta x\Delta y\left[(\rho w) + \tfrac{\partial(\rho w)}{\partial z}\Delta z - (\rho w)\right] = \Delta x\Delta y\left[\tfrac{\partial(\rho w)}{\partial z}\Delta z\right]$$

The mass inside control volume is

$$Mass\ inside\ CV = \rho\dot{\forall} = -\frac{\partial}{\partial t}(\rho\,\Delta x\,\Delta y\,\Delta z)$$

We now substitute the above expressions into mass balance equation

$$\Delta y\Delta z\left[\frac{\partial(\rho u)}{\partial x}\Delta x\right] + \Delta x\Delta z\left[\frac{\partial(\rho v)}{\partial y}\Delta y\right] + \Delta x\Delta y\left[\frac{\partial(\rho w)}{\partial z}\Delta z\right] = -\frac{\partial}{\partial t}(\rho\,\Delta x\,\Delta y\,\Delta z)$$

This leads to form

$$\frac{\partial(\rho u)}{\partial x} + \frac{\partial(\rho v)}{\partial y} + \frac{\partial(\rho w)}{\partial z} = -\frac{\partial\rho}{\partial t}$$

Conservation of mass and momentum equations in Cartesian, cylindrical and spherical coordinates

Cartesian coordinates:

$$\frac{\partial \rho}{\partial t}+\frac{\partial(\rho u)}{\partial x}+\frac{\partial(\rho v)}{\partial y}+\frac{\partial(\rho w)}{\partial z}=0$$

Cylindrical coordinates:

$$\frac{1}{r}\frac{\partial(r u_r)}{\partial r}+\frac{1}{r}\frac{\partial u_\theta}{\partial \theta}+\frac{\partial u_z}{\partial z}=0$$

Spherical coordinates:

$$\frac{1}{r}\frac{\partial}{\partial r}\left(r^2 u_r\right)+\frac{1}{\sin\varphi}\left[\frac{\partial}{\partial \varphi}\left(\sin\varphi\cdot u_\varphi\right)\right]+\frac{1}{\sin\varphi}\left[\frac{\partial}{\partial \theta}\left(u_\theta\right)\right]=0$$

1.1.2 MOMENTUM CONSERVATION

The Navier-Stokes equations, fundamental in fluid mechanics, describe momentum conservation within fluid flows. Their origins trace back to Leonhard Euler, who in 1757 formulated equations governing ideal (inviscid) fluids. However, real fluids exhibit viscosity affecting motion. Incorporating viscous terms was accomplished independently by Claude-Louis Navier in 1822 and George Gabriel Stokes in 1845, giving rise to the celebrated Navier-Stokes equations. These equations expanded Euler's approach by explicitly introducing viscous stress, accurately capturing viscous flow behavior such as that encountered in boundary layers, and turbulence. In this section, we derive the Navier-Stokes equations, starting from the Newton's second law of motion applied onto a fluid body.

Due to the nonlinear and coupled nature, exact analytical solutions are only possible for very few specialized cases. For more challenging cases of industrial applications, computational approaches are used to solve this equation. The numerical treatment of Navier-Stokes equations has thus driven significant developments in computational fluid dynamics (CFD), enabling realistic modeling of diverse fluid phenomena.

Figure 1.2 shows the infinitesimal control volume used for the balance of forces and momentum transfer. The stresses are different on opposing faces. The number 2 in subscript shows the stress on the opposing faces of the control volume. Using Taylor series we can estimate the normal stresses in x,y and z direction:

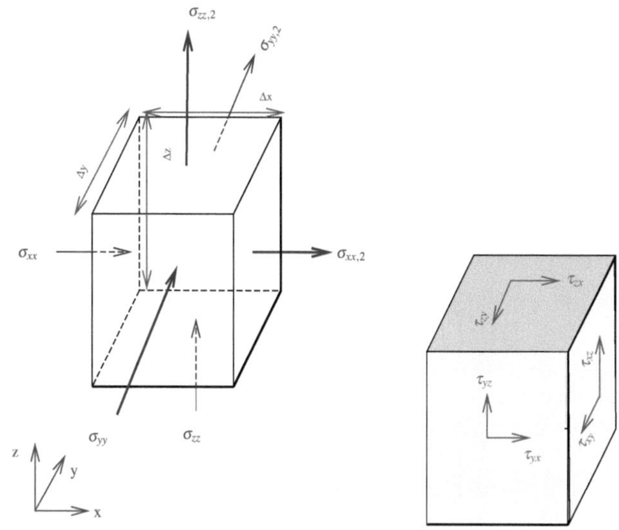

Figure 1.2: **Normal and shear stress distribution on an infinitesimal control volume. Normal stresses are represented Greek letter σ and shear stresses are represented by Greek letter τ.**

$$\sigma_{xx,2} = \sigma_{xx} + \left(\frac{\partial \sigma_{xx}}{\partial x}\right)\Delta x + \left(\frac{\partial^2 \sigma_{xx}}{\partial x^2}\right)\frac{\Delta x^2}{2} + \cdots$$

$$\sigma_{yy,2} = \sigma_{yy} + \left(\frac{\partial \sigma_{yy}}{\partial y}\right)\Delta y + \left(\frac{\partial^2 \sigma_{yy}}{\partial y^2}\right)\frac{\Delta y^2}{2} + \cdots$$

$$\sigma_{zz,2} = \sigma_{zz} + \left(\frac{\partial \sigma_{zz}}{\partial z}\right)\Delta z + \left(\frac{\partial^2 \sigma_{zz}}{\partial z^2}\right)\frac{\Delta z^2}{2} + \cdots$$

We can write the surface stress forces:

$$\begin{array}{l|l|l} F_{\sigma x} = (\sigma_{xx,2} - \sigma_{xx})\,\Delta z\Delta y & F_{\tau,yx} = (\tau_{yx,2} - \tau_{yx})\,\Delta x\Delta z & F_{\tau,zx} = (\tau_{zx,2} - \tau_{zx})\,\Delta x\Delta y \\ F_{\sigma y} = (\sigma_{yy,2} - \sigma_{yy})\,\Delta x\Delta z & F_{\tau,xy} = (\tau_{xy,2} - \tau_{xy})\,\Delta y\Delta z & F_{\tau,zy} = (\tau_{zy,2} - \tau_{zy})\,\Delta x\Delta y \\ F_{\sigma z} = (\sigma_{zz,2} - \sigma_{zz})\,\Delta x\Delta y & F_{\tau,xz} = (\tau_{xz,2} - \tau_{xz})\,\Delta y\Delta z & F_{\tau,yz} = (\tau_{yz,2} - \tau_{yz})\,\Delta x\Delta z \end{array}$$

From Newton's second law of motion we have $\sum F_x = m\,a_x$, $\sum F_y = m\,a_y$, and $\sum F_z = m\,a_z$:

For a control volume of size $\forall$ and having fluid with density ρ, the surface and body forces can be written as

$$F_{\sigma,xx} + F_{\tau,yx} + F_{\tau,zx} + F_{b,x} = m\,a_x = \rho\forall a_x$$

$$F_{\sigma,yy} + F_{\tau,xy} + F_{\tau,zy} + F_{b,y} = m\,a_y = \rho\forall a_y$$

$$F_{\sigma,zz} + F_{\tau,xz} + F_{\tau,yz} + F_{b,z} = m\,a_z = \rho\forall a_z$$

Substituting the normal and shear stresses, after neglecting H.O.T we have

$$\rho \forall a_x = (\sigma_{xx,2} - \sigma_{xx})\Delta y \Delta z + (\tau_{yx,2} - \tau_{yx})\Delta x \Delta z + (\tau_{zx,2} - \tau_{zx})\Delta x \Delta y + F_{b,x}$$

$$\rho \forall a_y = (\sigma_{yy,2} - \sigma_{yy})\Delta x \Delta z + (\tau_{xy,2} - \tau_{xy})\Delta y \Delta z + (\tau_{zy,2} - \tau_{zy})\Delta x \Delta y + F_{b,y}$$

$$\rho \forall a_z = (\sigma_{zz,2} - \sigma_{zz})\Delta x \Delta y + (\tau_{xz,2} - \tau_{xz})\Delta y \Delta z + (\tau_{yz,2} - \tau_{yz})\Delta x \Delta z + F_{b,z}$$

We divide the equations with volume $\forall$:

$$\rho a_x = (\sigma_{xx,2} - \sigma_{xx})\Delta z \Delta y + (\tau_{yx,2} - \tau_{yx})\Delta x \Delta z + (\tau_{zx,2} - \tau_{zx})\Delta x \Delta y + \frac{F_{b,x}}{\forall}$$

$$a_x = \frac{1}{\rho}\left\{\left(\frac{\partial \sigma_{xx}}{\partial x}\right) + \left(\frac{\partial \tau_{yx}}{\partial y}\right) + \left(\frac{\partial \tau_{zx}}{\partial z}\right)\right\} + f_{b,x}$$

Similarly we can cast the equations

$$a_y = \frac{1}{\rho}\left\{\left(\frac{\partial \sigma_{yy}}{\partial y}\right) + \left(\frac{\partial \tau_{xy}}{\partial x}\right) + \left(\frac{\partial \tau_{zy}}{\partial z}\right)\right\} + f_{b,y}$$

$$a_z = \frac{1}{\rho}\left\{\left(\frac{\partial \sigma_{zz}}{\partial z}\right) + \left(\frac{\partial \tau_{xz}}{\partial x}\right) + \left(\frac{\partial \tau_{yz}}{\partial y}\right)\right\} + f_{b,z}$$

where, $f_{b,x}$, $f_{b,x}$ and $f_{b,x}$ are body forces per unit mass in x,y and z directions. These equations are known as *Navier equations*.

For incompressible flow, Stoke proposed the constitutive relations which relate stresses with velocity gradients:

$$\tau_{ij} = 2\mu S_{ij} - \frac{2}{3}\mu S_{kk}\delta_{ij}$$

and

$$\sigma_{ij} = -P\delta_{ij} + 2\mu S_{ij} - \frac{2}{3}\mu S_{kk}\delta_{ij}$$

where, δ_{ij} is Kronecker delta, a unit second-order tensor. The strain-rate and rotation tensors are defined as

$$S_{ij} = \frac{1}{2}\left(\frac{\partial U_i}{\partial x_j} + \frac{\partial U_j}{\partial x_i}\right) \quad Strain-rate\, tensor$$

$$\Omega_{ij} = \frac{1}{2}\left(\frac{\partial U_i}{\partial x_j} - \frac{\partial U_j}{\partial x_i}\right) \quad Rotation\, tensor$$

in vector form, we can write for incompressible flows:

$$\sigma_{xx} = -P + 2\mu\frac{\partial u}{\partial x}, \sigma_{yy} = -P + 2\mu\frac{\partial v}{\partial y}, \sigma_{zz} = -P + 2\mu\frac{\partial w}{\partial z}$$

$$\tau_{xy} = \mu\left(\frac{\partial v}{\partial x} + \frac{\partial u}{\partial y}\right), \tau_{xz} = \mu\left(\frac{\partial w}{\partial x} + \frac{\partial u}{\partial z}\right), \tau_{yz} = \mu\left(\frac{\partial w}{\partial y} + \frac{\partial v}{\partial z}\right)$$

Substituting the Stoke's hypothesis into Navier equation, and introducing the Eulerian formulation for accelerations in x, y and z directions, we have

$$\frac{Du}{Dt}=\frac{\partial u}{\partial t}+u\frac{\partial u}{\partial x}+v\frac{\partial u}{\partial y}+w\frac{\partial u}{\partial z}=-\frac{1}{\rho}\frac{\partial P}{\partial x}+\frac{\mu}{\rho}\nabla^2 u+f_x \quad (1.1)$$

$$\frac{Dv}{Dt}=\frac{\partial v}{\partial t}+u\frac{\partial v}{\partial x}+v\frac{\partial v}{\partial y}+w\frac{\partial v}{\partial z}=-\frac{1}{\rho}\frac{\partial P}{\partial y}+\frac{\mu}{\rho}\nabla^2 v+f_y \quad (1.2)$$

$$\frac{Dw}{Dt}=\frac{\partial w}{\partial t}+u\frac{\partial w}{\partial x}+v\frac{\partial w}{\partial y}+w\frac{\partial w}{\partial z}=-\frac{1}{\rho}\frac{\partial P}{\partial z}+\frac{\mu}{\rho}\nabla^2 w+f_z \quad (1.3)$$

The above set of equations are called the Navier-Stokes equations.

Conservation of mass and momentum equations in cylindrical coordinates

Navier-Stokes Equations:

r-direction:

$$\rho\left(\frac{\partial u_r}{\partial t}+u_r\frac{\partial u_r}{\partial r}+\frac{u_\theta}{r}\frac{\partial u_r}{\partial \theta}-\frac{u_\theta^2}{r}+u_z\frac{\partial u_r}{\partial z}\right)$$
$$=-\frac{\partial P}{\partial r}+f_r+\mu\left[\frac{1}{r}\frac{\partial}{\partial r}\left(r\frac{\partial u_r}{\partial r}\right)-\frac{u_r}{r^2}+\frac{1}{r^2}\frac{\partial^2 u_r}{\partial \theta^2}-\frac{2}{r^2}\frac{\partial u_\theta}{\partial \theta}+\frac{\partial^2 u_r}{\partial z^2}\right]$$

θ-direction:

$$\rho\left(\frac{\partial u_\theta}{\partial t}+u_r\frac{\partial u_\theta}{\partial r}+\frac{u_\theta}{r}\frac{\partial u_\theta}{\partial \theta}+\frac{u_r u_\theta}{r}+u_z\frac{\partial u_\theta}{\partial z}\right)$$
$$=-\frac{1}{r}\frac{\partial P}{\partial \theta}+f_\theta+\mu\left[\frac{1}{r}\frac{\partial}{\partial r}\left(r\frac{\partial u_\theta}{\partial r}\right)-\frac{u_\theta}{r^2}+\frac{1}{r^2}\frac{\partial^2 u_\theta}{\partial \theta^2}-\frac{2}{r^2}\frac{\partial u_r}{\partial \theta}+\frac{\partial^2 u_\theta}{\partial z^2}\right]$$

z-direction:

$$\rho\left(\frac{\partial u_z}{\partial t}+u_r\frac{\partial u_z}{\partial r}+\frac{u_\theta}{r}\frac{\partial u_z}{\partial \theta}+u_z\frac{\partial u_z}{\partial z}\right)$$
$$=-\frac{\partial P}{\partial z}+f_z+\mu\left[\frac{1}{r}\frac{\partial}{\partial r}\left(r\frac{\partial u_z}{\partial r}\right)+\frac{1}{r^2}\frac{\partial^2 u_z}{\partial \theta^2}+\frac{\partial^2 u_z}{\partial z^2}\right]$$

1.1.3 ENERGY EQUATION

The **law of conservation of energy** is also known as **First Law of Thermodynamics** says that

> *Energy cannot be created or destroyed. It can transfer from one object to another and may change forms.*

Ther are various forms of energies, like sound energy, wave energy, solar energy, wind energy, nuclear energy etc.The energy E is a conserved quantity and will remain same. Energy can exist in several forms such as chemical, mechanical, kinetic, potential, electric, nuclear and magnetic etc. The total energy of a system (E) on a unit mass basis is denoted by e.

1.1.4 LUMPED ENERGY ANALYSIS

Let say there is a thermodynamic quantity of interest Φ, which got transported, convected or diffused due to sources or sinks. Mathematically one can write the balance of quantity Φ as:

$$\begin{aligned}\frac{D\Phi_{system}}{Dt} &= \underbrace{\sum_i (Convection)_i}_{at\ System\ boundaries} + \underbrace{\sum_j (Diffusion)_j}_{at\ System\ boundaries} \\ &+ \underbrace{\sum_k (Source)_k + \sum_\ell (Sink)_\ell}_{in\ system} \\ &+ \underbrace{\sum_m (Field)_m}_{acting\ on\ System\ volume}\end{aligned} \tag{1.4}$$

Figure 1.3 shows the schematic representation of energy transport processes.

Convection represents the macroscopic transport of the thermodynamic quantities with flow or mass. In a close system there would be no convection.

Diffusion is a phenomenon which happens due to gradient of the potential. It happens with the molecular movements. Any aspect of diffusion inside system will be an aspect of internal energy and covered in it. In case of mass transfer the potential is the concentration difference. In case of heat transfer the potential is temperature difference.

Souce and sink terms represent the transport due to effects internal to a system. For example, when the chemicals energy present in the explosives converts the point souce of thermal energy is generated. The internal heat generations in the nuclear fuel rods is also an example of same category.

Field effect represent the transport which arises due to fields affecting directly the volume of the system. Transports that are induced due to gravity or magnetic fields come under this category.

$$\left(\frac{DE}{Dt}\right)_{System} \underbrace{- \sum (e\dot{m})_{in} + \sum (e\dot{m})_{out}}_{at\ System\ boundaries} = \sum \dot{Q} + \sum \dot{W} - \underbrace{\sum p\frac{dV}{dt}}_{at\ System\ boundaries} \tag{1.5}$$

In a process the system volume may increase or decrease in time (dV $<$ 0 or dV $>$ 0). Since $\dot{V} = \dot{m}\upsilon$ The term P(dV/dt) is called the flow work can be written as $\dot{W}_{flow} = P\dot{m}\upsilon$ and Eq. can be written as:

$$\left(\frac{DE}{Dt}\right)_{System} \underbrace{- \sum (e\dot{m})_{in} + \sum (e\dot{m})_{out}}_{at\ System\ boundaries} = \sum \dot{Q} + \sum \dot{W} + \underbrace{\sum_{in} (P\upsilon\dot{m}) - \sum_{out} (P\upsilon\dot{m})}_{at\ System\ boundaries} \tag{1.6}$$

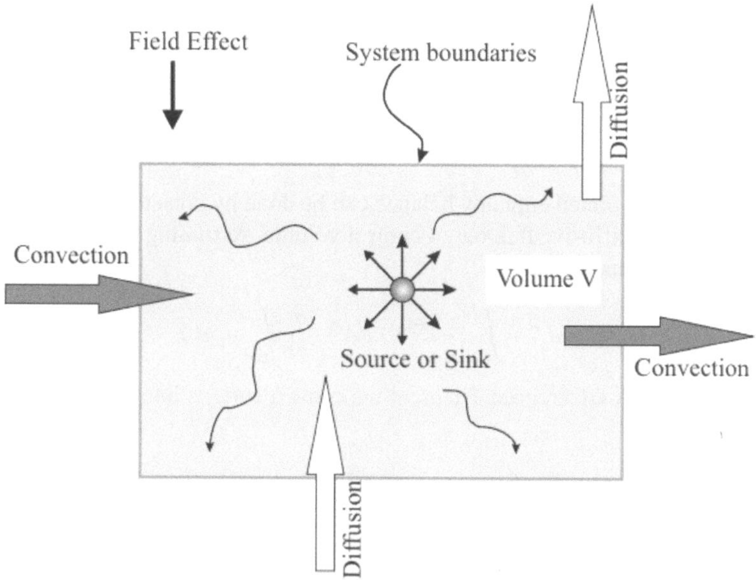

Figure 1.3: **The energy transfer through control volume.**

The Energy E per unit mass is defined as:

$$e = \left(u + gz + \frac{\tilde{u}^2}{2} \right) \tag{1.7}$$

We can define a property enthalpy(h) as u+Pυ.

$$\left(\frac{DE}{Dt}\right)_{System} - \underbrace{\sum((e+P\upsilon)\dot{m})_{in} + \sum((e+P\upsilon)\dot{m})_{out}}_{at\ System\ boundaries} = \sum\dot{Q} + \sum\dot{W} \tag{1.8}$$

$$\left(\frac{DE}{Dt}\right)_{open} - \sum_{in}\dot{m}\left(h + gz + \frac{\tilde{u}^2}{2}\right) + \sum_{out}\dot{m}\left(h + gz + \frac{\tilde{u}^2}{2}\right) = \sum\dot{Q} + \sum\dot{W} \tag{1.9}$$

This is First Law of Thermodynamics for Open Systems where energy and mass can flow through system. However we are not interested in distribution of energy inside system. We are interested in overall bulk energy balance.

1.2 INTEGRAL FORMULATION OF LUMPED ENERGY EQUATION

$$\frac{dN}{dt} = \int_{CS} \eta\rho\mathbf{V}.dA + \frac{\partial}{\partial t}\int_{CV} \eta\rho d\vartheta \tag{1.10}$$

Now as η = N/mass we have η as:
note that E = e.m = e. ρ ϑ
so η = E/mass = e

$$\frac{dE}{dt} = \int_{CS} e\rho \mathbf{V}.dA + \frac{\partial}{\partial t}\int_{CV} e\rho d\vartheta - \sum \dot{\mathtt{Q}} - \sum \dot{\mathtt{W}} \tag{1.11}$$

A more elaborated enthalpy balance can be done by considering enthalpy generation term and diffusive flux out of control volume. Assuming in absence of interactions with surrounding:

$$\int_{CS} h\rho \mathbf{V}.dA + \int_{CS}(-k\nabla T).dA - \frac{\partial}{\partial t}\int_{CV} u\rho d\vartheta - \int_{CV} \dot{q}_g d\vartheta \tag{1.12}$$

Using Gauss-Divergence Theorem we convert surface integral into volume integrals:

$$\int_{CV}\left[\rho\frac{\partial u}{\partial t} + \rho\nabla.(\mathbf{V}.h) - \nabla.(k\nabla T) - \dot{q}_g\right] d\vartheta = 0 \tag{1.13}$$

Here u is internal energy and **V** is velocity. Note that energy inside system is called internal energy so in volumetric balance only internal energy is relevant. Assuming k is independent of temperature and $h.\nabla\mathbf{V} = 0$ due to continuity equation we have:

$$\rho\left(\frac{\partial u}{\partial t} + \mathbf{V}.\nabla h\right) - k\nabla^2 T - \dot{q}_g = 0 \tag{1.14}$$

As h= c_pT and assuming u=h we write:

$$\boxed{c_p\rho\left(\frac{\partial T}{\partial t} + \mathbf{V}.\nabla T\right) - k\nabla^2 T - \dot{q}_g = 0} \tag{1.15}$$

This equation is suffice for heat diffusion analysis without convection currents.

1.2.1 DIFFERENTIAL FORMULATION OF ENERGY EQUATION

Differential formulation Using Divergence Theorem we replace area integral with volumetric integral as:

$$\sum \dot{\mathtt{Q}} + \sum \dot{\mathtt{W}} = \int_{CV} \mathbf{div}\left[e\rho\mathbf{V}\right] d\vartheta + \frac{\partial}{\partial t}\int_{CV} e\rho d\vartheta \tag{1.16}$$

$$\sum \dot{\mathtt{Q}} + \sum \dot{\mathtt{W}} = \int_{CV}\left[\mathbf{div}\left[e\rho\mathbf{V}\right] + \frac{\partial(e\rho)}{\partial t}\right] d\vartheta \tag{1.17}$$

$$\mathbf{div}\left[e\rho\mathbf{V}\right] + \frac{\partial(e\rho)}{\partial t} = \frac{\dot{\mathtt{Q}}}{d\vartheta} + \frac{\dot{\mathtt{W}}}{d\vartheta} \tag{1.18}$$

This is the conservation form

$$\frac{\partial(e\rho)}{\partial t} = \rho\frac{\partial(e)}{\partial t} + e\frac{\partial(\rho)}{\partial t} \tag{1.19}$$

$$\mathbf{div}\left[e\rho\mathbf{V}\right] = \rho\mathbf{V}.\nabla e + e\nabla.\left[\rho\mathbf{V}\right] \tag{1.20}$$

$$e\left[\nabla.\left[\rho\mathbf{V}\right] + \frac{\partial(\rho)}{\partial t}\right] + \rho\left[\mathbf{V}.\nabla e + \frac{\partial(e)}{\partial t}\right] = \frac{\dot{\mathrm{Q}}}{d\vartheta} + \frac{\dot{\mathrm{W}}}{d\vartheta} \tag{1.21}$$

This is the conservation form
Here

$$\left[\nabla.\left[\rho\mathbf{V}\right] + \frac{\partial(\rho)}{\partial t}\right]$$

is continuity equation so its zero. and term

$$\left[\mathbf{V}.\nabla e + \frac{\partial(e)}{\partial t}\right]$$

can be written as substantial derivative De/Dt

$$\rho\frac{De}{Dt} = \frac{\dot{\mathrm{Q}}}{d\vartheta} + \frac{\dot{\mathrm{W}}}{d\vartheta} \tag{1.22}$$

Heat can enter due to diffusion also by chemical energy be converted into thermal energy.

$$\dot{\mathrm{Q}} = -\int_{CS}\mathbf{q}.dA + \int_{CV}\tilde{q}d\vartheta \tag{1.23}$$

here $\tilde{q}$ is rate of thermal energy liberation in volume and $\mathbf{q}$ is diffusive flux.
Using Gauss-Dvergence Theorm:

$$\dot{\mathrm{Q}} = \int_{CV}\left[\tilde{q} - \nabla.\mathbf{q}\right]d\vartheta \tag{1.24}$$

For infinitesimal control volume:

$$\frac{\dot{\mathrm{Q}}}{d\vartheta} = \tilde{q} - \nabla.\mathbf{q} \tag{1.25}$$

Introducing this into equation we have:

$$\rho\frac{De}{Dt} = \frac{\dot{\mathrm{Q}}}{d\vartheta} + \frac{\dot{\mathrm{W}}}{d\vartheta} \tag{1.26}$$

$$\rho\frac{De}{Dt} = \tilde{q} - \nabla.\mathbf{q} + \frac{\dot{\mathrm{W}}}{d\vartheta} \tag{1.27}$$

$$\rho\frac{De}{Dt} - \tilde{q} + \nabla.\mathbf{q} = \frac{\dot{\mathrm{W}}}{d\vartheta} \tag{1.28}$$

For three-dimensional control volume in Cartesian coordinates we have:

$$\frac{De}{Dt} = \frac{\partial e}{\partial t} + u\frac{\partial e}{\partial x} + v\frac{\partial e}{\partial y} + w\frac{\partial e}{\partial z} \tag{1.29}$$

and energy equation is

$$\rho\left(\frac{\partial e}{\partial t}+u\frac{\partial e}{\partial x}+v\frac{\partial e}{\partial y}+w\frac{\partial e}{\partial z}\right)+\left(\frac{\partial q_x}{\partial x}+\frac{\partial q_y}{\partial y}+\frac{\partial q_z}{\partial z}\right)-\tilde{q}=\frac{\dot{\mathtt{W}}}{dxdydz} \tag{1.30}$$

The equation conceals that the total energy flux comprises two separate contributions: convective transport with the bulk flow and diffusive transport measured relative to that bulk motion. The thermal diffusion form of energy equation developed above is not enough to know the energy change between fluid layers due to fluid bulk movement (i.e., convection or advection) as energy transport is disconnected with stress distribution inside fluid body. $\mathtt{W}$ here is now not the shaft work but the work interaction between fluid surrounding and fluid. $\dot{\mathtt{W}}$ here is the rate of work interaction arise due to surrounding surface and body forces:

$$Work = force \cdot displacement$$

$$Work = Stress.area \cdot displacement$$

$$Work \approx Stress \cdot (Three\ Length\ dimension)$$

$$Stress = Work/Volume$$

The *Rate* can be introduced by *velocity/length* dimension $\approx$ 1/s:

$$\dot{\mathtt{W}}/\upsilon = \frac{(Stress \cdot velocity)}{length}$$

$$\begin{aligned}\frac{\dot{\mathtt{W}}}{dxdydz} &= \frac{\partial(u\sigma_{xx})}{\partial x}+\frac{\partial(u\tau_{yx})}{\partial y}+\frac{\partial(u\tau_{zx})}{\partial z}\\ &+\frac{\partial(v\tau_{xy})}{\partial x}+\frac{\partial(v\sigma_{yy})}{\partial y}+\frac{\partial(v\tau_{zy})}{\partial z}\\ &+\frac{\partial(w\tau_{xz})}{\partial x}+\frac{\partial(w\tau_{yz})}{\partial y}+\frac{\partial(w\sigma_{zz})}{\partial z}\\ &+uf_x+vf_y+wf_z\end{aligned}$$

$$\begin{aligned}\frac{\dot{\mathtt{W}}}{dxdydz} = uf_x+vf_y+wf_z &+u\frac{\partial(\sigma_{xx})}{\partial x}+\sigma_{xx}\frac{\partial(u)}{\partial x}+u\frac{\partial(\tau_{yx})}{\partial y}+\tau_{yx}\frac{\partial(u)}{\partial y}\\ &+u\frac{\partial(\tau_{zx})}{\partial z}+\tau_{zx}\frac{\partial(u)}{\partial z}+v\frac{\partial(\tau_{xy})}{\partial x}+\tau_{xy}\frac{\partial(v)}{\partial x}\\ &+v\frac{\partial(\sigma_{yy})}{\partial y}+\sigma_{yy}\frac{\partial(v)}{\partial y}+v\frac{\partial(\tau_{zy})}{\partial z}+\tau_{zy}\frac{\partial(v)}{\partial z}\\ &+w\frac{\partial(\tau_{xz})}{\partial x}+\tau_{xz}\frac{\partial(w)}{\partial x}+w\frac{\partial(\tau_{yz})}{\partial y}+\tau_{yz}\frac{\partial(w)}{\partial y}\\ &+w\frac{\partial(\sigma_{zz})}{\partial z}+\sigma_{zz}\frac{\partial(w)}{\partial z}\end{aligned} \tag{1.31}$$

$$
\begin{aligned}
\frac{\dot{W}}{dxdydz} &= uf_x + u\frac{\partial(\sigma_{xx})}{\partial x} + u\frac{\partial(\tau_{yx})}{\partial y} + u\frac{\partial(\tau_{zx})}{\partial z} \\
&+ vf_y + v\frac{\partial(\tau_{xy})}{\partial x} + v\frac{\partial(\sigma_{yy})}{\partial y} + v\frac{\partial(\tau_{zy})}{\partial z} \\
&+ wf_z + w\frac{\partial(\tau_{xz})}{\partial x} + w\frac{\partial(\tau_{yz})}{\partial y} + w\frac{\partial(\sigma_{zz})}{\partial z} \\
&+ \sigma_{xx}\frac{\partial(u)}{\partial x} + \tau_{yx}\frac{\partial(u)}{\partial y} + \tau_{zx}\frac{\partial(u)}{\partial z} \\
&+ \tau_{xy}\frac{\partial(v)}{\partial x} + \sigma_{yy}\frac{\partial(v)}{\partial y} + \tau_{zy}\frac{\partial(v)}{\partial z} \\
&+ \tau_{xz}\frac{\partial(w)}{\partial x} + \tau_{yz}\frac{\partial(w)}{\partial y} + \sigma_{zz}\frac{\partial(w)}{\partial z}
\end{aligned} \tag{1.32}
$$

$$
uf_x + u\frac{\partial(\sigma_{xx})}{\partial x} + u\frac{\partial(\tau_{yx})}{\partial y} + u\frac{\partial(\tau_{zx})}{\partial z} \tag{1.33}
$$

is actually ρuDu/Dt and likewise we can write for terms:

$$
\rho u\frac{Du}{Dt} + \rho v\frac{Dv}{Dt} + \rho w\frac{Dw}{Dt} = \rho\frac{D\mathbf{V}^2/2}{Dt} \tag{1.34}
$$

$$
\begin{aligned}
\frac{\dot{W}}{dxdydz} = \rho\frac{D\mathbf{V}^2/2}{Dt} &+ \sigma_{xx}\frac{\partial(u)}{\partial x} + \tau_{yx}\frac{\partial(u)}{\partial y} + \tau_{zx}\frac{\partial(u)}{\partial z} \\
&+ \tau_{xy}\frac{\partial(v)}{\partial x} + \sigma_{yy}\frac{\partial(v)}{\partial y} + \tau_{zy}\frac{\partial(v)}{\partial z} \\
&+ \tau_{xz}\frac{\partial(w)}{\partial x} + \tau_{yz}\frac{\partial(w)}{\partial y} + \sigma_{zz}\frac{\partial(w)}{\partial z}
\end{aligned} \tag{1.35}
$$

We introduced it ino the equation

$$
\begin{aligned}
&\rho\left(\frac{\partial e}{\partial t} + u\frac{\partial e}{\partial x} + v\frac{\partial e}{\partial y} + w\frac{\partial e}{\partial z}\right) \\
&+ \left(\frac{\partial q_x}{\partial x} + \frac{\partial q_y}{\partial y} + \frac{\partial q_z}{\partial z}\right) - \tilde{q} \\
&= \frac{\dot{W}}{dxdydz}
\end{aligned} \tag{1.36}
$$

After some simplifications, the equation will take the form (Uddin 2024):

$$
\rho c_p\left(\frac{DT}{Dt}\right) = k\left(\nabla^2.T\right) + \tilde{q} \tag{1.37}
$$

This is Energy equation we use for convective heat transfer.

$$\boxed{\rho c_p\left(\frac{\partial T}{\partial t}+u\frac{\partial T}{\partial x}+v\frac{\partial T}{\partial y}+w\frac{\partial T}{\partial z}\right)=k\left(\nabla^2.T\right)+\tilde{q}} \tag{1.38}$$

Equation 1.38 is valid for incompressible flows only.

Incompressible, constant-property energy equation

$$\frac{\partial T}{\partial t}+u\frac{\partial T}{\partial x}+v\frac{\partial T}{\partial y}+w\frac{\partial T}{\partial z}=\alpha\nabla^2T+\frac{\mu}{\rho c_v}\Phi+\frac{\tilde{q}}{\rho c_v}$$

$$\Phi=2\left[\left(\frac{\partial u}{\partial x}\right)^2+\left(\frac{\partial v}{\partial y}\right)^2+\left(\frac{\partial w}{\partial z}\right)^2\right]$$

$$+\left[\left(\frac{\partial u}{\partial y}+\frac{\partial v}{\partial x}\right)^2+\left(\frac{\partial v}{\partial z}+\frac{\partial w}{\partial y}\right)^2+\left(\frac{\partial w}{\partial x}+\frac{\partial u}{\partial z}\right)^2\right]$$

$$-\frac{2}{3}\left[\frac{\partial u}{\partial x}+\frac{\partial v}{\partial y}+\frac{\partial w}{\partial z}\right]^2$$

1.3 CONTINUUM TO DISCRETIZATION

Discretization is a fundamental concept in numerical analysis and computational modeling, where continuous mathematical equations are transformed into discrete representations for computation. Many physical phenomena, especially in fields like fluid mechanics, heat transfer, and electromagnetics, are governed by partial differential equations (PDEs) that describe changes over an infinite set of points in space and time. These equations inherently rely on the concept of limits to define smooth variations. However, in practical applications, solving such equations directly is often impossible due to their complexity. This is where discretization plays a crucial role.

The core idea behind discretization is breaking down a continuous domain into a finite number of discrete points, effectively replacing infinite variability with a finite set of representative values. This process eliminates the dependency on infinitesimally small changes, making calculations feasible for computers. Instead of considering continuous fields, variables such as velocity, temperature, or pressure are evaluated at specific locations, forming a grid or mesh that approximates the original system. One of the most significant consequences of discretization is the removal of limits. In a continuous formulation, limits define the behavior of derivatives and integrals, ensuring smooth transitions and infinitely small increments. However, once a system is discretized, limits are no longer explicitly present, as derivatives are approximated using finite differences between discrete points. This shift transforms PDEs into algebraic equations that relate values at distinct points, enabling numerical solutions through iterative or direct methods.

The concept of limits is fundamental in converting continuum equations to their discrete counterparts. In calculus, a limit is defined as:

$$\lim_{x \to a} f(x) = L \tag{1.39}$$

if, for every $\varepsilon > 0$, there exists a $\delta > 0$ such that for all x satisfying $0 < |x - a| < \delta$, we have:

$$|f(x) - L| < \varepsilon. \tag{1.40}$$

The derivative of a function is defined as the limit of the difference quotient:

$$f'(x) = \lim_{\Delta x \to 0} \frac{f(x + \Delta x) - f(x)}{\Delta x}. \tag{1.41}$$

In numerical methods, differentiation is approximated using small but finite values of Δx. This concept is critical in discretization since numerical schemes inherently involve approximating limits by small but nonzero increments.

Mathematically, we can cast the equation is an approximate discrete representation using a finite difference in time as

$$\frac{\Delta f}{\Delta t} \approx \frac{f_j^{n+1} - f_j^n}{\Delta t}$$

and

$$\frac{\partial f}{\partial t} = \lim_{\Delta t \to 0} \frac{f_j^{n+1} - f_j^n}{\Delta t}$$

where f represents a general scalar field, u is the velocity, Δt is the time step, and the superscripts n and $n+1$ denote the time levels.

The transition from continuum equations to discrete approximations is fundamental in computational fluid dynamics (CFD). The concept of limits provides the theoretical foundation for this transformation, ensuring that numerical approximations converge to the true physical solution. By refining the grid spacing (Δx) and time steps (Δt), we can achieve higher accuracy in solving complex fluid mechanics and energy equations.

While discretization makes problem-solving computationally feasible, it introduces challenges such as approximation errors and stability concerns. The choice of resolution, meaning the number of discrete points used, determines the accuracy of the solution. A finer resolution improves precision but increases computational cost, requiring a careful balance between efficiency and accuracy.

1.4 FLUID AND ENERGY TRANSPORT MODELS

The concept of transport is fundamental in the study of Fluid Mechanics. In nature, many animals naturally know how to carry important things like their food, babies, and materials for building nests or homes. In a similar way, we see and understand movement in liquids and in living things as a type of transport. For example, blood flowing in our bodies or water moving in a river are both forms of transport. There are different kinds of transport, each with its own characteristics and purposes which we identify as transport of mass, momentum and energy. This helps us study and use these various forms of transport more effectively in fields like biology, physics, and engineering.

When we differentiate currents and vortices in the context of fluid transport, we are, in fact, categorizing phenomena according to their magnitude and significance. Small vortices may appear less significant in comparison with bulk fluid movement. We need to understand and distinguish between advection, which involves the bulk movement of fluid, and diffusion, which involves the mingling of substances within the fluid.

Perhaps the reader is already acquainted with the kinetic molecular theory concerning gases. However, in order to represent fluid phenomena such as friction and drag, more precise equations such as the Navier-Stokes equations are required. For fluid flow analysis the concept of continuum is crucial. Continuum approaches break down when dealing with processes that occur at molecular scales, such as shock waves in a gas. In such cases, defining a continuous fluid velocity is impossible. Whenever, the continuum model fails to capture a local phenomenon accurately, we refer to it as a discontinuous flow.

When the fluid moves the fluid is not only bulk advected with the velocity, but it is also diffused in fluid layers. There will be diffusion present even if fluid flow is laminar. At the macroscopic level, molecules function as agents that facilitate the transportation and dispersion of diverse macroscopic properties via advection and diffusion. Taking into account the fact that individual molecule trajectories deviate from the mean path, which corresponds to the direction of the fluid's velocity, it becomes apparent that molecules or fluid particles are capable of transferring momentum in opposite directions to the mean. The aforementioned occurrence provides insight into the macroscopic nature of momentum and thermal diffusion.

Considering both advection and diffusion processes, the advection-diffusion equation for momentum per unit mass of fluid:

$$\frac{\partial u}{\partial t}+u\frac{\partial u}{\partial x}-D_m\frac{\partial^2 u}{\partial x^2}=0$$

and for the advection-diffusion equation for enthalpy or energy can be written as

$$\frac{\partial T}{\partial t}+u\frac{\partial T}{\partial x}-D_e\frac{\partial^2 T}{\partial x^2}=0$$

here ρ indicates density, u represents velocity, T represents the temperature, the parameter D_m is the diffusion coefficient for momentum, and the parameter D_e is the diffusion coefficient for energy. These equations describe the time evolution of momentum, and temperature in a fluid.

If we consider only advection phenomenon, then we can write the continuous advection equation:

$$\frac{\partial f}{\partial t}+u\frac{\partial f}{\partial x}=0$$

The advection equation models the transport of a quantity due to a bulk flow. It describes how substances like smoke, pollutants, or heat move with a velocity field. By studying advection, we understand how waves, shocks, and conserved properties propagate without spreading or dissipating. This is crucial in meteorology, oceanography, and engineering systems where convection plays a role.

The diffusion equation,

$$\frac{\partial f}{\partial t} - D_m \frac{\partial^2 f}{\partial x^2} = 0$$

in contrast, describes how substances spread due to random motion. This is governed by gradients in concentration, leading to smoothing and eventual uniform distribution over time. Studying diffusion helps in understanding processes like heat conduction, molecular diffusion, and biological transport mechanisms.

Modeling the advection equation, diffusion equation, and their combination the advection-diffusion equation provides critical insights into how physical quantities such as heat, pollutants, and fluids move and disperse in a medium. Each equation represents a fundamental transport mechanism in physics and engineering. When combined, the advection-diffusion equation balances transport by flow with spreading due to diffusion. This combined model is essential in environmental science (e.g., pollutant dispersion), chemical engineering (e.g., reactor design), and fluid mechanics. The numerical study of these equations teaches fundamental lessons about error analysis, stability, convergence, and computational efficiency, which are crucial in real-world simulations across physics, engineering, and environmental science. This will be discussed further in detail in next chapter.

1.5 COMPUTING POWER

Computing power, measured in FLOPS (Floating Point Operations Per Second), has grown exponentially over the last century, following Moore's Law and advancements in semiconductor technology. Early mechanical computers in the 1930s had negligible computing power, but by 1945, ENIAC performed around 0.004 FLOPS. The first supercomputers, like the CDC 6600 (1964) and Cray-1 (1976), achieved megaflops. By 1997, Intel's ASCI Red reached a teraflop, and in 2008, IBM Roadrunner surpassed a petaflop. Today, supercomputers like Frontier exceed an exaflop, enabling breakthroughs in AI, simulations, and quantum research. By 2025, experts predict computing power may reach 10 exaflops.

1.6 SOME IMPORTANT CONCEPTS

In the numerical solution of partial differential equations (PDEs), several core concepts underpin the reliability and accuracy of numerical methods. Before delving into the specifics of any discretization scheme, it is crucial to understand how consistency, stability, boundedness, convergence, and conservation form the foundation of a sound and physically realistic numerical methodology for PDEs.

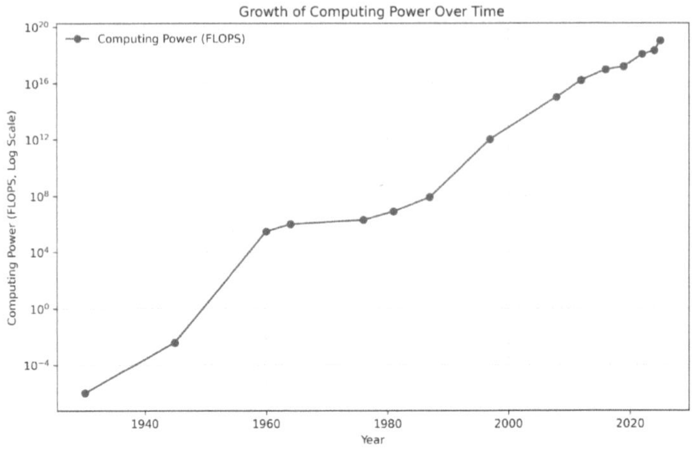

Figure 1.4: **Number of floating operations (indicate the computing speed) vs years.**

☞ **Consistency**: In finite difference methods, consistency means that the discrete finite difference equation correctly represents the underlying partial differential equation (PDE) as the grid spacing (in both time and space) tends to zero. Formally, one can show consistency by examining the local truncation error, i.e., how much error arises in a single step when substituting the exact (continuous) solution of the PDE into the finite difference scheme. If this local truncation error vanishes in the limit $\Delta x, \Delta t \to 0$, the scheme is said to be consistent with the PDE. If the mesh is very fine, the finite difference operator mimics the continuous differential operator accurately. for example: For a first-order PDE $u_t + a u_x = 0$, a forward-in-time, backward-in-space finite difference approximation is typically consistent if the difference quotient for u_t and u_x approximates the PDE's derivatives with $\mathscr{O}(\Delta x + \Delta t)$ accuracy.

☞ **Stability**: Stability concerns how errors, whether arising from initial conditions, boundary conditions, or round-off propagate (or grow) as the numerical solution advances in time (or iteration steps). A stable scheme does not amplify small perturbations unboundedly; instead, it keeps errors at a controlled or diminishing level. To assess stability is to insert a trial solution of the form e^{ikx} (or its discrete analog) and check if the amplification factor is ≤ 1. The amplification factor is

$$\text{Amplification Factor} = \frac{\text{Amplitude at next step}}{\text{Amplitude at current step}}$$

Therefore, when the amplification factor equals one, the amplitude remains unchanged; values greater than one indicate growth, while values less than one indicate decay. This procedure is called Von Neumann analysis.

☞ **Boundedness**: In many practical PDEs (especially linear or linearized problems), stability can be rephrased as the requirement that the discrete solution remain bounded (in some norm) over time. A scheme that allows the solution to blow up with small Δt or Δx is deemed unstable.

☞ **Convergence**: A scheme is convergent if the numerical solution u^n at each time (or iteration) converges to the exact PDE solution $u(x,t)$ in an appropriate norm as $\Delta x, \Delta t \to 0$. Convergence essentially states that the discrete solution matches the true continuous solution in the limit of infinite resolution. Convergence is usually expressed by an estimate of the form $\|u^n - U(\cdot,t_n)\| \leq \text{constant} \times (\Delta x^p + \Delta t^q)$, showing that as the mesh is refined, the solution error goes to zero.

Lax's Equivalence Theorem

Lax's Equivalence Theorem: tells us that for linear well-posed initial value problems, convergence happens if and only if the scheme is both stable and consistent. For a linear, well-posed initial value problem, Lax's Equivalence Theorem states that consistency plus stability is equivalent to convergence. Concretely:

$$(\text{Consistency}) \wedge (\text{Stability}) \quad \Longleftrightarrow \quad (\text{Convergence}).$$

This is a foundational result in numerical analysis of PDEs, as it simplifies the process of proving convergence. One shows that a scheme accurately reproduces the PDE (consistency) and that it does not blow up (stability), from which it follows that the discrete solution converges to the exact PDE solution.

☞ **Conservation:** For flow problems, formulating the governing equations in conservative form ensures that fluxes of mass, momentum, or energy are preserved across the computational domain. A typical conservative form is written as

$$\frac{\partial u}{\partial t} + \nabla \cdot F(u) = 0,$$

where $F(u)$ represents the flux of the conserved quantity u. In finite volume or conservative finite difference methods, these fluxes are integrated across cell boundaries, guaranteeing that the discrete solution maintains conservation laws such as mass conservation, momentum conservation, and energy conservation even in the presence of discontinuities or shocks.

The fundamental equations of fluid dynamics can be systematically cast into conservative form:

Continuity Equation (Mass Conservation):

The mass conservation principle requires that the rate of change of mass inside a control volume equals the net mass flux through its surface. In conservative form:

$$\frac{\partial \rho}{\partial t} + \nabla \cdot (\rho \vec{v}) = 0,$$

where ρ is the fluid density and $\vec{v}$ is the velocity vector.

Momentum Equations (Navier-Stokes Equations):

In conservative form, the Navier-Stokes equations for incompressible or compressible flow are written as

$$\frac{\partial(\rho\vec{v})}{\partial t}+\nabla\cdot(\rho\vec{v}\otimes\vec{v})=-\nabla p+\nabla\cdot\tau+\rho\vec{f},$$

where $\rho\vec{v}$ is the conserved momentum density, $\rho\vec{v}\otimes\vec{v}$ is the convective momentum flux tensor, p is the pressure, τ is the viscous stress tensor, and $\vec{f}$ represents body forces like gravity.

Energy Equation (First Law of Thermodynamics):

The conservation of total energy (internal energy plus kinetic energy) can be expressed in divergence form:

$$\frac{\partial(\rho E)}{\partial t}+\nabla\cdot[(\rho E+p)\vec{v}]=\nabla\cdot(k\nabla T)+\Phi,$$

where

$$E=e+\frac{|\vec{v}|^2}{2}$$

is the total energy per unit mass (internal energy e plus kinetic energy), k is the thermal conductivity, T is temperature, and Φ denotes viscous dissipation effects.

The left-hand side represents the temporal change and flux of total energy, while the right-hand side accounts for heat conduction and viscous dissipation.

We can cast these equations

$$\frac{\partial(\text{quantity})}{\partial t}+\nabla\cdot(\text{flux})=(\text{source terms}).$$

where, Quantity being conserved is $\rho\Phi$, $\rho v_j\Phi$ is convective flux, $-\Gamma\frac{\partial\Phi}{\partial x_j}$ is a diffusive flux, and S_Φ is a source term.

Thus, the complete conservation form is:

$$\frac{\partial(\rho\Phi)}{\partial t}+\frac{\partial}{\partial x_j}\left(\rho v_j\Phi-\Gamma\frac{\partial\Phi}{\partial x_j}\right)=S_\Phi.$$

or using divergence notation:

$$\frac{\partial(\rho\Phi)}{\partial t}+\nabla\cdot(\rho\Phi\vec{v}-\Gamma\nabla\Phi)=S_\Phi.$$

where $\vec{v}$ is the velocity vector.

☞ **Boundedness**: Boundedness of the numerical solution is often discussed alongside stability. In simple terms, a scheme is bounded if it does not produce unphysical oscillations or blow-up in the computed solution. Boundedness can be viewed as a stronger or more explicit form of stability. A stable scheme may allow some growth within a certain range, but a bounded scheme explicitly prevents the numerical solution from exceeding realistic or physically consistent bounds.

Maximum principle arguments: For parabolic or elliptic PDEs, one might show that the numerical solution satisfies a discrete maximum principle, thereby ensuring it remains bounded between the minimum and maximum of the boundary/initial conditions.

☞ **Total Variation Bounds**: In hyperbolic conservation laws, total variation diminishing (TVD) methods maintain boundedness in a sense of limited oscillations, and this property of numerical scheme is important for capturing shocks without spurious overshoot.

☞ **Monotonicity** in numerical schemes refers to the property that a discretization preserves the non-increasing or non-decreasing character of an initial data sequence at every subsequent time step, thereby prohibiting the introduction of new local extrema. Monotone schemes are valued in computational fluid dynamics because they prevent the occurrence of spurious oscillations when a sharp gradient or discontinuity is present in domain. Higher-order non-monotone methods generate dispersive ripples, seen as "undershoots" and "overshoots," whereas a monotone scheme damps these short spurious waves and maintains positivity of physically non-negative quantities such as density or concentration.

PROBLEMS

1P-1 Clarify the concept of consistency, convergence and stability for a numerical scheme.

1P-2 A partial differential equation has two independent variables x and t. What are the two conditions which are required for the solution of the difference equation to converge to the exact solution of the differential equation as $\Delta x \to 0$ and $\Delta t \to 0$?

1P-3 Consider the one–dimensional unsteady advection–diffusion equation

$$\frac{\partial \phi}{\partial t} + U_0 \frac{\partial \phi}{\partial x} = D \frac{\partial^2 \phi}{\partial x^2}$$

where U_0 and D are constants.

i Introduce the dimensionless variables $\hat{x} = x/L$, $\hat{t} = tU_0/L$, and $\hat{\phi} = \phi/\phi_0$. Derive the non-dimensional form of the equation.

ii Identify the dimensionless group that appears and state its physical interpretation.

iii Discuss the dominant transport mechanism when this group is (a) much smaller than unity and (b) much larger than unity.

1P-4 Consider the first-order partial differential equation

$$a(\xi)\frac{\partial u}{\partial x} + b(\xi)\frac{\partial u}{\partial y} = c(\xi)$$

where ξ is a function of certain variables. Classify the equation as linear, quasi-linear, or nonlinear in each of the following cases:

i a, b, c are constants or $\xi = \xi(x, y)$

ii $\xi = \xi(u)$

iii $\xi = \xi(u, \frac{\partial u}{\partial x}, \frac{\partial u}{\partial y}, u^n)$, where $n > 1$

1P-5 Consider a fluid with density ρ, velocity $\vec{u} = (u, v, w)$, and temperature T governed by the energy transport equation:

$$\rho c_p \left(\frac{\partial T}{\partial t} + u\frac{\partial T}{\partial x} + v\frac{\partial T}{\partial y} + w\frac{\partial T}{\partial z} \right) = k\nabla^2 T + \dot{q}_g$$

i Explain the physical meaning of each term in the equation.

ii What assumptions lead to the simplification of this equation into the classical heat equation?

iii For steady, one-dimensional conduction with no heat generation, reduce the above equation into a simpler form.

iv Discuss the impact of increasing k on the temperature profile.

REFERENCES

C. L. Navier, Mémoire sur les lois du mouvement des fluides, *Mémoires de l'Académie Royale des Sciences de l'Institut de France*, Année 1823, pp. 389-440, 1827.

G. G. Stokes, On the effect of the internal friction of fluids on the motion of pendulums, Transactions of the Cambridge Philosophical Society, 9 (Part II), 8-106, 1851.

N. Uddin, Fluid Mechanics, A Problem Solving Approach, CRC Press, 2023.

N. Uddin, Heat Transer, A Systematic Learning Approach, CRC Press, 2024.

R. W. MacCormack, The effect of viscosity in hyper velocity impact cratering, AIAA, 1969.

P.M. Roache, Computational fluid dynamics, Hermosa Publishers, 1972.

2 Classification of Fluid Flow Problems and Boundary Conditions

In this chapter, we introduce a phenomenological classification of fluid flow problems based on the physical behavior of the governing partial differential equations (PDEs). After completing this chapter, you should be able to:

- Distinguish between elliptic, parabolic, and hyperbolic PDEs based on their physical and mathematical behavior in fluid flow contexts.
- Identify how different types of PDEs model steady, diffusive, or wave-like transport phenomena.
- Understand the role of Initial Value Problems (IVPs) and Boundary Value Problems (BVPs) in determining flow evolution.
- Analyze the Navier–Stokes equations from the viewpoint of mixed parabolic–hyperbolic character and identify how each component governs transport mechanisms.
- Apply this classification framework to assess well-posedness and boundary condition requirements in canonical CFD problems.

Fluid flow problems are ubiquitous in both natural and industrial processes, ranging from weather systems and ocean currents to aircraft design and pipeline flows. Understanding and solving these problems require a deep knowledge of fluid dynamics and the mathematical principles that govern them. This chapter provides an introduction to the classification of fluid flow problems, focusing on the types of partial differential equations (PDEs) that describe them and how these are tackled through mathematical and numerical techniques.

The governing equations for fluid flow, such as the Navier-Stokes equations, can be classified based on their behavior into elliptic, parabolic, and hyperbolic PDEs. The type of equation influences the nature of the problem, whether it is time-dependent or steady-state, and guides the choice of numerical methods used to solve it. Fluid problems are further categorized into initial value problems (IVPs) and boundary value problems (BVPs), depending on how the solution domain is defined.

In this chapter, we will explore these classifications, discussing both theoretical and practical aspects. We will also introduce the concept of boundary layers and provide examples of typical fluid flow problems encountered in engineering and physics. This foundation will lead to a deeper understanding of the numerical techniques applied in later sections, such as Runge-Kutta methods and their use in solving fluid dynamics problems.

DOI: 10.1201/9781003663171-2

Table 2.1
Classification of Partial Differential Equations

Condition	Type	Example
$b^2 - 4ac > 0$	Hyperbolic	$\frac{\partial^2 u}{\partial t^2} - c^2 \frac{\partial^2 u}{\partial x^2} = 0$
$b^2 - 4ac = 0$	Parabolic	$\frac{\partial u}{\partial t} = c \frac{\partial^2 u}{\partial x^2}$
$b^2 - 4ac < 0$	Elliptic	$\frac{\partial^2 u}{\partial x^2} + \frac{\partial^2 u}{\partial y^2} = 0$

2.1 PHENOMENOLOGICAL CLASSIFICATION OF PDE TYPES IN CFD

As discussed in Chapter 1, the second-order partial differential equations in two dimensions:

$$a\frac{\partial^2 u}{\partial x^2} + b\frac{\partial^2 u}{\partial x \partial y} + c\frac{\partial^2 u}{\partial y^2} + d\frac{\partial u}{\partial x} + e\frac{\partial u}{\partial y} + fu + g = 0 \tag{1.8}$$

can be classified as follows.

Table 2.1 shows the classification of partial differential equations. Classification is based on the discriminant of the quadratic form associated with the second-order derivatives. If the discriminant is positive, the equation is of hyperbolic type. If zero, it is of parabolic type. If negative, it is of elliptic type. Each type dictates the behavior and properties of solutions to the differential equation.

Once the fundamental conservation equations governing fluid flows are derived, the next critical step is to consider the initial and boundary conditions required to complete a well-posed mathematical model. These conditions, when coupled with the governing equations, serve as the foundation for determining the evolution of the physical system. The proper specification of these conditions is essential, as they define how the system behaves over time and space, ensuring that the problem is both mathematically well-posed and physically meaningful. The significance of initial and boundary conditions cannot be overstated, as they directly influence the accuracy and stability of numerical solutions in computational fluid dynamics (CFD) simulations.

In fluid dynamics, problems typically fall into one of two broad categories: equilibrium problems and marching problems. Each of these categories requires different approaches to solve the governing equations and apply the appropriate conditions.

Equilibrium problems are those that describe steady-state scenarios, where the system does not change with time. Examples include the temperature distribution in a solid object or the stress field within a loaded structure. These problems are governed by elliptic partial differential equations, such as Laplace's equation, which describes steady, irrotational flow in incompressible fluids or steady heat transfer in conductive materials. For these types of problems, the governing equations do not involve time-dependent variables, and the solution typically seeks a stable configuration where all changes have settled.

On the other hand, marching problems involve time-dependent phenomena, where the system evolves dynamically over time. These problems require the solution of hyperbolic or parabolic partial differential equations, such as the Navier-Stokes equations that govern the motion of fluids. In such cases, initial conditions specify the state of the system at the beginning of the simulation, and boundary conditions define how the system interacts with its surroundings. These conditions must account for the flow's evolution, ensuring the solution captures the dynamic processes at each time step.

Equilibrium problems address steady-state situation, such as the distribution of temperature in a solid rod or the deformation of a loaded structure. Steady fluid flows also often belong to this category. These systems are governed by elliptic equations, such as:

$$\frac{\partial^2 \phi}{\partial x^2} + \frac{\partial^2 \phi}{\partial y^2} = 0,$$

which is the two-dimensional Laplace equation encountered in steady, irrotational, incompressible flow or steady heat conduction. Since elliptic problems represent boundary-value problems, one must specify conditions everywhere along the domain boundary. In such systems, disturbances propagate in all directions and solutions remain smooth even if boundary data contain jumps.

A simpler one-dimensional analog for steady heat conduction in a rod is:

$$k\frac{d^2T}{dx^2} = 0,$$

subject to boundary temperatures T_0 and T_L.

Marching problems capture unsteady phenomena and typically involve parabolic or hyperbolic equations. In these situations, one coordinate often plays the role of time (real or abstract) along which information "marches" forward. Parabolic equations describe diffusion-dominated processes (for example, unsteady heat conduction), while hyperbolic equations govern wave-like systems (for instance, acoustic or vibrational waves).

Parabolic equations:

$$\frac{\partial \phi}{\partial t} = \alpha\frac{\partial^2 \phi}{\partial x^2},$$

as in transient heat conduction, rely on initial conditions plus boundary conditions.

Hyperbolic equations:

$$\frac{\partial^2 \phi}{\partial t^2} = c^2\frac{\partial^2 \phi}{\partial x^2} \quad \text{or} \quad \frac{\partial \phi}{\partial t} = c\frac{\partial \phi}{\partial x}$$

as in wave propagation, similarly require initial conditions (e.g., initial displacement and velocity) and boundary conditions. In contrast to parabolic systems, hyperbolic solutions may exhibit sharp transitions or discontinuities.

A key challenge in CFD is selecting appropriate boundary and initial conditions for complex, real-world problems. The wrong choice can lead to incorrect or unstable solutions. For instance, poorly defined boundary conditions can cause numerical artifacts such as spurious oscillations or unrealistic flow patterns. Therefore, great care must be taken when applying these conditions, often requiring both physical insight and numerical experimentation to determine the most suitable approach.

Table 2.2 gives the classification of the partial differential equations (PDEs) as per physical behavior. Parabolic and hyperbolic equations, by contrast, may be formulated on either bounded or unbounded domains. In addition to initial data they demand appropriate outflow or radiation conditions to avoid artificial reflections. In computational fluid dynamics a bounded spatial region denotes internal-flow or confined-field situations, for instance cavity potential flow or solving the pressure-Poisson equation inside a finite box, whereas an exterior domain applies to external-aerodynamics or groundwater-seepage problems in which the physical region extends to infinity; here the solution of $\nabla^2\phi = 0$ (or its inhomogeneous form) must satisfy far-field decay requirements such as $\phi \to 0; \nabla\phi \to 0$ as $r \to \infty$ to guarantee uniqueness and represent the undisturbed free stream. This classification highlights the importance of selecting appropriate initial and boundary conditions, in conjunction with the governing equations, as they dictate the evolution of physical systems. It also determines whether the resulting problem is of the equilibrium (boundary-value) type or the marching (initial-boundary-value) type.

2.2 NATURE OF GOVERNING EQUATIONS

The basic flow equations are continuity equation

$$\frac{\partial \rho}{\partial t} + \frac{\partial}{\partial x_k}(\rho\, u_k) = 0 \tag{2.1}$$

and, the Navier-Stokes equations

$$\rho\left(\frac{\partial u_j}{\partial t} + u_i \frac{\partial u_j}{\partial x_i}\right) = \rho\, f_j - \frac{\partial P}{\partial x_j} + \frac{\partial \tau_{ij}}{\partial x_i} \tag{2.2}$$

where

$$s_{ij} = \tfrac{1}{2}\left(\frac{\partial u_i}{\partial x_j} + \frac{\partial u_j}{\partial x_i} - \tfrac{1}{3}\frac{\partial u_k}{\partial x_k}\,\delta_{ij}\right), \qquad \tau_{ij} = 2\mu\, s_{ij}, \qquad \Phi_{ij} = \tau_{ij}\, s_{ij}.$$

The Navier-Stokes equation is exhibiting parabolic and hyperbolic character simultaneously. The left hand side of equation 2.2

$$\rho\left(\frac{\partial u_j}{\partial t} + u_i \frac{\partial u_j}{\partial x_i}\right)$$

represents the hyperbolic character and this represent the bulk fluid motion which is generally called advection. Inserting the viscous-stress terms leads to the form

$$\rho\,\frac{Du_j}{Dt} = -\frac{\partial P}{\partial x_j} + \mu\,\frac{\partial}{\partial x_i}\left(\frac{\partial u_j}{\partial x_i} + \frac{\partial u_i}{\partial x_j} - \frac{2}{3}\,\delta_{ij}\,\frac{\partial u_k}{\partial x_k}\right) + \rho\, f_j$$

$$\rho\,\frac{Du_j}{Dt} - \mu\,\frac{\partial}{\partial x_i}\left(\frac{\partial u_j}{\partial x_i} + \frac{\partial u_i}{\partial x_j} - \frac{2}{3}\,\delta_{ij}\,\frac{\partial u_k}{\partial x_k}\right) = +\rho\, f_j - \frac{\partial P}{\partial x_j}$$

Table 2.2
Classification of Partial Differential Equations (PDEs) by Physical Behavior and Mathematical Characteristics

Attribute	Elliptic (Equilibrium)	Parabolic (Dissipative)	Hyperbolic (Wave-like)
Representative PDE	$\nabla^2\phi = 0$	$\frac{\partial\phi}{\partial t} = \alpha\nabla^2\phi$	$\frac{\partial\phi}{\partial t} + c\frac{\partial\phi}{\partial x} = 0$
Time Dependence	None (steady)	First-order in time	First-order in time
Domain Type	Bounded spatial region or exterior domain with decay conditions	Any spatial region; problem evolves from $t = 0$ onward	Any spatial region; may be finite or unbounded, evolves from $t = 0$ onward with radiation/outflow conditions
Conditions Required	Boundary conditions on all sides	Initial + boundary conditions	Initial + boundary conditions
Discontinuities	No discontinuities (smooth)	Smooth (no shocks)	May develop discontinuities (e.g., shocks)
Physical Phenomena	Steady heat conduction, potential flow	Unsteady heat diffusion, slow viscous flows	Acoustic waves, fluid shock waves, string vibrations
Mathematical Nature	Well-posed only with full boundary data	Well-posed with IC + BC	Hyperbolic system; requires characteristic treatment

where,

$$\rho\frac{Du_j}{Dt} = \rho\frac{\partial u_j}{\partial t} + \rho u_i\frac{\partial u_j}{\partial x_i}$$

The Navier-Stokes can be rearranged as

$$\rho\frac{\partial u_j}{\partial t} - \mu\frac{\partial}{\partial x_i}(\frac{\partial u_j}{\partial x_i} + \frac{\partial u_i}{\partial x_j} - \frac{2}{3}\delta_{ij}\frac{\partial u_k}{\partial x_k}) + \rho u_i\frac{\partial u_j}{\partial x_i} = +\rho f_j - \frac{\partial P}{\partial x_j}$$

Here we can see that the equation is exhibiting both parabolic and hyperbolic character. We can write it either

$$\underbrace{\rho\frac{\partial u_j}{\partial t} - \mu\frac{\partial}{\partial x_i}(\frac{\partial u_j}{\partial x_i} + \frac{\partial u_i}{\partial x_j} - \frac{2}{3}\delta_{ij}\frac{\partial u_k}{\partial x_k})}_{Parabolic\ Diffusion} + \rho u_i\frac{\partial u_j}{\partial x_i} = \underbrace{+\rho f_j - \frac{\partial P}{\partial x_j}}_{Sources}$$

or like

$$\underbrace{\rho\frac{\partial u_j}{\partial t} + \rho u_i\frac{\partial u_j}{\partial x_i}}_{Hyperbolic\ Advection} - \mu\frac{\partial}{\partial x_i}(\frac{\partial u_j}{\partial x_i} + \frac{\partial u_i}{\partial x_j} - \frac{2}{3}\delta_{ij}\frac{\partial u_k}{\partial x_k}) = \underbrace{+\rho f_j - \frac{\partial P}{\partial x_j}}_{Sources}$$

The Navier-Stokes equation in its mixed hyperbolic-parabolic form motivated a century of model studies. Between 1900 - 1930 fluid dynamicists such as Oseen (1901), Prandtl (1904) and Taylor (1923) observed that the inviscid part

$$\rho\Big(\frac{\partial u_j}{\partial t} + u_i\frac{\partial u_j}{\partial x_i}\Big) = 0$$

propagates disturbances along characteristics while the viscous stresses

$$\mu\frac{\partial}{\partial x_i}\Big(\frac{\partial u_j}{\partial x_i} + \frac{\partial u_i}{\partial x_j} - \tfrac{2}{3}\delta_{ij}\frac{\partial u_k}{\partial x_k}\Big)$$

diffuse momentum in a manner analogous to the heat equation. The interplay is evident in Oseen's linearised approximation and in Prandtl's boundary layer approximation where convective inertia forces balances viscous forces within boundary layer. To isolate the roles of advection and diffusion, researchers introduced simplified one-dimensional analogues:

2.3 MODEL EQUATIONS IN NUMERICAL FLUID MECHANICS

In this book, we shall examine the numerical treatment of four prototype partial differential equations so as to understand their stability limits, time-marching strategies and grid-spacing requirements. We will introduce a variety of discretization schemes, ranging from simple explicit upwind and central-difference formulas to explicit, implicit, and flux-limiter methods—and demonstrate how the stability and analogous diffusion constraints dictate the maximum allowable time step. Emphasis will be placed on the trade-off between accuracy and stability, the role of numerical (or artificial) viscosity in shock-capturing, and the proper choice of mesh resolution to resolve steep gradients without undue computational cost.

☞ Advection equation (purely hyperbolic)

$$\boxed{u_t + c\,u_x = 0}$$

The linear advection equation $u_t + c\,u_x = 0$ is a core model in CFD for understanding convective transport in both incompressible and compressible flows. For incompressible flows, it isolates the advection of quantities like velocity or temperature, helping to test numerical schemes without the added complexity of pressure coupling. It is widely used to study stability (e.g., CFL condition), numerical dispersion, and accuracy of discretization methods. In compressible flows, the equation represents wave propagation. It is essential for validating shock-capturing schemes and understanding how numerical errors affect wave motion. In both cases, it serves as a benchmark for assessing the performance of CFD solvers.

☞ Diffusion equation (purely parabolic)

$$\boxed{u_t = \nu\,u_{xx}}$$

The diffusion equation $u_t = \nu\,u_{xx}$ is fundamental in CFD for modeling heat diffusion, viscous diffusion, and scalar transport in both incompressible and compressible flows. As a purely parabolic equation, it provides a clean framework to study how information spreads smoothly over time. It helps test numerical schemes for stability, particularly implicit and Crank Nicolson methods, and is used to derive time-step restrictions for explicit schemes. The equation helps in understanding numerical diffusion, error propagation, and convergence behavior.

In CFD, this equation acts as model for viscous terms (parabolic part) in the Navier–Stokes equations. The analytical solution is often used for code verification and grid refinement studies.

☞ Advection-diffusion equation

$$\boxed{u_t + c\,u_x = \nu\,u_{xx}}$$

The advection–diffusion equation $u_t + c\,u_x = \nu\,u_{xx}$ combines both hyperbolic and parabolic behavior, making it a key model in CFD for studying convection-diffusion processes in incompressible fluid flows. It describes how a quantity is simultaneously transported by a flow (advection) and smoothed by viscosity or diffusion. It is used to evaluate the performance of numerical schemes under mixed transport conditions.

☞ Burgers equation (nonlinear counterpart of Navier-Stokes)

$$\boxed{u_t + u u_x = \nu u_{xx}}$$

The Burgers equation, proposed by Burgers in 1939, retains the quadratic convective term of the Euler equations while adding a linear viscous term. It demonstrates shock formation when viscosity is small and smooth diffusive behavior when viscosity dominates, mirroring the competing hyperbolic and parabolic tendencies in Navier-Stokes equation.

2.4 BOUNDARY CONDITIONS

Boundary conditions play a crucial role in numerical analysis by ensuring that a partial differential equation (PDE) problem is mathematically well-posed and physically meaningful. They define how the system interacts with its surroundings and control the behavior of the solution near the domain boundaries. Incorrect or inconsistent boundary conditions can lead to unphysical results, numerical instability, or non-convergence issues.

Figure 2.1 shows the backward-facing step configuration. Normally a fully developed or uniform velocity profile is imposed at the inlet with zero transverse and spanwise components, while the pressure gradient is assumed zero in the streamwise direction. No-slip and impermeability conditions are enforced along all wall boundaries, setting all velocity components to zero, while the pressure gradient normal to the wall is taken to vanish. At the outlet, a convective boundary condition is typically applied, specifying zero streamwise derivatives for all velocity components, while the pressure is set to a fixed value to anchor the solution and ensure well-posedness of the pressure field. This setup supports recirculation, reattachment, and shear layer development typical of separated internal flows.

$$\begin{array}{llllll}
\textbf{Inlet} & u = u_{\text{in}}(y), & v = 0, & w = 0, & \dfrac{\partial p}{\partial x} = 0 & \\
\textbf{Outlet} & \dfrac{\partial u}{\partial x} = 0, & \dfrac{\partial v}{\partial x} = 0, & \dfrac{\partial w}{\partial x} = 0, & p = p_{\text{out}} & \\
\textbf{Wall} & u = 0, & v = 0, & w = 0, & \dfrac{\partial p}{\partial n} = 0 &
\end{array}$$

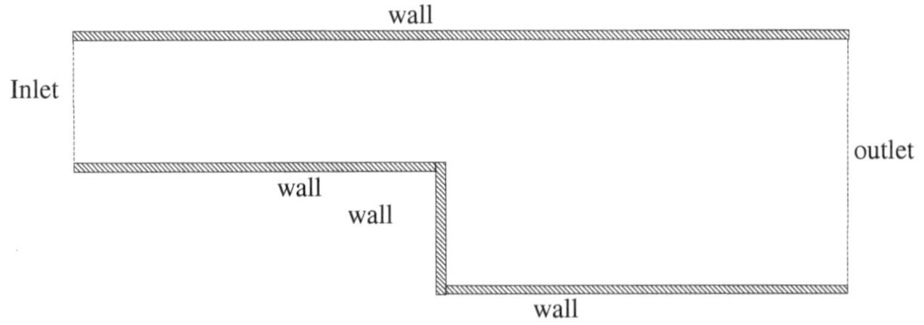

Figure 2.1: **Boundary conditions in case of flow over a backward-facing step.**

Periodic Boundary Conditions: Sometimes we use the periodic boundary conditions in flow problem. Periodic boundary conditions are used when the physical domain exhibits translational symmetry i.e., the flow field exiting one boundary re-enters through the opposite boundary with identical properties. Mathematically, if a variable $\phi(x,y)$ is periodic in the x-direction over a domain of length L, then:

$$\phi(0,y) = \phi(L,y), \quad \frac{\partial \phi}{\partial x}(0,y) = \frac{\partial \phi}{\partial x}(L,y)$$

This implies both the value and derivative are continuous across the boundaries. Periodicity is commonly applied in channels, pipes, and boxes to simulate fully developed or statistically repeating flows without modeling the entire geometry. For a 2D channel flow with periodic boundary conditions in the streamwise direction, the mathematical boundary conditions are:

Walls (y = 0, y = H):

$$u = 0, \quad v = 0, \quad \frac{\partial p}{\partial y} = 0$$

Inlet/Outlet Periodicity (for x = 0, and x = L) :

$$u(0,y) = u(L,y), \quad v(0,y) = v(L,y), \quad p(0,y) = p(L,y)$$

In this formulation, periodic boundary conditions imply that the flow exiting the domain at $x = L$ re-enters at $x = 0$ with identical velocity and pressure fields. This is commonly used in simulations of fully developed or statistically homogeneous flows, where streamwise gradients are negligible or artificially suppressed to study core flow physics without end effects. The no-slip and impermeability conditions along the top and bottom walls enforce viscous boundary layers and shear development.

Non-reflective boundary conditions (NRBCs): These conditions are applied to let waves such as pressure, velocity, or acoustic disturbances exit the computational domain without creating artificial reflections. These reflections can interfere with the internal flow field, degrade solution accuracy, or trigger instability in unsteady or compressible simulations. NRBCs approximate the behavior of an open or infinite domain, allowing information to pass through the boundary naturally. They are often applied at outlet boundaries in problems involving jet exhausts, atmospheric flows, acoustic propagation, or shock waves. One common method is to use characteristic-based formulations that separate incoming and outgoing wave components and only retain physically valid inflow signals. Another approach uses sponge zones or damping layers near the boundary that gradually attenuate the outgoing disturbances. By minimizing reflections, non-reflective boundaries help preserve the accuracy of wave structures and transient features, making them vital for simulations that involve open boundaries interacting with dynamic or wave-like phenomena.

For non-reflective outlet boundaries, a simplified advection equation is used to approximate wave-like behavior and allow outgoing information to leave the domain smoothly:

$$\frac{\partial \phi}{\partial t} + c\frac{\partial \phi}{\partial n} = 0$$

where, ϕ is a flow variable (e.g., pressure or velocity), c is the convection speed (often set to the local flow speed or speed of sound), and $\partial/\partial n$ is the derivative normal to the boundary. This is known as the Sommerfeld radiation condition and represents a one-way wave equation. It allows outward-traveling waves to pass through the boundary without reflection. This approach is especially effective for linearized acoustic waves and low-amplitude disturbances and is widely used in compressible flow simulations, aeroacoustics, and unsteady CFD involving open boundaries.

The Orlanski condition is a type of non-reflective (radiation) boundary condition introduced by I. Orlanski in 1976 for numerical weather prediction and is widely used in unsteady and wave-propagation problems. The basic form of the Orlanski condition is a convective (or wave-like) boundary condition:

$$\frac{\partial \phi}{\partial t} + c\frac{\partial \phi}{\partial x} = 0$$

where, ϕ is a flow variable (e.g., pressure, temperature, velocity component), c is the local phase speed (estimated from interior values), and x is the coordinate normal to the outflow boundary. Instead of prescribing a fixed value or extrapolating ϕ, the Orlanski method advects the variable out of the domain at a speed c, which is often computed as:

$$c = \left.\frac{\partial \phi/\partial t}{\partial \phi/\partial x}\right|_{\text{interior}}$$

This makes the method adaptive and more accurate than constant-speed radiation models. It is especially effective in flows with unsteady waves or vortices where the local wave speed varies.The Orlanski condition improves numerical stability and physical realism in simulations where disturbances must freely exit the domain common in open flows, atmospheric models, and compressible CFD.

Sponge Layer: A sponge layer (also called a damping layer or absorbing zone) is a numerical technique used near the boundaries of a computational domain to gradually absorb outgoing waves or flow disturbances, preventing artificial reflections from contaminating the interior solution. Sponge layers are common in compressible flow, aeroacoustics, weather modeling, and wave propagation, where physical boundaries are far from the region of interest and minimizing reflections is critical for accuracy. In a sponge layer, additional source or damping terms are added to the governing equations. These terms act to relax the solution smoothly toward a reference (often quiescent or freestream) state. A typical form is:

$$\frac{\partial \phi}{\partial t} + [\text{governing terms}] = -\sigma(x)\,(\phi - \phi_{\text{ref}})$$

where, ϕ is the flow variable (e.g., velocity, pressure), ϕ_{ref} is the target or ambient value (often zero or inflow state), and $\sigma(x)$ is a spatially varying damping coefficient, usually increasing gradually toward the boundary.

Symmetry boundary conditions are suitable for laminar and regular Reynolds-Averaged Navier-Stokes equations (RANS) based turbulence simulations. However,

Table 2.3
Mathematical boundary conditions for 2D steady subsonic flow

Boundary	$\vec{U}$	T	p	k	ω	ε
Wall	$\vec{U}=0$	$T=T_{\text{wall}}$	$\frac{\partial p}{\partial n}=0$	Wall-function	Wall-function	Wall-function
Symmetry	$\vec{U}\cdot\hat{n}=0, \quad \frac{\partial \vec{U}_{\parallel}}{\partial n}=0$	$\frac{\partial T}{\partial n}=0$	$\frac{\partial p}{\partial n}=0$	$\frac{\partial k}{\partial n}=0$	$\frac{\partial \omega}{\partial n}=0$	$\frac{\partial \varepsilon}{\partial n}=0$
Inlet	$\vec{U}=\vec{U}_{\text{in}}$	$T=T_{\text{in}}$	$\frac{\partial p}{\partial n}=0$	$k=\frac{3}{2}(I\lvert\vec{U}\rvert)^2$	$\omega=\frac{\sqrt{k}}{C_\mu^{1/4}L}$	$\varepsilon=C_\mu^{3/4}\frac{k^{3/2}}{L}$
Outlet	$\frac{\partial \vec{U}}{\partial n}=0$	$\frac{\partial T}{\partial n}=0$	$p=p_{\text{out}}$	$\frac{\partial k}{\partial n}=0$	$\frac{\partial \omega}{\partial n}=0$	$\frac{\partial \varepsilon}{\partial n}=0$

symmetry boundary conditions are not suitable for use in high-fidelity (Direct and Large Eddy Simulations) because these high-fidelity methods aim to resolve or model the full spectrum of turbulence, including its stochastic and three-dimensional nature. Symmetry conditions artificially constrain the solution by enforcing zero normal velocity and zero normal gradients of all quantities at the plane of symmetry, which suppresses turbulence fluctuations across that boundary. Therefore, even if the geometry is symmetric, it is important to model the full domain or apply periodic or physical boundary conditions that preserve turbulent coherent structures freedom.

Table 2.3 summarizes typical boundary conditions for subsonic, steady, two-dimensional flows, including turbulence quantities such as k, ω, and ε. Note that in Table 2.3, U is bulk inlet velocity, I is turbulent intensity, k is turbulent kinetic energy, ε is turbulence dissipation rate, ω is the specific dissipation rate of turbulence, defined as $\omega = \varepsilon/\beta^* k$. Both β^*, and C_μ are constant having value 0.09. Turbulence modeling will be discussed in Chapter 20. However, here we would like to introduce all important boundary conditions most often required to setup a numerical simulation.

Immersed Boundary Approach: Finite difference methods are not limited to simple geometries. The Immersed Boundary Method (IBM) introduced by Charles S. Peskin in the early 1970s demonstrated that it can be used for irregular boundaries. In his groundbreaking 1972 paper, Peskin showed how a finite difference-based approach could successfully simulate the interaction between moving heart valve leaflets and blood flow. This is done for an inherently complex, and dynamic geometry. His subsequent 1977 work extended this analysis to full cardiac fluid dynamics. These studies established the foundational idea of coupling Eulerian fluid equations with Lagrangian representations of moving boundaries using distributed body forces. This coupling, central to IBM, effectively enabled finite difference solvers on regular Cartesian grids to handle intricate, irregular, and time-dependent geometries—eliminating the need for body-fitted meshes and affirming FD's viability for complex flow problems.

PROBLEMS

2P-1 Consider the general partial differential equation given by:

$$a\frac{\partial^2 u}{\partial x^2}+b\frac{\partial^2 u}{\partial x\partial y}+c\frac{\partial^2 u}{\partial y^2}=f$$

i Analyze the nature of the equation in terms of the coefficients a, b, and c by drawing parallels to conic sections. What types of PDE characteristics can arise based on different values of these coefficients?

ii Describe the typical features of these equation types, including their dependence on solutions, and initial/boundary conditions.

2P-2 Identify the Eulers equation for inviscid flows as elliptic, parabolic or hyperbolic.

$$\frac{\partial \mathbf{U}}{\partial t}+\nabla\cdot\mathbf{F}(\mathbf{U})=0,$$

where, $\mathbf{U}=(\rho,\rho u,\rho v,\rho w,E)^T$ is the vector of conserved variables (density, momentum, and energy), and $\mathbf{F}$ represents the flux terms.

2P-3 In the case of a steady, two-dimensional, incompressible flow over a surface, the Prandtl boundary-layer approximation leads to a reduced set of partial differential equations under the assumptions $\delta/L \ll 1$ and large Reynolds number. The governing mass and momentum equations inside the boundary layer are given as follows:

$$\frac{\partial u}{\partial x}+\frac{\partial v}{\partial y}=0$$

$$u\frac{\partial u}{\partial x}+v\frac{\partial u}{\partial y}=-\frac{1}{\rho}\frac{dP}{dx}+\nu\frac{\partial^2 u}{\partial y^2}$$

Pressure in the normal direction is assumed to be constant across the boundary layer, i.e., $\frac{\partial P}{\partial y}=0$. Here, $u(x,y)$ and $v(x,y)$ are the velocity components in the streamwise and normal directions respectively, $P(x)$ is the pressure (constant in the y-direction), ρ is the fluid density, and ν is the kinematic viscosity. Based on the structure of these equations, identify the mathematical nature (elliptic, parabolic, or hyperbolic) of the governing partial differential equations within the boundary layer.

REFERENCES

H. Poincare, Sur les équations aux dérivées partielles de la physique mathématique, American Journal of Mathematics 12 (1890), 211–294, 1890.

R. Courant, K. O. Friedrichs (1948) - Supersonic Flow and Shock Waves, Interscience Publishers, 1948.

C. W. Oseen, Bidrag till teorien f"or vagr"orelse i str"ommar (Contribution to the theory of wave motion in streams), Lund, 1901. Published in Lunds Universitets rsskrift (Ny f"oljd, Avd. 2, Vol. 37, no. 7) and as Kungl. Fysiografiska S"allskapets i Lund Handlingar (Ny f"oljd, Vol. 12, no. 7), 34 pages, 1901.

J. M. Burgers, Application of a model system to illustrate turbulence and shock wave formation, Proceedings of the Koninklijke Nederlandse Akademie van Wetenschappen, 42, pp. 293-378, 1939.

G. I. Taylor, Stability of a viscous liquid contained between two rotating cylinders, Philosophical Transactions of the Royal Society of London, Series A, 223, pp. 289-343, 1923.

A. Sommerfeld, Die mathematische Theorie der Diffraction, Mathematische Annalen, 47, 317–374, 1912.

I. Orlanski, A simple boundary condition for unbounded hyperbolic flows, Journal of Computational Physics, 21(3), 251-269, 1976.

C. S. Peskin, Flow patterns around heart valves: A numerical method. Journal of Computational Physics, 10(2), 252-271, 1972.

C. S. Peskin, Numerical analysis of blood flow in the heart. Journal of Computational Physics, 25(3), 220-252, 1977.

3 Finite Difference Method

In this chapter, we will cover the foundational concepts of Computational Fluid Dynamics (CFD) and its application in solving fluid flow problems. After completing this chapter, you should be able to:

- Understand the concept of discretization.
- Learn the basic concepts of the Finite Difference Method (FDM) and its application in solving fluid flow equations.
- Develop an understanding of compact finite difference (KFD) method.
- Gain familiarity with the wavenumber.
- Understand the concept of numerical stability.

Computational Fluid Dynamics (CFD) is a branch of fluid mechanics that uses numerical algorithms to solve and analyze problems involving fluid flows. By discretizing the governing conservations equations Navier-Stokes equations, CFD allows the prediction of flow fields, temperature distributions, and scalar transport in complex geometries and boundary conditions. Its applications span a wide range of natural and engineered systems.

CFD techniques are broadly classified based on how they discretize and solve the governing equations of fluid flow. The most common approaches include:

Finite Difference Method (FDM): Uses structured grids and approximates derivatives using differences between neighboring points. It is efficient and easy to implement but limited in handling complex geometries.

Finite Volume Method (FVM): Integrates conservation laws over control volumes and ensures local conservation of mass, momentum, and energy. It is widely used in engineering applications and well-suited for unstructured grids and irregular domains.

Finite Element Method (FEM): Employs variational techniques and divides the domain into elements with interpolation functions. It offers flexibility in geometry and is preferred in structural-fluid interaction and porous media flows.

Spectral Methods: Expand the solution in terms of global basis functions (e.g., Fourier or Chebyshev polynomials). They provide very high accuracy for smooth problems but are less effective for complex geometries or discontinuities.

Lattice Boltzmann Method (LBM): A mesoscopic approach that models fluid behavior using particle distribution functions on a lattice. It is increasingly used for multiphase flows and flows in complex porous structures.

Each technique has strengths suited to specific flow regimes, geometries, and accuracy requirements. In this book we focus on FDM and FVM only.

3.1 FINITE DIFFERENCE METHOD

Finite difference methods (FDM) are a fundamental class of numerical techniques used to approximate solutions to differential equations, particularly in the fields of

DOI: 10.1201/9781003663171-3

fluid mechanics, heat transfer, and structural analysis. The core idea behind FDM is to replace continuous derivatives with discrete approximations based on values of the function at a set of grid points. These approximations are typically derived from Taylor series expansions and can achieve various orders of accuracy depending on the stencil used.

In one dimension, for example, the derivative of a function at a point can be approximated using forward, backward, or central difference formulas. For higher accuracy and improved resolution of gradients, compact schemes and high-order methods are often employed. The finite difference approach is most effective when applied to problems defined on structured grids, where spacing between points is uniform or smoothly varying.

FDM is widely used due to its simplicity, ease of implementation, and clear physical interpretation. It is especially suited for solving problems involving simple geometries. However, for flow with discontinuities (such as shocks), finite volume methods are more appropriate, and discussed in Chapter 17, and on wards.

We may use Taylor-series to develop approximation for the derivative. Developing f(x+Δx) around f(x) we have:

$$f(x+\Delta x) = f(x) + \frac{\partial f}{\partial x}\Delta x + \frac{1}{2}\frac{\partial^2 f}{\partial x^2}(\Delta x)^2 + \frac{1}{3!}\frac{\partial^3 f}{\partial x^3}(\Delta x)^3 + \cdots \tag{3.1}$$

Taylor expansion tells us is that we can approximate the function at an arbitrary distance far away from point x.

$$\frac{f(x+\Delta x) - f(x)}{\Delta x} = f_x(x) + \underbrace{\frac{\Delta x}{2}f_{xx}(x) + \frac{(\Delta x)^2}{6}f_{xxx}(x) + \cdots}_{Truncation\ error} \tag{3.2}$$

This shows that the right-hand side of above equation is an approximation to the first derivative fx in point x. The remaining terms on the right hand side represent the error associated to this approximation. If we restrict the truncation error to its dominant term, that is to the lower power in Δx, we have the first derivative f_x:

$$(f_x)_j = \left(\frac{\partial f}{\partial x}\right)_j = \frac{f(x+\Delta x) - f(x)}{\Delta x} = f_x(x) + \frac{\Delta x}{2}f_{xx}(x) = f_x(x) + O(\Delta x) \tag{3.3}$$

In terms of station position indicated as j we have

$$(f_x)_j = \left(\frac{\partial f}{\partial x}\right)_j = \frac{f_{j+1} - f_j}{\Delta x} \underbrace{+\frac{\Delta x}{2}(f_{xx})_j + \frac{\Delta x^2}{6}(f_{xxx})_j + \cdots}_{Truncation\ error} \tag{3.4}$$

and after truncation error approximated as

$$\boxed{(f_x)_j = \left(\frac{\partial f}{\partial x}\right)_j = \frac{f_{j+1} - f_j}{\Delta x}} \tag{3.5}$$

The approximation takes into account the point $(j+1)$, succeeding point j, hence it is termed the first-order forward difference approximation.

This is certainly not the only formula we can think of, as we can re-apply the relations by replacing everywhere Δx by (-Δx). This leads to the relation:

$$(f_x)_j = \left(\frac{\partial f}{\partial x}\right)_j = \frac{f_j - f_{j-1}}{\Delta x} \underbrace{+\frac{\Delta x}{2}(f_{xx})_j - \frac{\Delta x^2}{6}(f_{xxx})_j + \cdots}_{\textit{Truncation error}} \tag{3.6}$$

$$\boxed{(f_x)_j = \left(\frac{\partial f}{\partial x}\right)_j = \frac{f_j - f_{j-1}}{\Delta x} + O(\Delta x)} \tag{3.7}$$

The approximation takes into account the point $(j-1)$, preceding point j, hence it is termed the first-order backward difference approximation.

We can also formulate the derivative in another way. Using the above two Taylor series approximations for the first order derivative (Eqs. 3.5 and 3.7), we can formulate the second derivative of a function f with respect to x, using points $j+1$, $j-1$, and j. By summing the Taylor series expansions for f_j and f_{j+1}, we get:

$$f_{j+1} + f_{j-1} = 2f_j + \left(\frac{\partial^2 f}{\partial x^2}\right)(\Delta x)^2 + \left(\frac{\partial^4 f}{\partial x^4}\right)\frac{(\Delta x)^4}{12} \tag{3.8}$$

Upon solving for the second derivative, we find:

$$\boxed{\left(\frac{\partial^2 f}{\partial x^2}\right) = \frac{f_{j+1} + f_{j-1} - 2f_j}{(\Delta x)^2} + O(\Delta x)^2} \tag{3.9}$$

Example 3.1

Develop the second order accurate approximation for the derivative $(\partial u/\partial x)$.

Solution

We can develop an approximation using Taylor series and finite differences: The Taylor series expansion for u_{j+1} about u_j:

$$\begin{aligned} u_{j+1} \quad = \quad & u_j + \Delta x\left(\frac{\partial u}{\partial x}\right)_j + \frac{(\Delta x)^2}{2!}\left(\frac{\partial^2 u}{\partial x^2}\right)_j + \frac{(\Delta x)^3}{3!}\left(\frac{\partial^3 u}{\partial x^3}\right)_j \\ & + \frac{(\Delta x)^4}{4!}\left(\frac{\partial^4 u}{\partial x^4}\right)_j + \ldots \end{aligned}$$

The Taylor series expansion for u_{j-1} about u_j:

$$\begin{aligned} u_{j-1} \quad = \quad & u_j - \Delta x\left(\frac{\partial u}{\partial x}\right)_j + \frac{(\Delta x)^2}{2!}\left(\frac{\partial^2 u}{\partial x^2}\right)_j - \frac{(\Delta x)^3}{3!}\left(\frac{\partial^3 u}{\partial x^3}\right)_j \\ & + \frac{(\Delta x)^4}{4!}\left(\frac{\partial^4 u}{\partial x^4}\right)_j - \ldots \end{aligned}$$

Subtracting the second expansion from the first and divide by $2\Delta x$:

$$\frac{u_{j+1} - u_{j-1}}{2\Delta x} = \left(\frac{\partial u}{\partial x}\right)_j + \frac{(\Delta x)^2}{3!}\left(\frac{\partial^3 u}{\partial x^3}\right)_j + \frac{(\Delta x)^4}{5!}\left(\frac{\partial^5 u}{\partial x^5}\right)_j + \dots$$

Here, the even-order derivative terms cancel out because they have the same magnitude but opposite signs in the expansions for u_{j+1} and u_{j-1}. The resulting finite difference approximation is then:

$$\frac{u_{j+1} - u_{j-1}}{2\Delta x} = \left(\frac{\partial u}{\partial x}\right)_j + \frac{1}{6}\Delta x^2\left(\frac{\partial^3 u}{\partial x^3}\right)_j + \frac{1}{120}\Delta x^4\left(\frac{\partial^5 u}{\partial x^5}\right)_j + \dots$$

The second-order accuracy of the centered difference approximation described in this equation has been confirmed. Reducing the grid spacing, Δx, by half decreases the principal error component by a factor of four. Therefore, reducing the value of Δx increases the accuracy of the second-order approximation more rapidly than the first-order approximation.
The approximation for derivative is sometimes expressed as an operator:

$$\frac{\partial u}{\partial x} \approx (\delta_x u)_j = \frac{u_{j+1} - u_{j-1}}{2\Delta x}$$

Example 3.2

Formulate the Central Difference Approximation for Second Derivative $\partial^2 u/\partial x^2$ at a discrete grid point i.

Solution
To approximate the second derivative of a function $u(x)$ with respect to x at a discrete grid point i, we begin by considering the Taylor series expansions of u at points $i+1$ and $i-1$:

$$u_{i+1} = u_i + \left(\frac{\partial u}{\partial x}\right)_i \Delta x + \left(\frac{\partial^2 u}{\partial x^2}\right)_i \frac{(\Delta x)^2}{2!} + \left(\frac{\partial^3 u}{\partial x^3}\right)_i \frac{(\Delta x)^3}{3!} + \cdots \quad (3.10)$$

$$u_{i-1} = u_i - \left(\frac{\partial u}{\partial x}\right)_i \Delta x + \left(\frac{\partial^2 u}{\partial x^2}\right)_i \frac{(\Delta x)^2}{2!} - \left(\frac{\partial^3 u}{\partial x^3}\right)_i \frac{(\Delta x)^3}{3!} + \cdots \quad (3.11)$$

By summing these expansions, we can isolate terms involving the second derivative, and we notice that the terms involving the first and third derivatives cancel each other out, yielding:

$$u_{i+1} + u_{i-1} = 2u_i + \left(\frac{\partial^2 u}{\partial x^2}\right)_i (\Delta x)^2 + \left(\frac{\partial^4 u}{\partial x^4}\right)_i \frac{(\Delta x)^4}{12} + \cdots \tag{3.12}$$

Solving for the second derivative, we find:

$$\left(\frac{\partial^2 u}{\partial x^2}\right)_i = \frac{u_{i+1} + u_{i-1} - 2u_i}{(\Delta x)^2} - \left(\frac{\partial^4 u}{\partial x^4}\right)_i \frac{(\Delta x)^2}{12} + \cdots \tag{3.13}$$

Neglecting higher-order terms, the central finite difference approximation for the second derivative with respect to x evaluated at grid point i is:

$$\left(\frac{\partial^2 u}{\partial x^2}\right)_i \approx \frac{u_{i+1} + u_{i-1} - 2u_i}{(\Delta x)^2} \tag{3.14}$$

From the error term, $O(\Delta x)^2$, we conclude that this central difference approximation is of second-order accuracy.

Example 3.3

Formulate a 4th order approximation for Second Derivative $\partial^2 u/\partial x^2$ at a discrete grid point i.

Solution

To derive the fourth-order accurate finite difference approximation for the second derivative $\frac{\partial^2 u}{\partial x^2}$ at a grid point (i,j), we start by considering the Taylor series expansions of u at points surrounding (i, j) in the x-direction.

Given a uniform grid spacing Δx, the Taylor series expansions of u at points $(i+1, j)$, $(i+2, j)$, $(i-1, j)$, and $(i-2, j)$, about the point (i, j), are as follows:

$$\begin{aligned} u_{i+1,j} &= u_{i,j} + \left(\frac{\partial u}{\partial x}\right)_{i,j} \Delta x + \left(\frac{\partial^2 u}{\partial x^2}\right)_{i,j} \frac{(\Delta x)^2}{2!} + \left(\frac{\partial^3 u}{\partial x^3}\right)_{i,j} \frac{(\Delta x)^3}{3!} \\ &\quad + \left(\frac{\partial^4 u}{\partial x^4}\right)_{i,j} \frac{(\Delta x)^4}{4!} + \cdots \end{aligned}$$

$$\begin{aligned} u_{i-1,j} &= u_{i,j} - \left(\frac{\partial u}{\partial x}\right)_{i,j} \Delta x + \left(\frac{\partial^2 u}{\partial x^2}\right)_{i,j} \frac{(\Delta x)^2}{2!} - \left(\frac{\partial^3 u}{\partial x^3}\right)_{i,j} \frac{(\Delta x)^3}{3!} \\ &\quad + \left(\frac{\partial^4 u}{\partial x^4}\right)_{i,j} \frac{(\Delta x)^4}{4!} - \cdots \end{aligned}$$

$$\begin{aligned} u_{i+2,j} &= u_{i,j} + 2\left(\frac{\partial u}{\partial x}\right)_{i,j} \Delta x + 2^2\left(\frac{\partial^2 u}{\partial x^2}\right)_{i,j} \frac{(\Delta x)^2}{2!} + 2^3\left(\frac{\partial^3 u}{\partial x^3}\right)_{i,j} \frac{(\Delta x)^3}{3!} \\ &\quad + 2^4\left(\frac{\partial^4 u}{\partial x^4}\right)_{i,j} \frac{(\Delta x)^4}{4!} + \cdots \end{aligned}$$

$$u_{i-2,j} = u_{i,j} - 2\left(\frac{\partial u}{\partial x}\right)_{i,j}\Delta x + 2^2\left(\frac{\partial^2 u}{\partial x^2}\right)_{i,j}\frac{(\Delta x)^2}{2!} - 2^3\left(\frac{\partial^3 u}{\partial x^3}\right)_{i,j}\frac{(\Delta x)^3}{3!} + 2^4\left(\frac{\partial^4 u}{\partial x^4}\right)_{i,j}\frac{(\Delta x)^4}{4!} - \cdots$$

To isolate the second derivative term, we multiply the $(i+1, j)$ and $(i-1, j)$ expansions by 16, the $(i+2, j)$ and $(i-2, j)$ expansions by -1, and then sum them. The first and third derivative terms will cancel out, and we can solve for the second derivative term:

$$-u_{i+2,j} + 16u_{i+1,j} - 30u_{i,j} + 16u_{i-1,j} - u_{i-2,j} = 12\left(\frac{\partial^2 u}{\partial x^2}\right)_{i,j}(\Delta x)^2 + O((\Delta x)^4)$$

Dividing by $12(\Delta x)^2$, we obtain the fourth-order accurate central difference approximation for the second derivative:

$$\left(\frac{\partial^2 u}{\partial x^2}\right)_{i,j} = \frac{-u_{i+2,j} + 16u_{i+1,j} - 30u_{i,j} + 16u_{i-1,j} - u_{i-2,j}}{12(\Delta x)^2} + O((\Delta x)^4)$$

The error term $O((\Delta x)^4)$ indicates that this approximation is accurate to the fourth order in Δx, which means the error decreases rapidly as Δx becomes smaller.

Incorporating more data points typically enhances the precision of numerical approximations. Herein, a fourth-order formula for the first derivative of a function f at point j is provided, which utilizes the function values at points $j-2$, $j-1$, $j+1$, and $j+2$. The development of this formula via Taylor series expansions involves expanding the function values at adjacent points to f_j and resolving for the derivative f'_j.

Considering the Taylor series expansions of f_{j-2}, f_{j-1}, f_{j+1}, and f_{j+2} around f_j, we obtain:

$$f_{j-2} = f_j - 2\Delta x f'_j + \frac{(2\Delta x)^2}{2!}f''_j - \frac{(2\Delta x)^3}{3!}f'''_j + \frac{(2\Delta x)^4}{4!}f''''_j + \mathcal{O}(\Delta x^5)$$

$$f_{j-1} = f_j - \Delta x f'_j + \frac{\Delta x^2}{2!}f''_j - \frac{\Delta x^3}{3!}f'''_j + \frac{\Delta x^4}{4!}f''''_j + \mathcal{O}(\Delta x^5)$$

$$f_{j+1} = f_j + \Delta x f'_j + \frac{\Delta x^2}{2!}f''_j + \frac{\Delta x^3}{3!}f'''_j + \frac{\Delta x^4}{4!}f''''_j + \mathcal{O}(\Delta x^5)$$

$$f_{j+2} = f_j + 2\Delta x f'_j + \frac{(2\Delta x)^2}{2!}f''_j + \frac{(2\Delta x)^3}{3!}f'''_j + \frac{(2\Delta x)^4}{4!}f''''_j + \mathcal{O}(\Delta x^5)$$

To isolate f'_j, we construct a linear combination of these expansions:

$$af_{j-2} + bf_{j-1} + cf_{j+1} + df_{j+2} = f'_j + \mathcal{O}(\Delta x^4)$$

Next, by substituting the Taylor series expansions into the linear combination and solving for the coefficients a, b, c, and d to nullify the terms involving f_j, f_j'', and f_j''', ensuring the coefficient in front of f_j' is unity:

$$\begin{aligned} a(-2\Delta x) + b(-\Delta x) + c\Delta x + d(2\Delta x) &= 1 \\ a + b + c + d &= 0 \\ a(2\Delta x)^2 + b(\Delta x)^2 + c(\Delta x)^2 + d(2\Delta x)^2 &= 0 \\ -a(2\Delta x)^3 + b(\Delta x)^3 - c(\Delta x)^3 + d(2\Delta x)^3 &= 0 \end{aligned}$$

Upon solving the system, the coefficients are determined as:

$$a = \frac{1}{12}, b = -\frac{2}{3}, c = \frac{2}{3}, d = -\frac{1}{12}$$

The ensuing finite difference formula is then:

$$f_j' = \frac{f_{j-2} - 8f_{j-1} + 8f_{j+1} - f_{j+2}}{12\Delta x} + \mathcal{O}(\Delta x^4)$$

Table 3.0 lits the finite difference schemes of various orders for approximating from first to fourth derivatives.

Table 3.1 summarizes finite difference schemes for first and second derivatives using 2-, 3-, and 5-point stencils. The 2-point upwind scheme offers first-order accuracy and stability in advective flows. Central schemes provide higher accuracy—second-order for 3-point and fourth-order for 5-point—making them suitable for symmetric and diffusion-dominated problems.

3.2 SPURIOUS ERRORS

In computational fluid dynamics (CFD), numerical schemes can introduce errors. These errors stem from the discretization and approximation of the governing equations, and they can alter the accuracy and physical fidelity of simulations. Below we explain each type of error, their causes, their effects on simulations, and ways to mitigate them.

Dissipation Error (Numerical Diffusion): Dissipation error is a numerical error that causes artificial damping of solution amplitude. It arises when the leading term of the truncation error in a scheme's modified equation is an even order spatial derivative, effectively adding numerical viscosity into the solution. In simpler terms, the scheme behaves like it has extra diffusion—sharp gradients or wave amplitudes get smoothed out over time. First order upwind schemes are a classic example which introduce strong numerical diffusion, greatly smearing steep fronts or discontinuities as the solution is convected. This inherent dissipation in upwind schemes results in a smoothing effect on the solution, as higher wavenumbers are damped. Thus, selecting an appropriate finite difference scheme requires balancing accuracy and dissipation across the wavenumber spectrum, particularly when modeling wave phenomena or transport equations.

Table 3.0: Finite difference schemes of various orders (1st, 2nd, 4th) for approximating the first to fourth derivatives

(a) Forward differences $O(\Delta x)$

	f_j	f_{j+1}	f_{j+2}	f_{j+3}	f_{j+4}
$\Delta x \frac{\partial f}{\partial x}$	-1	1			
$(\Delta x)^2 \frac{\partial^2 f}{\partial x^2}$	1	-2	1		
$(\Delta x)^3 \frac{\partial^3 f}{\partial x^3}$	-1	3	-3	1	
$(\Delta x)^4 \frac{\partial^4 f}{\partial x^4}$	1	-4	6	-4	1

(b) Backward differences $O(\Delta x)$

	f_{j-4}	f_{j-3}	f_{j-2}	f_{j-1}	f_j
$\Delta x \frac{\partial f}{\partial x}$				-1	1
$(\Delta x)^2 \frac{\partial^2 f}{\partial x^2}$			1	-2	1
$(\Delta x)^3 \frac{\partial^3 f}{\partial x^3}$		-1	3	-3	1
$(\Delta x)^4 \frac{\partial^4 f}{\partial x^4}$	1	-4	6	-4	1

(c) Forward differences $O(\Delta x)^2$

$2\Delta x \frac{\partial f}{\partial x}$	f_j	f_{j+1}	f_{j+2}	f_{j+3}	f_{j+4}	f_{j+5}
	-3	4	-1			
$(\Delta x)^2 \frac{\partial^2 f}{\partial x^2}$	2	-5	4	-1		
$2(\Delta x)^3 \frac{\partial^3 f}{\partial x^3}$	-5	18	-24	14	-3	
$(\Delta x)^4 \frac{\partial^4 f}{\partial x^4}$	3	-14	26	-24	11	-2

(d) Backward differences $O(\Delta x)^2$

	f_{j-5}	f_{j-4}	f_{j-3}	f_{j-2}	f_{j-1}	f_j
$2\Delta x \frac{\partial f}{\partial x}$				1	-4	3
$(\Delta x)^2 \frac{\partial^2 f}{\partial x^2}$			-1	4	-5	2
$2(\Delta x)^3 \frac{\partial^3 f}{\partial x^3}$		3	-14	24	-18	5
$(\Delta x)^4 \frac{\partial^4 f}{\partial x^4}$	-2	11	-24	26	-14	3

(e) Central differences $O(\Delta x)^2$

	f_{j-2}	f_{j-1}	f_j	f_{j+1}	f_{j+2}
$2\Delta x \frac{\partial f}{\partial x}$		-1	0	1	
$(\Delta x)^2 \frac{\partial^2 f}{\partial x^2}$		1	-2	1	
$2(\Delta x)^3 \frac{\partial^3 f}{\partial x^3}$	-1	2	0	-2	1
$(\Delta x)^4 \frac{\partial^4 f}{\partial x^4}$	1	-4	6	-4	1

(f) Central differences $O(\Delta x)^4$

	f_{j-3}	f_{j-2}	f_{j-1}	f_j	f_{j+1}	f_{j+2}	f_{j+3}
$12\Delta x \frac{\partial f}{\partial x}$		1	-8	0	8	-1	
$12(\Delta x)^2 \frac{\partial^2 f}{\partial x^2}$		-1	16	-30	16	-1	
$8(\Delta x)^3 \frac{\partial^3 f}{\partial x^3}$	1	-8	13	0	-13	8	-1
$6(\Delta x)^4 \frac{\partial^4 f}{\partial x^4}$	-1	12	-39	56	-39	12	-1

Table 3.1
Finite Difference Stencils for First and Second Derivative Approximations

Stencil	First Derivative	Second Derivative
2-Point Stencil (Upwind)	$\frac{\phi_i-\phi_{i-1}}{\Delta x}$	–
3-Point Stencil (Central)	$\frac{\phi_{i+1}-\phi_{i-1}}{2\Delta x}$	$\frac{\phi_{i+1}-2\phi_i+\phi_{i-1}}{(\Delta x)^2}$
5-Point Stencil (Central)	$\frac{-\phi_{i+2}+8\phi_{i+1}-8\phi_{i-1}+\phi_{i-2}}{12\Delta x}$	$\frac{-\phi_{i+2}+16\phi_{i+1}-30\phi_i+16\phi_{i-1}-\phi_{i-2}}{12(\Delta x)^2}$

The main effect of dissipation error is a loss of amplitude and blurring of features. Peaks, vortices, or shock fronts in the flow become more diffuse and lower in magnitude than they should be. For instance, a sharp pulse being transported will steadily flatten and diminish in height due to excessive dissipation. Important small scale flow structures may vanish prematurely because the numerical scheme damps them out. In the extreme, a shock wave captured on a coarse grid with a very diffusive scheme will appear as a gradual slope rather than a sharp jump. The upside is that dissipative schemes tend to be very stable (they suppress oscillations), but the downside is a loss of accuracy in predicting the true strength or detail of flow features.

To mitigate dissipation errors in CFD simulations, one can use higher order or less dissipative schemes. Higher order discretizations (e.g., 2nd order or 4th order) have smaller truncation error and typically introduce less artificial viscosity than first order methods, preserving more of the true amplitude. For example, switching from a first order upwind to a second order scheme will reduce numerical diffusion. Finer spatial resolution and/or smaller time steps reduce truncation error, which in turn diminishes numerical diffusion. With more grid points resolving a gradient, the scheme doesn't need to smear it as much to remain stable.

Many CFD solvers introduce explicit artificial diffusion to ensure stability, but it should be set to the minimum level required. Excessive smoothing or filtering operations should be avoided unless absolutely necessary. Implementing flux limiters or adaptive blending (switching to a lower-order scheme only in regions where needed) helps manage dissipation, applying it selectively near steep gradients rather than uniformly across the domain.

Dispersion Error (Phase Error): Dispersion error is a numerical phase error that causes waves of different wavelengths to travel at incorrect speeds. It occurs when the leading truncation error terms are of odd order in space or time, which makes the numerical phase velocity differ from the physical velocity. In a non-dispersive physical system (like linear advection), all Fourier components of a wave should propagate at the same speed. However, a dispersive numerical scheme causes different frequency components to propagate at different speeds. This results in phase distortion—the shape of a wave packet changes because some parts lag or lead relative to where they should be. Numerically, one often sees small oscillatory wiggles in the solution, especially in the wake of a moving steep gradient, due to dispersion error.

The primary effect of dispersion error is distortion of wave position and shape. The error does not directly damp the wave (amplitude stays nearly correct), but the wave will be in the wrong place or oscillate. For example, if we simulate a propagating sine wave, dispersion error may cause its crests to shift relative to the true solution (a phase lag or lead). If the solution contains a sharp change, dispersive schemes often produce a wavering undershoot/overshoot pattern trailing the front—oscillations that are not present in the real physics. These oscillations are typically smaller in magnitude than Gibbs oscillations and arise from phase mismatches rather than instability. Over long times, dispersion error can seriously degrade accuracy: multiple wave components de synchronize, so a pulse might spread out or deform even if no physical dispersion is present. In engineering terms, dispersion error can mess up wave timing (important in acoustics, for instance) or cause non physical resonances due to phase lag accumulation.

To reduce dispersion errors, one can refine the mesh or time step. Ensuring a sufficient number of grid points per wavelength is crucial. Dispersion errors tend to become smaller as $k\Delta x \to 0$ (where k is wavenumber)—i.e., a finer grid resolves the wave better, reducing phase lag. Smaller time steps (for time marching schemes) can also improve phase accuracy, especially if the dispersion error comes from time discretization. Some numerical schemes are specifically designed to minimize dispersion. For instance, increasing the order of a centered scheme can reduce phase error (a 4th order scheme has less dispersion error than a 2nd order scheme for the same grid spacing). There are also dispersion relation preserving (DRP) schemes or optimized finite difference stencils that match the numerical wave speed to the true wave speed over a range of wavelengths. Choosing such a scheme helps waves propagate with the correct phase. If minor oscillations (ringing) appear due to dispersion, a light smoothing filter or slight artificial damping can be applied to the highest frequency components of the solution. This sacrifices a bit of amplitude accuracy to remove the most spurious wiggles. The filtering should be minimal so as not to introduce too much dissipation—just enough to damp out the unphysical high frequency noise while preserving the bulk of the wave.

Aliasing Error: The Nyquist wavenumber is a concept from signal processing and numerical methods, particularly relevant to the study of aliasing in wave-based systems. It represents the highest wavenumber (or frequency) that can be accurately represented on a discrete grid or sampled system without incurring aliasing errors.

The Nyquist wavenumber is defined as:

$$k_{\text{Nyquist}} = \frac{\pi}{\Delta x}$$

where, Δx is the grid spacing or the spatial resolution of the discrete system, and k_{Nyquist} is the maximum spatial frequency (or wavenumber) that can be represented without aliasing. The associated Nyquist wavelength is $\lambda_{\min} = 2\Delta x$, since $k = 2\pi/\lambda$, we have:

$$\lambda = \frac{2\pi}{k} \Rightarrow \lambda_{\min} = \frac{2\pi}{k_{\text{Nyquist}}} = 2\Delta x$$

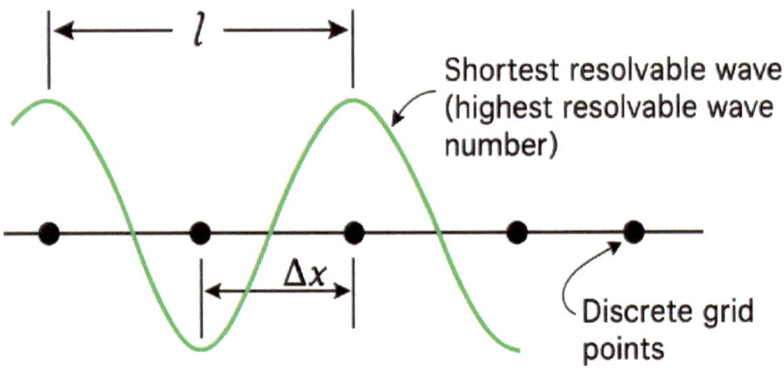

Figure 3.1: **Schematic representation of Nyquist wave.**

When a continuous function (such as a wave) is discretized on a grid, the grid imposes a limit on the highest spatial frequency that can be accurately captured. If the wavelength of a wave is too small relative to the grid spacing, the discrete representation can fold or alias higher frequencies back into the lower frequencies. This aliasing causes the wave to be misrepresented, resulting in incorrect or distorted results. The Nyquist wavenumber corresponds to the highest wavenumber that can be resolved with a given grid spacing. It corresponds to half the spatial sampling rate, analogous to how the Nyquist frequency defines half the sampling rate in one-dimensional signal processing. In physical terms, the Nyquist wavenumber corresponds to the shortest possible wavelength that can be distinguished on the grid, which is two grid points per wavelength (i.e., the shortest wavelength that can be represented by the grid).

When two waves, each characterized by the wave number k^*, interact:

$$e^{ik^*j} \cdot e^{ik^*j} = e^{i2k^*j}$$

they *generate* or *breed* a new wave component with a wave number $2k^*$ exactly twice of the original. This process, known as wave-number doubling, results in the continuous production of increasingly shorter waves. Eventually, these shorter waves surpass the numerical grid's maximum resolvable limit—the Nyquist frequency. Waves beyond this limit become aliased, appearing erroneously as longer waves, leading to numerical instability and inaccuracies in simulations. This means that waves with higher frequencies are incorrectly represented by lower frequencies, leading to errors in simulations or signal processing. To avoid aliasing and represent higher frequencies correctly, the grid resolution needs to be sufficiently fine. This means reducing Δx to allow the Nyquist wavenumber to increase, thus allowing the system to capture higher frequencies.

In numerical methods, such as finite difference schemes or spectral methods, the Nyquist wavenumber helps define the range of spatial frequencies that can be accurately modeled. It is a key factor in determining the appropriate grid spacing and time step when solving wave equations or other PDEs. Consider a 1D grid where the spacing is $\Delta x = 0.1$. The Nyquist wavenumber is:

$$k_{\text{Nyquist}} = \frac{\pi}{0.1} \approx 31.42 \text{ rad/m}$$

This means that any wavelength shorter than $2 \times \Delta x$ (or roughly 0.2 m) will be subject to aliasing if it is not properly handled by the numerical scheme.

3.2.1 ERRORS ASSOCIATED WITH DISCRETIZATION

The errors associated with finite difference discretization are as follows:

i Numerical Diffusion and Dispersion: When approximating solutions to PDEs, numerical schemes can introduce artificial diffusion (smearing of sharp gradients) and dispersion (phase errors in wave propagation), which may need to be corrected for accurate solutions.

ii Truncation Errors: These are errors made when a continuous mathematical process is represented discretely. For instance, when derivatives are approximated using finite differences, the omitted higher-order terms lead to truncation errors. These errors are intrinsic to the discretization process and cannot be entirely eliminated.

iii Information Loss at Sub-grid Scales: In the context of numerical simulations, especially those involving wave propagation, any fluctuations at scales smaller than the grid resolution (sub-grid scales) cannot be resolved and are therefore lost. This is analogous to the concept of the Nyquist frequency in signal processing, which states that for a signal to be properly sampled without aliasing, it must be sampled at least at twice the highest frequency present in the signal. The the non-dimensional wave number defined as

$$GPPW = \frac{2\pi}{k\Delta x} = \left(\frac{2\pi}{k^*}\right) \quad (k = 1, 2, 3, \ldots)$$

3.3 FOURIER SERIES ANALYSIS

Mathematically, functions can be represented through Fourier series. We can learn a lot by comparing the outcome of discretization in comparison with the sine functions. As theoretically the discretzed function must in principal be able to capture the sine and cosine functions.

The linear stability method utilizes the characteristic that functions can be articulated through a Fourier series, representable by a limited collection of real numbers. This mathematical method incrementally introduces additional terms to refine the precision of fitting a curve. Commonly, the series incorporates variations of sine and cosine functions, with each term scaled by a specific weighting factor. Functions with gentle, gradual changes demand fewer terms, whereas those with abrupt fluctuations or 'flashy' characteristics, along with functions that exhibit sudden breaks such as 'step-functions', generally need a greater quantity of terms to achieve an exact depiction.

The Fourier series can be represented using other special function forms in addition to sines and cosines, as long as they meet specific orthogonality requirements. Nevertheless, the complex exponential form is considered one of the more practical forms as it encompasses both the sine and cosine functions. The function can be expressed using Euler's relation, which combines the real and imaginary components to yield the representation in the complex plane.

The complex exponential $e^{i\alpha}$ can be expressed in terms of cosine and sine, which is shown in Euler's relation:

$$e^{i\alpha} = \cos(\alpha) + i\sin(\alpha)$$

This equation allows for a representation of functions in the complex plane, where the real part (cosine term) can be visualized as the horizontal coordinate and the imaginary part (sine term) as the vertical coordinate. The complex plane representation is particularly useful as it captures both the amplitude and phase of the wave, which can be visualized as the radius and angle in a polar coordinate system, respectively.

It is assumed that the solution of any finite-difference scheme may be expressed as a Fourier series using complex exponential notation at every given time level, n, in the following form:

$$h_j^n = \sum_{k=1}^{k=kk} \xi_k^n exp(ik\alpha_j), \quad k = 1, 2, 3, \ldots, kk$$

where ξ_k^n is the Fourier coefficient for wave number k at time-level n, and α is the dimensionless wave number as previously defined.

The longest wave that can be represented within a computational domain is described as having a wavelength that is twice the domain's length, which means only half of its wavelength fits within the domain. This is a fundamental concept in numerical methods, as it sets the limit for the lowest frequency (or longest wavelength) that can be modeled without aliasing. In the realm of numerical analysis, when modeling real cosine waves within a computational domain, the longest wave is characterized by having just half of its wavelength span the entire domain length. This means the full wave cycle is spread over double the domain's extent, which can be mathematically represented as $L/2$ if L is the domain's length. The description also extends to higher wave numbers, going up to the maximal wave number corresponding to the Nyquist frequency. At this point, the wave number k equals the number of grid points jj within the domain. A key aspect of linear stability analysis in numerical methods is to predict the behavior of each wave's Fourier coefficient over time, determining whether it will grow, diminish, or remain steady for any given wave number index k.

When a wave with a wavelength shorter than twice the domain length (or a frequency higher than the Nyquist frequency) is sampled, aliasing occurs. This means that these higher frequencies are misrepresented as lower frequencies in the computational model, leading to inaccuracies.

Let say $k = 1$ giving

$$h_j^n = \xi^n e^{i\alpha j}$$

then insert this solution into the finite-difference scheme. The solutions for any wave number at any space and time points can be represented as such.

$$h_j^{n+1} = \xi^{n+1} e^{i\alpha j}$$

$$h_{j+1}^n = \xi^n e^{i\alpha(j+1)}$$

When these terms are inserted into the F-D scheme and the resulting equation is reorganized into the form

$$\xi^{n+1} = A\xi^n$$

It is observed that for each time step and for each wave number, the Fourier coefficient on the right-hand side is multiplied by a factor A, known as the amplification factor. For stable finite-difference systems, the Fourier coefficients, denoted by ξ, must not exhibit unbounded growth.

$$|A| \leq 1$$

When applying finite difference methods (FDs) for spatial discretization, one can analyze the numerical stability by looking at the amplification factor, which describes how errors or oscillations are propagated with each time step. For stability, the magnitude of the amplification factor must be less than or equal to 1.

We have developed:

1. First order forward difference approximation:

$$f'_j = \frac{f_{j+1} - f_j}{\Delta x} + O(\Delta x)$$

   This is used to approximate the first derivative using the value at point j and the next point $j+1$.

2. Second order central difference approximation:

$$f'_j = \frac{f_{j+1} - f_{j-1}}{2\Delta x} - \frac{\Delta x^2}{6} f'''_j + \ldots$$

   This approximation uses the values at point $j+1$ and the previous point $j-1$ and is more accurate than the first order approximation.

3. Fourth order central difference approximation:

$$f'_j = \frac{f_{j-2} - 8f_{j-1} + 8f_{j+1} - f_{j+2}}{12\Delta x} + O(\Delta x^4)$$

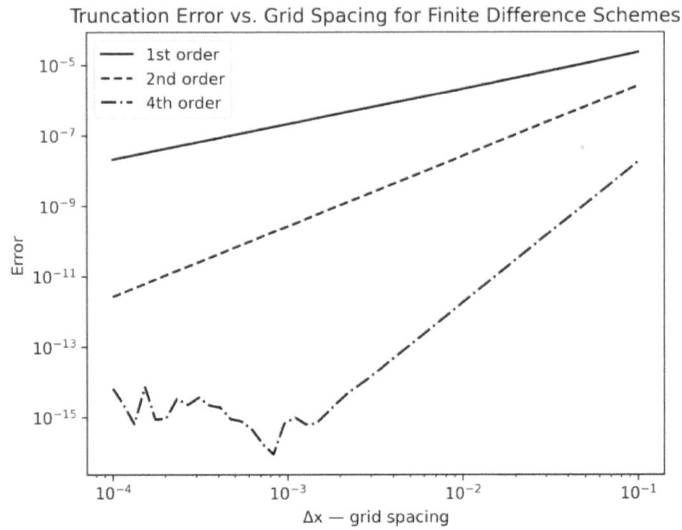

Figure 3.2: **Comparison of truncation error and grid spacing for three finite difference methods.**

Now if we have a function

$$f(x) = \frac{\sin x}{x^4}$$

we can estimate the apprximate drivative of this function using finite differences. For x=5, we can plot the error between the exact and finite-difference formulation in log-log scale as shown in Figure 3.2.

An alternative approach for evaluating the order of accuracy, which can be more revealing, is the modified wavenumber technique. This method probes the efficiency of a finite difference scheme in differentiating specific types of functions, particularly sinusoidal ones. These functions are a fundamental choice for analysis as they form the basis of Fourier series, which are commonly employed to approximate arbitrary functions. To properly represent and precisely differentiate high-frequency sinusoidal functions, a greater number of points is necessary. With a predetermined grid resolution, our focus is to determine the effectiveness of a finite difference scheme in differentiating the more complex high-frequency sinusoidal functions. It is generally anticipated that most differencing schemes will perform adequately with low-frequency functions that vary gradually. In solving non-linear differential equations, which typically comprise multiple frequencies, the modified wavenumber method provides a means to evaluate the accuracy with which different components of the solution are depicted.

To illustrate the concept, we consider an exponential function $u = e^{ikx}$. The exact derivative of the exponential function e^{ikx} is given by:

$$\frac{\partial e^{ikx}}{\partial x} = ike^{ikx}$$

Here, i is the imaginary unit, and k is called the wavenumber. To approximate the derivative of the function $u_j = e^{ikx_j}$, we apply the centered difference operator derived earlier:

$$\frac{\partial u}{\partial x} \approx (\delta_x u)_j = \frac{u_{j+1} - u_{j-1}}{2\Delta x}$$

where $x_j = j\Delta x$. From this, we can derive the modified wavenumber.

Substituting $u_j = e^{ikx_j}$ and expanding, we get:

$$(\delta_x u)_j = \frac{e^{ik\Delta x(j+1)} - e^{ik\Delta x(j-1)}}{2\Delta x}$$

This expression simplifies to:

$$(\delta_x u)_j = \frac{(e^{ik\Delta x} - e^{-ik\Delta x})e^{ikx_j}}{2\Delta x}$$

Using Euler's formula $e^{i\theta} - e^{-i\theta} = 2i\sin(\theta)$, we can further simplify:

$$(\delta_x u)_j = \frac{i\sin(k\Delta x)e^{ikx_j}}{\Delta x}$$

This equation represents the centered difference approximation of the derivative of e^{ikx} at the point x_j and includes the modified wavenumber term $\sin(k\Delta x)/\Delta x$, which modifies the exact wavenumber k due to discretization.

Thus, we can rewrite the expression as:

$$(\delta_x u)_j = ik^*_{\text{mod}} e^{ikx_j}$$

where $k^*_{\text{mod}} = \frac{\sin(k\Delta x)}{\Delta x}$ is the modified wavenumber.

The discrepancy between the modified wavenumber k^*_{mod} and the true wavenumber k arises from discretization errors in the numerical method. The closeness of the modified wavenumber to the true wavenumber determines the accuracy of the finite difference approximation. An approximation is considered accurate if the modified wavenumber k^*_{mod} is close to the true wavenumber k over a wide range of values. In contrast, large discrepancies indicate that the numerical approximation may not accurately represent wave behavior, especially for higher wavenumbers (shorter wavelengths).

The accuracy of the numerical approximation can be evaluated by analyzing the dispersion relation of the numerical scheme, which relates frequency and wavenumber. The closer the numerical dispersion relation is to the exact dispersion relation, the more accurate the finite difference scheme is in representing wave propagation. This accuracy is crucial in simulations where wave propagation characteristics are important, such as in computational fluid dynamics, acoustics, and other fields involving wave phenomena.

The numerical phase speed c_c is defined as:

$$c_c = \frac{\omega_c}{k} = \frac{ck^*_{\text{mod}}}{k}$$

Substitute the expression for k^*_{mod} into c_c:

$$c_c = c\frac{\sin(k\Delta x)}{k\Delta x}$$

To further analyze this expression, assume $k\Delta x \ll 1$, and using approximation:

$$\sin x \approx x - \frac{x^3}{6}$$

which allows us to use the Taylor series expansion for $\sin(k\Delta x)$:

$$\sin(k\Delta x) \approx k\Delta x - \frac{(k\Delta x)^3}{6}$$

Substitute this approximation into the expression for c_c:

$$c_c \approx c\frac{k\Delta x - \frac{(k\Delta x)^3}{6}}{k\Delta x}$$

Separate the terms:

$$c_c \approx c\left(1 - \frac{(k\Delta x)^2}{6}\right)$$

Thus, we have:

$$c_c \approx c\left[1 - \frac{1}{6}(k\Delta x)^2\right]$$

This shows that the numerical phase speed c_c deviates from the true phase speed c by a factor that depends on $(k\Delta x)^2$, indicating a second-order error in $k\Delta x$. The numerical phase speed of different schemes is critical in evaluating their accuracy for wave propagation problems. A scheme with a numerical phase speed close to the true wave speed preserves the wave shape and minimizes dispersion errors, which are distortions due to phase differences between numerical and exact solutions. By comparing numerical phase speeds, one can assess how well each scheme captures the physical wave behavior across various wavelengths. A good scheme will exhibit minimal phase-speed error, especially for well-resolved waves, indicating reliable performance in simulations.

Group Velocity: Group velocity, in numerical schemes for solving dispersive partial differential equations, represents the speed at which energy or wave packets propagate through a medium. Even when the underlying equation is non-dispersive, numerical discretization can introduce dispersion, causing variations in group velocity with respect to k. This numerical dispersion leads to inaccuracies, where different components of a wave packet travel at different speeds, resulting in phase errors and distortion of the wave structure. Understanding group velocity in numerical methods is crucial to capturing accurate energy propagation, as it ensures that the computational model closely mimics the physical behavior of the system being simulated.

The group velocity or wave celerity is defined as the derivative of the angular frequency with respect to the wavenumber:

$$v_g = \frac{\partial \omega_c}{\partial k}$$

Starting with the expression for the numerical angular frequency ω_c, given by

$$\omega_c = c\,k^*_{\text{mod}}$$

where k^*_{mod} is the modified wavenumber defined as

$$k^*_{\text{mod}} = \frac{\sin(k\Delta x)}{\Delta x}$$

we aim to determine the derivative $\frac{\partial \omega_c}{\partial k}$.

Substitute the expression for k^*_{mod} into ω_c:

$$\omega_c = c \cdot \frac{\sin(k\Delta x)}{\Delta x}$$

To find the group velocity, differentiate ω_c with respect to k:

$$\frac{\partial \omega_c}{\partial k} = c \cdot \frac{1}{\Delta x} \cdot \frac{\partial}{\partial k}(\sin(k\Delta x))$$

Applying the chain rule, the derivative of $\sin(k\Delta x)$ with respect to k becomes

$$\frac{\partial \omega_c}{\partial k} = c \cdot \frac{1}{\Delta x} \cdot \cos(k\Delta x) \cdot \frac{\partial}{\partial k}(k\Delta x)$$

Since $\frac{\partial}{\partial k}(k\Delta x) = \Delta x$, substitute this back:

$$\frac{\partial \omega_c}{\partial k} = c \cdot \frac{1}{\Delta x} \cdot \cos(k\Delta x) \cdot \Delta x$$

Now the Δx terms cancel out, yielding

$$\frac{\partial \omega_c}{\partial k} = c\cos(k\Delta x)$$

This expression represents the group velocity, showing that the speed of wave packet propagation depends on both the true wave speed c and the grid spacing Δx, with the dependence modulated by the cosine function.

In numerical schemes, the group velocity is essential for understanding how accurately energy or information is transported by a wave packet. The group velocity indicates the speed at which the wave packet's amplitude propagates, making it crucial for simulating dispersive waves, where different frequencies travel at different speeds. A good numerical scheme should have a group velocity close to the true physical group velocity to avoid artificial dispersion, which can lead to incorrect energy transport and phase distortion over time. By comparing the group velocities of different schemes, one can select a scheme that better preserves wave packet shape and ensures more realistic energy propagation in simulations.

Modified wavenumber k_{mod} for the second-order finite difference scheme: To find the modified wavenumber k_{mod} as a function of $k\Delta x$ for the second-order finite difference scheme for the first derivative $\frac{df}{dx} \approx \frac{f_{i+1}-f_{i-1}}{2\Delta x}$, let's analyze the scheme in the frequency domain by considering the effect of the finite difference scheme on a plane wave $f_j = e^{ikx_j}$, where $x_j = j\Delta x$.

Given $f_j = e^{ikx_j}$, we can represent the values at neighboring grid points as:

$$f_{i+1} = e^{ik(x_i+\Delta x)} = e^{ikx_i}e^{ik\Delta x} = f_i e^{ik\Delta x}$$
$$f_{i-1} = e^{ik(x_i-\Delta x)} = e^{ikx_i}e^{-ik\Delta x} = f_i e^{-ik\Delta x}$$

Substitute these into the finite difference formula:

$$\frac{df}{dx} \approx \frac{f_{i+1}-f_{i-1}}{2\Delta x} = \frac{e^{ikx_i}e^{ik\Delta x} - e^{ikx_i}e^{-ik\Delta x}}{2\Delta x}$$

This simplifies to:

$$\frac{df}{dx} \approx \frac{e^{ik\Delta x} - e^{-ik\Delta x}}{2\Delta x} e^{ikx_i} = \frac{2i\sin(k\Delta x)}{2\Delta x} e^{ikx_i}$$

$$\frac{df}{dx} \approx \frac{i\sin(k\Delta x)}{\Delta x} e^{ikx_i}$$

The ideal derivative of a plane wave $f(x) = e^{ikx}$ is $ikf(x)$, hence:

$$ikf(x) = ike^{ikx} \implies ik$$

Compare this with the approximate derivative:

$$ik_{\text{mod}} = \frac{i\sin(k\Delta x)}{\Delta x}$$

Remove the imaginary unit i and solve for k_{mod}:

$$k_{\text{mod}} = \frac{\sin(k\Delta x)}{\Delta x}$$

Since $k^*_{\text{mod}} = \frac{\sin(k\Delta x)}{\Delta x}$, we can write:

$$k^*_{\text{mod}}\Delta x = \sin(k\Delta x)$$

We plot this relationship in Figure 3.3. This modified wavenumber equation illustrates the accuracy with which each wavenumber component of the solution is represented over the range $0 \le k\Delta x \le \pi$. The plot shows that the approximation remains accurate as long as the grid spacing Δx is small.

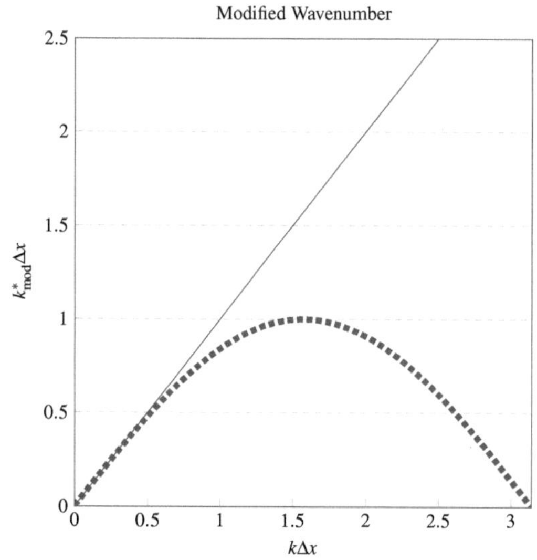

Figure 3.3: **Modified wavenumber for 2nd-order schemes.**

The wave number for first derivative with FDO4: The fourth-order finite difference approximation given is:

$$\frac{\partial f}{\partial x} \approx \frac{-f_{j-2} + 8f_{j-1} - 8f_{j+1} + f_{j+2}}{12\Delta x}$$

Substituting the plane wave $f_j = e^{ikx_j} = e^{ikj\Delta x}$ into the finite difference formula, we get the terms:

$$f_{j-2} = e^{ik(j-2)\Delta x} = e^{ikj\Delta x} e^{-2ik\Delta x} = f_j e^{-2ik\Delta x}$$
$$f_{j-1} = e^{ik(j-1)\Delta x} = e^{ikj\Delta x} e^{-ik\Delta x} = f_j e^{-ik\Delta x}$$
$$f_{j+1} = e^{ik(j+1)\Delta x} = e^{ikj\Delta x} e^{ik\Delta x} = f_j e^{ik\Delta x}$$
$$f_{j+2} = e^{ik(j+2)\Delta x} = e^{ikj\Delta x} e^{2ik\Delta x} = f_j e^{2ik\Delta x}$$

Plugging these expressions into the finite difference approximation:

$$\frac{\partial f}{\partial x} \approx \frac{1}{12\Delta x} \left(-e^{-2ik\Delta x} + 8e^{-ik\Delta x} - 8e^{ik\Delta x} + e^{2ik\Delta x} \right)$$

Utilizing Euler's formulas for the sine and cosine expansions, we convert this to:

$$\frac{\partial f}{\partial x} \approx \frac{1}{12\Delta x} (-2i\sin(2k\Delta x) + 16i\sin(k\Delta x))$$

Further simplifying the terms:

$$\frac{\partial f}{\partial x} \approx \frac{i}{12\Delta x} (16\sin(k\Delta x) - 2\sin(2k\Delta x))$$

Setting this equal to the ideal derivative of the plane wave ikf:

$$ik_{\text{mod}} = \frac{i}{12\Delta x}(16\sin(k\Delta x) - 2\sin(2k\Delta x))$$

Thus, solving for k_{mod}:

$$k_{\text{mod}} = \frac{1}{12\Delta x}(16\sin(k\Delta x) - 2\sin(2k\Delta x))$$

The wave number for first derivative with FDO6: For a sixth-order finite difference scheme for the first derivative, the approximation can typically involve more grid points to achieve higher accuracy. Let's consider a commonly used sixth-order finite difference stencil:

$$\frac{\partial f}{\partial x} \approx \frac{1}{60\Delta x}(-f_{i-3} + 9f_{i-2} - 45f_{i-1} + 45f_{i+1} - 9f_{i+2} + f_{i+3})$$

To derive k_{mod} for this scheme, we'll use the expression in the Fourier space by applying it to a plane wave $f(x) = e^{ikx}$, just like in the previous examples:

$$f_{i+n} = e^{ik(x_i + n\Delta x)} = e^{ikx_i}e^{ikn\Delta x} = f_i e^{ikn\Delta x}$$

Substitute these into the finite difference formula:

$$\frac{df}{dx} \approx \frac{1}{60\Delta x}\left(-e^{ik(-3\Delta x)} + 9e^{ik(-2\Delta x)} - 45e^{ik(-\Delta x)} + 45e^{ik\Delta x} - 9e^{ik(2\Delta x)} + e^{ik(3\Delta x)}\right)$$

Simplify using Euler's formula:

$$\frac{df}{dx} \approx \frac{1}{60\Delta x}\left(-e^{-3ik\Delta x} + 9e^{-2ik\Delta x} - 45e^{-ik\Delta x} + 45e^{ik\Delta x} - 9e^{2ik\Delta x} + e^{3ik\Delta x}\right)$$

Factor out e^{ikx_i} (the common term):

$$\begin{aligned}
\frac{df}{dx} &\approx \frac{e^{ikx_i}}{60\Delta x}\left(-e^{-3ik\Delta x} + 9e^{-2ik\Delta x} - 45e^{-ik\Delta x} + 45e^{ik\Delta x} - 9e^{2ik\Delta x} + e^{3ik\Delta x}\right) \\
&= \frac{ie^{ikx_i}}{60\Delta x}(-\sin(3k\Delta x) + 9\sin(2k\Delta x) - 45\sin(k\Delta x) + 45\sin(k\Delta x) \\
&\quad -9\sin(2k\Delta x) + \sin(3k\Delta x)) \\
&= \frac{ie^{ikx_i}}{\Delta x}\left(\frac{2}{60}(\sin(3k\Delta x) - 9\sin(2k\Delta x) + 45\sin(k\Delta x))\right) \\
k_{\text{mod}} &= \frac{1}{\Delta x}\left(\frac{1}{60}(2\sin(3k\Delta x) - 18\sin(2k\Delta x) + 90\sin(k\Delta x))\right)
\end{aligned}$$

Modified Wavenumber and Resolution Characteristics of Finite Difference Schemes

For First order derivative:

FDO2:

$$k_{\text{mod}}\Delta x = \sin(k\Delta x)$$

$$FDO2 \quad scheme: \quad k^*_{\text{mod, max}} = 1 \quad and \quad k^*(k^*_{\text{mod, max}}) = \frac{\pi}{2}$$

FDO4:

$$k_{\text{mod}}\Delta x = \frac{1}{12}\left(16\sin(k\Delta x) - 2\sin(2k\Delta x)\right)$$

$$FDO4 \quad scheme: \quad k^*_{\text{mod, max}} = 1.372 \quad and \quad k^*(k^*_{\text{mod, max}}) = 1.797$$

FDO6:

$$k_{\text{mod}}\Delta x = \frac{1}{60}(2\sin(3k\Delta x) - 18\sin(2k\Delta x) + 90\sin(k\Delta x))$$

$$FDO6 \quad scheme: \quad k^*_{\text{mod, max}} = 1.586 \quad and \quad k^*(k^*_{\text{mod, max}}) = 1.936$$

Note: Numerical instability typically initiates at the highest resolvable wavenumber, corresponding to $k^* = k^*_{\text{mod,max}}$. The relationship between wavelength λ and wavenumber k is given by:

$$\lambda = \frac{2\pi}{k}$$

A larger wavenumber k corresponds to a shorter wavelength λ, meaning the wave oscillates more frequently in space. A smaller wavenumber k corresponds to a longer wavelength λ, meaning the wave oscillates less rapidly in space.

Referring to Figure 3.4, one obtains the grid points per wavelength (GPPW) for each scheme by substituting the corresponding $k\Delta x$ values into the formula.

For FDO2: till $k\Delta x = 0.5$ the scheme matches with exact solution:

$$\text{GPPW}_{FDO2} = \frac{2\pi}{0.5} = 12.57 \approx 13$$

For FDO4: $k\Delta x = 1$

$$\text{GPPW}_{FDO4} = \frac{2\pi}{1} = 6.283 \approx 7$$

For FDO6: $k\Delta x = 1.3$

$$\text{GPPW}_{FDO6} = \frac{2\pi}{1.3} = 4.833 \approx 5$$

The quantity $k^*(k^*_{\text{mod,max}})$ guides us how far into the wavenumber spectrum a given finite difference scheme can resolve before significant phase error or instability arises.

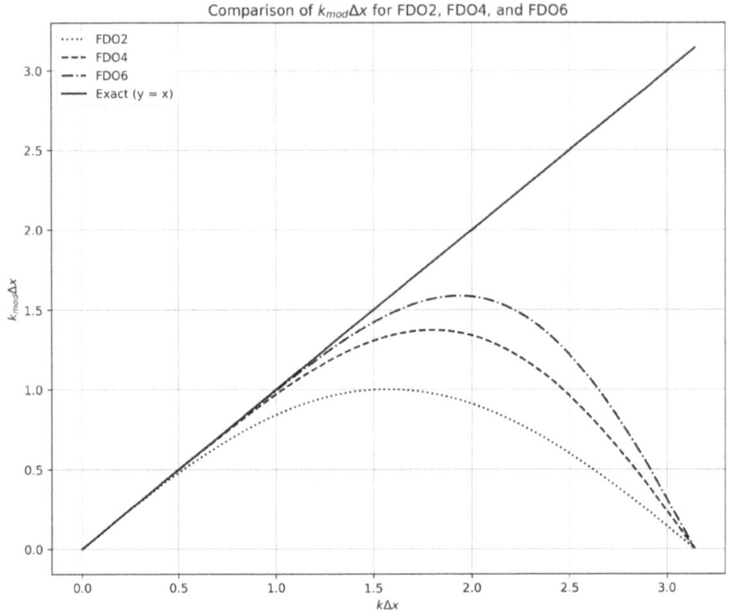

Figure 3.4: **Comparison of FD-O2, O4 and O6 schemes for first derivative.**

The Grid Points Per Wavelength (GPPW) is a critical measure in numerical simulations, especially when discussing the resolution of different finite difference (FD) schemes for accurately resolving wave phenomena. The GPPW indicates how many grid points are used to resolve one wavelength of the wave in the simulation. The choice of GPPW value significantly affects the accuracy and stability of a numerical scheme. It is especially relevant in contexts where wave dispersion and phase errors are concerns, such as in computational fluid dynamics, electromagnetics, and seismic imaging.

Figure 3.4 illustrates a comparative analysis of second-, fourth-, and sixth-order finite difference (FD-O2, FD-O4, and FD-O6) schemes for computing the first derivative of a function. The plot highlights the increasing accuracy with higher-order discretizations, as evidenced by the improved agreement with the analytical derivative and reduced numerical error. Such comparisons are essential in selecting appropriate schemes for simulations requiring a balance between accuracy and computational cost. In the realm of numerical simulations, particularly those involving wave phenomena, understanding the relationship between Grid Points Per Wavelength (GPPW) and the order of finite difference schemes (FDO2, FDO4, FDO6) is crucial.

When k^*_{mod} reaches its maximum for a given $k\Delta x$, it marks the shortest wavelength that the scheme can resolve before dispersion errors rapidly increase. Understanding this cutoff wavenumber provides several key insights: one can determine the minimum grid resolution needed to capture critical physical scales, assess whether a higher-order scheme allows for coarser grids without sacrificing accuracy, estimate numerical phase-speed errors to choose stable time-step sizes, and identify

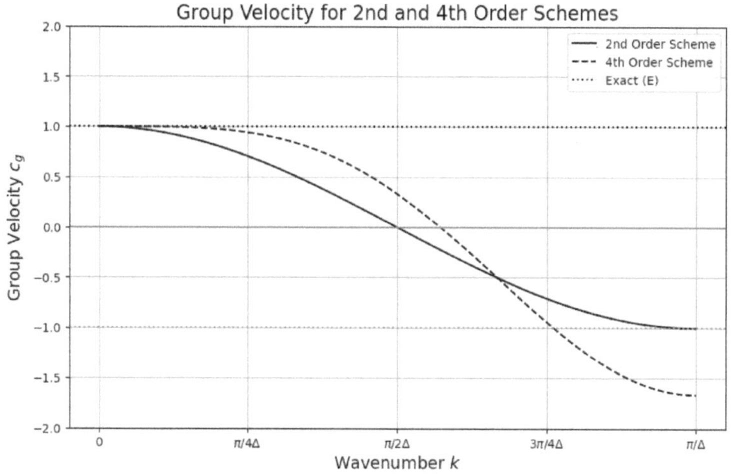

Figure 3.5: **Group velocity c_g for the second-order and fourth-order numerical schemes.**

where subgrid-scale modeling or filtering becomes necessary to avoid aliasing. Collectively, these considerations directly influence simulation accuracy, computational efficiency, and the ability to model complex, multiscale systems effectively.

When implementing these schemes, adjust the grid spacing Δx relative to the wavelength λ of the highest frequency we need to resolve as follows:

$$\Delta x \leq \frac{\lambda}{\text{GPPW}}$$

Adjusting Δx ensures that the wave is sampled adequately to capture its characteristics without introducing significant numerical errors.

Higher-order schemes such as FDO4 and FDO6 enable more efficient wave resolution with fewer grid points per wavelength, critical for reducing computational demands and enhancing the feasibility of simulating large or complex domains. The observed increase in maximum $k_{\text{mod}}\Delta x$ with higher-order schemes suggests improved accuracy and reduced numerical dispersion. These schemes are adept at simulating higher frequencies without aliasing, capturing finer details and sharper gradients of the modeled phenomena. For engineers and scientists, discerning the correlation between GPPW and finite difference orders aids in selecting the most appropriate numerical method. This selection is based on specific needs for accuracy, computational resources, and the physical characteristics of the problem, including the frequency content of the waves involved.

Overall, choosing the correct finite difference order and understanding its implications on GPPW is indispensable for optimizing numerical simulations across a broad spectrum of scientific and engineering applications.

Figure 3.5 shows that the fourth-order scheme generally provides a more accurate estimation of group velocity across a larger range of wavenumbers compared to the second-order scheme, making it preferable for problems involving higher-frequency waves. However, as k approaches the Nyquist limit, both schemes introduce significant dispersion, indicating they are unsuitable for accurately resolving very short-wavelength waves. The short-wavelength waves often have large negative group velocities in some numerical schemes. A negative group velocity implies that the wave energy or information propagates in the opposite direction of the wave vector, causing disturbances to spread away from the source of the short-wavelength waves. In the absence of dissipation (a mechanism that reduces wave amplitude over time, see section 3.2), this effect can cause short-wavelength noise to spread rapidly across the computational domain. Such noise can be problematic because it introduces unwanted oscillations or artifacts that interfere with the accurate representation of the simulated system. Essentially, without dissipation, the numerical scheme might inadvertently amplify and spread this noise, impacting the overall stability and accuracy of the solution.

3.3.1 SECOND ORDER DERIVATIVE

In general, it has been observed that second-derivative finite difference approximations exhibit better accuracy for high wavenumbers than first-derivative formulations.

The wave number for second derivative with FDO2: The second derivative using the central difference formula in a three-point stencil (FD-O2) is given by:

$$\left.\frac{d^2 f}{dx^2}\right|_j \approx \frac{f_{j-1} - 2f_j + f_{j+1}}{h^2}$$

where h is the grid spacing.

Assume a wave-like solution for $f(x)$ of the form:

$$f_j = e^{ikx_j} = e^{ikjh}$$

where $x_j = jh$. Here, k is the true wavenumber.

Substituting the wave solution $f_j = e^{ikjh}$ into the finite difference formula:

$$\left.\frac{d^2 f}{dx^2}\right|_j \approx \frac{e^{ik(j-1)h} - 2e^{ikjh} + e^{ik(j+1)h}}{h^2}$$

Expanding each term:

$$= \frac{e^{-ikh} - 2 + e^{ikh}}{h^2}$$

Using Euler's formula $e^{\pm ikh} = \cos(kh) \pm i\sin(kh)$:

$$\left.\frac{d^2 f}{dx^2}\right|_j = \frac{\cos(kh) - 2 + \cos(kh)}{h^2}$$

$$= \frac{2\cos(kh) - 2}{h^2}$$

Factoring out the 2:

$$= \frac{2(\cos(kh) - 1)}{h^2}$$

Now, introduce the dimensionless wavenumber $k^* = kh$, which means $kh = k^*$. Substitute $kh = k^*$ into the expression:

$$\left.\frac{d^2 f}{dx^2}\right|_j = \frac{2(\cos(k^*) - 1)}{h^2}$$

The true second derivative for the wave function $f(x) = e^{ikx}$ is:

$$\frac{d^2 f}{dx^2} = -k^2 e^{ikx}$$

In the numerical approximation, we aim to represent this as:

$$\left.\frac{d^2 f}{dx^2}\right|_j \approx -k_{\text{mod}}^2 e^{ikx}$$

where k_{mod} is the modified wavenumber.

Comparing coefficients, we have:

$$-k_{\text{mod}}^2 = \frac{2(\cos(k^*) - 1)}{h^2}$$

To remove h from the expression, multiply both sides by h^2:

$$-k_{\text{mod}}^{*2} = 2(\cos(k^*) - 1)$$

Thus, we find:

$$k_{\text{mod}}^{*2}|_{FDO2} = 2 \cdot (1 - \cos(k^*))$$

Using the trigonometric identity $1 - \cos(k^*) = 2\sin^2\left(\frac{k^*}{2}\right)$, we can rewrite k_{mod}^{*2} as:

$$k_{\text{mod}}^{*2} = 4\sin^2\left(\frac{k^*}{2}\right)$$

The wave number for second derivative with FDO4: The fourth-order central difference formula for the second derivative at grid point f_i is:

$$\left.\frac{d^2 f}{dx^2}\right|_i \approx \frac{-f_{i-2} + 16f_{i-1} - 30f_i + 16f_{i+1} - f_{i+2}}{12h^2}$$

where h is the grid spacing.

Assume a wave-like solution for $f(x)$:

$$f_j = e^{ikx_j} = e^{ikjh}$$

where $x_j = jh$ and k is the true wavenumber.

Substitute this solution into the finite difference formula:

$$\left.\frac{d^2f}{dx^2}\right|_i \approx \frac{-e^{ik(x_i-2h)}+16e^{ik(x_i-h)}-30e^{ikx_i}+16e^{ik(x_i+h)}-e^{ik(x_i+2h)}}{12h^2}$$

Since $x_i = ih$, this becomes:

$$\left.\frac{d^2f}{dx^2}\right|_i \approx \frac{-e^{-2ikh}+16e^{-ikh}-30+16e^{ikh}-e^{2ikh}}{12h^2}$$

Using Euler's formula, $e^{\pm ikh} = \cos(kh) \pm i\sin(kh)$, we can rewrite the expression in terms of cosine functions. The imaginary terms will cancel out due to symmetry:

$$\left.\frac{d^2f}{dx^2}\right|_i \approx \frac{1}{12h^2}\left[-2\cos(2kh)+32\cos(kh)-30\right]$$

Define the dimensionless wavenumber $k^* = kh$, so that $kh = k^*$. Substituting $kh = k^*$ into the expression, we get:

$$\left.\frac{d^2f}{dx^2}\right|_i \approx \frac{1}{12h^2}\left[-2\cos(2k^*)+32\cos(k^*)-30\right]$$

The true second derivative of $f(x) = e^{ikx}$ is:

$$\frac{d^2f}{dx^2} = -k^2e^{ikx}$$

In the numerical approximation, we want to represent this as:

$$\left.\frac{d^2f}{dx^2}\right|_i \approx -k_{\text{mod}}^2 e^{ikx}$$

where k_{mod} is the modified wavenumber.

Equating the coefficients, we get:

$$-k_{\text{mod}}^2 = \frac{-2\cos(2k^*)+32\cos(k^*)-30}{12h^2}$$

Multiplying both sides by h^2 to eliminate h^2:

$$-k_{\text{mod}}^{*2} = \frac{-2\cos(2k^*)+32\cos(k^*)-30}{12}$$

Thus:

$$k_{\text{mod}}^{*2}\big|_{FDO4} = \frac{2\cos(2k^*)-32\cos(k^*)+30}{12}$$

The wave number for second derivative with FDO6: We drive now k_{mod}^{*2} for the 6th-order finite difference approximation (FDO6) of the second derivative. For a 6th-order finite difference approximation of the second derivative, the finite difference operator in the Fourier domain is given by:

$$L(kh) = \frac{1}{180h^2}\left[4\cos(3kh)-54\cos(2kh)+540\cos(kh)-490\right]$$

This operator approximates the second derivative, so we can relate it to the modified squared wavenumber:

$$-k_{\text{mod}}^{*2} = L(kh)$$

Thus,

$$k_{\text{mod}}^{*2} = -L(kh)$$

To express $L(kh)$ in terms of $\sin^2\left(\frac{kh}{2}\right)$, we use trigonometric identities to expand each cosine term:

$$\cos(2kh) = 2\cos^2(kh) - 1$$

$$\cos(3kh) = 4\cos^3(kh) - 3\cos(kh)$$

$$\cos(kh) = 1 - 2\sin^2\left(\frac{kh}{2}\right)$$

Define $s = \sin^2\left(\frac{kh}{2}\right)$ to simplify notation.

Substitute $\cos(kh) = 1 - 2s$ into the expressions for $\cos(2kh)$ and $\cos(3kh)$ and then into $L(kh)$.

$$\cos(kh) = 1 - 2s$$

$$\cos^2(kh) = (1-2s)^2 = 1 - 4s + 4s^2$$

$$\cos^3(kh) = (1-2s)^3 = 1 - 6s + 12s^2 - 8s^3$$

Substitute these into $L(kh)$:

$$L(kh) = \frac{1}{180h^2}\left[4\cos(3kh) - 54\cos(2kh) + 540\cos(kh) - 490\right]$$

$$L(kh) = \frac{1}{180h^2}\left[4(1-6s+12s^2-8s^3) - 54(1-4s+4s^2) + 540(1-2s) - 490\right]$$

$$L(kh) = \frac{1}{180h^2}\left[4 - 24s + 48s^2 - 32s^3 - 54 + 216s - 216s^2 + 540 - 1080s - 490\right]$$

$$L(kh) = \frac{1}{180h^2}\left[-720s - 240s^2 - 128s^3\right]$$

Since $k_{\text{mod}}^{*2} = -L(kh)$:

$$k_{\text{mod}}^{*2} = \frac{1}{180h^2}(720s + 240s^2 + 128s^3)$$

Now, simplify the coefficients:

$$k_{\text{mod}}^{*2} = \frac{1}{h^2}\left[\frac{720}{180}s + \frac{240}{180}s^2 + \frac{128}{180}s^3\right]$$

$$k_{\text{mod}}^{*2}\big|_{FDO6} = \frac{1}{h^2}\left[4s + \frac{4}{3}s^2 + \frac{32}{45}s^3\right]$$

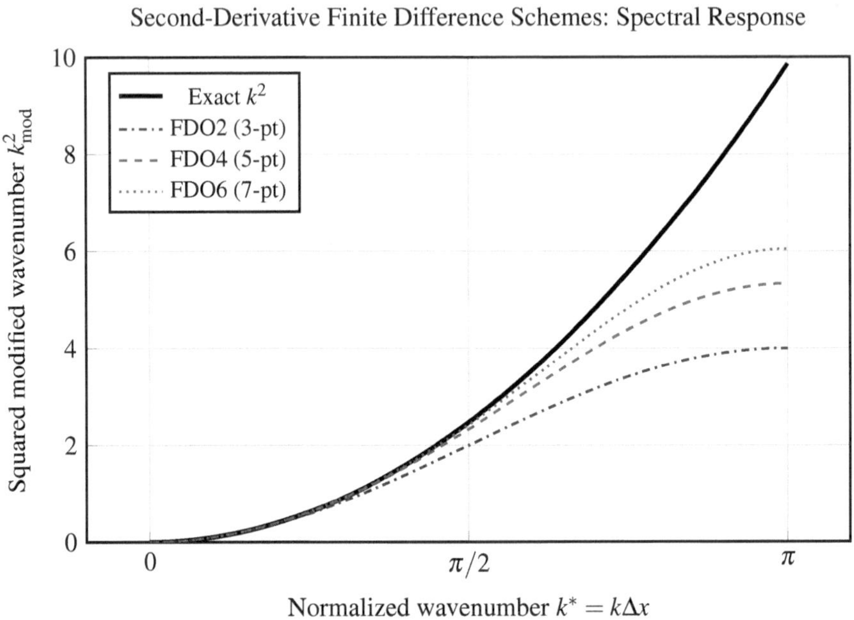

Figure 3.6: **Spectral response of second-derivative finite difference schemes. The plot compares the squared modified wavenumber k^2_{mod} of second-order (FDO2), fourth-order (FDO4), and sixth-order (FDO6) central difference schemes against the exact Fourier response k^2. Higher-order schemes maintain better agreement with the exact solution across more of the wavenumber domain.**

This expression is valuable for analyzing the dispersion properties of the 6th-order finite difference approximation.

Figure 3.6 compares the dispersion characteristics of second-, fourth-, and sixth-order central finite difference schemes by plotting the modified wavenumber k^2_{mod} as a function of the true wavenumber k^*. The closer the curve follows the ideal Fourier line, the more accurately the scheme resolves high-frequency content. The maximum resolvable wavenumber $k^2_{\texttt{mod,max}}$ increases with the scheme's order, demonstrating the effectiveness of higher-order methods in capturing finer flow structures with reduced numerical dispersion.

Table 3.2 presents the maximum dimensionless wavenumber squared k^{*2}_{mod} for several central finite-difference schemes of different orders. The FD-O2, FD-O4, and FD-O6 schemes progressively approach the spectral (Fourier) value of π^2, reflecting improved resolution properties with increasing order. Higher-order schemes yield better approximation to the exact derivative over a broader wavenumber range, thereby enhancing accuracy for shorter wavelengths. This demonstrates the superior spectral fidelity of higher-order finite-difference schemes.

Generalized Compact Finite Difference Formulas: Table 3.3 presents generalized finite-difference schemes for approximating first (Φ') and second (Φ'')

Table 3.2
Maximum dimensionless wave number k^{*2}_{mod} for some central difference finite-difference schemes

	Central FD			
	FD-O2	FD-O4	FD-O6	Fourier
$k^{*2}_{\text{mod, max}}$	4	5.3	6.04	π^2

derivatives using various stencil widths and orders of accuracy. The coefficients a through g define the weights assigned to the function values Φ at symmetric stencil points, while the parameters α, β, and γ describe the implicit coupling between derivatives at adjacent nodes. The normalization factor q scales the denominator to ensure consistency with the grid spacing and desired accuracy.

The general compact finite-difference formula for the first derivative Φ' is:

$$\alpha \Phi'_{i-1} + \beta \Phi'_i + \gamma \Phi'_{i+1} = \frac{1}{q}\left(a\Phi_{i-3} + b\Phi_{i-2} + c\Phi_{i-1} + d\Phi_i + e\Phi_{i+1} + f\Phi_{i+2} + g\Phi_{i+3}\right)$$

The general formula for the second derivative Φ'' scaled by Δ^2:

$$\alpha \Phi''_{i-1} + \beta \Phi''_i + \gamma \Phi''_{i+1} = \frac{1}{q}\left(a\Phi_{i-3} + b\Phi_{i-2} + c\Phi_{i-1} + d\Phi_i + e\Phi_{i+1} + f\Phi_{i+2} + g\Phi_{i+3}\right)$$

3.4 HIGHER ORDER AND COMPACT FINITE DIFFERENCE SCHEMES (KFD)

Classical finite-difference formulas arise by matching successive terms in a Taylor expansion of a smooth function about each node. The approach is transparent yet rapidly widens the stencil as higher-order accuracy is demanded. An alternative is Hermite interpolation, which fits a single polynomial that reproduces both the function and a selected set of its derivatives at the chosen nodes. For one-dimensional uniform grids, the Hermite representation of a function $u(x)$ can be expressed as

$$u(x) = \sum_{k=1}^{n} a_k(x)\,u_k + \sum_{k=1}^{m} b_k(x)\,(u_x)_k,$$

where the weights $a_k(x)$ and $b_k(x)$ depend only on the relative position within the cell. Because derivative information is embedded in the reconstruction, Hermite interpolation delivers the target order with a markedly smaller stencil than a pure Taylor expansion.

Pade'-type compact schemes translate this Hermite idea into a discrete operator for derivatives. A symmetric first-derivative template is

$$\beta u'_{i-2} + \alpha u'_{i-1} + u'_i + \alpha u'_{i+1} + \beta u'_{i+2} = \frac{a}{2\Delta x}(u_{i+1} - u_{i-1}) + \frac{b}{4\Delta x}(u_{i+2} - u_{i-2}) + \frac{c}{6\Delta x}(u_{i+3} - u_{i-3}),$$

with coefficients chosen so that the truncation error begins at $\mathscr{O}(\Delta x^p)$. If $\beta = 0$, the unknown derivatives couple only to their immediate neighbors, producing a tridiagonal system that is solved efficiently with the Thomas algorithm; when $\beta \neq 0$, the matrix is pentadiagonal, and the same banded-solver strategy applies.

Table 3.3
Generalized finite-difference schemes for approximating first (Φ') and second (Φ'') derivatives with uniform grid spacing

Generalized Finite Difference Stencils (First and Second Derivatives)												
α	β	γ	a	b	c	d	e	f	g	q	Order	Scheme Type
	Φ'				-1	1				Δ	O1	Explicit
	Φ'				-1	0	1			2Δ	O2	Explicit
	Φ'			1	-4	3				2Δ	O2	Explicit
	Φ'		-2	9	-18	11				6Δ	O3	Explicit
	Φ'			1	-6	3	2			6Δ	O3	Explicit
$1\Phi'$	$2\Phi'$	0			-5	4	1			2Δ	O3	Implicit
0	$2\Phi'$	$1\Phi'$			-1	-4	5			2Δ	O3	Implicit
	Φ'			1	-8	0	8	-1		12Δ	O4	Explicit
	0	Φ'	3	-16	36	-48	25			12Δ	O4	Implicit
	Φ'		-1	6	-18	10	3			12Δ	O4	Explicit
$1\Phi'$	$4\Phi'$	$1\Phi'$			-6	0	6			2Δ	O4	Implicit
$1\Phi'$	$3\Phi'$	0		-17	9	9	-1			6Δ	O4	Implicit
	Φ'		-1	9	-45	0	45	-9	1	60Δ	O6	Explicit
	Φ''				1	-2	1			Δ^2	O2	Explicit
	Φ''			-1	16	-30	16	-1		$12\Delta^2$	O4	Explicit
	Φ''		-1	4	-5	2				Δ^2	O2	Explicit
	Φ''		-1	4	6	-20	11			$12\Delta^2$	O3	Explicit
	0	Φ''	11	-56	114	-104	35			$12\Delta^2$	O3	Implicit
$1\Phi''$	$11\Phi''$	0			13	-27	15	-1		Δ^2	O3	Implicit
0	$11\Phi''$	$1\Phi''$		-1	15	-27	13			Δ^2	O3	Implicit
$1\Phi''$	$10\Phi''$	$1\Phi''$			12	-24	12			Δ^2	O4	Implicit
	Φ''		2	-27	270	-490	270	-27	2	$180\Delta^2$	O6	Implicit
	$3\Phi''$	Φ'/Δ		-2	27	-54	29			$6\Delta^2$	O3	Implicit
$-\Phi'/\Delta$	$3\Phi''$				29	-54	27	-2		$6\Delta^2$	O3	Implicit
	$18\Phi'' - 66\Phi'/\Delta$		4	-27	108	-85				Δ^2	O3	Implicit
	$18\Phi'' + 66\Phi'/\Delta$					-85	108	-27	4	Δ^2	O3	Implicit

Lele's 1992 study formalized this construction and presented fourth-, sixth-, and eighth-order operators together with matching second-derivative forms, demonstrating near-spectral dispersion properties in turbulent and aeroacoustic tests.

Because every compact stencil can be written as a Pade' rational approximation, the two terms are often used interchangeably. Table 3.4 summarizes the canonical symmetric family; the same matrices serve both the Hermite-interpolation and Pade' viewpoints.

Compact Pade' operators supply high-order accuracy, low dispersion, and minimal storage without the wide stencils of explicit formulas, and their flexibility allows coefficient sets tailored for enhanced modified-wavenumber fidelity.

Compact finite-difference schemes based on Pade' rational approximation provide high spectral resolution with narrow stencils, making them attractive for computational fluid dynamics and aeroacoustics. In the seminal 1992 study by Lele, tridiagonal Pade' operators were derived for fourth-, sixth-, and eighth-order accuracy, showing that coupling the derivative at a node with its two immediate neighbours yields dispersion characteristics rivaling spectral methods while retaining efficiency with the Thomas algorithm. Lele also presented compact second-derivative forms,

Table 3.4
Canonical symmetric Pade' Finite Difference Schemes

Scheme	Stencil Width	Nominal Order
Pade' (1,1)	3 points	4th
Pade' (2,2)	5 points	6th
Pade' (3,3)	7 points	8th
Pade' (4,4)	9 points	9th

stability analysis, and demonstrated advantages in turbulent and aeroacoustic simulations.

Higher Order Accuracy for the First Order Derivative: Lele (1992) provides a unified parametric formulation for the compact finite difference scheme coefficients for the first derivative. This formulation allows us to derive schemes of various formal orders (4th, 6th, 8th, 10th, etc.) by solving for the coefficients using Taylor series expansion constraints.

The general form for first derivative as per compact scheme is:

$$\beta f'_{i-2}+\alpha f'_{i-1}+f'_i+\alpha f'_{i+1}+\beta f'_{i+2}=\frac{a}{2h}(f_{i+1}-f_{i-1})+\frac{b}{4h}(f_{i+2}-f_{i-2})+\frac{c}{6h}(f_{i+3}-f_{i-3})$$

Lele derives a system of algebraic equations to match the Taylor series expansions on both sides of the above equation, leading to constraints like:

For second-order accuracy:

$$a+b+c=1+2\alpha+2\beta$$

For fourth-order:

$$a+2^2b+3^2c=2(\alpha+4\beta)$$

For sixth-order:

$$a+4^2b+9^2c=2(\alpha+16\beta)$$

For eighth-order:

$$a+6^2b+15^2c=2(\alpha+36\beta)$$

These allow a multi-parameter family of compact schemes. By choosing specific values for α, β, or fixing the stencil size (tridiagonal or pentadiagonal), one can solve the above equations to determine the corresponding a,b,c. Thus, the compact schemes stem from a systematic, unified framework that generates a broad class of high-order schemes with spectral-like resolution.

Higher Order Accuracy for the Second Order Derivative: The generalized Compact finite difference approximation for second derivative based on Lele approach is

$$\beta u''_{i-2}+\alpha u''_{i-1}+u''_i+\alpha u''_{i+1}+\beta u''_{i+2}$$
$$=\frac{a}{\Delta x^2}(u_{i+1}-2u_i+u_{i-1})+\frac{b}{4\Delta x^2}(u_{i+2}-2u_i+u_{i-2})+\frac{c}{9\Delta x^2}(u_{i+3}-2u_i+u_{i-3}) \tag{3.15}$$

For higher-order accuracy in approximating the second derivative, the following coefficient sets may be incorporated:

Sixth Order Accuracy

$$\alpha = \frac{2}{11}, \quad \beta = 0, \quad a = \frac{12}{11}, \quad b = \frac{3}{11}, \quad c = 0$$

and

Eighth Order Accuracy

$$\alpha = \frac{344}{1179}, \quad \beta = \frac{38\alpha - 9}{214},$$

$$a = \frac{696 - 1191\alpha}{428}, \quad b = \frac{2454\alpha - 294}{535}, \quad c = \frac{1179\alpha - 344}{2140}$$

Compact finite difference schemes, such as those developed by Lele (1992), are widely used for their high spectral-like accuracy and efficiency in resolving smooth solutions. However, these schemes rely on centered stencils and implicit formulations that become ill-posed or undefined near physical boundaries, where the necessary symmetric points are unavailable. As a result, compact schemes often produce spurious oscillations and loss of accuracy at or near the boundaries in non-periodic domains. It is common practice to employ lower-order one-sided or biased stencils near the boundaries to ensure numerical stability and consistency. These specially designed boundary closures may sacrifice some local accuracy but prevent non-physical behavior and allow seamless integration with the high-order interior scheme. Techniques such as buffer zones, ghost points, or explicit damping are also used to further suppress boundary-induced artifacts. Proper treatment of boundaries remains a critical aspect in the practical application of compact schemes for computational fluid dynamics and wave propagation problems.

Fornberg (1988) developed a systematic algorithm for generating finite-difference stencils of arbitrary order, including one-sided forward and backward formulas for second derivatives.

Compact finite-difference Stencils based on Lele's Approach

Compact finite-difference coefficients for first derivative (interior stencil)

$$\beta f'_{i-2} + \alpha f'_{i-1} + f'_i + \alpha f'_{i+1} + \beta f'_{i+2} = \frac{a}{2h}(f_{i+1} - f_{i-1}) + \frac{b}{4h}(f_{i+2} - f_{i-2}) + \frac{c}{6h}(f_{i+3} - f_{i-3})$$

Scheme	α	β	a	b	c
4th-order	1/4	0	3/2	0	0
6th-order	1/3	0	14/9	1/9	0
8th-order	3/8	0	25/16	1/5	−1/80

Boundary Stencils — First Derivative (2nd-Order Accuracy)

Boundary stencils for 4th order accurate interior scheme:
Left boundary (right-biased):

$$a_0 f_0' + a_1 f_1' = \frac{1}{h}(b_0 f_0 + b_1 f_1 + b_2 f_2)$$

Right boundary (left-biased):

$$a_0 f_n' + a_1 f_{n-1}' = \frac{1}{h}(b_0 f_n + b_1 f_{n-1} + b_2 f_{n-2})$$

Boundary stencils for 6th order accurate interior scheme:
i = 1 (second point from the left) and
i = n - 1 (second point from the right):

$$a_0 f_i' + a_1 f_{i-1}' + a_2 f_{i+1}' = \frac{b}{h}(f_{i+1} - f_{i-1})$$

Point	a_0	a_1	a_2	b_0	b_1	b_2
i = 0 (left boundary)	1	2	–	-5	4	1
i = n (right boundary)	1	2	–	5	-4	-1
i = 1 (second point from the left)	1	0.25	0.25	–	–	– uses $\frac{3}{2}(f_2 - f_0)$
i = n - 1 (second point from the right)	1	0.25	0.25	–	–	– uses $\frac{3}{2}(f_{n-1} - f_{n-3})$

Compact finite-difference coefficients for second derivative (interior stencil)

$$\beta f_{i-2}'' + \alpha f_{i-1}'' + f_i'' + \alpha f_{i+1}'' + \beta f_{i+2}'' = \frac{a}{h^2}(f_{i+1} - 2f_i + f_{i-1}) + \frac{b}{4h^2}(f_{i+2} - 2f_i + f_{i-2}) + \frac{c}{9h^2}(f_{i+3} - 2f_i + f_{i-3})$$

Scheme	α	β	a	b	c
4th-order	1/10	0	6/5	0	0
6th-order	2/11	0	12/11	3/11	0

Boundary Stencils — Second Derivative (Explicit, Non-Periodic)

Boundary Explicit Forward/Backward Stencils: The first and last m points are computed using explicit finite-difference approximations.
Forward-biased stencil (for left boundary):

$$f_i'' = \frac{1}{h^2}\sum_{k=0}^{m} c_k f_{i+k}$$

Backward-biased stencil (for right boundary):

$$f_i'' = \frac{1}{h^2}\sum_{k=0}^{m} c_k f_{i-k}$$

Order	m	Coefficients $[c_0, c_1, c_2, \ldots]$
4th	4	$\left[\frac{35}{12}, -\frac{26}{3}, \frac{19}{2}, -\frac{14}{3}, \frac{11}{12}\right]$
6th	5	$\left[\frac{15}{4}, -\frac{77}{6}, \frac{107}{6}, -13, \frac{61}{12}, -\frac{5}{6}\right]$

These explicit stencils are used at:
$i = 0, 1, \ldots, m-1$ (forward-biased)
$i = n-m, \ldots, n-1$ (backward-biased)

Example 3.4

Fourth-order compact derivative on a uniform periodic grid
Using Lele's symmetric compact finite-difference scheme, derive a fourth-order accurate central approximation for the first derivative $u'(x)$ on a uniform periodic grid. Begin from the implicit five-point template

$$\beta u'_{i-2} + \alpha u'_{i-1} + u'_i + \alpha u'_{i+1} + \beta u'_{i+2} = a\frac{u_{i+1} - u_{i-1}}{2\Delta x} + b\frac{u_{i+2} - u_{i-2}}{4\Delta x} + c\frac{u_{i+3} - u_{i-3}}{6\Delta x}$$

determine the coefficient set α, β, a, b, c that yields fourth-order accuracy, write the resulting interior stencil, and formulate the cyclic tridiagonal system for the unknown derivative vector u', indicating how a rank-2 Sherman–Morrison update gives an $O(N)$ solution.

Solution
Lele's symmetric implicit template

$$\beta u'_{i-2} + \alpha u'_{i-1} + u'_i + \alpha u'_{i+1} + \beta u'_{i+2} = a\frac{u_{i+1} - u_{i-1}}{2\Delta x} + b\frac{u_{i+2} - u_{i-2}}{4\Delta x} + c\frac{u_{i+3} - u_{i-3}}{6\Delta x}$$

yields fourth–order accuracy with

$$\alpha = \frac{1}{4}, \beta = 0, a = \frac{3}{2}, b = 0, c = 0.$$

Interior stencil

$$\frac{1}{4}u'_{i-1} + u'_i + \frac{1}{4}u'_{i+1} = \frac{3}{2}\frac{u_{i+1} - u_{i-1}}{2\Delta x}$$

This gives

$$\mathbf{A}\mathbf{u}' = \mathbf{r},$$

where

$$A = \text{tridiag}(\alpha, 1, \alpha) + \alpha\left(\mathbf{e}_1\mathbf{e}_N^{\mathsf{T}} + \mathbf{e}_N\mathbf{e}_1^{\mathsf{T}}\right),$$

and

$$r_i = a\frac{u_{i+1} - u_{i-1}}{2\Delta x}.$$

A is a tridiagonal matrix with α on the sub- and super-diagonals and 1 on the main diagonal. Plus a rank-2 correction $\alpha(\mathbf{e}_1\mathbf{e}_N^{\mathsf{T}} + \mathbf{e}_N\mathbf{e}_1^{\mathsf{T}})$, where $\mathbf{e}_1$ and

$\mathbf{e}_N$ are the first and N-th standard basis vectors, respectively. This adds α to the $(1,N)$ and $(N,1)$ entries, making the matrix circulant (or periodic). $\mathbf{r}$ is a vector whose i-th component is given by a finite difference approximation: $r_i = a\frac{u_{i+1}-u_{i-1}}{2\Delta x}$, which resembles a central difference approximation for a first derivative.

This system appears in numerical methods for solving differential equations, possibly with periodic boundary conditions. The matrix A is symmetric and has a structured form that can be exploited for efficient solving.

See `Python` code `Ch3-Ex34` for solution of this problem.

The problem involves solving a linear system $A\mathbf{u}' = \mathbf{r}$ at each time step where the derivative is required. The matrix A combines a tridiagonal core with periodic boundary conditions.

Running code with $N = 128$:

```
L2 error            = 2.281e-08
Maximum point error = 3.227e-08
```

is in agreement with the expected $C\Delta x^4$ behaviour and close to machine precision, confirming both correctness and $O(N)$ efficiency.

Example 3.5

Express the general expression for the compact finite-difference scheme for the second derivative based on the sixth-order accuracy (Lele approach).

Solution The finite difference approximation for the second derivative based on Lele's approach (sixth-order accuracy) is given by:

$$\beta u''_{i-2} + \alpha u''_{i-1} + u''_i + \alpha u''_{i+1} + \beta u''_{i+2}$$
$$= \frac{a}{\Delta x^2}(u_{i+1} - 2u_i + u_{i-1}) + \frac{b}{4\Delta x^2}(u_{i+2} - 2u_i + u_{i-2}) + \frac{c}{9\Delta x^2}(u_{i+3} - 2u_i + u_{i-3})$$

For the sixth-order accuracy, the coefficients are:

$$\alpha = \frac{2}{11}, \quad \beta = 0, \quad a = \frac{12}{11}, \quad b = \frac{3}{11}, \quad c = 0$$

Since $\beta = 0$ and $c = 0$, the terms involving u''_{i-2}, u''_{i+2}, u_{i+3}, and u_{i-3} drop out. The expression simplifies to:

$$\frac{2}{11}u''_{i-1} + u''_i + \frac{2}{11}u''_{i+1}$$
$$= \frac{12}{11\Delta x^2}(u_{i+1} - 2u_i + u_{i-1}) + \frac{3}{44\Delta x^2}(u_{i+2} - 2u_i + u_{i-2})$$

This is the sixth-order compact finite difference scheme for the second derivative based on Lele's approach.

3.4.1 MODIFIED WAVENUMBER ANALYSIS FOR COMPACT FINITE DIFFERENCE SCHEMES

Compact finite difference schemes achieve high accuracy by coupling derivatives across grid points via implicit systems. A critical tool for evaluating their performance is modified wavenumber analysis, which compares the numerical approximation to the exact spectral derivative across Fourier modes. This analysis reveals how well the scheme resolves different wavelengths, particularly at high wavenumbers where dispersion errors typically dominate. By deriving the effective wavenumber-dependent amplification factor, we quantify resolution limits and artificial dissipation. This section applies the analysis to the proposed tridiagonal compact scheme, emphasizing the impact of its periodic correction terms on spectral accuracy and demonstrating its superiority over explicit counterparts for smooth problems.

In this section, we will do the modified wavenumber analysis for Lele-type compact finite difference schemes of different order.

The modified wave number for first derivative: To derive the modified wavenumber k_{mod} for an eighth-order compact finite difference (FD) scheme, we start with the general form of the FD equation and proceed through the logical steps to express k_{mod} as a function of the discrete wavenumber k.

The eighth-order central compact finite difference schemes developed by Lele (1992) is

$$\beta u'_{i-2} + \alpha u'_{i-1} + u'_i + \alpha u'_{i+1} + \beta u'_{i+2} = \frac{a}{2\Delta x}(u_{i+1} - u_{i-1}) + \frac{b}{4\Delta x}(u_{i+2} - u_{i-2})$$

where:
u'_i represents the derivative at grid point i,
Δx is the grid spacing,
$\alpha, \beta, a,$ and b are coefficients determined by the desired accuracy order.

For eighth-order accuracy:

$$\alpha = \frac{4}{9}, \beta = \frac{1}{36}, a = \frac{40}{27}, b = \frac{25}{54}$$

To analyze the modified wavenumber, assume a plane wave solution:

$$u_i = e^{ikx_i} = e^{iki\Delta x}$$

Then, the derivative u'_i should ideally be iku_i, but due to the numerical scheme, the actual computed derivative will be scaled by k_{mod}, where k_{mod} is the modified wavenumber.

Left Side (compact FD terms for u'): Substitute $u'_i = ik_{\text{mod}}u_i$ and similarly for shifted indices:

$$\beta e^{-2ik\Delta x} ik_{\text{mod}}u_i + \alpha e^{-ik\Delta x} ik_{\text{mod}}u_i + ik_{\text{mod}}u_i + \alpha e^{ik\Delta x} ik_{\text{mod}}u_i + \beta e^{2ik\Delta x} ik_{\text{mod}}u_i$$

Factor out $ik_{\text{mod}}u_i$:

$$ik_{\text{mod}}u_i \left(\beta e^{-2ik\Delta x} + \alpha e^{-ik\Delta x} + 1 + \alpha e^{ik\Delta x} + \beta e^{2ik\Delta x}\right)$$

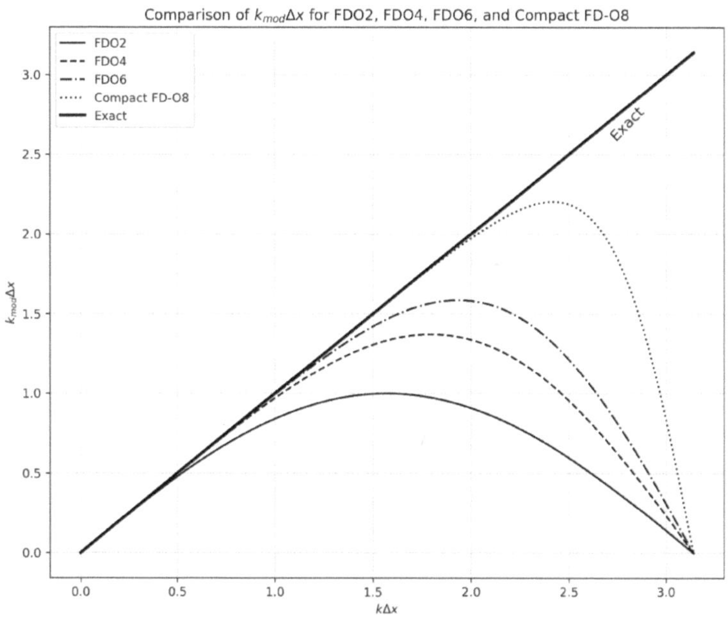

Figure 3.7: **Comparison of Compact Finite difference scheme with regular FD schemes.**

Define $G(k\Delta x) = \beta e^{-2ik\Delta x} + \alpha e^{-ik\Delta x} + 1 + \alpha e^{ik\Delta x} + \beta e^{2ik\Delta x}$.

Right Side (finite differences of u_i): Substitute $u_i = e^{iki\Delta x}$ into the finite difference terms:

$$\frac{a}{2\Delta x}(e^{ik\Delta x} - e^{-ik\Delta x})u_i + \frac{b}{4\Delta x}(e^{2ik\Delta x} - e^{-2ik\Delta x})u_i$$

Simplify using the sine function:

$$u_i\left(\frac{a}{\Delta x}\sin(k\Delta x) + \frac{b}{2\Delta x}\sin(2k\Delta x)\right)$$

Equate the terms from the left and right sides of the equation:

$$ik_{\text{mod}}G(k\Delta x) = i\left(\frac{a}{\Delta x}\sin(k\Delta x) + \frac{b}{2\Delta x}\sin(2k\Delta x)\right)$$

Dividing both sides by $iu_iG(k\Delta x)$, we get:

$$k_{\text{mod}} = \frac{\frac{a}{\Delta x}\sin(k\Delta x) + \frac{b}{2\Delta x}\sin(2k\Delta x)}{G(k\Delta x)}$$

where $G(k\Delta x) = \beta e^{-2ik\Delta x} + \alpha e^{-ik\Delta x} + 1 + \alpha e^{ik\Delta x} + \beta e^{2ik\Delta x}$.

Figure 3.7 presents a comparison of dispersion relations for classical finite difference schemes (FD-O2, FD-O4, FD-O6) and a compact finite difference scheme (FD-O8). It shows the normalized modified wavenumber $k_{\text{mod}}\Delta x$ as a function of the actual wavenumber $k\Delta x$, with the ideal (exact) line representing the continuous Fourier case. The FD-O2 scheme exhibits substantial dispersion error and deviates early from the exact line. The FD-O4 and FD-O6 schemes show progressively better accuracy but still fall short at higher wavenumbers. In contrast, the compact FD-O8 scheme closely follows the exact line over a wide range of $k\Delta x$, demonstrating its superior spectral resolution. This indicates that compact schemes are better suited for capturing high-frequency phenomena with reduced dispersion, making them attractive for high-fidelity simulations in computational fluid dynamics and wave propagation problems.

Figure 3.8 presents the dispersion characteristics of various finite difference schemes by plotting the phase ratio $k^*\Delta x/k\Delta x$ against the normalized wavenumber $k\Delta x$. This ratio reflects the accuracy with which a numerical scheme captures the phase speed of a propagating wave. The exact solution, represented by a horizontal line at unity, denotes perfect phase accuracy across all wavenumbers. The second-order central scheme exhibits significant dispersion error, as the phase ratio declines sharply with increasing wavenumber. The fourth-order scheme improves this behavior but still deviates notably at higher $k\Delta x$ values. In contrast, the sixth-order compact Padé scheme closely follows the exact line over a broader range, demonstrating its superior ability to resolve wave phases accurately. This comparison underscores the advantages of high-order and compact schemes in minimizing dispersion error, making them more suitable for simulations involving wave propagation and long-range accuracy requirements.

In numerical analysis of partial differential equations, the concept of effective wavenumber is essential for understanding the accuracy of finite difference schemes in approximating wave propagation. The highest resolvable wavenumber k on a grid with spacing Δx is given by $\pi/\Delta x$, so the range of interest for k is from 0 to $\pi/\Delta x$. This range reflects the Nyquist limit, beyond which aliasing errors occur, causing high-frequency components to be misrepresented. For finite difference operators, such as the second-order (FD02), fourth-order (FD04), and higher-order schemes, the effective wavenumber k_{mod} differs from the actual wavenumber k, especially as k approaches the Nyquist limit. The Taylor series expansion approximation remains accurate only for $k < \pi/2\Delta x$, the lower half of the wavenumber range, where the modified wavenumber aligns closely with the exact wavenumber. Higher-order finite difference schemes, like FD04 and FD06, improve accuracy over a larger portion of this range but still diverge as k increases. Compact finite difference schemes, such as the Compact FD-08 shown in the plot, exhibit better accuracy near the Nyquist limit, reducing phase errors. However, the choice of scheme affects the nature of the error: while centered schemes approximate k_{mod} with real values, upwind discretizations introduce complex effective wavenumbers, reflecting dissipative behavior.

We now derive the modified wavenumber k_{mod} for the second derivative using the compact finite difference scheme. For a generalized compact finite difference (FD)

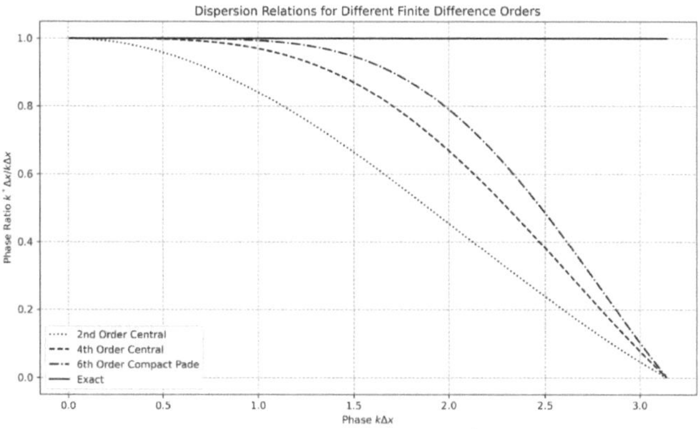

Figure 3.8: **Dispersion relation comparison between different finite difference schemes (2nd order, 4th order, 6th order) and the exact solution.**

approximation of the second derivative, we have:

$$\begin{aligned}\beta u''_{i-2}+\alpha u''_{i-1}+u''_{i}+\alpha u''_{i+1}+\beta u''_{i+2} &= \frac{a}{\Delta x^2}(u_{i+1}-2u_i+u_{i-1}) \\ &\quad +\frac{b}{4\Delta x^2}(u_{i+2}-2u_i+u_{i-2}) \\ &\quad +\frac{c}{9\Delta x^2}(u_{i+3}-2u_i+u_{i-3}).\end{aligned}$$

Assume a harmonic solution of the form $u_i = e^{jki\Delta x}$, where j is the imaginary unit, k is the wavenumber, and i is the index of the grid point.

Then the second derivative u''_i in the harmonic form can be expressed as:

$$u''_i = -k^2_{\text{mod}} u_i.$$

For each term on the left side, substitute the assumed harmonic solution:

$$u''_{i\pm1} = -k^2_{\text{mod}} e^{\pm jk\Delta x} u_i,$$

$$u''_{i\pm2} = -k^2_{\text{mod}} e^{\pm 2jk\Delta x} u_i.$$

Thus, the left side of the equation becomes:

$$\beta(-k^2_{\text{mod}} e^{-2jk\Delta x})+\alpha(-k^2_{\text{mod}} e^{-jk\Delta x})+(-k^2_{\text{mod}})+\alpha(-k^2_{\text{mod}} e^{jk\Delta x})+\beta(-k^2_{\text{mod}} e^{2jk\Delta x}).$$

Factoring out $-k^2_{\text{mod}}$, the left side becomes:

$$-k^2_{\text{mod}}\left(\beta e^{-2jk\Delta x}+\alpha e^{-jk\Delta x}+1+\alpha e^{jk\Delta x}+\beta e^{2jk\Delta x}\right).$$

Define $G(k\Delta x)$ as:

$$G(k\Delta x) = \beta e^{-2jk\Delta x}+\alpha e^{-jk\Delta x}+1+\alpha e^{jk\Delta x}+\beta e^{2jk\Delta x}.$$

Then, the left side becomes:

$$-k_{\text{mod}}^2 G(k\Delta x) u_i.$$

Similarly, for the right side terms:

$$u_{i+1} - 2u_i + u_{i-1} = (e^{jk\Delta x} - 2 + e^{-jk\Delta x}) u_i,$$

$$u_{i+2} - 2u_i + u_{i-2} = (e^{2jk\Delta x} - 2 + e^{-2jk\Delta x}) u_i,$$

$$u_{i+3} - 2u_i + u_{i-3} = (e^{3jk\Delta x} - 2 + e^{-3jk\Delta x}) u_i.$$

Thus, the right side becomes:

$$\frac{a}{\Delta x^2}(e^{jk\Delta x} - 2 + e^{-jk\Delta x}) + \frac{b}{4\Delta x^2}(e^{2jk\Delta x} - 2 + e^{-2jk\Delta x}) + \frac{c}{9\Delta x^2}(e^{3jk\Delta x} - 2 + e^{-3jk\Delta x}).$$

Define the right side term $R(k\Delta x)$ as:

$$\begin{aligned} R(k\Delta x) &= \frac{a}{\Delta x^2}(e^{jk\Delta x} - 2 + e^{-jk\Delta x}) + \frac{b}{4\Delta x^2}(e^{2jk\Delta x} - 2 + e^{-2jk\Delta x}) \\ &\quad + \frac{c}{9\Delta x^2}(e^{3jk\Delta x} - 2 + e^{-3jk\Delta x}). \end{aligned}$$

Since the harmonic solution holds for all u_i, we can equate the left and right sides:

$$-k_{\text{mod}}^2 G(k\Delta x) = R(k\Delta x).$$

Thus, the expression for the modified wavenumber k_{mod} becomes:

$$k_{\text{mod}}^2 = -\frac{R(k\Delta x)}{G(k\Delta x)}.$$

where:

$$G(k\Delta x) = \beta e^{-2jk\Delta x} + \alpha e^{-jk\Delta x} + 1 + \alpha e^{jk\Delta x} + \beta e^{2jk\Delta x},$$

$$\begin{aligned} R(k\Delta x) &= \frac{a}{\Delta x^2}(e^{jk\Delta x} - 2 + e^{-jk\Delta x}) + \frac{b}{4\Delta x^2}(e^{2jk\Delta x} - 2 + e^{-2jk\Delta x}) \\ &\quad + \frac{c}{9\Delta x^2}(e^{3jk\Delta x} - 2 + e^{-3jk\Delta x}). \end{aligned}$$

Table 3.5

Maximum dimensionless wave numbers for simple finite-difference and Lele-type compact finite difference schemes

	FD-O2	FD-O4	FD-O6	KFD-O4	KFD-O6	Fourier
$k^*_{mod,max}$	1	1.37	1.59	1.73	1.99	π
$k^*(k^*_{mod,max})$	1.57	1.79	1.93	2.10	2.24	π

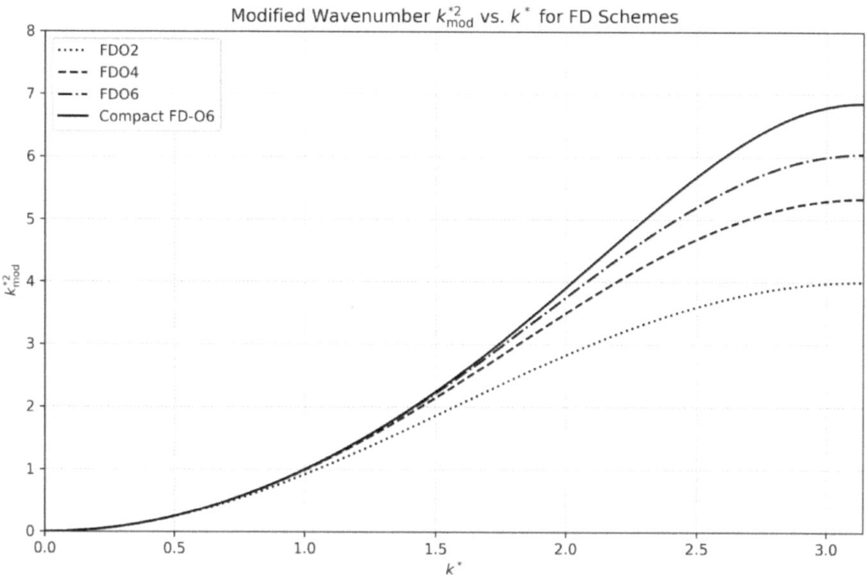

Figure 3.9: k^2_{mod} **for various KFD schemes.**

3.4.2 PENTADIAGONAL COMPACT FINITE-DIFFERENCE SCHEME

Building on Lele's foundational work, Kim and Lee in 1996 introduced an optimised family of compact schemes that preserved the narrow-band structure but tuned the free coefficients by minimising integrated phase error over the resolvable wavenumber band. Their eighth-order variant keeps $\beta \neq 0$ yet avoids the longer ± 3 stencil of Lele, delivering superior resolution on coarse meshes and smoother error near the Nyquist limit. Comparative studies show that Lele operators maximise formal accuracy per node, whereas Kim and Lee operators maximise practical spectral accuracy, so practitioners choose according to conservation requirements, resolution needs, and allowable computational overhead. Both schemes fit within the general Pade' template and extend naturally to curvilinear grids.

The optimized 8th-order pentadiagonal compact finite-difference scheme developed by Kim and Lee (1996) offers a high-resolution method for computing first derivatives, particularly well-suited for direct numerical simulations (DNS) and wave propagation problems:

$$\beta f'_{i-2} + \alpha f'_{i-1} + f'_i + \alpha f'_{i+1} + \beta f'_{i+2}$$
$$= \frac{a}{2\Delta x}(f_{i+1} - f_{i-1}) + \frac{b}{4\Delta x}(f_{i+2} - f_{i-2}) + \frac{c}{6\Delta x}(f_{i+3} - f_{i-3})$$

This scheme (often called an OHOC – Optimized High-Order Compact scheme) has different coefficient values, because it prioritizes minimizing dispersion error up to the Nyquist limit. The scheme employs a seven-point stencil and constructs a pentadiagonal system that implicitly relates the derivatives at five consecutive nodes. Its general form includes weighted combinations of function values and their derivatives, with coefficients optimized to minimize dispersive (phase) errors in the wavenumber domain. This optimization is performed analytically using a weighted integral of the phase error over a restricted spectral range, typically up to 90% of the Nyquist limit. The resulting coefficients:

$$a = \frac{379}{276}, b = \frac{575}{706}, c = \frac{7}{419}, \alpha = \frac{173}{322}, \beta = \frac{59}{909}$$

yield significantly improved resolution characteristics when compared to standard schemes. In contrast, Lele's (1992) widely-used 4th-order pentadiagonal compact scheme is designed to provide spectral-like resolution by matching the numerical dispersion relation to the exact one across the full wavenumber domain. While Lele's approach delivers smoother global accuracy, Kim and Lee's scheme offers superior phase accuracy within a practical and physically relevant wavenumber band. As a result, the optimized 8th-order scheme proves more effective in capturing fine-scale features in simulations where minimal numerical dispersion is critical. This makes it particularly advantageous in computational aeroacoustics and DNS of turbulence, where both high order and targeted resolution are essential.

The modified wavenumber is computed using the scheme's coefficients (Kim and Lee, 1996) with the formula:

$$k_{\text{mod}}(k) = \frac{a \sin k + \frac{b}{2} \sin 2k + \frac{c}{3} \sin 3k}{1 + 2\alpha \cos k + 2\beta \cos 2k}$$

matches the exact k with relative error below 0.5% up to $k_f \approx 0.79\pi$. The corresponding resolving efficiency is $e = 0.7896$; thus, a mere three grid points per wavelength suffice for spectral-like accuracy.

We define resolving efficiency ε of 0.005 for 0.5% error, and thus the relative error is

$$\text{relative\_error}(k) = \frac{|k_{\text{mod}}(k) - k|}{k}.$$

The critical wavenumber k_f is the point where this error first exceeds ε:

$$k_f = \min\{k \mid \text{relative\_error}(k) > \varepsilon\}.$$

The resolving efficiency $e(\varepsilon)$ and minimum grid points per wavelength ($GPPW$) are derived as

$$e(\varepsilon) = \frac{k_f}{\pi}, \quad GPPW = \frac{2}{e(\varepsilon)}.$$

We observe from Figure 3.10 that for the eighth-order scheme,

$$k_f \approx 2.4805 \approx 0.7896\pi,$$

$$e(\varepsilon) = 0.7896, \quad GPPW = \frac{2}{0.7896} = 2.53 \approx 3.$$

Therefore, only three grid points are required for spectral-like resolution, allowing shorter wavelengths to be resolved with fewer points.

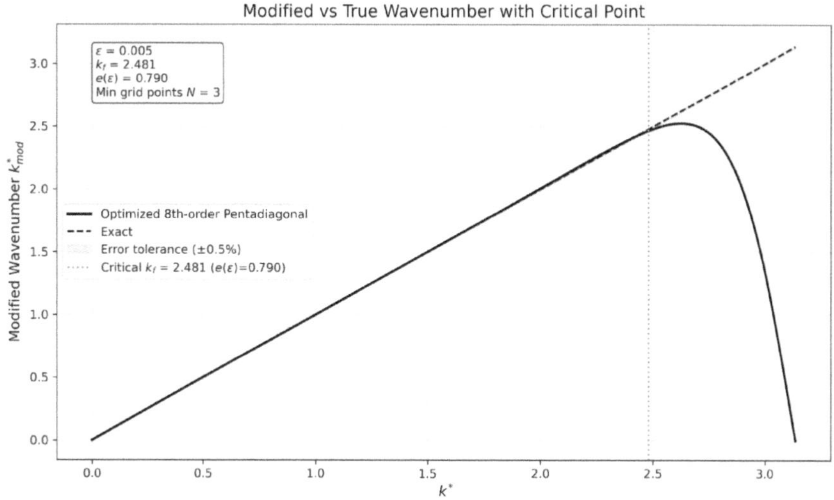

Figure 3.10: **Optimized eighth-order pentadiagonal compact scheme for first derivative.**

This analysis demonstrates how high-order optimized schemes, though more complex to implement, substantially reduce the number of grid points required to achieve the same resolution as lower-order schemes. The eighth-order compact scheme, optimized by Kim and Lee (1996), achieves a higher resolving efficiency by minimizing phase errors over a targeted spectral range, allowing it to resolve shorter wavelengths accurately. This advantage becomes especially significant in simulations involving wave propagation, turbulence, and acoustics, where phase accuracy directly impacts solution fidelity.

Example 3.6

Write the eighth-order compact finite difference scheme based on optimized 8th-order compact finite difference scheme developed by Kim and Lee (1996) for the first derivative based on the coefficients:

$$\alpha = \frac{4}{9}, \quad \beta = \frac{1}{36}, \quad a = \frac{40}{27}, \quad b = \frac{25}{54}, \quad c = 0$$

Solution

Given:

$$\alpha = \frac{4}{9}, \quad \beta = \frac{1}{36}, \quad a = \frac{40}{27}, \quad b = \frac{25}{54}, \quad c = 0$$

Substitute these into the general compact finite difference scheme:

$$\frac{1}{36}u'_{i-2}+\frac{4}{9}u'_{i-1}+u'_i+\frac{4}{9}u'_{i+1}+\frac{1}{36}u'_{i+2}=\frac{40}{27}\frac{u_{i+1}-u_{i-1}}{2\Delta x}+\frac{25}{54}\frac{u_{i+2}-u_{i-2}}{4\Delta x}$$

This equation represents the eighth-order compact finite difference scheme for approximating the first derivative u' using central differences and the given coefficients.

3.5 FINITE DIFFERENCE ON NON-UNIFORM GRIDS

In many practical problems the spatial resolution requirements vary across the domain, making non-uniform grids an attractive choice for finite-difference discretization. By clustering grid points in regions where solution gradients are steep—such as mixing layers or boundary layers, one can achieve higher accuracy without a prohibitive increase in overall grid size.

In this section, we introduce the construction of non-uniform meshes via hyperbolic-tangent stretching and demonstrate how standard central-difference formulas for first and second derivatives must be adapted when the grid spacing is not constant. We then apply these ideas first to the hyperbolic-tangent mixing-layer profile and subsequently to the one-seventh power-law velocity distribution in a channel with coordinate transformation, illustrating how point clustering near walls enhances the fidelity of numerical approximations.

Mixing Layers: Mixing layers form when two parallel streams with different velocities come into contact, producing a thin region of intense shear where vorticity is generated and fluid from each stream is entrained into the other. The velocity profile across such a layer is often well represented by a hyperbolic-tangent function, whose steep gradients demand high resolution near the layer center.

Employing a non-uniform grid that clusters points where $\left|\frac{du}{dy}\right|$ is largest allows accurate computation of derivatives without an impractically fine mesh everywhere. In this section we examine the hyperbolic-tangent mixing-layer profile, derive its analytical first and second derivatives, and then show how to apply central-difference formulas on a non-uniform grid to approximate those derivatives with high fidelity.

The function u based on the hyperbolic tangent, for $y<0$:

$$u(y)=-1+\frac{\tanh(\alpha(y+1))}{\tanh(\alpha)}$$

and for $y\geq 0$:

$$u(y)=1+\frac{\tanh(\alpha(y-1))}{\tanh(\alpha)}$$

The values of α=3 can be set as a first choice to have a balanced trade-off between resolving steep gradients and maintaining smooth grid spacing.

Flow between walls: In simulating fluid flow between two parallel walls of a planar channel, a coordinate transformation is employed to enhance resolution near the boundaries $x = \pm 1$. This transformation strategically clusters computational points close to these walls to effectively capture the dynamics of thin vorticity layers. The transformation is defined as:

$$x = \frac{\tanh(a(\xi - 0.5))}{\tanh\left(\frac{a}{2}\right)}$$

Choosing $a = 3$ generally achieves an optimal balance between accurately capturing the thin vorticity layers and preserving smoothly varying grid spacing. Should gradients remain under-resolved, increase a to 6; conversely, if the clustering becomes excessively tight—resulting in unnecessarily small cells—reduce a to 1.

Metric-Based Finite Difference Formulation on Stretched Grids: In practical CFD applications, non-uniform grids are often used to cluster points in regions of sharp gradients, such as near walls or mixing layers. To maintain second-order accuracy in such cases, the physical coordinate x can be mapped to a uniformly spaced computational coordinate ξ using a transformation $x = f(\xi)$. Derivatives of any function $g(x)$ then transform as:

$$\frac{\partial g}{\partial x} = \frac{1}{x_\xi}\frac{\partial g}{\partial \xi}, \quad \frac{\partial^2 g}{\partial x^2} = \frac{1}{x_\xi}\left(\frac{1}{x_\xi}\frac{\partial g}{\partial \xi}\right)_\xi$$

On a uniform computational grid ξ_i with spacing $\Delta\xi = \frac{1}{n_x - 1}$, the first derivative becomes:

$$g_x = \frac{1}{x_\xi} \cdot \frac{g_{i+1} - g_{i-1}}{2\Delta\xi}, \quad \text{with} \quad x_\xi = \frac{x_{i+1} - x_{i-1}}{2\Delta\xi}$$

The second derivative is written as:

$$g_{xx} = \frac{1}{x_\xi \Delta\xi}\left(\frac{1}{x_\xi^+} g_\xi^+ - \frac{1}{x_\xi^-} g_\xi^-\right)$$

with:

$$g_\xi^+ = \frac{g_{i+1} - g_i}{\Delta\xi}, \quad g_\xi^- = \frac{g_i - g_{i-1}}{\Delta\xi}, \quad x_\xi^+ = \frac{x_{i+1} - x_i}{\Delta\xi}, \quad x_\xi^- = \frac{x_i - x_{i-1}}{\Delta\xi}$$

This leads to a second-order accurate nodal expression:

$$\boxed{g_{xx} = \frac{1}{x_\xi \Delta\xi^2}\left(\frac{1}{x_\xi^+} g_{i+1} - \left(\frac{1}{x_\xi^+} + \frac{1}{x_\xi^-}\right) g_i + \frac{1}{x_\xi^-} g_{i-1}\right)}$$

This formulation applies to any scalar function on a non-uniform structured mesh.

For convective terms such as $(U_n u)_x$, a discrete formulation that ensures both symmetry and kinetic energy conservation is given by:

$$(U_n u)_x = \frac{1}{x_\xi}\left(\frac{U_i u_{i+1} - U_{i-1}u_{i-1}}{2\Delta\xi}\right)$$

This second-order scheme avoids spurious growth of discrete kinetic energy and is widely used in high-fidelity simulations such as DNS and LES of incompressible flows on body-fitted or stretched meshes.

Remark: On a uniform grid where $x_\xi = 1$ and $\Delta\xi = \Delta x$, the transformed expressions reduce to classical central difference formulas. The transformation framework thus extends finite difference methods naturally to stretched and curvilinear grids.

At the domain boundaries (i.e., $i = 0$ and $i = N-1$), the central difference formulas become undefined due to the absence of symmetric neighboring points. In such cases, second-order accurate one-sided finite difference approximations must be used instead. These should be combined with the appropriate metric terms to maintain consistency with the coordinate transformation. For instance, the second derivative at the left boundary can be approximated using a forward-difference formulation scaled by local x_ξ values, while a backward-difference form is applied at the right boundary. This ensures stable and accurate evaluation of derivatives even on non-uniform meshes.

PROBLEMS

3P-1 Following data is obtained from the experimental data.

y in mm	u in m/s
0	0
2.54	45.892
5.08	87.375
7.62	124.984

Find velocity gradient at the wall using:

Explicit O2 (Forward)

Explicit O4 (Forward)

Simple Forward Difference (O1)

3P-2 The third-order accurate finite difference formula for a derivative is

$$\left(\frac{\partial u}{\partial y}\right)_{j=1} = \frac{-11u_{j=1} + 18u_{j=2} - 9u_{j=3} + 2u_{j=4}}{6\Delta y}$$

Demonstrate that the equation derived from the polynomial approach is consistent with a Taylor series expansion.

3P-3 Develop the second order accurate derivative:

$$\left(\frac{\partial u}{\partial x}\right)_i = \frac{-3u_i + 4u_{i+1} - u_{i+2}}{2\Delta x}$$

Demonstrate that the equation derived from the polynomial approach is consistent with a Taylor series expansion.

3P-4 Consider the fourth-order central difference approximation for the second derivative, given by

$$\left(\frac{\partial^2 u}{\partial x^2}\right)_x = \frac{-u_{x+2}+16u_{x+1}-30u_x+16u_{x-1}-u_{x-2}}{12(\Delta x)^2} + \mathscr{O}(\Delta x^4),$$

which is commonly used to discretize spatial derivatives in wave equations or similar linear PDEs. This leads to the finite difference scheme

$$\tilde{u}_{x+2} - 16\tilde{u}_{x+1} + \left(30 - \frac{12\omega^2(\Delta x)^2}{a^2}\right)\tilde{u}_x - 16\tilde{u}_{x-1} + \tilde{u}_{x-2} = 0,$$

a fourth-order accurate difference equation. Given only two physical boundary conditions, such as $\tilde{u}_0 = 0$, $\tilde{u}_M = 0$, analyze the implications of this higher-order scheme: determine the nature and number of admissible solutions, discuss the potential for nonphysical or spurious modes, and explain why additional boundary conditions may be necessary. Identify the challenges involved in ensuring a unique and physically meaningful solution, and propose strategies to address them. Furthermore, demonstrate that this finite difference scheme not only retains the exact eigenfunction structure of the continuous operator but also provides a substantially more accurate approximation of eigenfrequencies compared to second-order methods. To this end, assume an eigenfunction of the form

$$\tilde{u}_x = \sin\left(\frac{n\pi x}{M}\right), \quad \text{where } \theta = \frac{n\pi}{M}.$$

3P-5 Sengupta and Dey (2004) devised an optimized high-accuracy upwind compact scheme. Its interior stencil is given by

$$p_{j-1}u'_{j-1} + u'_j + p_{j+1}u'_{j+1} = \frac{1}{h}\sum_{n=-2}^{2} q_n u_{j+n} \quad (3.62)$$

where

$$p_{j\pm1} = D \pm \frac{\eta_*}{60}, \quad q_{\pm2} = \pm\frac{F}{4} + \frac{\eta_*}{300}, \quad q_{\pm1} = \pm\frac{E}{2} + \frac{\eta_*}{30}, \quad q_0 = -\frac{11\eta_*}{150}.$$

The coefficients are designated as $D = 0.3793894912$, $E = 1.5755379$, and $F = 0.183205192$. The term η_* acts as the upwind parameter, and setting $\eta_* = 0$ converts the scheme to a central type. These parameters D, E, and F are specifically optimized to enhance resolution in the spectral domain. Plot k*/k vs k.Δx, for η^*=0,1,3,5. Comment if the ratio k^*/k exceeds 1 on what is it indicating.

3P-6 In numerical simulations of wave phenomena, aliasing can occur when high-frequency waves are incorrectly represented as lower-frequency waves due to insufficient resolution on the grid. Consider a one-dimensional grid with spacing Δx, and a wave described by the function $u(x) = \sin(kx)$, where k is the wave number. The shortest wavelength that can be represented on this grid is $2\Delta x$, corresponding to the wave number $k_{\max} = \frac{\pi}{\Delta x}$. In a simulation, if a wave with a wave number larger than $k_{\max}$ is generated, it will lead to aliasing. Specifically, when the advection term $u\frac{\partial u}{\partial x}$ is applied to the wave, the derivative of $\sin(kx)$ generates a new wave with a wave number of $2k_{\max}$, corresponding to a wavelength shorter than $2\Delta x$. This new wave is too small to be represented on the grid and will be aliased to a lower-frequency wave. Explain how aliasing occurs in this case and the consequences it has on the accuracy and stability of the simulation. Discuss the role of the numerical period and how the aliasing of high-frequency waves results in spurious solutions in the simulation.

3P-7 In numerical simulations, waves shorter than $2\Delta x$ can be generated due to nonlinear interactions, even when the initial waves have wavelengths greater than or equal to $2\Delta x$. Consider a function $u = \sum u_n$ that contains harmonic components, and a nonlinear interaction term of the form $\sin(k_1 x)\sin(k_2 x)$. Show that this interaction generates two new waves with wave numbers $k_1 \pm k_2$. Also, explain the aliasing phenomenon when the wave number k exceeds the maximum resolvable wave number $k_{\max} = \frac{\pi}{\Delta x}$, and how this results in the misrepresentation of high-frequency waves as lower-frequency ones. Discuss how the wave number k gets aliased to $k^* = 2k_{\max} - k$, and explain the implications for the numerical solution, including energy buildup and spurious wave generation.

3P-8 Consider a situation in which a physical quantity u depends on a spatial coordinate x that is a function of a computational coordinate ξ, so that

$$u = u\big(x(\xi)\big).$$

By successively applying the chain rule and the product rule, one can show that the second derivative of u with respect to x transforms according to

$$\frac{\partial^2 u}{\partial x^2} = \frac{\frac{\partial^2 u}{\partial \xi^2}\left(\frac{d\xi}{dx}\right)^2 - \frac{\partial u}{\partial \xi}\frac{d^2 x}{d\xi^2}}{\left(\frac{dx}{d\xi}\right)^3}.$$

i Derive an expression for $\frac{\partial u}{\partial x}$ in terms of $\frac{\partial u}{\partial \xi}$ and $\frac{d\xi}{dx}$.

ii Apply the product rule to $\frac{d}{dx}\left(\frac{\partial u}{\partial \xi}\frac{d\xi}{dx}\right)$, transforming all x-derivatives into ξ-derivatives.

iii Combine terms to obtain the final formula for $\frac{\partial^2 u}{\partial x^2}$ under the change of variables.

3P-9 Consider the non-uniform grid defined by

$$y_i = \frac{\tanh(\alpha x_i)}{\tanh(\alpha)},$$

where $\alpha = 2$.

i Derive expressions for the first and second derivatives using the central-difference approximations.

ii Evaluate u'_{11} and u''_{11} by the formula derived in [i], where 11 are number of grid points.

iii Discuss how the choice of stretching parameter $\alpha = 2$ influences the accuracy of the derivative approximations near $y = 0$, and suggest whether increasing α would improve or degrade the results.

3P-10 Consider one-seventh power-law velocity profile in a planar channel of half-height $R = 1$ and maximum centreline velocity $u_{\max} = 2$ is discretized on uniform grids of size $N = 50, 100, 1000, 10000$:

$$u(x) = u_{\max}\left(1 - |x/R|^{1/7}\right),$$

i Derive the expressions for $\frac{du}{dx}$ and $\frac{d^2u}{dx^2}$.

ii Write central-difference formulas for the first and second derivatives on a uniform grid.

iii For $N = 100$, compute numerically d^2u/dx^2 at the grid point nearest $x = 0.5$ and evaluate its percentage error against the analytical value.

iv Comment on how increasing N affects the error and whether there is any indication of an optimal grid size beyond which further refinement yields diminishing returns.

3P-11 Consider function:

$$u(x) = -\sin(x) + \sin^2(2x)\cos(x)$$

i Derive expression for first derivative using compact KFD-O4 scheme.

ii On $0 \leq x \leq 2\pi$, compute numerical u'_i for grid points N=32, 64, 128, 256, 512, 1024, 5000.

iii Evaluate the L_2-error $||E||_2 = \sqrt{\frac{1}{N}\Sigma(u'_i - u_x(x_i))^2}$ and tabulate versus Δx.

iv Discuss which scheme best balances accuracy and computational cost.

3P-12 Consider the smooth function $f(x) = \tanh(x)$, defined over the domain x in [-3, 3] discretized uniformly using different grid points. The first and second derivatives of this function are known analytically:

$$f'(x) = \frac{1}{\cosh^2(x)}, \quad f''(x) = -2\tanh(x) \cdot \frac{1}{\cosh^2(x)}.$$

Compute the first and second derivatives using the second-order central difference scheme (FDO2), and the sixth-order compact finite-difference scheme (KFD-O6) with the following interior stencil for first and second derivative.

i Experiment with grid points 5,10, 50 and estimate the absolute errors for both schemes at all interior points x_i, excluding boundary points.

ii Which scheme gives better accuracy? Justify answer based on the plotted results and the error behavior.

3P-13 Compare the grid resolution requirements of two numerical schemes—the standard second-order central difference (FD-O2) scheme and the optimized eighth-order pentadiagonal compact scheme (Kim and Lee, 1996) for resolving a wave with a phase error tolerance of $\varepsilon = 0.005$. Use the resolving efficiency $e(\varepsilon)$, defined as the fraction of the wavenumber spectrum accurately resolved within the specified tolerance, to compute the critical wavenumber $k_f = e(\varepsilon) \cdot \pi$, where $k = \omega \Delta x$, and the minimum number of grid points per wavelength $N = \lceil 2/e(\varepsilon) \rceil$. The error tolerance is defined as $\varepsilon = |k_{\text{mod}}(k) - k|/k$, representing the relative dispersion error. Using following information, determine the grid resolution requirements.

Scheme	**Error Tolerance** ε	**Resolving Efficiency** $e(\varepsilon)$
FD-O2 (Second-order central)	0.005	0.055
Optimized 8th-order pentadiagonal	0.005	0.788

3P-14 Consider a flow adjacent to a solid wall where velocity data is available only within the flow field. Using a second-degree polynomial representation of the function $u(y)$ near the wall, derive a second-order accurate forward finite difference expression for the first derivative $\left(\frac{\partial u}{\partial y}\right)_1$ at the wall (point 1). Use three equidistant points located at $y = 0$, $y = \Delta y$, and $y = 2\Delta y$, as shown in Figure 3.11.

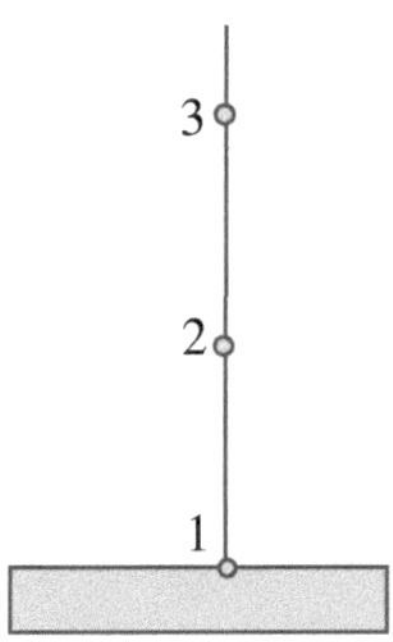

Figure 3.11: Equidistant grid points with Δy spacing.

REFERENCES

M. B. Abbott, D. R. Basco, Computational fluid dynamics. Longman Scientific and Technical, New York, USA, 1989.

L. Collatz, The Numerical Treatment of Differential Equations, Springer-Verlag, 1966.

F. B. Hildebrand, Introduction to Numerical Analysis. New York: McGraw-Hill, 1956.

S. K. Lele, Compact finite difference schemes with spectral-like resolution. J. Comp. Phys., 183, 16-42, 1992.

C. W. Hirt, Heuristic Stability Theory for Finite-Difference Equations, Journal of Computational Physics, 1968.

T. K. Sengupta, S. Dey, Proper orthogonal decomposition of direct numerical simulation data of by-pass transition. Comput. Struct. 82, 2693-2703, 2004.

T.K. Sengupta; G. Ganeriwal; S. De. Analysis of central and upwind compact schemes. , 192(2), 677-694, 2003

K. A. Hoffmann, S. T. Chiang, Computational Fluid Dynamics, Vol. 1, Engineering Education System, Wichita, Kansas, 1998.

P. Orlandi, Fluid Flow Phenomena: A Numerical Toolkit, Springer Science+Business Media, originally published by Kluwer Academic Publishers, 2000.

J. W. Kim, D. J. Lee, Optimized Compact Finite Difference Schemes with Maximum Resolution, AIAA Journal, Vol. 34, No. 5, pp. 887-893, 1996.

B. Fornberg, Generation of finite difference formulas on arbitrarily spaced grids, Mathematics of Computation, 51(184):699-706, 1988

K. Mahesh, G. Constantinescu, P. Moin, A numerical method for largeeddy simulation in complex geometries. Journal of Computational Physics, 197, 215-240, 2004.

4 Time-Stepping Techniques

In this chapter, we will explore time integration techniques for solving ordinary differential equations (ODEs), focusing on methods such as Euler's explicit and implicit methods, as well as the Runge-Kutta family. Understanding these methods is essential for solving both simple and stiff ODE systems, which appear in numerous fluid dynamics applications. After finishing this chapter, you should be able to:

- Understand the fundamental principles of time integration techniques for ODEs.
- Learn the key concepts and differences between explicit and implicit time-stepping methods.
- Develop an understanding of the Runge-Kutta methods and their applications for solving ODEs.
- Be familiar with the advantages and limitations of the Euler and Runge-Kutta methods in both stiff and non-stiff problems.
- Analyze the stability and accuracy of different time integration methods for ODEs.

Initial Value Problems (IVPs) and Boundary Value Problems (BVPs) are two fundamental classes of problems that arise in the mathematical modeling of physical systems, including fluid dynamics, heat transfer, and structural analysis. Understanding the distinction between these two problem types is essential for selecting appropriate solution methods, whether analytical or numerical. An Initial Value Problem (IVP) is characterized by a system of differential equations where the solution is determined by initial conditions provided at a specific point in time. These problems typically govern time-dependent phenomena, such as the evolution of a fluid flow over time or the motion of a particle under a force. The solution progresses from the initial state using time-stepping methods, such as Runge-Kutta or Euler's method. On the other hand, a Boundary Value Problem (BVP) requires the solution to satisfy conditions at multiple points in space, often at the boundaries of the domain. These problems commonly arise in steady-state systems, where the distribution of temperature, pressure, or velocity is governed by spatial conditions. Solving BVPs involves finding a solution that satisfies the differential equation throughout the domain while meeting the boundary conditions at the edges.

In the study of numerical methods for solving differential equations, time integration techniques are crucial for understanding how a system evolves over time. While partial differential equations (PDEs) are commonly used to model complex fluid dynamics, simpler ordinary differential equations (ODEs) often arise in idealized or reduced systems, providing insight into key behaviors such as chemical

DOI: 10.1201/9781003663171-4

reactions or trajectories of individual fluid parcels. These simpler equations, though more straightforward, serve as valuable stepping stones for understanding the time-stepping methods needed for more complex systems.

This chapter introduces fundamental time integration techniques such as Euler's explicit and implicit methods and the more sophisticated Runge-Kutta methods. These methods form the foundation of many numerical solvers, including those used for solving the Navier-Stokes equations, and understanding them helps in selecting the right approach for different types of ODE problems. Euler's explicit method, known for its simplicity, is particularly useful for problems where computational efficiency is prioritized, but it suffers from stability limitations when applied to stiff systems. The Euler implicit method, though computationally more intensive, offers better stability for such systems and is often preferred in cases where the system exhibits rapid changes over different time scales. Additionally, we introduce the Runge-Kutta family of methods, which are popular for their balance of accuracy and computational efficiency. These methods, particularly the fourth-order Runge-Kutta method (RK4), are widely used for solving non-stiff problems where high precision is required. For more challenging problems, especially stiff ODEs, implicit Runge-Kutta methods such as Backward Euler and the Radau IIA methods provide the necessary stability while allowing larger time steps, though they come with increased computational demands. The chapter also explores the concept of stiffness, a property of certain systems where large disparities exist between fast and slow time scales. For these systems, the usual explicit methods may require prohibitively small time steps to maintain stability. Thus, implicit methods become indispensable for efficient integration. By comparing the stability and accuracy of these methods, this chapter provides the tools necessary to select the most appropriate time integration technique for a given ODE system.

While the equations that govern the fluid flow are partial differential equations, a lot can be learned by analyzing the ordinary differential equations (ODEs) that arise in various situations. For example, the trajectories of individual fluid parcels in inviscid flow are governed by simple ODEs, and ODE systems can describe chemical reactions or highly idealized dynamical systems. Since the numerical methods for solving ODEs are generally simpler than those for partial differential equations, and since the time-stepping methods used in partial differential equations are closely related to those in ODEs, this chapter focuses on analyzing methods for approximating the solutions of ODEs. The lessons learned will helps us later in adjusting the time step for the progress of numerical solution of Navier-Stokes equation.

4.1 STIFFNESS OF DIFFERENTIAL EQUATION

A system of differential equations is said to be stiff when it contains components that evolve on widely varying timescales. This typically occurs when the ratio of the largest to the smallest (in magnitude) eigenvalue of the system's Jacobian matrix is very large, leading to a high condition number κ. In linear systems of the form

$$\frac{d\mathbf{y}}{dt} = A\mathbf{y}$$

the stiffness is directly linked to the spectral properties of the matrix A. When some eigenvalues have large negative real parts, the corresponding solution components decay rapidly, while others may evolve slowly. To accurately resolve the dynamics without numerical instability, especially with explicit time-stepping methods, one must choose a very small time step Δt to satisfy the stability criteria, often dictated by the fastest decaying modes.

Stiffness is not inherently related to the physical timescales of interest but rather to the numerical limitations imposed by stability constraints. For example, even if the fast-decaying modes are not crucial to the solution's long-term behavior, they still restrict the maximum allowable time step in explicit schemes. Implicit time integration methods are preferred for stiff systems because they are unconditionally stable for many problems, allowing the use of larger time steps without compromising stability.

A system of inherently stable ODEs is considered stiff if

$$|\lambda_{min}| \ll |\lambda_{max}|,$$

where λ_{min} and λ_{max} represent the smallest and largest eigenvalues (in magnitude), respectively. Based on stiffness ratio κ, stiffness can be categorized as follows:

Category	**Stiffness Ratio κ**
Mildly stiff	$\kappa < 10^2$
Strongly stiff	$10^3 < \kappa < 10^5$
Extremely stiff	$10^6 < \kappa < 10^8$
Pathologically stiff	$\kappa > 10^9$

Example 4.1

Consider a differential equation

$$y''(x) + 2002y'(x) + 2001\,y(x) = 0$$

and identify it as stiff or non-stiff.

Solution To determine stiffness, we first find the characteristic roots of the equation. The equation can be rewritten as a second-order linear homogeneous ODE with constant coefficients. Assuming a solution of the form $y(x) = e^{\lambda x}$, we substitute into the ODE and get:

$$\lambda^2 + 2002\lambda + 2001 = 0$$

Solving for λ gives the characteristic roots:

$$\lambda = \frac{-2002 \pm \sqrt{2002^2 - 4 \cdot 2001}}{2}$$

Simplifying:

$$\lambda = \frac{-2002 \pm \sqrt{4008004 - 8004}}{2} = \frac{-2002 \pm \sqrt{4000000}}{2}$$

$$\lambda = \frac{-2002 \pm 2000}{2}$$

This gives two roots:

$$\lambda_1 = -1 \quad \text{and} \quad \lambda_2 = -2001$$

The two eigenvalues $\lambda_1 = -1$ and $\lambda_2 = -2001$ have a large difference in magnitude. Since $|\lambda_2| \gg |\lambda_1|$, the system has a high condition number (ratio of the largest to smallest eigenvalues), which is a characteristic of stiffness. The large eigenvalue λ_2 causes rapid decay, while the smaller eigenvalue λ_1 represents a slower decay component.
The condition number κ for this system is calculated as the ratio of the magnitudes of the largest and smallest eigenvalues, which are given as:

$$\lambda_1 = -1, \quad \lambda_2 = -2001$$

Thus,

$$\kappa = \frac{|\lambda_{\max}|}{|\lambda_{\min}|} = \frac{2001}{1} = 2001$$

Since $\kappa = 2001$, which is a large value, this system has a high condition number, indicating stiffness. This large condition number confirms that the system requires a small time step for stable numerical integration, as typical for stiff systems.
Stiff ODEs require very small time steps for stable numerical integration, especially to handle the fast-decaying component (corresponding to λ_2). Even though the solution associated with λ_1 decays slowly, the large negative eigenvalue λ_2 forces the use of a much smaller time step, making standard numerical methods inefficient unless a stiff solver is used.
A differential equation (DE) system $\dot{z} = Az$ is stiff if $\lambda_{\max} \gg \lambda_{\min}$, indicating a large condition number. Here, $\lambda_{\max} := \max_i |\lambda_i|$ and $\lambda_{\min} := \min_i |\lambda_i|$.
Using the eigenvectors of A, the system can be transformed and decoupled into DEs of the form $\dot{z}_i = \lambda_i z_i$, each with its individual time step limit $h_{\max,i} = \frac{s_r}{|\lambda_i}|$ (assuming real eigenvalues $\lambda_i < 0$).
If $\kappa = 2001$ is the condition number of our system, then for stable integration of the entire system, a time step

$$\text{If } \kappa = \frac{|\lambda_{\max}|}{|\lambda_{\min}|}, \text{ then } h_{\max} = \frac{s_r}{|\lambda_{\max}|} = \frac{1}{\kappa} \cdot \frac{s_r}{|\lambda_{\min}|}$$

4.2 SINGLE-STEP METHODS

One-step methods include the explicit and implicit Euler schemes, various Runge-Kutta methods, and the Crank-Nicolson method.

4.2.1 EULER METHODS: EXPLICIT AND IMPLICIT

Euler Explicit Method (Forward Euler) Euler's explicit method, often called the Forward Euler method, is among the simplest numerical methods for solving ordinary differential equations (ODEs). It provides a straightforward, first-order approach for initial value problems (IVPs). The core principle is to use the known slope of the solution at the current time step to predict the solution at the next time step. Although its simplicity makes it easy to implement and understand, it suffers from stability issues and limited accuracy if the step size is not sufficiently small.

Consider an ODE of the form

$$\frac{dy}{dt} = f(t,y), \quad y(t_0) = y_0.$$

Here, $y(t)$ is the unknown function, $f(t,y)$ governs the rate of change, and t_0 is the initial time where the solution is known to be y_0. The Forward Euler method approximates the solution by stepping forward in discrete increments of size h. At each step, the slope $f(t_n, y_n)$ is treated as constant over the interval from t_n to $t_{n+1} = t_n + h$. The approximate solution value is then given by

$$y_{n+1} = y_n + h f(t_n, y_n).$$

As a typical illustration, consider the first-order decay problem

$$\frac{dy}{dt} = -ky, \quad y(0) = y_0,$$

where k is a positive constant. The continuous solution exhibits exponential decay. In the Forward Euler scheme, one applies

$$y_{n+1} = y_n - hky_n.$$

If $y_0 = 1$, $k = 0.1$, and $h = 0.5$, then the solution at each step is computed by substituting the previous value. The first step yields

$$y_1 = 1 - 0.5 \times 0.1 \times 1 = 0.95,$$

and the second step becomes

$$y_2 = 0.95 - 0.5 \times 0.1 \times 0.95 = 0.9025.$$

Repeating this procedure generates successive approximations that decay as expected, though the Forward Euler method's first-order accuracy implies that the global error grows proportionally to the time step size.

One of the principal advantages of Euler's explicit method is its simplicity; it is easy to implement and computationally inexpensive per step for non-stiff problems. However, it is only accurate to first order in h. Consequently, if high accuracy is desired, a very small time step is required, which can increase the overall computational cost. Furthermore, Forward Euler can become unstable when solving stiff problems unless h is extremely small, making the method impractical in such cases.

Euler Implicit Method (Backward Euler) The Euler implicit method, also called the Backward Euler method, is another first-order technique for solving ODEs, but it differs significantly in stability properties. Unlike the explicit version, the Backward Euler approach is implicit, requiring that one solve an algebraic equation at each time step. Its principal benefit is greater stability, making it particularly suitable for stiff problems where the Forward Euler method would demand prohibitively small time steps.

With the ODE

$$\frac{dy}{dt} = f(t,y), \quad y(t_0) = y_0,$$

the Backward Euler method approximates the solution via the relation

$$y_{n+1} = y_n + h f(t_{n+1}, y_{n+1}).$$

Because y_{n+1} appears on both sides, one typically must rearrange or solve a nonlinear equation (in the general case) to find y_{n+1}. This extra step adds computational effort, but grants much better stability characteristics.

Returning to the decay problem

$$\frac{dy}{dt} = -ky, \quad y(0) = y_0,$$

the Backward Euler formula becomes

$$y_{n+1} = y_n + h(-k y_{n+1}).$$

Rearranging to solve for y_{n+1}, one obtains

$$y_{n+1} = \frac{y_n}{1+hk}.$$

As a result, when $y_0 = 1$, $k = 0.1$, and $h = 0.5$, the first computed step is

$$y_1 = \frac{1}{1+0.5\times 0.1} = \frac{1}{1.05} \approx 0.9524,$$

and the next step is

$$y_2 = \frac{0.9524}{1.05} \approx 0.9077.$$

This method also captures the decay trend, and its implicit nature makes it less sensitive to large time steps when the system is stiff.

A central advantage of the Backward Euler method is its unconditional stability for linear problems, allowing it to handle larger time steps more effectively. This feature is essential in stiff systems, where explicit methods would require impractically small steps. On the other hand, Backward Euler is still only first order in accuracy, so reducing the time step is necessary when seeking high precision. Moreover, because the method is implicit, one must solve an algebraic equation at each step, which is more computationally expensive than in the explicit approach, especially for nonlinear problems.

4.2.2 CRANK-NICOLSON METHOD

The Crank-Nicolson method is a time integration scheme widely used in solving ordinary and partial differential equations. It is also referred to as the **trapezoidal method** because it is based on applying the trapezoidal rule for numerical integration in time. This connection becomes clear when considering a general ordinary differential equation of the form

$$\frac{du}{dt} = f(t,u),$$

and integrating from t^n to t^{n+1} using the trapezoidal rule. The result is

$$u^{n+1} = u^n + \frac{\Delta t}{2}\left[f(t^n,u^n) + f(t^{n+1},u^{n+1})\right]$$

which is the update form used in the Crank-Nicolson method. This approach takes the average of the derivative evaluated at the current and future time levels, and the integral corresponds geometrically to the area under a trapezoid, giving rise to the alternative name. For linear problems, the Crank-Nicolson scheme coincides with the θ-method with $\theta = 0.5$, making it time-centered.

The method is particularly popular in the numerical solution of parabolic partial differential equations, such as the diffusion equation, where it is referred to more commonly by its Crank-Nicolson name. In contrast, when used in the context of ordinary differential equations or integral approximations, it may simply be called the trapezoidal rule. It is second-order accurate in time, implicit in nature, and provides a good balance between accuracy and stability.

Importantly, the Crank-Nicolson method is a single-step time-stepping method. A time-stepping method progresses the solution forward in time by computing values at successive time levels. In a one-step method, the solution at the new time level t^{n+1} depends only on values from the current time level t^n, and not on any earlier time history like t^{n-1}, t^{n-2}, etc. The Crank-Nicolson scheme uses the discretized form

$$\frac{U^{n+1} - U^n}{\Delta t} = \frac{1}{2}\left(\mathscr{L}U^n + \mathscr{L}U^{n+1}\right),$$

where $\mathscr{L}$ is a spatial differential operator. This formulation highlights that only U^n and U^{n+1} are involved, confirming its single-step character.

Crank-Nicolson method is a second-order accurate, implicit, single-step, time-stepping method. It is widely appreciated for its ability to provide unconditionally stable solutions for linear problems while retaining temporal accuracy.

4.2.3 RUNGE-KUTTA METHODS

Runge-Kutta methods are a family of iterative techniques widely used for solving ordinary differential equations (ODEs). They are popular due to their balance of accuracy and computational efficiency, making them suitable for many types of problems, including non-stiff initial value problems. The classical Runge-Kutta method is an explicit method, known as RK4 (fourth-order), which calculates the solution at each time step using intermediate evaluations of the derivative. This provides a good trade-off between accuracy and ease of implementation. RK method is ideal for numerically solving IVP problems. Many BVPs, especially in fluid flow, heat transfer, or structural mechanics, are nonlinear in nature. The RK method, especially the higher-order versions like RK4, can efficiently handle these nonlinearities with good accuracy.

For stiff problems, implicit Runge-Kutta methods such as the backward Euler method or the Radau IIA methods are preferred. These methods ensure stability for large time steps and allow efficient handling of stiff systems. This section will delve into the derivation of various Runge-Kutta methods, their applications to both stiff and non-stiff problems, and their advantages over other time-stepping methods.

The explicit versions of Runge-Kutta methods are widely used for solving ordinary differential equations (ODEs) where the next value of the solution is computed directly using the known values at the current time step, without requiring the solution of implicit equations. Some common Explicit Runge-Kutta Schemes are

RK2 (Second-order Runge-Kutta) or Heun Method or Improved Euler Method

The Heun method is a second-order accurate, single-step numerical integration technique used to solve initial value problems of the form:

$$\frac{dy}{dt} = f(t,y), \quad y(t_0) = y_0.$$

This method is an explicit member of the Runge-Kutta (RK) family and is also known as the Improved Euler method or the Modified Euler method. Unlike the basic Euler method, which uses only the slope at the beginning of the interval, Heun's method improves accuracy through a two-stage predictor-corrector approach.

Algorithm: Given a current solution point (t_n, y_n) and a step size h, the method proceeds as follows:

1. Predictor Step (Euler Approximation): An initial estimate for the next point is computed using the slope at the current point:

$$y^* = y_n + h \cdot f(t_n, y_n).$$

2. Corrector Step (Averaged Slope): The final approximation is obtained by averaging the slopes at the beginning and the predicted end of the interval:

$$y_{n+1} = y_n + \frac{h}{2}\left[f(t_n, y_n) + f(t_n + h, y^*)\right].$$

This averaging reduces error, making the method second-order accurate.

Order of Accuracy: The mthod has local truncation error of order $\mathcal{O}(h^3)$, and has global error of order $\mathcal{O}(h^2)$

Heun method can be summarized by its Butcher tableau:

$$\begin{array}{c|cc} 0 & 0 & 0 \\ 1 & 1 & 0 \\ \hline & \frac{1}{2} & \frac{1}{2} \end{array}$$

This confirms that Heun's method is a two-stage explicit method (since entries below the diagonal are zero), and is a part of the Runge-Kutta family (specifically, an RK2 method). The method is suitable for moderately accurate integration of non-stiff ODEs. Due to its balance between simplicity and improved accuracy over Euler's method, Heun's method is widely preferred.

RK3 (Third-order Runge-Kutta): A third-order method that provides a midpoint between RK2 and RK4 in terms of accuracy and efficiency.
General Form:

$$k_1 = f(t_n, u_n)$$

$$k_2 = f(t_n + \frac{1}{2}\Delta t, u_n + \frac{1}{2}k_1\Delta t)$$

$$k_3 = f(t_n + \Delta t, u_n - k_1\Delta t + 2k_2\Delta t)$$

$$u_{n+1} = u_n + \frac{1}{6}(k_1 + 4k_2 + k_3)\Delta t$$

RK4 (Fourth-order Runge-Kutta): The most popular explicit RK method, offering a good balance between accuracy and computational cost. It uses four intermediate calculations (slopes) to achieve a fourth-order accuracy.
General Form:

$$k_1 = f(t_n, u_n)$$

$$k_2 = f(t_n + \frac{1}{2}\Delta t, u_n + \frac{1}{2}k_1\Delta t)$$

$$k_3 = f(t_n + \frac{1}{2}\Delta t, u_n + \frac{1}{2}k_2\Delta t)$$

$$k_4 = f(t_n + \Delta t, u_n + k_3\Delta t)$$

$$u_{n+1} = u_n + \frac{1}{6}(k_1 + 2k_2 + 2k_3 + k_4)\Delta t$$

Explicit RK schemes are widely used for solving ODEs in many applications, especially when high accuracy and efficiency are needed for non-stiff problems.

Merson's Runge-Kutta Method (Order 4 with 5 Stages): Merson's method is a 5-stage RK method that produces a 4th-order solution but also computes a 5th-order estimate for error control (adaptive step-size adjustment).

The general form is:

$$y_{n+1} = y_n + h\sum_{i=1}^{s} b_i k_i,$$

where the slopes k_i are computed as:

$$k_i = f\left(t_n + c_i h, y_n + h\sum_{j=1}^{i-1} a_{ij}k_j\right).$$

Coefficients for Merson's Method: The Butcher tableau for Merson's method is

$$
\begin{array}{c|ccccc}
0 & 0 & 0 & 0 & 0 & 0 \\
\frac{1}{3} & \frac{1}{3} & 0 & 0 & 0 & 0 \\
\frac{1}{3} & \frac{1}{6} & \frac{1}{6} & 0 & 0 & 0 \\
\frac{1}{2} & \frac{1}{8} & 0 & \frac{3}{8} & 0 & 0 \\
1 & \frac{1}{2} & 0 & -\frac{3}{2} & 2 & 0 \\
\hline
\text{4th order} & \frac{1}{6} & 0 & 0 & \frac{2}{3} & \frac{1}{6} \\
\text{5th order} & \frac{1}{10} & 0 & \frac{3}{10} & \frac{2}{5} & \frac{1}{5}
\end{array}
$$

The c_i values (nodes) are: $c = [0, \frac{1}{3}, \frac{1}{3}, \frac{1}{2}, 1]$. The a_{ij} coefficients (from the lower triangular matrix in the tableau) define how the slopes k_i depend on previous slopes. The b_i coefficients for the 4th-order solution are $[\frac{1}{6}, 0, 0, \frac{2}{3}, \frac{1}{6}]$. The alternative b_i^* coefficients for the 5th-order estimate (used for error control) are $[\frac{1}{10}, 0, \frac{3}{10}, \frac{2}{5}, \frac{1}{5}]$.

Using the Butcher tableau, the slopes are computed as:

$$
\begin{aligned}
k_1 &= f(t_n, y_n), \\
k_2 &= f\left(t_n + \frac{1}{3}h, y_n + \frac{1}{3}hk_1\right), \\
k_3 &= f\left(t_n + \frac{1}{3}h, y_n + \frac{1}{6}hk_1 + \frac{1}{6}hk_2\right), \\
k_4 &= f\left(t_n + \frac{1}{2}h, y_n + \frac{1}{8}hk_1 + \frac{3}{8}hk_3\right), \\
k_5 &= f\left(t_n + h, y_n + \frac{1}{2}hk_1 - \frac{3}{2}hk_3 + 2hk_4\right).
\end{aligned}
$$

4th Order Solution:

$$
y_{n+1} = y_n + h\left(\frac{1}{6}k_1 + 0 \cdot k_2 + 0 \cdot k_3 + \frac{2}{3}k_4 + \frac{1}{6}k_5\right).
$$

5th Order Estimate (for Error Control):

$$
\hat{y}_{n+1} = y_n + h\left(\frac{1}{10}k_1 + 0 \cdot k_2 + \frac{3}{10}k_3 + \frac{2}{5}k_4 + \frac{1}{5}k_5\right).
$$

4.3 AMPLIFICATION FACTOR AND STABILITY DIAGRAM

We analyze the behavior of numerical time-stepping methods by using a linear test equation

$$
\frac{dy}{dt} = \alpha y,
$$

where, $\alpha \in \mathbb{C}$ is a complex constant which represents the growth or decay rate (e.g., an eigenvalue of the system or Jacobian matrix)[1].

[1] In some literature, the notation $\sigma = \lambda h$ is used, where λ denotes an eigenvalue of the system. Under this convention, the equivalence $\alpha \equiv \lambda$ and $p \equiv \sigma$ holds. Thus, the analysis remains consistent regardless of nomenclature. The polynomial form of $A(p)$ becomes a powerful tool for studying the stability, accuracy, and qualitative behavior of numerical methods when solving ordinary or partial differential equations. Amplification factor can also be written as $A(\sigma) = 1 + \sigma + a_2\sigma^2 + \ldots$ in the $\lambda - \sigma$ framework.

Amplification factor (A) is a central concept in numerical methods, which characterizes how a numerical scheme propagates the solution over each time step. Thus, it represents the ratio of successive numerical solutions:

$$A = \frac{y_{n+1}}{y_n}$$

indicating how the solution at one step relates to the previous. For many methods, A can be expressed as a polynomial in the dimensionless parameter $p = \alpha h$, where α is the system's eigenvalue and h is the time step. A typical form is

$$A(p) = 1 + p + a_2 p^2 + a_3 p^3 + a_4 p^4 + a_5 p^5 + \mathrm{R6},$$

where $a_2, a_3, \ldots$ are method-dependent coefficients, and R6 denotes higher-order terms.

The parameter $p = \alpha h$ is crucial because it combines the stiffness of the system (through α) and the size of the time step h. It serves as the input to the function $A(p)$, which determines the stability behavior of the numerical method. The behavior of the method depends on the modulus $|A(p)|$:

i If $|A(p)| < 1$, the method damps the solution.

ii If $|A(p)| > 1$, the method amplifies the solution, which may lead to instability.

iii If $|A(p)| = 1$, the method is neutrally stable, which is ideal for oscillatory problems.

Table 4.1 lists the characteristic polynomials corresponding to various time-integration methods, written in terms of the dimensionless parameter $p = \alpha h$, where α represents the eigenvalue of the linearized system. In some textbooks the formulation $p = \lambda \Delta t$ is used, where λ is the eigenvalue of the spatial discretization operator and Δt is the time step used in the numerical scheme. Each polynomial is expressed in terms of the amplification factor A, which determines how perturbations in the numerical solution grow or decay over time. This shows that $\alpha = \lambda$ and $h = \Delta t$, aligning the notation with the alternate form $p = \lambda \Delta t$.

For a method to be stable, all roots of the characteristic polynomial, that is, the values of A, must satisfy $|A| \leq 1$. The table includes both explicit and implicit schemes, covering first- to higher-order methods such as Forward Euler, Leapfrog, the trapezoidal rule, Adams-Bashforth, Adams-Moulton, ABM, and Gazdag methods. These characteristic polynomials are fundamental to linear stability analysis, where the combined effect of the eigenvalue λ and time step Δt is captured by the parameter p. The location of p in the complex plane determines whether a given method remains stable when applied to stiff, oscillatory, or diffusive systems.

The parameter $p = \alpha h$ plays a central role in the stability analysis of numerical integration methods. It encapsulates both the eigenvalue α of the system and the time step h, and serves as the key variable in determining the region of stability for a given scheme. The amplification factor $A(p)$, as a function of this parameter, reveals how the method responds to different modes in the solution. In particular, stability properties are classified based on the behavior of $A(p)$ in the complex plane:

☞ A-stability: the method is stable for all $\Re(p) \leq 0$

Table 4.1
Characteristic polynomials (where, $p = \alpha h = \lambda\Delta t$)

A-polynomial (in p)	Method
$A - 1 - p = 0$	Explicit Euler-O1
$A^2 - 2pA - 1 = 0$	Leapfrog-O2
$A^2 - (1 + \frac{3}{2}p)A + \frac{1}{2}p = 0$	AB-O2
$A^3 - \left(1 + \frac{23}{12}p\right)A^2 + \frac{16}{12}pA - \frac{5}{12}p = 0$	AB-O3
$A\left(1 - p\right) - 1 = 0$	Implicit Euler-O1
$A\left(1 - \frac{1}{2}p\right) - \left(1 + \frac{1}{2}p\right) = 0$	Trapezoidal-O2
$A^2\left(1 - \frac{2}{3}p\right) - \frac{4}{3}A + \frac{1}{3} = 0$	Backward-O2
$A^2\left(1 - \frac{5}{12}p\right) - A\left(1 + \frac{8}{12}p\right) + \frac{p}{12} = 0$	AM-O3
$A^3 - \left(1 + \frac{13p}{12} + \frac{115p^2}{144}\right)A^2 + \left(\frac{p}{12} + \frac{80p^2}{144}\right)A - \frac{25p^2}{144} = 0$	ABM-O3
$A^3 - (1 + 2p)A^2 + \frac{2}{3}pA - \frac{1}{2}p = 0$	Gazdag-O3

☞ L-stability: the method is A-stable and additionally satisfies $A(p) \to 0$ as $p \to \infty$

These criteria ensure that stiff modes decay appropriately and that the method remains stable regardless of step size, which is essential in the integration of stiff differential equations.

Methods are classified as Zero-stability, A-stability, B-stability, L-stability, G-stability and Strong stability.

Zero-stability (0-stability) applies to multi-step methods and is a prerequisite for their convergence. A method is 0-stable if small perturbations in initial data or round-off errors do not grow without bound as the computation proceeds. This ensures that the numerical method is well-posed as the timestep approaches zero. BDF1 (backward Euler) and BDF2 are 0-stable, but BDF methods beyond order 2 lose this property.

A-stability refers to the property of a time integration method where the numerical solution remains bounded for all problems with eigenvalues lying in the left-half complex plane, i.e., for all complex numbers p such that $\mathrm{Re}(p) \leq 0$. This means the method can handle stiff decay without introducing instability. A-stable methods are particularly useful in stiff systems, as they guarantee unconditional stability for linear problems. Examples include the implicit Euler method and the trapezoidal rule.

B-stability is a concept that extends the idea of A-stability to nonlinear problems. A method is B-stable if it preserves contractivity: that is, if two solutions starting from different initial data get closer under the continuous dynamics, the same should hold under the numerical method. B-stability is important for ensuring the robustness of numerical integration in nonlinear and stiff systems. The implicit midpoint method is an example of a B-stable scheme.

L-stability is a stronger condition than A-stability. An L-stable method is A-stable and, in addition, its amplification factor $A(p)$ tends to zero as $p \to \infty$. This ensures that fast-decaying transients (associated with very stiff components of the system) are heavily damped in the numerical solution. L-stability is crucial for problems where high-frequency components must be suppressed quickly. A classic example of an L-stable method is the implicit Euler scheme.

Strong Stability Preserving (SSP) methods are designed to maintain monotonicity properties (such as total variation diminishing, TVD) of the spatial discretization when coupled with explicit time stepping. These methods prevent the formation of new oscillations, especially important in hyperbolic problems and shock capturing. The SSP property is often associated with Runge-Kutta methods, such as SSPRK(3,3), which are tailored to work with upwind or TVD spatial schemes.

A numerical method is said to possess G-stability if it satisfies a generalization of the B-stability condition, specifically tailored to preserve energy-like norms in time integration. This concept is particularly relevant for implicit Runge-Kutta methods applied to nonlinear or energy-conserving problems. G-stable schemes are especially valuable in the analysis of systems such as structural dynamics and Hamiltonian mechanics, where maintaining bounded or conserved energy is crucial for long-term numerical stability.

Algebraic stability is a technical condition used in the analysis of implicit Runge-Kutta methods. It ensures that the method has favorable stability properties even when applied to certain classes of stiff nonlinear problems. Algebraic stability helps prove that the method does not increase solution energy or violate contractive properties, and it is closely related to the Butcher tableau of the method.

It is important to note that

1 All L-stable methods are also A-stable.

2 Trapezoidal rule is A-stable but not L-stable (does not damp stiff modes).

3 Gauss-Legendre methods are A-stable but not L-stable (they are symplectic and suitable for conservative systems).

4 L-stable methods are preferred for stiff dissipative problems.

Table 4.2 lists the Stability classifications of time-integration methods with representative examples.

Figure 4.1 compares the stability regions of ten fundamental numerical integration methods, visualized through their characteristic behavior in the complex p-plane where $p = \alpha h$. The shaded blue areas represent stable regions where the amplification factor A satisfies $|A| \leq 1$ for all roots of each method's characteristic polynomial, ensuring bounded solutions. The blue boundary curves trace $|A| = 1$, demarcating stable and unstable zones. Key observations include the bounded stability regions of explicit methods (e.g., Euler's circular region) versus the expansive or unbounded stable zones of implicit methods (e.g., Implicit Euler's entire left half-plane), highlighting critical trade-offs between computational efficiency and stability in numerical integration. This comprehensive visualization enables direct comparison of stability constraints across classical temporal discretization techniques.

Table 4.2
Stability classifications of time-integration methods

Stability Type	Key Purpose	Examples
A-stability	Stable for stiff linear problems	Implicit Euler; Trapezoidal rule (Crank Nicolson)
L-stability	Damps fast transients in very stiff systems	Radau IIA(3); two-stage SDIRK
B-stability	Contractive behavior for nonlinear ODEs	Implicit midpoint method
0-stability	Necessary for convergence of multistep schemes	Backward Differentiation Formula of Order 1 ad 2
Strong Stability Preserving	Maintains monotonicity for hyperbolic/shock problems	Three-stage third-order RK; Five-stage fourth-order RK
G-stability	Preserves discrete energy norms	Implicit midpoint; Gauss-Legendre Runge Kutta
Algebraic stability	Extends B-stability to stiff nonlinear cases	Gauss-Legendre Runge Kutta family

4.4 MULTISTEP METHODS

Adams-Bashforth Methods: Adams-Bashforth method can be described in a general form as an explicit multistep method:

$$y_{n+1} = y_n + h\left(a_1 f(x_n, y_n) + a_2 f(x_{n-1}, y_{n-1}) + a_3 f(x_{n-2}, y_{n-2}) + \dots\right)$$

Where,

i y_{n+1} is the solution at the next time step.

ii h is the time step size.

iii $f(x_k, y_k)$ represents the derivative evaluated at past time levels.

iv $a_1, a_2, a_3, \dots$ are coefficients specific to the order of the Adams-Bashforth scheme.

This scheme uses only known information from previous steps (i.e., it is explicit) to compute y_{n+1}, which is characteristic of Adams-Bashforth methods.

Adams-Moulton Method : The Adams-Moulton (AM) method is an implicit linear multistep method used for solving ordinary differential equations (ODEs) of the form:

$$\frac{dy}{dx} = f(x,y), \quad y(x_0) = y_0$$

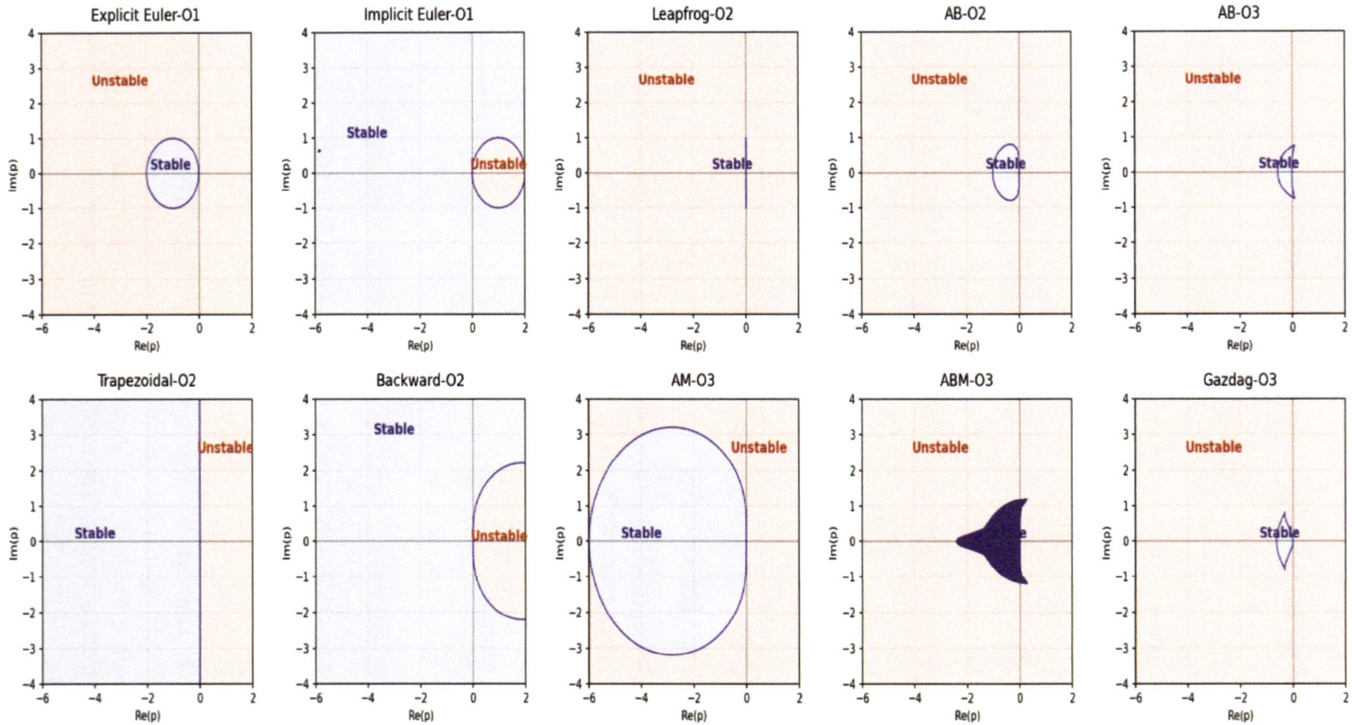

Figure 4.1: **Stability regions (shaded blue) in the complex p-plane for various numerical integration methods. The boundary $|A| = 1$ is shown in blue. Methods arrangement: Top row—Explicit Euler (O1), Leapfrog (O2), Adams-Bashforth (AB-O2, AB-O3), and Implicit Euler (O1); Bottom row—Trapezoidal (O2), Backward Differentiation (O2), Adams-Moulton (AM-O3), Adams-Bashforth-Moulton predictor-corrector (ABM-O3), and Gazdag (O3). Stable regions satisfy $|A| \leq 1$ for all roots of the characteristic polynomial.**

Table 4.3
Adams-Bashforth methods (AB-ON), where N denotes the order of truncation error. Listed are the coefficients a_k in the formula and the corresponding imaginary (s_i) and real (s_r) parts of the amplification factor roots.

N	a_1	a_2	a_3	a_4	a_5	s_i	s_r
1	1					0	2.00
2	3/2	-1/2				0	1.00
3	23/12	-16/12	5/12			0.72	0.54
4	55/24	-59/24	37/24	-9/24		0.43	0.30
5	1901/720	-2774/720	2616/720	-1274/720	251/720	0	0.16

It is part of the Adams family of predictor-corrector methods , where:
The Adams-Bashforth (AB) method (explicit) provides a predicted value.
The Adams-Moulton (AM) method (implicit) refines this prediction using a corrector step.

The Adams-Moulton (AM) methods are implicit multistep schemes that require solving an equation at each time step because the unknown future value y_{n+1} appears on both sides of the formula. A general AM method has the form

$$y_{n+1} = y_n + h \sum_{k=0}^{s} \beta_k f(x_{n+1-k}, y_{n+1-k})$$

where the evaluation at x_{n+1} involves y_{n+1} itself, necessitating iterative procedures such as fixed-point iteration or Newton's method. This implicit nature increases computational effort but yields significantly greater stability, particularly for stiff equations. In terms of accuracy, AM methods exhibit smaller local truncation errors than their explicit counterparts (e.g., Adams-Bashforth methods) of the same order. For instance, AM2—equivalent to the trapezoidal rule—is second-order accurate and more precise than the second-order Adams-Bashforth method; similarly, AM3 achieves third-order accuracy. These methods are preferred when stability is critical, as their stability regions are substantially larger than those of explicit multistep schemes, making them suitable for a wide range of ODE problems.

The s-step Adams-Moulton method is given by:

$$y_{n+1} = y_n + h(\beta_0 f(x_{n+1}, y_{n+1}) + \beta_1 f(x_n, y_n) + \cdots + \beta_s f(x_{n+1-s}, y_{n+1-s}))$$

where, h = step size, β_k = coefficients determined by polynomial interpolation, s = number of steps (order depends on s).

Tables 4.7 and 4.8 summarize the numerical stability properties of various time integration methods by listing the imaginary (s_i) and real (s_r) parts of the roots of their amplification factors, which indicate a method's ability to resolve oscillatory

Table 4.4
Common Adams-Moulton Methods

Method Name	Formula	Stability
Backward Euler (AMO1)	$y_{n+1} = y_n + hf(x_{n+1}, y_{n+1})$	A-stable
Trapezoidal Rule (AMO2)	$y_{n+1} = y_n + \frac{h}{2}[f(x_n, y_n) + f(x_{n+1}, y_{n+1})]$	A-stable
AMO3	$y_{n+1} = y_n + h(\frac{5}{12}f(x_{n+1}, y_{n+1}) + \frac{8}{12}f(x_n, y_n) - \frac{1}{12}f(x_{n-1}, y_{n-1}))$	Conditionally stable
AMO4	$y_{n+1} = y_n + h(\frac{9}{24}f(x_{n+1}, y_{n+1}) + \frac{19}{24}f(x_n, y_n) - \frac{5}{24}f(x_{n-1}, y_{n-1}) + \frac{1}{24}f(x_{n-2}, y_{n-2}))$	Conditionally stable

Table 4.5
Adams-Moulton Coefficients and Stability Limits

N	β_0	β_1	β_2	β_3	β_4	β_5	s_i	s_r
1	1	–	–	–	–	–	∞	∞
2	$\frac{1}{2}$	$\frac{1}{2}$	–	–	–	–	∞	∞
3	$\frac{5}{12}$	$\frac{8}{12}$	$-\frac{1}{12}$	–	–	–	0	6
4	$\frac{9}{24}$	$\frac{19}{24}$	$-\frac{5}{24}$	$\frac{1}{24}$	–	–	0	3
5	$\frac{251}{720}$	$\frac{646}{720}$	$-\frac{264}{720}$	$\frac{106}{720}$	$-\frac{19}{720}$	–	1.21	1.83
6	$\frac{475}{1440}$	$\frac{1427}{1440}$	$-\frac{798}{1440}$	$\frac{482}{1440}$	$-\frac{173}{1440}$	$\frac{27}{1440}$	1.37	1.18

behavior and control numerical damping. Table 4.7 presents data for Runge-Kutta (RK) methods, showing that higher-order schemes generally exhibit larger values of s_i and s_r, reflecting improved resolution of wave-like solutions and better stability. Notably, the classical fourth-order RK method and Merson's fifth-order method provide a good balance between accuracy and damping. Table 4.8 focuses on Adams Predictor-Corrector (APK) methods, where an explicit Adams-Bashforth (AB) predictor is followed by an implicit Adams-Moulton (AM) corrector. The stability of these schemes is largely governed by the AM step, as evidenced by the favorable values of s_r and s_i. For comparison, the final row shows the performance of a two-step AB-only scheme, which lacks the stabilizing influence of the corrector and thus shows weaker damping. These values help quantify the trade-offs in accuracy and stability across explicit, implicit, and mixed methods.

If the imaginary root s_i of a Runge-Kutta (RK) scheme is zero, this implies that the stability region of the method does not extend into the imaginary axis, or more

Table 4.6
$A = 1 + p + a_2p^2 + a_3p^3 + a_4p^4 + a_5p^5 +$ **H.O.T where** $p = \alpha h$

Scheme	a_2	a_3	a_4	a_5
RK2-O2	1/2			
RK3-O3	1/2	1/6		
RK4-O4	1/2	1/6	1/24	
RK5-O4-Merson	1/2	1/6	1/24	1/144
RK6-O5	1/2	1/6	1/24	1/120

Table 4.7
The imaginary and real parts of the roots of the amplification factor in Runge-Kutta methods with two to six stages

Method	s_i	s_r
RK2-O2	0	2
RK3-O3	1.73	2.51
RK4-O4	2.83	2.79
RK4-O5 - Merson	3.46	3.55
RK6-O5	0	3.21

Table 4.8
Imaginary (s_i) and real (s_r) parts of the roots of the amplification factor for Adams Predictor-Corrector (APK) schemes. Stability is primarily governed by the corrector (AM) step. The last row compares an AB-only scheme over two steps.

Predictor / Corrector Scheme	s_i	s_r
AB-O2 / AM-O2	1.28	2.00
AB-O2 / AM-O3	1.20	2.40
AB-O3 / AM-O4	1.17	1.93
AB-O3 / AB-O3 (2-step comparison)	1.46	1.08

specifically, the method does not handle purely imaginary eigenvalues effectively (see Figure 4.2).

In practical terms:

☞ Limited Stability for Oscillatory Systems: A zero imaginary root s_i indicates that the method has no stability along the imaginary axis, which suggests it is not suitable for solving problems with purely oscillatory behavior (e.g.,

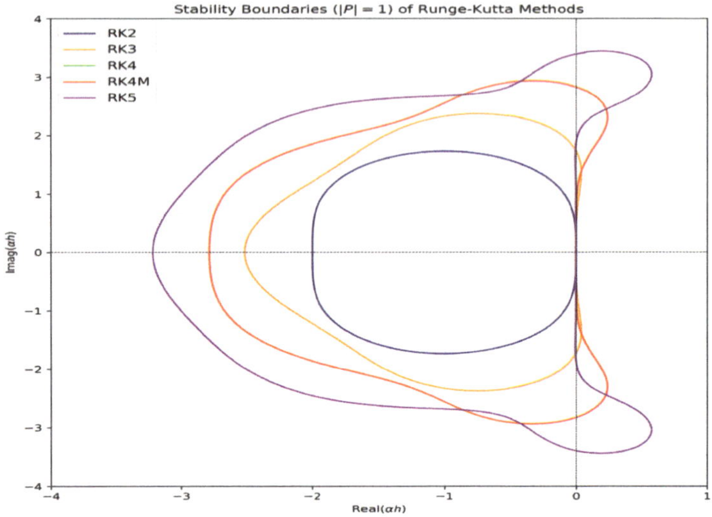

Figure 4.2: **Stability region for RK-03 and O4 schemes**

Table 4.9

The imaginary and real parts of the roots of the amplification factor in Adam-Bashforth methods

Order	s_i	s_r
AB-O1	0	2
AB-O2	0	1
AB-O3	0.72	0.54
AB-O4	0.43	0.30
AB-O5	0	0.16

systems with eigenvalues that lie on the imaginary axis). This is commonly encountered in wave equations or harmonic oscillators, where eigenvalues are purely imaginary.

☞ Better Suited for Dissipative Systems: Such an RK method would be more effective for systems with negative real eigenvalues (dissipative systems) rather than oscillatory systems. For example, it would work well for problems like diffusion or decay processes, where solutions tend to zero over time rather than oscillate.

☞ Reduced Accuracy in Time-Dependent Oscillations: If an RK scheme has no stability region in the imaginary direction, applying it to oscillatory systems may lead to numerical instability or inaccurate results over long time steps, as the method fails to appropriately handle the oscillations.

Stiff Problems and Implicit RK schemes: Stiffness refers to a particular behavior in differential equations, particularly in systems involving multiple time scales. Stiff systems are characterized by the presence of rapid changes in some components of the solution, while other components evolve more slowly. The disparity in these rates can make the numerical integration of stiff systems challenging because small time steps are required to accurately capture the fast dynamics, even when the solution's slow dynamics are the primary interest. In the context of fluid dynamics, chemical kinetics, and structural analysis, stiffness often arises in equations describing diffusion, chemical reactions, or flows involving high Reynolds numbers.

For solving stiff systems, explicit numerical methods like standard Runge-Kutta schemes can be inefficient or unstable because they require extremely small time steps for stability. This is where specialized methods like implicit Runge-Kutta methods or stiff solvers become useful. These methods allow for stable integration with larger time steps, even for stiff systems.

Unlike explicit methods, implicit RK schemes involve solving a system of equations at each time step, making them more computationally intensive but more stable for larger time steps. These methods are particularly suited for stiff equations, where explicit schemes would require prohibitively small time steps for stability.

4.5 PATHLINES AND FLOW TRAJECTORIES

Pathline equations are crucial in flow visualization techniques such as Particle Image Velocimetry (PIV) and other flow measurement methods. These techniques are used to observe and analyze the motion of fluid particles by tracking tracer particles that are seeded within the flow. PIV, for example, captures images of the particle positions at different time intervals, allowing the determination of instantaneous velocity fields and the tracking of particle trajectories. The pathlines, which represent the trajectories of individual particles, are used to visualize the flow patterns and obtain detailed information about the velocity distribution in the fluid. By knowing the pathlines, researchers can gain insights into the flow characteristics, such as turbulence, vortices, and the behavior of the fluid under different conditions. Also, pathlines are important in flow visualization methods like Lagrangian tracking, where the motion of individual particles is traced over time to study the evolution of the flow. They play a significant role in understanding the dynamics of complex flow phenomena, such as mixing, transport of particles, and interaction of fluid with obstacles or boundaries. By analyzing the pathlines, engineers and scientists can optimize designs in fluid systems, predict flow behavior in different geometries, and enhance the understanding of transport phenomena in various applications, from aerodynamics to environmental fluid dynamics. Thus, pathline equations derived from the fundamental fluid mechanics PDEs are indispensable tools in both theoretical and applied fluid dynamics.

The pathline equations are obtained by solving the velocity components u and v in the respective directions, resulting in the parametric equations that describe the motion of a particle over time. In the two-dimensional flow field, the trajectory of a fluid particle can be expressed through the pathline equations:

$$\frac{dx}{dt} = u(x,y,t), \quad \frac{dy}{dt} = v(x,y,t),$$

where $x(t)$ and $y(t)$ represent the position of the particle at time t, and u and v are the velocity components in the x and y directions, respectively. These equations are solved for the specific velocity field that describes the flow, yielding the trajectory of the fluid particle.

The pathline equations represent an Initial Value Problem (IVP) as they are defined by initial conditions rather than boundary conditions. In this case, the initial conditions are the position of the fluid particle at time $t = 0$, i.e., (x_p, y_p), and the initial velocity components at that time. The solution to these equations provides the trajectory of the fluid particle as it evolves with time.

An IVP is characterized by the specification of initial conditions at a single point (typically at $t = 0$) and the goal is to find the evolution of the system over time. In the case of fluid flow, the trajectory or pathline is determined by solving the equations from the initial position, and the velocity field dictates how the particle moves through the flow domain.

Pathlines, on the other hand, describe the actual trajectory of a fluid particle as it moves over time. They show where a fluid particle travels from a specific starting point under the influence of the velocity field. In many CFD simulations, especially those involving steady flow, the streamlines are often plotted to represent the fluid's motion. But infact they are the pathlines rather than streamlines. Streamline is a hypothetical line where velocity is tangent and they are not the flow trajectories. The confusion arises because in some cases the flow, streamlines, pathlines, and streaklines all coincide. This is because the velocity field does not change over time, so the trajectory of a particle (pathline) follows the same direction as the streamline at each point. However, in unsteady (time-varying) flow, streamlines and pathlines become distinctly different.

In flow visualization techniques like Particle Image Velocimetry (PIV), the pathlines of tracer particles are often tracked to obtain information about the velocity field. Here, the trajectory of the particles over time is much more relevant than the instantaneous direction of flow. Using the term streamlines for the pathlines in such cases would misrepresent the concept of what is being visualized.

Example 4.4

A two-dimensional flow field is described by the following velocity components:

$$u = 3x(t+1), \quad v = 3y(t-1)$$

Determine the trajectory of the fluid particle that passes through the point (x_p, y_p) at $t = 0$.

Solution

Because u and v represent the velocities in the x and y directions respectively, write:

$$u = \frac{dx}{dt} = 3x(t+1) \quad \Rightarrow \quad \frac{dx}{dt} = 3x(t+1).$$

Separate the variables:

$$\frac{dx}{x} = 3(t+1)\,dt.$$

Integrate both sides:

$$\int \frac{dx}{x} = 3\int (t+1)\,dt \quad \Rightarrow \quad \ln(x) = \frac{3}{2}t^2 + 3t + C_1.$$

Hence,

$$x(t) = C_1 \exp\left(\tfrac{3}{2}t^2 + 3t\right).$$

Similarly, for $y(t)$:

$$v = \frac{dy}{dt} = 3y(t-1) \quad \Rightarrow \quad \frac{dy}{dt} = 3y(t-1).$$

Separate the variables:

$$\frac{dy}{y} = 3(t-1)\,dt.$$

Integrate both sides:

$$\int \frac{dy}{y} = 3\int (t-1)\,dt \quad \Rightarrow \quad \ln(y) = \frac{3}{2}t^2 - 3t + C_2.$$

Thus,

$$y(t) = C_2 \exp\left(\tfrac{3}{2}t^2 - 3t\right).$$

Apply the initial condition (x_p, y_p) at $t = 0$:

$$x(0) = x_p = C_1 \exp\left(\tfrac{3}{2} \times 0^2 + 3 \times 0\right) \quad \Rightarrow \quad C_1 = x_p,$$

$$y(0) = y_p = C_2 \exp\left(\tfrac{3}{2} \times 0^2 - 3 \times 0\right) \quad \Rightarrow \quad C_2 = y_p.$$

Therefore, the trajectory of the fluid particle that passes through (x_p, y_p) at $t = 0$ is given by:

$$x(t) = x_p \exp\left(\tfrac{3}{2}t^2 + 3t\right), \quad y(t) = y_p \exp\left(\tfrac{3}{2}t^2 - 3t\right).$$

These parametric equations describe the pathline of the fluid particle.
See `Python` code `Ch4-Ex44` for solution of this problem.
We compute the L_2-norm as

$$\mathrm{L}_2^{(x)} = \sqrt{\frac{1}{N}\sum_{i=1}^{N}\left(x_i^{\text{num}} - x_i^{\text{exact}}\right)^2}, \qquad \mathrm{L}_2^{(y)} = \sqrt{\frac{1}{N}\sum_{i=1}^{N}\left(y_i^{\text{num}} - y_i^{\text{exact}}\right)^2}$$

L_2 norm errors in x and y for different time-stepping methods.

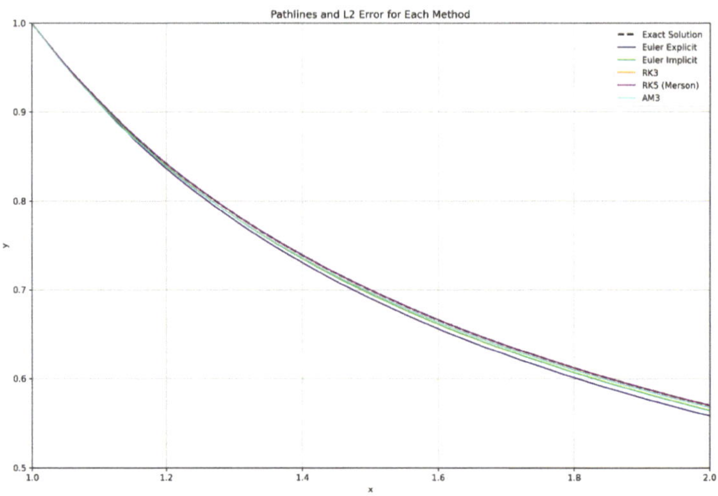

Figure 4.3: **Pathlines according to exact solution and different numerical methods.**

Method	$L_2^{(x)}$	$L_2^{(x)}$
Euler Explicit	2.6452	6.1808×10^{-3}
Euler Implicit	1.9899	1.6851×10^{-3}
RK3	4.4708×10^{-4}	1.5090×10^{-7}
RK5 (Merson)	1.25739×10^{-6}	$5.73500e \times 10^{-10}$
AM3	7.2611×10^{-2}	1.0393×10^{-3}

Figure 4.3 and list shows that RK5 (Merson) performed best compared to other methods.

4.6 TIME STEPPING IN NAVIER-STOKES SOLVERS

Efficient time stepping is critical for the practical simulation of fluid flows, as it directly impacts computational cost and accuracy. The numerical solution of the Navier-Stokes equations presents unique challenges in temporal discretization, requiring careful consideration of stability, accuracy, and computational efficiency. These governing equations for fluid flow exhibit nonlinear coupling between velocity and pressure fields. They also span a wide range of temporal scales, particularly in turbulent or high Reynolds number flows. Two fundamental strategies emerge for temporal discretization: the direct (monolithic) approach and the method of lines, each offering distinct advantages for different classes of problems.

The direct discretization approach simultaneously treats spatial and temporal derivatives, yielding a fully coupled nonlinear system at each time level. This approach naturally accommodates implicit time integration schemes, which prove

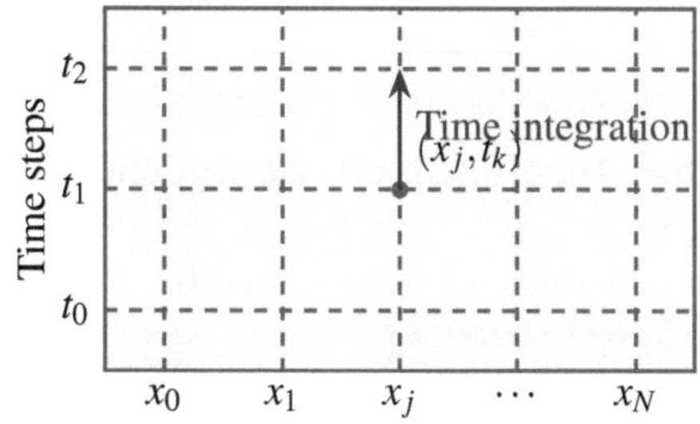

Figure 4.4: **Method of lines approach for numerically solving partial differential equations.**

essential for stiff problems characteristic of viscous-dominated flows or fine spatial discretizations. The simplest such method, the first-order Euler implicit (Backward Euler) scheme, provides unconditional stability at the cost of numerical diffusion and first-order temporal accuracy. For improved accuracy while maintaining stability, the second-order Crank-Nicolson method offers an attractive compromise through its time-centered differencing. However, its semi-implicit nature can sometimes lead to weak instability in nonlinear problems, prompting the use of more robust alternatives like BDF2 (second-order Backward Differentiation) in production codes. The monolithic approach proves particularly valuable when solving coupled multiphysics problems or when employing algebraic splitting methods for pressure-velocity coupling. The resulting large nonlinear systems typically require Newton-type iterations coupled with advanced linear solvers, making the computational cost per time step substantial but often necessary for stability.

The method of lines provides an alternative paradigm by first discretizing spatial derivatives, reducing the original PDE system to a set of coupled ODEs in time (see Figure 4.4). This separation of concerns allows flexible pairing of spatial discretization methods (finite difference, volume, or element) with specialized ODE integrators. For explicit time marching, the forward Euler method serves as the simplest option, though its severe CFL stability restriction makes it impractical for most Navier-Stokes applications. Higher-order explicit Runge-Kutta methods can improve efficiency for convection-dominated flows, but remain constrained by stability limits. In contrast, implicit time integrators like the theta-method (which generalizes both Euler implicit and Crank-Nicolson) overcome these restrictions at the expense of solving nonlinear systems. The method of lines particularly shines when combined with operator splitting techniques, allowing different physical processes (convection, diffusion, pressure projection) to be integrated with optimal time steps. Modern implementations often employ adaptive time stepping based on error estimation, dynamically balancing accuracy requirements with computational effort across different flow regimes.

Table 4.10
Comparison of Time Discretization Approaches for Navier-Stokes Solvers

Aspect	Method of Lines (Semi-Discretization)	Full Discretization
Discretization	Spatial derivatives discretized first, leaving time continuous.	Spatial and temporal derivatives discretized simultaneously.
Resulting Equations	System of ODEs in time.	System of nonlinear algebraic equations.
Solution Method	Adaptive ODE solvers (e.g., Runge-Kutta, Adams-Moulton)	Implicit time integrators (e.g., Crank-Nicolson, BDF2) with nonlinear solvers (e.g., Newton-type)
Flexibility	High flexibility in time-step choice and adaptive stepping.	Less flexible; time step often constrained by nonlinear solver.
Use Cases	Unsteady flows with wide time-scale ranges (e.g., turbulence).	Steady-state or stiff problems (e.g., creeping flows).

The choice between these approaches ultimately depends on problem characteristics and computational objectives. Monolithic methods excel when strong coupling between variables mandates simultaneous solution, while the method of lines offers superior flexibility for problems where different physical processes exhibit disparate time scales. Practical implementations frequently hybridize these strategies, using implicit treatment of stiff terms while handling non-stiff components explicitly. This delicate balance between accuracy, stability, and efficiency forms the core challenge in developing robust Navier-Stokes solvers, with time integration strategy playing a pivotal role in determining overall solver performance.

The two approaches for time discretization are compared in Table 4.10.

Lessons Learned

☞ Time-stepping methods for ODEs (e.g., Euler, Runge-Kutta) form the computational backbone for time integration in both ODEs and PDEs, especially in fluid dynamics where they govern the evolution of flow variables.

☞ Explicit vs. implicit methods present a trade-off: explicit methods (like

Forward Euler, RK4) are easy to implement but require very small time steps for stiff systems, while implicit methods (like Backward Euler) are unconditionally stable for many cases but involve solving algebraic systems at each step.

☞ Stiffness is characterized by a large spread in eigenvalues of the system's Jacobian. When the condition number is high, explicit schemes become impractical, and stiff solvers (typically implicit) are necessary for efficiency and stability.

☞ Amplification factor analysis helps understand how numerical methods propagate or damp signals. The parameters s_i (imaginary root) and s_r (real root) describe the method's ability to handle oscillations and decay respectively.

☞ Temporal stability (via s_i) along with spatial resolution limits (via k_{mod}), give practical constraints on the allowable time step in simulations.

☞ Numerical accuracy depends not only on the formal order of the method but also on its stability and dispersion properties. Higher-order methods are not always better if their amplification behavior poorly represents the underlying dynamics.

☞ Stability diagrams, such as the $|A(p)| \leq 1$ regions in the complex plane, provide intuitive visual tools to assess and compare the stability properties of various time integration schemes.

☞ Pathlines, governed by IVPs, represent true particle trajectories and are crucial in Lagrangian flow visualization. They are distinct from streamlines, especially in unsteady flows where velocity fields evolve over time.

PROBLEMS

4P-1 Consider the linear initial value problem

$$\frac{\mathrm{d}y}{\mathrm{d}x} = ay, \quad y(0) = y_0.$$

Its exact solution is

$$y(x) = y_0 \exp(ax).$$

Using a forward difference approximation with a uniform step Δx, the difference scheme becomes

$$y_{i+1} = \left(1 + a\Delta x\right) y_i.$$

Solve this problem using Euler Forward method. Compares with the exact exponential solution.

4P-2 Consider the challenges that arise in solving ordinary differential equations of the form $y' = Ay$, where matrix A has eigenvalues λ_j.

1. A system is deemed stiff if the largest absolute eigenvalue significantly exceeds the smallest one: $\lambda_{\max} \gg \lambda_{\min}$, where $\lambda_{\max} = \max_j |\lambda_j|$ and $\lambda_{\min} = \min_j |\lambda_j|$.

2. The maximal feasible integration step size, Δt, is governed by $\lambda_{\max}$, and is typically approximated by $\Delta t_{\max} \approx \frac{1}{\lambda_{\max}}$. For stiff systems, this limitation mandates a Δt that is often smaller than necessary for the majority of the solution components, especially when the magnitude of the component associated with $\lambda_{\max}$ is relatively small.

4P-3 Consider the linear equation

$$\frac{dy}{dt} = -5y, \qquad y(0) = 1$$

i Apply the Forward Euler and Backward Euler methods with time step $\Delta t = 0.1$ for $t \in [0, 1]$.

ii Compare the numerical solutions with the exact solution.

iii Plot the absolute error $|y_{\text{num}} - y_{\text{exact}}|$ versus time.

iv Explain why the Backward Euler method performs better in terms of stability for this stiff decay problem.

4P-4 Given the velocity field

$$u(x,t) = 2x(t+1), \quad \frac{dx}{dt} = u(x,t),$$

i Use RK2 (Heun method) and RK4 to compute the trajectory of a particle from $x(0) = 1$ to $t = 1$ using step size $h = 0.1$.

ii Derive the exact solution analytically and compare with numerical solutions.

iii Tabulate the global error for both methods and verify the expected convergence order.

4P-5 A system of ODEs is given by

$$\frac{d\mathbf{y}}{dt} = A\mathbf{y}, \quad A = \begin{bmatrix} -1 & 0 \\ 0 & -1000 \end{bmatrix}, \quad \mathbf{y}(0) = \begin{bmatrix} 1 \\ 1 \end{bmatrix}.$$

i Show analytically that this system is stiff and compute the condition number κ.

ii Using step size $h = 0.01$, apply Forward Euler and Backward Euler methods for $t \in [0, 0.1]$.

iii Discuss which components of the solution cause the stiffness and how each method behaves in resolving them.

iv Explain why implicit methods are preferred for such systems and derive a step-size restriction for Forward Euler to remain stable.

4P-6 Solve the following IVP using the classical RK4 (4th-order Runge-Kutta) method:

$$\frac{dy}{dt} = t - y^2, \quad y(0) = 1.$$

i Compute the solution up to $t = 2$ with $h = 0.2$.

ii Compare with the Heun (RK2) method using the same step size.

iii Tabulate the errors at $t = 1$ for both methods and discuss the trade-offs between accuracy and computational cost.

4P-7 The velocity field $u(x,y,t) = 2x(t+1)$, $v(x,y,t) = -2y(t-1)$ and initial position $(x_0, y_0) = (1,1)$.

i Derive exact pathline equations $x(t)$ and $y(t)$, and plot the trajectory for $t \in [0,1]$.

ii Solve the pathline equations using RK4 using Heun's method (RK2) and Forward Euler, and compare with the exact solution and compute the L_2 error.

iii Analyze how the choice of method changes if $u = -1000x$. Justify.

REFERENCES

G. Hellwig, Partial Differential Equations, An Introduction, Blaisdell, New York, USA, 1964.

M. J. Kloker, Numerische Loser (Zeitintegrationsverfahren) für die gewöhnliche Mod-ëlld-ifferentialgleichung $y' = \alpha y$, Institut fur Aerodynamik und Gasdynamik der Universität Štuttgart, 1996.

5 Preserving Oscillatory Accuracy in Numerical Time-Stepping

In this chapter, we will explore the importance of amplitude and phase in the numerical solutions of oscillatory differential equations. These concepts are crucial for evaluating the stability and accuracy of numerical methods, particularly in systems involving wave motion or oscillations. After finishing this chapter, you should be able to:

- Understand the role of amplitude and phase in numerical solutions.
- Learn how Euler's explicit and implicit methods affect amplitude and phase.
- Quantify and calculate amplitude and phase errors.
- Analyze the stability and accuracy of Euler's methods for oscillatory systems.

In numerical solutions of oscillatory differential equations, amplitude and phase are crucial for ensuring accurate approximations of real-world phenomena, such as wave propagation. This chapter focuses on how numerical methods, particularly Euler's explicit and implicit schemes, affect these two characteristics. We will explore the impact of amplitude errors, which indicate instability, and phase errors, which can lead to incorrect physical behavior. The chapter highlights the importance of controlling these errors, especially in oscillatory systems, and uses examples like the equation $y' = i\pi y$ to demonstrate the application of these concepts. The comparison between Euler's explicit and implicit methods will showcase their ability to handle oscillatory behavior, with implicit methods providing better stability.

In the context of numerical solutions for differential equations, particularly those involving oscillatory systems (like the equation $y' = i\pi y$), amplitude and phase play crucial roles in evaluating the accuracy and stability of the numerical methods used. Here's how these concepts relate to numerical solutions.

5.1 INFLUENCE OF TRUNCATION ERROR IN NUMERICAL SCHEMES

To understand the role of truncation errors in numerical schemes, it is essential to examine how spatial derivatives influence the temporal evolution of solutions in partial differential equations. Spatial derivatives appear in mathematical models to represent physical processes like diffusion, convection, and wave propagation. These derivatives, when discretized, can introduce numerical artifacts depending on the order and sign of the terms involved. By analyzing a simple linear equation in the

DOI: 10.1201/9781003663171-5

Fourier domain, we can gain insight into the effect of each derivative's order on solution behavior—particularly in terms of damping, amplification, and phase speed. This knowledge is foundational for assessing the stability and accuracy of numerical methods, especially when approximating higher-order spatial derivatives in schemes such as finite difference or spectral methods.

To analyze the effect of truncation errors in numerical schemes, let us begin with a general linear partial differential equation:

$$\frac{\partial u}{\partial t} = \alpha \frac{\partial^n u}{\partial x^n}$$

A natural way to explore how solutions behave under such an equation is to assume a Fourier-mode representation. Suppose the solution takes the form:

$$u(x,t) = V(t)e^{ikx}$$

Substituting this form into the original equation gives:

$$\frac{\partial}{\partial t}\left(V(t)e^{ikx}\right) = \alpha \frac{\partial^n}{\partial x^n}\left(V(t)e^{ikx}\right)$$

Since the exponential function does not depend on time, the left-hand side becomes $\frac{dV(t)}{dt}e^{ikx}$. On the right-hand side, applying the n-th spatial derivative to the exponential yields $(ik)^n V(t)e^{ikx}$, leading to:

$$\frac{dV(t)}{dt}e^{ikx} = \alpha(ik)^n V(t)e^{ikx}$$

Canceling the common exponential term, we obtain a simple ordinary differential equation:

$$\frac{dV(t)}{dt} = \alpha(ik)^n V(t)$$

This equation is separable. Dividing both sides and integrating:

$$\frac{1}{V(t)}\frac{dV(t)}{dt} = \alpha(ik)^n \quad \Rightarrow \quad \ln V(t) = \alpha(ik)^n t + C$$

Exponentiating both sides:

$$V(t) = V(0)e^{\alpha(ik)^n t}$$

Thus, the complete solution in space and time becomes:

$$u(x,t) = V(0)e^{ikx+\alpha(ik)^n t}$$

This general result tells us that the behavior of the solution depends critically on the exponent n. Let's explore a few cases:

i If $n = 1$, we get:

$$u(x,t) = V(0)e^{ik(x+\alpha t)}$$

which describes pure translation at a constant phase speed α.

ii For $n = 2$:

$$u(x,t) = V(0)e^{ikx}e^{-\alpha k^2 t}$$

If $\alpha > 0$, this describes damping or diffusion of the mode without translation.

iii For $n = 3$:

$$u(x,t) = V(0)e^{ik(x-\alpha k^2 t)}$$

The solution still translates, but now the velocity depends on k^2, introducing dispersion.

iv For $n = 4$:

$$u(x,t) = V(0)e^{\alpha k^4 t}e^{ikx}$$

Here, the solution grows or decays very rapidly, depending on the sign of α, and is highly sensitive to wavenumber.

This Fourier-based analysis reveals that the order of spatial derivatives in a numerical scheme directly governs how different wavenumber components evolve in time. First- and third-order derivatives contribute to phase shifts or dispersion, altering the speed at which wave modes propagate, while even-order derivatives like second and fourth introduce amplitude modulation, resulting in diffusion or artificial amplification.

These insights are crucial when interpreting truncation errors from Taylor expansions in finite-difference schemes. In practice, higher-order truncation terms—though seemingly small—can introduce unintended damping, dispersion, or instability, particularly at high wavenumbers. Therefore, understanding their effect is essential not only for predicting the accuracy of a scheme but also for ensuring its physical fidelity and numerical robustness. This underscores why numerical analysts often perform Fourier or von Neumann stability analysis: to diagnose and control the behavior of discretized PDEs across the entire spectral domain.

5.2 AMPLITUDE IN NUMERICAL SOLUTIONS

Amplitude in numerical solutions refers to the maximum value of the numerical approximation of the oscillatory solution, akin to the concept of amplitude in the exact solution.

In numerical methods, particularly for oscillatory problems, the amplitude must be monitored to ensure that numerical solutions do not grow unbounded. An increasing amplitude indicates instability in the numerical method. The amplitude of the numerical solution can be compared to the exact solution's amplitude to compute the amplitude error. This error quantifies how closely the numerical method replicates the behavior of the true solution. The amplitude error is given by:

$$\Delta A = |A_{\text{num}} - A_{\text{exact}}|$$

In practical scenarios, like simulating wave motion or electrical signals, the amplitude of the solution reflects physical quantities like energy or intensity. Ensuring the numerical amplitude matches the exact amplitude is critical for accurate modeling.

5.3 PHASE IN NUMERICAL SOLUTIONS

Phase in numerical solutions refers to the argument (or angle) of the complex number representing the numerical solution at a given time. It indicates the temporal position of the oscillation within its cycle. When using numerical methods, particularly those like the explicit Euler method, the phase of the computed solution can differ from the exact phase. This phase error can lead to misrepresentations of the physical behavior being modeled. The phase error quantifies the discrepancy between the numerical phase and the exact phase:

$$\Delta\Theta = |\Theta_{\text{num}} - \Theta_{\text{exact}}|$$

This measurement is essential for understanding how well the numerical solution mimics the actual behavior of the system, especially in oscillatory systems where phase differences can have significant implications. Interference Effects : In systems where multiple waves interact, phase differences between numerical solutions can lead to constructive or destructive interference. This aspect is crucial in fields such as acoustics and electromagnetic theory.

5.4 AMPLITUDE AND PHASE OF EULER METHOD

Amplitude and phase are fundamental concepts in the study of oscillatory differential equations, and understanding their behavior in numerical simulations is vital for ensuring that solutions reflect the true dynamics of a system. Throughout this chapter, the focus will be on analyzing how numerical methods, particularly the explicit (Forward) Euler and implicit (Backward) Euler approaches, capture the amplitude and phase of solutions to an ordinary differential equations. These two aspects, amplitude and phase, serve as metrics for gauging the fidelity of a numerical approximation. Amplitude represents the peak or characteristic magnitude of an oscillatory response, highlighting whether the solution remains bounded and whether it matches expected physical quantities like energy or wave intensity. Phase, on the other hand, designates the position of the oscillation within its cycle. When solving problems that involve periodic or oscillatory phenomena, any appreciable deviation in phase can translate into large qualitative errors, especially in contexts where the interaction of multiple wave components leads to interference patterns.

i **Explicit Euler Method** : The explicit Euler method updates the solution based on the current state and can lead to issues with stability if the time step h is too large. This instability can result in non-physical amplitudes or phase shifts that diverge from the expected behavior. -As the amplitude is computed based on the update rules, the errors can accumulate, leading to significant deviations from the expected exact solution.

ii **Implicit Euler Method** : The implicit Euler method often provides better stability for stiff equations and oscillatory systems. This method tends to control the amplitude growth more effectively and provides phase accuracy, leading to solutions that stay closer to the expected oscillatory behavior. - By inherently dampening the growth of amplitudes, the implicit method helps maintain accuracy in both amplitude and phase over time.

5.5 LESSONS FROM A MODEL ODE $Y' = \alpha Y$

To illustrate these properties, we use a model equation, $y' = \alpha y$, featuring a complex constant α that can have both real and imaginary parts. In cases where α is purely imaginary, neutral oscillations occur, while purely real α can produce exponential decay or growth, depending on its sign. The interplay between real and imaginary components leads to damped or unstable oscillations, revealing how time-stepping schemes must carefully handle both amplitude and phase in order to remain stable and accurate.

The exact solution to this differential equation

$$y' = \alpha y$$

is given as

$$y = y_0 \exp(\alpha t)$$

where α is a complex constant defined as:

$$\alpha = \alpha_r + i\alpha_i$$

Here, α_r is the real part, and α_i is the imaginary part, which leads to different behaviors in the solution based on its components.

5.5.1 SPECIAL CASES:

☞ Neutral Oscillation : When $\alpha = i\alpha_i$ (where $\alpha_r = 0$), this indicates neutral oscillation with a wave number α_i. The complex ODE corresponds to a system of two real differential equations:

$$y_r' = -\alpha_i y_i, \quad y_i' = \alpha_i y_r$$

Here, y_r and y_i represent the real and imaginary parts of the solution, respectively. An example given is when $\alpha_i = \pi$:

$y_r = y_{r0}\cos(\pi t) - y_{i0}\sin(\pi t)$ with initial conditions $y_{r0} = 1, y_{i0} = 0$. This results in $y_r = \cos(\pi t)$.

☞ Exponential Decay : When $\alpha = \alpha_r$ (where $\alpha_i = 0$), this indicates pure exponential decay ($\alpha_r < 0$). For example: If $\alpha_r = -1$:

$$y = y_r e^{-x}, \quad \text{where } y_0 = 1 \implies y = e^{-x}$$

The stability condition is expressed mathematically as:

$$\left|\frac{y_{n+1}}{y_n}\right| = |A(p)| \leq 1 \quad \text{for } \alpha_r < 0,\ \alpha_i \text{ arbitrary}$$

where:

$A(p)$ is the amplification factor of the numerical method, $p = \alpha h$, with $\alpha = \alpha_r + i\alpha_i$, α_r is the real part of the eigenvalue, and α_i is the imaginary part.

The Taylor series for e^x around $x = 0$ is given by:

$$e^x = \sum_{n=0}^{\infty} \frac{x^n}{n!} = 1 + \frac{x}{1!} + \frac{x^2}{2!} + \frac{x^3}{3!} + \frac{x^4}{4!} + \ldots$$

So the exact amplification factor P is

$$P_{\text{exact}}(p) = e^p = 1 + \frac{p}{1!} + \frac{p^2}{2!} + \frac{p^3}{3!} + \frac{p^4}{4!} + \cdots \quad (5.1)$$

where $p = \alpha h$. This series expansion arises from the Taylor series of the exponential function.

5.6 EULER EXPLICIT SCHEME

We consider a first-order differential equation given by:

$$\frac{dy}{dt} = \alpha y$$

According to explicit Euler method, the update rule for the solution at the next time step is

$$y_{n+1} = y_n + hf(y_n)$$

where:
y_n is the value of y at the current time step,
h is the time step size,
$f(y_n)$ is the function evaluated at the current value of y.
For equation $\frac{dy}{dt} = \alpha y$, we can substitute $f(y)$ with αy:

$$y_{n+1} = y_n + h(\alpha y_n)$$

This can be rearranged to give:

$$y_{n+1} = y_n(1 + \alpha h)$$

The expression can be recognized as the iterative relation for the solution at each step. The ratio of consecutive values can be written as:

$$\frac{y_{n+1}}{y_n} = 1 + \alpha h$$

This leads to the characteristic polynomial of the method:

$$A(p) = 1 + p \quad \text{with } p = \alpha h \quad (5.2)$$

The comparison between the exact polynomial P_{exact} (equation 5.1) and the characteristic polynomial $A(p)$ (equation 5.2) illustrates how the numerical method captures the dynamics of the true solution. The term $O(h)$ indicates that the explicit Euler method is first-order accurate, meaning it reproduces the exact series up to the term of order h^1.

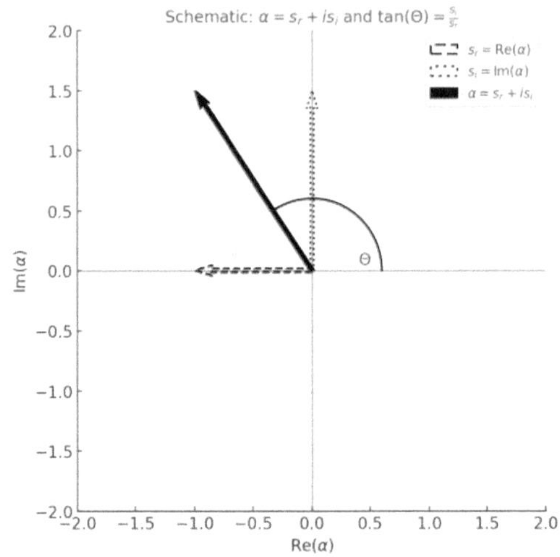

Figure 5.1: **Geometric interpretation of the complex eigenvalue $\alpha = s_r + is_i$ in the complex plane. The real part $s_r = \mathrm{Re}(\alpha)$ is shown as a dashed vector, and the imaginary part $s_i = \mathrm{Im}(\alpha)$ is shown as a dotted vector. The full complex eigenvalue α is represented by the solid vector. The angle Θ between the real axis and α satisfies $\tan(\Theta) = s_i/s_r$, and it plays a central role in stability limits for time-integration schemes involving complex eigenvalues.**

Figure 5.1 shows the schematic diagram of the complex eigenvalue $\alpha = s_r + is_i$ as a vector in the complex plane. The dashed horizontal vector represents the real part $s_r = \mathrm{Re}(\alpha)$, which controls exponential growth or decay. The dotted vertical vector represents the imaginary part $s_i = \mathrm{Im}(\alpha)$, responsible for oscillations. The solid diagonal vector is the full eigenvalue α, combining both effects. The angle Θ between α and the real axis is defined by

$$\tan(\Theta) = \frac{s_i}{s_r}$$

This angle is important for stability analysis. It appears in the general stability limit

$$h_{\max} = \frac{s(\Theta)}{\sqrt{s_r^2 + s_i^2}}$$

where $s(\Theta)$ is a method-specific stability bound. The diagram helps visualize how eigenvalue components and their direction affect the stability of time integration schemes.

Different Cases for α

1. For $\alpha = i\alpha_i$: Here, α is purely imaginary. The stability condition can be represented as $s_i = s$ (with $\Theta = 90°$), which leads to:

$$h_{\max} = \frac{s_i}{|\alpha_i|}$$

This means that the maximum allowable time step size is inversely proportional to the magnitude of the imaginary part of α.

2. For $\alpha = \alpha_r$: Here, α is purely real. The stability condition can be represented as $s_r = s$ (with $\Theta = 0°$), which leads to:

$$h_{\max} = \frac{s_r}{|\alpha_r|}$$

In this case, the maximum time step size is inversely proportional to the magnitude of the real part of α.

3. General Stability Limit : For cases where α has both real and imaginary parts, the stability condition can be expressed as:

$$\tan(\Theta) = \frac{|\alpha_i|}{|\alpha_r|}$$

Rearranging gives:

$$h_{\max} = \frac{s(\Theta)}{\sqrt{|\alpha_i|^2 + |\alpha_r|^2}}$$

This expression provides a general limit for the maximum step size in terms of the angle Θ and the magnitudes of the real and imaginary parts of α.

Figure 5.2 shows the plot of following equations using Euler Explicit approach.

$$y' = i\pi y \quad (y' = \alpha y, \alpha_r = 0, \alpha_i = \pi;\ h\text{-Step size})$$
$$y' = -y \quad (y' = \alpha y, \alpha_r = -1, \alpha_i = 0;\ h\text{-Step size})$$

The expression

$$y_{n+1} = P y_n$$

describes the evolution of a solution y from the current time step n to the next time step $n+1$. Here, P represents a transformation that incorporates both growth and oscillatory behavior.

$P = A\exp(i\theta)$: This equation defines P as a complex number.

A is the amplitude factor, which reflects how much the solution's magnitude changes from one time step to the next.
θ is the phase angle, which determines the phase shift of the wave.

$P_{\text{exact}} = \exp(\alpha h)$
$A_{\text{exact}} = \exp(\alpha_r h)$
$\theta_{\text{exact}} = (\alpha_i h)$

In the numerical implementation, such as with the explicit Euler method, these exact relationships must be approximated. The process involves:

1. Substituting the Model Equation : By inserting the governing differential equations (model equations) into the numerical integration formulas, we derive modified factors that account for the numerical method's behavior.

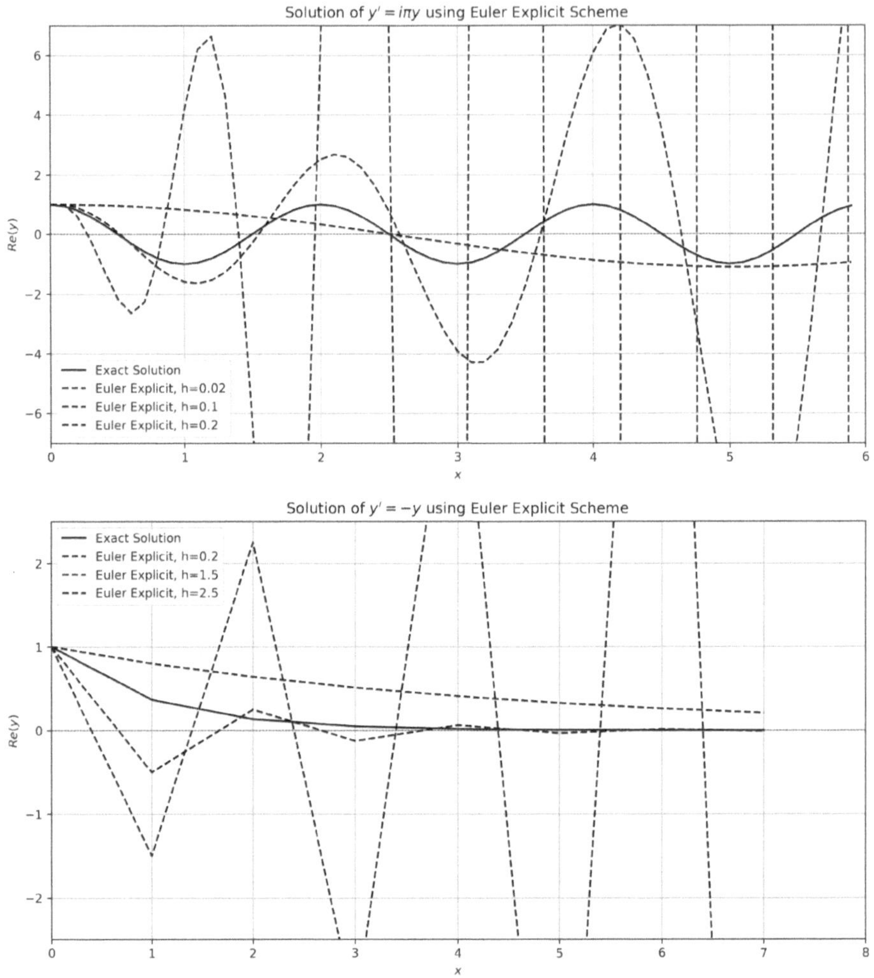

Figure 5.2: **Euler Explicit scheme.**

2. Calculating Modified Amplitude Factors and Wave Numbers : Through the numerical approach, we obtain factors that adjust the amplitudes and wave numbers to reflect the computational method's inherent errors and stability constraints.

Numerical methods are subject to errors, primarily due to rounding and discretization. This means that while the method aims to provide a close approximation to the true solution, various factors can introduce discrepancies. Understanding these errors helps refine the numerical technique and ensures more accurate results.

A_{mod} : This is the modified amplitude decay factor per step in the integration process.

AP_{mod} : Represents the adjustment factor for exact wavelengths, ensuring that the method's response aligns with the theoretical expectations. The expression $AP_{\text{mod}} =$

$A_{\text{mod}}^{(2\pi/|\alpha_i|h)}$ indicates how this factor scales with the step size and the properties of the equation.

$(\alpha_i h)_{\text{mod}}$: This reflects how the dimensionless wave number is adjusted for numerical simulations.

Cases: 1. For $\alpha = i\alpha_i$: In this case, when α is purely imaginary, the amplitude factor $A_{\text{mod}} = \sqrt{1+(\alpha_i h)^2} > 1$ indicates that there is growth in the solution. The relationship $(\alpha_i h)_{\text{mod}} = \tan^{-1}(\alpha_i h)$ helps understand how the phase evolves, which can influence stability.

2. For $\alpha = \alpha_r$: Here, when α is real and negative, the modified amplitude factor decreases the magnitude of the solution. This reflects damping in the system, which is often desirable for stability.

Example 5.1

Solve the first-order differential equation:

$$y' = i\pi y \quad \text{for} \quad 0 \leq x \leq 4, \quad \text{with the initial condition} \quad y(0) = 1.$$

1. Find the real part of the exact solution and compute the amplitude and phase as a function of x.

2. Derive the analytical expression for the solution using the explicit Euler method.

3. Compute the error in y, amplitude, and phase for $x = 4$ using two different step sizes.

4. Provide the equations for the numerical solution using the explicit Euler method.

Solution

To solve this analytically:

1. Exact Solution: Separate variables and integrate:

$$\frac{dy}{y} = i\pi dx$$

Integrating both sides:

$$\ln(y) = i\pi x + C$$

Exponentiate both sides:

$$y(x) = e^{i\pi x + C}$$

Since $y(0) = 1$, we can determine the constant C:

$$y(0) = e^{C} = 1 \quad \Rightarrow \quad C = 0$$

Thus, the exact solution is:

$$y_{\text{ex}}(x) = e^{i\pi x}$$

2. Real and Imaginary Parts of the Exact Solution: Using Euler's formula $e^{i\theta} = \cos(\theta) + i\sin(\theta)$, the exact solution becomes:

$$y_{\text{ex}}(x) = \cos(\pi x) + i\sin(\pi x)$$

So, the real part and imaginary parts are:

$$\text{Re}(y_{\text{ex}}(x)) = \cos(\pi x)$$

$$\text{Im}(y_{\text{ex}}(x)) = \sin(\pi x)$$

3. Amplitude and Phase of the Exact Solution: Amplitude $A_{\text{mod, ex}}$: The amplitude of the solution is given by the magnitude of the complex number:

$$A_{\text{mod, ex}} = |y_{\text{ex}}(x)| = \sqrt{\cos^2(\pi x) + \sin^2(\pi x)} = 1$$

The amplitude remains constant at 1 for all values of x.
Phase $\Theta_{\text{mod, ex}}$: The phase is given by the argument (or arctangent) of the imaginary part divided by the real part:

$$\Theta_{\text{mod, ex}} = \arg(y_{\text{ex}}(x)) = \arctan\left(\frac{\sin(\pi x)}{\cos(\pi x)}\right) = \pi x$$

The explicit Euler method is a first-order numerical method for solving ordinary differential equations. For the equation $y' = i\pi y$, the update formula is given by:

$$y_{n+1} = y_n + h \cdot f(x_n, y_n)$$

where $f(x_n, y_n) = i\pi y_n$ is the right-hand side of the equation. Substituting this into the update formula:

$$y_{n+1} = y_n + h \cdot i\pi y_n = y_n(1 + i\pi h)$$

Thus, the numerical solution evolves according to:

$$y_{n+1} = y_n(1 + i\pi h)$$

where h is the step size and $y_0 = 1$.
We now compute the error in the solution at $x = 4$ using two different step sizes: $h_1 = 0.2$ and $h_2 = 0.1$.
1. Step Size $h_1 = 0.2$: Number of steps :

$$N_1 = \frac{4}{h_1} = \frac{4}{0.2} = 20 \quad \text{steps}$$

Numerical Solution : The numerical solution is:

$$y_n = (1 + i\pi h_1)^n y_0 = (1 + i\pi \cdot 0.2)^{20} \cdot 1$$

Compute the numerical solution for $y_{n=20}$.
2. Step Size $h_2 = 0.1$: Number of steps :

$$N_2 = \frac{4}{h_2} = \frac{4}{0.1} = 40 \quad \text{steps}$$

Numerical Solution : Similarly, for $h_2 = 0.1$:

$$y_n = (1 + i\pi h_2)^{40} \cdot 1$$

Compute the numerical solution for $y_{n=40}$.
3. Errors in y, Amplitude, and Phase:
Exact Solution : The exact solution at $x = 4$ is:

$$y_{\text{ex}}(4) = e^{i\pi \cdot 4} = 1$$

Error in y : Compute the error in the numerical solution compared to the exact solution:

$$\Delta y = |y_{\text{num}}(4) - y_{\text{ex}}(4)|$$

Amplitude Error : The amplitude error is computed as:

$$\Delta A = |A_{\text{num}}(4) - A_{\text{ex}}(4)|$$

where $A_{\text{ex}}(4) = 1$.
Phase Error : The phase error is computed as:

$$\Delta\Theta = |\Theta_{\text{num}}(4) - \Theta_{\text{ex}}(4)|$$

Amplitude Modulus Calculation A_{mod}: The amplitude modulus A_{mod} represents the magnitude of the numerical solution at the given point. It is computed from the amplification factor applied n-times (steps) in the numerical solution:

$$A_{\text{mod}} = |y_{\text{num}}| = |A^n_{\text{mod}} e^{i\Theta_{\text{mod}} n}| = A^n_{\text{mod}}$$

The explicit Euler formula for solving $y' = f(x,y)$ is given by:

$$y_{n+1} = y_n + hf(x_n, y_n)$$

For the given differential equation $y' = i\pi y$, this becomes:

$$y_{n+1} = y_n + h \cdot i\pi y_n$$

Since $f(x_n, y_n) = i\pi y_n$, the update equation becomes:

$$y_{n+1} = y_n(1 + i\pi h)$$

We are starting with the initial condition $y(0) = 1$, and at each step, the value of y is updated using the formula above. The general form after n steps is:

$$y_n = y_0(1 + i\pi h)^n$$

Given $y_0 = 1$ and $h = 0.2$, we can write:

$$y_n = (1 + i\pi \cdot 0.2)^n = (1 + 0.2i\pi)^n$$

To find the amplitude (magnitude) of y_n, we need to compute the modulus of the complex number:

$$|y_n| = |(1 + 0.2i\pi)^n|$$

First, we compute the modulus of $(1 + 0.2i\pi)$:

$$|1 + 0.2i\pi| = \sqrt{1^2 + (0.2\pi)^2} = \sqrt{1 + 0.04\pi^2}$$

Approximating $\pi \approx 3.1416$, we get:

$$\sqrt{1 + 0.04 \times 9.8696} = \sqrt{1 + 0.39478} = \sqrt{1.39478} \approx 1.181$$

Now, after $n = 20$ steps, the amplitude will be:

$$A_{\text{mod}}^{n_1} = (1.181)^{20}$$

Using a calculator, we find:

$$A_{\text{mod}}^{n_1} \approx 27.866$$

This indicates how much the solution has grown or decayed after performing the Euler steps.
Amplitude Error ΔA: The amplitude error ΔA is computed as the difference between the numerical amplitude A_{mod} and the exact amplitude A_{ex} (which is known to be 1, as the exact solution oscillates but does not grow):

$$\Delta A = A_{\text{mod}}^{n_1} - A_{\text{ex}} = 27.866 - 1 = 26.866$$

This value shows a large amplitude error, which is indicative of how much the numerical method is deviating from the true oscillatory behavior at this point.
Phase Modulus Θ_{mod}: The phase modulus Θ_{mod} is the argument (angle) of the complex number representing the numerical solution. It is calculated by the expression:

$$\Theta_{\text{mod}} = \arg(y_{\text{num}}(x_n)) = \arg(A_{\text{mod}}^n e^{i\Theta_{\text{mod}} n}) = n \cdot \Theta_{\text{mod}}$$

We know the differential equation is $y' = i\pi y$ with the initial condition $y(0) = 1$, and the explicit Euler method formula is:

$$y_{n+1} = y_n + h \cdot i\pi y_n$$

This can be simplified to:

$$y_{n+1} = y_n(1 + i\pi h)$$

For step size $h = 0.2$, this becomes:

$$y_{n+1} = y_n(1 + 0.2i\pi)$$

Thus, the general solution after n steps is:

$$y_n = (1 + 0.2i\pi)^n$$

To find the phase of y_n, we need the argument of the complex number $(1 + 0.2i\pi)$. The argument of a complex number $z = a + ib$ is given by:

$$\arg(z) = \tan^{-1}\left(\frac{b}{a}\right)$$

For $1 + 0.2i\pi$, the real part $a = 1$, and the imaginary part $b = 0.2\pi$.
The argument of a complex number $z = a + ib$ is defined as the angle θ between the positive real axis and the vector representing z in the complex plane. Mathematically, it's given by:

$$\theta = \tan^{-1}\left(\frac{b}{a}\right)$$

where: - b is the imaginary part - a is the real part
For $z = 1 + 0.2i\pi$, this becomes:

$$\theta = \tan^{-1}\left(\frac{0.2\pi}{1}\right) = \tan^{-1}(0.62832)$$

Using a calculator to find the arctangent:

$$\theta = \tan^{-1}(0.62832) \approx 0.5586 \text{radians}$$

After $n = 20$ steps, the total phase shift is the argument of $(1 + 0.2i\pi)^n$, which is simply n times the argument of $(1 + 0.2i\pi)$:

$$\Theta_{\text{mod}}^{n_1} = 20 \times \arg(1 + 0.2i\pi)$$

Substitute the calculated argument:

$$\Theta_{\text{mod}}^{n_1} = 20 \times 0.5586 = 11.172 \text{ radians}$$

The phase angle after 20 steps is approximately:

$$\Theta_{\text{mod}}^{n_1} \approx 11.22 \text{ radians}$$

The value $\Theta_{\text{mod}}^{n_1} = 11.22$ radians is obtained by multiplying the argument of $1 + 0.2i\pi$ (which is about 0.5586 radians) by the number of steps $n = 20$, resulting in a total phase shift after 20 steps.
Phase Error $\Delta\Theta$: The phase error $\Delta\Theta$ is the difference between the numerical phase Θ_{mod} and the exact phase Θ_{ex}, which is computed by the exact solution for $x = 4$:

$$\Delta\Theta = \Theta_{\text{mod}}^{n_1} - \Theta_{\text{ex}} = 11.22 \text{ rad} - \pi = -1.34 \text{ rad}$$

This value represents how much the numerical method has shifted the phase compared to the exact oscillatory solution.
Error in y, Δy: Finally, the error in y at $x = 4$ is the difference between the numerical and exact solutions, which is given as:

$$\Delta y = |y_{\text{num}}(x = 4) - y_{\text{ex}}(x = 4)|$$

From the notes:

$$y(x = 4) = A_{\text{mod}} \cdot \text{Re}(e^{in_1\Theta_{\text{mod}}}) = 6.192$$

And the error Δy is computed as:

$$\Delta y = 6.192 - 1 = 5.192$$

This error shows a significant deviation from the exact solution.
Equation for the Numerical Solution with the Explicit Euler Method For the Euler method, the update equation is:

$$y_{n+1} = y_n + h \cdot f(x_n, y_n) = y_n(1 + i\pi h)$$

Thus, the solution at any point $x_n = n \cdot h$ can be approximated as:

$$y_n = (1 + i\pi h)^n y_0$$

where $y_0 = 1$ is the initial condition.
Stability Criterion : For the Euler method, the stability is governed by the amplification factor $|A_{\text{mod}}|$. The method is stable if $|A_{\text{mod}}| \leq 1$.

$$|A_{\text{mod}}| = \sqrt{1 + (\pi h)^2}$$

If h is too large, the method will become unstable as $|A_{\text{mod}}| > 1$. This leads to unbounded growth in the numerical solution, which is not consistent with the exact solution.

Conclusion:
The exact solution is purely oscillatory with a constant amplitude of 1 and a linearly increasing phase.
The explicit Euler method gives an approximation for small step sizes, but larger step sizes introduce phase and amplitude errors.
The stability of the Euler method is contingent on keeping the step size small enough to ensure that the amplification factor remains bounded.

5.7 EULER IMPLICIT FORMULATION

The explicit Euler method was shown to update solutions by taking a step proportional to the function's current slope. Although it is straightforward to implement, this scheme can display significant amplitude growth if the time step h is not kept sufficiently small, especially when α includes an imaginary component. The error in amplitude compounds with each step, and the phase often lags or leads relative to the exact solution. This shortcoming underscores the importance of verifying that the ratio $\left|1+\alpha h\right|$ stays within a stable region. The implicit Euler method addresses part of this issue by basing each new step on the value of the function evaluated at the future time. This requires solving an implicit equation each time, but the added computational cost can pay off for stiff or highly oscillatory systems because the method naturally provides stronger damping of unphysical growth in amplitude, as well as better phase consistency over many steps.

To discuss it further, we consider the differential equation

$$y' = i\pi y$$

with the initial condition $y(0) = 1$, for $0 \leq x \leq 4$, and we need to:

1. Give the real part of the exact solution and discuss the amplitude and phase over x. 2. Derive the analytical solution using the explicit Euler method with grid step h and steps $x_k = k \cdot h$, $k = 0, 1, 2, \ldots, N$. 3. Estimate the error in y, the amplitude error, and phase error for specific values of h. 4. Provide the equation for the numerical solution using the Euler implicit method.

Solution

The implicit Euler method is based on the discretization of the differential equation:

$$y' = f(t, y)$$

The backward Euler method approximates this by:

$$y_{n+1} = y_n + hf(t_{n+1}, y_{n+1})$$

Here, h is the time step, and $t_{n+1} = t_n + h$.

Rearranging this gives:

$$y_{n+1} - hf(t_{n+1}, y_{n+1}) = y_n$$

This shows that the next value y_{n+1} depends on itself, making it an implicit method.

Since y_{n+1} appears on both sides of the equation, it often requires an iterative method to solve for y_{n+1}, such as Newton's method or fixed-point iteration.

The differential equation we are dealing with is:

$$y' = i\pi y$$

with the initial condition $y(0) = 1$ for $0 \leq x \leq 4$.

Exact Solution: To solve this equation, we can separate the variables:

$$\frac{dy}{y} = i\pi\, dx$$

Integrating both sides gives:

$$\ln|y| = i\pi x + C$$

Exponentiating yields:

$$y = e^C e^{i\pi x}$$

Applying the initial condition $y(0) = 1$:

$$1 = e^C \Rightarrow C = 0 \implies y = e^{i\pi x}$$

Using Euler's formula, we can express this as:

$$y = \cos(\pi x) + i \sin(\pi x)$$

Amplitude : The amplitude of y is given by the magnitude of y:

$$|y| = \sqrt{\cos^2(\pi x) + \sin^2(\pi x)} = 1$$

This indicates that the amplitude is constant over time.

Phase : The phase θ is given by:

$$\theta = \tan^{-1}\left(\frac{\sin(\pi x)}{\cos(\pi x)}\right) = \pi x$$

The phase varies linearly with x.

For the computation of the modified amplitude A_{mod}:

$$A_{\text{mod}} = \sqrt{\frac{1}{1 + (\pi h)^2}}$$

The reference amplitude A_{ex}:

$$A_{\text{ex}} = 1$$

Table 5.1
Results of the Analysis

Index	h	A^n_{mod}	A_{ex}	$A^n_{\text{mod}} - A_{\text{ex}}$	$n \cdot \theta_{\text{mod}} - \theta_{\text{ex}}$
0	0.2	0.035886	1	-0.964114	-1.405229
1	0.1	0.152212	1	-0.847788	-1.405271

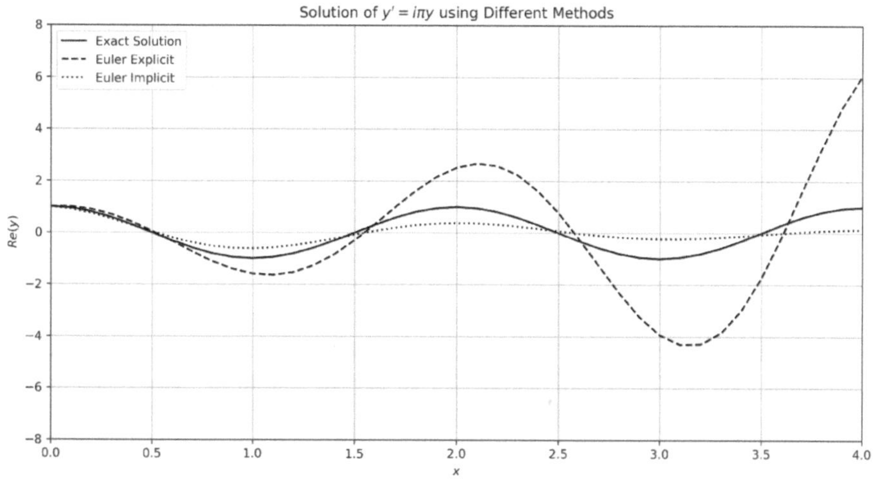

Figure 5.3: **Comparison of Euler Explicit and Euler Implicit schemes.**

For the computation of A^n_{mod}:

$$A^n_{\text{mod}} = (A_{\text{mod}})^n$$

To calculate the difference between the modified amplitude raised to n and the reference amplitude:

$$A^n_{\text{mod}} - A_{\text{ex}}$$

For the derived phase θ_{mod}:

$$\theta_{\text{mod}} = \tan^{-1}\left(-\frac{h}{\pi^2}\right)$$

To compute the difference in phase scaled by n:

$$n \cdot \theta_{\text{mod}} - \theta_{\text{ex}}$$

Where θ_{ex} is the expected phase value (in this case set to 1).

Preserving both amplitude and phase fidelity is paramount when numerically solving oscillatory differential equations like $y' = i\pi y$. The explicit Euler scheme, although appealing because of its simplicity, can introduce significant numerical discrepancies if the chosen step size is not sufficiently small. One of the clearest manifestations of these discrepancies is in the amplitude, which can grow unboundedly for marginally large steps. This phenomenon arises from the way the method updates the solution solely based on the derivative at the current time, failing to incorporate any feedback from the next step. The result is a characteristic polynomial $P = 1 + \alpha h$ that needs to lie within the unit circle in the complex plane. In purely imaginary cases, slight variations in h can push this polynomial outside the stable region, quickly leading to spurious oscillatory growth. Phase errors, albeit separate from amplitude errors, carry their own significance, as a divergence in the phase angle can drastically alter how multiple waves interact, leading to erroneous interference patterns or timing mismatches in models that rely on precise oscillation tracking.

Conversely, the implicit Euler (or Backward Euler) method highlights the advantages of an approach that evaluates the function at the future time step. While this method demands solving an implicit relation for y_{n+1} at each step, which may be costly or cumbersome, its payoff in stability can be substantial. For problems that are stiff, or for those that require long time integration of oscillations, the damping characteristic of implicit Euler tends to preserve the amplitude within acceptable bounds. This is particularly relevant for cases where α has a large imaginary part because explicit Euler schemes risk unbounded error if they are not operated with a very fine step size. Furthermore, when α also has a negative real part, implicit Euler can more faithfully reproduce the exponential decay portion of the solution while keeping the oscillatory component under control. The net effect is an improved balance of amplitude and phase accuracy over time, but one pays for it with increased computational overhead and the need for iterative solvers.

Examples comparing the numerical solutions from each method to the exact solution $\exp(i\pi x)$ laid bare these strengths and weaknesses. Quantitative measures for error—whether in the solution itself (Δy), the amplitude (ΔA), or the phase ($\Delta\Theta$)—highlight how crucial it is to tune the time step and, more broadly, to select the correct method for the problem's characteristics. In some scenarios, explicit Euler's simplicity might suffice if high-frequency oscillations or stiff behavior are not a concern, or if one can afford a very small h. In others, implicit Euler's robustness and capacity to stabilize the solution are more compelling. Ultimately, a thorough understanding of amplitude and phase dynamics helps practitioners choose or develop better numerical methods, guiding them to refine step sizes, incorporate more advanced schemes (like higher-order Runge-Kutta or implicit-explicit hybrids), or focus on domain-specific optimizations. This understanding confirms that the synergy between theory—through concepts like characteristic polynomials, stability regions, and Taylor expansions—and empirical verification is indispensable. By applying these principles to progressively complex equations, one can ensure that the numerical treatment of oscillatory systems remains reliable, stable, and physically meaningful over time.

5.8 PHASE AND AMPLITUDE ACCURACY IN OSCILLATORY AND DECAYING ODES

We will now investigate the numerical behavior of the RK2-02 method for two distinct ordinary differential equations (ODEs). In the first case, we consider the oscillatory and decaying ODEs:

Oscillatory ODE:

$$y'(t) = i\alpha y(t),\ \alpha > 0$$

For the oscillatory equation, we define the non-dimensional parameter

$$z = i\alpha h,$$

where h is the time step. The local amplification factor of the RK2-02 method is

$$A(z) = 1 + z + \frac{z^2}{2},$$

so that $A(z)$ is, in general, a complex number. From this amplification factor, we extract three key parameters:

$$A_{\text{mod}} = |A(z)|, \quad (\alpha h)_{\text{mod}} = \arg\big(A(z)\big), \quad AP_{\text{mod}} = \big|A(z)\big|^{\frac{2\pi}{\alpha h}}.$$

Here, $(\alpha h)_{\text{mod}}$ measures the phase error introduced per time step (since $\arg(A(z))$ differs from the exact phase αh), A_{mod} is the per-step amplitude factor (which ideally should be unity), and AP_{mod} captures how amplitude errors accumulate over one exact period of the continuous solution.

Because the quantity αh measures the size of the time step relative to the frequency, $(\alpha h)_{\text{mod}}$ directly reflects how much the method's effective frequency deviates from the true frequency. Meanwhile, $A_{\text{mod}} \neq 1$ indicates the presence of artificial growth or decay of amplitude per step, and raising this value to the number of steps per period provides AP_{mod}, the effective amplification over one full cycle.

In Figure 5.4, the behavior of $(\alpha h)_{\text{mod}}$ (dashed line), A_{mod} (dash-dotted line), and AP_{mod} (dotted line) is plotted versus the non-dimensional parameter αh. The exact amplification factor, $A_{\text{ex}} = 1$, serves as a reference.

Decaying ODE:

$$y'(t) = \alpha y(t),\ \alpha < 0$$

For the purely decaying ODE, we define the non-dimensional parameter

$$z = \alpha h$$

(with $\alpha < 0$), which is now purely real. The local amplification factor is again

$$A(z) = 1 + z + \frac{z^2}{2},$$

but because z is real, the phase of $A(z)$ is zero and the only relevant measure is its magnitude. Consequently, we track two parameters:

$$A_{\text{mod}} = |A(z)|, \quad AP_{\text{mod}} = |A(z)|^{\frac{2\pi}{|\alpha h|}}.$$

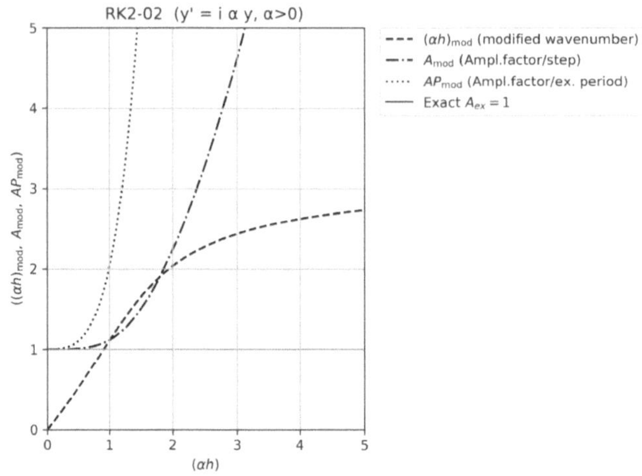

Figure 5.4: **Numerical parameters for ODE with oscillatory behavior using RK-O2.**

In this context, $\alpha < 0$ gives an exact solution that decays exponentially; hence the primary concern is whether $A(z)$ remains below unity (so that the method also decays) or deviates in a way that introduces spurious growth or an incorrect rate of decay.

In Figure 5.5, the per-step amplitude factor A_{mod} (dash-dotted line) and the per-period amplitude factor AP_{mod} (dotted line) are shown versus αh. Once more, $A_{\text{ex}} = 1$ stands for the exact solution's decay rate (i.e., no artificial growth).

In Figure 5.4 (oscillatory ODE), phase errors appear as deviations in $(\alpha h)_{\text{mod}}$, meaning the numerical solution oscillates with a frequency different from the exact one. Likewise, if $A_{\text{mod}} \neq 1$, an incorrect amplitude results. In Figure 5.5 (decaying ODE), there is no phase component, so only amplitude errors are plotted; again, a value of $A_{\text{mod}} \neq 1$ indicates unwanted growth or an incorrect decay rate. The quantity AP_{mod} in both cases demonstrates how these per-step discrepancies accumulate over one exact period of the underlying continuous solution.

These parameters thus reveal the characteristic errors introduced by the RK2-02 method. For the oscillatory ODE, a deviation in the modified wavenumber $(\alpha h)_{\text{mod}}$ signals phase errors, altering the effective oscillation frequency. For both the oscillatory and the decaying problems, any deviation of $|A(z)|$ from unity indicates how the method may erroneously amplify or damp the solution. By examining AP_{mod}, one can see the total effect of these discrepancies across a full cycle of the continuous solution.

To maintain accuracy when using the second-order Runge-Kutta method (RK2-O2) for solving both oscillatory and decaying ordinary differential equations, careful control of the time step h is critical.

Figure 5.4 shows that for the oscillatory case, where the equation is $y' = i\alpha y$ with $\alpha > 0$, the figure shows that both the amplitude factor per step A_{mod} and the ampli-

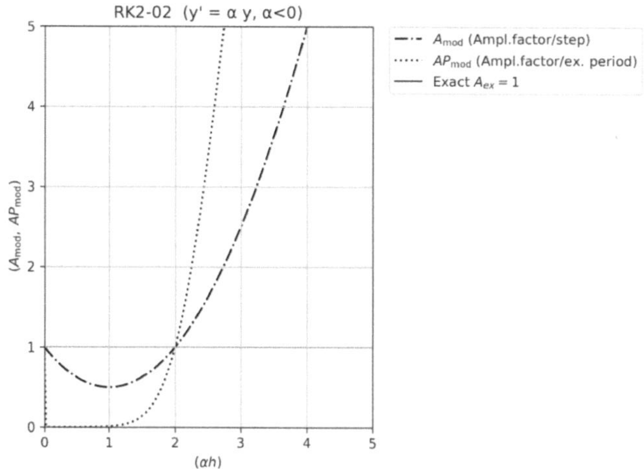

Figure 5.5: **Numerical parameters for ODE with decaying behavior using RK-O2.**

fication over a numerical period AP_{mod} increase rapidly as αh grows. Additionally, the modified wavenumber $(\alpha h)_{\text{mod}}$, which reflects phase accuracy, significantly deviates from the exact value. These distortions lead to incorrect amplitude and phase behavior, which compromise the fidelity of the numerical solution. To minimize such errors, the time step should be kept sufficiently small so that $\alpha h \lesssim 1$. This constraint ensures that the amplification remains close to unity and the phase error is controlled, preserving both the oscillation amplitude and wave propagation characteristics.

Figure 5.5 shows that for the decaying case, where the equation is $y' = \alpha y$ with $\alpha < 0$, the RK2-O2 method similarly fails to capture correct decay when αh becomes large. As shown in the corresponding plot, the amplification factor A_{mod} initially remains below 1—indicating decay—but eventually exceeds 1, introducing artificial growth. The per-period amplification AP_{mod} also grows rapidly, indicating accumulated numerical error. To maintain accurate decay behavior, the time step must be small enough to ensure that $|A(z)| < 1$ and that the decay rate is preserved. Typically, this requires that αh stays below 2, with better accuracy achieved for smaller values.

Classical 4th Order Runge Kutta Method: The algorithm advances the solution from (t_n, y_n) to $(t_{n+1} = t_n + h, y_{n+1})$ in four stages:

$$
\begin{aligned}
k_1 &= \lambda\, y_n, \\
k_2 &= \lambda \left(y_n + \tfrac{h}{2} k_1 \right), \\
k_3 &= \lambda \left(y_n + \tfrac{h}{2} k_2 \right), \\
k_4 &= \lambda \left(y_n + h k_3 \right), \\
y_{n+1} &= y_n + \frac{h}{6} \left(k_1 + 2k_2 + 2k_3 + k_4 \right).
\end{aligned}
$$

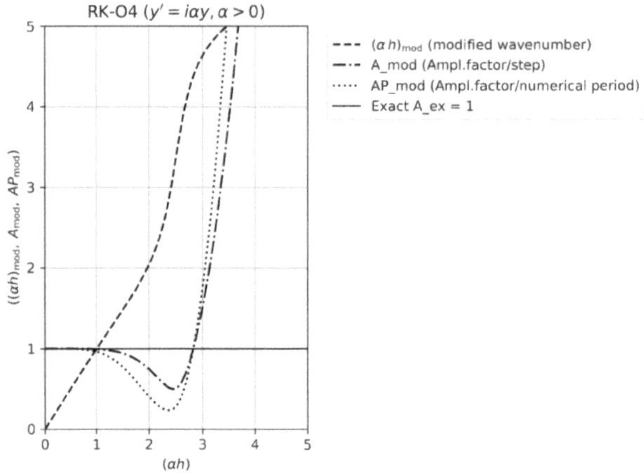

Figure 5.6: **Numerical parameters for ODE with oscillatory behavior using RK-O4.**

We use the local amplification factor, defined as

$$z = \lambda h.$$

Since the ODE is linear, we can factor out y_n from each stage and write

$$y_{n+1} = A(z)\, y_n,$$

where

$$A(z) = 1 + z + \frac{z^2}{2} + \frac{z^3}{6} + \frac{z^4}{24}$$

is the local amplification factor (LAF).

Oscillatory Case: If $\lambda = i\alpha$ with $\alpha > 0$, then $z = i\alpha h$ is purely imaginary. The factor $A(z)$ will generally have a magnitude different from 1 and a nonzero phase. We can define

$$A_{\text{mod}} = |A(z)|, \quad (\alpha h)_{\text{mod}} = \arg\big(A(z)\big), \quad AP_{\text{mod}} = |A(z)|^{\frac{2\pi}{\arg(A(z))}},$$

where the exponent $\frac{2\pi}{\arg(A(z))}$ corresponds to amplification over one numerical period.

To maintain accuracy when solving an oscillatory ordinary differential equation of the form $y' = i\alpha y$, where $\alpha > 0$, using the classical fourth-order Runge-Kutta (RK4) method, it is important to monitor how the numerical scheme distorts both amplitude and phase. As shown in Figure 5.6, the amplification factor per step $A_{\text{mod}} = |A(z)|$ deviates from unity as αh increases, leading to artificial amplification or damping of the oscillations. Simultaneously, the modified wavenumber $(\alpha h)_{\text{mod}} = \arg\big(A(z)\big)$ diverges from the exact value αh, introducing phase errors that result in inaccurate wave propagation speeds. The per-period amplification AP_{mod},

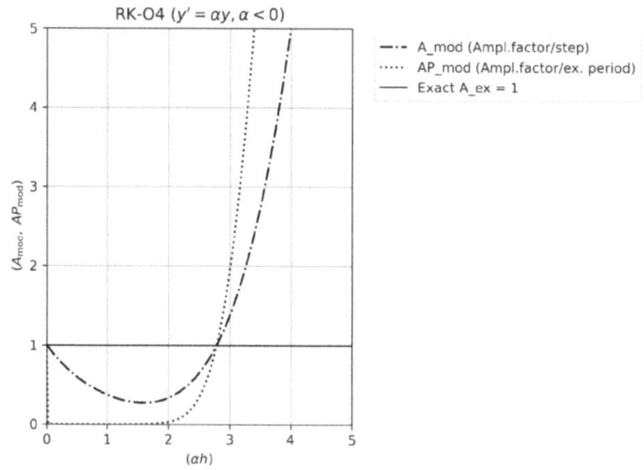

Figure 5.7: **Numerical parameters for ODE with decaying behavior using RK-O4.**

which reflects the cumulative effect over one full numerical period, also deviates significantly for large αh, further indicating loss of fidelity. To ensure both amplitude and phase accuracy, the time step h must be chosen such that αh remains small—typically $\alpha h \lesssim 2$. This ensures the numerical method stays within a regime where the magnitude and phase of $A(z)$ closely approximate those of the exact solution. Therefore, for oscillatory problems, reducing h not only improves stability but also preserves the correct wave characteristics in the numerical simulation.

Decaying Case: If $\lambda = \alpha < 0$ is real and negative, then $z = \alpha h$ is real. In this case, $\arg\big(A(z)\big) = 0$, and $|A(z)|$ determines whether the method preserves decay or introduces spurious growth. One may also define an amplification over a chosen period by

$$|A(z)|^{\frac{2\pi}{|\alpha h|}},$$

if desired.

Figures 5.6 and 5.7 show the numerical parameters for the oscillating and decaying behavior of ODE. Plots show that, while the RK4-O4 method generally performs well in preserving amplitude and phase, errors accumulate as the step size increases. These errors manifest as changes in amplitude (growth or decay) and phase shifts in the oscillatory case. The method is more accurate for smaller h, and its accuracy diminishes as αh increases.

If AP_{mod} is close to zero then it means that the method is introducing a significant error in amplitude, either by over-damping the oscillatory solution or decaying the solution too quickly. For oscillatory problems, this is undesirable, as it suggests the oscillations are being damped out incorrectly. For decaying problems, this suggests the decay rate is being exaggerated or the solution is decaying too rapidly compared to the true behavior.

To maintain accuracy when solving a decaying ordinary differential equation of the form $y' = \alpha y$, with $\alpha < 0$, using the classical fourth-order Runge-Kutta (RK4)

Table 5.2
General expressions for amplitude and phase behavior in damped and oscillatory model ODE $y' = \alpha y$ using numerical methods. The amplification factor $A(z)$ depends on the specific scheme (e.g., Euler, RK2, RK4 etc.).

Quantity	Damped Case	Oscillatory Case
	$(\alpha = \alpha_r < 0)$	$(\alpha = i\alpha_i)$
Exact Solution		
$y(t)$	$y_0 e^{\alpha_r t}$	$y_0 e^{i\alpha_i t}$
		$= y_0[\cos(\alpha_i t) + i\sin(\alpha_i t)]$
A_{ex}	$e^{\alpha_r t}$	1
θ_{ex}	0 or π	$\alpha_i t$
Numerical Method (General)		
Amplification factor $A(z)$	$z = \alpha_r h \in \mathbb{R}$	$z = i\alpha_i h \in i\mathbb{R}$
$A_{\text{mod}} = \lvert A(z)\rvert$	$\lvert A(z)\rvert$	$\lvert A(z)\rvert$
$\theta_{\text{mod}} = \arg(A(z))$	0 or π	$\arg(A(z))$
A^n_{mod}	$(A_{\text{mod}})^n$	$(A_{\text{mod}})^n$
AP_{mod}	- *	$A_{\text{mod}}^{2\pi/\theta_{\text{mod}}}$

*Not defined for damped systems, since no oscillation period exists.

method, it is essential to carefully control the time step h. Figure 5.7 illustrates that for small values of the product αh, the numerical amplification factor per step $A_{\text{mod}} = |A(z)|$ remains below one, which correctly reflects the decaying nature of the exact solution. However, as αh increases, the amplification factor begins to exceed one, indicating that the numerical solution may start exhibiting unphysical growth rather than decay. This behavior is further amplified when observing the per-period amplification AP_{mod}, which grows rapidly for large αh, suggesting the accumulation of error over time. To ensure both stability and fidelity to the true decaying dynamics, the time step h should be selected such that αh remains within the stable region—typically less than approximately 2.5. A conservative and practical guideline is to choose $h \ll \frac{1}{|\alpha|}$, which helps ensure that the numerical method faithfully reproduces the decay of the exact solution without introducing spurious amplification.

In general, it is essential to select time-stepping methods that accurately preserve both amplitude and phase characteristics. Small errors in numerical amplitude can artificially damp or exaggerate instability, while phase errors distort wave propagation and interaction. These inaccuracies lead to incorrect predictions. To mitigate such errors, the time step must be sufficiently small, and the numerical scheme should exhibit low dispersion and dissipation.

Table 5.2 summarizes the essential results governing amplitude and phase behavior in the damped and oscillatory regimes of the model ordinary differential equation $y' = \alpha y$. It outlines how one-step numerical methods represent the temporal evolution of disturbances, including the effects of amplification, phase shift, and per-period growth, based on the nature of the complex coefficient α.

PROBLEMS

5P-1 Consider the differential equation

$$y' = \left(-1 + i\pi\right) y, \quad y(0) = 1, \quad 0 \leq x \leq 4.$$

i Show that this ODE has an exact solution

$$y(x) = \exp\left((-1 + i\pi)x\right).$$

ii Find the exact amplitude $\left|y(x)\right|$ and the phase $\arg\left(y(x)\right)$ for $0 \leq x \leq 4$.

iii Apply the Euler Explicit and Euler Implicit methods with two different step sizes (for instance, $h = 0.1$ and $h = 0.2$) and compare:

a The net amplitude $\left(A_{\text{mod}}\right)^n$ after $n = \frac{4}{h}$ steps.

b The amplitude error relative to the exact amplitude at $x = 4$.

c The numerical phase and its deviation from the exact phase.

iv Conclude which method provides greater stability and accuracy under different conditions (such as larger h) and how the interplay between the negative real part -1 and the imaginary part π affects the oscillatory decay in the solution.

5P-2 Consider the following methods for solving the ODE $y' = \alpha \cdot y$, where $\alpha = -1 + 2i$, $y \in \mathbb{C}$, with initial condition $y(0) = 1$ and step sizes $h = \{0.1, 0.2, 0.3\}$:

i Implicit Euler

ii Heun's method (RK2)

iii RK4-O4

Comment on the results.

5P-3 Given the oscillation equation $\frac{dU}{dt} = i\omega U$, apply the von Neumann method to analyze the stability of different numerical schemes.

i Define the amplification factor A and explain its role in stability analysis.

ii Compare the stability of the Euler, Heun, and Backward schemes based on their amplification factors.

iii Discuss the stability condition for each scheme and identify which one approximates the true solution accurately.

iv Explain how the von Neumann method can predict and control instability in numerical simulations.

5P-4 Consider a fluid system where the wave frequency ω is real, and the perturbation is given by the equation $\hat{V} = v(y)e^{i\alpha x - i\omega t}$, where $\alpha = \alpha_r + i\alpha_i$ is the complex wave number. Here, α_r represents the real part, while α_i represents the imaginary part. The term ω is the frequency. Explain how the imaginary part of the wave number, α_i, governs the spatial growth rate of the disturbance. Discuss the significance of $\alpha_i < 0$ and $\alpha_i > 0$, and how these conditions determine whether the system experiences spatial instability or stability. Specifically, describe the physical meaning of a positive and negative α_i in terms of the growth or decay of disturbances in the system. Finally, derive the criterion for spatial instability based on the spatial growth rate and explain under which conditions the system becomes unstable.

5P-5 The Orr-Sommerfeld equation governs the linear evolution of infinitesimal disturbances in viscous shear flows and is expressed as:

$$\left(D^2 - \alpha^2\right)^2 \hat{v} = i\alpha Re\left[U\left(D^2 - \alpha^2\right)\hat{v} - U''\hat{v}\right],$$

where $\hat{v}(y,t)$ is the wall-normal disturbance velocity, $\alpha = \alpha_r + i\alpha_i$ is the complex streamwise wavenumber, $U(y)$ is the base flow profile, $D = d/dy$, and Re is the Reynolds number. After spatial discretization, the equation reduces to a system of ODEs for the modal state vector $\mathsf{V}(t)$, of the form $\dot{\mathsf{V}} = \mathsf{L}\mathsf{V}$, where, L is the discrete Orr-Sommerfeld operator matrix. Assuming a time-stepping approach is used to integrate this system such as explicit or implicit Euler, or Runge-Kutta RKO2 and RKO4 methods, explain how the numerical amplitude factor A_{mod} and phase $\theta_{\text{mod}} = \arg(A(z))$, derived from the method-specific amplification factor $A(z)$, influence the predicted stability characteristics. How do discrepancies in amplitude (e.g., $A_{\text{mod}} < A_{\text{ex}}$) or phase error $\Delta\Theta = |\theta_{\text{mod}} - \alpha_i h|$ distort the physical growth rate and wave speed of unstable modes? Discuss the implications of these numerical errors on identifying neutral stability curves and transition thresholds in the Orr-Sommerfeld equation.

5P-6 The following techniques for the solution of a differential equation $y' = \alpha y$ with $\alpha, y \in \mathbb{C}$ are considered:

ABO3

AMO3

RKO4

i Provide the order of the methods and the maximum extent of the method's stability region along the real and imaginary axes (s_r, s_i) for all three methods:

ii Plot the accuracy diagram for decaying solutions ($\alpha = \alpha_r < 0$, A_{mod} over $-\alpha_r h$, step size) for all three methods and include the stability limits s_r.

5P-7 The following techniques for the solution of a differential equation $y' = \alpha y$ with $\alpha, y \in \mathbb{C}$ are considered:

RK4-O4

Trapezoidal rule

Explicit Euler

i Provide the order of the methods and the stability limits (s_r, s_i) for all three methods:

ii Plot the accuracy diagram for decaying solutions ($\alpha = \alpha_r < 0$, A_{mod} over $-\alpha_r h$, step size) for all three methods and include the stability limits s_r.

REFERENCES

M. J. Kloker, Numerische Löser (Zeitintegrationsverfahren) für die gewöhnliche Modelldifferentialgleichung $y' = \alpha y$, Institut für Aerodynamik und Gasdynamik der Universität Stuttgart, 1996.

Evans, Partial Differential Equations (2nd ed.), American Mathematical Society, 2010.

Strauss, Partial Differential Equations: An Introduction (2nd ed.), John Wiley & Sons.

6 Boundary Value Problems

In this chapter, we will delve into the concept of Boundary Value Problems (BVPs), focusing on their application in solving physical systems with boundary conditions specified at different points in the domain. Boundary value problems arise in a wide range of fields, from fluid dynamics to heat transfer, and their solutions often require numerical methods due to the complexity of the governing differential equations. After finishing this chapter, you should be able to:

- Understand the theory behind Boundary Value Problems and how they differ from Initial Value Problems.
- Learn how to apply the shooting method to solve nonlinear boundary value problems, such as the Blasius equation.
- Be familiar with how boundary conditions are incorporated into the numerical solution of fluid dynamics problems.

Boundary Value Problems (BVPs) are a class of problems in which the solution to a differential equation must satisfy specified conditions at different points, usually at the boundaries of the domain. Unlike Initial Value Problems (IVPs), where the solution is determined by initial conditions at a single point in time, BVPs are governed by conditions at multiple points, making them more complex and challenging to solve. These types of problems are ubiquitous in engineering and physics, particularly in fluid dynamics, heat transfer, and structural analysis. The solutions to BVPs often describe steady-state systems or systems that evolve under steady conditions over a spatial domain.

This chapter focuses on the methods used to solve BVPs, with a particular emphasis on the shooting method. The shooting method is a powerful technique that transforms a boundary value problem into an initial value problem, making it solvable with standard numerical integration methods like Runge-Kutta. By guessing the initial condition for the unknown function, solving the resulting IVP, and iteratively adjusting the guess, the correct solution that satisfies the boundary conditions at both ends of the domain can be obtained. This method is particularly useful for solving nonlinear BVPs, such as the Blasius equation, which describes the velocity profile in a laminar boundary layer over a flat plate. The Blasius equation itself serves as a key example, demonstrating the application of the shooting method in solving a third-order nonlinear ordinary differential equation. By transforming this equation into a system of first-order ODEs, we can apply numerical methods like Runge-Kutta to solve it. The shooting method iteratively adjusts the guess for the unknown boundary condition until the solution converges, ensuring that the boundary conditions at both the plate surface and far from the surface are satisfied. In addition to discussing the shooting method, this chapter explores the concept of pathlines in fluid flow, a

DOI: 10.1201/9781003663171-6

fundamental aspect of fluid dynamics. Pathlines describe the trajectories followed by individual fluid particles over time. We will work through an example of a two-dimensional flow field and derive the parametric equations for the pathline of a fluid particle, illustrating the application of separation of variables and integration techniques.

6.1 APPLICATION: SOLUTION OF BLASIUS PROBLEM

The Blasius equation is a third-order nonlinear boundary value problem because it requires solving a differential equation with boundary conditions specified at different points (in this case, at $\eta = 0$ and $\eta \to \infty$). The Blasius equation describes the steady, incompressible laminar boundary layer over a flat plate. The equation is a third-order nonlinear ordinary differential equation (ODE) derived from the Navier-Stokes equations for boundary layer flow. It governs the velocity profile in the boundary layer, where the velocity components in the boundary layer are functions of a similarity variable η, defined as:

$$\eta = \frac{y}{\sqrt{\frac{\nu x}{U_\infty}}}$$

where y is the distance from the plate, x is the distance along the plate, U_∞ is the free-stream velocity, and ν is the kinematic viscosity.

The Blasius equation is:

$$\frac{d^3 f}{d\eta^3} + \frac{1}{2} f \frac{d^2 f}{d\eta^2} = 0$$

where: - $f(\eta)$ is a dimensionless stream function related to the velocity components. - $f'(\eta)$ represents the dimensionless velocity profile in the boundary layer, with $f'(0) = 0$ and $f'(\eta \to \infty) = 1$. - $f''(\eta)$ represents the velocity gradient at the wall, which is linked to the shear stress.

The boundary conditions for the Blasius equation are as follows:

$$f(0) = 0, \quad f'(0) = 0, \quad f'(\eta \to \infty) = 1$$

These conditions represent the no-slip condition at the plate surface ($f(0) = 0$ and $f'(0) = 0$) and the free-stream velocity far from the surface ($f'(\eta \to \infty) = 1$).

The Blasius equation is a nonlinear boundary value problem. To solve this, the shooting method is applied, which converts the boundary value problem into an initial value problem. The method involves:
Guessing the unknown initial condition for $f''(0)$, denoted as $f''(0) = \text{guess}$.
Solving the resulting initial value problem using a numerical integration scheme, typically a Runge-Kutta method.
Iteratively adjusting the guessed value of $f''(0)$ until the boundary condition at $\eta \to \infty$, $f'(\eta \to \infty) = 1$, is satisfied.

The shooting method uses root-finding techniques (such as the bisection method) to find the correct value of $f''(0)$ that satisfies the boundary conditions.

The Blasius equation is transformed into a system of three first-order ODEs:

$$\frac{df}{d\eta} = f'$$

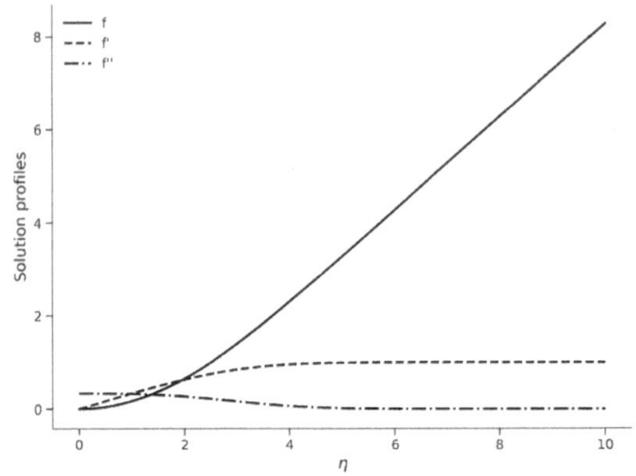

Figure 6.1: **Blasius Solution.**

$$\frac{df'}{d\eta} = f''$$

$$\frac{df''}{d\eta} = -\frac{1}{2} f f''$$

With initial conditions:

$$f(0) = 0, \quad f'(0) = 0, \quad f''(0) = \text{guess}$$

The goal is to integrate these equations numerically using the guessed value of $f''(0)$ and then adjust the guess until the condition $f'(\eta \to \infty) = 1$ is met.

The function `solve_ivp` in `SciPy` integrates an ordinary differential equation as an initial value problem. Laminar flat plate flow is a boundary value problem,

$$2f''' + f f'' = 0, \quad f(0) = 0, \; f'(0) = 0, \; f'(\eta \to \infty) = 1,$$

and, we can apply the shooting method to turn it into a sequence of initial value problems (IVPs):

Assume a trial value for the missing initial condition (e.g., $f''(0) = \alpha$).

Use `solve_ivp` to integrate

$$\frac{dy}{d\eta} = F(\eta, y), \quad y(0) = [f(0), f'(0), \alpha],$$

from $\eta = 0$ to a large η.

Evaluate the residual $R(\alpha) = f'(\eta_{\max}; \alpha) - 1$.

Adjust α (via a root-finding routine such as bisection or secant) until $R(\alpha) = 0$.

In this way, each "shot" turns the boundary value problem (BVP) into an IVP solved by `solve_ivp`, and the outer root finder ensures the far-field boundary condition is met.

Pseudo algorithm for `solve_ivp`

1. Define the ODE system $\frac{dy}{dt} = f(t,y)$, the interval $[t_0, t_f]$ and the initial condition $y(t_0) = y_0$.

2. Choose initial time step h, absolute and relative tolerances $(atol, rtol)$, and points t_{eval} at which to store the solution.

3. Initialize
$$t \leftarrow t_0, \quad y \leftarrow y_0, \quad \mathscr{S} \leftarrow \{(t_0, y_0)\}.$$

4. While $t < t_f$ do

   (a) If $t + h > t_f$, set $h \leftarrow t_f - t$.

   (b) Compute Runge-Kutta stage increments
   $$k_i = f\big(t + c_i h,\ y + \sum_{j<i} a_{ij} h k_j\big) \quad \text{for } i = 1, \ldots, s.$$

   (c) Form the fourth and fifth order solutions
   $$y^{(4)} = y + h \sum_i b_i^{(4)} k_i, \quad y^{(5)} = y + h \sum_i b_i^{(5)} k_i.$$

   (d) Estimate the local error
   $$\varepsilon = \left\| y^{(5)} - y^{(4)} \right\|.$$

   (e) Compute the allowed error tolerance
   $$\varepsilon_{\text{tol}} = atol + rtol \max\big(\|y\|, \|y^{(5)}\|\big).$$

   (f) If $\varepsilon \leq \varepsilon_{\text{tol}}$ then i. Accept the step:
   $$t \leftarrow t + h, \quad y \leftarrow y^{(5)}, \quad \mathscr{S} \leftarrow \mathscr{S} \cup \{(t, y)\}.$$

   (g) Adjust the time step
   $$h \leftarrow 0.9\, h \left(\tfrac{\varepsilon_{\text{tol}}}{\varepsilon}\right)^{1/5}.$$

   (h) If step was rejected, do not advance t; repeat from 4(b.) with the new h.

5. Interpolate or extract the solution at the specified t_{eval} from $\mathscr{S}$.

6. Return a solution object containing $\{t, y\}$, solver status, and diagnostics.

Python code for Blasius Solution

See `Python` code `Ch6-Balsius` for solution of Blasius problem.

PROBLEMS

6P-1 Consider the boundary value problem:

$$y'' + 2002y' + 2001y = 0$$

If we want to solve it using shooting method then does the resulting system of ODEs have any property that makes the numerical solution difficult to achieve?

6P-2 Consider the second-order differential equation

$$y'' + 2y' - 13y = 0.$$

Transform this boundary value problem into an equivalent system of first-order ordinary differential equations suitable for numerical solution via the shooting method.

Discuss whether the resulting system exhibits any properties that could pose challenges to numerical integration. If such properties exist, identify and explain them.

6P-3 Given the ordinary differential equation:

$$2y'' - 204i \cdot y' - 400y = 0$$

i Rearrange the ODE so that it can be solved using a shooting method.

ii Does the resulting system have properties that make the numerical solution a difficult task? if so, what are these properties called?

6P-4 Solve the boundary value problem using the shooting method:

$$y'' + 4y' - 5y = 0$$

Assess whether the system of ordinary differential equations derived from this reformulation possesses characteristics that could challenge the numerical solution.

6P-5 Consider a steady two-dimensional laminar boundary layer developing over a semi-infinite flat plate in a uniform free stream of velocity U_∞ and temperature T_∞. Through the similarity transformation one obtains the Blasius equation

$$2f'''(\eta) + f(\eta)f''(\eta) = 0$$

subject to the boundary conditions: $f(0) = 0, \quad f'(0) = 0, \quad f'(\eta) \to 1 \quad \text{as } \eta \to \infty$, and the thermal boundary layer equation

$$\theta''(\eta) + \frac{\Pr}{2} f(\eta)\theta'(\eta) = 0$$

with boundary conditions: $\theta(0) = 0, \quad \theta(\eta) \to 1 \quad$ as $\eta \to \infty$. This is known as Polhausen problem. Solve this problem for Prandtl number Pr=0.72, using `solve_ivp` and find the values of $f'(\eta)$ and $\theta(\eta)$. Determine the local skin-friction coefficient $C_f(x) = 2f''(0)/\sqrt{\mathrm{Re}_x}$, and prove that the local Nusselt number $Nu_x = \theta'(0)\sqrt{\mathrm{Re}_x}$, and discuss how varying the Prandtl number Pr influences both the velocity and thermal boundary layer thicknesses as well as the rate of wall heat transfer. Prove that $\theta'(0) = 0.29561$ at η=0.

REFERENCES

L. Prandtl, Über Flüssigkeitsbewegung bei sehr kleiner Reibung (On fluid motion with very small friction), presented at the III. Internationalen Mathematiker-Kongress, Heidelberg, 1904. Published in Verhandlungen des III. Internationalen Mathematiker-Kongresses, Heidelberg 1904, B. G. Teubner, Leipzig, pp. 484-491, 1905.

P. J. Roache, Computational fluid dynamics. Hermosa, Albuquerque, USA, 1977.

7 Advection Equation

In this chapter, we will explore the one-dimensional linear advection equation, which models the transport of a quantity in a given direction. After completing this chapter, you should be able to:

- Understand the advection equation and its significance in modeling transport phenomena.
- Analyze the stability of numerical schemes for the advection equation using linear stability analysis.
- Evaluate phase errors and phase ratios in various numerical discretizations of the advection equation.
- Implement and compare numerical schemes such as Leapfrog (LF) and Crank-Nicolson (CN) for the advection equation.
- Evaluate group velocity, phase error, and amplitude response in different schemes.

The linear advection equation models the transport of a scalar quantity, such as heat or pollutants, and is central to many physical processes. This chapter focuses on solving this equation numerically, with an emphasis on stability, accuracy, and the challenges introduced by numerical methods like finite differences. The Courant-Friedrichs-Lewy (CFL) condition plays a critical role in ensuring stability by constraining the time step relative to the spatial grid and wave speed. We also explore the effects of numerical errors, such as dispersion and diffusion, on wave propagation. By analyzing different numerical schemes, including Euler Explicit and Crank-Nicolson methods, we evaluate their performance in terms of amplitude and phase accuracy.

The linear advection equation can be generally written as:

$$\frac{\partial u}{\partial t} + a\frac{\partial u}{\partial x} = 0$$

where:

u is the quantity being advected (transported),
a is the advection speed, which is a constant in the simplest case,
t represents time,
x represents space.

This equation states that the rate of change of u with respect to time plus the advection speed times the spatial derivative of u equals zero. The implication is that the quantity u is being transported along the x-direction at a speed a. The solution to this equation represents the propagation of waveforms moving without alteration in shape at the speed a. If a is positive, the wave moves to the right; if negative, it moves to the left.

DOI: 10.1201/9781003663171-7

The linear advection equation is a hyperbolic equation. For the linear advection equation, the characteristics are straight lines along which the initial data is transported. In this case, the characteristics are given by $x - at =$ constant, which means any disturbance initially at position x_0 moves to the right (if $a > 0$) along the line determined by its initial position at a speed a.

In numerical simulations, capturing the behavior of the linear advection equation accurately requires careful consideration of the discretization methods, especially because numerical diffusion and dispersion can distort the waveform: Some numerical schemes introduce artificial diffusion which can cause the wave to spread or dissipate, which isn't a characteristic of the ideal advection equation. The *Courant-Friedrichs-Lewy* (CFL) condition is crucial in the numerical solution to ensure stability. It relates the time step size Δt, the spatial grid size Δx, and the wave speed a to maintain stability.

7.1 LINEAR STABILITY ANALYSIS FOR ADVECTION PROBLEM

In numerical solutions of partial differential equations, stability analysis is essential to ensure that errors introduced during computation do not grow over time, compromising the accuracy and reliability of the solution. This section focuses on the stability characteristics of the one-dimensional advection equation, a fundamental equation that models the transport of quantities in a given direction. For this purpose, we examine a discretized version of the advection equation, which uses a finite difference approach to approximate the time and spatial derivatives. A key parameter in this analysis is the Courant number, a dimensionless factor that represents the ratio between the numerical wave propagation speed and the actual physical wave speed. The Courant number dictates the stability of the scheme: if it exceeds a certain threshold, the solution becomes unstable, leading to amplified errors over time. By conducting a linear stability analysis, we evaluate how different values of the Courant number affect the behavior of the numerical solution. This includes identifying the range of Courant numbers for which the numerical method remains stable, as well as assessing both amplitude and phase accuracy, which are crucial for faithfully capturing wave propagation. This comprehensive analysis not only ensures stability but also provides insights into the accuracy and fidelity of the numerical solution in representing the underlying physical phenomenon.

$$\frac{\partial h}{\partial t} + c\frac{\partial h}{\partial x} = 0$$

7.1.1 VON NEUMANN PROCEDURE

Von Neumann analysis is widely regarded as the most commonly used approach for determining stability criteria due to its simplicity, straightforward application, and reliability. However, it is limited to providing necessary and sufficient conditions for the stability of linear initial value problems with constant coefficients. Using finite difference discretization of a 1D advection equation with a forward difference in time and a backward difference in space, characteristic of an explicit upwind scheme, we have

$$\frac{h_j^{n+1}-h_j^n}{\Delta t}+c\frac{h_j^n-h_{j-1}^n}{\Delta x}=0$$

$$h_j^{n+1}=h_j^n-\frac{c\Delta t}{\Delta x}h_j^n+\frac{c\Delta t}{\Delta x}h_{j-1}^n$$

$$h_j^{n+1}=(1-CFL)h_j^n+CFLh_{j-1}^n$$

where,

$$CFL=\frac{\text{Physical distance a wave travels in one time step with highest speed}}{\text{Minimal numerical grid spacing}}=\frac{a_{\max}\Delta t}{\Delta x_{\min}}$$

It quantifies whether information can move across more than one numerical cell per step.

introducing

$$h_j^{n+1}=\xi^{n+1}e^{ik^*j}$$

$$h_{j+1}^n=\xi^n e^{ik^*(j+1)}$$

we have

$$\xi_{ij}^{n+1}=(1-CFL)\xi_{ij}^n+CFL\xi_{i(j-1)}^n$$

$$\xi^{n+1}e^{ik^*j}=(1-CFL)\xi^n e^{ik^*j}+CFL\xi^n e^{ik^*(j-1)}$$

$$\xi^{n+1}=\left[(1-CFL)+CFLe^{-ik^*}\right]\xi^n$$

$$A=1-CFL+CFLe^{-ik^*}$$

Figure 7.1 presents the solution computed using the Euler explicit method with the 2-point finite difference stencil. It is important to note that the 2-point scheme is simply the first-order upwind method. As the wave propagates in time, the solution exhibits noticeable damping, a characteristic effect of the numerical diffusion inherent in this low-order scheme.

Figure 7.2 shows the stability regions in the complex plane for different values of CFL. The circles represent the loci of the amplification factor A for CFL values of $1/4$, $1/2$, $3/4$, 1, and $5/4$, plotted against the real and imaginary axes. The dashed line is the unit circle which marks the boundary of $|A|\leq 1$, indicating stability. The shaded region inside the unit circle shows the stable region where the condition $|A|\leq 1$ is satisfied.

The Courant number (CFL), which is also called Courant-Friedrichs-Lewy (CFL) condition, is important for ensuring stability in computational fluid mechanics.

1. Stability Condition for CFL ≤ 1 : When the Courant number CFL is less than or equal to 1, the modulus of the amplification factor $|A|$ is less than or equal to 1. This means that for any wavenumber k^*, the Fourier coefficients of the numerical method are not growing in time, ensuring stability.

2. Instability for CFL > 1 : If CFL exceeds 1, then $|A|>1$, meaning the Fourier coefficients are amplified at every time step, leading to instability. In other words, when CFL is greater than 1, the method becomes unstable because errors in the numerical solution grow over time.

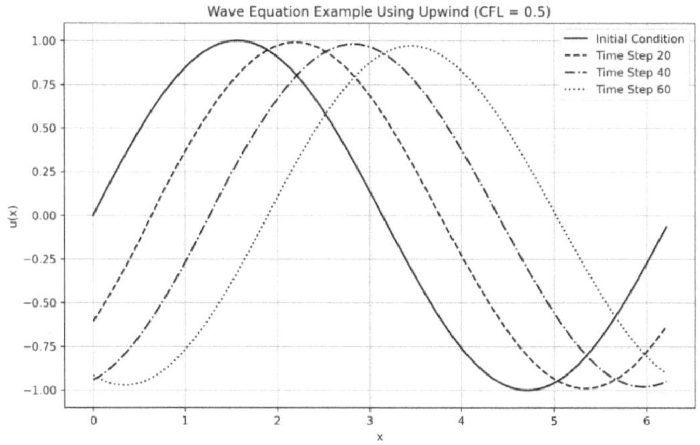

Figure 7.1: **Solution of linear advection equation using Euler-explicit 2pt-FD stencil.**

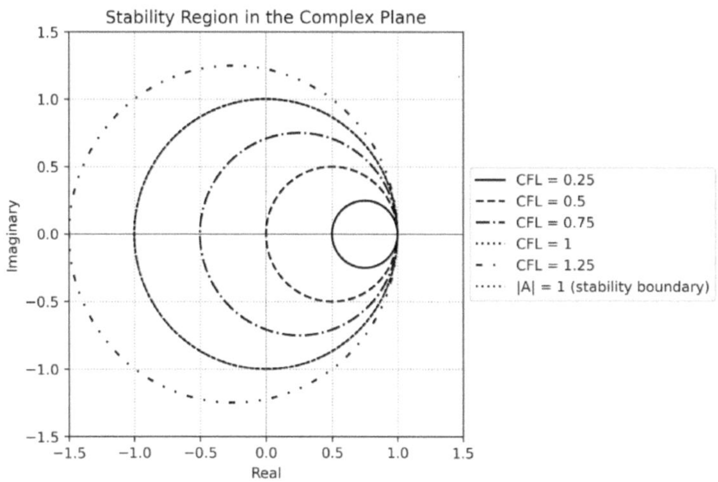

Figure 7.2: **Argand diagram of amplification factor.**

3. Stability Condition : For stability, the condition CFL ≤ 1 must be satisfied. This is directly related to the CFL condition, which defines the maximum allowable time step size in relation to the grid spacing for a numerical method to remain stable. This is critical for solving partial differential equations (PDEs) numerically, particularly in time-dependent problems.

The CFL ≤ 1 is the key stability condition, identical to the CFL condition, ensuring the numerical method does not amplify errors over time. Additionally, the analysis of amplitude and phase accuracy offers a more comprehensive evaluation of the method's performance.

The Courant-Friedrichs-Lewy (CFL) number plays a critical role in the stability and accuracy of numerical schemes for solving partial differential equations, particularly in the context of the one-dimensional linear advection equation. In numerical simulations, the CFL number governs the relationship between the time step, spatial step, and the speed of wave propagation, directly affecting the performance of discretization methods. For explicit schemes such as Euler-Explicit and Upwind, adherence to a suitable CFL condition is essential to prevent numerical instabilities and oscillations. This following problem examines the influence of the CFL number on the stability and accuracy of these schemes, illustrating how variations in the CFL number impact error propagation, solution smoothness, and overall fidelity to the underlying physics of advection.

Example 7.1

Problem: One-Dimensional Linear Advection—Explicit Upwind Scheme Analysis

Consider the one-dimensional linear advection equation given by

$$\frac{\partial u}{\partial t} + a\frac{\partial u}{\partial x} = 0$$

on the spatial domain $0 \leq x \leq 5$, with a constant wave speed $a = 1.0$. The initial condition is defined as a step function:

$$u(x,0) = \begin{cases} 1, & \text{if } 1.8 \leq x \leq 2.2 \\ 0, & \text{otherwise} \end{cases}$$

Assume a fixed boundary condition at the left boundary and simulate the system using a first-order **Euler explicit upwind scheme**, with the spatial derivative approximated as

$$\left.\frac{\partial u}{\partial x}\right|_j \approx \frac{u_j - u_{j-1}}{\Delta x}$$

and the time derivative discretized as

$$\left.\frac{\partial u}{\partial t}\right|_j \approx \frac{u_j^{n+1} - u_j^n}{\Delta t}.$$

i Implement the Euler explicit upwind scheme for the initial step condition described above using a uniform grid.

ii Investigate till final time $T = 3.0$ using three different Courant numbers: $CFL = 0.5$, 1.0, and 2.0.

iii Analyze and discuss the behavior of the numerical solution for each CFL case. Comment on damping, numerical diffusion, dispersion, and instability.

iv Identify which CFL values lead to acceptable solutions and which result in spurious oscillations or instabilities.

Solution

We may code this problem in python. See `Python` code `Ch7-Ex71` for solution of this problem.

Figure 7.3 shows the influence of CFL and dt/dx over the solution progress. At CFL=1/2, the solution damps out as it progress in time. With CFL=1, the solution just shift, and in case of CFL=2, the solutions grows and damps out with spurious oscillations. The solution is stable as long as CFL remains less than 1.

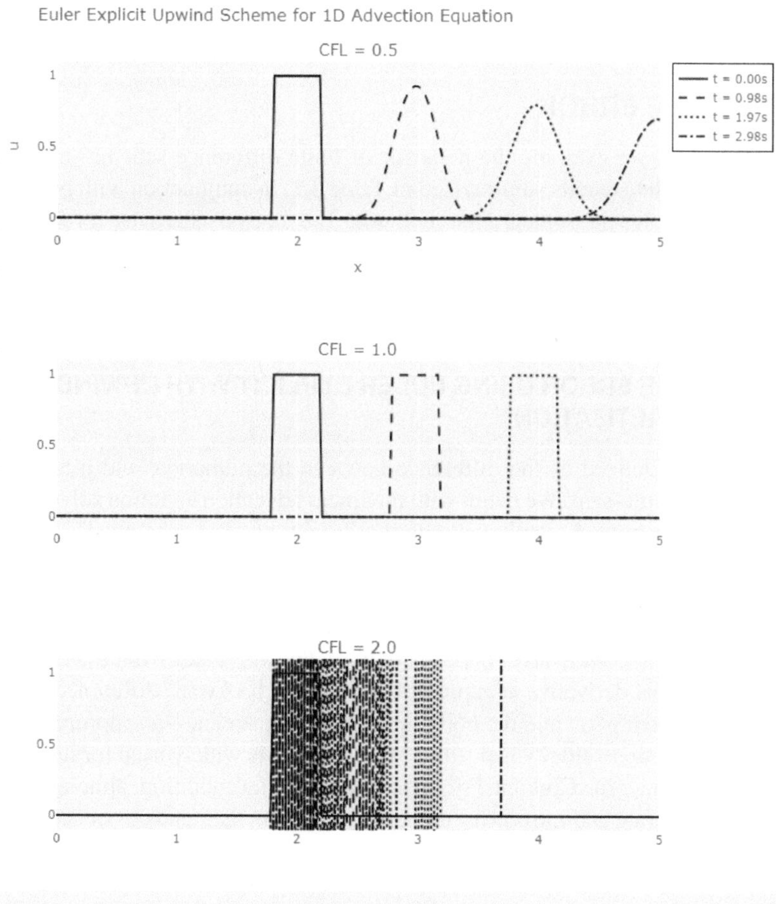

Figure 7.3: **Euler Explicit with Upwind Scheme: The solution is unstable at CFL=2.**

Lessons:

☞ The explicit upwind scheme is conditionally stable and only valid for CFL ≤ 1.

☞ At CFL = 0.5, the scheme is stable but exhibits numerical diffusion.

☞ At CFL = 1.0, the wave propagates correctly but accuracy depends on resolution.

☞ At CFL = 2.0, the scheme becomes unstable and develops spurious oscillations.

☞ Low-order upwind schemes smear sharp gradients significantly over time.

7.2 PHASE ERROR

In this section, we examine the behavior of finite difference schemes using the 2-point and 3-point stencils summarized in Table 3.1, in conjunction with both explicit and implicit Euler time integration methods. The analysis starts by formulating the discrete numerical scheme, followed by an investigation of the resulting solution characteristics—particularly how the numerical wave's amplitude and phase compare to those of the exact solution.

7.2.1 PHASE ERROR USING EULER EXPLICIT WITH UPWIND SPATIAL DISCRETIZATION

Phase error is defined as the difference between the numerical and true phase angles after one time-step. We begin with the linear advection equation solved using an Euler-explicit approach combined with the 2-point upwind scheme. The discretization of the one-dimensional linear advection equation using a combination of forward Euler time integration and a two-point upwind spatial scheme has already been presented in section 7.1.1. In this scheme, the time is advanced explicitly, meaning the solution at the next time level depends only on known values from the current time level. The spatial derivative is approximated using a backward difference, which relies on the current point and the point immediately upstream—an appropriate choice when the advection velocity is positive. This scheme is widely used for its simplicity and stability under the Courant-Friedrichs-Lewy (CFL) condition, although it is only first-order accurate and introduces numerical diffusion that tends to smear sharp gradients over time. Earlier, we have developed the expression for amplification factor as

$$A = 1 - \text{CFL}(1 - e^{-ik\Delta x})$$

where $\text{CFL} = \frac{a\Delta t}{\Delta x}$ is the Courant-Friedrichs-Lewy number.

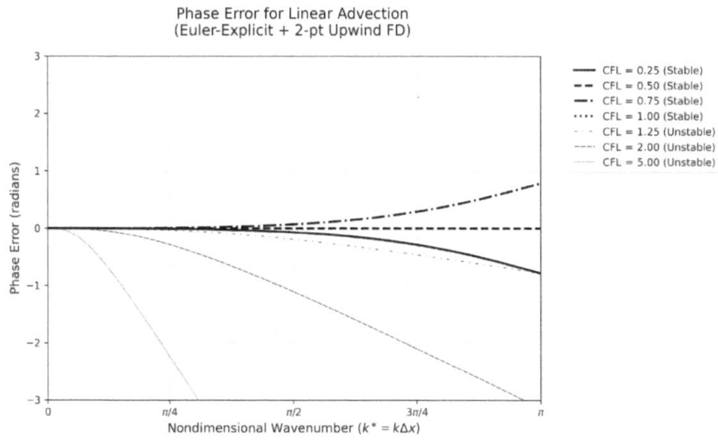

Figure 7.4: **Phase error associated with Euler-explicit 2pt-FD stencil for linear advection equation.**

From the above:

$$A = 1 - \text{CFL}(1 - \cos(k\Delta x) + i\sin(k\Delta x))$$

$$A = 1 - \text{CFL} + \text{CFL}\cos(k\Delta x) - i\text{CFL}\sin(k\Delta x)$$

The amplitude A gives the rate at which the amplitude of the numerical solution changes relative to the true solution:

$$|A| = \sqrt{(1 - \text{CFL} + \text{CFL}\cos(k\Delta x))^2 + (\text{CFL}\sin(k\Delta x))^2}$$

The phase error measures the shift in phase angle per time step compared to the exact solution. The phase error, incorporating the negative numerical phase shift explicitly, is

$$\phi|_{num} = \arg(A) = \tan^{-1}\left(\frac{-\text{CFL}\sin(k\Delta x)}{1 - \text{CFL} + \text{CFL}\cos(k\Delta x)}\right)$$

$$er_{\omega} = \phi|_{exact} - \phi|_{num}$$

$$er_{\omega} = k^* \cdot CFL - \phi|_{num}$$

This phase error is plotted in Figure 7.4. The phase error shows the error in the wave's speed introduced by the numerical scheme, while the amplitude error shows how the scheme affects the wave's amplitude, potentially leading to numerical damping or growth. The accuracy of both phase and amplitude in numerical simulations is critical in preserving the physical properties of the wave over time.

Example 7.2

Problem: One-Dimensional Linear Advection—Analysis of Numerical Schemes

Consider the one-dimensional linear advection equation

$$\frac{\partial u}{\partial t} + a\frac{\partial u}{\partial x} = 0$$

on a spatially periodic domain $0 \leq x \leq 2\pi$, where the wave speed $a = 1.0$ is constant. This problem is to be numerically solved using Forward-Time Central-Space (FTCS) scheme finite-difference schemes to study their accuracy properties under different Courant-Friedrichs-Lewy (CFL) numbers. The initial condition is prescribed as a smooth sinusoidal wave $u(x,0) = \sin x$. Apply the Forward-Time Central-Space (FTCS) scheme, where spatial derivatives are approximated using a centered second-order finite difference, and time integration is via Euler-explicit method. Observe how the numerical solution evolves as the wave circulates around the periodic domain and analyze the impact of the instability inherent in this scheme. Present results using several CFL numbers to illustrate how increasing the time step affects the phase error and amplitude preservation.

Assess the development of spurious oscillations, new extrema, and dispersion artifacts as the wave propagates. Discuss how the behavior changes at CFL = 0.5, 1.0, and 2.0. Comment on the inability of this scheme to maintain boundedness and monotonicity in the presence of discontinuities or steep gradients.

Solution

Advection equation with smooth initial condition

We will solve the one-dimensional linear advection equation

$$\frac{\partial u}{\partial t} + a\frac{\partial u}{\partial x} = 0$$

on a uniform periodic domain $0 \leq x \leq 2\pi$, with constant wave speed a=1.0. Spatial differentiation is performed with the central finite-difference stencil, and the solution advances in time by an explicit forward-Euler step whose size follows $CFL = 0.1$. This forward-time central-space (FTCS) scheme keeps all variables on the same grid and is linearly unstable for pure advection. Starting from the smooth wave $u(x,0) = sinx$, the solution evolution is investigated.

Figure 7.5 demonstrates how the wave, as it repeatedly traverses the periodic domain, suffers from a growing phase error and a growth in amplitude. This behavior is a direct result of the linear instability inherent in the Euler explicit scheme when coupled with a central difference stencil for pure advection. The plot illustrates how error accumulates with time, producing a visible phase shift and distortion in wave shape, even at a low CFL number.

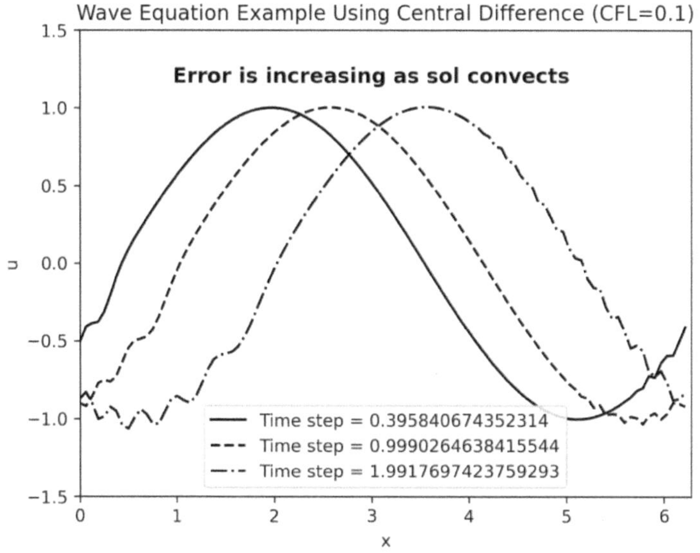

Figure 7.5: **Phase error for advection equation (Euler Explicit 3 point FD stencil).**

Lessons:

☞ FTCS is linearly unstable for pure advection at any Courant-Friedrichs-Lewy number; its amplification factor always exceeds unity except at $\theta = 0$.

☞ Instability appears as unbounded oscillations and rapid amplitude growth, with the highest-frequency modes diverging fastest.

☞ Phase error accumulates each timestep, producing an increasing phase lag that distorts the wave long before blow-up.

☞ Because of these defects, FTCS is inappropriate for convection-dominated problems; use a stable upwind or Lax-Wendroff-type scheme instead (see Chapter 16).

7.2.2 EULER IMPLICIT WITH UPWIND SPATIAL DISCRETIZATION

In this approach, the one-dimensional linear advection equation is solved using an **implicit upwind scheme**. Time discretization is performed implicitly, meaning that the solution at the future time level $n+1$ is used in the formulation. For the spatial derivative, an upwind finite difference is applied, which naturally involves the value at the previous spatial location, also at the future time step. Specifically, the time derivative is approximated as

$$\frac{\partial u}{\partial t} \approx \frac{u_j^{n+1} - u_j^n}{\Delta t},$$

while the spatial derivative is represented using the backward difference form

$$\frac{\partial u}{\partial x} \approx \frac{u_j^{n+1} - u_{j-1}^{n+1}}{\Delta x}.$$

Substituting into the advection equation $\frac{\partial u}{\partial t} + a\frac{\partial u}{\partial x} = 0$, the resulting discretized equation is

$$\frac{u_j^{n+1} - u_j^n}{\Delta t} + a\frac{u_j^{n+1} - u_{j-1}^{n+1}}{\Delta x} = 0.$$

This leads to a linear system involving the unknown values u_j^{n+1}, which must be solved at each time step.

Temporal Discretization :

$$\frac{\partial u}{\partial t} \approx \frac{u_j^{n+1} - u_j^n}{\Delta t}$$

Spatial Discretization (Upwind):

$$\frac{\partial u}{\partial x} \approx \frac{u_j^{n+1} - u_{j-1}^{n+1}}{\Delta x}$$

Advection Equation Setup : Plugging these into the advection equation $\frac{\partial u}{\partial t} + a\frac{\partial u}{\partial x} = 0$ gives:

$$\frac{u_j^{n+1} - u_j^n}{\Delta t} + a\frac{u_j^{n+1} - u_{j-1}^{n+1}}{\Delta x} = 0$$

Rearranging : Multiplying through by Δt and rearranging terms to isolate u_j^{n+1} on one side:

$$u_j^{n+1} - u_j^n + a\frac{\Delta t}{\Delta x}(u_j^{n+1} - u_{j-1}^{n+1}) = 0$$

$$u_j^{n+1} = u_j^n - a\frac{\Delta t}{\Delta x}(u_j^{n+1} - u_{j-1}^{n+1})$$

$$u_j^{n+1} = u_j^n - CFL(u_j^{n+1} - u_{j-1}^{n+1})$$

Example 7.3

Consider the one-dimensional linear advection equation

$$\frac{\partial u}{\partial t} + a\frac{\partial u}{\partial x} = 0$$

on a spatially periodic domain $0 \leq x \leq 2\pi$, where the wave speed $a = 1.0$ is constant. This problem is to be numerically solved using Euler implicit-upwind finite difference scheme to study their accuracy properties under different Courant-Friedrichs-Lewy (CFL) numbers. Consider a sharp initial pulse condition:

$$u(x,0) = \begin{cases} 1, & x_p \leq x < x_p + w \\ 0, & \text{otherwise} \end{cases}$$

where, $x_p = L/3$ is the starting position of the pulse (approximately one-third into the domain of length L), and $w = \texttt{pulse\_width} \cdot \Delta x$ is the width of the pulse. Assess the development of spurious oscillations, new extrema, and dispersion artifacts as the wave propagates. Discuss how the behavior changes at CFL = 0.5.

Solution

Advection equation with sharp initial condition

We solve the problem for the sharp initial condition using Euler-implicit and Upwind Schemes. Figure 7.6 shows the time evolution of a localized square pulse governed by the linear advection equation using the Euler-Implicit scheme combined with a 2-point upwind finite difference method at a CFL number of 0.5. The initial condition is a sharp pulse located at one-third of the domain. As time progresses, the pulse moves to the right, consistent with the positive advection speed. However, the solution exhibits significant numerical diffusion, evident from the loss of amplitude and the broadening of the pulse over time. This behavior is a known limitation of first-order upwind schemes, particularly when used with implicit time integration. Although the method is stable at this CFL, it fails to preserve the sharpness of the initial wave profile, making it unsuitable for problems where high fidelity in capturing steep gradients is important.

lessons:

☞ The Euler-implicit upwind scheme is unconditionally stable for linear advection.

☞ It is only first-order accurate and introduces significant numerical diffusion that intensifies with larger CFL, quickly eroding sharp fronts.

☞ Because it is monotone, it avoids spurious oscillations but cannot preserve amplitude or steep gradients.

☞ The method is best reserved for stiff or diffusion-dominated problems where robustness outweighs accuracy.

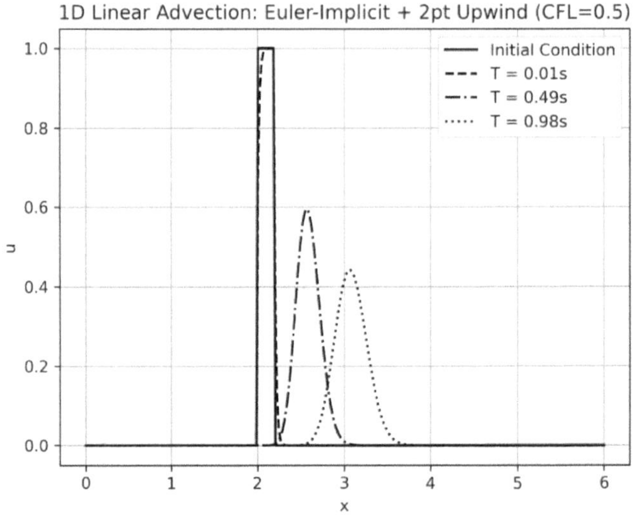

Figure 7.6: **Advection equation solved using Euler-Implicit and FD Upwind schemes at different CFL numbers.**

Amplification Factor : The Euler-Implicit scheme with 2-point upwind spatial discretization for the advection equation is written as:

$$u_j^{n+1} = u_j^n - \text{CFL}\left(u_j^{n+1} - u_{j-1}^{n+1}\right)$$

Assuming a Fourier mode solution:

$$u_j^n = \hat{u}^n e^{ikx_j}, \quad u_j^{n+1} = \hat{u}^{n+1} e^{ikx_j}, \quad u_{j-1}^{n+1} = \hat{u}^{n+1} e^{ikx_{j-1}}$$

Using $x_{j-1} = x_j - \Delta x$, we have:

$$e^{ikx_{j-1}} = e^{ikx_j} e^{-ik\Delta x}$$

Substitute into the finite difference scheme:

$$\hat{u}^{n+1} e^{ikx_j} = \hat{u}^n e^{ikx_j} - \text{CFL}\left(\hat{u}^{n+1} e^{ikx_j} - \hat{u}^{n+1} e^{ikx_j} e^{-ik\Delta x}\right)$$

Cancel e^{ikx_j} on both sides:

$$\hat{u}^{n+1} = \hat{u}^n - \text{CFL}\left(\hat{u}^{n+1} - \hat{u}^{n+1} e^{-ik\Delta x}\right)$$

Isolate $\hat{u}^{n+1}$:

$$\hat{u}^{n+1}\left(1 + \text{CFL} - \text{CFL}e^{-ik\Delta x}\right) = \hat{u}^n$$

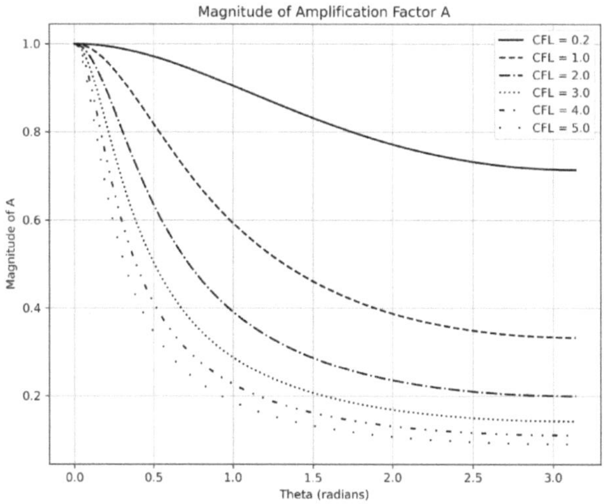

Figure 7.7: **Amplitude response of the Euler-Implicit scheme with 2-point upwind stencil for various wavenumbers k^* and CFL values.**

The amplification factor A is then:

$$A = \frac{\hat{u}^{n+1}}{\hat{u}^n} = \frac{1}{1 + \text{CFL} - \text{CFL}e^{-ik\Delta x}}$$

Let $k^* = k\Delta x$, so the expression becomes:

$$A = \frac{1}{1 + \text{CFL}(1 - e^{-ik^*})}$$

This complex amplification factor governs both the damping and phase shifting of each Fourier mode due to the numerical scheme. The modulus and argument of A control the dissipation and dispersion, respectively.

Phase Error Analysis: To analyze phase distortion, write the denominator of A using Euler's identity:

$$e^{-ik^*} = \cos k^* - i \sin k^*$$

So the amplification factor becomes:

$$A = \frac{1}{1 + \text{CFL} - \text{CFL}(\cos k^* - i \sin k^*)} = \frac{1}{1 + \text{CFL}(1 - \cos k^*) + i\text{CFL} \sin k^*}$$

Let the real and imaginary parts of the denominator be:

$$D_R = 1 + \text{CFL}(1 - \cos k^*), \quad D_I = \text{CFL} \sin k^*$$

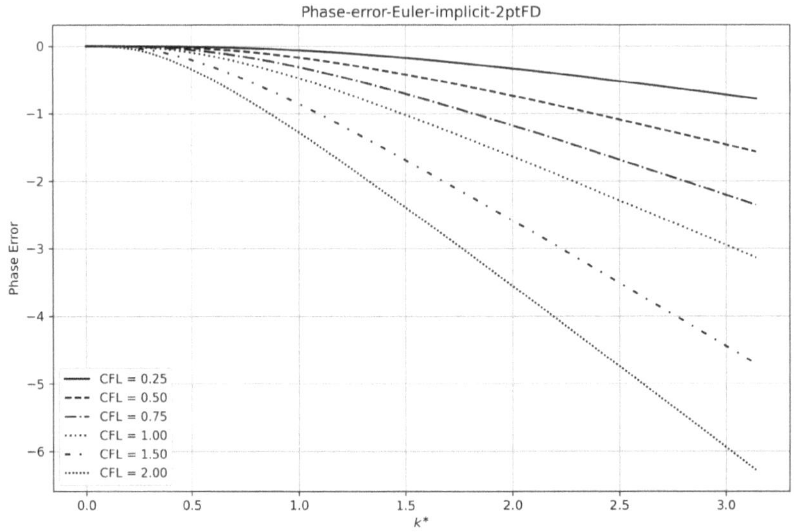

Figure 7.8: **Phase error of Euler-Implicit and 2-point upwind scheme for the advection equation. Numerical dispersion leads to cumulative lag in wave propagation.**

Then the phase angle ϕ is:

$$\phi = \arctan\left(\frac{D_I}{D_R}\right) = \arctan\left(\frac{\text{CFL}\sin k^*}{1+\text{CFL}(1-\cos k^*)}\right)$$

The exact phase shift per time step is $\phi_{\text{true}} = -ak\Delta t = -\text{CFL}\cdot k^*$, so the phase error is:

$$er_\omega = \phi|_{\text{exact}} - \phi|_{num} = -\text{CFL}\cdot k^* - \arctan\left(\frac{\text{CFL}\sin k^*}{1+\text{CFL}(1-\cos k^*)}\right)$$

7.3 PHASE RATIO

The phase ratio is a key metric in advection-dominated flows because it quantifies how accurately a numerical scheme reproduces the propagation speed of different wave components. The phase ratio is defined as

$$\text{Phase Ratio} = \frac{c_{\text{num}}}{c_{\text{phys}}}$$

where, c_{num} is the numerical phase velocity, and c_{phys} is the physical phase velocity

A phase ratio equal to unity indicates that all wave components are transported at the correct speed. However, when the phase ratio deviates from one, the scheme introduces phase errors, causing different Fourier modes to travel at incorrect velocities. This leads to artificial dispersion, distortion, and loss of accuracy in sharp gradients or wavefronts, which is especially problematic in advection-dominated problems. Maintaining a phase ratio close to one is therefore essential to ensure the correct propagation of transported features and to preserve the physical fidelity of numerical simulations.

Consider an The advection equation, given by $\frac{\partial u}{\partial t} + a\frac{\partial u}{\partial x} = 0$ (where $a > 0$ is the constant speed of propagation), describes the transport of a quantity u without change in shape in the direction of flow. The Fourier transform approach provides a significant insight into solving this partial differential equation (PDE).

The general solution for $u(x,t)$ using Fourier methods is expressed as:

$$u(x,t) = f(t)e^{ikx}$$

Here, $f(t)$ represents the temporal part of the solution, and e^{ikx} represents the spatial part where k is the wave number.

From the advection equation:

$$f_t' e^{ikx} + af_k' ike^{ikx} = 0$$

Solving this, we obtain:

$$f_t' = -af_k' ik$$

This implies that $f(t) = e^{-iakt}$, assuming $f(0) = 1$.

The wave function becomes:

$$u(x,t) = e^{-iakt}e^{ikx} = e^{ik(x-at)}$$

This represents a wave moving to the right with speed a, maintaining its initial shape.

When discretizing in space, if the exact continuous solution to the PDE with spatial discretization (like an ordinary differential equation) is considered:

$$u_{numerical}(x,t) = f(0)e^{ik(x-a^*t)}$$

Here, a^* represents the numerical phase speed which differs from the actual speed a due to numerical dispersion.

Phase Speed and Dispersion: The phase ratio $\frac{a^*}{a} = \frac{k^*}{k}$ determines how accurately the wave speed is represented in the numerical model. If $\frac{a^*}{a} < 1$, the wave propagates too slowly, and if $\frac{a^*}{a} > 1$, it propagates too fast.

Dispersion occurs when different frequency components of the solution travel at different velocities, leading to phase errors. The imaginary portion of a^* (and hence k^*) results in errors in the amplitude (either damping or growth):

$$u_{numerical}(x,t) = f(0)e^{ka_i^*t}e^{ik(x-a^*t)}$$

This factor $e^{ka_i^*t}$ modifies the amplitude, demonstrating either growth or decay of the wave amplitude over time, which is an artifact of the numerical scheme.

7.3.1 EULER EXPLICIT AND FD-O2 SCHEME

To derive the phase ratio expression for the advection equation using the Euler explicit time-stepping scheme and the second order finite difference spatial discretization (FDO2), we need to set up and discretize the advection equation:

$$\frac{\partial u}{\partial t} + a\frac{\partial u}{\partial x} = 0$$

where a is the constant advection speed.

The Euler explicit method approximates the time derivative, and the second order central finite difference method approximates the spatial derivative. We define the grid points as u_i^n where i is the spatial index and n is the time index.

$$\frac{u_i^{n+1} - u_i^n}{\Delta t}$$

$$\frac{u_{i+1}^n - u_{i-1}^n}{2\Delta x}$$

By substituting these into the advection equation, we rearrange it to get:

$$u_i^{n+1} = u_i^n - a\frac{\Delta t}{2\Delta x}(u_{i+1}^n - u_{i-1}^n)$$

To examine the numerical dispersion and phase error, consider a plane wave solution:

$$u(x,t) = e^{i(kx-\omega t)}$$

Substituting into our discretized scheme and simplifying, we expect:

$$e^{-i\omega\Delta t} = 1 - i\frac{a\Delta t}{\Delta x}\sin(k\Delta x)$$

Solving for ω, the angular frequency, gives:

$$\omega = -\frac{i}{\Delta t}\log\left(1 - i\frac{a\Delta t}{\Delta x}\sin(k\Delta x)\right)$$

Assuming Δt is small, we can expand the logarithm for small arguments:

$$\omega \approx \frac{a\sin(k\Delta x)}{\Delta x}$$

The numerical phase speed a^* is derived from the relationship:

$$\omega = a^* k$$

Thus:

$$a^* \approx \frac{a\sin(k\Delta x)}{k\Delta x}$$

The phase ratio $\frac{a^*}{a}$ indicates how the numerical method modifies the wave speed:

$$\frac{a^*}{a} \approx \frac{\sin(k\Delta x)}{k\Delta x}$$

$$\left(\frac{a^*}{a}\right)_{FD-O2} \approx \frac{\sin(k^*)}{k^*}$$

This expression captures how the second order central difference scheme used in conjunction with Euler explicit time-stepping affects the advection wave propagation. This shows that for small $k\Delta x$, the phase speed is close to the actual speed, but as $k\Delta x$ increases, numerical dispersion introduces significant phase errors. This effect is a critical consideration in the stability and accuracy of numerical simulations for hyperbolic PDEs like the advection equation.

7.3.2 EULER EXPLICIT AND FD-O4 SCHEME

Usin explicit Euler time-stepping alongside the fourth-order central difference for spatial discretization:

$$\frac{\partial u}{\partial t} + a\frac{\partial u}{\partial x} = 0$$

$$\frac{u_i^{n+1} - u_i^n}{\Delta t} + a\frac{-u_{i+2}^n + 8u_{i+1}^n - 8u_{i-1}^n + u_{i-2}^n}{12\Delta x} = 0$$

Rearranged for the Next Time Step:

$$u_i^{n+1} = u_i^n - \frac{a\Delta t}{12\Delta x}\left(-u_{i+2}^n + 8u_{i+1}^n - 8u_{i-1}^n + u_{i-2}^n\right)$$

Assuming a plane wave solution:

$$u(x,t) = e^{i(kx-\omega t)}$$

Substituting into the discretized scheme:

$$e^{-i\omega\Delta t} = 1 - \frac{ia\Delta t}{12\Delta x}\left(-e^{2ik\Delta x} + 8e^{ik\Delta x} - 8e^{-ik\Delta x} + e^{-2ik\Delta x}\right)$$

Let's simplify the expression inside the parentheses:

$$-e^{2ik\Delta x} + 8e^{ik\Delta x} - 8e^{-ik\Delta x} + e^{-2ik\Delta x}$$

Group Similar Terms:

$$= 8e^{ik\Delta x} - 8e^{-ik\Delta x} - e^{2ik\Delta x} + e^{-2ik\Delta x}$$

Apply Euler's Formula:

$$e^{ik\Delta x} - e^{-ik\Delta x} = 2i\sin(k\Delta x)$$

$$e^{2ik\Delta x} - e^{-2ik\Delta x} = 2i\sin(2k\Delta x)$$

Substitute Back:

$$8e^{ik\Delta x} - 8e^{-ik\Delta x} = 16i\sin(k\Delta x)$$
$$-e^{2ik\Delta x} + e^{-2ik\Delta x} = -2i\sin(2k\Delta x)$$

Combine Terms:

$$-e^{2ik\Delta x} + 8e^{ik\Delta x} - 8e^{-ik\Delta x} + e^{-2ik\Delta x} = 16i\sin(k\Delta x) - 2i\sin(2k\Delta x)$$

Substitute into the Original Equation:

$$e^{-i\omega\Delta t} = 1 - \frac{ia\Delta t}{12\Delta x}\left[16i\sin(k\Delta x) - 2i\sin(2k\Delta x)\right]$$

$$= 1 - \frac{a\Delta t}{12\Delta x}\left[16\sin(k\Delta x) - 2\sin(2k\Delta x)\right]$$

Assuming Δt is small, we can use the approximation $e^{-i\omega\Delta t} \approx 1 - i\omega\Delta t$. Thus:

$$1 - i\omega\Delta t \approx 1 - \frac{a\Delta t}{12\Delta x}\left[16\sin(k\Delta x) - 2\sin(2k\Delta x)\right]$$

Equate the Terms:

$$-i\omega\Delta t \approx -\frac{a\Delta t}{12\Delta x}\left[16\sin(k\Delta x) - 2\sin(2k\Delta x)\right]$$

$$\omega \approx \frac{a}{12\Delta x}\left[16\sin(k\Delta x) - 2\sin(2k\Delta x)\right]$$

From the relationship $\omega = a^*k$:

$$a^*k \approx \frac{a}{12\Delta x}\left[16\sin(k\Delta x) - 2\sin(2k\Delta x)\right]$$

Solving for a^*:

$$a^* \approx a\left(\frac{16\sin(k\Delta x) - 2\sin(2k\Delta x)}{12k\Delta x}\right)$$

$$\frac{a^*}{a} \approx \frac{16\sin(k\Delta x) - 2\sin(2k\Delta x)}{12k\Delta x} = \frac{8\sin(k\Delta x) - \sin(2k\Delta x)}{6k\Delta x}$$
$$\left(\frac{a^*}{a}\right)_{FDO4} \approx \frac{8\sin(k^*) - \sin(2k^*)}{6k^*}$$

7.3.3 EULER EXPLICIT AND FD-O6 SCHEME

Using the explicit Euler time-stepping and the sixth-order central difference for spatial discretization:

$$\frac{u_i^{n+1}-u_i^n}{\Delta t}+a\frac{u_{i-3}^n-9u_{i-2}^n+45u_{i-1}^n-45u_{i+1}^n+9u_{i+2}^n-u_{i+3}^n}{60\Delta x}=0$$

Substituting the plane wave solution $u(x,t)=e^{i(kx-\omega t)}$ into the discretized equation, we arrive at:

$$e^{-i\omega\Delta t}=1-\frac{ia\Delta t}{60\Delta x}\left(-e^{-3ik\Delta x}+9e^{-2ik\Delta x}-45e^{-ik\Delta x}+45e^{ik\Delta x}-9e^{2ik\Delta x}+e^{3ik\Delta x}\right)$$

Expanding the exponentials using Euler's formulas and simplifying, we obtain

$$\omega\Delta t\approx\frac{a\Delta t}{60\Delta x}\left(-2\sin(3k\Delta x)+18\sin(2k\Delta x)-90\sin(k\Delta x)\right)$$

From the relationship $\omega=a^*k$, we express the numerical phase speed as:

$$a^*\approx a\left(\frac{-2\sin(3k\Delta x)+18\sin(2k\Delta x)-90\sin(k\Delta x)}{k\Delta x}\right)$$

$$\left(\frac{a^*}{a}\right)_{FDO6}\approx\frac{-2\sin(3k^*)+18\sin(2k^*)-90\sin(k^*)}{60\cdot k^*}$$

7.3.4 EULER EXPLICIT AND KFD-O4 SCHEME

The compact fourth-order finite difference approximation for the first derivative is given by:

$$\frac{1}{4}u'_{i-1}+u'_i+\frac{1}{4}u'_{i+1}=\frac{3}{4}\frac{u_{i+1}-u_{i-1}}{\Delta x}$$

Here, u'_i is the numerical approximation of the derivative at point i, and Δx is the spatial grid spacing.

Assume a plane wave solution of the form:

$$u_i=e^{i(kx_i-\omega t)}$$

Then the derivatives are:

$$u'_i=ika^*u_i$$

where a^* is the numerical phase speed we want to find.

Substitute u_i and u'_i into the compact scheme:

$$\frac{1}{4}(ika^*u_{i-1})+ika^*u_i+\frac{1}{4}(ika^*u_{i+1})=\frac{3}{4}\frac{u_{i+1}-u_{i-1}}{\Delta x}$$

Simplify the left side:

$$ika^* \left(\frac{1}{4} u_{i-1} + u_i + \frac{1}{4} u_{i+1} \right)$$

Simplify the right side:

$$\frac{3}{4} \frac{u_{i+1} - u_{i-1}}{\Delta x}$$

Using $u_i = e^{i(kx_i)}$:
- $u_{i+1} = e^{i(kx_i + k\Delta x)}$ - $u_{i-1} = e^{i(kx_i - k\Delta x)}$
Left side:

$$ika^* u_i \left(\frac{1}{4} e^{-ik\Delta x} + 1 + \frac{1}{4} e^{ik\Delta x} \right)$$

Right side:

$$\frac{3}{4} \frac{e^{ik\Delta x} - e^{-ik\Delta x}}{\Delta x} u_i$$

Left side:

$$ika^* u_i \left(1 + \frac{1}{2} \cos(k\Delta x) \right)$$

Right side:

$$\frac{3}{4} \frac{2i \sin(k\Delta x)}{\Delta x} u_i = \frac{3i \sin(k\Delta x)}{2\Delta x} u_i$$

Cancel u_i and i from both sides:

$$ka^* \left(1 + \frac{1}{2} \cos(k\Delta x) \right) = \frac{3 \sin(k\Delta x)}{2\Delta x}$$

Recall that $k\Delta x = k^*$:

$$k = \frac{k^*}{\Delta x}$$

Substitute k back into the equation:

$$\frac{\theta}{\Delta x} a^* \left(1 + \frac{1}{2} \cos(k^*) \right) = \frac{3 \sin(k^*)}{2\Delta x}$$

Cancel Δx:

$$\theta a^* \left(1 + \frac{1}{2} \cos(k^*) \right) = \frac{3 \sin(k^*)}{2}$$

Solve for a^*:

$$a^* = \frac{3 \sin(k^*)}{2k^* \left(1 + \frac{1}{2} \cos(k^*) \right)}$$

Simplify the denominator:

$$1 + \frac{1}{2} \cos(k^*) = \frac{2 + \cos(k^*)}{2}$$

So the expression becomes:

$$a^* = \frac{3\sin(k^*)}{2k^*\left(\frac{2+\cos(k^*)}{2}\right)} = \frac{3\sin(k^*)}{k^*(2+\cos(k^*))}$$

Thus, the numerical phase speed ratio is:

$$\left(\frac{a^*}{a}\right)_{KFDO4} = \frac{3\sin(k^*)}{k^*(2+\cos(k^*))}$$

7.3.5 COMPACT FD-O6 (KFD-O6)

A typical compact sixth-order finite difference scheme is:

$$\frac{1}{3}u'_{i-1} + u'_i + \frac{1}{3}u'_{i+1} = \frac{14}{9}\frac{u_{i+1}-u_{i-1}}{2\Delta x} + \frac{1}{9}\frac{u_{i+2}-u_{i-2}}{4\Delta x}$$

Using the same plane wave solution:

$$u_i = e^{i(kx_i-\omega t)}, \quad u'_i = ika^* u_i$$

Left side:

$$ika^*\left(\frac{1}{3}u_{i-1} + u_i + \frac{1}{3}u_{i+1}\right)$$

Right side:

$$\frac{14}{9}\frac{u_{i+1}-u_{i-1}}{2\Delta x} + \frac{1}{9}\frac{u_{i+2}-u_{i-2}}{4\Delta x}$$

Compute each term:
- $u_{i\pm n} = e^{i(kx_i \pm nk\Delta x)}$
Left side:

$$ika^* u_i\left(\frac{1}{3}e^{-ik\Delta x} + 1 + \frac{1}{3}e^{ik\Delta x}\right)$$

Simplify:

$$ika^* u_i\left(1 + \frac{2}{3}\cos(k\Delta x)\right)$$

Right side:
First term:

$$\frac{14}{9}\frac{e^{ik\Delta x}-e^{-ik\Delta x}}{2\Delta x} = \frac{14}{9}\frac{2i\sin(k\Delta x)}{2\Delta x} = \frac{14i\sin(k\Delta x)}{9\Delta x}$$

Second term:

$$\frac{1}{9}\frac{e^{2ik\Delta x}-e^{-2ik\Delta x}}{4\Delta x} = \frac{1}{9}\frac{2i\sin(2k\Delta x)}{4\Delta x} = \frac{i\sin(2k\Delta x)}{18\Delta x}$$

Combine:

$$iu_i\left(\frac{14\sin(k\Delta x)}{9\Delta x}+\frac{\sin(2k\Delta x)}{18\Delta x}\right)$$

Equating Both Sides
Cancel iu_i:

$$ka^*\left(1+\frac{2}{3}\cos(k\Delta x)\right)=\frac{14\sin(k\Delta x)}{9\Delta x}+\frac{\sin(2k\Delta x)}{18\Delta x}$$

Substitute $k=\frac{k^*}{\Delta x}$:

$$\frac{\theta}{\Delta x}a^*\left(1+\frac{2}{3}\cos(k^*)\right)=\frac{14\sin(k^*)}{9\Delta x}+\frac{\sin 2(k^*)}{18\Delta x}$$

Multiply both sides by Δx:

$$\theta a^*\left(1+\frac{2}{3}\cos(k^*)\right)=\frac{14\sin(k^*)}{9}+\frac{\sin 2(k^*)}{18}$$

Simplify right side:

$$\frac{14\sin(k^*)}{9}+\frac{\sin 2k^*}{18}=\frac{28\sin k^*}{18}+\frac{\sin 2k^*}{18}$$

Recall that $\sin 2k^*=2\sin k^*\cos k^*$:

$$\frac{28\sin k^*}{18}+\frac{2\sin k^*\cos k^*}{18}=\frac{\sin k^*(28+2\cos k^*)}{18}$$

Simplify numerator:

$$28+2\cos k^*=2(14+\cos k^*)$$

So the right side becomes:

$$\frac{\sin k^*\cdot 2(14+\cos k^*)}{18}=\frac{\sin k^*(14+\cos k^*)}{9}$$

Now, equate both sides:

$$\theta a^*\left(1+\frac{2}{3}\cos k^*\right)=\frac{\sin k^*(14+\cos k^*)}{9}$$

Simplify the left side denominator:

$$1+\frac{2}{3}\cos k^*=\frac{3+2\cos k^*}{3}$$

Therefore:

$$\theta a^*\left(\frac{3+2\cos k^*}{3}\right)=\frac{\sin k^*(14+\cos k^*)}{9}$$

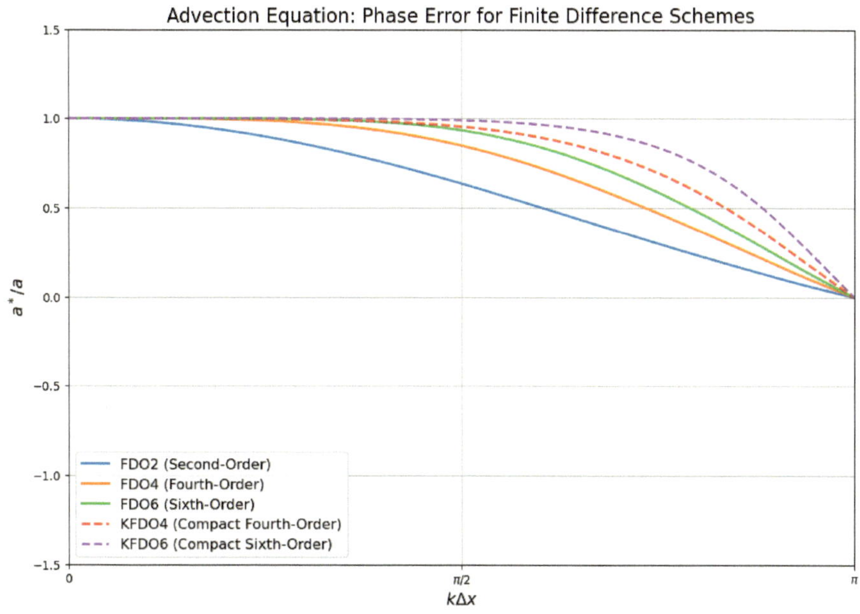

Figure 7.9: **Phase ratio for advection equation using different FD schemes.**

Solve for a^*:

$$a^* = \frac{\sin k^*(14+\cos k^*)}{9k^*} \cdot \frac{3}{3+2\cos k^*}$$

Simplify:

$$a^* = \frac{\sin k^*(14+\cos k^*)}{3k^*(3+2\cos k^*)}$$

Thus, the numerical phase speed ratio is:

$$\left(\frac{a^*}{a}\right)_{KFDO6} = \frac{\sin k^*(14+\cos k^*)}{3k^*(3+2\cos k^*)}$$

Figure 7.9 compares the phase accuracy of various finite difference (FD) schemes for the linear advection equation. The phase error er_ω is plotted against the nondimensional wavenumber $k^* = k\Delta x$. The second-order central difference scheme (FD02) exhibits the largest phase lag, particularly at higher wavenumbers. In contrast, higher-order schemes such as FD04 and FD06 demonstrate significantly lower phase error, while their compact counterparts (KFD04 and KFD06) show even greater accuracy across the spectrum. The compact schemes preserve near-zero phase error over a broader wavenumber range, making them highly suitable for wave-resolving problems that demand high spectral fidelity.

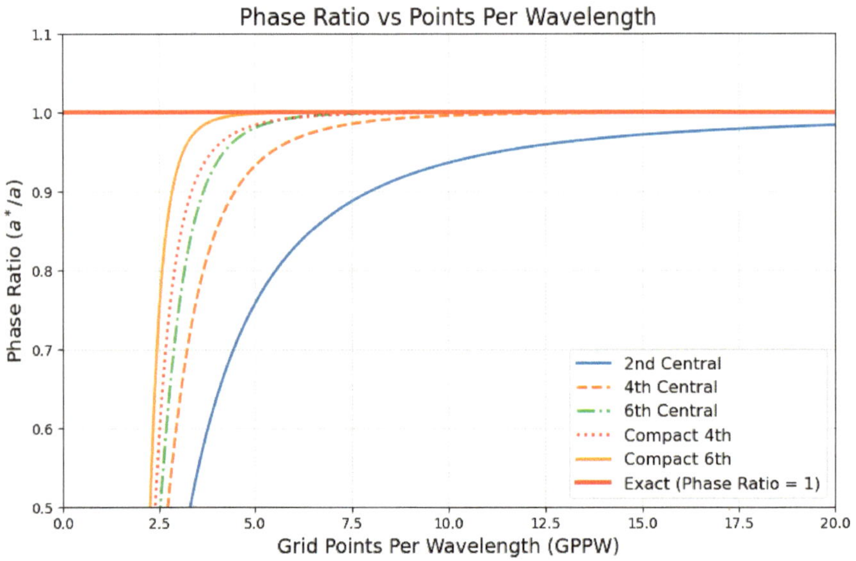

Figure 7.10: **Grid points per wavelength. Note that just 8 points are needed for FDO6 scheme.**

Figure 7.10 illustrates how different finite difference schemes perform in terms of phase ratio accuracy as a function of grid points per wavelength (GPPW). A phase ratio of one indicates perfect phase propagation. The second-order scheme requires more than 12 points per wavelength to achieve moderate accuracy, and even more for high-fidelity wave resolution. In contrast, the sixth-order compact scheme achieves near-exact phase propagation with only 8 points. This highlights the efficiency of higher-order and compact schemes for wave-dominated problems, as they reduce numerical dispersion and computational cost while maintaining accuracy with fewer grid points.

Table summarizes the grid points per wavelength for some FD schemes. Moderate accuracy corresponds to a phase error within 5%, while high accuracy corresponds to a phase error within 1%. Compact schemes (KFD) achieve higher spectral resolution and require fewer grid points to maintain accuracy.

7.4 LEAPFROG SCHEME (LF)

The Leapfrog scheme uses values from two previous time steps (u_j^{n-1} and u_j^{n+1}), hence the name 'Leapfrog', as it 'leaps' over the intermediate time step (u_j^n). It calculates the spatial derivative using a central difference, which involves the points u_j^{n+1} and u_j^{n-1}, resulting in a second-order accuracy in space. The value at the next time step u_j^{n+1} is calculated based on the value two steps back in time (u_j^{n-1}) and the

Table 7.1
Estimated grid points per wavelength (GPPW) required to achieve moderate and high phase accuracy for various finite difference schemes applied to the linear advection equation

Scheme	GPPW (Moderate Accuracy)	GPPW (High Accuracy)
FD02 (2nd-order)	≥12	≥25
FD04 (4th-order)	≥8	≥16
FD06 (6th-order)	≥6	≥12
KFD04 (Compact 4th-order)	≥6	≥10
KFD06 (Compact 6th-order)	≥4	≥8

spatial derivative at the current time step (u_j^n). The scheme is typically expressed as:

$$u_j^{n+1} = u_j^{n-1} - \frac{\Delta t}{\Delta x}\left(u_{j+1}^n - u_{j-1}^n\right)$$

The Leapfrog method is explicit and straightforward to implement. It provides good phase accuracy and minimal numerical dispersion, making it particularly suitable for wave propagation problems. The Leapfrog scheme can be unstable if not used under appropriate conditions of spatial and temporal discretization. It can also exhibit computational modes or ”2Δt oscillations” due to its central nature in time.

The Leapfrog scheme requires u^{n-1} to compute u^{n+1}, which presents a problem at the very beginning of the simulation where u^{n-1} does not exist. To address this, another method must be employed for the first time step. A first-order method like Forward Euler is used to compute u^1 from u^0 due to its simplicity and direct approach:

$$u_j^1 = u_j^0 - \frac{\Delta t}{\Delta x}\left(u_{j+1}^0 - u_{j-1}^0\right)$$

or

$$u_j^{n+1} = u_j^{n-1} - \frac{a\Delta t}{\Delta x}(u_{j+1}^n - u_{j-1}^n)$$

Leapfrog scheme (explicit scheme second order in both space and time) is very sensitive to initial conditions. In case of non-smooth initial conditions the suprious oscillations are generated as solution progresses (see Figure 7.11).

The given Leapfrog scheme in the code is defined by:

$$u_j^{n+1} = u_j^{n-1} - \frac{a\Delta t}{\Delta x}\left(u_{j+1}^n - u_{j-1}^n\right)$$

To perform Fourier analysis, we assume a solution of the form:

$$u_j^n = A^n e^{ikj\Delta x}$$

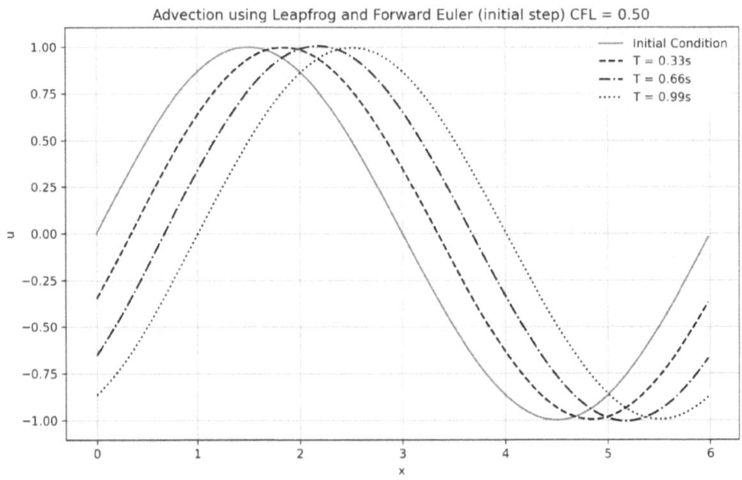

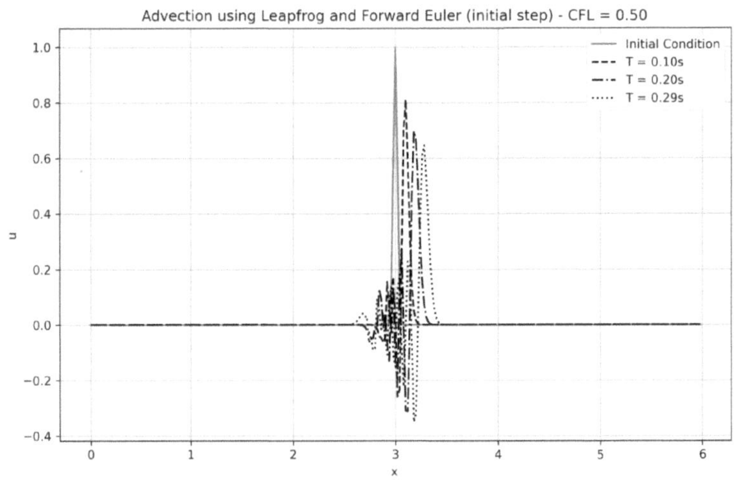

Figure 7.11: **1D Advection equation solved using Leapfrog scheme with smooth and sharp initial conditions.**

where: A is the amplification factor we wish to find, k is the wave number, and j and n are the spatial and temporal indices, respectively.

Substitute $u_j^n = A^n e^{ikj\Delta x}$ into the Leapfrog scheme:

$$A^{n+1} e^{ikj\Delta x} = A^{n-1} e^{ikj\Delta x} - \frac{a\Delta t}{\Delta x}\left(A^n e^{ik(j+1)\Delta x} - A^n e^{ik(j-1)\Delta x}\right)$$

We can factor out $e^{ikj\Delta x}$ and A^n from both sides:

$$A^2 = A^{-1} - \frac{a\Delta t}{\Delta x}\left(e^{ik\Delta x} - e^{-ik\Delta x}\right)$$

Using the identity $e^{ix} - e^{-ix} = 2i\sin(x)$, we get:

$$A^2 = A^{-1} - \frac{2ia\Delta t}{\Delta x}\sin(k\Delta x)$$

This gives us a quadratic equation for A:

$$A^2 - i2C\sin(k\Delta x)A - 1 = 0$$

where $C = \frac{a\Delta t}{\Delta x}$ is the Courant number.

Using the quadratic formula to solve for A, we get:

$$A = -iC\sin(k\Delta x) \pm \sqrt{1 - C^2\sin^2(k\Delta x)}.$$

We label these two roots as A_+ and A_-, corresponding to the $+$ and $-$ signs in the square root.

For the scheme to be stable, the amplification factor A must have a magnitude $|A| \leq 1$. This ensures that the solution does not grow unbounded over time.

The stability condition here is:

$$1 - C^2\sin^2(k\Delta x) \geq 0,$$

which simplifies to:

$$C \leq 1.$$

This is known as the Courant-Friedrichs-Lewy (CFL) condition for stability, and represented often as CFL. When $CFL \leq 1$, the amplification factor satisfies $|A| = 1$, meaning there is no amplification or damping of the solution over time. This property is desirable for accurately preserving wave amplitudes.

The amplification factor A can be rewritten in polar form as:

$$A = |A|e^{i\beta},$$

where β represents the phase shift per time step associated with the propagation of the wave. Since $|A| = 1$ when $CFL \leq 1$, we focus on the phase behavior.

For the positive root A_+:

$$A_+ = -iCFL\sin(k\Delta x) + \sqrt{1 - CFL^2\sin^2(k\Delta x)}.$$

We can express A_+ in terms of a real and imaginary part, then compute β as:

$$\beta = \arg(A_+) = \arctan\left(\frac{\text{Imaginary Part}}{\text{Real Part}}\right).$$

Let:

$$a = CFL \sin(k\Delta x).$$

Then:

$$A_+ = -ia + \sqrt{1-a^2},$$

so the phase β_+ is:

$$\beta_+ = \arctan\left(\frac{-a}{\sqrt{1-a^2}}\right).$$

The Leapfrog scheme produces two roots, A_+ and A_-, for the amplification factor. Each root corresponds to a different mode in the solution:
Physical Mode (A_+): This represents the actual wave propagation with phase angle β_+, describing the true behavior of the wave solution.
Computational Mode (A_-): This is a secondary solution with a different phase angle, often seen as a non-physical artifact of the Leapfrog scheme. This mode typically has a phase shift of π relative to the physical mode, leading to oscillations that do not correspond to the physical wave.

The general solution for a single wave number k can be expressed as a combination of both modes:

$$u_j^n = A_+^n e^{ikj\Delta x} + A_-^n e^{ikj\Delta x}.$$

This linear combination includes both the physical and computational modes, and the computational mode may oscillate in an alternating pattern due to the phase difference.

The key takeaways for this analysis are:
Stability Condition : The scheme is stable if $CFL \leq 1$.
Amplitude Preservation : When $CFL \leq 1$, $|A| = 1$, meaning there is no amplification or decay, which is desirable for accurate wave propagation.
Phase Behavior : The phase β determines the speed and direction of wave propagation. The physical mode has a phase angle that correctly represents the wave, while the computational mode introduces an artificial oscillation.
Computational Mode : The Leapfrog scheme inherently includes a computational mode that does not represent the physical wave solution. This mode is periodic with an alternating phase shift and does not affect the stability as long as $CFL \leq 1$.

This analysis shows that the Leapfrog scheme is effective for wave propagation as it maintains amplitude and stability under the CFL condition. However, care should be taken with the computational mode, as it can introduce non-physical oscillations in the numerical solution.

7.4.1 GROUP VELOCITY IN LF SCHEME

Analyzing group velocity is crucial when studying the advection equation, as it indicates how the envelope of a wave packet or a localized disturbance travels through the domain. While phase velocity describes the motion of individual wave components, it is the group velocity that governs the overall transport of information, energy, and physical features in the solution. In numerical modeling, if the computed group velocity does not accurately represent the true propagation speed, the result can be artificial dispersion or distortion of wave packets, leading to physically incorrect solutions. Therefore, understanding group velocity helps in evaluating and improving numerical schemes, ensuring that the key characteristics of advected profiles—such as sharp fronts or gradients—are preserved and advected at the correct speed, which is particularly important in convection-dominated problems.

Consider the one-dimensional linear advection equation is given by

$$u_t = -u_x$$

which describes a wave propagating in a specific direction with constant speed. To analyze its numerical properties using the Leapfrog (LF) scheme, we aim to derive the dispersion relation, group velocity, and phase speed step by step.

First, the Leapfrog scheme discretizes the temporal and spatial derivatives as follows:

$$u_j^{n+1} - u_j^{n-1} = -\frac{\Delta t}{\Delta x}(u_{j+1}^n - u_{j-1}^n),$$

where u_j^n is the value of u at spatial index j and time step n, with Δt and Δx representing the time and space steps, respectively. To analyze the wave propagation, assume a plane wave solution $u_j^n = Ae^{i(kj\Delta x - \omega n\Delta t)}$, where k is the wavenumber and ω is the angular frequency. Substituting this plane wave into the Leapfrog equation, we compute the derivatives and simplify the terms.

For u_j^{n+1} and u_j^{n-1}, substitute the plane wave expression to get:

$$u_j^{n+1} = Ae^{i(kj\Delta x - \omega(n+1)\Delta t)} = Ae^{i(kj\Delta x - \omega n\Delta t)}e^{-i\omega\Delta t},$$

$$u_j^{n-1} = Ae^{i(kj\Delta x - \omega(n-1)\Delta t)} = Ae^{i(kj\Delta x - \omega n\Delta t)}e^{i\omega\Delta t}.$$

Similarly, for u_{j+1}^n and u_{j-1}^n:

$$u_{j+1}^n = Ae^{i(k(j+1)\Delta x - \omega n\Delta t)} = Ae^{i(kj\Delta x - \omega n\Delta t)}e^{ik\Delta x},$$

$$u_{j-1}^n = Ae^{i(k(j-1)\Delta x - \omega n\Delta t)} = Ae^{i(kj\Delta x - \omega n\Delta t)}e^{-ik\Delta x}.$$

Substituting these into the Leapfrog scheme cancels the common factor $Ae^{i(kj\Delta x - \omega n\Delta t)}$, leaving:

$$e^{-i\omega\Delta t} - e^{i\omega\Delta t} = -\frac{\Delta t}{\Delta x}\left(e^{ik\Delta x} - e^{-ik\Delta x}\right).$$

Using $e^{-i\theta} - e^{i\theta} = -2i\sin(\theta)$, this simplifies to:

$$-2i\sin(\omega\Delta t) = -\frac{\Delta t}{\Delta x}\cdot 2i\sin(k\Delta x).$$

Canceling $-2i$, the dispersion relation becomes:

$$\sin(\omega\Delta t) = CFL\sin(k\Delta x),$$

where $CFL = \frac{\Delta t}{\Delta x}$ is the Courant number. This relation connects the temporal and spatial frequencies of the wave in the numerical scheme and reflects how the discretization introduces numerical dispersion, especially for high wavenumbers.

The group velocity is defined as $v_g = \frac{\partial \omega}{\partial k}$. To compute it, differentiate the dispersion relation implicitly:

$$\cos(\omega\Delta t) \cdot \frac{\partial(\omega\Delta t)}{\partial k} = CFL\cos(k\Delta x) \cdot \Delta x.$$

Simplifying

$$\Delta t \cdot \frac{\partial \omega}{\partial k} = CFL\Delta x\cos(k\Delta x)/\cos(\omega\Delta t),$$

$$c_g = \frac{\partial \omega}{\partial k} = \frac{\cos(k\Delta x)}{\cos(\omega\Delta t)}.$$

The phase speed v_p is defined as $v_p = \frac{\omega}{k}$. From the dispersion relation:

$$\sin(\omega\Delta t) = CFL\sin(k\Delta x),$$

solve for ω:

$$\omega = \frac{1}{\Delta t}\arcsin(CFL\sin(k\Delta x)).$$

Thus, the phase speed becomes:

$$v_p = \frac{\omega}{k} = \frac{1}{k\Delta t}\arcsin(CFL\sin(k\Delta x)).$$

The group velocity reflects the speed at which energy propagates, while the phase speed determines the speed of individual wave components. The dependence on k in the Leapfrog scheme reveals numerical dispersion, where different wavenumbers propagate at different speeds.

Example 7.4

To analyze the group velocity v_g, consider a wave packet in the spatial interval $[0,3]$, discretized using a uniform grid with spacing $\Delta x = \frac{1}{160}$. The initial condition is a sine wave modulated by a Gaussian envelope, defined as:

$$u(x,0) = e^{-16(x-\frac{1}{2})^2}\sin(kx),$$

where the Gaussian is centered at $x = \frac{1}{2}$, and k is the wavenumber of the sine wave. The wavenumber is chosen such that there are 8 grid points per wavelength, satisfying $k\Delta x = k^* = \frac{2\pi}{8} \approx 0.79$, resulting in $k \approx 125.7$. This ensures that the wave packet is well-resolved on the numerical grid.

Solution

For the one-dimensional advection equation $u_t = -u_x$, the exact analytical solution predicts that the wave packet propagates to the right at a constant speed $v = 1$, maintaining its shape. However, when the Leapfrog (LF) scheme is used to solve the equation numerically with a Courant-Friedrichs-Lewy (CFL) number CFL $= \frac{\Delta t}{\Delta x} = 0.4$, numerical dispersion is introduced. This dispersion modifies the propagation speed of the wave packet, causing deviations from the exact solution.

The group speed is based on the movement of the wave packet's peak position over time.

Initial Peak Position: The position of the peak of the initial wave packet ($t = 0$). This identifies the index of the maximum value in the wave packet, which is then converted to a physical position by multiplying it by the grid spacing, Δx.

Final Peak Position: The position of the peak of the wave packet at the end of the simulation ($t = T$) is converted to a physical position.

Group Speed Calculation: The group speed is computed as the rate of change of the peak position over the total simulation time:

$$c_g = \frac{\text{final\_peak\_position} - \text{initial\_peak\_position}}{T}$$

Here, T is the total simulation time.

This method provides an estimate of the group velocity by observing the actual movement of the wave packet in the simulation. The group velocity c_g for a linear advection equation $\partial u/\partial t + c\partial u/\partial x = 0$ is equal to the wave speed c, which is 1.0 in this case. At $t = 0$, the wave packet is centered around $x = 0.5$. Since the group velocity is 1.0, the wave packet moves a distance equal to $c_g \times t = 1.0 \times 2.0 = 2.0$. Adding the displacement to the initial position:

$$x_{\text{final}} = x_{\text{initial}} + c_g \times t = 0.5 + 2.0 = 2.5$$

Thus, the exact solution shows that the wave packet, under ideal conditions, should travel to $x = 2.5$ after $t = 2$ seconds.

The observed speed of the wave packet is approximately $v_g \approx 0.759$, which corresponds to the theoretical group velocity for the given $k^* = \frac{2\pi}{8}$ under the LF scheme. This demonstrates the dispersive nature of the numerical method, where different wavenumbers propagate at different speeds. Understanding and computing the group velocity is essential to characterize numerical dispersion and its impact on wave propagation simulations.

To derive the group velocity c_g under the Leapfrog (LF) scheme in terms of the Courant-Friedrichs-Lewy (CFL) number, we begin with the dispersion relation. For the Leapfrog scheme applied to the 1D advection equation, the dispersion relation is given by:

$$\sin(\omega\Delta t) = \text{CFL} \cdot \sin(k^*),$$

where ω is the angular frequency, $k^* = k\Delta x$ is the non-dimensional wavenumber, Δt is the time step size, Δx is the grid spacing, and $\text{CFL} = c\Delta t/\Delta x$ is the Courant number.

To find the group velocity c_g, defined as the derivative of ω with respect to k, we first isolate ω in the dispersion relation. For small time steps, the argument of arcsin remains valid, leading to the expression:

$$\omega = \frac{1}{\Delta t} \arcsin(\text{CFL} \cdot \sin(k^*)).$$

The group velocity c_g is defined as:

$$c_g = \frac{\partial \omega}{\partial k}.$$

Using the chain rule, the derivative is expressed as:

$$\frac{\partial \omega}{\partial k} = \frac{1}{\Delta t} \cdot \frac{1}{\sqrt{1-(\text{CFL} \cdot \sin(k^*))^2}} \cdot \frac{\partial}{\partial k}[\text{CFL} \cdot \sin(k^*)].$$

The term $\frac{\partial}{\partial k}[\text{CFL} \cdot \sin(k^*)]$ simplifies to:

$$\frac{\partial}{\partial k}[\text{CFL} \cdot \sin(k^*)] = \text{CFL} \cdot \Delta x \cdot \cos(k^*).$$

Substituting this back, the group velocity becomes:

$$c_g = \frac{\text{CFL} \cdot \Delta x \cdot \cos(k^*)}{\Delta t \cdot \sqrt{1-(\text{CFL} \cdot \sin(k^*))^2}}.$$

Using the definition $\text{CFL} = c\Delta t/\Delta x$, substitute $\Delta t = \text{CFL} \cdot \Delta x/c$:

$$c_g = \frac{\cos(k^*) \cdot c}{\sqrt{1-(\text{CFL} \cdot \sin(k^*))^2}}.$$

For small k^*, approximate $\cos(k^*) \approx 1 - \frac{1}{2}(k^*)^2$ and $\sin(k^*) \approx k^*$. The denominator becomes:

$$\sqrt{1-\text{CFL}^2 \sin^2(k^*)} \approx \sqrt{1-\text{CFL}^2 (k^*)^2} \approx 1 - \frac{1}{2}\text{CFL}^2(k^*)^2.$$

Substituting these approximations, we get:

$$c_g \approx \frac{1-\frac{1}{2}(k^*)^2}{1-\frac{1}{2}\text{CFL}^2(k^*)^2}.$$

For small $(k^*)^2$, use the first-order approximation $\frac{1-x}{1-y} \approx 1-(x-y)$:

$$c_g \approx 1 - \frac{1}{2}(k^*)^2 + \frac{1}{2}\text{CFL}^2(k^*)^2.$$

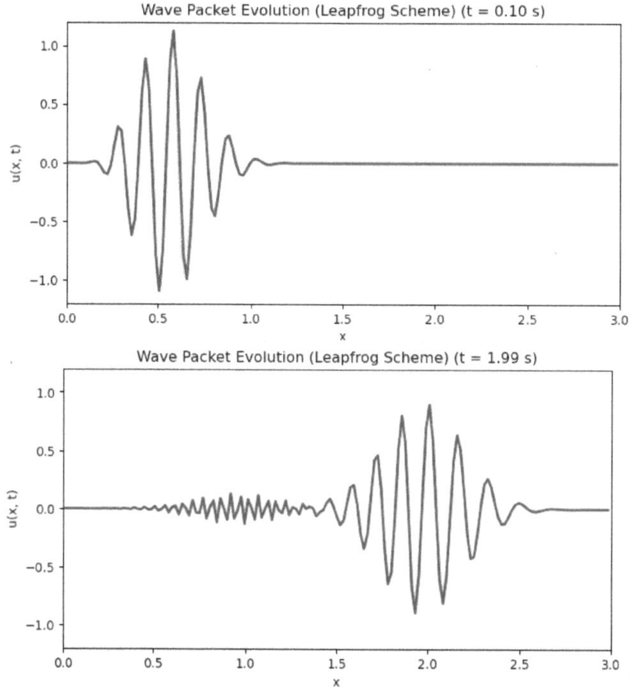

Figure 7.12: **Propagation of a wave packet with k*=0.79 and CFL=0.4. The packet was expected move with speed 1 as in advection equation but its moving at approximate group speed of 0.74.**

Simplifying:

$$c_g \approx 1 - \frac{1}{2}(1 - \text{CFL}^2)(k^*)^2.$$

The group velocity c_g in terms of the CFL number and the non-dimensional wavenumber is:

$$c_g \approx 1 - \frac{1}{2}(1 - \text{CFL}^2)(k^*)^2.$$

This result shows that for small k^*, the group velocity approximates the physical velocity ($c = 1$ for the advection equation). As k^* increases, numerical dispersion reduces c_g, with the deviation influenced by the factor $(1 - \text{CFL}^2)$. For k*=2 π/8 and CFL=0.4 the c_g is 0.74.

This example highlights the importance of analyzing group velocity errors in difference schemes. It shows that inaccuracies in a difference scheme are not solely tied to truncation errors. Although the wave deviates significantly from the exact solution in a pointwise sense, the qualitative behavior remains accurate. If one were to rely solely on accumulated truncation error estimates, the conclusion might be that the computation is unreliable. However, systematic effects like dispersions and dissipations, rather than random disturbances, drive the observed errors.

Movie: `wave_packet_evolution_LF.gif`: This movie illustrates the evolution of a wave packet under advection. The animation demonstrates how a wave packet, initially localized in space, evolves as it propagates in x-direction.

Lessons:

☞ Leapfrog is second-order accurate in time and space and conserves amplitude for CFL ≤ 1.

☞ At CFL $= 1$ the group velocity is exact for well-resolved waves ($k^* \lesssim \pi/2$), but dispersive error grows for higher wavenumbers and the most poorly resolved modes even propagate backward.

☞ The scheme suffers from computational (parasitic) modes and requires careful initial step treatment.

☞ Suitable for smooth solutions; not recommended for discontinuous profiles.

By understanding group velocity errors, one can achieve several benefits. For instance, it helps identify when a mesh is too coarse for the nature of the numerical solution, guides the selection of a suitable difference scheme and mesh size to minimize such errors, and allows making the most of an imperfect solution when further mesh refinement would be computationally prohibitive.

For meteorological problems that do not involve shocks, Kreiss and Oliger (1973) recommend using the leapfrog scheme, which is second-order in time and fourth-order in space. This scheme is both convenient to implement and provides high accuracy.

7.5 CRANK-NICOLSON SCHEME (CN)

We now use the numerical discretization that combines elements of both the forward and backward time differences and central spatial differences. This method is known as the Crank-Nicolson scheme, which is a specific type of the trapezoidal rule applied to partial differential equations, particularly for advection or diffusion equations.

Time Discretization: The scheme uses a centered time difference for u_j, approximating the derivative by $(u_j^{n+1} - u_j^n)/\Delta t$. This represents the average of forward and backward time derivatives.

Space Discretization: The spatial derivative is approximated using a central difference, not just at the current time step but also averaged across the n and $n+1$ time steps, specifically

$$\frac{u_{j+1}^{n+1} - u_{j-1}^{n+1} + u_{j+1}^{n} - u_{j-1}^{n}}{4\Delta x}$$

This central differencing in space also offers second-order spatial accuracy.

Trapezoidal Rule (Crank-Nicolson): This method averages the fluxes or derivatives between the current and next time levels, following the principle of the trapezoidal rule for numerical integration. The averaging helps suppress numerical oscillations and enhances stability, rendering the scheme unconditionally stable for linear problems such as diffusion.

The advection equation is

$$\frac{\partial u}{\partial t} = -c\frac{\partial u}{\partial x}$$

Now, discretize the time derivative:

$$\frac{u_j^{n+1} - u_j^n}{\Delta t} = -c\left(\frac{u_{j+1}^{n+1} - u_{j-1}^{n+1} + u_{j+1}^n - u_{j-1}^n}{4\Delta x}\right)$$

Multiply both sides by Δt:

$$u_j^{n+1} - u_j^n = -\frac{c\Delta t}{4\Delta x}\left(u_{j+1}^{n+1} - u_{j-1}^{n+1} + u_{j+1}^n - u_{j-1}^n\right)$$

Now, bring all terms involving u_j^{n+1} to the left side:

$$u_j^{n+1} + \frac{c\Delta t}{4\Delta x}\left(u_{j+1}^{n+1} - u_{j-1}^{n+1}\right) = u_j^n - \frac{c\Delta t}{4\Delta x}\left(u_{j+1}^n - u_{j-1}^n\right)$$

Again, define $CFL = \frac{c\Delta t}{\Delta x}$:

$$u_j^{n+1} = u_j^n - \frac{\text{CFL}}{2}\left((u_{j+1}^{n+1} - u_{j-1}^{n+1}) + (u_{j+1}^n - u_{j-1}^n)\right)$$

This form shows how the Crank-Nicolson scheme computes the new time step value by incorporating both the current and the next values of u in its spatial derivative calculations, which is characteristic of the trapezoidal rule.

The typical Crank-Nicolson scheme for advection (or similarly structured problems) leads to a tridiagonal matrix when written in matrix form. This tridiagonal nature arises due to the central differencing in space, which involves neighboring grid points on both sides. However, in the case of a pure implicit upwind discretization, the system's matrix can take on a lower or upper bidiagonal form depending on the direction of the wind or flow.

In a bidiagonal system, each equation in the linear system involves only the current and either the previous or next spatial point at the new time level, depending on whether the matrix is lower or upper bidiagonal. This setup allows for a recursive calculation where the solution at one grid point is computed based on the solution at an adjacent point:

$$u_j^{n+1} = f(u_{j-1}^{n+1}, \text{known terms})$$

where f represents some function derived from the discretization. In the case of an advection equation handled with an implicit upwind method, this recursive relationship helps in efficiently solving the system using straightforward forward or backward substitutions.

Given the bidiagonal (or tridiagonal in general Crank-Nicolson implementations) structure, we can solve these systems efficiently using direct methods. For a lower bidiagonal system, we would use forward substitution, and for an upper bidiagonal system, backward substitution.

The Crank-Nicolson method, being a second-order accurate scheme in both time and space, typically provides good accuracy for linear advection problems at lower CFL numbers. However, as seen in Figure 7.13, its performance can degrade if the CFL condition is too high, leading to instabilities.

At times, efforts to enhance the accuracy of numerical solutions can actually result in poorer outcomes. Additionally, numerical schemes that perform well for smooth solutions might not be suitable for handling discontinuous solutions. In such cases, these schemes can generate unphysical oscillations and create new, unintended extrema.

We now drive the amplitude equation for the 1D linear advection equation is given by:

$$\frac{\partial u}{\partial t} + c\frac{\partial u}{\partial x} = 0$$

where c is the advection speed. The Crank-Nicolson discretization for this equation can be written as:

$$\frac{u_j^{n+1} - u_j^n}{\Delta t} + c\frac{u_{j+1}^{n+1} - u_{j-1}^{n+1} + u_{j+1}^n - u_{j-1}^n}{4\Delta x} = 0$$

Assuming $u_j^n = A^n e^{ikj\Delta x}$ where A is the amplification factor, k is the wave number, j is the spatial index, and n is the time step index, the discrete Fourier mode substitution into the Crank-Nicolson scheme provides:

$$Ae^{ikj\Delta x} - e^{ikj\Delta x} + \frac{c\Delta t}{4\Delta x}\left(Ae^{ik(j+1)\Delta x} - Ae^{ik(j-1)\Delta x} + e^{ik(j+1)\Delta x} - e^{ik(j-1)\Delta x}\right) = 0$$

Simplifying the equation by canceling out $e^{ikj\Delta x}$ and using the identity $e^{ik\Delta x} = \cos(k\Delta x) + i\sin(k\Delta x)$, we get:

$$A - 1 + \frac{c\Delta t}{4\Delta x}\left(A(e^{ik\Delta x} - e^{-ik\Delta x}) + (e^{ik\Delta x} - e^{-ik\Delta x})\right) = 0$$

$$A - 1 + \frac{c\Delta t}{2\Delta x}(Ai\sin(k\Delta x) + i\sin(k\Delta x)) = 0$$

$$A - 1 + i\frac{c\Delta t}{2\Delta x}\sin(k\Delta x)(A+1) = 0$$

Rearranging and factoring out and solving for A:

$$A(1 + i\frac{c\Delta t}{2\Delta x}\sin(k\Delta x)) = 1 - i\frac{c\Delta t}{2\Delta x}\sin(k\Delta x)$$

$$A = \frac{1 - i\frac{c\Delta t}{2\Delta x}\sin(k\Delta x)}{1 + i\frac{c\Delta t}{2\Delta x}\sin(k\Delta x)}$$

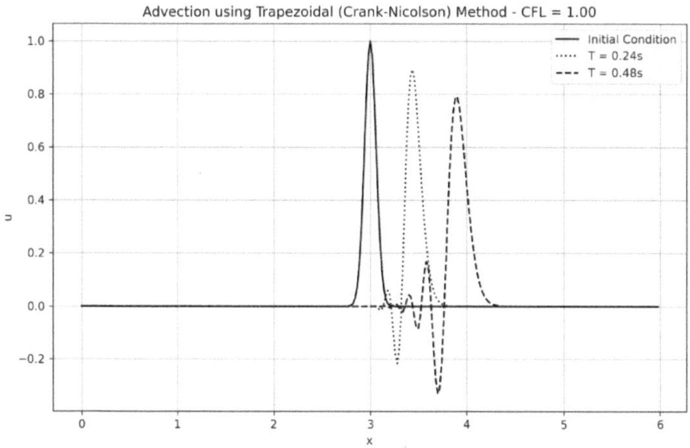

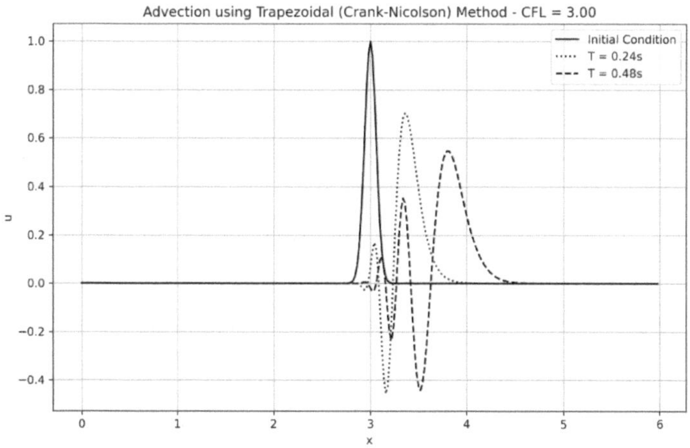

Figure 7.13: **1D Advection equation solved using Crank Nicolson schemes.**

This A can be further simplified to get the modulus (amplitude) $|A|$:

$$|A| = \sqrt{\frac{(1 - \frac{c\Delta t}{2\Delta x}\sin(k\Delta x))^2}{(1 + \frac{c\Delta t}{2\Delta x}\sin(k\Delta x))^2}} = 1$$

$$|A| = \left|\frac{1 - \frac{\mu}{2} i \sin(k\Delta x)}{1 + \frac{\mu}{2} i \sin(k\Delta x)}\right|$$

Let $a = 1$, $b = \frac{\mu}{2}\sin(k\Delta x)$:

$$A = \frac{a - ib}{a + ib}$$

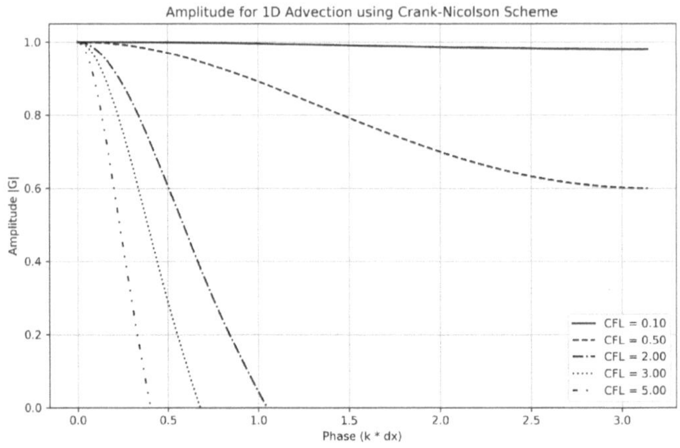

Figure 7.14: **Amplitude for 1D advection equation using CN Scheme.**

Compute the modulus squared:

$$|A|^2 = \frac{(a-ib)(a+ib)}{(a+ib)(a-ib)} = 1$$

Simplify:

$$|A|^2 = \frac{a^2+b^2}{a^2+b^2} = 1$$

Therefore:

$$|A| = 1$$

The modulus of the amplification factor A is 1, indicating that the scheme is neutrally stable in terms of amplitude. This means the Crank-Nicolson scheme conserves the amplitude of the wave for each Fourier component. The Crank-Nicolson scheme for the 1D advection equation is stable for this linear advection problem as long as CFL is less than 1.

Phase Error: The amplification factor A for the Crank-Nicolson scheme is a complex number with unit modulus $|A| = 1$, and can be expressed as:

$$A = e^{-i\theta}$$

where θ is the numerical phase shift per time step. We compute θ as the argument of A. Define:

$$\xi = \frac{\text{CFL}}{2}\sin(k\Delta x)$$

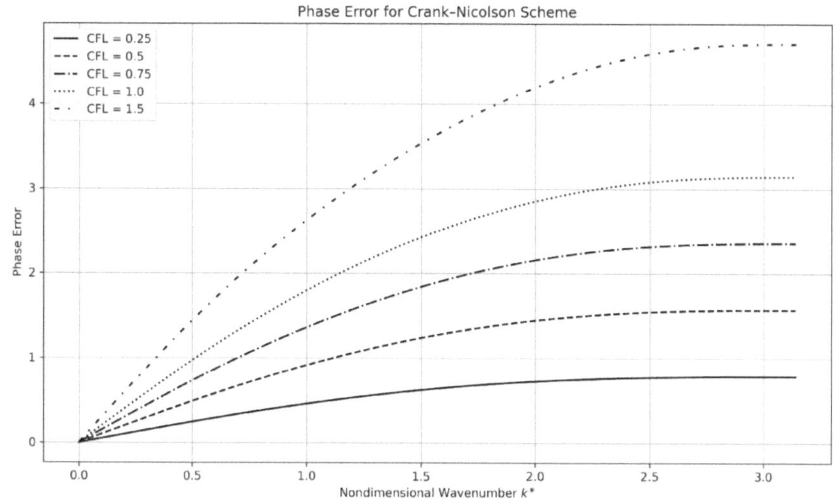

Figure 7.15: **Phase error for the CN Scheme in case of 1D Advection equation.**

and

$$A = \frac{1 - i\xi}{1 + i\xi}$$

The argument of a quotient is the difference of the arguments:

$$\theta = \arg(A) = \arg(1 - i\xi) - \arg(1 + i\xi)$$

$$\theta = -\arctan(\xi) - \arctan(\xi) = -2\arctan\left(\frac{\text{CFL}}{2}\sin(k\Delta x)\right)$$

The exact phase shift per time step is:

$$\phi_{\text{exact}} = ak\Delta t = \text{CFL} \cdot k\Delta x$$

So, the phase error is:

$$\text{Phase Error} = \phi_{\text{exact}} - \theta = \text{CFL} \cdot k\Delta x + 2\arctan\left(\frac{\text{CFL}}{2}\sin(k\Delta x)\right)$$

Introducing the nondimensional wavenumber $k^* = k\Delta x$, the expression becomes:

$$\text{Phase Error} = \text{CFL} \cdot k^* + 2\arctan\left(\frac{\text{CFL}}{2}\sin(k^*)\right)$$

Figure 7.15 shows the phase error introduced by the Crank-Nicolson scheme for various CFL numbers as a function of nondimensional wavenumber k^*. The phase error increases with both k^* and CFL, indicating that higher frequencies and larger time steps lead to greater phase distortion. At low k^*, all curves are close to zero, confirming high accuracy for long wavelengths. However, at higher k^*, the numerical wave significantly lags the exact solution. Although Crank-Nicolson is amplitude-preserving, this plot highlights its dispersive nature, making it less suitable for advection-dominated problems at high CFL values.

While CN is attractive for its unconditional stability and second-order accuracy, its poor phase accuracy at high CFL makes it ill-suited for convection-dominated problems. Schemes that explicitly manage dispersion and dissipation such as Leapfrog, Lax-Wendroff, or high-order compact schemes with RK integration—are more appropriate for preserving waveforms and phase fidelity.

Phase Velocity Error: The phase error affects the numerical phase velocity v_p^{num}, which is also called numerical wave celerity, and can be computed as:

$$v_p^{\text{num}} = \frac{\text{Numerical Angular Frequency}}{k} = \frac{\theta}{k\Delta t}$$

Comparison with Physical Phase Velocity:

$$v_p^{\text{exact}} = c$$

Phase Velocity Error :

$$\text{Phase Velocity Error} = v_p^{\text{num}} - v_p^{\text{exact}} = \left(\frac{\theta}{k\Delta t}\right) - c$$

Impact on Numerical Simulations: A non-zero phase error can lead to numerical dispersion, causing waves to propagate at incorrect speeds, which may distort the solution over time.

Figure 7.15 shows that the Crank-Nicolson scheme is non-dissipative, meaning it doesn't naturally dampen oscillations. Instead, it tends to introduce oscillations at high CFL conditions or for high-frequency components. The plot shows that for large CFL numbers, the phase error not only grows but also shifts to negative values (indicating phase lag), which corresponds to a slower wave propagation in the numerical solution. In practice, this behavior is why the Crank-Nicolson scheme is generally favored for diffusion-dominated problems, where stability and second-order accuracy are prioritized, rather than advection-dominated problems where minimizing phase error is crucial. For advection-dominated equations, phase errors due to high CFL conditions can significantly impact the accuracy and quality of the solution.

Note that in solving the Navier-Stokes equations, the Crank-Nicolson scheme works well due to the diffusive (viscous) terms present in the equations. These terms naturally help in damping out high-frequency oscillations, making the Crank-Nicolson scheme more stable for these types of problems. The Navier-Stokes equations are a combination of advection and diffusion. While the CN scheme has issues with advection alone (due to phase errors, as shown in the plot), the presence of diffusion mitigates this problem. The diffusion terms counteract some of the oscillations and phase errors introduced by the CN scheme's handling of advection.

For high Reynolds number flows, where advection dominates and phase error can be problematic, practitioners sometimes use a hybrid approach, combining CN with other methods (e.g., implicit-explicit schemes or fractional step methods) to handle advection more effectively. The CN method performs best when the CFL number is kept within a reasonable range (typically less than 1 or 2 for Navier-Stokes applications). In practical simulations, this is often achieved by refining the grid or adjusting the time step, which helps maintain accuracy without introducing significant phase errors. For flows with high Reynolds numbers or where advection strongly dominates, the Crank-Nicolson method alone may not be sufficient, as it can lead to non-physical oscillations and phase errors. In such cases, methods with more explicit control of advection terms, such as the upwind or total variation diminishing (TVD) schemes, are often used

7.5.1 GROUP VELOCITY IN CN SCHEME

To derive the dispersion relation for the advection equation $u_t = -u_x$ using the Crank-Nicolson (CN) scheme, we start by expressing the numerical scheme. The Crank-Nicolson method discretizes the equation as

$$u_j^{n+1} - u_j^n = -\frac{\text{CFL}}{2}\left[\left(u_{j+1}^{n+1} - u_{j-1}^{n+1}\right) + \left(u_{j+1}^n - u_{j-1}^n\right)\right],$$

where $\text{CFL} = c\Delta t/\Delta x$ is the Courant number. Next, we substitute a trial solution of the form $u_j^n = e^{i(kj\Delta x - \omega n\Delta t)}$, where k is the wavenumber and ω is the angular frequency. Substituting this expression into the CN scheme, the terms u_{j+1}^{n+1}, u_{j-1}^{n+1}, u_{j+1}^n, and u_{j-1}^n transform using the relationships $u_{j+1}^{n+1} = u_j^{n+1}e^{ik\Delta x}$, $u_{j-1}^{n+1} = u_j^{n+1}e^{-ik\Delta x}$, and so on. By substituting these into the scheme and dividing through by u_j^n, we simplify the expression using Euler's formulas, where $e^{ik\Delta x} - e^{-ik\Delta x} = 2i\sin(k\Delta x)$ and $e^{ik\Delta x} + e^{-ik\Delta x} = 2\cos(k\Delta x)$. After further algebraic manipulations, the dispersion relation is derived as

$$2\tan\left(\frac{\omega\Delta t}{2}\right) = \text{CFL}\cdot\sin(k\Delta x)$$

This relation connects the angular frequency ω, the Courant number CFL, and the wavenumber k, describing the influence of numerical dispersion introduced by the Crank-Nicolson scheme.

To derive the group speed expression for the Crank-Nicolson (CN) scheme:

The CN scheme applied to the 1D advection equation:

$$\frac{\partial u}{\partial t} + c\frac{\partial u}{\partial x} = 0,$$

is discretized as:

$$\frac{u_j^{n+1} - u_j^n}{\Delta t} + \frac{c}{2}\left(\frac{u_{j+1}^n - u_{j-1}^n}{2\Delta x} + \frac{u_{j+1}^{n+1} - u_{j-1}^{n+1}}{2\Delta x}\right) = 0,$$

where c is the wave speed, Δt is the time step, and Δx is the grid spacing. Rearranging gives:

$$u_j^{n+1} - \frac{\text{CFL}}{2}\left(u_{j+1}^{n+1} - u_{j-1}^{n+1}\right) = u_j^n + \frac{\text{CFL}}{2}\left(u_{j+1}^n - u_{j-1}^n\right),$$

where $\text{CFL} = c\Delta t/\Delta x$ is the Courant number.

We assume a wave-like solution:

$$u_j^n = \hat{u}e^{i(kj\Delta x - \omega n\Delta t)},$$

where k is the wavenumber, ω is the angular frequency, and $\hat{u}$ is the amplitude. Substituting this into the discretized CN scheme gives:

$$e^{-i\omega\Delta t} - \frac{\text{CFL}}{2}\left(e^{ik\Delta x} - e^{-ik\Delta x}\right) = 1 + \frac{\text{CFL}}{2}\left(e^{ik\Delta x} - e^{-ik\Delta x}\right).$$

Simplify the terms using $\sin(k\Delta x) = \frac{e^{ik\Delta x} - e^{-ik\Delta x}}{2i}$:

$$e^{-i\omega\Delta t} = \frac{1 - i\frac{\text{CFL}}{2}\sin(k\Delta x)}{1 + i\frac{\text{CFL}}{2}\sin(k\Delta x)}.$$

Taking the logarithm of both sides to isolate ω:

$$\omega\Delta t = -2i\tan^{-1}\left(\frac{\text{CFL}}{2}\sin(k\Delta x)\right).$$

Thus, the angular frequency is:

$$\omega = -\frac{2i}{\Delta t}\tan^{-1}\left(\frac{\text{CFL}}{2}\sin(k\Delta x)\right).$$

To find the group speed c_g, compute the derivative of ω with respect to k:

$$c_g = \frac{\partial\omega}{\partial k}.$$

The group speed simplifies to:

$$c_g = \frac{\cos(k\Delta x)}{1 + \frac{\text{CFL}^2}{4}\sin^2(k\Delta x)},$$

where $k^* = k\Delta x$ is the dimensionless wavenumber. For small values of k, expand $\cos(k\Delta x)$ and $\sin^2(k\Delta x)$ using Taylor series expansions:

$$\cos(k\Delta x) \approx 1 - \frac{1}{2}(k\Delta x)^2,$$

$$\sin^2(k\Delta x) \approx (k\Delta x)^2.$$

Substituting these approximations into $c_g(k)$:

$$c_g \approx 1 - \frac{1}{2}(k\Delta x)^2 - \frac{\text{CFL}^2}{4}(k\Delta x)^2.$$

Combining terms:

$$c_g \approx 1 - \frac{1}{4}(2 + \text{CFL}^2)(k\Delta x)^2.$$

This is the approximate group speed for the Crank-Nicolson scheme, showing the numerical dispersion effect as $k\Delta x$ increases.

Example 7.5

Derive the Crank-Nicolson (CN) dispersion relation for the initial condition *Gaussian-modulated sine wave*:

$$u(x,0) = e^{-16(x-0.5)^2}\sin(kx),$$

Solution

We start with the Crank-Nicolson discretization scheme applied to the one-dimensional linear advection equation:

$$u_t + cu_x = 0.$$

The semi-discrete Crank-Nicolson formulation in time involves averaging the temporal and spatial terms at the current and future time steps. The discretized form is given as:

$$\frac{u_j^{n+1} - u_j^n}{\Delta t} + \frac{c}{2}\left(\frac{u_{j+1}^n - u_{j-1}^n}{2\Delta x} + \frac{u_{j+1}^{n+1} - u_{j-1}^{n+1}}{2\Delta x}\right) = 0,$$

where u_j^n represents the solution at the j-th spatial grid point and the n-th time step, c is the wave speed, Δt is the time step size, and Δx is the spatial grid spacing.

Rearranging terms gives:

$$u_j^{n+1} - \frac{\text{CFL}}{4}\left(u_{j+1}^{n+1} - u_{j-1}^{n+1}\right) = u_j^n + \frac{\text{CFL}}{4}\left(u_{j+1}^n - u_{j-1}^n\right),$$

where $\text{CFL} = \frac{c\Delta t}{\Delta x}$ is the Courant-Friedrichs-Lewy number.

We assume a wave-like solution of the form:

$$u_j^n = \hat{u}e^{i(kj\Delta x - \omega n\Delta t)},$$

where k is the wavenumber, ω is the angular frequency, and $\hat{u}$ is the amplitude. Substituting this solution into the discretized Crank-Nicolson equation yields:

$$e^{-i\omega\Delta t} - \frac{\text{CFL}}{4}\left(e^{ik\Delta x} - e^{-ik\Delta x}\right) = 1 + \frac{\text{CFL}}{4}\left(e^{ik\Delta x} - e^{-ik\Delta x}\right).$$

Simplify the terms involving exponential functions. Using $\sin(k\Delta x) = \frac{e^{ik\Delta x} - e^{-ik\Delta x}}{2i}$ and rearranging, we obtain:

$$2\tan\left(\frac{\omega\Delta t}{2}\right) = \text{CFL}\sin(k\Delta x).$$

This is the Crank-Nicolson dispersion relation. To express the phase speed $c_g(k^*)$, we rewrite ω as:

$$\omega = \frac{2}{\Delta t}\tan^{-1}\left(\frac{\text{CFL}}{2}\sin(k^*\Delta x)\right),$$

where $k^* = k\Delta x$. The group speed $c_g(k^*)$ is defined as the derivative of ω with respect to k^*. For small values of k^*, expand $\tan^{-1}(x) \approx x - \frac{x^3}{3}$ and $\sin(x) \approx x - \frac{x^3}{6}$. Substituting these approximations into the expression for ω, we find:

$$c_g \approx 1 - \frac{2 + \text{CFL}^2}{4}(k^*)^2.$$

This expression shows the effect of numerical dispersion introduced by the Crank-Nicolson scheme. For small values of k^*, the numerical group speed closely approximates the physical wave speed. However, as k^* increases, dispersion errors grow, leading to a deviation between the numerical group speed and the physical speed.

The observed group speed is much smaller ($v_g = 0.0187$) than the theoretical group speed ($v_g = 1$). This discrepancy suggests that the numerical implementation or damping boundary conditions significantly slowed the propagation of the wave packet. This could be caused by numerical dispersion or errors introduced during the simulation that slow down the wave packet's propagation.

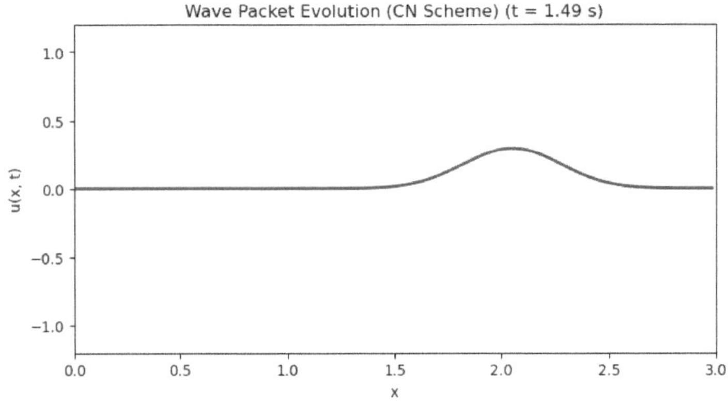

Figure 7.16: **Group Speed using CN scheme at CFL=1. The expected and observed group velocities are same.**

Movie: `wave_packet_evolution_CN.gif` : This movie illustrates the evolution of a wave packet under advection. The animation demonstrates how a wave packet, initially localized in space, evolves as it propagates in x-direction.
Lessons:

☞ Crank-Nicolson with the centred spatial stencil is second-order accurate in both time and space and neutrally stable, so every Fourier mode keeps its amplitude ($|AI = 1$).

☞ Phase and group-velocity errors rise with both nondimensional wavenumber k^* and CFL; larger CFL bends the numerical dispersion curve farther from the exact line.

☞ For poorly resolved waves ($k^* > \pi/2$) the computed group velocity becomes negative, causing the highest-frequency content to propagate in the wrong direction.

☞ The scheme works well for diffusion-dominated problems where amplitude accuracy matters more than phase accuracy, but on convection-dominated flows it can distort wave packets unless the grid is sufficiently fine.

☞ In high-speed wave applications Crank-Nicolson should be combined with filters or replaced by a dispersion-optimised integrator to limit spurious dispersion.

Figure 7.17 compares the approximate group velocity c_g of the Crank-Nicolson (CN) and Leapfrog (LF) schemes for the linear advection equation as a function of nondimensional wavenumber k^*, across different CFL numbers. At low wavenumbers, all curves begin near $c_g = 1$, indicating accurate propagation speed for long waves. However, as k^* increases, CN exhibits a steep drop in group velocity, particularly at higher CFL values. For CFL = 1.0, CN's group velocity becomes negative at relatively small k^*, which implies that high-frequency components propagate in the wrong direction (an unphysical result caused by severe numerical dispersion). In contrast, the Leapfrog scheme shows much better behavior: at CFL = 1.0, the group velocity remains flat and accurate across the entire k^* range, confirming that LF is dispersion-free at leading order in this configuration. Even at lower CFL values, LF maintains better phase and group speed accuracy than CN. This analysis reinforces that while CN is unconditionally stable, its dispersion error renders it unsuitable for wave propagation problems. Leapfrog, though conditionally stable, is much more accurate in preserving wave characteristics, making it preferable in convection- and wave-dominated flows.

Table 7.2 summarizes the approximate group velocity expressions and dispersion behavior for the Crank-Nicolson (CN) and Leapfrog (LF) schemes applied to the linear advection equation. The CN scheme shows a group velocity that decreases

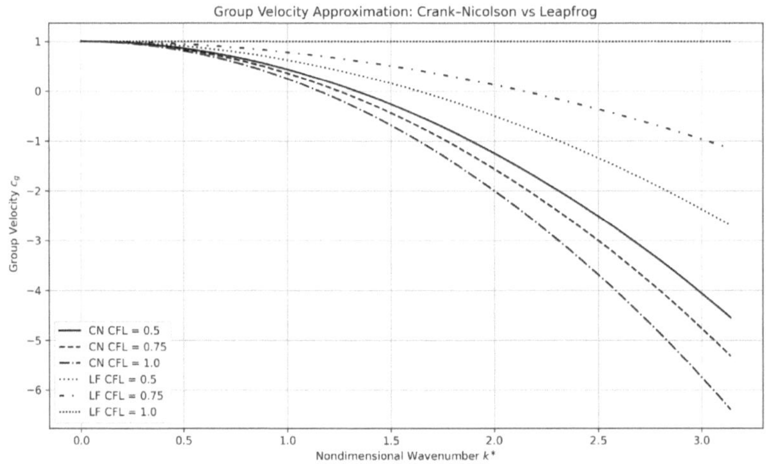

Figure 7.17: **Approximate group velocity c_g as a function of nondimensional wavenumber k^* for Crank-Nicolson (CN) and Leapfrog (LF) schemes at different CFL numbers. CN exhibits significant dispersion errors at high k^*, especially for large CFL values, with group velocity even becoming negative. In contrast, LF preserves group velocity much more accurately, particularly when CFL = 1.0, making it better suited for wave-resolving problems.**

rapidly with increasing non-dimensional wavenumber k^*, and the error worsens with larger CFL numbers. At high wavenumbers, the group velocity can even become negative, leading to severe phase lag and unphysical wave reversal. In contrast, the Leapfrog scheme maintains significantly better group velocity accuracy. When CFL = 1, LF yields exact group velocity at leading order, making it ideal for wave propagation problems. Even for CFL < 1, LF introduces only moderate dispersion compared to CN. This comparison highlights that, although CN is stable and non-dissipative, its poor phase accuracy limits its usefulness in advection-dominated simulations. Leapfrog, while conditionally stable, offers much better dispersion properties and is generally preferred for accurately resolving wave speed and waveform over time.

Errors introduced by finite differencing are not random fluctuations but rather the result of a systematic combination of dispersion and, potentially, dissipation at different orders. Gaining insight into the behavioral characteristics, such as group velocity errors, can provide several advantages. It can help identify when a mesh is too coarse based on the behavior of the numerical solution, aid in selecting an appropriate difference scheme and mesh size to minimize these errors, and assist in making the most out of an imperfect solution when further mesh refinement is too costly.

Table 7.2
Approximate group velocity expressions and dispersion characteristics for Crank-Nicolson and Leapfrog schemes

Scheme	Approximate Group Velocity c_g	Dispersion Characteristics
Crank-Nicolson (CN)	$1-\frac{2+\text{CFL}^2}{4}(k^*)^2$	High dispersion; $c_g<0$ for large k^*; severe phase lag
Leapfrog (LF)	$1-\frac{1}{2}(1-\text{CFL}^2)(k^*)^2$	Low dispersion; exact group velocity at $\text{CFL}=1$

7.6 EIGENVALUE ANALYSIS OF THE ADVECTION EQUATION

In the analysis of a convection-dominant problem employing a 3-point central difference scheme, the treatment of boundary conditions significantly affects the computation within the entire domain. The mathematical descriptions provided below detail how each node, including boundaries, is addressed in this setup:

Discretized Equations for Interior Points: Within the domain's interior, the convection term for a node i, where $1 \leq i \leq M-2$, is approximated using a central difference scheme. This discretization for the convection term at node i is expressed as:

$$\frac{du_i}{dt}+a\frac{u_{i+1}-u_{i-1}}{2\Delta x}=0$$

Here, u_i denotes the function value at node i, a represents the advection speed, and Δx is the spatial grid spacing.

Boundary Conditions: At the First Node ($i=0$):

The condition at the first node is modeled similar to an interior point, using a central difference to the right, which effectively mimics a zero-gradient condition at the boundary:

$$\frac{du_0}{dt}+a\frac{u_1-u_{-1}}{2\Delta x}=0$$

For practical purposes, since u_{-1} is outside the computational domain, it is typically approximated such that $u_{-1}=u_1$, simplifying the equation to:

$$\frac{du_0}{dt}+a\frac{u_1-u_1}{2\Delta x}=0 \quad \Rightarrow \quad \frac{du_0}{dt}=0$$

At the Last Node ($i=M-1$):

The last node utilizes an upwind scheme to effectively handle the outflow condition, described by:

$$\frac{du_{M-1}}{dt}+a\frac{u_{M-1}-u_{M-2}}{\Delta x}=0$$

This approach models the characteristics exiting the domain, where the upwind difference accounts for the flow direction and enhances numerical stability.

Matrix Representation Summary:
The matrix A, representing these discretizations, is structured as follows for a system where $\frac{d\mathbf{u}}{dt} = A\mathbf{u}$:

$$A = \begin{bmatrix} 0 & -\frac{a}{2\Delta x} & 0 & \cdots & 0 \\ \frac{a}{2\Delta x} & 0 & -\frac{a}{2\Delta x} & \cdots & 0 \\ 0 & \frac{a}{2\Delta x} & 0 & \cdots & 0 \\ \vdots & \vdots & \vdots & \ddots & \vdots \\ 0 & \cdots & 0 & \frac{a}{\Delta x} & -\frac{a}{\Delta x} \end{bmatrix}$$

This matrix accounts for each node's discretized convection equation, including special considerations at the boundaries. The choice of numerical scheme and how boundary conditions are applied crucially affects the solution's stability and accuracy, particularly critical in flows dominated by convection.

The eigenvalues from this matrix setup will feature both real and imaginary components. The real components are influenced by the asymmetry from the upwind scheme at the last boundary, while the imaginary components reflect the characteristics of wave propagation modeled by the central differences. This configuration and the resulting eigenvalues demonstrate the effects of boundary treatments on solution behavior, highlighting how these conditions manage inflow and outflow dynamics within the computational framework. The eigenvalue distribution is not symmetrically aligned along the imaginary axis, indicating the nuanced effects of boundary conditions and the stability traits of the adopted numerical approach.

> Real Parts: The real parts of all the eigenvalues are negative, which is crucial for the stability of the numerical scheme. Negative real parts indicate that any perturbations in the numerical solution will exponentially decay over time, contributing to the stability of the system. This is a desired characteristic in numerical simulations of physical systems, as it prevents the amplification of errors.
> Imaginary Parts: The imaginary parts of the eigenvalues represent the oscillatory nature of the solution. In physical terms, these components relate to wave propagation phenomena in the system. The presence of significant imaginary parts suggests that the system supports wave-like solutions, which are typical in convection-dominated problems.
> Singular Real Eigenvalue: The eigenvalue -0.13053446 with zero imaginary part stands out as it is solely real and has the largest magnitude. This is often called bulge.

3 point Stencil with Periodic BC: The eigenvalue distribution indicates that the numerical scheme exhibits marginal stability under periodic boundary conditions. This means the system's solution is predominantly oscillatory due to the dominance of imaginary components in the eigenvalues. The real parts of the eigenvalues are negligible (approximately zero), which ensures there is no exponential growth ($\text{Re}(\lambda) > 0$) or decay ($\text{Re}(\lambda) < 0$). This behavior is typical for wave-like systems modeled using central difference schemes, where physical diffusion or artificial damping is minimal.

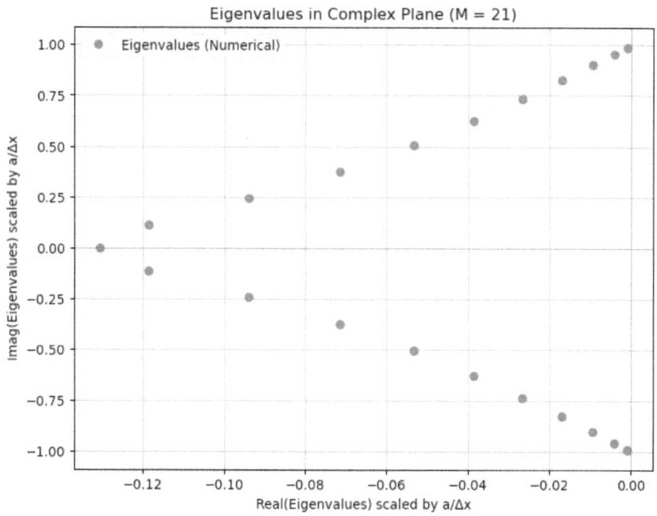

Figure 7.18: **Eigenvalues of convection matrix with 3-point stencil and open boundaries.**

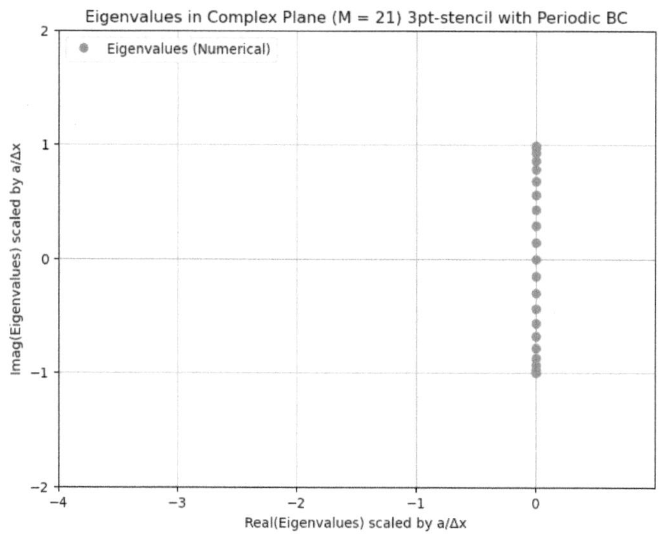

Figure 7.19: **Eigenvalues of convection matrix with 3-point stencil and periodic boundaries.**

While the absence of positive real parts confirms that there is no intrinsic instability in the system, the scheme is not unconditionally stable. It is conditionally stable, meaning that the numerical stability of the solution also depends on the choice of the time-stepping method and adherence to its associated stability criterion (e.g., the Courant-Friedrichs-Lewy (CFL) condition for explicit methods):

$$\text{CFL} = \frac{a\Delta t}{\Delta x} \leq 1$$

where a is the advection speed, Δt is the time step size, and Δx is the spatial grid spacing.

The marginal stability of the eigenvalues ensures that, in the absence of numerical errors or inappropriate time-stepping, the system remains bounded and preserves the oscillatory wave-like nature of convection phenomena. The analysis confirms that the central difference scheme with periodic boundary conditions is well-suited for accurately modeling periodic convection dynamics, as it maintains the physical properties of the system without introducing artificial dissipation or instability.

7.7 TIME-STEP CONSTRAINTS IN ADVECTION PROBLEMS

When solving the advection equation numerically, the choice of finite difference scheme significantly influences the maximum allowable time step size. With central finite differences, the time step is directly limited by the convective Courant-Friedrichs-Lewy (CFL) condition, explicitly tied to the maximum modified wavenumber and the discretization parameters. The equation is coming from the Courant-Friedrichs-Lewy (CFL) condition, which states that the time step Δt must be small enough to satisfy:

$$\Delta t < \frac{\text{Characteristic Length}}{\text{Wave Speed}}$$

In numerical methods for time integration, s_i and s_r represent the imaginary and real parts of the roots of the amplification factor, respectively on stability plot. The real part s_r indicates how the method manages damping or amplification: negative s_r values suggest damping, which promotes stability, while positive values may amplify oscillations and risk instability. In contrast, the imaginary part s_i reflects the method's capacity to handle oscillatory behavior; higher s_i values mean the method can capture more complex oscillations, making it ideal for wave-like or transient phenomena. Methods with zero s_i, such as RK2-O2 and RK6-O5, lack oscillatory behavior, focusing instead on stable growth or decay. Thus, understanding s_i and s_r helps in selecting an appropriate RK method depending on whether accurate oscillation resolution or stable real-part damping is the priority.

For the advection equation with central FDs, the following holds: (7.1)

$$\Delta t_{\text{max}} = \frac{s(\Theta = 90^\circ)}{\frac{c}{\Delta x} \cdot k_{\text{mod, max}}} = \frac{s_i}{k_{\text{mod, max}}} \frac{\Delta x}{|c|} \quad \text{("Convective Limit")}$$

The term s_i represents the imaginary part of the amplification factor root for the time integration method (e.g., RK4-O4). Higher s_i values allow for greater stability with larger oscillations, while $k^*_{\text{mod, max}}$ represents the maximum effective wavenumber that can be captured by the spatial discretization scheme.

With central finite differences, the time step size must be controlled for the advection equation

$$\Delta t_{\max} = \frac{s(\Theta = 90^\circ)}{c/\Delta x \cdot k^*_{\text{mod,max}}} = \frac{s_i}{k^*_{\text{mod,max}}} \frac{\Delta x}{c} = CFL_{\max} \frac{\Delta x}{c} \quad \text{(Convective limit)}$$

With any other finite difference scheme, the CFL number must be controlled for advection equation solution:

$$CFL_{\max} \cong \frac{s(\Theta)}{\sqrt{\left(k^*_{\text{mod,i,max}}\right)^2 + \left(k^*_{\text{mod,r,max}}\right)^2}}, \qquad \Theta = \arctan \frac{k^*_{\text{mod,r,max}}}{-k^*_{\text{mod,i,max}}}$$

For a three-dimensional advection problem with uniform spatial discretization in all directions, the time step should be one-third of the $\Delta t_{\max}$.

Example 7.6

In an advection problem the speed is $c = 1$ and the uniform size of numerical grid is$\Delta x = 0.5$. Calculate maximum timestep $\Delta t_{\max}$ required for the stable run using RK4-O4 method?

Solution

To calculate $\Delta t_{\max}$ for the RK4-O4 method, we use the following information:

$$\Delta t_{\max} = \frac{s_i}{k^*_{\text{mod, max}}} \cdot \frac{\Delta x}{|c|}$$

For the FD-O2 scheme: $k^*_{\text{mod, max}} = 1$
For RK4-O4: $s_i \approx 2.83$
Spatial step size: $\Delta x = 0.5$
Wave speed: $|c| = 1$
Substituting these values into the formula:

$$\Delta t_{\max} = \frac{2.83}{1} \cdot \frac{0.5}{1} = 2.83 \times 0.5 = 1.415$$

$$\Delta t_{\max} = 1.415s$$

Example 7.7

We want to solve 1D advection equation, and for spatial discretization FD-O2, central, $\Delta x = 0.5$ is used. For time integration methods, third order Adams-Bashforth method is considered. How large can the time step Δt be chosen for this method?
What problem is encountered in practice when using AB methods if, as here, the integration is intended to begin at $t = 0$?
What is k^* for the exact solution on the chosen grid, and which wave becomes unstable first when increasing the time step (in terms of wavelength)?

Solution

$$\Delta t_{\text{max}} = \frac{s_i}{k^*_{\text{mod,max}}} \cdot \frac{\Delta x}{|c|}$$

$$\Delta t_{\text{max}} = \frac{0.72}{1} \cdot \left(\frac{0.5}{|0.5}\right) = 0.72$$

In practice, there is an initialization problem because previous values are always required (even for the first time step). A remedy is to calculate the initial steps successively with increasing order or with other methods. Additionally, the previous slopes (spatial derivatives) must be stored, which significantly increases the memory requirements for higher-order methods.
We know the k so

$$k^* = k\Delta x = \frac{2\pi}{3}(0.5) = \frac{\pi}{3} = 1.046$$

The relationship between k and wavelength is

$$\lambda = \frac{2\pi}{k}$$

This gives

$$k^* = k\Delta x = \left(\frac{2\pi}{\lambda}\right)\Delta x$$

and for FD-O2 schemes for first derivative

$$\lambda = \left(\frac{2\pi}{k^*}\right)\Delta x = \left(\frac{2\pi}{k^*_{unstable}}\right)0.5 = \left(\frac{2\pi}{\pi/2}\right)0.5 = 4 \times 0.5 = 2$$

Table 7.3 compares schemes commonly used for advection term modeling in Incompressible Flow Applications.

Table 7.3
Comparison of Schemes for $u_t + cu_x = 0$ in Incompressible Flow Applications.

Name	Amplification factor	Stability Comments
Exact Solution	$A_{exact} = \exp(i\,\mathrm{CFL}\,k^*)$	Always stable
Forward Euler	$A = 1 - \mathrm{CFL}(1 - e^{-ik^*})$	Stability depends on CFL number
FTCS (Forward Time, Central Space)	$A = 1 - i\mathrm{CFL}\sin(k^*)$	Unstable
Backward Euler	$A = \dfrac{1}{1 + \mathrm{CFL} - \mathrm{CFL}e^{-ik^*}}$	Unconditionally stable
Leapfrog	$A = -i\,\mathrm{CFL}\sin(k^*) \pm \sqrt{1 - \mathrm{CFL}^2 \sin^2(k^*)}$	Stable for $\mathrm{CFL} \leq 1$
Crank-Nicolson	$A = \dfrac{1 - i\,\frac{\mathrm{CFL}}{2}\sin(k^*)}{1 + i\,\frac{\mathrm{CFL}}{2}\sin(k^*)}$	Neutrally stable ($\lvert A \rvert = 1$) (dispersive for large k^*) not preferred

Lessons Learned

☞ The Von Neumann method is the most effective tool for analyzing stability in linear, periodic problems. The matrix method offers a more general, though complex, alternative suitable for non-periodic or variable-coefficient systems.

☞ Numerical schemes may exhibit conditional stability, requiring a time-step restriction for convergence.

☞ The Courant-Friedrichs-Lewy (CFL) condition ensures that the numerical domain of dependence fully contains the physical domain of dependence in hyperbolic problems. This check is critical for stability of explicit schemes.

☞ In hyperbolic, convection-dominated problems, two major numerical errors arise: numerical diffusion (amplitude damping) and numerical dispersion (phase distortion). These errors depend on frequency content and scheme characteristics.

☞ A key factor affecting accuracy in time-dependent wave propagation is the number of grid points per wavelength. Insufficient resolution leads to phase errors and poor dispersion representation.

☞ First-order implicit methods, such as backward Euler, are unconditionally stable but exhibit large numerical diffusion and low temporal accuracy, making them unsuitable for unsteady advection problems.

☞ Implicit schemes allow large time steps but introduce excessive numerical damping at high CFL. While useful for steady-state convergence, this property degrades accuracy in unsteady simulations unless CFL is kept small.

☞ The Leapfrog scheme conserves amplitude and is non-dissipative, but suffers from decoupled parasitic modes unless filtered or paired with staggered updates.

☞ Crank-Nicolson is non-dissipative and second-order accurate, but may exhibit phase errors at high CFL due to symmetric averaging in time.

☞ Compact schemes offer improved phase accuracy and allow accurate wave propagation with fewer grid points compared to standard finite-difference methods.

PROBLEMS

7P-1 Consider the unsteady partial differential equation (PDE):

$$\frac{\partial u}{\partial t} + c\frac{\partial u}{\partial x}, \quad u = u(x,t)$$

where $c \in \mathbb{R}$ is a constant.

a) What type of PDE is this, and which types of flow problems does it model?

b) This PDE is to be integrated in time using central finite differences of second order for the spatial discretization. The following time integration methods are applied:

   (i) Euler implicit

   (ii) Heun's method

   (iii) Runge-Kutta 4th order (RK4-O4)

   What is the maximum allowable CFL number for stable integration, beyond which the method becomes unstable? (For FD-O2: $k^*_{\text{mod,max}} = 1$, and for FD-O4: $k^*_{\text{mod,max}} = 1.37$)

c) Can we suggest a modification to the spatial discretization (of the same order) that would allow us to use an absolutely stable method for solving this problem?

7P-2 Consider the one-dimensional linear advection equation:

$$\frac{\partial u}{\partial t} + c\frac{\partial u}{\partial x} = 0$$

with the following initial and boundary conditions:

Initial Condition:

$$u(x,0) = e^{-16(x-0.5)^2}\sin(\xi x)$$

Boundary Condition at $x = 0$:

$$u(0,t) = A\sin(\omega t)$$

where, $A = 1$ is the amplitude, $\omega = 100$ rad/s is the forcing frequency, ξ is a given wavenumber, c is the wave speed. Using the Explicit Euler method for time discretization and the Second-Order Central Finite Difference (FDO2) for spatial discretization, derive:

i The dispersion relation for the numerical scheme.

ii The group velocity based on the dispersion relation.

7P-3 Solve the pure advection equation using forward-in-time, backward-in-space (FTBS) explicit scheme:

$$\frac{\partial h}{\partial t} + c\frac{\partial h}{\partial x} = 0$$

The initial and boundary conditions in terms of $h(x,t)$ are

$$h(x,0) = \begin{cases} 6.0\,\mathrm{m} & \text{for}\, 0 \leq x < 6000\,\mathrm{m} \\ 7.5\,\mathrm{m} & \text{for}\, 6000\,\mathrm{m} \leq x < 12000\,\mathrm{m} \\ 6.0\,\mathrm{m} & \text{for}\, 12000\,\mathrm{m} \leq x < 18000\,\mathrm{m} \\ 0.0\,\mathrm{m} & \text{for}\, x \geq 18000\,\mathrm{m} \end{cases}$$

$$h(0,t) = 6.0\,\mathrm{m}$$

$$h(L,t) = 24000\,\mathrm{m}$$

Take $\Delta x = 3000\,\mathrm{m}$ and timesteps: $\Delta t = 1000\,\mathrm{s}$, $\Delta t = 2000\,\mathrm{s}$, and $\Delta t = 3000\,\mathrm{s}$, with wave speed $c = 1.5\,\mathrm{m/s}$.

i Compute values of h at grid points for the given initial and boundary data using forward-in-time, forward-in-space (FTFS) explicit scheme.

ii Determine the computationally best solution considering stability and accuracy.

iii Experiment with FTFS scheme and comment on why it might fail for pure advection problem.

7P-4 Solve the linear advection equation on a periodic domain using the Compact Finite Difference Operator of sixth order (FDO6) for spatial discretization. Implement time integration using three schemes: Explicit Euler, RK2 (second-order Runge-Kutta), and RK4 (fourth-order Runge-Kutta).

The initial condition is given by:

```
u_initial = -np.sin(x) + (np.sin(2 * x) ** 2) * np.cos(x)
```

Compare the performance of the time integration schemes using the same spatial operator and observe the evolution and accuracy over time.

7P-5 For the partial differential equation

$$u_t + uu_x = 0$$

i Determine the order of truncation error τ for the discretization of uu_x, if the discretization of the convective term uu_x as: is done through difference term

$$uu_x = \frac{u_i}{6\Delta x}(-11u_i + 18u_{i+1} - 9u_{i+2} + 2u_{i+3}) + \tau$$

ii Setup the discretized equation using Euler explicit scheme.

iii Using the CFL condition estimate the maximum permissible time step Δt.

7P-6 Consider the unsteady partial differential equation (PDE)

$$\frac{\partial u}{\partial t} + c \cdot \frac{\partial u}{\partial x} = 0, \quad u = u(x,t), \quad c \geq 0, \text{ constant.}$$

i What is the type of this PDE, and for what types of flow problems can it be considered as a representative model equation?

ii The PDE is to be integrated using as time stepping methods, along with a central FD-O2 for the spatial derivatives. What is the maximum CFL number that can be used in the integration without making the method unstable?

iii What spatial discretization approaches are possible that can stabilize the potentially unstable time integration scheme for the PDE?

iv Solve this using Leapfrog scheme using Python.

7P-7 Consider the model equation for convection processes in the range $0 \leq x \leq 1$ with $c = 1$:

$$\frac{\partial u}{\partial t} + c\frac{\partial u}{\partial x} = 0, \quad u = u(x,t)$$

The following initial and boundary conditions hold:

Initial Boundary: $u(x,0) = 2\sin(kx), \quad k = \frac{\pi}{2}$

Right Boundary: $u(0,t) = 2\sin(-ckt)$

The spatial discretization is carried out inside the integration region using central finite differences of second order (FD-02), and at the outflow boundary, one-sided finite differences of first order are used. For the equidistant grid, a spatial step size of $\Delta x = 0.5$ is used. The time integration uses the classical fourth-order Runge-Kutta method (RK4-O4).

i Sketch the integration region with the initial condition and the expected evolution of the exact solution at a later time $t > t_0$. How many points per exact wavelength is the wave resolved spatially? Also, based on central FD in the interior, give the largest time step at which a stable solution is expected.

ii What effects would the use of higher-order finite differences (e.g., FD-04) have, disregarding boundary effects, on the given spatial resolution? Discuss qualitatively the impact on the spatial solution behavior and the time integration.

iii Calculate the exact solution of PDE for the initial condition with the ansatz:

$$F(x,t) = A(t) \cdot e^{ikx}, \quad A, F \in \mathbb{C}.$$

7P-8 Consider the model equation for convection processes in the range $0 \leq x \leq 1$ with $c = 1$:

$$\frac{\partial u}{\partial t} + c\frac{\partial u}{\partial x} = 0, \quad u = u(x,t)$$

Initial Boundary: $u(x,0) = 2\sin(kx), \quad k = \frac{\pi}{2}$

Right Boundary: $u(0,t) = 2\sin(-ckt)$

(a) Solve this equation numerically using Runge-Kutta RKO2 or RKO4 method with time step restricted for domain $0 \leq x \leq 1$ and time $0 \leq t \leq 1$.

(b) Plot both the exact and numerical solutions vs time and discuss the deviations/errors that occur due to respective spatial discretization, or time stepping.

7P-9 Consider the partial differential equation:

$$\frac{\partial u}{\partial t} + c\frac{\partial u}{\partial x} = 0, \quad u = u(x,t), \quad c, u \in \mathbb{R}$$

i Employing a spatial solution approach of the form $u(x,t) = u_k(t)e^{ikx}$ transforms the original equation into a differential equation described by $\dot{y} = \beta y$. Determine the expression for the coefficient β.

ii For the RK-O4 method, compute the maximum time step size $\Delta t_{\max}$ that maintains stable integration when used as the time integrator for this equation.

iii Predict the behavior of the numerical solution when $\Delta t > \Delta t_{\max}$ is used, where $\Delta t_{\max}$ is the time step required for stable solution via RK-O4 method.

7P-10 Consider the unsteady partial differential equation (PDE):

$$\frac{\partial u}{\partial t} + c\frac{\partial u}{\partial x} = 0, \quad u = u(x,t), \quad c, u \in \mathbb{R}$$

The exact solution for a specified spatial wave with wave number k is:

$$u(x,t) = \sin(k(x - ct)), \quad \text{where } k = \frac{\pi}{2}, \text{ and } c = 0.5.$$

This PDE is numerically solved using the Euler explicit method in both time and space using finite difference (FD) approach. The spatial step size Δx is 1, and the time step size Δt is set to 0.5.

i Prove that the amplification factor per step can be expressed as:

$$A_{\text{mod}}(CFL, k^*) = \sqrt{(1 + CFL \cdot k_{\text{mod},i})^2 + (CFL \cdot k_{\text{mod},r})^2}$$

where $CFL = c\Delta t / \Delta x$.

1. Calculate the amplification factor A_{mod} for the specified wave and discretization if the finite difference is applied as $u_{x,n} = \frac{1}{2\Delta x}(u_{n-2} - 4u_{n-1} + 3u_n)$.

2. Evaluate how large the time step $\Delta t = 1$ can be to maintain numerical stability for short-term integration.

3. Assess whether the employed spatial discretization leads to an amplitude error in the exact time integration.

7P-11 Consider the equation for a non-stationary partial differential equation (PDE):

$$\frac{\partial u}{\partial t} + c\frac{\partial u}{\partial x} = 0, \quad u = u(x,t), \quad c, u \in \mathbb{R}$$

over the domain $-3 \leq x \leq 3$. The solution is modeled for a specified wave with:

$$u(x,t) = \sin\left(\frac{\pi}{3}(x - ct)\right), \quad c = 1$$

The problem is treated as spatially periodic, implying:

$$u, u_x, \ldots (x,t) = u, u_x, \ldots (x+6,t).$$

i Identify the general type of Equation and the fundamental PDE category it belongs to. Calculate the exact solution at location points $x = -3, -2, ..., 3$ for $t = 0$ and $t = 0.5$.

ii Implement the Euler explicit method for temporal and second-order centered finite differences for spatial discretization to solve the equation in the domain $0 \leq x \leq 3$ with $\Delta x = 1$.

iii Compute and plot the numerical solution for $t = 0$ and $t = 0.5$, deriving spatial derivatives at grid points. Assess the stability of the method and explain the behavior based on the CFL condition.

7P-12 The partial differential equation

$$\frac{\partial u}{\partial t} + c\frac{\partial u}{\partial x} = 0, \quad u = u(x,t), \quad c = 0.5$$

is considered in the region $\{0 \leq x \leq 1.5\}$. The spatial step size for numerical integration is $\Delta x = 0.5$. The following initial/boundary conditions apply:

Initial Condition :

$$u(x,t=0) = \sum_{j=1}^{2} u_j = \hat{u}_1 \cdot \cos(k_1 \cdot x) + \hat{u}_2 \cdot \cos(k_2 \cdot x)$$

Boundary Condition :

$$u(x=0,t) = \sum_{j=1}^{2} \hat{u}_j \cdot \cos(-k_j \cdot c \cdot t)$$

with $\hat{u}_1 = 1$ and $\hat{u}_2 = 0.2$ as well as $k_1 = \pi/4$ and $k_2 = 2\pi$. The exact solution is thus

$$u(x,t) = \sum_{j=1}^{2} \hat{u}_j \cdot \cos\left(k_j \cdot (x - c \cdot t)\right).$$

Solve it using RK-O4 and first order finite-difference scheme.

7P-13 On the domain $385 \leq \mathrm{x} \leq 415$ solve the linear advection equation

$$u_t + c u_x = 0, \qquad c = 1$$

using N = 300 grid points and a nine-point central difference for u_x. Advance in time with fourth-order Runge-Kutta and run three cases with Δt = 1.60, 1.13, 0.56 (T =10s). The initial condition is

$$u(x,0) = 0.5\,\exp\left[-(x-400)^2\right]$$

and homogeneous Dirichlet values are enforced at both boundaries. Integrate to t = T for each Δt and overlay the numerical profile on the exact translation of the Gaussian. Explain how the error trends correlate with the corresponding CFL numbers and with the dispersion properties of the nine-point stencil.

7P-14 Simulate wave propagation in a domain using the Crank-Nicolson method, with a sinusoidal forcing function applied at the left boundary. The wave equation $u_t + u_x = 0$ governs the system, and the boundary condition is given by $u(0,t) = \sin(100t)$. Explain how the boundary condition influences wave propagation and describe the numerical stability of the solution, taking into account the Courant-Friedrichs-Lewy (CFL) condition. Discuss the behavior of the wavefront, including how it moves and how damping at the right boundary affects the solution. Finally, compute the observed wave speed and compare it with the expected value.

7P-15 Investigate the dispersion of various wave numbers in a signal. A pulse propagated through the spatial domain $x \in [0,3]$, with a grid resolution $\Delta x = \frac{1}{160}$, and $CFL = 0.4$. Use Leapfrog scheme used for the numerical method. The initial distribution of the wave is given by

$$u(x,0) = e^{-3200(x-1/2)^2}$$

Discuss how the dispersion of the wave can be understood in terms of its Fourier components, and explain how the numerical method affects the dispersion behavior and the speed of wave propagation across different frequencies.

REFERENCES

J. Crank, P. Nicolson, A practical method for numerical evaluation of solutions of partial differential equations of the heat-conduction type. Proceedings, Cambridge Philosophical Society, 43(50): 50-67, 1947.

E. C. DuFort, S. P. Frankel, Stability conditions in the numerical treatment of parabolic differential equations. Math Tables and Other Aids to Computation 7: 135-52, 1953.

L. N. Trefethen, Group velocity in finite difference schemes. SIAM Rev., 24, 113-136, 1982.

H. O. Kreiss, J. Oliger, Methods for the Approximate Solution of Time-Dependent Problems, CARP Publ. No. 10. World Meteorol. Organ., Geneva, 1973.

R. F. Warming and B. J. Hyett, The modified equation approach to the stability and accuracy of finite-difference methods, J. Comput. Phys., 14(2):159 - 179, 1974.

8 Diffusion Problem

In this chapter, we will explore the concept of diffusion as a physical and numerical phenomenon. We will analyze how diffusion is modeled in partial differential equations and investigate various numerical methods for solving diffusion-dominated problems. Special emphasis will be given to the stability, accuracy, and performance of explicit and implicit schemes, supported by Fourier analysis and matrix-based techniques.

- Understand the distinction between physical and numerical diffusion.
- Learn about Fourier analysis and continuum response functions.
- Develop and analyze explicit and implicit diffusion schemes.
- Become familiar with linear stability analysis and amplification factors.
- Analyze the role of eigenvalue decomposition in diffusion solvers.

In advection-dominated problems, diffusion only arises due to numerical errors, such as numerical amplitude errors, which result from the discretization of the equation. These errors introduce artificial diffusion, known as numerical diffusion, that is not part of the actual physical process. However, when diffusion is a physical phenomenon that needs to be resolved within the system, the partial differential equation (PDE) itself must account for it.

A one dimensional diffusion equation can be written as

$$\frac{\partial u}{\partial t} = \mathscr{D}\frac{\partial^2 u}{\partial x^2}$$

where $\mathscr{D}$ is the diffusion coefficient. If the diffusion process is averaged over a spatial scale, the coefficient $\mathscr{D}$ changes to $\mathscr{D}_{\text{avg}}$, reflecting the extent of averaging.

The main idea is that when diffusion is an inherent part of the physical system being modeled, it is essential to ensure that the diffusion coefficient in the partial differential equation accurately reflects the scale of the underlying physical process. If not appropriately chosen, the diffusion coefficient may become unphysical or misleading in numerical simulations. This contrasts with purely advective problems, where physical diffusion is absent, but numerical diffusion can emerge due to discretization errors. Since diffusion is inherently time-dependent, it necessitates the use of time-stepping methods for its numerical treatment.

8.1 FOURIER ANALYSIS OF DIFFUSION EQUATION

Consider the diffusion equation:

$$\frac{\partial \phi}{\partial t} = \mathscr{D}\frac{\partial^2 \phi}{\partial x^2}$$

DOI: 10.1201/9781003663171-8

where, $\phi(x,t)$ represents the concentration or some diffused quantity, $\mathscr{D}$ is the diffusion coefficient, and the equation describes how ϕ evolves over time due to diffusion. We assume a solution of the form of a Fourier mode:

$$\phi(x,t) = \exp(ikx) \cdot T(t)$$

where,

k is the wave number, and $k = \frac{2\pi n}{L}$ (for a domain of length L).

$T(t)$ is a time-dependent function that we aim to determine.

Substituting this assumed form into the original diffusion equation

$$\frac{\partial}{\partial t}[\exp(ikx)T(t)] = \mathscr{D}\frac{\partial^2}{\partial x^2}[\exp(ikx)T(t)]$$

The left-hand side becomes

$$\exp(ikx)\frac{dT(t)}{dt}$$

The right-hand side involves differentiating $\exp(ikx)$ twice with respect to x. Each differentiation brings down a factor of ik, so

$$\mathscr{D}\frac{\partial^2}{\partial x^2}[\exp(ikx)T(t)] = -\mathscr{D}k^2\exp(ikx)T(t)$$

Cancelling $\exp(ikx)$ from both sides of the equation

$$\frac{dT(t)}{dt} = -\mathscr{D}k^2T(t)$$

This is a simple first-order linear differential equation. The solution is

$$T(t) = \exp(-\mathscr{D}k^2t)$$

Substituting $T(t)$ back into the assumed form for $\phi(x,t)$, we get

$$\phi(x,t) = \exp(ikx)\exp(-\mathscr{D}k^2t)$$

Combining the exponentials

$$\phi(x,t) = \exp(ikx - \mathscr{D}k^2t)$$

This equation represents the diffusion of a Fourier mode over time. The wave number k describes the spatial variation, while the term $-\mathscr{D}k^2t$ describes the decay of the mode due to diffusion.

To compare this analytical solution to discrete numerical methods (finite difference, for example), we introduce some non-dimensional parameters, which are used to standardize the analysis:

i Dimensionless Wave Number $k^* = k\Delta x$, which shows how the wave number k relates to the grid spacing Δx. This non-dimensional form helps in comparing how well a numerical method resolves different spatial frequencies.

ii Dimensionless Diffusion number (σ) is defined as

$$\sigma = \frac{\mathscr{D}\Delta t}{\Delta x^2}$$

where, $\mathscr{D}$ is the diffusion coefficient. Δt and Δx are the time step and spatial step sizes, respectively. This dimensionless number σ represents how diffusion evolves over time in relation to the numerical grid. It gives a way to compare the strength of diffusion relative to the numerical resolution (the grid spacing Δx). If σ is too large, the diffusion might be overestimated, causing excessive smoothing of the solution.

These dimensionless parameters are critical for comparing the behavior of the analytical solution (the exact solution of the diffusion equation) with the behavior of numerical methods. When solving PDEs numerically (e.g., using finite difference or finite element methods), these quantities help to assess how well the numerical grid and time steps resolve the physical processes like diffusion or advection.

Introduce the mesh quantities

$$k^* = k\,\Delta x, \qquad \sigma = \frac{\mathscr{D}\,\Delta t}{\Delta x^2}.$$

Because $k = k^*/\Delta x$ and $\mathscr{D} = \sigma\,\Delta x^2/\Delta t$, the exact Fourier mode

$$\phi(x,t) = \exp\big(ikx - \mathscr{D}k^2 t\big)$$

can be rewritten entirely in terms of k^*, σ, Δt:

$$\phi(x,t) = \exp\Big(i\frac{k^*}{\Delta x}x - \sigma k^{*2}\frac{t}{\Delta t}\Big).$$

For discrete time levels $t = n\,\Delta t\,(n = 0,1,2,\ldots)$ this reduces to

$$\phi^n(x) = \exp\big(i\tfrac{k^*}{\Delta x}x - \sigma k^{*2} n\big),$$

showing explicitly that each step multiplies the mode amplitude by the continuum response factor $\exp(-\sigma k^{*2})$. The diffusion equation does not transport a wave; it damps every Fourier component without introducing any phase shift. Therefore, only the real part is important and we define the continuum response function $Re(k^*)$ per step as

$$R_e(k^*) \approx [\phi^n(x)]^{real} = \exp\big(-\sigma k^{*2}\big)$$

This represents the exact decay of a Fourier mode due to diffusion.

8.2 EXPLICIT SCHEME

We start analyzing the diffusion equation with the Euler explicit scheme for time:

$$\frac{\partial \phi}{\partial t} = \mathscr{D} \frac{\partial^2 \phi}{\partial x^2}$$

For the time derivative $\frac{\partial \phi}{\partial t}$, we use the forward difference approximation in time:

$$\frac{\partial \phi}{\partial t} \approx \frac{\phi_j^{n+1} - \phi_j^n}{\Delta t}$$

where

ϕ_j^n is the value of ϕ at spatial index j and time level n;

ϕ_j^{n+1} is the value of ϕ at the same location j one time step later;

Δt is the time-step size.

For the second spatial derivative we employ the centred (second-order) difference

$$\frac{\partial^2 \phi}{\partial x^2} \approx \frac{\phi_{j+1}^n - 2\phi_j^n + \phi_{j-1}^n}{\Delta x^2},$$

where

ϕ_{j+1}^n and ϕ_{j-1}^n are the values at the neighbouring nodes $j+1$ and $j-1$ at time level n;

Δx is the uniform spatial grid spacing.

We now substitute the finite difference approximations into the original diffusion equation:

$$\frac{\phi_j^{n+1} - \phi_j^n}{\Delta t} = \mathscr{D} \frac{\phi_{j+1}^n - 2\phi_j^n + \phi_{j-1}^n}{\Delta x^2}$$

To solve for ϕ_j^{n+1}, multiply both sides of the equation by Δt and rearrange:

$$\phi_j^{n+1} = \phi_j^n + \mathscr{D} \frac{\Delta t}{\Delta x^2} \left(\phi_{j+1}^n - 2\phi_j^n + \phi_{j-1}^n\right)$$

Introduce the dimensionless diffusion coefficient σ, defined as:

$$\sigma = \frac{\mathscr{D} \Delta t}{\Delta x^2}$$

This parameter measures how the diffusion term scales with the time step and spatial grid size.

Substituting σ into the equation, we get the explicit scheme:

$$\phi_j^{n+1} = \sigma \phi_{j+1}^n + (1 - 2\sigma)\phi_j^n + \sigma \phi_{j-1}^n$$

This is the explicit finite difference scheme for the diffusion equation.

To substitute the expression

$$\phi_j^n = \xi^n e^{ik^* j}$$

into the explicit finite difference scheme:

$$\phi_j^{n+1} = \sigma\phi_{j+1}^n + (1-2\sigma)\phi_j^n + \sigma\phi_{j-1}^n$$

We now perform the stability analysis for this scheme. We substitute $\phi_j^n = \xi^n e^{ik^* j}$ into the explicit scheme

For each term in the scheme, substitute the expression $\phi_j^n = \xi^n e^{ik^* j}$:

$$\phi_j^n = \xi^n e^{ik^* j}$$

$$\phi_{j+1}^n = \xi^n e^{ik^*(j+1)} = \xi^n e^{ik^* j} e^{ik^*}$$

$$\phi_{j-1}^n = \xi^n e^{ik^*(j-1)} = \xi^n e^{ik^* j} e^{-ik^*}$$

To derive the amplification factor, substitute these expressions into the explicit scheme:

$$\xi^{n+1} e^{ik^* j} = \sigma\xi^n e^{ik^*(j+1)} + (1-2\sigma)\xi^n e^{ik^* j} + \sigma\xi^n e^{ik^*(j-1)}$$

$$\xi^{n+1} e^{ik^* j} = \sigma\xi^n e^{ik^* j} e^{ik^*} + (1-2\sigma)\xi^n e^{ik^* j} + \sigma\xi^n e^{ik^* j} e^{-ik^*}$$

Since $e^{ik^* j}$ appears in every term, we can factor it out and cancel from both sides of the equation:

$$\xi^{n+1} = \sigma\xi^n e^{ik^*} + (1-2\sigma)\xi^n + \sigma\xi^n e^{-ik^*}$$

Now simplify the equation:

$$\xi^{n+1} = \xi^n \left(\sigma e^{ik^*} + (1-2\sigma) + \sigma e^{-ik^*}\right)$$

Using Euler's formula $e^{ik^*} + e^{-ik^*} = 2\cos(k^*)$, this simplifies to:

$$\xi^{n+1} = \xi^n ((1-2\sigma) + 2\sigma\cos(k^*))$$

Thus, the equation for ξ^{n+1} becomes:

$$\xi^{n+1} = \xi^n (1 - 2\sigma(1-\cos(k^*)))$$

This is the final expression after substituting $\phi_j^n = \xi^n e^{ik^* j}$ into the explicit finite difference scheme.

We derived the expression:

$$\xi^{n+1} = \xi^n (1 - 2\sigma(1-\cos(k^*)))$$

Recall the identity:

$$1-\cos(k^*) = 2\sin^2\left(\frac{k^*}{2}\right)$$

Substitute this identity into the equation:

$$\xi^{n+1} = \xi^n\left(1-2\sigma\cdot 2\sin^2\left(\frac{k^*}{2}\right)\right)$$

This simplifies to:

$$\xi^{n+1} = \xi^n\left(1-4\sigma\sin^2\left(\frac{k^*}{2}\right)\right)$$

Thus, we have the desired form:

$$\boxed{\xi^{n+1} = \left(1-4\sigma\sin^2\left(\frac{k^*}{2}\right)\right)\xi^n}$$

$$\xi^{n+1} = A\xi^n$$

This equation describes how the Fourier coefficient ξ^n evolves from one time step n to the next time step $n+1$. The factor A is the amplification factor, which governs the growth or decay of the Fourier coefficient over time.

The Amplification Factor A for Euler explicit scheme for diffusion equation :

$$\boxed{A = 1-4\sigma\sin^2\left(\frac{k^*}{2}\right)}$$

where, σ is the dimensionless diffusion coefficient, which relates to the time step Δt and spatial grid size Δx, and k^* is the dimensionless wave number. The term $\sin^2\left(\frac{k^*}{2}\right)$ represents how the numerical scheme handles different wave numbers (spatial frequencies). The factor A tells us whether the solution grows, decays, or remains stable at each time step. For the Euler explicit scheme to be stable, the Fourier coefficient ξ must not grow unbounded over time. This implies that the amplification factor must satisfy:

$$|A| \leq 1$$

This condition ensures that the numerical method does not amplify errors or oscillations over time. If $|A| > 1$, then the scheme is unstable, as errors will grow exponentially with each time step. Conversely, if $|A| \leq 1$, the errors remain bounded, and the scheme is considered stable. The amplification factor A controls whether the numerical solution grows, decays, or remains steady over time. In diffusion-dominated problems, we expect the solution to decay (since diffusion tends to smooth out variations), and thus, we desire $A \leq 1$ for stability.

The specific form of A shows that stability depends on both the diffusion number (σ) and the spatial wave number (k^*). The trigonometric function $\sin^2\left(\frac{k^*}{2}\right)$ introduces a dependence on the spatial frequency, meaning that higher-frequency components (larger k^*) may be more prone to instability unless σ is small enough to ensure $|A| \leq 1$.

For the Forward-Time-Centered-Space (FTCS) scheme the amplification factor, stability requires that the modulus of the factor does not exceed unity for every admissible wavenumber:

$$|A(k^*)| \leq 1 \qquad \text{for all } k^* \in [0, \pi].$$

Because the sine term is non-negative, the largest possible value of $|A|$ is $A_{\max} = 1$ (obtained when $\sin^2 = 0$). The smallest value occurs when $\sin^2 = 1$:

$$A_{\min} = 1 - 4\sigma.$$

Demanding $|A_{\min}| \leq 1$ gives

$$-1 \leq 1 - 4\sigma \leq 1.$$

The upper bound is always satisfied. The lower bound yields the only restriction:

$$1 - 4\sigma \geq -1 \Longrightarrow -4\sigma \geq -2 \Longrightarrow 4\sigma \leq 2 \Longrightarrow \sigma \leq 0.5$$

Hence the mesh Fourier (diffusion) number must satisfy

$$\sigma = \frac{\mathscr{D}\Delta t}{\Delta x^2} \leq \frac{1}{2}$$

to keep $|A| \leq 1$ for every Fourier component, guaranteeing stability of the explicit diffusion scheme.

We have proved that for a one-dimensional diffusion equation expressed the continuum (exact) response function is

$$R_e(k^*) = \exp\left(-\sigma k^{*2}\right).$$

The $R_e(k^*)$ represents the ideal continuous decay rate, while $A(k^*)$ quantifies the additional numerical damping introduced by the explicit finite-difference scheme.

i Stability requires $\sigma \leq \frac{1}{2}$; beyond that bound $|A| > 1$ and high frequency modes grow.

ii For small k^* the Taylor expansion of $A(k^*)$ recovers $R_e(k^*)$ up to second order, so long wave behaviour is reproduced correctly.

The Linear Stability Analysis provides a more comprehensive understanding of the behavior of numerical schemes than simply relying on the amplification factor for stability. By plotting amplitude portrait, one can assess the performance of a numerical method for various wave numbers and determine its accuracy. While the

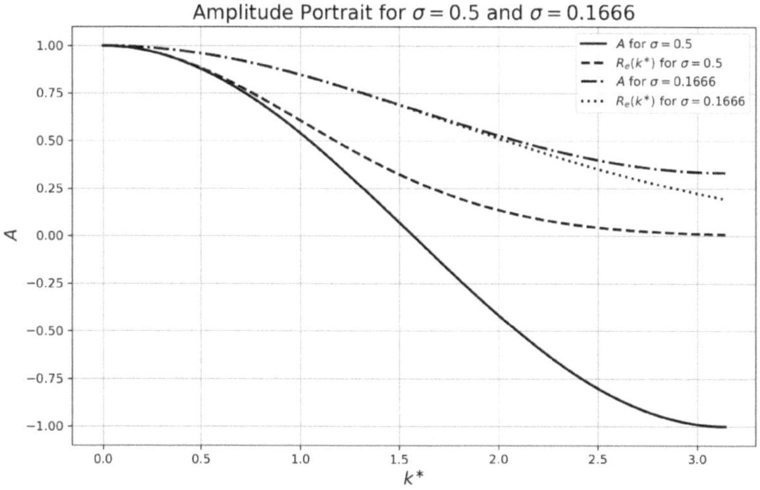

Figure 8.1: **Amplification factor for explicit scheme.**

lagged explicit scheme is analyzed in this context, more sophisticated schemes might offer better accuracy. This analysis is especially valuable for understanding how well short waves (high-frequency components) are handled in numerical solutions.

Figure 8.2 (in polar coordinates) shows the angular component (θ or k^*) ranges from 0° to 180°, and the radial component represents the amplification factor A. The shaded (hatched) area represents the stability zone for $\sigma \leq 0.5$. This zone is considered stable, as the amplification factor $A \leq 1$ in this region. The stability zone is shaded gray, highlighting that within this range, the numerical scheme is stable. The curves that fall inside the shaded region (for $\sigma \leq 0.5$) correspond to stable configurations of the numerical scheme. For values of $\sigma > 0.5$, such as $\sigma = 0.625$ and $\sigma = 1.0$, the curves extend outside the stability region, indicating that the scheme is unstable for those values of σ. The outermost curve corresponds to the boundary for maximum amplification factor, beyond which the scheme becomes unstable.

8.3 LEAPFROG SCHEME

The Leapfrog scheme is a second-order time integration method that uses values from two previous time levels to compute the next. Applied to the one-dimensional diffusion equation:

$$\frac{\partial \phi}{\partial t} = \mathscr{D}\frac{\partial^2 \phi}{\partial x^2}$$

the leapfrog discretization uses a centered difference in time over two steps and a second-order central difference in space. The discrete form is:

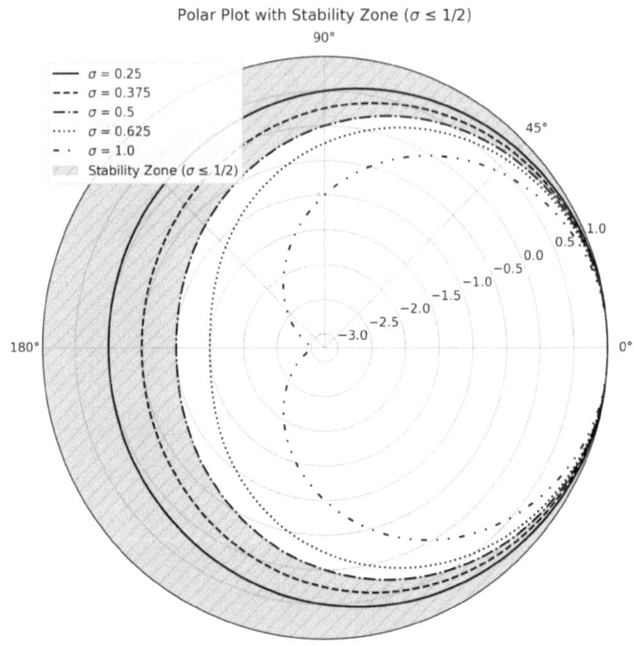

Figure 8.2: **Polar plot for amplification factor for Explicit scheme with indicated stability zone.**

$$\frac{\phi_j^{n+1} - \phi_j^{n-1}}{2\Delta t} = \mathscr{D}\frac{\phi_{j+1}^n - 2\phi_j^n + \phi_{j-1}^n}{\Delta x^2}$$

Multiplying both sides by $2\Delta t$, we obtain the explicit leapfrog update formula:

$$\phi_j^{n+1} = 2\sigma\left(\phi_{j+1}^n - 2\phi_j^n + \phi_{j-1}^n\right) + \phi_j^{n-1}$$

where $\sigma = \dfrac{\mathscr{D}\Delta t}{\Delta x^2}$ is the dimensionless diffusion number.

To analyze the stability of this scheme, we perform a von Neumann (Fourier) analysis by assuming a solution of the form:

$$\phi_j^n = A^n e^{ik^* j}$$

where A is the amplification factor, $k^* = k\Delta x$ is the nondimensional wave number, and j is the spatial index.

We substitute this into the leapfrog update formula. Using:

$$\phi_j^{n+1} = A^{n+1} e^{ik^* j}, \quad \phi_j^{n-1} = A^{n-1} e^{ik^* j}, \quad \phi_j^n = A^n e^{ik^* j}$$

$$\phi_{j+1}^n = A^n e^{ik^*(j+1)} = A^n e^{ik^* j} e^{ik^*}, \quad \phi_{j-1}^n = A^n e^{ik^*(j-1)} = A^n e^{ik^* j} e^{-ik^*}$$

Substituting into the scheme gives:

$$A^{n+1} e^{ik^* j} = 2\sigma A^n e^{ik^* j} (e^{ik^*} - 2 + e^{-ik^*}) + A^{n-1} e^{ik^* j}$$

Dividing both sides by $e^{ik^* j}$:

$$A^{n+1} = 2\sigma A^n (e^{ik^*} - 2 + e^{-ik^*}) + A^{n-1}$$

Using the identity:

$$e^{ik^*} + e^{-ik^*} = 2\cos(k^*) \quad \Rightarrow \quad e^{ik^*} - 2 + e^{-ik^*} = -4\sin^2\left(\frac{k^*}{2}\right)$$

we have

$$A^{n+1} = -8\sigma \sin^2\left(\frac{k^*}{2}\right) A^n + A^{n-1}$$

This is a linear, constant-coefficient recurrence relation:

$$A^{n+1} + 8\sigma \sin^2\left(\frac{k^*}{2}\right) A^n - A^{n-1} = 0$$

Assume a solution of the form $A^n = \alpha^n$, where α is a complex scalar. Substituting into the recurrence:

$$\alpha^{n+1} + 8\sigma \sin^2\left(\frac{k^*}{2}\right) \alpha^n - \alpha^{n-1} = 0$$

Divide through by α^{n-1}:

$$\alpha^2 + 8\sigma \sin^2\left(\frac{k^*}{2}\right) \alpha - 1 = 0$$

This is a quadratic equation in α:

$$\alpha^2 + b\alpha - 1 = 0 \quad \text{with} \quad b = 8\sigma \sin^2\left(\frac{k^*}{2}\right)$$

The roots are given by:

$$\alpha = \frac{-b \pm \sqrt{b^2+4}}{2} = -4\sigma \sin^2\left(\frac{k^*}{2}\right) \pm \sqrt{16\sigma^2 \sin^4\left(\frac{k^*}{2}\right) + 1}$$

Therefore, the two amplification factors are:

$$\boxed{A = -4\sigma \sin^2\left(\frac{k^*}{2}\right) \pm \sqrt{16\sigma^2 \sin^4\left(\frac{k^*}{2}\right) + 1}}$$

To assess the stability, we examine the magnitude $|A|$ of these roots. Since $\sigma > 0$ and $\sin^2\left(\frac{k^*}{2}\right) > 0$, the square root is always slightly greater than the absolute value of the first term. Therefore, one of the roots always has magnitude greater than 1:

$$|A| = \left| -4\sigma \sin^2\left(\frac{k^*}{2}\right) \pm \sqrt{16\sigma^2 \sin^4\left(\frac{k^*}{2}\right) + 1} \right| > 1$$

This proves that at least one amplification factor exceeds unity in magnitude for any nonzero σ, meaning the leapfrog scheme for the diffusion equation is unconditionally unstable. Errors will grow exponentially with time regardless of the grid spacing or time step.

Let us define:

$$\alpha = \sin^2\left(\frac{k^*}{2}\right), \quad \text{with } 0 \leq \alpha \leq 1$$

Then the amplification factors simplify to:

$$A = -4\sigma\alpha \pm \sqrt{16\sigma^2\alpha^2 + 1}$$

We now examine the magnitude of both roots to assess stability.

Case 1: Root with positive sign

$$|A_1| = \left| -4\sigma\alpha + \sqrt{16\sigma^2\alpha^2 + 1} \right|$$

Since the square root is strictly greater than $4\sigma\alpha$, this expression is positive:

$$|A_1| = \sqrt{16\sigma^2\alpha^2 + 1} - 4\sigma\alpha$$

We test whether $|A_1| \leq 1$. Squaring both sides:

$$\left(\sqrt{16\sigma^2\alpha^2 + 1} - 4\sigma\alpha\right)^2 \leq 1$$

$$16\sigma^2\alpha^2 + 1 - 8\sigma\alpha\sqrt{16\sigma^2\alpha^2 + 1} + 16\sigma^2\alpha^2 \leq 1$$

$$32\sigma^2\alpha^2 + 1 - 8\sigma\alpha\sqrt{16\sigma^2\alpha^2 + 1} \leq 1$$

$$32\sigma^2\alpha^2 \leq 8\sigma\alpha\sqrt{16\sigma^2\alpha^2 + 1}$$

Divide both sides by $8\sigma\alpha$ (assuming $\sigma > 0$, $\alpha > 0$):

$$4\sigma\alpha \leq \sqrt{16\sigma^2\alpha^2+1}$$

This inequality is always satisfied because:

$$(4\sigma\alpha)^2 = 16\sigma^2\alpha^2 < 16\sigma^2\alpha^2+1$$

Hence,

$$|A_1| < 1$$

Case 2: Root with negative sign

$$|A_2| = \left|-4\sigma\alpha - \sqrt{16\sigma^2\alpha^2+1}\right| = 4\sigma\alpha + \sqrt{16\sigma^2\alpha^2+1}$$

Since both terms are positive, we immediately conclude:

$$|A_2| > \sqrt{16\sigma^2\alpha^2+1} > 1$$

for any $\sigma > 0$ and $k^* \neq 0$, which proves that the leapfrog scheme is **unconditionally unstable** for the diffusion equation.

Leapfrog scheme is unstable for diffusion because $|A| > 1$ for at least one root at any $\sigma > 0$.

Note: Only one root lies inside the unit circle; the other always exceeds 1 in magnitude, making the scheme unstable for pure diffusion equation.

8.4 THE DUFORT-FRANKEL SCHEME

Starting from the diffusion equation:

$$\frac{\partial \phi}{\partial t} = \mathscr{D}\frac{\partial^2 \phi}{\partial x^2}$$

The DuFort-Frankel scheme uses the following approximations: 1. Time derivative: central difference

$$\frac{\partial \phi}{\partial t} \approx \frac{\phi_j^{n+1} - \phi_j^{n-1}}{2\Delta t}$$

2. Spatial derivative: central difference with ϕ_j^n replaced by the average of ϕ_j^{n+1} and ϕ_j^{n-1}:

$$\frac{\partial^2 \phi}{\partial x^2} \approx \frac{\phi_{j+1}^n - (\phi_j^{n+1} + \phi_j^{n-1}) + \phi_{j-1}^n}{\Delta x^2}$$

Substituting these into the diffusion equation gives:

$$\frac{\phi_j^{n+1} - \phi_j^{n-1}}{2\Delta t} = \mathscr{D}\frac{\phi_{j+1}^n - \phi_j^{n+1} - \phi_j^{n-1} + \phi_{j-1}^n}{\Delta x^2}$$

Now, let $\sigma = \frac{\mathscr{D}\Delta t}{\Delta x^2}$ and rearrange the equation to solve for ϕ_j^{n+1}:

$$\phi_j^{n+1} - \phi_j^{n-1} = 2\sigma\left(\phi_{j+1}^n - \phi_j^{n+1} - \phi_j^{n-1} + \phi_{j-1}^n\right)$$

Collect terms involving ϕ_j^{n+1} on the left:

$$\phi_j^{n+1} + 2\sigma\phi_j^{n+1} = \phi_j^{n-1} - 2\sigma\phi_j^{n-1} + 2\sigma(\phi_{j+1}^n + \phi_{j-1}^n)$$

Factorize:

$$(1+2\sigma)\phi_j^{n+1} = (1-2\sigma)\phi_j^{n-1} + 2\sigma(\phi_{j+1}^n + \phi_{j-1}^n)$$

Finally, solve for ϕ_j^{n+1}:

$$\phi_j^{n+1} = \frac{1-2\sigma}{1+2\sigma}\phi_j^{n-1} + \frac{2\sigma}{1+2\sigma}(\phi_{j+1}^n + \phi_{j-1}^n)$$

This is the DuFort-Frankel scheme.
Stability Analysis: Substitute the Fourier mode $\phi_j^n = u^n e^{ikj\Delta x}$ into the scheme:

$$u^{n+1}e^{ikj\Delta x} = \frac{1-2\sigma}{1+2\sigma}u^{n-1}e^{ikj\Delta x} + \frac{2\sigma}{1+2\sigma}u^n\left(e^{ik(j+1)\Delta x} + e^{ik(j-1)\Delta x}\right)$$

Divide through by $e^{ikj\Delta x}$:

$$u^{n+1} = \frac{1-2\sigma}{1+2\sigma}u^{n-1} + \frac{2\sigma}{1+2\sigma}u^n\left(e^{ik\Delta x} + e^{-ik\Delta x}\right)$$

Using $e^{ik\Delta x} + e^{-ik\Delta x} = 2\cos(k\Delta x)$, we get:

$$u^{n+1} = \frac{1-2\sigma}{1+2\sigma}u^{n-1} + \frac{4\sigma\cos(k\Delta x)}{1+2\sigma}u^n$$

This is a two-step recurrence relation. To analyze stability, assume $u^{n+1} = Au^n$ and $u^n = Au^{n-1}$, which gives:

$$A^2u^{n-1} = \frac{1-2\sigma}{1+2\sigma}u^{n-1} + \frac{4\sigma\cos(k\Delta x)}{1+2\sigma}Au^{n-1}$$

Divide through by u^{n-1}:

$$A^2 = \frac{1-2\sigma}{1+2\sigma} + \frac{4\sigma\cos(k\Delta x)}{1+2\sigma}A$$

Rearrange into a quadratic equation:

$$A^2 - \frac{4\sigma\cos(k\Delta x)}{1+2\sigma}A - \frac{1-2\sigma}{1+2\sigma} = 0$$

This is the quadratic equation for the amplification factor A. The scheme is unconditionally stable if $|A| \leq 1$ for all $k\Delta x$ and $\sigma > 0$.

Solving the Quadratic Equation: The quadratic equation is:

$$A^2 - \left(\frac{4\sigma\cos\theta}{1+2\sigma}\right)A - \frac{1-2\sigma}{1+2\sigma} = 0$$

where $k^* = k\Delta x$.

The discriminant D is:

$$\mathtt{D} = \left(\frac{4\sigma\cos(k^*)}{1+2\sigma}\right)^2 + 4\frac{1-2\sigma}{1+2\sigma}$$

Simplify $\mathscr{D}$:

$$\mathtt{D} = \frac{4(1-4\sigma^2\sin^2(k^*))}{(1+2\sigma)^2}$$

The roots are:

$$A = \frac{2\sigma\cos(k^*) \pm \sqrt{1-4\sigma^2\sin^2(k^*)}}{1+2\sigma}$$

For stability, we require $|A| \leq 1$. It can be shown that for all $\sigma > 0$ and all (k^*), the magnitude of A is indeed bounded by 1, confirming unconditional stability.

8.5 DAMPED LEAPFROG SCHEME

The damped leapfrog scheme is a variant of the leapfrog method, which is a second-order accurate scheme in both space and time. The leapfrog method itself can suffer from numerical oscillations and instability, especially when applied to stiff problems or when the time step size is not well chosen.

To counteract this, a damping term (sometimes also called an artificial viscosity term) is introduced to the standard leapfrog method to suppress these oscillations. The damped leapfrog scheme is thus a modified version that reduces or prevents oscillations at the cost of some accuracy, while maintaining second-order accuracy in time.

Key characteristics of the damped leapfrog scheme:

1. Second-order accuracy: Both in time and space.
2. Damping term: Introduced to control oscillations and improve stability.
3. Simplicity: It retains the simplicity of the leapfrog method but sacrifices a bit of accuracy due to the artificial damping.

The modified leapfrog, a damped leapfrog scheme (a combination of Leapfrog and Lax-Wendroff) could be employed, as follows:

$$u_j^{n+1} = u_j^{n-1} + 2\frac{\Delta t}{\Delta x^2}\left(u_{j+1}^n - 2u_j^n + u_{j-1}^n\right) - \varepsilon(u_{j+1}^n - 2u_j^n + u_{j-1}^n)$$

Where ε is a small damping factor to control instability while maintaining second-order accuracy.

To find the amplification factor A, we need to substitute a Fourier mode into the equation, perform a von Neumann stability analysis, and simplify the resulting expression.

Assume a Fourier mode solution of the form:

$$u_j^n = \hat{u}^n e^{ikj\Delta x}$$

where $\hat{u}^n$ is the amplitude of the mode, k is the wavenumber, and j indexes the spatial position.

Substitute this Fourier mode into the given difference equation:

$$u_j^{n+1} = u_j^{n-1} + 2\frac{\Delta t}{\Delta x^2}\left(u_{j+1}^n - 2u_j^n + u_{j-1}^n\right) - \varepsilon\left(u_{j+1}^n - 2u_j^n + u_{j-1}^n\right)$$

Now we use the fact that:

$$u_{j+1}^n = \hat{u}^n e^{ik(j+1)\Delta x} = \hat{u}^n e^{ik\Delta x} e^{ikj\Delta x}$$

$$u_{j-1}^n = \hat{u}^n e^{ik(j-1)\Delta x} = \hat{u}^n e^{-ik\Delta x} e^{ikj\Delta x}$$

Substituting these into the difference equation, we get:

$$\begin{aligned}\hat{u}^{n+1} e^{ikj\Delta x} \quad = \quad & \hat{u}^{n-1} e^{ikj\Delta x} + 2\frac{\Delta t}{\Delta x^2}\left(\hat{u}^n e^{ik(j+1)\Delta x} - 2\hat{u}^n e^{ikj\Delta x} + \hat{u}^n e^{ik(j-1)\Delta x}\right) \\ & -\varepsilon\left(\hat{u}^n e^{ik(j+1)\Delta x} - 2\hat{u}^n e^{ikj\Delta x} + \hat{u}^n e^{ik(j-1)\Delta x}\right)\end{aligned}$$

Factor out the common terms $e^{ikj\Delta x}$:

$$\hat{u}^{n+1} = \hat{u}^{n-1} + 2\frac{\Delta t}{\Delta x^2}\hat{u}^n\left(e^{ik\Delta x} - 2 + e^{-ik\Delta x}\right) - \varepsilon\hat{u}^n\left(e^{ik\Delta x} - 2 + e^{-ik\Delta x}\right)$$

The expression $e^{ik\Delta x} - 2 + e^{-ik\Delta x}$ simplifies using Euler's formula:

$$e^{ik\Delta x} + e^{-ik\Delta x} = 2\cos(k\Delta x)$$

Thus:

$$e^{ik\Delta x} - 2 + e^{-ik\Delta x} = 2(\cos(k\Delta x) - 1)$$

Substitute this into the equation:

$$\hat{u}^{n+1} = \hat{u}^{n-1} + 4\frac{\Delta t}{\Delta x^2}\hat{u}^n(\cos(k\Delta x) - 1) - 2\varepsilon\hat{u}^n(\cos(k\Delta x) - 1)$$

Now, define the amplification factor A as the ratio of $\hat{u}^{n+1}$ to $\hat{u}^n$:

$$A = \frac{\hat{u}^{n+1}}{\hat{u}^n}$$

We rewrite the equation in terms of A:

$$A\hat{u}^n = \hat{u}^{n-1} + \left[4\frac{\Delta t}{\Delta x^2}(\cos(k\Delta x) - 1) - 2\varepsilon(\cos(k\Delta x) - 1)\right]\hat{u}^n$$

Dividing by $\hat{u}^n$:

$$A = \frac{\hat{u}^{n-1}}{\hat{u}^n} + 4\frac{\Delta t}{\Delta x^2}(\cos(k\Delta x) - 1) - 2\varepsilon(\cos(k\Delta x) - 1)$$

Since $\hat{u}^{n-1} = A^{-1}\hat{u}^n$, we substitute this:

$$A = A^{-1} + 4\frac{\Delta t}{\Delta x^2}(\cos(k\Delta x) - 1) - 2\varepsilon(\cos(k\Delta x) - 1)$$

Multiply both sides by A:

$$A^2 = 1 + A\left(4\frac{\Delta t}{\Delta x^2}(\cos(k\Delta x) - 1) - 2\varepsilon(\cos(k\Delta x) - 1)\right)$$

This is a quadratic equation for A:

$$A^2 - A\left(4\frac{\Delta t}{\Delta x^2}(\cos(k\Delta x) - 1) - 2\varepsilon(\cos(k\Delta x) - 1)\right) - 1 = 0$$

We can solve this quadratic equation to find the amplification factor A:

$$A = \frac{\left(4\frac{\Delta t}{\Delta x^2}(\cos(k\Delta x)-1)-2\varepsilon(\cos(k\Delta x)-1)\right) \pm \sqrt{\left(4\frac{\Delta t}{\Delta x^2}(\cos(k\Delta x)-1)-2\varepsilon(\cos(k\Delta x)-1)\right)^2+4}}{2}$$

8.6 IMPLICIT SCHEME: BTCS

The one-dimensional diffusion equation

$$\frac{\partial u}{\partial t} = \mathscr{D}\frac{\partial^2 u}{\partial x^2}$$

is discretised with a backward (implicit) difference in time and a central-difference in space, giving

$$u_i^{n+1} = u_i^n + \frac{\mathscr{D}\Delta t}{\Delta x^2}\left(u_{i+1}^{n+1} - 2u_i^{n+1} + u_{i-1}^{n+1}\right).$$

Fourier (von Neumann) stability analysis: Assume a single Fourier mode

$$u_i^n = \xi^n e^{ik\,i\Delta x},$$

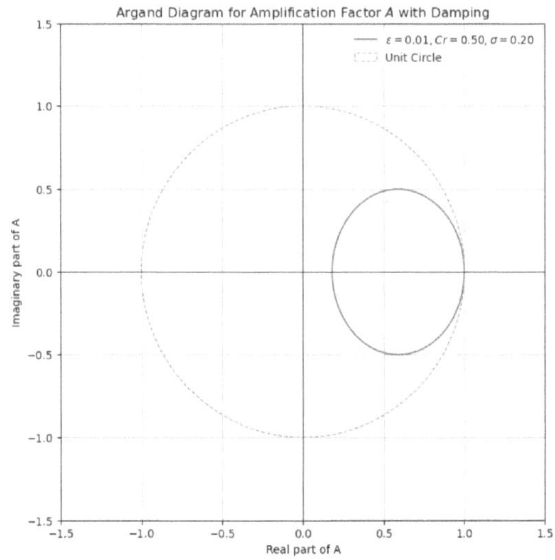

Figure 8.3: **Modified Leap Frog scheme with damping.**

where k is the real wave number. At the new time level $n+1$

$$u_i^{n+1} = \xi^{n+1} e^{iki\Delta x}, \qquad u_{i\pm1}^{n+1} = \xi^{n+1} e^{iki\Delta x} e^{\pm ik\Delta x}.$$

Insert these expressions into the discrete scheme:

$$\xi^{n+1} e^{iki\Delta x} = \xi^n e^{iki\Delta x} + \frac{\mathscr{D}\Delta t}{\Delta x^2} \xi^{n+1} e^{iki\Delta x} \left(e^{ik\Delta x} - 2 + e^{-ik\Delta x}\right).$$

Divide by $e^{iki\Delta x}$ and rearrange:

$$\xi^{n+1} \left[1 - \frac{2\mathscr{D}\Delta t}{\Delta x^2} (1 - \cos(k\Delta x))\right] = \xi^n.$$

Define the diffusion number

$$\sigma = \frac{\mathscr{D}\Delta t}{\Delta x^2},$$

then the amplification factor $A = \xi^{n+1}/\xi^n$ is

$$A = \frac{1}{1 + 2\sigma\left(1 - \cos(k\Delta x)\right)}.$$

Using $1 - \cos(k^*) = 2\sin^2((k^*)/2)$ gives

$$A = \frac{1}{1 + 4\sigma \sin^2\left(k\Delta x/2\right)}.$$

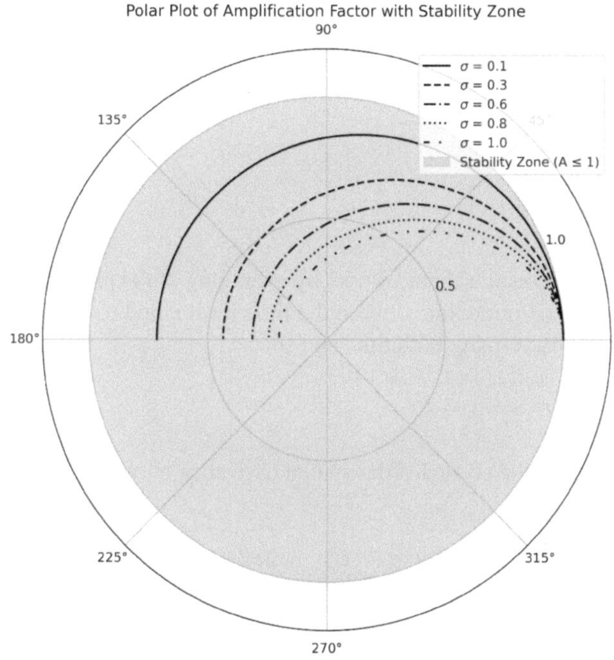

Figure 8.4: **Stability limits for Implicit BTCS scheme under different diffusion numbers.**

$$\boxed{A_{\text{BTCS}} = \frac{1}{1 + 4\sigma \sin^2(k\Delta x/2)}}$$

Because $\sigma > 0$ and $\sin^2(\cdot) \geq 0$, the denominator satisfies $1 + 4\sigma \sin^2(\cdot) \geq 1$. Hence $|A_{BTCS}| \leq 1$ for every $\Delta t > 0$ and all wave numbers k. The implicit BTCS scheme for the diffusion equation is therefore unconditionally stable (see Figure 8.4).

i For small σ: The scheme is stable for most wave numbers, with A close to 1.

ii For large σ: Stability is still maintained, but the scheme becomes dissipative, with high wave numbers experiencing significant damping.

The stability condition is met as long as $A \leq 1$, meaning that the numerical solution remains bounded. This is reflected in the shaded stability zone on the polar plot, where the amplification factor remains below or equal to 1 for stable configurations.

8.7 IMPLICIT: CRANK-NICOLSON SCHEME

The Crank-Nicolson scheme is a finite difference method used to numerically solve partial differential equations, specifically for time-dependent problems. The method is an implicit scheme and is known for being second-order accurate in both space and time.

Given the equation:

$$\frac{\partial \phi}{\partial t} = \mathscr{D}\frac{\partial^2 \phi}{\partial x^2}$$

The Crank-Nicolson scheme works by averaging the explicit and implicit methods, i.e., it takes the midpoint between the time level n and $n+1$.

We start by discretizing both time and space:

Time step: $t_n \to t_{n+1} = t_n + \Delta t$

Space step: $x_i \to x_{i+1} = x_i + \Delta x$

Now let's write the second-order spatial derivative at two different time levels (at n and $n+1$):

$$\frac{\partial^2 \phi}{\partial x^2} \approx \frac{\phi_{i+1}^n - 2\phi_i^n + \phi_{i-1}^n}{\Delta x^2}$$

at time level n, and similarly at $n+1$:

$$\frac{\partial^2 \phi}{\partial x^2} \approx \frac{\phi_{i+1}^{n+1} - 2\phi_i^{n+1} + \phi_{i-1}^{n+1}}{\Delta x^2}$$

The Crank-Nicolson method averages these two spatial derivatives:

$$\frac{\partial \phi}{\partial t} \approx \frac{\phi_i^{n+1} - \phi_i^n}{\Delta t}$$

$$\frac{\partial^2 \phi}{\partial x^2} \approx \frac{1}{2}\left(\frac{\phi_{i+1}^{n+1} - 2\phi_i^{n+1} + \phi_{i-1}^{n+1}}{\Delta x^2} + \frac{\phi_{i+1}^n - 2\phi_i^n + \phi_{i-1}^n}{\Delta x^2}\right)$$

Now, using the original equation and substituting the expressions above:

$$\frac{\phi_i^{n+1} - \phi_i^n}{\Delta t} = \mathscr{D} \cdot \frac{1}{2}\left(\frac{\phi_{i+1}^{n+1} - 2\phi_i^{n+1} + \phi_{i-1}^{n+1}}{\Delta x^2} + \frac{\phi_{i+1}^n - 2\phi_i^n + \phi_{i-1}^n}{\Delta x^2}\right)$$

Multiplying through by Δt and rearranging terms:

$$\phi_i^{n+1} - \frac{\mathscr{D}\Delta t}{2\Delta x^2}\left(\phi_{i+1}^{n+1} - 2\phi_i^{n+1} + \phi_{i-1}^{n+1}\right) = \phi_i^n + \frac{\mathscr{D}\Delta t}{2\Delta x^2}\left(\phi_{i+1}^n - 2\phi_i^n + \phi_{i-1}^n\right)$$

This is the Crank-Nicolson scheme for the given equation. The method is implicit because the unknown future values ϕ_i^{n+1}, ϕ_{i+1}^{n+1}, and ϕ_{i-1}^{n+1} appear on the left-hand side, and solving it requires solving a system of equations at each time step.

To develop the amplification factor A for the Crank-Nicolson scheme, we need to analyze the behavior of the scheme in terms of a Fourier mode. This process involves assuming a solution of the form:

$$\phi_i^n = A^n e^{ikx_i}$$

where A is the amplification factor, k is the wave number, and x_i is the position in space.

We start with the Crank-Nicolson discretization of the diffusion equation:

$$\frac{\phi_i^{n+1} - \phi_i^n}{\Delta t} = \mathscr{D} \cdot \frac{1}{2} \left(\frac{\phi_{i+1}^{n+1} - 2\phi_i^{n+1} + \phi_{i-1}^{n+1}}{\Delta x^2} + \frac{\phi_{i+1}^n - 2\phi_i^n + \phi_{i-1}^n}{\Delta x^2} \right)$$

Substituting the assumed form $\phi_i^n = A^n e^{ikx_i}$ for each ϕ, we get:
For ϕ_i^{n+1}:

$$\phi_i^{n+1} = A^{n+1} e^{ikx_i}$$

For ϕ_{i+1}^{n+1} and ϕ_{i-1}^{n+1}:

$$\phi_{i+1}^{n+1} = A^{n+1} e^{ikx_{i+1}} = A^{n+1} e^{ik(x_i+\Delta x)} = A^{n+1} e^{ikx_i} e^{ik\Delta x}$$

$$\phi_{i-1}^{n+1} = A^{n+1} e^{ikx_{i-1}} = A^{n+1} e^{ik(x_i-\Delta x)} = A^{n+1} e^{ikx_i} e^{-ik\Delta x}$$

Similarly, for time level n:

$$\phi_{i+1}^n = A^n e^{ik(x_i+\Delta x)} = A^n e^{ikx_i} e^{ik\Delta x}$$

$$\phi_{i-1}^n = A^n e^{ik(x_i-\Delta x)} = A^n e^{ikx_i} e^{-ik\Delta x}$$

Now, substitute these into the Crank-Nicolson scheme:

$$\frac{A^{n+1} e^{ikx_i} - A^n e^{ikx_i}}{\Delta t} = \mathscr{D} \cdot \frac{1}{2\Delta x^2} \Big((A^{n+1} e^{ikx_i} e^{ik\Delta x} + A^{n+1} e^{ikx_i} e^{-ik\Delta x} - 2A^{n+1} e^{ikx_i})$$
$$+ (A^n e^{ikx_i} e^{ik\Delta x} + A^n e^{ikx_i} e^{-ik\Delta x} - 2A^n e^{ikx_i}) \Big)$$

We can cancel e^{ikx_i} from both sides and simplify:

$$\frac{A^{n+1} - A^n}{\Delta t} = \mathscr{D} \cdot \frac{1}{2\Delta x^2} \left(A^{n+1} (e^{ik\Delta x} + e^{-ik\Delta x} - 2) + A^n (e^{ik\Delta x} + e^{-ik\Delta x} - 2) \right)$$

Using the trigonometric identity $e^{i\theta} + e^{-i\theta} = 2\cos\theta$, this becomes:

$$\frac{A^{n+1} - A^n}{\Delta t} = \mathscr{D} \cdot \frac{1}{2\Delta x^2} \left(A^{n+1} (2\cos(k\Delta x) - 2) + A^n (2\cos(k\Delta x) - 2) \right)$$

Factor out the common terms:

$$\frac{A^{n+1}-A^n}{\Delta t} = \mathscr{D}\cdot\frac{1}{\Delta x^2}\left((\cos(k\Delta x)-1)(A^{n+1}+A^n)\right)$$

Multiply both sides by Δt and solve for A^{n+1}:

$$A^{n+1}-A^n = \frac{\mathscr{D}\Delta t}{\Delta x^2}(\cos(k\Delta x)-1)(A^{n+1}+A^n)$$

Let $\sigma = \frac{\mathscr{D}\Delta t}{\Delta x^2}$, the diffusion number, and solve for the amplification factor A:

$$A^{n+1}-A^n = \sigma(\cos(k\Delta x)-1)(A^{n+1}+A^n)$$

Now, collect terms involving A^{n+1} on the left and A^n on the right:

$$A^{n+1}(1-\sigma(\cos(k\Delta x)-1)) = A^n(1+\sigma(\cos(k\Delta x)-1))$$

Finally, the amplification factor A is given by:

$$A = \frac{1+\sigma(\cos(k\Delta x)-1)}{1-\sigma(\cos(k\Delta x)-1)}$$

This is the amplification factor for the Crank-Nicolson scheme, which determines the stability of the scheme. The scheme is stable when $|A|\leq 1$. Since Crank-Nicolson is unconditionally stable, $|A|\leq 1$, and in fact, $|A|=1$ for all wavenumbers, indicating no artificial damping (pure diffusion) for any value of the time step Δt.

We can cast this amplification factor into another form, using the identity:

$$\cos(k\Delta x) = 1-2\sin^2\left(\frac{k\Delta x}{2}\right)$$

Substitute this into the expression for A:

$$A = \frac{1+\sigma(1-2\sin^2\left(\frac{k\Delta x}{2}\right)-1)}{1-\sigma(1-2\sin^2\left(\frac{k\Delta x}{2}\right)-1)}$$

Simplifying:

$$A_{CN} = \frac{1-2\sigma\sin^2\left(\frac{k^*}{2}\right)}{1+2\sigma\sin^2\left(\frac{k^*}{2}\right)}$$

The Crank-Nicolson scheme is unconditionally stable because all amplification factors lie within the unit circle. This makes the scheme highly suitable for solving parabolic PDEs like the diffusion equation without concern for instability, even at larger time steps. Additionally, the lack of significant imaginary components indicates minimal dispersion error.

As an implicit scheme, Crank-Nicolson can handle larger time steps without introducing instability (unconditionally stable for parabolic problems). This makes it

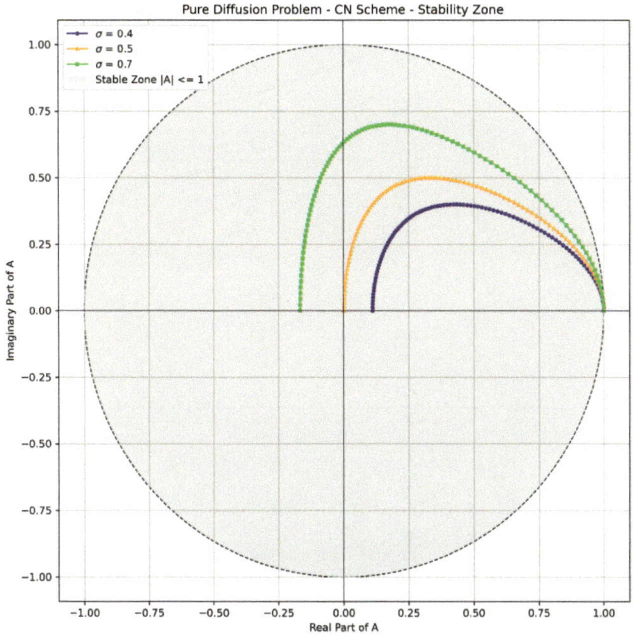

Figure 8.5: **Crank-Nicolson scheme.**

more accurate in practical applications, where larger time steps are used to speed up computations, while still maintaining stability.

The Crank-Nicolson scheme, which is centered in time, strategically balances numerical dissipation and dispersion. This balanced approach inherently minimizes, but does not eliminate, numerical artifacts such as excessive damping and oscillations, making it well-suited for a wide range of continuous problems. Particularly in diffusion-type scenarios where the solution is expected to be smooth, this method ensures that diffusive effects are accurately modeled without introducing artificial oscillations, enhancing physical fidelity in simulations like heat transfer or contaminant dispersion.

However, in cases involving sharp discontinuities or non-smooth solutions, the Crank-Nicolson scheme, despite its advantages, can still manifest small oscillations due to the Gibbs phenomenon. These oscillations, while generally less pronounced compared to those produced by explicit second-order schemes like Lax-Wendroff, may still necessitate the use of filtering techniques or limiters to maintain numerical accuracy and stability near these discontinuities. This need arises because, although the scheme effectively handles the amplitude of solution features by avoiding phase

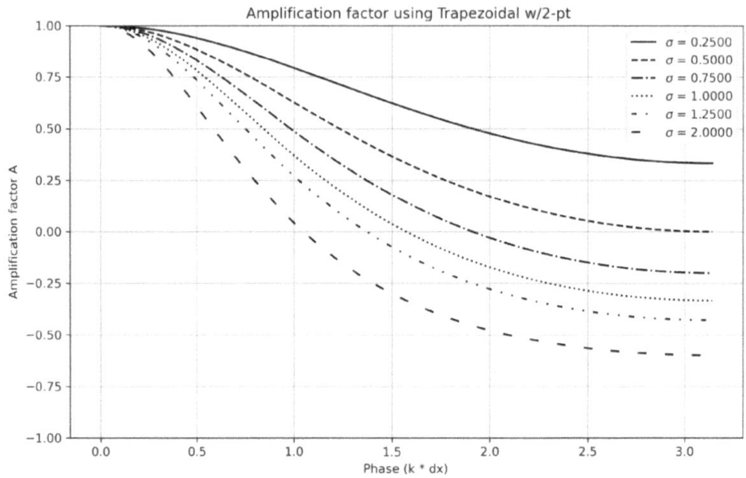

Figure 8.6: **Amplitude vs phase for Crank-Nicolson scheme. This shows the scheme is unconditionally stable.**

shifts (thanks to the real nature of the amplification factor-A), the presence of discontinuities can still challenge its capacity to deliver completely smooth results

8.8 LAASONEN SCHEME

To derive the Laasonen Scheme, we start with the diffusion equation and apply implicit discretization in time and central differencing in space. The general form of the diffusion equation is:

$$\frac{\partial \phi}{\partial t} = \mathscr{D} \frac{\partial^2 \phi}{\partial x^2}$$

Time Stepping: backward Euler at t^{n+1}

$$\left.\frac{\partial \phi}{\partial t}\right|_i^{n+1} \approx \frac{\phi_i^{n+1} - \phi_i^n}{\Delta t}$$

Space discretization: centered second order at the same new level

$$\left.\frac{\partial^2 \phi}{\partial x^2}\right|_i^{n+1} \approx \frac{\phi_{i+1}^{n+1} - 2\phi_i^{n+1} + \phi_{i-1}^{n+1}}{\Delta x^2}$$

Insert these in the PDE and multiply by Δt:

$$\phi_i^{n+1} - \phi_i^n = \sigma\left(\phi_{i+1}^{n+1} - 2\phi_i^{n+1} + \phi_{i-1}^{n+1}\right).$$

Re-arrange to the standard tridiagonal form

$$-\sigma\,\phi_{i-1}^{n+1} \;+\; (1+2\sigma)\phi_i^{n+1} \;-\; \sigma\,\phi_{i+1}^{n+1} \;=\; \phi_i^n. \tag{1}$$

Von Neumann (Fourier-mode) analysis: Assume periodic domain and insert a single Fourier component

$$\phi_i^n = A^n\, e^{iki\Delta x} = A^n\, e^{ik^* i}.$$

Then

$$\phi_{i\pm1}^{n+1} = A^{n+1}\, e^{ik^*(i\pm1)} = A^{n+1} e^{ik^* i} e^{\pm ik^*}.$$

Substitute these three expressions into eq. and divide by the common factor $e^{ik^* i}$:

$$-\sigma A^{n+1} e^{-ik^*} \;+\; (1+2\sigma)A^{n+1} \;-\; \sigma A^{n+1} e^{ik^*} \;=\; A^n.$$

Factor out A^{n+1} on the left and use $\cos\theta = \frac{1}{2}(e^{i\theta}+e^{-i\theta})$:

$$A^{n+1}\Big[1+2\sigma-2\sigma\cos(k^*)\Big] = A^n.$$

Finally, the amplification factor

$$A \;\equiv\; \frac{\phi^{n+1}}{\phi^n} = \frac{1}{1+2\sigma\big[1-\cos(k^*)\big]} = \frac{1}{1+4\sigma\sin^2\!\big(k^*/2\big)}$$

Because the denominator in the above equation is at least 1 for every $\sigma > 0$ and every k^*, the amplification factor

$$A = \frac{1}{1+4\sigma\sin^2\!\big(k^*/2\big)}$$

is real, strictly positive, and satisfies $0 < A \le 1$.

$$0 < A \le 1$$

hence the Laasonen scheme is unconditionally stable (decays or leaves amplitudes unchanged, never grows them).

Table 8.1 lists amplification factor for some numerical schemes used for pure diffusion equation.

8.9 TIME STEP FOR DIFFUSION PROBLEMS

The viscous time step limit is

$$\Delta t_{\max} = \frac{s_r}{k^{*2}_{\text{mod, max}}} \cdot \frac{\Delta x^2}{\mathscr{D}} = \sigma_{\max} \cdot \frac{\Delta x^2}{\mathscr{D}}$$

The dimensionless diffusion number σ has an upper limit

$$\sigma \le 0.5$$

Table 8.1
Implicit Numerical Schemes and their Amplification Factors for the Diffusion Equation

Scheme	Amplification Factor $A(k^*)$
Crank-Nicolson (CN)	$A = \dfrac{1+\sigma(\cos k^* - 1)}{1-\sigma(\cos k^* - 1)}$
Laasonen (Backward-Euler)	$A = \dfrac{1}{1+4\sigma\sin^2(k^*/2)}$

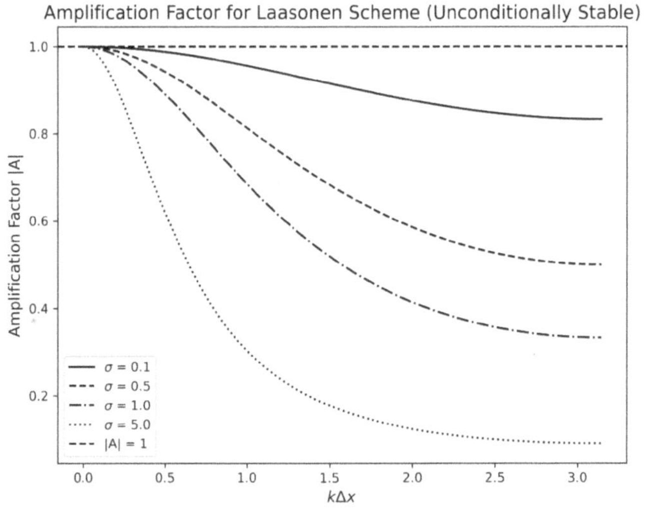

Figure 8.7: Laasonen Scheme.

The equation above is arrived from the stability analysis of numerical schemes for solving diffusion (or parabolic) equations. This equation sets an upper limit on the time step Δt to maintain numerical stability. It is specifically relevant for explicit time integration schemes applied to problems involving second-order spatial derivatives, such as the diffusion equation. When using finite differences of a certain order, we can define a modified wavenumber $k^*_{\text{mod, max}}$ to characterize how well the numerical scheme approximates the true spatial derivative. The term $k^{*2}_{\text{mod, max}}$ represents the squared maximum modified wavenumber of the scheme, which relates to the highest frequency that can be stably resolved by the spatial discretization.

For diffusion-like equations, stability requires that the time step Δt be small enough to prevent numerical instability. In explicit time-stepping schemes (e.g., Forward Euler), stability is governed by the diffusion number, analogous to the CFL condition for advection equations. The diffusion number states that

$$\Delta t < \frac{\Delta x^2}{\mathscr{D}} \frac{1}{k^{*2}_{\text{mod, max}}}$$

Here, s_r represents a stability factor specific to the time integration method, indicating how well it handles the real (non-oscillatory) parts of the solution. This criterion helps ensure that the numerical solution remains stable and free of oscillations or exponential growth.

Example 8.1

For unsteady Couette flow, the governing differential equation is derived from the unsteady Navier-Stokes equations, taking into account that there is no pressure gradient and the flow is incompressible:

$$\frac{\partial u}{\partial t} = \nu \frac{\partial^2 u}{\partial y^2}$$

where,
$u = u(y,t)$ is the velocity in the x-direction as a function of y and t,
ν is the kinematic viscosity of the fluid,
t is the time variable.

Boundary and initial conditions:

$u(0,t) = u_0 = 40\ \text{m s}^{-1}$	(impulsively moved lower wall),
$u(h,t) = 0$	(gap between walls),
$u(y,0) = 0 \quad (0 < y \leq h)$	(fluid initially at rest).

Grid and parameters: $h = 0.04$ m, $\Delta y = 0.001$ m, $\nu = 2.17 \times 10^{-4}\,\text{m}^2/\text{s}$.
Time integration: Consider Forward-time centered-space (explicit Euler) with two different time steps so that the mesh Fourier number $\sigma = \nu\Delta t/\Delta y^2$ = 0.217 and 0.5034 respectively.

Solution The time step is computed as

$$\Delta t_{\text{max,viscous}} = \frac{\sigma \Delta y^2}{\nu} = \frac{0.434 \times \Delta y^2}{\nu}$$

See `Python` code `Ch8-Ex81` for solution of this problem. We repeat this procedure for the second diffusion number with different time step.

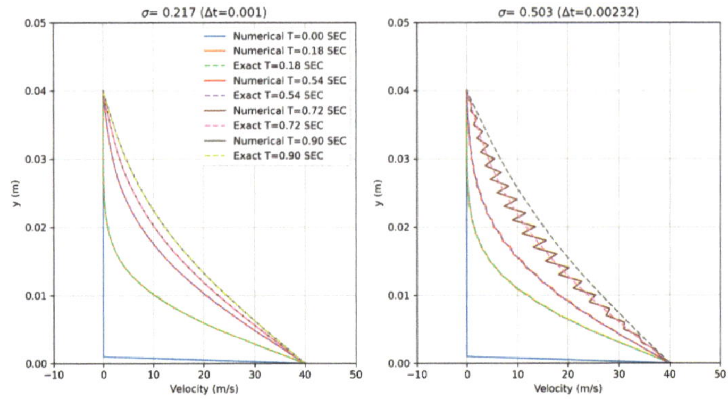

Figure 8.8: **Numerical solution of the unsteady Couette flow problem using FD-O2 and Euler explicit.**

Figure 8.8 shows the velocity distributions for the unsteady Couette flow problem using FD-O2 and Euler explicit schemes.

8.10 DIFFUSION EQUATION WITH MATRIX DECOMPOSITION

The general form of the diffusion equation is:

$$\frac{\partial u}{\partial t} = \mathscr{D}\frac{\partial^2 u}{\partial x^2}$$

Where:
$u(x,t)$ is the unknown function (e.g., temperature, concentration, etc.).
$\mathscr{D}$ is the diffusion coefficient.
$\frac{\partial^2 u}{\partial x^2}$ is the second spatial derivative.

For simplicity, we assume $\mathscr{D} = 1$, which means we focus on the equation:

$$\frac{\partial u}{\partial t} = \frac{\partial^2 u}{\partial x^2}$$

We discretize the spatial and temporal domains to solve this equation numerically. The second-order finite difference approximation (FD-02) for the second spatial derivative is given by:

$$\frac{\partial^2 u}{\partial x^2} \approx \frac{u_{j+1} - 2u_j + u_{j-1}}{\Delta x^2}$$

Where: u_j represents the value of the function at the grid point j.
Δx is the grid spacing in space.
The notation u_{j+1}, u_{j-1} represent the function values at adjacent points.

Similarly, the time derivative $\frac{\partial u}{\partial t}$ can be approximated by the forward difference in time:

$$\frac{\partial u}{\partial t} \approx \frac{u_j^{n+1} - u_j^n}{\Delta t}$$

Where:
u_j^n is the value of u at time step n.
u_j^{n+1} is the value at the next time step $n+1$.
Δt is the time step.

Substitute the spatial derivative approximation into the diffusion equation:

$$\frac{u_j^{n+1} - u_j^n}{\Delta t} = \frac{u_{j+1}^n - 2u_j^n + u_{j-1}^n}{\Delta x^2}$$

Rearranging to solve for u_j^{n+1}:

$$u_j^{n+1} = u_j^n + \frac{\Delta t}{\Delta x^2}\left(u_{j+1}^n - 2u_j^n + u_{j-1}^n\right)$$

This is the explicit scheme (FD-02) for solving the diffusion equation.

To represent this equation in matrix form, let's consider the values u_j^n as a vector u^n:

$$u^n = \begin{pmatrix} u_1^n \\ u_2^n \\ \vdots \\ u_{N-1}^n \end{pmatrix}$$

Now, we need to express the equation

$$u_j^{n+1} = u_j^n + \frac{\Delta t}{\Delta x^2}\left(u_{j+1}^n - 2u_j^n + u_{j-1}^n\right)$$

in matrix form. The term $\frac{\Delta t}{\Delta x^2}\left(u_{j+1}^n - 2u_j^n + u_{j-1}^n\right)$ can be represented by a tridiagonal matrix A. The matrix A corresponds to the discrete second derivative operator:

$$A = \frac{1}{\Delta x^2}\begin{pmatrix} -2 & 1 & 0 & \cdots & 0 \\ 1 & -2 & 1 & \cdots & 0 \\ 0 & 1 & -2 & \cdots & 0 \\ \vdots & \vdots & \vdots & \ddots & 1 \\ 0 & 0 & 0 & 1 & -2 \end{pmatrix}$$

The numerical scheme can then be written in matrix form as:

$$\frac{\partial u}{\partial t} = \frac{\mathscr{D}}{\Delta x^2} A u$$

Where:
A is the matrix that represents the finite difference approximation for the second spatial derivative.
u is the vector containing the values of u_j at each grid point.
$\mathscr{D}$ is diffusion coefficient.

This allows us to solve for the next time step u^{n+1} by iterating over the time steps using matrix operations, which can be computationally efficient, especially for large systems.

Symmetric and Orthogonal Matrices: A matrix S is said to be *symmetric* if $S = S^T$, meaning it is equal to its transpose—its entries are symmetric with respect to the main diagonal. A matrix is *orthogonal* if $S^{-1} = S^T$, which implies that its columns and rows are orthonormal vectors.

Any real symmetric matrix A can be diagonalized by an orthogonal matrix S as

$$A = S\Lambda S^T,$$

where Λ is a diagonal matrix whose entries are the eigenvalues of A. The corresponding eigenvectors can be chosen to be orthonormal, forming the columns of S.

Transformation of Vectors: Given a vector $\mathbf{u}$ in the original coordinate system, the corresponding modal vector in the eigenbasis is obtained via

$$\mathbf{q} = S^T\mathbf{u}.$$

This transformation simplifies the system, since the matrix A becomes diagonal in the modal space.

The modal representation of the diffusion equation then becomes a system of uncoupled ordinary differential equations:

$$\dot{\mathbf{q}} = \frac{\nu}{\Delta x^2}\mathbf{q},$$

where $\mathbf{q}$ is the vector of modal amplitudes. The initial condition transforms as $\mathbf{q}_0 = S^T\mathbf{u}_0$, and the solution is

$$\mathbf{q}(t) = \mathbf{q}_0 \exp\left(\frac{\nu}{\Delta x^2}t\right).$$

Discretized Diffusion Equation: The continuous diffusion equation

$$\frac{\partial u}{\partial t} = \nu\frac{\partial^2 u}{\partial x^2}$$

is discretized using central finite differences as

$$\frac{d\mathbf{u}}{dt} = \nu A\mathbf{u},$$

where A is a symmetric tridiagonal matrix with -2 on the diagonal and 1 on the sub- and super-diagonals.

Solving the eigenvalue problem

$$A\phi_k = \lambda_k \phi_k$$

yields the modal basis. In this basis, each component $q_k(t)$ evolves independently as

$$\frac{dq_k}{dt} = \nu \lambda_k q_k \quad \Rightarrow \quad q_k(t) = q_{k0} \exp(\nu \lambda_k t).$$

Modal Expansion: The complete solution to the diffusion equation can be expressed as a superposition of eigenmodes:

$$u(x,t) = \sum_k q_{k0} \exp(\nu\lambda_k t)\, \phi_k(x),$$

where q_{k0} are the modal coefficients determined from the initial condition, λ_k are the eigenvalues of the discretized operator A, and $\phi_k(x)$ are the corresponding eigenfunctions.

In vector form, the solution evolves as

$$\mathbf{u}(t) = S\mathbf{q}(t),$$

where $\mathbf{q}(t)$ is the vector of modal amplitudes, and S is the matrix of eigenvectors of A. The diagonalization of A simplifies the time integration process by decoupling the system into independent scalar ODEs for each mode.

For diffusion problems, the eigenvalues λ_k are typically negative, implying that each mode decays exponentially over time. The higher the frequency (i.e., the larger the magnitude of λ_k), the faster the decay. If any λ_k were positive, the corresponding mode would grow nonphysically, indicating numerical instability. The matrix S serves as a transformation that maps the initial condition onto the eigenbasis, and the evolved modal components are then combined to reconstruct the full physical-space solution.

To solve the system, we need to find the eigenvalues of the matrix A. The matrix A is symmetric and tridiagonal, which means that its eigenvalues can be determined analytically. We now need to solve the equation

$$A\phi_k = \lambda_k \phi_k$$

for the eigenvalue problem for A, where λ_k are the eigenvalues and ϕ_k are the eigenvectors. The matrix A is a discrete version of the second derivative operator, and its eigenvalues are well-known for finite difference approximations to the diffusion equation. By applying Fourier analysis to the system, we find that the eigenvalues λ_k are given by:

$$k^{*2}_{mod} = 2(1 - \cos(k^*))$$

$$\lambda_k = 2\left(1 - \cos\left(\frac{k\pi}{N}\right)\right)$$

The matrix A represents a discretization of the second derivative operator in space, and the eigenfunctions of the second derivative operator are sinusoidal functions in the continuous case. This is because the second derivative of a sine function gives a negative multiple of the same sine function, which is a characteristic of the eigenfunctions for second-order differential operators. In the discrete case, the solution to the eigenvalue problem for the finite difference matrix involves solving for the modes of the system. These modes correspond to sine functions of the form $\sin\left(\frac{k\pi x}{N}\right)$, where k is the mode number. The eigenvalues associated with these modes are $\lambda_k = 2\left(1 - \cos\left(\frac{k\pi}{N}\right)\right)$, where the cosine term arises from the discretization of the second derivative.

This formula is derived by considering the spatial frequencies of the sine functions and using the properties of the finite difference approximation. The result shows that the eigenvalues λ_k are dependent on the spatial frequency k and the grid spacing Δx.

Each eigenvalue λ_k corresponds to the rate at which the corresponding spatial mode decays over time. Larger eigenvalues correspond to faster-decaying modes, and smaller eigenvalues correspond to slower-decaying modes. The spatial frequencies k determine the rate of decay, with lower frequencies (corresponding to larger wavelengths) decaying more slowly than higher frequencies.

The eigenvalues govern the decay rates of the spatial modes of the system when using a finite difference method for solving diffusion-like equations.

Minimum eigenvalue ($\lambda_{\min}$): For the second-difference matrix that results from a centred-difference (FD-O2) discretisation of $\partial^2 u/\partial x^2$ on a uniform grid with N interior nodes and homogeneous Dirichlet boundaries, the spectrum is known in closed form:

$$\lambda_k = -\frac{4}{\Delta x^2}\sin^2\!\left(\frac{k\pi}{2N}\right), \qquad k = 1, 2, \ldots, N-1$$

The smallest-magnitude (and therefore slowest-decaying) eigenvalue occurs at $k = 1$

$$\lambda_{\min} = -\frac{4}{\Delta x^2}\sin^2\!\left(\frac{\pi}{2N}\right)$$

Large-N (fine-mesh) approximation: When the grid is sufficiently fine, $\theta = \frac{\pi}{2N} \ll 1$ and $\sin\theta \approx \theta$; thus

$$\lambda_{\min} \approx -\frac{4}{\Delta x^2}\left(\frac{\pi}{2N}\right)^2 = -\frac{\pi^2}{N^2\,\Delta x^2}$$

Hence

$$\lambda_{\min} \propto -\frac{1}{N^2}$$

so, as the number of grid points N increases (i.e., the mesh is refined), $|\lambda_{\min}|$ becomes smaller and the slowest mode decays more slowly in time—exactly the trend expected in diffusion problems.

Maximum Eigenvalue λ_{max}: This eigenvalue corresponds to the fastest decaying mode in the system. It is given by:

$$\lambda_{\text{max}} = -4$$

The decay rate of the fastest mode is fixed and does not depend on the grid resolution. The negative sign shows that this mode decays over time.

In general λ will be complex. However, for a physically stable problem it will be necessary that the real part of λ_k is negative.

The ratio of the maximum eigenvalue to the minimum eigenvalue gives an indication of the system's stiffness. A stiff system means that it has modes with vastly different decay rates, which can cause numerical challenges during integration. The ratio is given by:

$$\frac{\lambda_{\text{max}}}{\lambda_{\text{min}}} \approx \frac{4N^2}{\pi^2}$$

As the grid resolution N increases, the ratio of λ_{max} to λ_{min} increases. This means that the system becomes stiffer, with much slower decaying modes compared to faster decaying ones. As the system becomes stiffer, it requires careful numerical treatment, especially when using explicit time-stepping methods, as large differences in decay rates can lead to numerical instability. This stiffness means that smaller time steps are needed for stability as the grid resolution increases.

$$\sigma_{\text{max}} = \frac{s_r}{k^{*2}_{\text{mod,max}}}$$

In this equation: σ_{max} represents the maximum stability factor for the system.
s_r is the stability factor specific to the time integration method being used, which indicates how well the method handles the real (non-oscillatory) parts of the solution.
$k^2_{\text{mod, max}}$ refers to the maximum wavenumber squared, which represents the highest frequency component of the solution. The wavenumber is tied to how quickly the solution changes in space.

This formula provides the relationship between the stability factor s_r and the spatial frequency component $k_{\text{mod, max}}$. As $k_{\text{mod, max}}$ increases (i.e., as the solution involves higher spatial frequencies), the maximum allowable time step Δt_{max} decreases to maintain stability. The wavenumber affects the time step in such a way that finer grids (with higher wavenumbers) require smaller time steps to avoid instability in the numerical solution.

The equation:

$$\Delta t_{\text{max}} = \frac{\sigma_{\text{max}} \frac{\Delta x^2}{v}}{2}$$

where,
Δt_{max} is the maximum allowable time step for the simulation.
σ_{max} is the maximum stability factor, which we derived earlier.

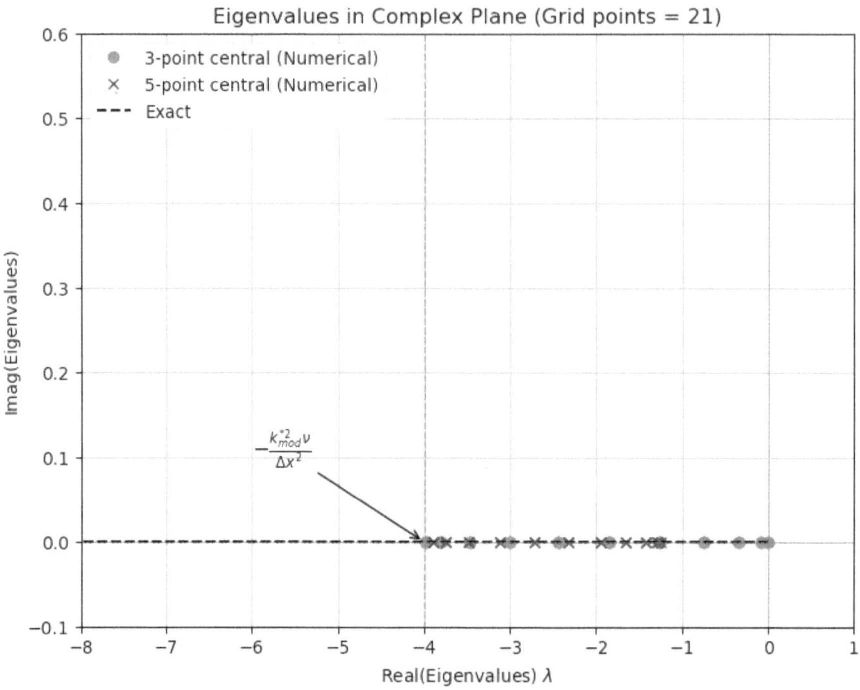

Figure 8.9: **Comparison of numerical eigenvalues computed using 3-point and 5-point central finite-difference schemes against the exact eigenvalues for the discretized diffusion operator. The eigenvalues are plotted in the complex plane, demonstrating purely real values aligned along the negative real axis, confirming numerical stability for the diffusion equation discretization (21 grid points).**

Δx^2 is the spatial grid spacing squared, which defines the spatial resolution of the numerical grid.
ν is the diffusion coefficient, which defines the rate of diffusion in the system.

This states that the maximum allowable time step $\Delta t_{\max}$ is proportional to the square of the grid spacing Δx^2 and inversely proportional to the diffusion coefficient ν. Essentially, as the grid becomes finer (smaller Δx), the maximum time step must decrease to maintain numerical stability. Similarly, if the diffusion coefficient ν is large, the system diffuses more quickly, allowing for a larger time step without compromising stability.

```
3-point scheme and N=100:
Max eigenvalue (real) = -0.0
Min eigenvalue (real) = -3.977661652450257
```

8.11 TIME STEP IN DIFFUSION EQUATION

For numerical solutions of the diffusion equation using central finite differences, stability imposes a strict upper bound on the allowable time step size. This limit arises from the viscous diffusion process, characterized by the diffusive Courant number or Fourier number, σ_max. The maximum permissible time step is determined explicitly by the spatial discretization size, the kinematic viscosity ν, and the maximum modified wavenumber associated with the discretization scheme. The resulting condition ensures numerical stability by constraining the rate at which diffusive effects propagate through the computational grid.

With central finite differences, the time step size must be controlled for the diffusion equation

$$\Delta t_{\text{max}} = \frac{s(\Theta = 0^\circ)}{\nu/\Delta x^2 \cdot k^{*2}_{\text{mod,max}}} = \frac{s_r}{k^{*2}_{\text{mod,max}}} \frac{\Delta x^2}{\nu} = \sigma_{\text{max}} \frac{\Delta x^2}{\nu}$$

For a three-dimensional diffusion problem with uniform spatial discretization in all directions, the time step should be one-third of the Δt_{max}.

Table 8.2 compares schemes commonly used for diffusion term modeling in flow applications.

Lessons Learned

☞ In advection-dominated problems, diffusion appears as a numerical artifact (numerical diffusion), whereas in diffusion-dominated problems, diffusion is explicitly part of the physical PDE.

☞ Diffusion equations must represent averaged or filtered quantities to reflect physical reality accurately, and the diffusion coefficient must match the scale of this averaging.

☞ Explicit finite difference schemes for diffusion require careful consideration of stability constraints, typically characterized by the amplification factor:$A = 1 - 4\sigma \sin^2\left(\frac{k^*}{2}\right)$.

☞ Stability criterion for explicit schemes requires: $|A| \leq 1, \quad \sigma \leq 0.5$.

☞ The Leapfrog scheme is second-order accurate but unconditionally unstable for diffusion because its amplification factor always exceeds unity

☞ The Damped Leapfrog Scheme introduces artificial viscosity to enhance stability and reduce oscillations, providing second-order accuracy with controlled damping.

☞ The explicit FTCS (Forward-Time Central-Space) scheme is conditionally stable, requiring $\sigma \leq 0.5$, whereas the implicit Laasonen (backward-Euler) scheme is unconditionally stable, allowing any time step size without compromising stability.

☞ Crank-Nicolson scheme is unconditionally stable, second-order accurate, and balances numerical dissipation and dispersion, minimizing artificial oscillations.

☞ The viscous stability limit (time step size limit) for explicit diffusion problems is given by:

$$\Delta t_{\text{max}} = \sigma_{\text{max}} \frac{\Delta x^2}{\nu}, \quad \sigma_{\text{max}} \leq 0.5$$

☞ Matrix decomposition methods (eigenvalue/eigenvector approaches) transform diffusion problems into decoupled modal systems.

☞Eigenvalues for discretized diffusion problems indicate decay rates for spatial modes, highlighting that larger eigenvalues (higher-frequency modes) decay faster:

☞Numerical stiffness arises from large ratios of eigenvalues (fastest to slowest decay modes), requiring careful selection of time steps in explicit methods.

Table 8.2
Schemes for one–dimensional diffusion equation: $u_t = \mathscr{D} u_{xx}$ (Diffusion number: $\sigma = \mathscr{D}\Delta t / \Delta x^2$)

Name	Amplification factor $A(k^*)$	Stability comment
Exact (Fourier)	$A_e = \exp\left(-\sigma k^{*2}\right)$	Always stable
Implicit Euler (Laasonen)	$A_{IE} = \dfrac{1}{1 + 4\sigma \sin^2\left(k^*/2\right)}$	Unconditionally stable
Crank–Nicolson	$A_{CN} = \dfrac{1 - 2\sigma \sin^2\left(k^*/2\right)}{1 + 2\sigma \sin^2\left(k^*/2\right)}$	Unconditionally stable (weak damping)
Leapfrog	$A_{LF} = -4\sigma \sin^2(k^*/2) \pm \sqrt{1 + 16\sigma^2 \sin^4(k^*/2)}$	Unstable for any $\sigma > 0$
DuFort–Frankel (3-level)	$A_{DF} = \dfrac{2\sigma \cos k^* \pm \sqrt{1 - 4\sigma^2 \sin^2 k^*}}{1 + 2\sigma}$	Unconditionally stable

PROBLEMS

8P-1 Consider a wall that is 1 ft thick and extends infinitely in the other directions. Initially, the temperature throughout the wall is uniform at 100°F. Suddenly, the temperatures at both surfaces of the wall are raised to 300°F and maintained at that level. The wall is made of nickel steel (40% Ni), with a thermal diffusivity of $\alpha = 0.1\,\text{ft}^2/\text{hr}$. The objective is to calculate how the temperature distribution inside the wall evolves over time.

The governing equation for this situation is the one-dimensional unsteady heat conduction equation in Cartesian coordinates:

$$\frac{\partial T}{\partial t} = \alpha \frac{\partial^2 T}{\partial x^2}$$

with the initial condition $T_i = 100°\text{F}$ and surface temperatures maintained at $T_s = 300°\text{F}$.

The numerical method should be used to solve the equation over a spatial step $\Delta x = 0.05$ ft and a time step $\Delta t = 0.01$ hr. The analytical solution provided is to be used to validate the numerical results. In all cases, the temperature distribution should be printed and plotted for all positions along the wall at intervals of 0.1 hours, up to a total of 0.5 hours. The analytical solution involves an infinite series:

$$T = T_s + 2(T_i - T_s) \sum_{m=1}^{\infty} e^{-(m\pi/L)^2 \alpha t} \frac{1-(-1)^m}{m\pi} \sin\left(\frac{m\pi x}{L}\right)$$

In practice, the sum is truncated after a finite number of terms. If too few terms are included. Investigate using DuFort-Frankel Scheme.

8P-2 The diffusion equation is given as

$$\frac{\partial u}{\partial t} = \nu \frac{\partial^2 u}{\partial y^2}$$

where ν is the kinematic viscosity of the fluid. The initial and boundary conditions are stated as follows:

Initial Condition:

$$t = 0, \quad u = U_0 \text{ for } y = 0, \quad u = 0 \text{ for } 0 < y \leq h$$

Boundary Conditions:

$$t \geq 0, \quad u = U_0 \text{ for } y = 0, \quad u = 0 \text{ for } y = h$$

The fluid is oil with a kinematic viscosity of $0.000217\,\text{m}^2/\text{s}$, and the spacing between plates is 40 mm. The velocity of the lower wall is specified as $U_0 = 40\,\text{m/s}$. A grid system with $\Delta y = 0.001\,\text{m}$ and various time steps is used to investigate the numerical schemes and the effect of time step on stability and accuracy. The number of spatial points in the y-direction is $JM = 41$, and the time steps are denoted by NM. Use numerical Schemes:

(a) FTCS Explicit Method:
(I) $\Delta t = 0.002$, $NM = 341$
(II) $\Delta t = 0.00232$, $NM = 341$

(b) DuFort-Frankel Explicit Method:
(I) $\Delta t = 0.002$, $NM = 341$
(II) $\Delta t = 0.003$, $NM = 341$

$$u = U_0 \left\{ \sum_{n=0}^{\infty} [\operatorname{erfc}(2n\eta_1 + \eta) - \operatorname{erfc}(2n\eta_1 - \eta)] \right\}$$

which simplifies as

$$u = U_0 \{\operatorname{erfc}(\eta) - \operatorname{erfc}(2\eta_1 - \eta) + \operatorname{erfc}(4\eta_1 + \eta) - \operatorname{erfc}(4\eta_1 - \eta) + \cdots\}$$

where: $\eta_1 = \frac{h}{2\sqrt{\nu t}}$. This series represents the solution to the Navier-Stokes equation subject to the given boundary conditions.

The error term for comparison is defined as

$$\text{ER} = \left(\frac{\text{Analytical value} - \text{Computed value}}{\text{Analytical value}} \right) \times 100$$

8P-3 Solve

$$\frac{\partial u}{\partial t} = \nu \frac{\partial^2 u}{\partial x^2}, \quad u = u(x,t)$$

with the real constant $\nu > 0$ in the domain $\{0 \le x \le 4\}$.

The problem is spatially periodic, i.e.,

$$u(x,t) = u(x+4,t), \quad u_{xx}(x,t) = u_{xx}(x+4,t)$$

Initial Condition:

$$u(x,t=0) = \sin(kx), \quad k = \frac{\pi}{2}$$

The exact solution is:

$$u(x,t) = \exp\left(-\nu \left(\frac{\pi}{2}\right)^2 t\right) \cdot \sin\left(\frac{\pi}{2}x\right)$$

(i) Characterize the equation and sketch the initial condition and the expected solution at a time $t > 0$. Determine the constant ν so that the amplitude is halved at time $t = 1.4$.

(ii) The equation is discretized with central finite differences of second order using five equidistant grid points. Time stepping is done using the explicit Euler method. Calculate the numerical solution at $t = 1$ with a time step $\Delta t = 1$. Provide the numerical and exact solution at $t = 1$, and plot both.

(iii) Use a second-order backward difference method for the spatial discretization. The second derivative is approximated by:

$$u_{xx,i} \approx \frac{-u_{i-3} + 4u_{i-2} - 5u_{i-1} + 2u_i}{\Delta x^2}$$

Perform time integration using the RK2 method. Calculate $u(x,t)$ at $t = 1$ and compare with the result from Task (ii).

8P-4 Consider two infinite parallel plates, separated by a distance of 0.5 cm. The upper plate remains stationary, while the lower plate oscillates according to

$$u(0,t) = u_0 \cos(1200t) = u_0 \cos(1200 n\Delta t)$$

where n represents the computational time level, with $n = 1$ at $t = 0.0$.

The governing equation is derived from the Navier-Stokes equation:

$$\frac{\partial u}{\partial t} = \nu \frac{\partial^2 u}{\partial y^2}$$

Assume that the kinematic viscosity ν is constant with a value of 0.000175 m^2/s, and $u_0 = 35$ m/s. Select 41 evenly spaced grid points, with $i = 1$ positioned at the lower plate. The time step is specified as $\Delta t = 0.000015$s. Use the Euler explicit method with the central differencing scheme FD-O2 (FTCS) to solve the equation within the domain up till time $t = 0.0084$s. Investigate the solution at the time levels: 0.0, 0.002, 0.004, 0.006, and 0.0084 seconds. Also, plot the velocity profiles at the indicated time levels.

8P-5 Test Laasonen and Crank-Nicolson schemes for

$$\frac{\partial \phi}{\partial t} = \mathscr{D} \frac{\partial^2 \phi}{\partial x^2}$$

The initial condition $\phi(x)$ can be expressed as:

$$\phi(x) = \begin{cases} 0 & \text{if } x \leq 0.9 \\ 10(x - 0.9) & \text{if } 0.9 \leq x \leq 1.0 \\ 10(1.1 - x) & \text{if } 1.0 \leq x \leq 1.1 \\ 0 & \text{if } x \geq 1.1 \end{cases}$$

This piecewise function defines $\phi(x)$ such that it rises from 0 to 1 between $0.9 \leq x \leq 1.0$ and falls from 1 to 0 between $1.0 \leq x \leq 1.1$, while it is zero outside these intervals. Take the number of spatial grid points: 51. The number of time steps: $n_t = 20$. The diffusion coefficient is 0.01. The spatial domain has $x_{\min} = 0$ and $x_{\max} = 2$.

8P-6 Consider the unsteady partial differential equation (PDE):

$$\frac{\partial u}{\partial t} = \nu \frac{\partial^2 u}{\partial x^2}, \quad u = u(x,t), \quad \nu \geq 0, \text{ constant.}$$

i What is the type of this PDE, and for which kind of flow problems can it be considered a representative model equation?

ii The PDE is to be integrated using time stepping methods, and with a central FD-O2 for the spatial derivatives. What is the maximum timestep that can be used during the integration, without making the method unstable?

8P-7 Given the partial differential equation

$$\frac{\partial u}{\partial t} = \nu \frac{\partial^2 u}{\partial x^2}, \qquad \nu = 1$$

The spatial discretization is performed using central finite differences of second or sixth order, while the explicit Euler method is used for time integration. For both spatial discretizations, determine the maximum diffusion number for which the method remains stable.

8P-8 Consider the unsteady partial differential equation:

$$\frac{\partial u}{\partial t} = \nu \frac{\partial^2 u}{\partial x^2}, \qquad u = u(x,t), \nu \geq 0, \, \nu \in \mathbb{R}$$

with the spatial solution assumption $u(x,t) = u_k(t)e^{ikx}$. The second derivative u_{xx} can then be numerically expressed as:

$$u_{xx} = -k_{mod}^2 \frac{u}{\Delta x^2}, \qquad \text{where } k^* = k\Delta x.$$

i The suggested discretisation methods are:

1 Fourier (exact)

2 Central FD-02

3 One-sided 3-point FD-01

Match the following three specifications of k_{mod}^2 with the spatial discretisations listed below:

a $k_{mod}^2 = -1 + 2\cos(k) - \cos(2k) + i\beta_1$

b $k_{mod}^2 = k^2$

c $k_{mod}^2 = 2\,(1 - \cos(k))$

ii What is the effect of using non-central differences for the spatial discretization in the integration of this PDE?

8P-9 Consider the unsteady one-dimensional momentum equation for the x-component of velocity $u(y,t)$ in a vertical channel flow, given by

$$\frac{\partial u}{\partial t} = -\frac{1}{\rho} \frac{\partial p}{\partial x} + \nu \frac{\partial^2 u}{\partial y^2}$$

where ρ is the fluid density, ν is the kinematic viscosity The equation is subject to the boundary conditions

$$u(y=0,t)=U_1(t), \qquad u(y=h,t)=U_2(t),$$

which enforce no-slip conditions at the channel walls. Assume $U_1(t)=U_2(t)=0$, and an initial condition $u(y,0)=0$. The pressure gradient $\partial p/\partial x$ is assumed constant. Ignore gravitational body force. Discretize the domain $0\le y\le h$ using a uniform grid with $N+1$ points. Implement a numerical solution to the governing equation using both the Euler explicit and Euler implicit finite-difference schemes. In each case:

i Use a first-order forward difference in time.

ii Use a second-order central difference for the spatial derivative.

iii Integrate the system up to a fixed final time T.

Compute using grid resolutions in the y-direction for grid $N=50$, $N=100$, and $N=500$, and assume that the channel flow is driven by a constant pressure gradient in the x-direction, having $\partial p/\partial x$:

$$\text{Case A:}\quad \frac{\partial p}{\partial x}=-10\ \text{Pa/m}, \qquad \text{Case B:}\quad \frac{\partial p}{\partial x}=-100\ \text{Pa/m}.$$

For the numerical computations, take the water kinematic viscosity as $\nu=1\times 10^{-6}\ \text{m}^2/\text{s}$, and the water density is $\rho=1000\ \text{kg/m}^3$. The channel height is set as $h=1.0$ m. Plot the velocity profiles, and investigate after how much time the flow will be fully developed.

The exact steady-state solution to the unsteady 1D momentum equation in a vertical channel flow under a constant pressure gradient and no-slip boundary conditions is given by the parabolic velocity profile:

$$u_{\text{exact}}(y)=\frac{1}{2\nu}\left(-\frac{1}{\rho}\frac{\partial p}{\partial x}\right)y(h-y),$$

and peak Velocity (Centerline, $y=h/2$) is

$$u_{\max}=\frac{h^2}{8\nu\rho}\left|\frac{\partial p}{\partial x}\right|$$

where, $u(y)$ is the streamwise velocity as a function of wall-normal coordinate y, h is the channel height, ν is the kinematic viscosity, ρ is the fluid density, $\partial p/\partial x$ is the constant pressure gradient in the streamwise direction.

8P-10 A gas turbine blade, made of Inconel 718 (a high-performance superalloy), is exposed to high-temperature gas flow. The blade has a length of 0.5 m and a thickness of 5 cm. The surface of the blade is subjected to a convective heat transfer coefficient of 3000 $\text{W/m}^2\text{K}$, and the temperature difference between the blade surface and the surrounding gas is 600 K. The blade material has a thermal conductivity of

20 W/m·K, a density of 8200 kg/m^3, and a specific heat capacity of 500 J/kg·K. Using the Crank-Nicolson method, determine the temperature distribution across the blade at steady-state for a given time step. Assume that the initial temperature of the blade is uniform and equal to the surrounding gas temperature (300 K). The solution should be computed for a time step of 0.001 s and a spatial grid resolution of 10 points. Solve for the temperature profile at multiple time steps, considering both the convective boundary condition at the surface and an insulated boundary condition at the other end of the blade. Present the temperature distribution along the length of the blade.

REFERENCES

J.C. Strikwerda, Finite Difference Schemes and Partial Differential Equations, 2nd ed., SIAM, 2004.

R.J. LeVeque, Finite Difference Methods for Ordinary and Partial Differential Equations, SIAM, 2007.

R. Haberman, Applied Partial Differential Equations with Fourier Series and Boundary Value Problems, 5th ed., Pearson, 2013.

G. H. Golub and C.F. Van Loan, Matrix Computations, 4th ed., Johns Hopkins University Press, 2013.

K.W. Morton and D.F. Mayers, Numerical Solution of Partial Differential Equations: An Introduction, 2nd ed., Cambridge University Press, 2005.

P. Laasonen, Über eine Methode zur Lösung der Wärmeleitungsgleichung, Acta Math. 81 pp. 309 - 317, 1949.

9 Stabilizing Explicit Schemes with Hyperviscosity and Filtering

Numerical simulations of fluid flows often face challenges in maintaining stability, particularly when resolving high-gradient regions or shocks. Hyperviscosity, a numerical technique involving higher-order derivative terms provides an effective stabilization mechanism by selectively damping spurious high-frequency oscillations without significantly altering large-scale physics. This chapter explores the mathematical foundation and practical implementation of hyperviscosity in explicit finite-difference schemes. We analyze its role in modifying the dispersion and dissipation characteristics of the governing equations, demonstrating how added hyperviscosity term suppress grid-scale instabilities while preserving solution accuracy.

After finishing this chapter, you should be able to:

- Understand the role of hyperviscosity in stabilizing numerical simulations of high-gradient flows.
- Learn about the mathematical formulation of the diffusion equation with hyperviscosity.

In computational fluid dynamics and other numerical simulations, managing numerical stability is crucial, especially when dealing with high-gradient flows or shock waves. One effective approach to addressing these challenges is the introduction of hyperviscosity into the governing equations. The concept of hyperviscosity, while not representing a physical property of the fluid, serves as a numerical technique to stabilize solutions by smoothing out discontinuities or high-frequency oscillations that can arise in simulations.

When designing numerical methods for solving partial differential equations, a fundamental concern is the accuracy of approximations used for spatial derivatives. These approximations inevitably introduce truncation errors due to the discrete representation of continuous functions. One common example is the central difference scheme for the second derivative, which, while second-order accurate, introduces a leading error proportional to the fourth derivative of the function. By examining this through Taylor series expansion, we observe that the numerical approximation not only deviates from the exact differential operator but also modifies the underlying

DOI: 10.1201/9781003663171-9

equation. This modification introduces artificial terms absent in the original physical model, thereby altering the behavior of the solution in subtle yet significant ways. To understand how numerical methods introduce errors, consider a typical second-order central difference approximation of the second derivative:

$$\frac{\partial^2 u}{\partial x^2} \approx \frac{u(x+\Delta x) - 2u(x) + u(x-\Delta x)}{\Delta x^2}$$

To evaluate the accuracy of this approximation, expand $u(x \pm \Delta x)$ using Taylor series:

$$u(x \pm \Delta x) = u(x) \pm \Delta x u' + \frac{\Delta x^2}{2} u'' \pm \frac{\Delta x^3}{6} u''' + \frac{\Delta x^4}{24} u^{(''''')} + \cdots$$

Adding the two expansions:

$$u(x+\Delta x) + u(x-\Delta x) = 2u(x) + \Delta x^2 u'' + \frac{\Delta x^4}{12} u^{('''')} + \cdots$$

Substituting back into the finite difference formula:

$$\frac{u(x+\Delta x) - 2u(x) + u(x-\Delta x)}{\Delta x^2} = u'' + \frac{\Delta x^2}{12} u^{('''')} + \cdots$$

So, the approximation introduces a leading truncation error term of the form:

$$\xi \frac{\partial^4 u}{\partial x^4}, \quad \text{where} \quad \xi = \frac{\Delta x^2}{12}$$

$$u'' \approx \frac{u(x+\Delta x) - 2u(x) + u(x-\Delta x)}{\Delta x^2} - \frac{\Delta x^2}{12} u^{('''')}$$

The inclusion of the truncation error term modifies the diffusion equation by introducing an additional fourth-order derivative term, leading to a form

$$\frac{\partial u}{\partial t} - \nu \frac{\partial^2 u}{\partial x^2} - \xi \frac{\partial^4 u}{\partial x^4} = 0$$

Assume a Fourier mode solution of the form:

$$u(x,t) = \mathsf{V}(t) e^{ikx}$$

We substitute the spatial derivatives into the PDE:

$$\frac{d\mathsf{V}}{dt} e^{ikx} + \nu k^2 \mathsf{V}(t) e^{ikx} - \xi k^4 \mathsf{V}(t) e^{ikx} = 0$$

Divide both sides by e^{ikx} gives

$$\frac{d\mathsf{V}}{dt} = -\nu k^2 \mathsf{V}(t) - \xi k^4 \mathsf{V}(t)$$

The evolution equation for the Fourier amplitude becomes

$$\frac{\partial u}{\partial t} = -\nu k^2 u - \xi k^4 u$$

This leads to an effective dissipation term

$$D_{\text{eff}} = \nu k^2 + \xi k^4$$

The first term represents physical diffusion, while the second term is purely numerical. As the wavenumber k increases, the artificial dissipation term dominates, disproportionately damping small-scale features.

This implies that small-scale features—those associated with high frequencies—are disproportionately damped by the numerical method. This unphysical smoothing effect must be considered in applications such as turbulent flows, where high-frequency content is essential.

This result shows that numerical schemes can introduce significant numerical damping of small-scale features, especially when k is large.

i Numerical dissipation increases rapidly with k, especially for higher-order truncation terms.

ii It is sensitive to grid resolution Δx, which must be chosen carefully.

iii For high-order schemes, the artificial damping may dominate physical diffusion, distorting the solution.

This analysis highlights how truncation errors in finite difference methods affect both the accuracy and the qualitative behavior of numerical solutions. The emergence of artificial dissipation, particularly prominent at high wavenumbers, illustrates the need for careful grid resolution and scheme selection. As numerical dissipation scales with $\Delta x^2 k^4$, its impact grows with finer features in the solution, often damping out high-frequency content. Consequently, in computational simulations where small-scale accuracy is critical, such as in turbulence or wave propagation problems, understanding the role and magnitude of truncation errors is essential.

9.1 STABILIZING EXPLICIT SCHEMES USING HYPERVISCOSITY

This section focuses on a modified diffusion equation that incorporates hyperviscosity through an additional fourth-order spatial derivative term. The goal is to explore how this modification enhances the stability and robustness of numerical schemes, particularly in the context of finite difference methods. We analyze the impact of the hyperviscosity term on the solution's behavior and investigate the conditions under which it effectively stabilizes high-gradient flows. By examining the stability of the resulting numerical schemes and considering the influence of the viscosity parameter, we aim to provide insights into the practical application of hyperviscosity in computational simulations, especially for problems involving shocks, discontinuities, and other challenging flow regimes.

Consider the PDE:

$$\frac{\partial u}{\partial t} = \nu \frac{\partial^2 u}{\partial x^2} - \varepsilon \frac{\partial^4 u}{\partial x^4}$$

The fourth-order term acts as a hyperviscosity term, selectively damping high-frequency oscillations.

Hyperviscosity is a numerical technique introduced into simulations to enhance stability and suppress non-physical high-wavenumber modes, particularly in regions with sharp gradients or discontinuities. Although it does not correspond to any physical fluid property, hyperviscosity mimics the dissipative effects of viscosity at a higher derivative order, without significantly altering the large-scale dynamics of the solution.

In this formulation, the coefficient ε has units of L^4/T, ensuring that the added term $\varepsilon\partial^4 u/\partial x^4$ is dimensionally consistent with the other terms in the equation. This fourth-order diffusion acts to smooth out rapidly varying components in the solution, particularly those near the Nyquist wavenumber. Such stabilization is crucial in finite-difference methods, where high-frequency numerical noise can otherwise cause instability.

Hyperviscosity is commonly applied in explicit schemes where stability constraints are more restrictive, especially in simulations involving turbulence, shocks, or fine spatial resolution. It allows the use of larger time steps by dampening unstable high-frequency content without introducing artificial dissipation across all scales. Proper tuning of the hyperviscosity coefficient is essential to preserve accuracy while achieving numerical robustness.

9.1.1 FDO2 AND EULER EXPLICIT SCHEME

We now analyze the modified wavenumber associated with the discretized hyperviscosity term. The standard central difference approximation for the fourth derivative at point i, using a uniform grid spacing Δx, is given by:

$$\frac{\partial^4 u}{\partial x^4} \approx \frac{u_{i+2} - 4u_{i+1} + 6u_i - 4u_{i-1} + u_{i-2}}{\Delta x^4}$$

This finite-difference stencil uses coefficients $[1, -4, 6, -4, 1]$ applied to the nodes at $i+2, i+1, i, i-1$, and $i-2$, respectively.

To determine how this operator acts on a Fourier mode, assume the solution takes the form:

$$u(x) = e^{ikx}$$

Then, the values at neighboring grid points become:

$$u_{i+2} = e^{ik(x_i+2\Delta x)} = e^{ikx_i} e^{2ik\Delta x},$$
$$u_{i+1} = e^{ik(x_i+\Delta x)} = e^{ikx_i} e^{ik\Delta x},$$
$$u_{i-1} = e^{ik(x_i-\Delta x)} = e^{ikx_i} e^{-ik\Delta x},$$
$$u_{i-2} = e^{ik(x_i-2\Delta x)} = e^{ikx_i} e^{-2ik\Delta x}.$$

Substituting these into the finite-difference formula gives:

$$\frac{\partial^4 u}{\partial x^4} \approx \frac{e^{ikx_i}\left(e^{2ik\Delta x} - 4e^{ik\Delta x} + 6 - 4e^{-ik\Delta x} + e^{-2ik\Delta x}\right)}{\Delta x^4}$$

Now apply Euler's identities to express the result in terms of trigonometric functions:

$$e^{2ik\Delta x} + e^{-2ik\Delta x} = 2\cos(2k\Delta x),$$
$$e^{ik\Delta x} + e^{-ik\Delta x} = 2\cos(k\Delta x)$$

Thus, the expression simplifies to:

$$\frac{\partial^4 u}{\partial x^4} \approx \frac{e^{ikx_i}\left(2\cos(2k\Delta x) - 8\cos(k\Delta x) + 6\right)}{\Delta x^4}$$

The factor multiplying e^{ikx_i} defines the **modified wavenumber** for the fourth derivative. Therefore, we define:

$$k^{*4}_{\text{mod}} = \frac{2\cos(2k\Delta x) - 8\cos(k\Delta x) + 6}{\Delta x^4}$$

This can also be written in a more conventional form as:

$$k^{*4}_{\text{mod}} = \frac{6 - 8\cos(k\Delta x) + 2\cos(2k\Delta x)}{\Delta x^4}$$

This expression quantifies how the discrete approximation deviates from the continuous operator in wavenumber space. It plays a central role in analyzing the spectral dissipation introduced by hyperviscosity in numerical simulations.

For stability considerations when using second-order finite differences (FDO2) with an explicit Euler time-stepping scheme, the upper bound for the hyperviscosity coefficient is determined by the stability constraint of the time integrator (see Figure 9.2):

$$\left|-\varepsilon k^{*4}_{\text{mod}}\right|_{\max} \approx s_r$$

Given that the maximum modified wavenumber for the fourth derivative is $k^{*4}_{\text{mod,max}} = 16$ (as shown in Figure 9.1), the upper limit for the hyperviscosity coefficient becomes:

$$\varepsilon_{\max} = \frac{s_r}{k^{*4}_{\text{mod,max}}} = \frac{2}{16} = \frac{1}{8}$$

In practice, typical values used for ε fall in the range:

$$0.05 \leq \varepsilon \leq 0.1$$

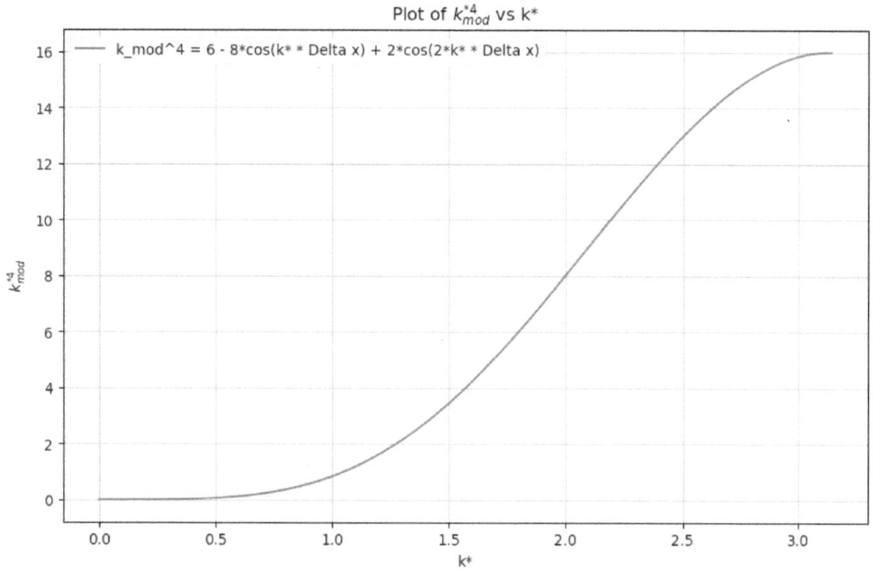

Figure 9.1: **Modified wavenumber for hyperviscosity.**

It is important to note that, under this formulation, the amount of damping per time step remains constant with respect to the stability condition, even though the actual damping depends on both Δt and Δx through the numerical formulation.

We now revisit the governing equation and analyze its behavior in wavenumber space. The modified partial differential equation, incorporating hyperviscosity in its discretized form, is:

$$\frac{\partial u}{\partial t} = \nu \frac{\partial^2 u}{\partial x^2} - \frac{\varepsilon \Delta x^4}{\Delta t} \frac{\partial^4 u}{\partial x^4}$$

This equation contains a second-order diffusion term u_{xx} and a fourth-order hyperviscosity term u_{xxxx}. It is a representative model for schemes that incorporate numerical stabilization while solving parabolic PDEs.

To analyze this equation in the frequency domain using an FDO2 scheme with explicit Euler time integration, we assume a trial solution of the form:

$$u(x,t) = e^{i(kx-\omega t)}$$

where k is the wavenumber, and ω is the frequency.

Using a central difference approximation for the second derivative:

$$u_{xx} \approx \frac{u(x+\Delta x,t) - 2u(x,t) + u(x-\Delta x,t)}{\Delta x^2}$$

Substituting the Fourier form:

$$u_{xx} \approx \frac{e^{ik\Delta x} - 2 + e^{-ik\Delta x}}{\Delta x^2} u(x,t) = \frac{2\cos(k\Delta x) - 2}{\Delta x^2} u(x,t)$$

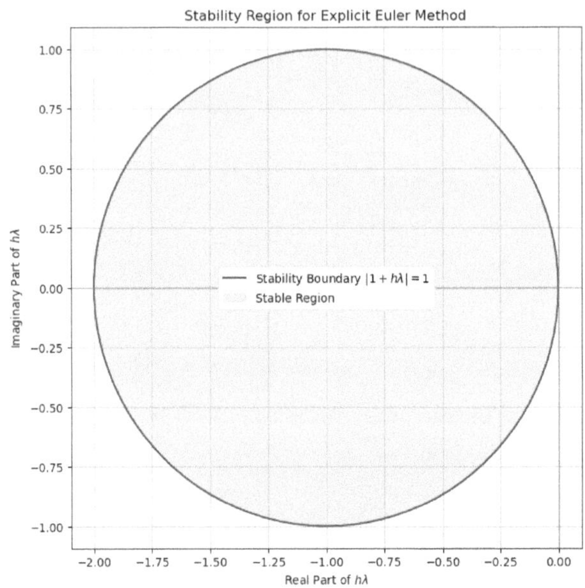

Figure 9.2: **Stability plot of Euler Explicit Scheme.**

For the fourth derivative, using a five-point central difference:

$$u_{xxxx} \approx \frac{e^{2ik\Delta x} - 4e^{ik\Delta x} + 6 - 4e^{-ik\Delta x} + e^{-2ik\Delta x}}{\Delta x^4} u(x,t) = \frac{2\cos(2k\Delta x) - 8\cos(k\Delta x) + 6}{\Delta x^4} u(x,t)$$

The explicit Euler scheme updates the solution as:

$$\frac{u(x,t+\Delta t) - u(x,t)}{\Delta t} = \left[\frac{2\cos(k\Delta x) - 2}{\Delta x^2} - \frac{\varepsilon \Delta x^4}{\Delta t} \cdot \frac{2\cos(2k\Delta x) - 8\cos(k\Delta x) + 6}{\Delta x^4}\right] u(x,t)$$

To simplify, introduce the diffusion number:

$$\sigma = \frac{\nu \Delta t}{\Delta x^2}$$

and define:

$$R = \frac{\varepsilon \Delta x^4}{\Delta t}$$

The update equation becomes:

$$\frac{u(x,t+\Delta t) - u(x,t)}{\Delta t} = \left[\frac{\sigma}{\Delta t}(2\cos(k\Delta x) - 2) - \frac{R}{\Delta t}(2\cos(2k\Delta x) - 8\cos(k\Delta x) + 6)\right] u(x,t)$$

Introduce the modified wavenumbers:

$$k_{\text{mod}}^{*2} = 2(1 - \cos(k\Delta x)), \quad k_{\text{mod}}^{*4} = 6 - 8\cos(k\Delta x) + 2\cos(2k\Delta x)$$

Then the update becomes:

$$\frac{u(x,t+\Delta t)-u(x,t)}{\Delta t}=\left[-\frac{\sigma}{\Delta t}k^{*2}_{\text{mod}}-\frac{R}{\Delta t}k^{*4}_{\text{mod}}\right]u(x,t)$$

Rearranging:

$$u(x,t+\Delta t)=u(x,t)+\left[-\sigma k^{*2}_{\text{mod}}-Rk^{*4}_{\text{mod}}\right]u(x,t)$$

or

$$u(x,t+\Delta t)=\left[1-\sigma k^{*2}_{\text{mod}}-Rk^{*4}_{\text{mod}}\right]u(x,t)$$

To achieve stability and match the analytical decay at the highest resolvable mode, we impose the following approximation at $k\Delta x=\pi$:

$$-\sigma k^{*2}_{\text{mod}}-Rk^{*4}_{\text{mod}}\approx-\sigma\pi^2$$

This gives:

$$-Rk^{*4}_{\text{mod}}\approx-\sigma\pi^2+\sigma k^{*2}_{\text{mod}}\Rightarrow R\approx\sigma\cdot\frac{\pi^2-k^{*2}_{\text{mod}}}{k^{*4}_{\text{mod}}}$$

At the Nyquist frequency $k\Delta x=\pi$, we have:

$$k^{*2}_{\text{mod}}=2(1-\cos(\pi))=4$$

$$k^{*4}_{\text{mod}}=6-8\cos(\pi)+2\cos(2\pi)=6+8+2=16$$

Substituting in:

$$\frac{R}{\sigma}=\frac{\pi^2-4}{16}\approx0.3668$$

Thus, the hyperviscosity coefficient required to ensure equivalent damping at $k\Delta x=\pi$ is:

$$\varepsilon=0.3668\,\sigma\,\frac{\Delta t}{\Delta x^4}$$

With this choice, the hyperviscosity term provides numerically optimal damping of high-wavenumber modes while preserving the stability structure of the second derivative term. It ensures that at the Nyquist limit, the discretized hyperviscous contribution mimics the spectral decay of the true Laplacian operator.

If instead ε is chosen arbitrarily, independent of σ and Δt, the resulting damping will be misaligned with the intended physical behavior—either overly diffusive or insufficient—leading to numerical inaccuracies or instability.

Therefore, the calibrated governing equation becomes:

$$\boxed{\frac{\partial u}{\partial t}=\nu\frac{\partial^2 u}{\partial x^2}-0.3668\,\sigma\frac{\Delta x^4}{\Delta t}\frac{\partial^4 u}{\partial x^4}}$$

This formulation guarantees that the numerical damping imposed by the hyperviscosity term is tuned to match the stability requirements of the explicit scheme while preserving the physical fidelity of the solution.

We will now investigate the one-dimensional diffusion equation using an explicit finite-difference method:

$$\frac{\partial u}{\partial t} = \nu \frac{\partial^2 u}{\partial x^2}$$

where $u(x,t)$ is the scalar field (e.g., temperature or concentration), and ν is the constant diffusion coefficient set to 1.0. The simulation is performed over a spatial domain $x \in [0,1]$ using a grid of 101 equally spaced points, giving a spatial resolution $\Delta x = 0.01$.

The initial condition is specified as a sine wave: $u(x,0) = \sin(\pi x)$, which satisfies homogeneous Dirichlet boundary conditions $u(0,t) = u(1,t) = 0$ for all t. These boundary conditions are implicitly enforced in the code by not updating the first and last elements of the solution array during time integration. The boundary values remain fixed, consistent with the analytical behavior of the sine function at the endpoints.

The diffusion equation is discretized using a second-order central difference scheme (FDO2) for the spatial second derivative, and the forward Euler method for time integration. In the stability analysis of explicit Euler schemes, the term s_r typically denotes the maximum allowed magnitude within the stability region of the time integration scheme. Specifically, for an explicit Euler scheme applied to purely dissipative equations, the numerical stability criterion requires:

$$|1 + \lambda \Delta t| \leq 1$$

where λ represents the eigenvalue of the spatial discretization operator. For purely real, negative eigenvalues (as in diffusion-type equations), this inequality simplifies to:

$$-2 \leq \lambda \Delta t \leq 0$$

Thus, the stability boundary $s_r = 2$ emerges naturally as the critical limit of stability for explicit Euler schemes. This value was chosen precisely because it represents the largest magnitude of the negative real axis interval within the explicit Euler stability region. The theoretical limit for stability is

$$\Delta t_{\max} = \frac{s_r}{k^{*2}_{\text{mod,max}}} \frac{\Delta x^2}{\nu} = \frac{2}{4} \frac{\Delta x^2}{\nu}$$

$$\Delta t \leq \frac{\Delta x^2}{2\nu}$$

so setting $\Delta t = 2.5 \cdot \frac{\Delta x^2}{2\nu}$ results in a diffusion number $\sigma = \frac{\nu \Delta t}{\Delta x^2} = 2.5$, well beyond the permissible limit $\sigma \leq 0.5$. This setup guarantees that the solution becomes unstable as time progresses. The result shows that violating the stability condition in explicit schemes leads to unbounded growth in the solution.

Next, we solve the modified one-dimensional diffusion equation with an hyperviscosity term, given by the partial differential equation:

$$\frac{\partial u}{\partial t} = \nu \frac{\partial^2 u}{\partial x^2} - 0.3668\,\sigma \frac{\Delta x^4}{\Delta t}\frac{\partial^4 u}{\partial x^4}$$

where,
$u(x,t)$ is the scalar field (e.g., temperature or concentration),
$\nu = 1.0$ is the physical diffusion coefficient,
$\sigma = \frac{\nu \Delta t}{\Delta x^2}$ is the diffusion number, and the coefficient 0.3668σ controls the strength of the hyperviscosity.

This PDE is a combination of second-order diffusion (physical) and fourth-order hyperviscosity (numerical stabilization). The second-order term diffuses the field, while the fourth-order term damps high-frequency numerical noise and prevents instability due to a large time step.

The initial condition is same as previously defined:

$$u(x,0) = \sin(\pi x)$$

The boundary conditions are homogeneous Dirichlet:

$$u(0,t) = 0, \quad u(1,t) = 0$$

Figures 9.3 compare the numerical solutions of the diffusion equations and illustrate the role and impact of hyperviscosity in stabilizing explicit finite-difference schemes. In the case where the pure diffusion equation is solved using second-order central differences (FDO2) and an explicit Euler time integration method, the parameters are selected such that the diffusion number σ exceeds the known stability limit for explicit schemes, which requires $\sigma \leq 0.5$ for stability. As a result, the numerical solution becomes unstable. This instability appears as rapidly growing high-frequency oscillations that corrupt the physical solution, eventually leading to an exponential blow-up.

The numerical results demonstrate the effectiveness of hyperviscosity in stabilizing an otherwise unstable explicit scheme for the diffusion equation. With the diffusion number σ = 1.25 (purposely exceeding the stability limit of 0.5 for pure diffusion), the solution remains stable when the fourth-order hyperviscosity term is included, as shown by the smooth blue curve matching the expected diffusive decay pattern. The hyperviscosity coefficient ε = 0.3668 $\sigma \Delta x^4/\Delta t$ selectively damps the high-frequency numerical oscillations that would normally cause instability, while preserving the physical behavior of the solution at larger scales. The numerical solution with hyperviscosity closely follows the analytical solution (red dashed line) of the pure diffusion equation, confirming that the stabilization does not significantly alter the physical diffusion process. The slight deviation between the numerical and analytical solutions arises because the hyperviscosity-modified equation being solved differs from the pure diffusion equation, particularly in its treatment of high-wavenumber components. This implementation validates the theoretical stability analysis showing that hyperviscosity allows stable computations with larger time steps than normally permitted by the standard diffusion equation discretization.

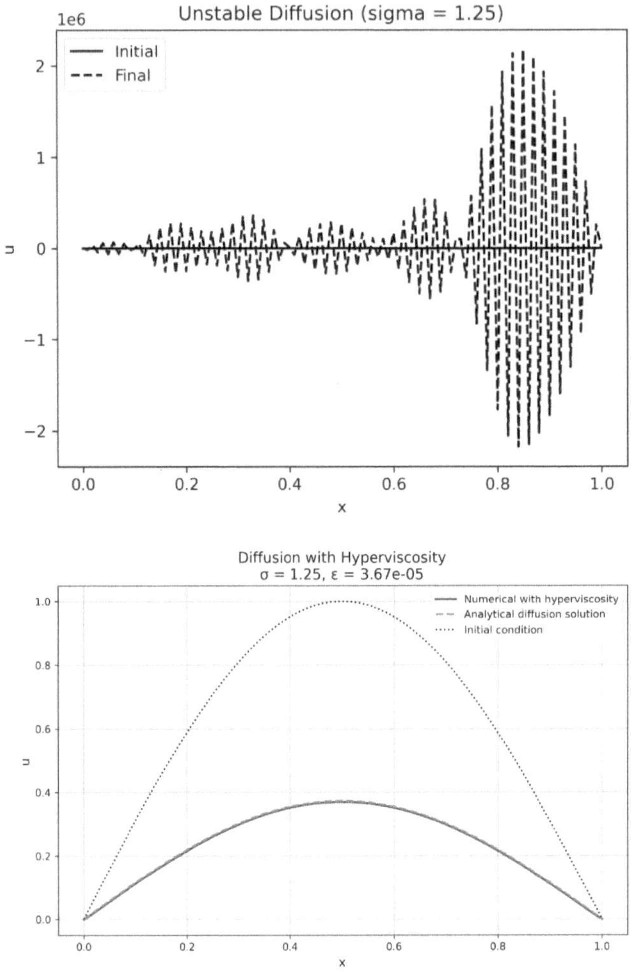

Figure 9.3: **Comparison of numerical solutions for the diffusion equation with and without hyperviscosity. The left plot shows the result of solving the pure diffusion equation using an explicit Euler scheme with a diffusion number $\sigma = 1.25$, exceeding the stability limit and resulting in rapidly growing, non-physical oscillations. The right plot includes a fourth-order hyperviscosity term $-0.3668\,\sigma\,u_{xxxx}$, which selectively damps unstable high-frequency modes and restores numerical stability under otherwise unstable time-stepping conditions.**

9.2 NUMERICAL STABILIZATION OF THE EXPLICIT DIFFUSION EQUATION USING GAUSSIAN FILTERING

We now investigate the stabilization of the explicit finite-difference solution of the one-dimensional diffusion equation using Gaussian filtering of the Laplacian term. The governing equation is the standard diffusion (heat) equation,

$$\frac{\partial u}{\partial t} = \nu \frac{\partial^2 u}{\partial x^2},$$

where $u(x,t)$ is the dependent variable (e.g., temperature or concentration) defined on the spatial domain $x \in [0,1]$, and ν is the diffusion coefficient. For all computations, the initial condition was taken as a single sine mode,

$$u(x,0) = \sin(\pi x),$$

which satisfies homogeneous Dirichlet boundary conditions,

$$u(0,t) = u(1,t) = 0 \qquad \forall t \geq 0.$$

Spatial discretization was performed using the second-order central finite-difference operator (FDO2) for the Laplacian. The explicit Euler method was used for time integration. Under these settings, the classical stability criterion for the explicit scheme requires the diffusion number,

$$\sigma = \frac{\nu \Delta t}{\Delta x^2} \leq 0.5,$$

to ensure stability. However, in this study, time steps deliberately exceeding the classical stability limit ($\sigma > 0.5$) were employed to demonstrate the effect of numerical filtering.

The key modification to the classical method involved the application of a Gaussian filter to the computed discrete Laplacian at each time step. Specifically, at every update, the standard Laplacian u_{xx} was convolved with a Gaussian kernel of width Δ_g (the filter width), such that the update equation became

$$u_i^{n+1} = u_i^n + \nu \Delta t \left[\mathcal{G}(u_{xx})\right]_i,$$

where $\mathcal{G}(u_{xx})$ denotes the Gaussian-filtered Laplacian evaluated at grid point i. This filtering operation acts as a high-frequency smoother, selectively suppressing spurious high-wavenumber components responsible for numerical instability in explicit schemes. The effect of different filter widths (Δ_g) was systematically tested. For sufficiently large Δ_g, the scheme remained stable even for diffusion numbers well above 0.5, demonstrating the filter's powerful stabilizing effect. For moderate filter values, the numerical solution remained in close agreement with the analytical solution,

$$u(x,t) = sin(\pi x) exp(-\nu \pi^2 t),$$

as quantified by a small maximum absolute error after integrating to time T.

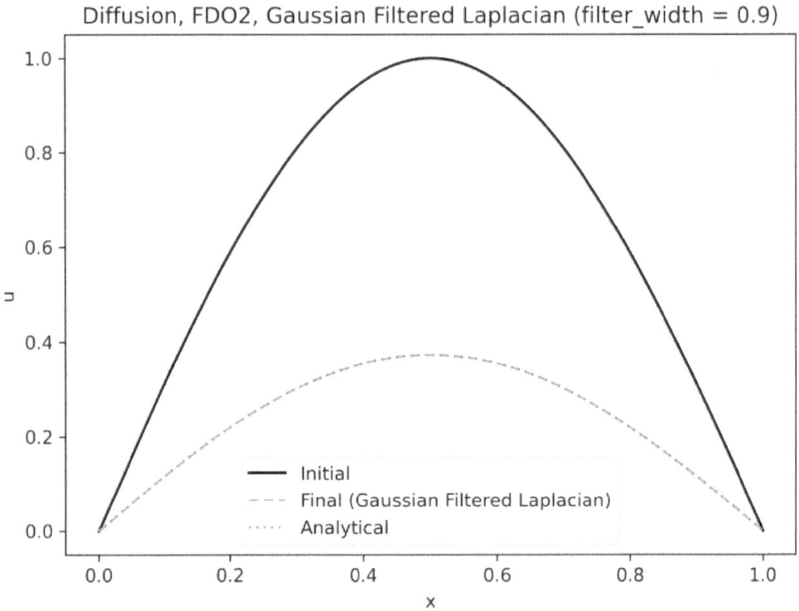

Figure 9.4: **Stabilization of the Explicit Diffusion Equation Using Gaussian Filtering with Δ_g=0.9.**

Figure 9.4 shows that Gaussian filtering of the Laplacian enables the use of much larger time steps in explicit finite-difference simulations of the diffusion equation, without incurring instability or significant loss of accuracy—provided the filter width is judiciously chosen. When compared to the analytical solution, the filtered scheme can achieve excellent agreement, with errors on the order of 10^{-5} or less, even for diffusion numbers as high as 1.25. However, small filtering width ($\Delta_g < 0.55$) can lead to stability issues and oscillations. The Gaussian-filtered Laplacian approach offers a flexible, practical method for accelerating explicit parabolic solvers by balancing stability and accuracy.

From a computational perspective, Gaussian filtering is often more localized and cheaper to implement than hyperviscosity, which requires discretizing fourth or even higher-order derivatives using wide stencils. These extended stencils increase both the memory footprint and computational cost, particularly in multi-dimensional simulations or when using implicit time-stepping methods. Filters, on the other hand, can be applied using compact kernels (e.g., 3- or 5-point symmetric stencils), and when implemented in spectral space via FFTs, they are computationally efficient and highly accurate. Thus, for applications where computational cost is critical and phase preservation is not paramount, filtering may offer a favorable trade-off compared to hyperviscosity.

Lessons Learned

☞ Hyperviscosity introduces artificial fourth-order dissipation that selectively targets high-wavenumber components, improving stability without significantly affecting the large-scale structure of the solution. Its effect is minimal on smooth modes but critical for suppressing grid-scale instabilities.

☞ A properly tuned hyperviscosity coefficient, scaled with the diffusion number, allows stable simulations even when the time step exceeds the classical stability limit. This expands the efficiency of explicit schemes in stiff problems.

☞ Filtering enhances the stability of explicit schemes by acting as a low-pass operator that damps non-physical high-frequency modes while preserving essential large-scale features. When carefully applied with high-order or spectral methods, it improves robustness without sacrificing accuracy, though excessive filtering can suppress meaningful dynamics.

PROBLEMS

9P-1 Solve the modified diffusion equation with hyperviscosity:

$$\frac{\partial u}{\partial t} = \nu \frac{\partial^2 u}{\partial x^2} - \varepsilon \frac{\partial^4 u}{\partial x^4}, \quad \varepsilon = 0.01,$$

using FD-O4 for space and explicit Euler for time. Use $u(x,0) = \sin(2\pi x)$ on $x \in [0,1]$ with Dirichlet BCs $u(0,t) = u(1,t) = 0$. The parameters ν and ε are dimensionally L^2/T and L^4/T, respectively.

9P-2 For the PDE:

$$\frac{\partial u}{\partial t} = \nu \frac{\partial^2 u}{\partial x^2} - \varepsilon \frac{\partial^4 u}{\partial x^4},$$

i Discretize with FD-O2 and explicit Euler (equidistant mesh).

ii Derive the truncation error.

iii Perform von Neumann stability analysis and derive the stability limit for ε.

9P-3 Consider the nonlinear Boussinesq equation given by:

$$\frac{\partial^2 u}{\partial t^2} - \frac{\partial^2 u}{\partial x^2} = \frac{\partial^2}{\partial x^2}\left(\frac{u^2}{2}\right) + \varepsilon \frac{\partial^4 u}{\partial x^4},$$

with the initial condition:

$$u(x,0) = 0.072\,\text{sech}^2\left[1.073(x-10)\right] + 0.048\,\text{sech}^2\left[0.876(x-20)\right],$$

and vanishing boundary conditions at infinity.

a) Solve the equation numerically using explicit Euler time integration and fourth-order finite difference (FD-O4) schemes for spatial derivatives. Apply a mild smoothing filter at each time step to stabilize the simulation. Observe and describe the effect of the dispersion term $\varepsilon\partial^4 u/\partial x^4$ on the evolution of the solitary wave pulses.

b) Repeat the simulation after removing the dissipative term $\varepsilon\partial^4 u/\partial x^4$, keeping all other numerical methods unchanged. Comment on how the wave behavior differs in the absence of dispersion-related damping. What role does the smoothing filter now play in stabilizing the numerical solution?

In both cases, compare the persistence, amplitude, and possible interactions of the solitary waves, and discuss the impact of the dissipation and filtering on the long-time accuracy of the numerical method.

9P-4 In the context of stabilizing explicit finite-difference schemes for parabolic equations, both hyperviscosity and Gaussian filtering are employed to suppress high-frequency numerical instabilities through artificial damping. Hyperviscosity introduces targeted dissipation via a fourth-order spatial derivative, with a spectral damping effect that grows rapidly with wavenumber and peaks at the Nyquist frequency. In contrast, Gaussian filtering acts as a smooth low-pass filter that attenuates a broader band of frequencies in a more gradual manner. Analyze the spectral damping profiles of both methods and argue, with reference to frequency response and numerical behavior, which technique offers more robust stabilization of the scheme and under what conditions.

REFERENCES

K. A. Hoffmann, S. T. Chiang, Computational Fluid Dynamics, Vol. 1, Engineering Education System, Wichita, Kansas, 1998.

D. W. Peaceman, H. H. Rachford, The Numerical Solution of Parabolic and Elliptic Differential Equations, Journal of the Society for Industrial and Applied Mathematics, Vol. 3, No. 1, pp. 28-41, 1955.

J. Von Neumann, R. D. Richtmyer, A method for the numerical calculation of hydrodynamic shocks. Journal of Applied Physics, 21(3), 232-237, 1950.

A. Jameson, W. Schmidt, E. Turkel, Numerical Solution of the Euler Equations by Finite Volume Methods Using Runge-Kutta Time-Stepping Schemes. AIAA Paper 81-1259, 1981.

J. Y. K. Cho, L. M. Polvani, The emergence of jets and vortices in freely evolving, shallow-water turbulence on a sphere. Physics of Fluids, 8(6), 1531-1552, 1996.

A. W. Cook, W. H. Cabot, Hyperviscosity for shock-turbulence interactions. Journal of Computational Physics, 203(2), 379-385. 2005.

A. G. Lamorgese, D. A. Caughey, S. B. Pope, Direct numerical simulation of homogeneous turbulence with hyperviscosity. Physics of Fluids, 17(1), 015106, 2005.

J. L. Guermond, S. Prudhomme, Mathematical analysis of a spectral hyperviscosity LES model for turbulent flows. M2AN (ESAIM), 37(6), 893-908, 2003.

10 2D Unsteady Parabolic Problems

In this chapter, we will explore the concept of 2D unsteady parabolic flow with a focus on numerical solutions of the heat equation using explicit and implicit schemes. We introduce the Gaussian pulse as an initial condition and study the evolution of temperature fields under time-dependent boundary conditions. We apply FDO2, ADI, and SOR schemes to understand stability, efficiency, and convergence behavior.

- Understand the formulation and behavior of 2D parabolic diffusion problems.
- Learn about FDO2, ADI, and SOR schemes for time-dependent 2D heat equations.
- Develop stable numerical solvers using explicit and implicit time integration.
- Be familiar with handling Gaussian pulse initial conditions and time-varying boundaries.
- Analyze the convergence and stability characteristics of numerical schemes.

The study of two-dimensional unsteady parabolic flows is critical in understanding a variety of time-dependent transport phenomena. These problems commonly arise in heat transfer, diffusion of solutes in porous media, and unsteady viscous flow fields. The 2D heat equation, a representative form of parabolic PDEs, characterizes how thermal energy diffuses within a material over time. Unlike its 1D counterpart, the 2D problem introduces increased complexity due to the interaction of diffusion processes in both spatial directions. Consequently, numerical approaches must handle the coupling between dimensions efficiently while maintaining stability and accuracy.

In this chapter, we begin by discretizing the 2D heat equation using second-order central difference schemes for the spatial derivatives and forward Euler for time. The limitations of explicit methods, such as restrictive time step constraints for stability, lead us to explore implicit alternatives. The Alternating Direction Implicit (ADI) method is introduced to address these challenges. ADI provides an efficient solution strategy by decoupling the 2D system into successive 1D implicit steps. Each sub-step involves solving a tridiagonal system, thereby significantly reducing computational cost compared to solving a fully coupled system. The stability properties of ADI make it well suited for stiff problems and long-time integration. To complement ADI, the Successive Over-Relaxation (SOR) method is also discussed. SOR accelerates convergence of iterative solvers by introducing a relaxation factor, balancing between past and updated values. Through systematic derivation, we examine

DOI: 10.1201/9781003663171-10

how the SOR method is constructed from the fully implicit discretization of the heat equation. The convergence history and numerical performance of ADI and SOR are then compared to assess their efficiency and stability in solving large systems.

We will solve a 2D unsteady parabolic equation with Gaussian pulse initial conditions, using FDO2 method, which involves discretizing the space. The heat equation in two spatial dimensions is given by:

$$\frac{\partial u}{\partial t} = \alpha \left(\frac{\partial^2 u}{\partial x^2} + \frac{\partial^2 u}{\partial y^2} \right)$$

where $u(x,y,t)$ is the temperature at position (x,y) and time t, and α is the thermal diffusivity of the material.

We'll use a Gaussian pulse as the initial condition:

$$u(x,y,0) = A \exp\left(-\frac{(x-x_0)^2 + (y-y_0)^2}{2\sigma^2} \right)$$

where A is the amplitude of the pulse, (x_0, y_0) is the center of the pulse, and σ is the standard deviation, which controls the spread of the pulse.

Let's say the boundary conditions are defined such that the temperature on the boundary increases linearly over time:

$$u(x,y,t) = t \quad \text{on all boundaries}$$

For the spatial part, the second derivatives in both the x and y directions are approximated using the central difference formula:

$$\frac{\partial^2 u}{\partial x^2} \approx \frac{u_{i+1,j} - 2u_{i,j} + u_{i-1,j}}{dx^2}$$

$$\frac{\partial^2 u}{\partial y^2} \approx \frac{u_{i,j+1} - 2u_{i,j} + u_{i,j-1}}{dy^2}$$

For the time derivative, the forward difference method is used:

$$\frac{\partial u}{\partial t} \approx \frac{u_{i,j}^{n+1} - u_{i,j}^{n}}{dt}$$

By substituting the spatial approximations into the heat equation and rearranging for $u_{i,j}^{n+1}$, the explicit formula for the temperature at the next time step $(n+1)$ at each grid point (i,j) in terms of the temperatures at the current time step (n) is:

$$u_{i,j}^{n+1} = u_{i,j}^{n} + \alpha \frac{dt}{dx^2}(u_{i+1,j}^{n} - 2u_{i,j}^{n} + u_{i-1,j}^{n}) + \alpha \frac{dt}{dy^2}(u_{i,j+1}^{n} - 2u_{i,j}^{n} + u_{i,j-1}^{n})$$

Stability Constraints:

$$\sigma_{\max} = \frac{S_r}{k_{\text{mod,max}}^2}$$

$$\Delta t_{diff} = \frac{1}{2} \frac{\sigma_{\max} \Delta x^2}{\alpha}$$

where, Δt_{diff} is diffusion related timestep, S_r=1 stability limit on real axis for for Euler Explicit scheme, $k^2_{\text{mod,max}}$ = 4 for FDO2, and σ_{max} is the maximum diffusion number. The factor 1/2 is used to account for the two-dimensional domain.

10.1 ALTERNATING DIRECTION IMPLICIT METHOD

The Alternating Direction Implicit (ADI) method is a numerical technique commonly used for solving partial differential equations (PDEs), particularly those that are parabolic or elliptic in nature. It is well suited for problems involving multidimensional PDEs, such as those arising in heat conduction, fluid flow, and diffusion processes.

The ADI method works by breaking a multi-dimensional problem into a series of simpler one-dimensional problems, solving each of these in an alternating sequence. In each time step or iteration, the method first solves for one direction (e.g., the x-direction) while treating the other direction (e.g., the y-direction) implicitly and keeping it fixed. In the next step, the roles are reversed. This splitting of directions allows for the efficient solving of the equations without needing to deal with a fully coupled system at each step.

We solve the two-dimensional diffusion equation:

$$\frac{\partial u}{\partial t} = \alpha \left(\frac{\partial^2 u}{\partial x^2} + \frac{\partial^2 u}{\partial y^2} \right)$$

where, $u = u(x,y,t)$ represents the temperature or concentration at position (x,y), t is time, and α is the diffusion coefficient.

The boundary conditions (BCs) used are Dirichlet boundary conditions where the boundaries are implicitly assumed to have fixed values. The values along the boundaries of the grid are fixed, and the internal points are updated during the iterations. The initial condition is a sine wave, which represents a non-zero, smooth distribution to start the diffusion process.

For the ADI (Alternating Direction Implicit) method, each time step is divided into two half-steps, alternating between treating the x and y directions implicitly. This ensures computational efficiency and stability. In the first half-step, the x-direction is solved implicitly, while the y-direction remains explicit. This means that for each row, a tridiagonal system of equations is solved in the x-direction:

$$\begin{aligned}
\frac{u_{i,j}^{(n+1/2)} - u_{i,j}^{(n)}}{\Delta t/2} &= \alpha \left(\frac{u_{i+1,j}^{(n+1/2)} - 2u_{i,j}^{(n+1/2)} + u_{i-1,j}^{(n+1/2)}}{h_x^2} \right) \\
&\quad + \alpha \left(\frac{u_{i,j+1}^{(n)} - 2u_{i,j}^{(n)} + u_{i,j-1}^{(n)}}{h_y^2} \right)
\end{aligned}$$

In the second half-step, the intermediate value $u_{i,j}^{(n+1/2)}$ is used to solve in the y-direction:

$$\frac{u_{i,j}^{(n+1)} - u_{i,j}^{(n+1/2)}}{\Delta t/2} = \alpha \left(\frac{u_{i+1,j}^{(n+1/2)} - 2u_{i,j}^{(n+1/2)} + u_{i-1,j}^{(n+1/2)}}{h_x^2} \right) + \alpha \left(\frac{u_{i,j+1}^{(n+1)} - 2u_{i,j}^{(n+1)} + u_{i,j-1}^{(n+1)}}{h_y^2} \right)$$

This alternation between the directions helps make the computation stable while reducing the effort of directly inverting a large matrix for both directions simultaneously.

The ADI method splits each full time step Δt into two equal half-steps ($\Delta t/2$) to decouple the spatial dimensions. In the first half-step, the x-direction is treated implicitly (solved via a tridiagonal system), while the y-direction terms are handled explicitly. The roles reverse in the second half-step: the y-direction becomes implicit, and the x-direction terms are updated explicitly. This alternating strategy avoids solving a fully coupled 2D system, reducing computational complexity while maintaining stability. By dividing the time step, ADI effectively balances accuracy and efficiency, as each half-step locally enforces stability through implicit discretization in one dimension, even for larger Δt. The splitting aligns with the method's core philosophy of dimensionally decoupled updates, as reflected in the half-step equations for $u^{(n+1/2)}$ and $u^{(n+1)}$.

10.2 SUCCESSIVE OVER-RELAXATION METHOD (SOR)

The Finite Difference Method (FDM) for the Successive Over-Relaxation (SOR) method is derived from the discretization of the diffusion equation. Consider the 2D steady-state diffusion equation (or heat equation in its implicit form):

$$\frac{\partial u}{\partial t} = \alpha \left(\frac{\partial^2 u}{\partial x^2} + \frac{\partial^2 u}{\partial y^2} \right)$$

The diffusion equation is discretized using central differences for the spatial derivatives and backward Euler for the time derivative. Let $u_{i,j}^n$ represent the solution at the grid point (i,j) at the n-th time step, and $u_{i,j}^{n+1}$ represent the updated solution at the next time step. The discretized equation becomes:

$$\frac{u_{i,j}^{n+1} - u_{i,j}^n}{\Delta t} = \alpha \left(\frac{u_{i+1,j}^{n+1} - 2u_{i,j}^{n+1} + u_{i-1,j}^{n+1}}{\Delta x^2} + \frac{u_{i,j+1}^{n+1} - 2u_{i,j}^{n+1} + u_{i,j-1}^{n+1}}{\Delta y^2} \right)$$

Rearranging for $u_{i,j}^{n+1}$:

$$u_{i,j}^{n+1} = \frac{\alpha \Delta t}{\Delta x^2} \left(u_{i+1,j}^{n+1} + u_{i-1,j}^{n+1} \right) + \frac{\alpha \Delta t}{\Delta y^2} \left(u_{i,j+1}^{n+1} + u_{i,j-1}^{n+1} \right) + u_{i,j}^n \cdot \left(1 - 2\frac{\alpha \Delta t}{\Delta x^2} - 2\frac{\alpha \Delta t}{\Delta y^2} \right)$$

Table 10.1
Comparison between ADI and SOR methods

ADI Pseudocode	SOR Pseudocode
Initialize u with the sine-based initial condition. Enforce boundary conditions (set domain edges to zero). For each time step: 1. Copy current u into a temporary array. 2. Perform the first half-step: solve in x-direction using a tridiagonal solver and treat y terms explicitly. 3. Perform the second half-step: solve in y-direction using a tridiagonal solver and treat x terms explicitly. 4. Re-apply boundary conditions at domain edges. 5. Compute error as the difference between new u and the temporary array for convergence tracking.	Initialize u with the sine-based initial condition. Enforce boundary conditions (set domain edges to zero). For each time step: 1. Copy current u into a temporary array named u_{prev}. 2. For a chosen number of iterations: (a) Create a copy of u for convergence checks. (b) For each interior cell, update u using the SOR relaxation formula that depends on r_x, r_y, and ω. (c) Re-apply boundary conditions at domain edges. (d) Compute error as the difference between U and the copy made at step (a). If below tolerance, stop.

Define $r_x = \alpha\Delta t/\Delta x^2$, and $r_y = \alpha\Delta t/\Delta y^2$. The equation simplifies to

$$\begin{aligned} u_{i,j}^{n+1} &= \frac{r_x}{1+2(r_x+r_y)}\left(u_{i+1,j}^{n+1}+u_{i-1,j}^{n+1}\right)+\frac{r_y}{1+2(r_x+r_y)}\left(u_{i,j+1}^{n+1}+u_{i,j-1}^{n+1}\right) \\ &\quad +\frac{1}{1+2(r_x+r_y)}u_{i,j}^{n} \end{aligned}$$

The SOR update rule introduces the relaxation factor ω, which combines the previous approximation $U_{i,j}^{\text{old}}$ and the new value computed from the FDM equation:

$$\begin{aligned} u_{i,j}^{\text{new}} &= (1-\omega)u_{i,j}^{\text{old}}+\frac{\omega}{1+2(r_x+r_y)}\Big[r_x\left(u_{i+1,j}^{\text{new}}+u_{i-1,j}^{\text{new}}\right)+r_y\Big(u_{i,j+1}^{\text{new}} \\ &\quad +u_{i,j-1}^{\text{new}}\Big)+u_{i,j}^{n}\Big] \end{aligned}$$

In this equation:

1. The first term, $(1-\omega)u_{i,j}^{\text{old}}$, retains a portion of the old value for stability.
2. The second term computes a weighted sum of neighboring grid points and the contribution from the previous time step.

This equation is used iteratively for each interior grid point until the solution converges, defined by the RMS error falling below a specified tolerance. Boundary values are enforced explicitly after each iteration.

Table 10.1 compares the computational frameworks of the Alternating Direction Implicit (ADI) and Successive Over-Relaxation (SOR) methods for solving the two-dimensional diffusion equation. The ADI method splits each time step Δt into two successive half-steps, alternating between implicit solves in the x- and y-directions while treating the other dimension explicitly. This decomposition enables efficient updates via tridiagonal solvers and ensures numerical stability, particularly for stiff problems. In contrast, the SOR method performs multiple point-wise iterative updates within a single Δt, guided by neighboring values and a relaxation factor ω. Whereas ADI assesses convergence once per time step, SOR evaluates residuals at each iteration, allowing adaptive early termination based on a specified tolerance.

The methods differ in their computational trade-offs. ADI's directional splitting avoids solving fully coupled systems, offering robustness for larger time steps; SOR, on the other hand, can achieve faster convergence through optimal tuning of ω, though it is more sensitive to parameter choices. These distinctions underscore the respective strengths of each approach—ADI for stable, large-scale simulations and SOR for problems amenable to fine-tuned iterative strategies. The table highlights how algorithmic design—dimensionally split versus iteratively relaxed—shapes performance in the numerical solution of parabolic PDEs.

Example 9.2

We will solve a two-dimensional diffusion equation with Dirichlet boundary conditions. The domain is defined by Lx = 1.0 and Ly = 1.0, with Nx = 50 and Ny = 50 points distributed along each axis. The grid spacing is determined by hx and hy, computed as Lx / (Nx - 1) and Ly / (Ny - 1). The time step dt is set to 0.0005, and the diffusion coefficient alpha is chosen to be 0.1. Two numerical methods, ADI and SOR, are employed. ADI uses a split approach, solving implicit equations along each dimension in consecutive half-steps. SOR applies iterative relaxation updates per time step, guided by a relaxation factor omega equal to 1.5. The code enforces zero Dirichlet boundary conditions on all sides by explicitly setting values to zero at the edges.
The initial condition is

$$u(x,y,0) = \sin(\pi x)\sin(\pi y)$$

which provides a nontrivial sinusoidal profile at t=0. At each time step, the solution arrays are updated by the ADI and SOR methods. ADI executes two implicit sweeps per step. The SOR method relies on iterative relaxation factor $\omega = 1.5$.

Solution

The RMS error measures how much the solution changes between time steps, indicating the rate at which the numerical method converges to the true solution. While the ADI method takes larger, stable time steps due to its implicit nature, the per-step changes in the solution can be small, leading to flatter convergence curves in residual plots. This may give the impression of slow convergence compared to the SOR method, which performs multiple iterations per time step to drive the residual down. Since ADI solves tridiagonal systems in each spatial direction separately, it maintains strong stability properties, especially for stiff problems. However, this stability comes at the cost of smaller solution updates per step, emphasizing robustness over rapid residual reduction.

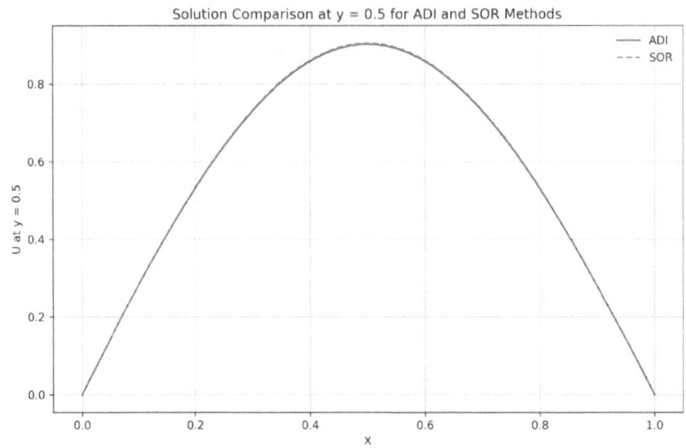

Figure 10.1: **Solution of 2D diffusion equation using ADI and SOR method.**

Case Study 10A: 2D Unsteady Heat Conduction with Explicit Schemes

Problem Definition

This case study addresses the two-dimensional unsteady heat conduction problem within a square domain of size 10×10. The case involves transient thermal diffusion originating from an initial Gaussian pulse centered at the domain's center. Thermal diffusivity is defined as $\alpha = 0.1$, and the temperature field evolves under a combination of spatially non-uniform Dirichlet boundary conditions. Specifically, constant values of 100°C are imposed on the top, bottom, and left boundaries, while the right boundary is held at 200°C. The initial temperature distribution is given by a Gaussian pulse with amplitude 150 and standard deviation 1.0, centered at (5, 5). The simulation goal is to track the temporal evolution of the temperature field up to $t = 9.96$ s.

Governing Equations

The governing equation is the unsteady two-dimensional heat conduction equation:

$$\frac{\partial T}{\partial t} = \alpha \left(\frac{\partial^2 T}{\partial x^2} + \frac{\partial^2 T}{\partial y^2} \right)$$

Here, $T(x,y,t)$ denotes the temperature field, and α is the constant thermal diffusivity. The simulation assumes a purely diffusive process, hence no convective terms are involved. The medium is isotropic and homogeneous, and the heat transfer is unsteady with no source term. The problem is solved under the assumption of constant spatial resolution and explicit time advancement.

Numerical Setup

The numerical implementation uses the Finite Difference Method (FDM) with a second-order central difference (FDO2) scheme for spatial discretization and the explicit Euler method for time integration. A uniform structured Cartesian grid is adopted with spacing $\Delta x = \Delta y = 0.1$, leading to a 101×101 node grid. The time step Δt is determined by the stability condition for the explicit scheme, with a diffusion number constraint $\sigma_{\text{max}} = 0.4$. The calculated value of Δt ensures both numerical stability and accuracy for the integration over time. Boundary conditions are applied as follows: all edges except the right are maintained at 100°C, while the right edge is set at 200°C throughout the simulation. The initial Gaussian pulse introduces a localized temperature gradient, initiating the diffusion process.

Results and Discussion

The simulation results illustrate the evolution of the Gaussian pulse under the effect of heat diffusion and boundary influence. At the beginning, the temperature is sharply peaked around the center of the domain. As time progresses, the pulse spreads out symmetrically while interacting with the fixed boundary temperatures. The most significant distortion is observed near the right boundary where the higher temperature (200°C) influences the gradient profile, producing an asymmetric field. The central high-temperature core gradually diminishes, with the domain tending toward a smooth, steady-state temperature distribution that reflects the imposed boundary values. The 3D surface plots clearly reveal this transition from localized heat to a more uniform temperature field. A horizontal slice at $y = 6$ demonstrates how the peak flattens and the profile shifts toward the right wall temperature, validating the diffusive transport.

Numerical Challenges

The explicit Euler method coupled with second-order spatial discretization proves effective for short-time simulations of diffusive problems, provided that the time step is chosen carefully to satisfy stability limits. The anisotropic boundary condition (with one wall hotter than the others) adds complexity to the symmetry of the solution, emphasizing the importance of physical boundary realism in diffusion problems. A critical numerical challenge is maintaining stability while resolving steep gradients from the initial Gaussian pulse. The choice of grid spacing and time step greatly influences the convergence and accuracy of results. Interested reader may further explore the improvements with adaptive time stepping, implicit schemes for time, or extension to variable thermal diffusivity cases for more realistic materials. A movie `animation-2DheatConduction` is provided to see the evolution of the temperature over time.

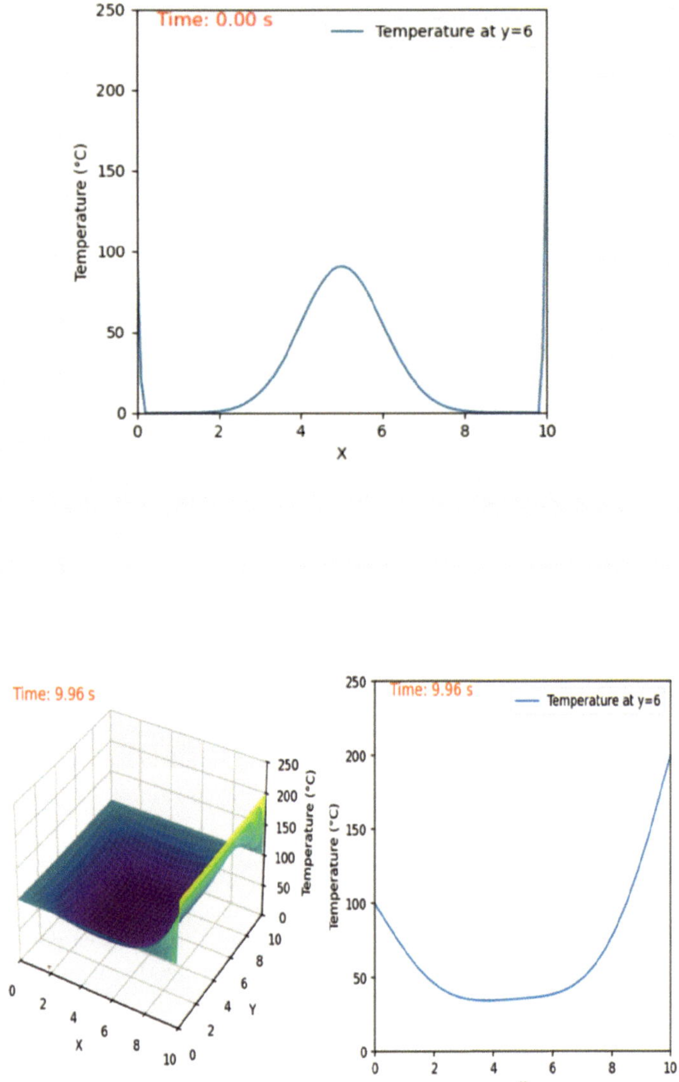

Figure 10.2: **Evolution of the temperature field in a 2D square domain from a central Gaussian pulse. Top: initial condition. Bottom: final distribution at $t = 9.96$ s.**

PROBLEMS

10P-1 Consider the two-dimensional unsteady heat conduction equation on a unit square domain $0 \leq x, y \leq 1$, subject to Dirichlet boundary conditions. The top boundary at $y = 1$ is maintained at a temperature of 1, while all other edges are held at zero temperature. The domain is discretized using a uniform grid, and the Alternating Direction Implicit (ADI) method is applied to compute the transient temperature field until a steady-state solution is reached. The ADI method alternates between solving tridiagonal systems along the x-direction (for fixed y) and the y-direction (for fixed x) within each time step. This approach enhances numerical stability while allowing efficient computation for parabolic PDEs.

1. Implement the ADI method using second-order central differences in space and an implicit discretization in time with alternating sweeps.
2. Use a stable time step proportional to the square of the grid spacing, and iterate until the solution converges to steady state within a specified tolerance.
3. Impose Dirichlet boundary conditions: $T(x, 1) = 1$ (top wall), and $T = 0$ on all other boundaries.
4. Visualize the resulting steady-state temperature distribution as a 3D surface plot using Plotly or equivalent visualization tools.
5. Discuss the convergence behavior and explain how the ADI method improves numerical efficiency and stability compared to a fully explicit scheme.

10P-2 Derive the stability condition for ADI method.

10P-3 Discuss the role of the relaxation parameter ω in the SOR method. How does the choice of ω affect convergence speed? What happens when $\omega < 1$, $\omega = 1$, and $\omega > 1$? Derive the optimal ω estimate for a rectangular domain based on spectral radius.

10P-4 Solve the 2D Laplace equation

$$\frac{\partial^2 T}{\partial x^2} + \frac{\partial^2 T}{\partial y^2} = 0$$

over a rectangular domain $[H = 1, L = 2]$ with the following boundary conditions:

At the top edge ($y = H$): $T = 0$

At the bottom edge ($y = 0$): $T = 200$

At the left edge ($x = 0$): $T = 0$

At the right edge ($x = L$): $T = 0$

1. Discretize using a 5-point stencil; handle Dirichlet and Neumann BCs.

2. Solve using SOR with $\omega = 1.75$; stop when

$$\frac{\|T^{(k+1)} - T^{(k)}\|_2}{\|T^{(k)}\|_2} < 10^{-6}$$

3. Plot $T(x,y)$; discuss boundary influence.
4. Comment on the role of ω in convergence.

Compare numerical result with an analytical solution:

$$T(x,y) = 400 \cdot \sum_{n=1}^{\infty} \frac{1-(-1)^n}{n\pi} \sin\left(\frac{n\pi x}{L}\right) \cdot \frac{\sinh\left(\frac{n\pi(H-y)}{L}\right)}{\sinh\left(\frac{n\pi H}{L}\right)}$$

REFERENCES

K. A. Hoffmann, S. T. Chiang, Computational Fluid Dynamics, Vol. 1, Engineering Education System, Wichita, Kansas, 1998.

D. W. Peaceman, H. H. Rachford, The Numerical Solution of Parabolic and Elliptic Differential Equations, Journal of the Society for Industrial and Applied Mathematics, Vol. 3, No. 1, pp. 28-41, 1955.

P. Laasonen, Über eine Methode zur Lösung der Wärmeleitungsgleichung, Acta Math. 81 pp. 309 - 317, 1949.

11 Linear Advection-Diffusion Equation

This chapter explores the concept of linear advection-diffusion (LAD) and its implications for numerical modeling. The advection-diffusion equation governs a wide range of transport phenomena in physics and engineering, describing how scalar quantities, such as temperature, concentration, or velocity, are transported by fluid motion (advection) and simultaneously dispersed through molecular or turbulent mixing. We analyze the structure of the governing equation, discretization strategies, and amplification factors using von Neumann analysis. Emphasis is placed on cell Reynolds number, and stability criteria. After finishing this chapter, you should be able to:

- Understand the role of advection and diffusion in scalar transport phenomena.
- Learn about numerical discretization schemes and stability criteria.
- Develop amplification factor expressions via von Neumann stability analysis.
- Be familiar with cell Reynolds number and its impact on solution stability.
- Analyze the response function and eigenvalue distribution in advection-diffusion systems.

In many natural and engineered systems, transport mechanisms are driven by both advection (bulk motion) and diffusion (molecular or turbulent mixing). The advection-diffusion equation captures these two effects and is central to modeling scalar field evolution in fluids, heat conduction with flow, pollutant dispersion, and other processes. This chapter introduces the one-dimensional form of the equation and builds toward its higher-dimensional analogs. We begin by formulating the linear advection-diffusion equation in 1D and analyze its behavior through finite difference approximations. Both central difference schemes and forward Euler methods are introduced to discretize space and time. The combined scheme is then subjected to von Neumann stability analysis, leading to a general expression for the amplification factor. This factor, expressed in terms of the Courant number (CFL) and diffusion number (σ), reveals how different modes propagate and dissipate over time. Key insight is gained by evaluating the amplification factor for special cases: pure advection, pure diffusion, and their combination. These analyses allow us to interpret how advection introduces phase shifts, while diffusion smooths gradients and damps high-frequency components.

In many physical systems, transport phenomena are governed by a combination of advection and diffusion processes. The advection-diffusion equation provides a

DOI: 10.1201/9781003663171-11

mathematical model to describe the evolution of a scalar quantity, such as heat, concentration, or velocity, as it is carried by a flow (advection) and spreads out over time (diffusion). Understanding how these two competing mechanisms interact is crucial for accurately predicting the behavior of such systems.

We shall now consider the 1-D advection-diffusion equation in the following form:

$$\frac{\partial f}{\partial t} + U\frac{\partial f}{\partial x} = \mathscr{D}\frac{\partial^2 f}{\partial x^2}$$

This equation combines the effects of advection (transport by the velocity U) and diffusion (spreading, controlled by the diffusion coefficient $\mathscr{D}$).

The finite difference form is derived from discretizing both time and space. For the time derivative and spatial derivatives, a combination of explicit schemes for advection and diffusion can be used.

1. Advection Term: Discretized using central difference scheme:

$$U\frac{\partial f}{\partial x} \approx U\frac{f_{j+1}^n - f_{j-1}^n}{2h}$$

2. Diffusion Term: Discretized using central difference:

$$\mathscr{D}\frac{\partial^2 f}{\partial x^2} \approx \mathscr{D}\frac{f_{j+1}^n - 2f_j^n + f_{j-1}^n}{h^2}$$

Combining these into the finite-difference form:

$$\frac{f_j^{n+1} - f_j^n}{\Delta t} + U\frac{f_{j+1}^n - f_{j-1}^n}{2h} = \mathscr{D}\frac{f_{j+1}^n - 2f_j^n + f_{j-1}^n}{h^2}$$

This is an explicit scheme, and we can solve for f_j^{n+1} directly given the values at the previous time step n. The CFL condition, or stability condition, will depend on the relative sizes of U, $\mathscr{D}$, Δt, and h.

Amplification factor: To derive the amplification factor A for the 1-D advection-diffusion equation, we will assume the finite-difference form of the equation and apply the von Neumann stability analysis. The governing finite-difference equation is:

$$\frac{f_j^{n+1} - f_j^n}{\Delta t} + U\frac{f_{j+1}^n - f_{j-1}^n}{2\Delta x} = \mathscr{D}\frac{f_{j+1}^n - 2f_j^n + f_{j-1}^n}{\Delta x^2}$$

Where:

Δt is the time step size.
Δx is the spatial grid size.
U is the velocity of advection.
$\mathscr{D}$ is the diffusion coefficient.

To perform the von Neumann stability analysis, we assume a solution of the form:

$$f_j^n = A^n e^{ikj\Delta x}$$

Where:

A is the amplification factor to be determined.
k is the wave number.
i is the imaginary unit.

We substitute this form into the finite-difference equation for each term:
- For f_j^{n+1}:

$$f_j^{n+1} = A^{n+1} e^{ikj\Delta x} = A f_j^n$$

- For f_{j+1}^n and f_{j-1}^n:

$$f_{j+1}^n = A^n e^{ik(j+1)\Delta x} = A^n e^{ikj\Delta x} e^{ik\Delta x}$$

$$f_{j-1}^n = A^n e^{ik(j-1)\Delta x} = A^n e^{ikj\Delta x} e^{-ik\Delta x}$$

Substitute the terms into the advection-diffusion equation:

$$\frac{A f_j^n - f_j^n}{\Delta t} + U \frac{f_{j+1}^n - f_{j-1}^n}{2\Delta x} = \mathscr{D} \frac{f_{j+1}^n - 2 f_j^n + f_{j-1}^n}{\Delta x^2}$$

Expanding each term:

$$\frac{A-1}{\Delta t} e^{ikj\Delta x} + U \frac{e^{ikj\Delta x} e^{ik\Delta x} - e^{ikj\Delta x} e^{-ik\Delta x}}{2\Delta x} = \mathscr{D} \frac{e^{ikj\Delta x} e^{ik\Delta x} - 2e^{ikj\Delta x} + e^{ikj\Delta x} e^{-ik\Delta x}}{\Delta x^2}$$

Simplify the terms:
- The advection term becomes:

$$U \frac{e^{ik\Delta x} - e^{-ik\Delta x}}{2\Delta x} = U \frac{2i\sin(k\Delta x)}{2\Delta x} = \frac{Ui\sin(k\Delta x)}{\Delta x}$$

- The diffusion term becomes:

$$D \frac{e^{ik\Delta x} - 2 + e^{-ik\Delta x}}{\Delta x^2} = \mathscr{D} \frac{2\cos(k\Delta x) - 2}{\Delta x^2} = -\mathscr{D} \frac{-2(1-\cos(k\Delta x))}{\Delta x^2}$$

Substitute back into the equation:

$$\frac{A-1}{\Delta t} + \frac{Ui\sin(k\Delta x)}{\Delta x} = \mathscr{D} \frac{-2(1-\cos(k\Delta x))}{\Delta x^2}$$

Multiply through by Δt to isolate A:

$$A - 1 = -\Delta t \left(\frac{Ui\sin(k\Delta x)}{\Delta x} + \mathscr{D} \frac{2(1-\cos(k\Delta x))}{\Delta x^2} \right)$$

Solving for A:

$$A = 1 - \Delta t \left(\frac{Ui\sin(k\Delta x)}{\Delta x} + \mathscr{D} \frac{2(1-\cos(k\Delta x))}{\Delta x^2} \right)$$

Expressed in terms of CFL and diffusion numbers:

$$A = 1 - CFL \cdot i\sin(k\Delta x) - 2\sigma(1 - \cos(k\Delta x))$$

Where:

$$CFL = \frac{U\Delta t}{\Delta x}, \quad \sigma = \frac{\mathscr{D}\Delta t}{\Delta x^2}$$

$$A = 1 - CFLi\sin(k\Delta x) - 2\sigma(1 - \cos(k\Delta x))$$

This expression contains both real and imaginary parts. The imaginary part corresponds to the wave propagation (due to advection), and the real part corresponds to the dissipation (due to diffusion).

Real part (diffusion):

$$\text{Re}(A) = 1 - 2\sigma(1 - \cos(k\Delta x))$$

Imaginary part (advection):

$$\text{Im}(A) = -CFL\sin(k\Delta x)$$

Thus, the full amplification factor A is:

$$A = [1 - 2\sigma(1 - \cos(k\Delta x))] - iCFL\sin(k\Delta x)$$

The real part $1 - 2\sigma(1 - \cos(k\Delta x))$ affects the stability related to diffusion. For stability, the real part should remain close to 1. The imaginary part $-CFL\sin(k\Delta x)$ affects the phase accuracy of the wave propagation.

The scheme satisfies von Neumann stability for sufficiently small values of CFL and σ, as the amplification factor A remains within the unit circle. However, as either parameter increases, the amplification factor deviates further from the real axis, and may eventually cross the stability boundary $|A| = 1$, indicating the onset of numerical instability.

Figure 11.1 shows the Argand diagrams of amplification factor. We explore three cases:

1. **Case** $CFL = 0.5, \sigma = 0.2$: The amplification factor forms a relatively small ellipse positioned near the right half of the complex plane, entirely within the stability zone ($|A| \leq 1$). The modest diffusion ($\sigma = 0.2$) slightly damps the amplitude of all wave modes, evident from the ellipse's shift toward values smaller than 1 on the real axis. The imaginary part, representing numerical dispersion, causes moderate phase shifts in wave modes. As highlighted by Pozrikidis (2008), an amplification factor ellipse within the unit circle indicates stable but dissipative behavior, with numerical diffusion smoothing the solution while numerical dispersion introduces phase errors.

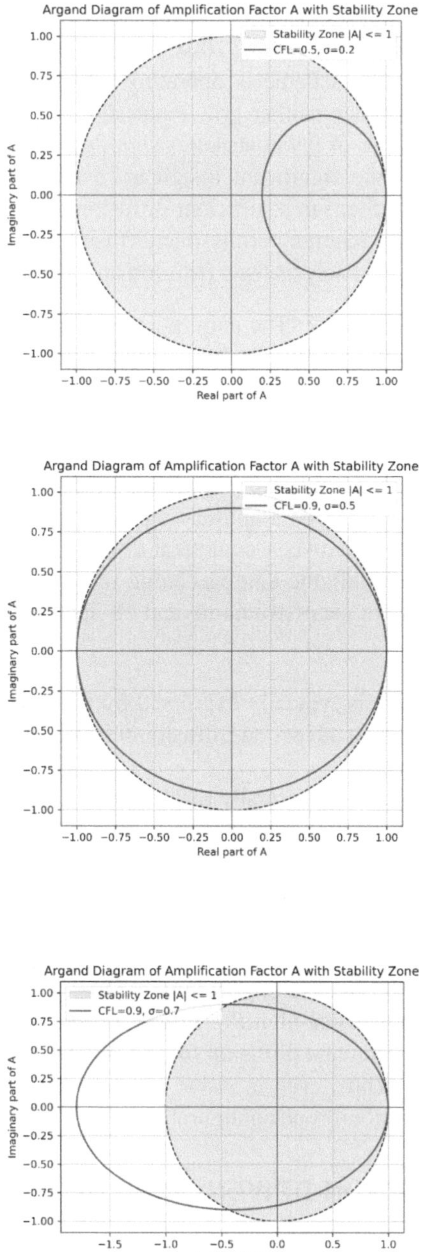

Figure 11.1: **Amplification factor for 1D Advection-Diffusion equation using central differencing scheme.**

2. **Case** $CFL = 0.9, \sigma = 0.5$: In this case, the ellipse nearly coincides with the boundary of the unit circle. This condition represents marginal stability, as noted by researchers (Pozrikidis, 2008; Morton and Mayers, 2005). Such parameters indicate the scheme is operating near its stability limit. The amplitude of wave modes neither grows nor significantly decays in each step. The Courant number of 0.9 contributes significantly to numerical dispersion, clearly visible in the significant imaginary extent of the ellipse. The relatively large $\sigma = 0.5$ ensures sufficient diffusion to maintain stability, though it approaches the maximum permissible diffusion for explicit schemes, as described in numerical analysis texts (Morton and Mayers, 2005).

3. **Case** $CFL = 0.9, \sigma = 0.7$: The amplification factor ellipse expands dramatically, crossing far beyond the stability circle, particularly extending into the negative-real and large-imaginary domains. This indicates numerical instability—modes with certain wavenumbers will amplify exponentially. The larger diffusion number ($\sigma = 0.7$), despite implying increased damping, does not necessarily guarantee stability because excessive numerical damping at higher Courant numbers shifts the amplification factor ellipse significantly, leading to complex interplay between numerical dispersion and diffusion. Pozrikidis (2008) explicitly highlights that exceeding recommended diffusion numbers leads to instability or excessive numerical dissipation that compromises accuracy.

The stability criterion is typically expressed by the conditions $CFL \leq 0.5$ and $\sigma \leq 0.5$ for practical explicit advection-diffusion schemes (Pozrikidis, 2008; Morton and Mayers, 2005)

Though the stability limit is formally reached around $\sigma = 0.5$, numerical results often show instability beginning near $\sigma = 0.7$, even when $CFL < 1$. This demonstrates that in the advection-diffusion system, the diffusion number plays a significant role in governing stability through the real part of the amplification factor.

It is important to distinguish between stability and oscillatory artifacts. A solution may be stable in the von Neumann sense (i.e., $|A| \leq 1$), yet still exhibit wiggles i.e., non-physical oscillations resulting from poor phase accuracy or insufficient numerical damping, particularly at high R_Δ. Wiggle suppression depends more on the interplay between advection and diffusion (quantified by the cell Reynolds number) than on stability alone. Thus, even in stable regimes, care must be taken to avoid under-resolved or dispersion-prone configurations that visually degrade the solution.

11.1 DISCRETIZATION-DRIVEN DIFFUSION IN ADVECTION-DIFFUSION MODELS

Numerical diffusion (also called artificial diffusion) arises when discretizing transport equations using finite difference methods. It introduces an additional term into the discrete approximation that mimics physical diffusion but originates solely from the numerical scheme. This phenomenon is particularly evident in the context of solving the 1D advection-diffusion equation :

$$\frac{\partial c}{\partial t} + u\frac{\partial c}{\partial x} - \mathscr{D}\frac{\partial^2 c}{\partial x^2} = 0$$

Let us consider the Forward Time Centered Space (FTCS) method, which uses a forward difference for time and central differences for space. When this scheme is applied to the advection-diffusion equation with $\theta = 0$, corresponding to a fully explicit scheme, the discretized form becomes:

$$\frac{c_j^{n+1} - c_j^n}{\Delta t} + u \cdot \frac{c_{j+1}^n - c_{j-1}^n}{2\Delta x} - \mathscr{D} \cdot \frac{c_{j+1}^n - 2c_j^n + c_{j-1}^n}{\Delta x^2} = 0$$

The FTCS scheme with central differencing for the advection term represents a classic explicit finite-difference formulation. It combines a forward Euler time step with a second-order central difference in space, leading to the following discretization for the linear advection equation:

$$\frac{f_j^{n+1} - f_j^n}{\Delta t} + u\frac{f_{j+1}^n - f_{j-1}^n}{2\Delta x} = 0$$

Although this scheme is first-order accurate in time and second-order in space, it is unconditionally unstable for pure advection problems. The central difference operator lacks directional bias and provides no numerical damping, allowing high-frequency (short-wavelength) errors to grow rapidly. This instability is clearly revealed through von Neumann analysis, where the amplification factor exceeds unity for certain wavenumbers, even when using arbitrarily small time steps.

In contrast, the same FTCS approach applied to the diffusion equation can be conditionally stable, provided the time step satisfies an appropriate upper bound. The key difference lies in the nature of the underlying PDE: advection requires directional accuracy to handle wave propagation, whereas diffusion inherently damps fluctuations.

Therefore, FTCS with central differencing for advection is best viewed as a pedagogical example of instability. It serves to illustrate why stabilizing mechanisms such as upwind differencing, artificial viscosity, or implicit time integration—are essential for accurate and robust simulation of convection-dominated flows.

To analyze the truncation error, we expand all terms in a Taylor series about the point (x_j, t_n). The exact values of c_{j+1}^n, c_{j-1}^n, etc., are expanded using:

$$c_{j\pm 1}^n = c \pm \Delta x\frac{\partial c}{\partial x} + \frac{\Delta x^2}{2}\frac{\partial^2 c}{\partial x^2} \pm \frac{\Delta x^3}{6}\frac{\partial^3 c}{\partial x^3} + \frac{\Delta x^4}{24}\frac{\partial^4 c}{\partial x^4} + \cdots$$

$$c_j^{n+1} = c + \Delta t\frac{\partial c}{\partial t} + \frac{\Delta t^2}{2}\frac{\partial^2 c}{\partial t^2} + \cdots$$

Substituting these into the FTCS scheme and simplifying yields:

$$\frac{\partial c}{\partial t} + u\frac{\partial c}{\partial x} - \mathscr{D}\frac{\partial^2 c}{\partial x^2} = -\frac{\Delta t}{2}\frac{\partial^2 c}{\partial t^2} - \frac{1}{6}u\Delta x^2\frac{\partial^3 c}{\partial x^3} + \frac{1}{12}\mathscr{D}\Delta x^2\frac{\partial^4 c}{\partial x^4} + \cdots$$

To interpret this error, note that the time derivative term $\frac{\partial^2 c}{\partial t^2}$ can be rewritten using the PDE itself. Differentiating the original equation with respect to time gives:

$$\frac{\partial^2 c}{\partial t^2} = \frac{\partial}{\partial t}\left(-u\frac{\partial c}{\partial x} + \mathscr{D}\frac{\partial^2 c}{\partial x^2}\right) = -u\frac{\partial^2 c}{\partial x \partial t} + \mathscr{D}\frac{\partial^3 c}{\partial x^2 \partial t}$$

Assuming constant u and $\mathscr{D}$, and replacing $\frac{\partial c}{\partial t}$ from the PDE again gives:

$$\frac{\partial^2 c}{\partial t^2} = u^2\frac{\partial^2 c}{\partial x^2} - 2u\mathscr{D}\frac{\partial^3 c}{\partial x^3} + \mathscr{D}^2\frac{\partial^4 c}{\partial x^4} \tag{12.2}$$

Substituting this back into equation

$$\frac{\partial c}{\partial t} + u\frac{\partial c}{\partial x} - \mathscr{D}\frac{\partial^2 c}{\partial x^2} = -\frac{\Delta t}{2}\frac{\partial^2 c}{\partial t^2} - \frac{1}{6}u\Delta x^2\frac{\partial^3 c}{\partial x^3} + \frac{1}{12}\mathscr{D}\Delta x^2\frac{\partial^4 c}{\partial x^4} + \cdots$$

Now focus on the first term:

$$-\frac{\Delta t}{2}\frac{\partial^2 c}{\partial t^2}$$

This is still a time derivative, but we want to interpret it using only spatial derivatives. So we substitute $\frac{\partial^2 c}{\partial t^2}$ using equation, which is derived by applying the original PDE again:

$$\frac{\partial^2 c}{\partial t^2} = u^2\frac{\partial^2 c}{\partial x^2} - 2u\mathscr{D}\frac{\partial^3 c}{\partial x^3} + \mathscr{D}^2\frac{\partial^4 c}{\partial x^4} \tag{12.2}$$

Substitute this into the error term:

$$-\frac{\Delta t}{2}\frac{\partial^2 c}{\partial t^2} = -\frac{\Delta t}{2}\left(u^2\frac{\partial^2 c}{\partial x^2} - 2u\mathscr{D}\frac{\partial^3 c}{\partial x^3} + \mathscr{D}^2\frac{\partial^4 c}{\partial x^4}\right)$$

Now distribute:

$$= -\frac{1}{2}u^2\Delta t\frac{\partial^2 c}{\partial x^2} + u\mathscr{D}\Delta t\frac{\partial^3 c}{\partial x^3} - \frac{1}{2}\mathscr{D}^2\Delta t\frac{\partial^4 c}{\partial x^4}$$

Now observe: the first term

$$-\frac{1}{2}u^2\Delta t\frac{\partial^2 c}{\partial x^2}$$

has the same mathematical form as the physical diffusion term:

$$\mathscr{D}\frac{\partial^2 c}{\partial x^2}$$

It looks like a diffusion term, but it does not come from the physics rather it comes from the discretization. So, we define the coefficient multiplying $\frac{\partial^2 c}{\partial x^2}$ in that term as the numerical diffusion coefficient, denoted:

$$D_{\text{num}} = -\frac{1}{2}u^2\Delta t$$

This numerical diffusion is not a physical phenomenon but rather a consequence of the discretization method. In this case, the numerical diffusion is negative, which has a destabilizing effect—it reduces the total effective diffusion present in the system. As a result, the natural damping of high-frequency components is weakened, potentially leading to numerical instability. Conversely, positive numerical diffusion enhances stability by artificially increasing damping, but this comes at the cost of accuracy, as it tends to smear sharp solution features. To ensure stability, the total diffusion should remain positive:

$$\mathscr{D} + D_{\text{num}} > 0 \Rightarrow \Delta t < \frac{2\mathscr{D}}{u^2}$$

This sets a time step restriction for the FTCS scheme.

For the generalized Crank-Nicolson scheme, the numerical diffusion term becomes:

$$D_{\text{num}} = (\theta - \tfrac{1}{2})u^2 \Delta t$$

This expression clearly reveals the role of θ:

i When $\theta = \frac{1}{2}$, $D_{\text{num}} = 0$: the method has no artificial diffusion.

ii When $\theta > \frac{1}{2}$, $D_{\text{num}} > 0$: stabilizing artificial diffusion is introduced.

iii When $\theta < \frac{1}{2}$, $D_{\text{num}} < 0$: the scheme may become unstable.

The parameter θ allows deliberate tuning of the numerical diffusion to balance between accuracy and stability. Numerical diffusion is a discretization artifact that may be beneficial (to damp high-frequency noise) or harmful (causing instability or unphysical behavior). Proper understanding and control are essential in designing stable and accurate numerical schemes. Note that $\theta = 0.5$ (Crank-Nicolson) introduces no artificial diffusion, while $\theta > 0.5$ adds stabilizing diffusion.

For 2D flow, the stability conditions come out to be:

$$\Delta t \leq \min\left(\frac{\Delta x^2}{4\mathscr{D}}, \frac{\Delta x}{|U|}, \frac{\Delta y}{|V|}\right)$$

This condition ensures that the modulus of the amplification factor $|A|$ does not exceed 1, which would cause the solution to grow unbounded (unstable). These conditions combine the requirements for stability due to diffusion and advection in both dimensions. Smaller Courant numbers CFL_x, CFL_y and larger diffusion numbers σ tend to stabilize the numerical scheme by keeping the amplification factor A closer to the real axis and ensuring $|A| \leq 1$. Larger Courant numbers and smaller diffusion numbers introduce more oscillations and can lead to instability, as seen in the more stretched contours with significant imaginary parts.

11.2 INFLUENCE OF CELL REYNOLDS NUMBER

One way to analyze the combined effect of advection and diffusion is through the *cell Reynolds number* R_Δ :

$$R_\Delta = \frac{\frac{U\Delta t}{\Delta x}}{\frac{\mathscr{D}\Delta t}{\Delta x^2}} = \frac{U\Delta x}{\mathscr{D}}$$

This is called the cell Reynolds number. Here, U is advection velocity, Δx is the grid cell size, $\mathscr{D}$ is the viscous diffusion coefficient.

The cell Reynolds number represents the relative importance of advection to diffusion on the discrete grid. Higher values of R_Δ correspond to advection-dominated behavior, which often leads to oscillatory solutions. The cell Reynolds number can be cast in terms of the Courant number (CFL) and diffusion (via σ):

$$R_\Delta = \frac{\text{CFL}}{\sigma}$$

The cell Reynolds number $R_\Delta = \frac{\text{CFL}}{\sigma}$ provides insight into the relative importance of advection and diffusion in a numerical scheme. When CFL dominates, indicating high advection relative to diffusion, R_Δ becomes large. Conversely, if diffusion dominates (i.e., σ is high), R_Δ becomes small. This ratio is particularly useful for assessing whether a numerical scheme is advection-dominated ($R_\Delta \gg 1$) or diffusion-dominated ($R_\Delta \ll 1$), which is critical for determining stability and accuracy in computational fluid dynamics. A large R_Δ suggests that the scheme must carefully handle sharp gradients and wave propagation, while a small R_Δ implies that diffusive effects are more significant, requiring appropriate treatment of viscous terms. Understanding this balance helps in selecting or designing numerical methods tailored to the dominant physical processes in a CFD problem.

11.3 EIGENVALUE ANALYSIS OF ADVECTION-DIFFUSION EQUATION

We analyze the eigenvalues of the one-dimensional advection-diffusion equation with periodic boundary conditions:

$$\frac{\partial u}{\partial t} + a\frac{\partial u}{\partial x} = \mathscr{D}\frac{\partial^2 u}{\partial x^2},$$

where, $u(x,t)$ is the scalar field (e.g., temperature or concentration), a is the advection velocity, D is the diffusion coefficient, $x \in [0,L]$ is the spatial domain.

Periodic boundary conditions imply:

$$u(0,t) = u(L,t).$$

The first-order derivative $\frac{\partial u}{\partial x}$ is discretized using the central difference scheme:

$$\frac{\partial u}{\partial x} \approx \frac{u_{i+1} - u_{i-1}}{2\Delta x}.$$

This is a second-order accurate scheme for spatial derivatives.

Diffusion (Second-Order Central Difference): The second-order derivative $\frac{\partial^2 u}{\partial x^2}$ is discretized as:

$$\frac{\partial^2 u}{\partial x^2} \approx \frac{u_{i+1} - 2u_i + u_{i-1}}{\Delta x^2}.$$

Combining the discretized advection and diffusion terms results in the following operator for each grid point i:

$$\text{Convection term: } \frac{a}{2\Delta x}\left(u_{i+1} - u_{i-1}\right),$$

$$\text{Diffusion term: } \frac{\mathscr{D}}{\Delta x^2}\left(u_{i+1} - 2u_i + u_{i-1}\right).$$

The coefficients of the resulting linear system are assembled into a sparse matrix (called the convection-diffusion matrix). The periodic boundary conditions are enforced by *wrapping around* the neighbors:

i For $i = 0$, $u_{-1} = u_{M-1}$,

ii For $i = M - 1$, $u_M = u_0$.

Convection-Diffusion Matrix: The matrix is constructed as:

$$A[i, (i-1) \mod M] = \frac{a}{2\Delta x} + \frac{\mathscr{D}}{\Delta x^2},$$

$$A[i, i] = -\frac{2\mathscr{D}}{\Delta x^2},$$

$$A[i, (i+1) \mod M] = -\frac{a}{2\Delta x} + \frac{\mathscr{D}}{\Delta x^2}.$$

This matrix describes the linear operator applied to the grid points.

Eigenvalue Analysis: The eigenvalues of the convection-diffusion matrix provide insights into the stability and dispersion properties of the numerical scheme.

For diffusion-only ($a = 0$), the eigenvalues lie along the negative real axis:

$$\lambda_k = -\frac{4\mathscr{D}}{\Delta x^2} \sin^2\left(\frac{k\pi}{M}\right), \quad k = 0, 1, \ldots, M-1.$$

When both advection and diffusion are present, the eigenvalues are generally complex, indicating a combination of damping (due to diffusion) and phase shift (due to advection).

1. $R_\Delta = 2.0$: The eigenvalues are well-distributed around a circular shape, which is typical for systems with balanced diffusion and advection. Indicates a relatively strong influence of diffusion smoothing the advection effect.

2. $R_\Delta = 5.0$: The eigenvalues compress toward the real axis, forming a stretched ellipse. This compression implies a dominance of advection relative to diffusion, leading to less smoothing and potentially more oscillatory behavior.

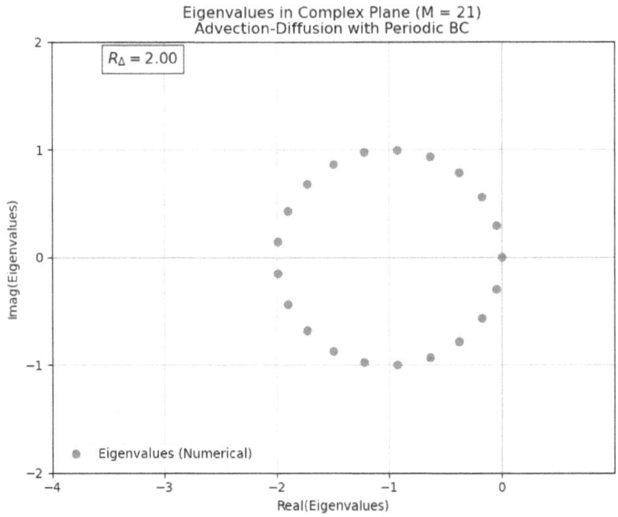

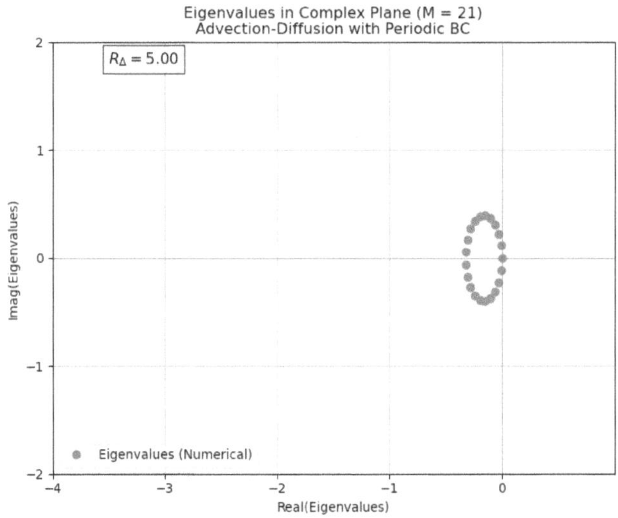

Figure 11.2: **Eigen values of advection-diffusion matrix using 3pt stencil and periodic boundary condition.**

Example 11.1

In this numerical experiment, we aim to solve the 1D advection-diffusion equation:

$$\frac{\partial f}{\partial t} + u\frac{\partial f}{\partial x} = \mathscr{D}\frac{\partial^2 f}{\partial x^2}$$

where, $f(x,t)$ is the field of interest (e.g., concentration of a substance), u is the constant advection speed, and $\mathscr{D}$ is the diffusion coefficient.
The equation is discretized using an explicit finite difference scheme on a uniform grid. The numerical simulation explores the effect of the *cell Reynolds number* (R_Δ) defined as:

$$R_\Delta = \frac{u\Delta x}{\mathscr{D}}$$

This parameter is also sometimes referred to as Peclet number.
The goal of the numerical experiment is to examine the behavior of the solution $f(x,t)$ as it evolves over time for a cell Reynolds number of $R_\Delta = 2.0$, and to investigate the interplay between advection and diffusion.
Initial and Boundary Conditions:

i The initial condition is a step-like profile for $f(x,0)$, with a sharp transition between $x = 0.25$ and $x = 0.5$.

ii The boundary conditions are $f(0,t) = f(1,t) = 0$, meaning the solution is fixed to zero at the boundaries.

iii The cell Reynolds number $R_\Delta = 2.0$ is used to ensure a balance between advection and diffusion.

iv The simulation runs for 100 time steps, and the solution is recorded at intervals of 20 time steps.

The finite difference discretization scheme used to solve the advection-diffusion equation is as follows:

$$\frac{f_j^{n+1} - f_j^n}{\Delta t} + u\frac{f_{j+1}^n - f_{j-1}^n}{2\Delta x} = \mathscr{D}\frac{f_{j+1}^n - 2f_j^n + f_{j-1}^n}{\Delta x^2}$$

Solution
This scheme is solved explicitly, and the stability of the solution is monitored via the cell Reynolds number.

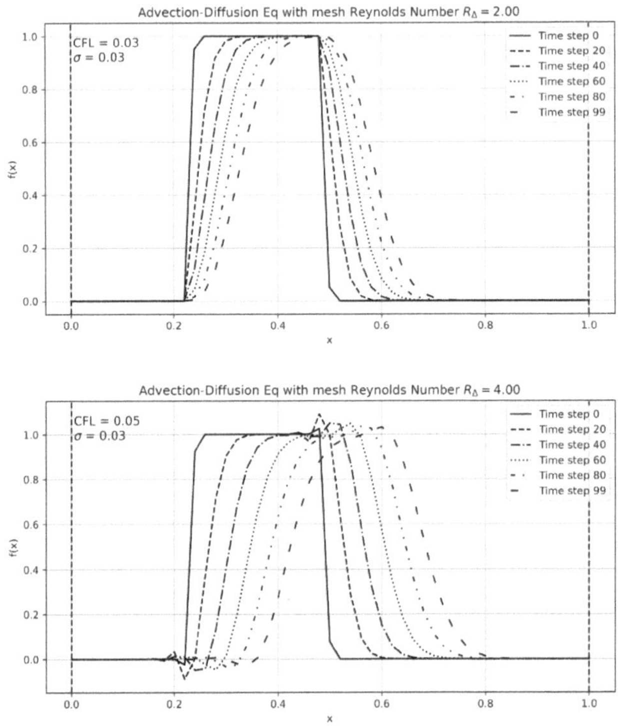

Figure 11.3: **The evolution of the solution f(x) over time for the advection-diffusion equation, with a cell Reynolds number.**

The results shown in Figure 11.3 the evolution of the solution $f(x,t)$ at different time steps (0, 20, 40, 60, 80, and 99). The solution begins as a sharp step function and gradually smooths out due to the effects of advection and diffusion. The step shifts to the right due to advection, while diffusion smooths out the steep gradients.
See `Python` code `Ch11-Ex111` for solution of this problem.

Example 11.2

Solve the convection-diffusion equation in two dimensions:

$$\frac{\partial f}{\partial t} + U\frac{\partial f}{\partial x} + V\frac{\partial f}{\partial y} = \mathscr{D}\left(\frac{\partial^2 f}{\partial x^2} + \frac{\partial^2 f}{\partial y^2}\right)$$

Where, f is the concentration field, U is the flow velocity in the x-direction, V is the flow velocity in the y-direction.

$\mathscr{D}$ is the diffusion coefficient, t is the time, and x and y are the spatial coordinates.

Initial and Boundary Conditions:

i Left Boundary : At $x = 0$ (the left edge of the domain), the concentration is fixed:

$$f(y,0,t) = 1 \quad \text{for } 0 \le y \le L_y$$

This represents a condition where a concentration of 1 enters the domain.

ii Right Boundary : At $x = L_x$ (the right edge of the domain), the concentration is set to zero:

$$f(y,L_x,t) = 0 \quad \text{for } 0 \le y \le L_y$$

This indicates that the concentration at the exit of the domain is zero.

iii Top Boundary : At $y = L_y$ (the top edge of the domain), the concentration is also zero:

$$f(L_y,x,t) = 0 \quad \text{for } 0 \le x \le L_x$$

This represents a no-flux condition at the top boundary.

iv Bottom Boundary : At $y = 0$ (the bottom edge of the domain), the concentration remains zero:

$$f(0,x,t) = 0 \quad \text{for } 0 \le x \le L_x$$

Similar to the top boundary, this condition also indicates a no-flux condition.

Solution
Figure 11.4 shows that solution is not exhibiting wiggles at cell Reynolds number less than 2. The solution will show wiggles at cell Reynolds number R_Δ >2 (see Figure 11.5). Also, the grid refinement can lead to issues with the wiggles.
The threshold $R_\Delta \le 2$ is relevant for central differencing schemes, but it is not a universal stability criterion—other schemes, such as upwind methods, behave differently and may remain stable even when $R_\Delta > 2$.

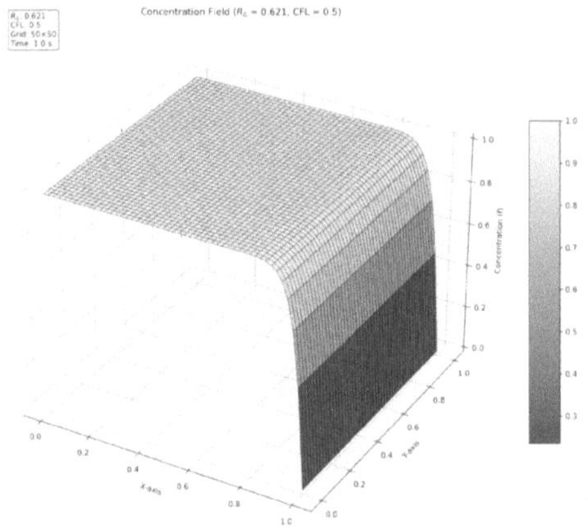

Figure 11.4: **Influence of cell Reynolds number (R_Δ=0.621) on Advection-Diffusion equation using 50 × 50 grid.**

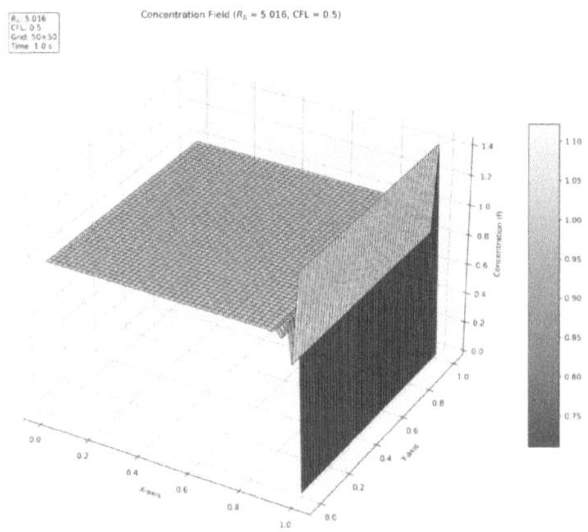

Figure 11.5: **Influence of cell Reynolds number (R_Δ=5) on Advection-Diffusion equation using 50 × 50 grid and CDS for convective term.**

This example shows why the cell Reynolds number R_Δ matters in two-dimensional simulations. It demonstrates how unwanted oscillations, commonly called wiggles, can appear when advection dominates over diffusion. Interestingly, even when we refine the grid (for instance, going from 32×32 to 40×40), which decreases the grid spacing Δx and lowers R_Δ, wiggles don't necessarily disappear. That's because central difference schemes can still struggle with capturing sharp gradients if there isn't enough numerical damping built into the method. It's also important to remember that just because a solution looks smooth doesn't mean it's numerically stable. The formal checks like von Neumann stability analysis are still needed. Overall, this example helps clarify the limits of central differencing in advection-dominated flows and suggests why upwind-biased or higher-order schemes, like QUICK or flux-limited methods, are often a better choice for maintaining stability and accuracy across a wide range of R_Δ.

11.4 APPROXIMATING THE CONVECTION TERM IN FINITE-DIFFERENCE SCHEMES

In the numerical simulation of advection–diffusion equations, spurious oscillations or wiggles typically arise when the convective terms are discretized using central difference schemes (CDS) and the mesh Reynolds number (local Peclet number) exceeds a critical value, often around 2. These wiggles are nonphysical and result from the dispersive nature of central differencing in convection-dominated regimes. However, when the convective terms are modeled using a first-order upwind scheme, such oscillations do not appear. The upwind scheme introduces numerical diffusion that stabilizes the solution and ensures monotonicity, even for high mesh Reynolds number.

Although implicit schemes, are generally stable, can have issues when the cell Reynolds number is high. In these cases, computed amplification factors can become negative, leading to instability. The numerical solutions are marred by *wiggles* typically caused in case of central difference schemes. These oscillations are more likely to occur when $R_\Delta > 2$, which signals instability in the numerical scheme. In numerical simulations, wiggles refer to oscillations that are not physically present in the actual system. They occur when a numerical scheme fails to correctly capture sharp gradients in the solution, leading to non-physical, oscillatory behavior in the computed results.

Central difference schemes compute derivatives by averaging neighboring points. This approach is effective when the solution varies smoothly but becomes problematic when the solution contains steep changes, such as shocks or sharp gradients, common in convection-diffusion problems. As a result, the scheme produces oscillations with a wavelength $2\Delta x$, which can grow or decay over time. The cell Reynolds number R_Δ quantifies the ratio of advective to diffusive transport in a system. When $P_\Delta > 2$, advection dominates, meaning the scheme has to handle fast-moving gradients. Central differencing, in particular, struggles in this regime because it tends to oscillate around steep changes in the solution, leading to the wiggles phenomenon. The numerical solution becomes unstable when R_Δ exceeds 2. The wiggles can grow uncontrollably, leading to a failure of the method unless corrective measures are introduced. This instability is a consequence of the numerical method's inability to resolve the sharp advective fronts accurately.

The cell Reynolds number $R_\Delta = \frac{CFL}{\sigma} = \frac{|u|\Delta x}{K}$ represents the ratio of advective transport to diffusive transport. If $R_\Delta \leq 2$, we are in a regime where diffusion dominates or balances advection. In this scenario, the central differencing scheme does not produce wiggles, which are high-frequency, non-physical oscillations that arise from the numerical method being unable to handle sharp gradients. Wiggles typically occur when advection is strong relative to diffusion, as central difference schemes struggle to represent sharp fronts in advection-dominated flows. For $R_\Delta > 2$, the scheme can produce these wiggles.

Von Neumann stability analysis evaluates whether numerical errors introduced in each time step grow, decay, or stay constant as the solution progresses. The von Neumann criterion assesses the growth factor of the errors, and for a method to be stable, the amplification factor $|A|$ must be less than or equal to 1 for all wavenumbers, meaning that errors do not grow uncontrollably. The condition $R_\Delta \leq 2$ ensures that wiggles are not present, which is related to the physical behavior of the solution. Specifically, it indicates that the central differencing scheme avoids generating oscillatory behavior due to strong advection. However, absence of wiggles does not automatically imply numerical stability. A method could still be numerically unstable in the von Neumann sense even without wiggles. This means that errors can still grow even if the solution looks physically reasonable (without wiggles). In many cases, the absence of wiggles suggests the solution is well-behaved for certain parameters, but it's not a formal guarantee that the method is stable in the von Neumann sense.

The condition $R_\Delta \leq 2$ serves as a heuristic guideline rather than a strict stability criterion, specifically indicating when non-physical oscillations (wiggles) may arise in central differencing schemes for the advection-diffusion equation. While the cell Reynolds number $R_\Delta = U\Delta x/\mathscr{D}$ quantifies the local dominance of advection over diffusion, the threshold $R_\Delta \leq 2$ is empirically derived to minimize wiggles in central differencing—where symmetric stencils are prone to dispersive errors when advection dominates. Importantly, exceeding this limit does not necessarily imply numerical instability (which is governed by von Neumann analysis requiring $|A| \leq 1$), but rather predicts degraded solution quality through spurious oscillations. Alternative schemes like upwind or QUICK can mitigate wiggles at higher R_Δ by introducing numerical diffusion or higher-order accuracy, respectively. Thus, while $R_\Delta \leq 2$ is a useful rule of thumb for central differencing, it is not a universal stability constraint, and practitioners must still verify stability through amplification factor analysis while selecting discretization methods tailored to their problem's advection-diffusion balance.

Table 11.1 lists three dimensionless parameters that needs to be controlled for the accurate and stable solution of an advection-diffusion equation.

Finite Difference Methods (FDM) for convection-diffusion and transport equations have evolved through key early works. Two fundamental discretization approaches are the central difference scheme (a centered, symmetric derivative approximation) and the upwind difference scheme (a one-sided derivative using the upstream value). Early researchers recognized a trade-off between the higher accuracy of central differencing and the greater stability (but added numerical diffusion) of upwind differencing.

Table 11.1
Dimensionless parameters appearing in the advection-diffusion analysis

Parameter	Definition	Physical meaning
CFL	$U\,\Delta t/\Delta x$	Non-dimensional advection time step
σ	$\mathscr{D}\,\Delta t/\Delta x^2$	Non-dimensional diffusion time step
R_Δ	$U\,\Delta x/\mathscr{D} = \text{CFL}/\sigma$	Ratio of advection to diffusion on the grid

11.4.1 FIRST-ORDER UPWIND SCHEME

In the context of the advection-diffusion equation, the first-order upwind scheme was introduced by Courant, Isaacson, and Rees (1952) to handle the advective transport term with stability and physical consistency. For a one-dimensional steady advection-diffusion problem with velocity u, diffusion coefficient $\mathscr{D}$, and uniform grid spacing Δx, the convective flux at the cell face $i+\frac{1}{2}$ is approximated as:

$$\phi_{i+1/2} = \begin{cases} \phi_i, & u > 0 \\ \phi_{i+1}, & u < 0 \end{cases}$$

This upwind selection means the spatial derivative of the convective flux uses a backward or forward difference aligned with the flow direction, ensuring the numerical domain of dependence matches the physical one. For positive velocity, the discrete convection derivative is

$$\frac{\partial(u\phi)}{\partial x} \approx \frac{u(\phi_i - \phi_{i-1})}{\Delta x}$$

This scheme, often called the donor-cell method, stabilizes the advection term by introducing numerical dissipation. The dimensionless ratio R_Δ, defined as

$$R_\Delta = \frac{U\Delta x}{\mathscr{D}} = \frac{\text{CFL}}{\sigma}$$

measures the relative strength of advection to diffusion on the grid. When R_Δ is large (advection-dominated regime), the upwind scheme maintains stability by damping spurious oscillations common in central differencing.

The scheme is unconditionally stable for linear advection when the Courant-Friedrichs-Lewy (CFL) number

$$\text{CFL} = \frac{u\Delta t}{\Delta x} \leq 1$$

but this comes at the cost of increased numerical diffusion, which tends to smear sharp gradients and reduce solution accuracy near steep fronts.

11.4.2 CENTRAL DIFFERENCE SCHEME

The central difference scheme approximates spatial derivatives in the advection-diffusion equation using values at the grid points, achieving second-order spatial accuracy. For uniform grid spacing, the convection term derivative is approximated as

$$\frac{\partial \phi}{\partial x} \approx \frac{\phi_{i+1} - \phi_{i-1}}{2\Delta x}$$

Central differencing has historically been favored for its simplicity and higher formal accuracy, especially when diffusion effects are significant. The stability of central differencing depends critically on the ratio R_Δ. When

$$R_\Delta = \frac{U\Delta x}{\mathscr{D}} < 2$$

meaning diffusion dominates or is comparable to advection on the mesh scale, the scheme delivers accurate and oscillation-free results without artificial dissipation.

11.4.3 HIGHER-ORDER CONVECTION SCHEMES: QUICK AND QUICKEST SCHEMES

The Quadratic Upstream Interpolation for Convective Kinematics (QUICK) scheme, introduced by Brian P. Leonard in 1979, represents a major advancement in computational fluid dynamics. Designed to address the limitations of classical finite difference methods—such as excessive numerical diffusion in first-order upwind schemes and oscillatory behavior in central differencing-QUICK employs a quadratic interpolation approach using two upstream nodes and one downstream node. This three-point weighted interpolation significantly reduces numerical diffusion and dispersion errors, achieving third-order spatial accuracy.

QUICK was originally formulated within the finite difference framework. It incorporates curvature terms to minimize non-physical oscillations at high grid Reynolds numbers, while also addressing the excessive numerical diffusion found in first-order schemes. By using a quadratic interpolation for the convective fluxes at grid points, QUICK achieves higher-order accuracy and better preserves steep gradients. The scheme's inherent convective stability allows reliable computations even on relatively coarse grids, reducing computational cost without sacrificing accuracy. These strengths have established QUICK as a widely used scheme for convection-dominated problems in aerodynamics, hydrodynamics, and environmental modeling.

Originally developed for incompressible flows, QUICK gained widespread adoption in the 1980s and 1990s within finite volume and finite difference approaches. Its success led to further refinements, such as the QUICKEST (QUICK with Estimated Streaming Terms) scheme by Leonard in 1991, which improved temporal accuracy for unsteady flows. Later variants, incorporate flux limiters to enhance stability near discontinuities. These developments solidified QUICK's role as a foundational method for high-accuracy CFD applications.

The QUICK scheme approximates the value of a scalar ϕ at a cell face using a quadratic interpolation between two upstream nodes and one downstream node. For a uniform grid with spacing Δx, the face value ϕ_f is computed as:

For flow to the right ($u_f > 0$):

$$\phi_f = \phi_{i-1} + \frac{1}{8}\left(3\phi_i - 2\phi_{i-1} - \phi_{i-2}\right)$$

For flow to the left ($u_f < 0$):

$$\phi_f = \phi_{i+1} + \frac{1}{8}\left(3\phi_i - 2\phi_{i+1} - \phi_{i+2}\right)$$

Compact Form (General Direction):

$$\phi_f = \phi_{\text{UP}} + \frac{1}{8}\left(3\phi_C - 2\phi_{\text{UP}} - \phi_{\text{DOWN}}\right)$$

where, ϕ_{UP} = immediate upstream node value, ϕ_C = central (current) node value,, ϕ_{DOWN} = second upstream node used in the quadratic stencil.

The scheme is third-order accurate in space but second-order overall when combined with standard temporal discretizations.

The QUICKEST (QUICK with Estimated Streaming Terms) scheme was developed by Brian P. Leonard in 1991 as an extension of the QUICK scheme to improve temporal accuracy and stability in unsteady flow simulations. While QUICK was primarily a spatial discretization method, QUICKEST introduced time-dependent corrections to reduce phase errors (numerical dispersion) in transient problems. This made it particularly valuable for wave propagation, vortex dynamics, and other time-accurate CFD applications. The scheme became widely used in large-eddy simulations (LES) and direct numerical simulations (DNS) where high-order accuracy is crucial. Modern variants, such as ULTIMATE-QUICKEST, incorporate flux limiters to suppress spurious oscillations near shocks.

The equation for ϕ is:

$$\begin{aligned}
\phi_j^{n+1} \quad = \quad & \phi_j^n + \left[\sigma(1-CFL) - \tfrac{CFL}{6}(CFL^2 - 3CFL + 2)\right]\phi_{j+1}^n \\
& - \left[\sigma(2-3CFL) - \tfrac{CFL}{2}(CFL^2 - 2CFL - 1)\right]\phi_j^n \\
& + \left[\sigma(1-3CFL) - \tfrac{CFL}{2}(CFL^2 - CFL - 2)\right]\phi_{j-1}^n \\
& + \left[\sigma(CFL) + \tfrac{CFL}{6}(CFL^2 - 1)\right]\phi_{j-2}^n
\end{aligned}$$

We substitute the Fourier mode representation $\phi_j^n = A^n e^{ikj\Delta x}$ in each term:

$$\phi_{j+2}^n = A^n e^{ikj\Delta x} e^{2ik\Delta x}$$

$$\phi_{j-1}^n = A^n e^{ikj\Delta x} e^{-ik\Delta x}$$

$$\phi_{j-2}^n = A^n e^{ikj\Delta x} e^{-2ik\Delta x}$$

For each of these exponential terms, we use Euler's formula:

$$e^{ik\Delta x} = \cos(k\Delta x) + i\sin(k\Delta x)$$

$$e^{-ik\Delta x} = \cos(k\Delta x) - i\sin(k\Delta x)$$

$$e^{2ik\Delta x} = \cos(2k\Delta x) + i\sin(2k\Delta x)$$

$$e^{-2ik\Delta x} = \cos(2k\Delta x) - i\sin(2k\Delta x)$$

Substituting these into the equation to obtain the real and imaginary parts.

The real part includes the cosine terms from $\cos(k\Delta x)$ and $\cos(2k\Delta x)$. Gathering terms, we get:

$$\begin{aligned}\text{Re}(A) \;=\; & 1 + \left(\sigma(1-CFL) + \sigma(CFL) - \tfrac{CFL}{6}(2CFL^2 - 2CFL + 1)\right)\cos(2k\Delta x) \\ & + \left(\sigma(1-3CFL) - \tfrac{CFL}{2}(CFL^2 - CFL - 2)\right)\cos(k\Delta x) \\ & - \left(\sigma(2-3CFL) - \tfrac{CFL}{2}(CFL^2 - 2CFL - 1)\right)\end{aligned}$$

This real part represents how diffusion and advection affect the wave amplitude.

The imaginary part arises from the sine terms $\sin(k\Delta x)$ and $\sin(2k\Delta x)$. Gathering terms, we get:

$$\begin{aligned}\text{Im}(A) \;=\; & \left(\sigma(1-CFL) + \sigma(CFL) - \tfrac{CFL}{6}(2CFL^2 - 2CFL + 1)\right)\sin(2k\Delta x) \\ & + \left(\sigma(1-3CFL) - \tfrac{CFL}{2}(CFL^2 - CFL - 2)\right)\sin(k\Delta x)\end{aligned}$$

The imaginary part reflects how the advection (through CFL) introduces phase shifts in the wave. Performing a von Neumann linear stability analysis yields the following expression for the amplification factor $A(k^*)$ (Leonard 1979):

$$\begin{aligned}A(k^*) = & 1 + 2\left(\sigma + \tfrac{1}{2}CFL^2\right)(\cos k^* - 1) \\ & + \tfrac{1}{6}CFL\left(1 - CFL^2 - 6\sigma\right)(4\cos k^* - \cos 2k^* - 3) \\ & - \mathrm{i}CFL\left[\sin k^* + \tfrac{1}{6}(1 - CFL^2 - 6\sigma)(2\sin k^* - \sin 2k^*)\right]\end{aligned}$$

The amplitude portraits for varying CFL and σ is plotted in Figure 11.6. Leonard identified the region of stability for the QUICKEST method in the (CFL, σ) plane. Although the scheme is explicit, it remains stable for $CFL > 1$ whenever $\sigma \neq 0$; the special case $\sigma = 0$ corresponds to the FTCS method, which is known to be unstable for pure advection.

QUICK: A third-order spatially accurate scheme based on quadratic upstream interpolation; it reduces numerical diffusion but performs best on smooth solution fields.

QUICKEST: An extension of QUICK with temporal correction terms; it improves phase accuracy and reduces dispersion errors in unsteady flow simulations.

The QUICK scheme marks a considerable evolution in numerical methods for convection-diffusion problems, effectively bridging the gap between the stability

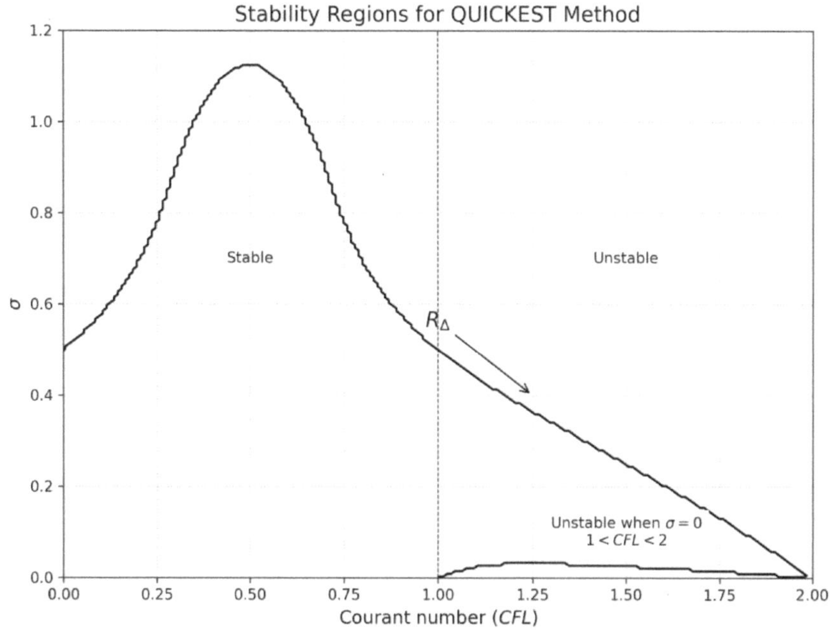

Figure 11.6: **The stability region in the (σ, CFL) plane for the QUICKEST method highlights its extended domain beyond $CFL = 1$. In the non-diffusive limit $\sigma = 0$ (R_Δ=∞), instability occurs when $1 < CFL < 2$ and again for $CFL > 2$.**

of upstream differencing and the accuracy of central differencing. Its sophisticated quadratic upstream interpolation strategy significantly reduces numerical errors, enabling highly accurate and stable solutions even at relatively coarse grid resolutions. This efficiency translates into reduced computational requirements, making QUICK particularly attractive for practical engineering simulations involving high cell Reynolds numbers. Its conservative nature further ensures the physical realism and reliability of simulation results, essential for predictive accuracy in fluid dynamics.

11.5 ALIASING, GRID RESOLUTION AND DIFFUSION SMOOTHING

In the advection-diffusion equation, the diffusion term $\mathscr{D}\partial^2 u/\partial x^2$ plays a crucial role in smoothing out sharp gradients by diffusing energy across the domain. This diffusive action effectively damps high-frequency components, limiting the formation of small-scale features that could otherwise lead to aliasing. In contrast, purely advective or inviscid nonlinear systems, such as the inviscid Burgers' equation, lack this diffusion-driven smoothing mechanism, making them far more susceptible to aliasing.

The convective term $U\partial u/\partial x$ in the advection-diffusion equation can still generate high-frequency components, especially when the initial condition contains sharp gradients or discontinuities. However, the diffusion term mitigates the effects of these high frequencies by smoothing out the steep gradients responsible for their generation. As a result, aliasing is generally less of a concern in the advection-diffusion equation compared to inviscid systems like the Burgers' equation, where the nonlinearity continuously produces high-frequency components without any inherent smoothing from diffusion.

Nonetheless, aliasing can occur in the advection-diffusion equation if the grid resolution is too coarse to capture the spatial scales present in the solution. However, the presence of diffusion reduces the severity of aliasing by naturally suppressing high-frequency components. If the grid resolution (spatial step Δx) is chosen sufficiently fine, and the diffusion coefficient ν is large enough to provide adequate smoothing, aliasing can generally be avoided.

The high-order procedures and schemes for computing convective fluxes in highly compressible flows that contain shocks or other discontinuities will be discussed in Chapter 19.

11.6 DISPERSION AND DISSIPATION IN NUMERICAL ADVECTION-DIFFUSION PROBLEM

In numerical PDE analysis, dispersion refers to phase errors (distortion of wave speeds) while dissipation refers to amplitude errors (numerical damping). These are quantified by comparing the scheme's numerical dispersion relation to the exact PDE dispersion relation. In computational fluid dynamics, we can decompose a field into wavenumber components. Each mode evolves according to the operator's dispersion and dissipation characteristics. For the one-dimensional advection-diffusion equation, the governing relation is

$$\frac{\partial \mathrm{T}}{\partial t} + u\frac{\partial \mathrm{T}}{\partial x} = \kappa\frac{\partial^2 \mathrm{T}}{\partial x^2}$$

so that in Fourier space each mode k satisfies

$$\omega(k) = uk \quad \delta(k) = -\kappa k^2$$

High-wavenumber modes ($k \gg 1$) incur strong dissipation through the $\delta(k)$ term, ensuring solution stability by damping grid-scale noise. Low-wavenumber modes carry the bulk of physical transport and must remain minimally distorted. Any numerical scheme is judged by how closely its discrete dispersion relation matches $\omega(k)$ and how its artificial dissipation approximates $-\kappa k^2$. Excess numerical dispersion generates spurious oscillations near discontinuities, while excessive numerical dissipation over-smooths important features. In order to design and tune high-order advection-diffusion solver, we quantifies error in each mode and balances fidelity against stability. By plotting and comparing analytic and discrete $\omega(k)$ and $\delta(k)$ curves, practitioners can predict scheme performance and avoid nonphysical artifacts.

$\omega(k)$ is the dispersion function for each spectral mode k. It relates temporal frequency to spatial variation, determining how fast waves of wavenumber k travel. The phase velocity of mode k is

$$c(k) = \frac{\omega(k)}{k}$$

and for pure advection,

$$\omega(k) = uk$$

so every mode moves at the same speed u. $\delta(k)$ is the dissipation rate for each mode k. It appears as an exponential factor $exp[\delta(k)t]$ in Fourier space and controls amplitude change over time. For physical diffusion,

$$\delta(k) = -\kappa k^2$$

so high k modes decay fastest, damping small-scale features and stabilizing the solution. In designing a CFD scheme that balances dispersion and dissipation, one first derives its discrete dispersion relation $\omega_{\text{disc}}(k)$ and dissipation rate $\delta_{\text{disc}}(k)$ and compares them to the analytic relations $\omega(k) = uk$ and $\delta(k) = -\kappa k^2$, focusing on the wavenumber range most relevant to the flow. Coefficients in high-order or compact finite-difference stencils are then tuned to minimize phase-speed errors at low and mid wavenumbers while providing controlled damping at high wavenumbers; dispersion-optimized operators (such as the KFD scheme) exemplify this calibration.

In regions of sharp gradients or discontinuities, hybrid central/upwind formulations and nonlinear limiters suppress spurious oscillations while avoiding excessive diffusion (discussed in Chapter 19).

Even when the cell Reynolds number ($R_\Delta = U\Delta x/\mathscr{D}$) is below 2, indicating a diffusion-dominated regime, numerical dispersion and dissipation can still occur. This is because these errors arise inherently from the numerical discretization itself and are influenced by grid spacing, time-stepping, and stencil choice, rather than solely by the relative dominance of advection or diffusion. Thus, even at $R_\Delta < 2$, a poorly selected numerical scheme or inadequate grid resolution can still generate phase errors (dispersion), causing distortions in wave propagation speeds, and amplitude errors (dissipation), leading to unwanted smoothing of the solution. Careful analysis and optimization of the discretization method remain essential irrespective of the cell Reynolds number regime.

Lessons Learned

☞ Understanding the dimensionless groups CFL and σ is essential, as they govern the stability limits of explicit schemes. The cell Reynolds number R_Δ, while not a stability criterion, provides insight into the relative dominance of advection versus diffusion and helps predict non-physical oscillations in central differencing schemes.

☞ The cell Reynolds number $R_\Delta = U\Delta x/\mathscr{D}$ offers a physical interpretation of numerical behavior; when $R_\Delta > 2$, oscillations (wiggles) can appear even if stability in the von Neumann sense is preserved. Central-difference schemes remain free of spurious oscillations only when the cell Reynolds number does not exceed about 2 locally. However, the absence of wiggles does not imply numerical stability as errors may still grow if the amplification factor exceeds unity. Thus, wiggle-free solutions must still be verified through formal stability analysis.

☞ Numerical diffusion is a discretisation artefact whose sign and magnitude can be tuned with implicitness parameter θ; choosing $\theta > 0.5$ adds stabilising diffusion, while $\theta < 0.5$ risks instability. Discretization-driven (artificial) diffusion can act both as a stabiliser and as a source of error; understanding its form and magnitude allows for deliberate control of accuracy and stability.

☞ Higher-order convection procedures such as QUICK and QUICKEST achieve third-order spatial accuracy, but their stability domain in the (CFL, σ) plane must be respected, especially for low-diffusion, advection-dominated problems.

☞ Eigenvalue analysis of the advection-diffusion matrix provides a global view of numerical dispersion and damping, especially under periodic boundary conditions.

☞ High-order schemes such as DRP, QUICK, and QUICKEST are specifically designed to minimize numerical dispersion and dissipation in advection-dominated problems, offering improved accuracy over standard differencing techniques.

☞ Second-order central differencing (FD-O2) remains a robust and efficient choice for discretizing diffusion terms, as it inherently captures the smoothing nature of diffusion with minimal complexity.

PROBLEMS

11P-1 A quantity of salt is released into a channel with a steady mean flow velocity of $u = 0.02$ m/s. The salt undergoes both convection and diffusion, governed by the 1D advection-diffusion equation:

$$\frac{\partial c}{\partial t} + u\frac{\partial c}{\partial x} - \mathscr{D}\frac{\partial^2 c}{\partial x^2} = 0,$$

where the diffusion coefficient is $\mathscr{D} = 1.6$ m^2/s. At a given instant, the salt forms a concentrated cloud extending 2000 m in length. Estimate the shape of this salt distribution when it reaches a location 12 km downstream from the release point. To do this, implement a numerical model using the Crank-Nicolson method applied to the discretized form of the governing equation:

$$\frac{c_j^{n+1}-c_j^n}{\Delta t}+\theta\left\{u\frac{c_{j+1}^{n+1}-c_{j-1}^{n+1}}{2\Delta x}-D\frac{c_{j+1}^{n+1}-2c_j^{n+1}+c_{j-1}^{n+1}}{\Delta x^2}\right\}$$

$$+(1-\theta)\left\{u\frac{c_{j+1}^{n}-c_{j-1}^{n}}{2\Delta x}-D\frac{c_{j+1}^{n}-2c_j^{n}+c_{j-1}^{n}}{\Delta x^2}\right\}=0$$

Use $\theta=0.55$, $\Delta x=250$ m, and $\Delta t=5000$ s in simulation. Simulate the transport of the salt cloud and describe or plot the concentration profile at $t=0$, $t=2.5\times10^5 s$, and $t=5\times10^5$ s. Note that θ=0.55 is a small amount of numerical diffusion to the scheme to enhance stability.

11P-2 Prove that for Crank-Nicolson method the numerical diffusion is:

$$D_{\text{num}}=(\theta-\tfrac{1}{2})u^2\Delta t$$

Interpret the influence of θ on numerical diffusion.

11P-3 Consider the one-dimensional linearized equation:

$$\frac{\partial T}{\partial t}+u\frac{\partial T}{\partial x}=\kappa\frac{\partial^2 T}{\partial x^2}-\beta\frac{\partial^3 T}{\partial x^3},$$

posed on a periodic domain $x\in[0,2\pi]$, where u is the advection velocity, $\kappa\geq0$ is the diffusivity, and $\beta\geq0$ is the dispersive coefficient. The initial condition is constructed as a superposition of odd harmonic modes:

$$T(x,0)=\Re\left[\sum_{m=1,3,5,\ldots}^{49}\frac{1}{m}e^{imx}\right],$$

which represents a sharp, non-symmetric pulse decaying as $1/m$ in Fourier space.

(a) Derive the general solution in Fourier space:

$$\widehat{T}(k,t)=\widehat{T}(k,0)\,\exp\left[\left(-\kappa k^2-i(uk-\beta k^3)\right)t\right].$$

Identify the dissipation rate $\sigma(k)=-\kappa k^2$ and the dispersion relation $\omega(k)=uk-\beta k^3$. How does $\sigma(k)$ affect wave amplitudes over time, and what does $\omega(k)$ imply about the phase velocity $c(k)=\omega(k)/k$?

(b) For the parameters:

Case I: $\kappa=0.05,\beta=0$ (pure diffusion),

Case II: $\kappa=0,\beta=0.05$ (pure dispersion),

Case III: $\kappa=0.05,\beta=0.05$ (combined effects),

investigate the evolution up to $t = 2$ and understand how small-scale features decay, how peaks broaden, and how total energy evolves. What is the smallest β value if a small amount of dispersion is unavoidable.

11P-4 For the PDE:

$$u_t + au_x = bu_{xxxx}, \quad a > 0,\ b > 0,$$

i Discretize the PDE using an explicit Euler time scheme, with fourth-order finite difference (FD-O4) approximations for both the advection and fourth-derivative terms.

ii Derive the local truncation error for the combined FD-O4 and Euler scheme.

iii Perform von Neumann stability analysis and derive a condition for the time step Δt in terms of Δx, a, and b.

iv Repeat the simulation by including an explicit low-pass spatial filter (e.g., a three-point filter) applied after each time step. Analyze and discuss the impact of filtering on the numerical stability and behavior of the solution, particularly in suppressing high-frequency instabilities.

REFERENCES

B. P. Leonard, A stable and accurate convective modelling procedure based on quadratic upstream interpolation. Computer Methods in Applied Mechanics and Engny 19: 59 - 98, 1979.

M. B. Abbott, D. R. Basco, Computational fluid dynamics. Longman Scientific and Technical, New York, USA, 1989.

R. Courant, E. Isaacson, M. Rees. On the Solution of Nonlinear Hyperbolic Differential Equations by Finite Difference. Commun. Pure Appl. Math., 5: 243-255, 1952.

J. von Neumann, R. D. Richtmyer. A Method for the Numerical Calculation of Hydrodynamic Shocks. J. Appl. Phys., 21 3: 232-237, 1950.

G. G. O'Brien, M. A. Hyman, S. Kaplan. A Study of the Numerical Solution of Partial Differential Equations. J. Math. Phys., 29: 223-251, 1950.

D. B. Spalding. A Novel Finite-Difference Formulation for Differential Expressions Involving Both First and Second Derivatives. Int. J. Numer. Methods Eng., 4 4: 551-559, 1972.

R. K. Runchal. Convergence and Accuracy of Three Finite Difference Schemes for a Two-Dimensional Conduction and Convection Problem. Int. J. Numer. Methods Eng., 4 4: 541-550, 1972.

J. Crank, P. Nicolson, A practical method for numerical integration of solutions of the heat-conduction equation," Proc. Camb. Phil. Soc. 43: 50-67, 1947.

C. Pozrikidis, Numerical Computation in Science and Engineering, Oxford University Press, Oxford, UK, 2008.

K. W. Morton, D. F. Mayers, Numerical Solution of Partial Differential Equations: An Introduction, 2nd ed., Cambridge University Press, UK, 2005.

12 The Viscous Burgers' Equation

In this chapter, we study the viscous Burgers equation, a nonlinear prototype equation widely used to explore convection-diffusion phenomena in fluid dynamics. Unlike the linear advection-diffusion equation, Burgers' equation exhibits nonlinear convection, which leads to richer dynamics. Despite its simplicity, it captures key features such as nonlinear wave steepening, shock formation, and viscous dissipation, making it a valuable model for understanding aspects of turbulence and energy transfer. We examine both analytical and numerical solutions under various boundary and initial conditions, analyze the energy spectra, investigate aliasing errors and modal decay, and assess the accuracy and stability of high-order numerical schemes applied to this equation. After finishing this chapter, you should be able to:

- Understand the physical meaning and mathematical structure of the viscous Burgers' equation.
- Learn about numerical solution strategies on uniform and non-uniform grids.
- Develop explicit and implicit finite difference schemes for nonlinear problems.
- Be familiar with energy spectrum analysis and the effects of numerical dispersion.

The Navier-Stokes equations were formulated over a period of time with significant contributions from two scientists: Claude-Louis Navier and George Gabriel Stokes. Navier introduced his formulation in 1822, and Stokes further refined the equations in 1845. Lewis Fry Richardson was a British scientist whose interdisciplinary work spanned mathematics, physics, meteorology, and psychology. Born in 1881, he is most renowned for pioneering numerical weather prediction (NWP) and making significant early contributions to computational fluid dynamics (CFD). His innovative approach laid the groundwork for modern meteorological forecasting and the numerical analysis of fluid flows. Richardson's most notable scientific endeavor was his attempt to apply mathematical techniques to predict weather conditions. In the early 20th century, weather forecasting relied mainly on empirical methods. Richardson sought a more systematic approach using the fundamental laws of physics to predict future weather states quantitatively.

Richardson focused on the equations of fluid motion, which are central to both meteorology and CFD. He was particularly interested in the Navier-Stokes equations, which describe the motion of fluid substances. These equations are central to

DOI: 10.1201/9781003663171-12

CFD and are used to model weather systems by considering atmospheric dynamics. Richardson attempted to solve a simplified version of these equations, incorporating both advection and diffusion processes:

$$\frac{\partial u}{\partial t} + u\frac{\partial u}{\partial x} = \nu\frac{\partial^2 u}{\partial x^2}$$

Here, u represents the fluid velocity, ν is the momentum diffusivity (representing viscous effects in the fluid), the term $u\frac{\partial u}{\partial x}$ accounts for advection, and $\nu\frac{\partial^2 u}{\partial x^2}$ represents diffusion.

To solve these equations numerically, Richardson used finite difference methods, discretizing time and space to transform the differential equations into algebraic ones that could be solved iteratively. This method required extensive calculations, far beyond what could be manually computed efficiently at the time. Richardson's work was ultimately unsuccessful in his lifetime due to the limitations of the computational resources available and the initial inaccuracy of the data used in his calculations. However, his conceptual framework and methodologies have since become foundational in the fields of numerical weather prediction and computational fluid dynamics. Modern supercomputers use sophisticated versions of Richardson's initial techniques to perform daily weather forecasts and complex fluid dynamics simulations across various industries.

Despite the visionary nature of his work, Richardson's initial forays into numerical weather prediction were met with limited success. The computational resources of his time were simply inadequate, and the data he had at his disposal were fraught with inaccuracies. Yet, undeterred by these practical failures, Richardson dreamed even bigger. He envisioned vast "forecast factories," filled with legions of human computers, all working in concert to predict the weather. This grand vision, though unrealized in his lifetime, presaged the modern era of parallel computing and remains a testament to his forward-thinking approach.

Burgers' equation is a fundamental mathematical model in fluid dynamics, utilized to explore various flow phenomena that involve both viscosity and shock waves. It is particularly valued for its ability to simulate the nonlinear convection and linear diffusion processes that are characteristic of turbulent flows and the propagation of shock waves in viscous media. This makes it applicable not only in studies of fluid mechanics but also in modeling traffic flows, where it describes the formation and dissipation of traffic jams in a manner analogous to shock waves. Additionally, it can be adapted to study wave propagation in shallow waters under certain conditions. The equation balances nonlinear advection with diffusion effects, providing a simplified yet insightful framework for analyzing complex fluid behaviors and their interactions in theoretical and applied physics contexts.

The Burgers' equation, a fundamental partial differential equation in fluid dynamics, is typically expressed as:

$$\text{Conservative form: } \frac{\partial u}{\partial t} + \frac{\partial}{\partial x}\left(\frac{u^2}{2}\right) = \nu\frac{\partial^2 u}{\partial x^2}$$

or

$$\text{Non-conservative form: } \frac{\partial u}{\partial t} + u\frac{\partial u}{\partial x} = \nu\frac{\partial^2 u}{\partial x^2}$$

Here, u represents the fluid velocity, t denotes time, x is the spatial coordinate, ν stands for the kinematic viscosity of the fluid, controlling the rate of diffusion. The two forms are mathematically equivalent for smooth solutions, but behave differently in the presence of discontinuities. The conservative form is preferred in capturing shocks accurately.

The key difference between the Burgers' equation and the advection-diffusion equation lies in their treatment of the advection term. The Burgers' equation is nonlinear, featuring an advection term $u\frac{\partial u}{\partial x}$, where the wave speed depends on the solution itself. This nonlinearity can lead to phenomena such as shock formation, especially when viscosity is low. In contrast, the advection-diffusion equation is linear, with a constant advection velocity c, making the wave speed independent of the solution. As a result, sharp gradients in the advection-diffusion equation are smoothed out by the diffusion term, and shocks do not form. While the Burgers' equation is often used to model complex fluid flows and nonlinear wave propagation, the advection-diffusion equation is applied in transport phenomena, such as pollutant dispersion or heat transfer, where advection and diffusion occur simultaneously in a linear system.

Aliasing is a key challenge in the numerical solution of the Burgers' equation, particularly when nonlinear interactions generate high-frequency components that surpass the Nyquist frequency of the grid. These components, with wave numbers $k > k_{\text{Nyquist}} = \pi/\Delta x$, cannot be properly resolved and are misrepresented as lower frequencies, causing aliasing. While viscosity does not eliminate aliasing, it damps high-frequency components, especially those near the Nyquist limit, reducing the visible impact of aliasing-related artifacts in the solution. The diffusion term $\nu\frac{\partial^2 u}{\partial x^2}$ plays a key role in attenuating unresolved high-wavenumber modes, effectively suppressing the spectral artifacts introduced by aliasing.

12.1 VISCOUS BURGERS' EQUATION SOLUTION

The viscous Burgers' equation is

$$\frac{\partial u}{\partial t} + u\frac{\partial u}{\partial x} = \nu\frac{\partial^2 u}{\partial x^2}$$

In this equation, $u(x,t)$ represents the velocity field as a function of space x and time t, and ν is the kinematic viscosity coefficient. The term $u\frac{\partial u}{\partial x}$ represents the nonlinear convective acceleration, while $\nu\frac{\partial^2 u}{\partial x^2}$ accounts for the diffusive effects due to viscosity.

Initial and Boundary Conditions:

The exact solution to Burgers' equation (with periodic boundary) is obtained by defining an auxiliary function $\phi(x,t)$ as the sum of Gaussians, used in Cole–Hopf transformation.

$$\phi(x,t) = \exp\left(-\frac{(x-4t)^2}{4\nu(t+1)}\right) + \exp\left(-\frac{(x-4t-2\pi)^2}{4\nu(t+1)}\right)$$

The initial velocity profile $u(x,0)$ is calculated using:

$$u(x,0) = -2\nu \frac{\partial}{\partial x} \ln \phi(x,0) + 4$$

This expression is derived by taking the spatial derivative of $\phi(x,t)$, dividing it by $\phi(x,t)$, and scaling appropriately. This results in a velocity profile derived from the exact solution via the Cole-Hopf transformation, which satisfies the Burgers' equation for all $t \geq 0$.

The boundary conditions are periodic over the domain $[0, 2\pi]$. This means that:

$$u(0,t) = u(2\pi,t)$$

Periodicity is enforced in the code by equating the first and last elements of the numerical solution array at each time step, ensuring the solution wraps around the spatial domain seamlessly.

To numerically solve the PDE, the code discretizes the spatial domain into nx grid points with spacing $\Delta x = \frac{2\pi}{nx-1}$ and time into nt steps with spacing Δt. The spatial and temporal derivatives in the PDE are approximated using finite difference methods.

For the convective term $u\frac{\partial u}{\partial x}$, the spatial derivative $\frac{\partial u}{\partial x}$ is approximated using a backward difference:

$$\left(\frac{\partial u}{\partial x}\right)_i \approx \frac{u_i - u_{i-1}}{\Delta x}$$

For the diffusive term $\nu\frac{\partial^2 u}{\partial x^2}$, the second spatial derivative $\frac{\partial^2 u}{\partial x^2}$ is approximated using a central difference:

$$\left(\frac{\partial^2 u}{\partial x^2}\right)_i \approx \frac{u_{i+1} - 2u_i + u_{i-1}}{\Delta x^2}$$

The temporal derivative $\frac{\partial u}{\partial t}$ is approximated using a forward difference:

$$\left(\frac{\partial u}{\partial t}\right)_i \approx \frac{u_i^{n+1} - u_i^n}{\Delta t}$$

Combining these approximations, the finite difference scheme for updating u_i^{n+1} becomes:

$$u_i^{n+1} = u_i^n - \frac{u_i^n \Delta t}{\Delta x}(u_i^n - u_{i-1}^n) + \nu \frac{\Delta t}{\Delta x^2}(u_{i+1}^n - 2u_i^n + u_{i-1}^n)$$

This equation is used for all interior grid points $i = 1$ to $nx - 2$. For the boundary points, the periodic boundary conditions are applied:

At $i = 0$:

$$u_0^{n+1} = u_0^n - \frac{u_0^n \Delta t}{\Delta x}(u_0^n - u_{nx-1}^n) + \nu \frac{\Delta t}{\Delta x^2}(u_1^n - 2u_0^n + u_{nx-1}^n)$$

At $i = nx - 1$:

$$u_{nx-1}^{n+1} = u_{nx-1}^{n} - \frac{u_{nx-1}^{n}\Delta t}{\Delta x}(u_{nx-1}^{n} - u_{nx-2}^{n}) + \nu\frac{\Delta t}{\Delta x^2}(u_0^n - 2u_{nx-1}^{n} + u_{nx-2}^{n})$$

These ensure that the solution remains continuous and periodic across the spatial domain.

The explicit finite difference scheme employed in the code is conditionally stable. The stability condition is determined by the Courant-Friedrichs-Lewy (CFL) criterion, which relates the time step Δt, the spatial step Δx, the maximum velocity $u_{\max}$, and the viscosity ν.

For the convective term, the CFL condition requires:

$$\frac{u_{\max}\Delta t}{\Delta x} \leq C_{\text{conv}}$$

For the diffusive term, the stability condition is:

$$\frac{\nu \Delta t}{\Delta x^2} \leq \frac{1}{2}$$

The time step Δt is carefully chosen to satisfy the more restrictive condition, which typically comes from the diffusion term due to the second-order spatial derivative.

Figure 12.1 presents the numerical solution of the Burgers' equation over time, computed using a second-order finite difference method in space and the explicit Euler method in time. Each curve represents the solution u as a function of spatial coordinate x at different time points t, ranging from 0.100 seconds to 0.500 seconds. The plot shows the time evolution of the velocity profile, capturing nonlinear steepening and eventual shock-like structures, especially noticeable between $t = 0.300$ and $t = 0.500$ seconds. This visualization helps in understanding the dynamics of nonlinear wave propagation and shock formation in fluid dynamics.

Figure 12.2 depicts the energy spectra E_k for viscous Burgers' equation at various times, showing how energy is distributed across different scales represented by the wavenumber k. The energy spectrum is generated from the velocity field $u(x,t)$ using its spatial Fourier transform. $u(x,t)$ is the solution of the Burgers' equation at a given time t, discretized over space. The energy spectrum is obtained by applying the Fourier transform to $u(x,t)$:

$$\hat{u}_k = \mathscr{F}[u(x,t)]$$

The energy in each Fourier mode k is given by:

$$E_k = \frac{1}{2}|\hat{u}_k|^2$$

where $k = \frac{2\pi n}{L}$, with $n = 0, 1, 2, \ldots$, and L is the domain length.

For the viscous Burgers' equation, which includes a diffusion term, the spectral energy representation provides valuable insight into how energy is distributed and

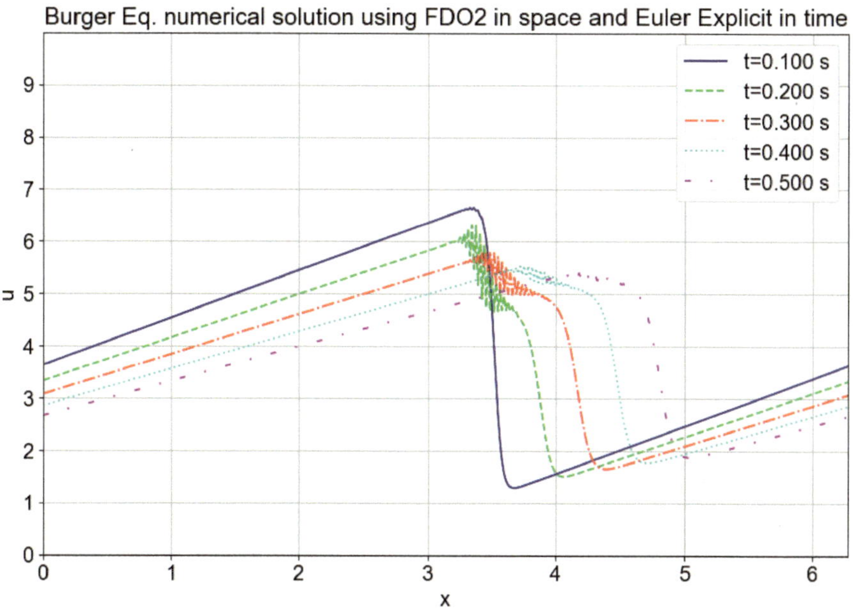

Figure 12.1: **Burgers' equation over time, computed using a second-order finite difference method in space and the explicit Euler method in time.**

dissipated across spatial scales. At the initial time $t = 0$, the energy is primarily concentrated at low wavenumbers, indicating that large-scale structures dominate the flow. These correspond to smoothly varying, long-wavelength features.

As time evolves, the effects of viscosity become increasingly prominent. Two key processes are observed:

1. Energy Cascade: Nonlinear convection transfers energy from large scales (low k) to smaller scales (higher k). This mechanism is evident from the progressive extension of the spectrum toward higher wavenumbers.

2. Viscous Dissipation: At later times (e.g., $t = 0.7$ and $t = 1.0$ s), energy at high wavenumbers rapidly decays due to the action of viscosity, which preferentially damps small-scale structures.

However, numerical artifacts can distort this ideal behavior. When spatial or temporal discretization is too coarse to resolve steep gradients or small-scale features, artificial energy accumulation occurs at high wavenumbers. This phenomenon is not a physical backflow of energy, but rather a result of aliasing and dispersion errors. These spurious energy peaks are common when numerical schemes under-resolve the gradient scales present in the solution.

Using higher-order schemes, such as fourth-order finite difference (FD-04) in space and fourth-order Runge-Kutta (RK-04) in time, significantly reduces such artifacts. These methods exhibit lower numerical dispersion and better spectral accuracy,

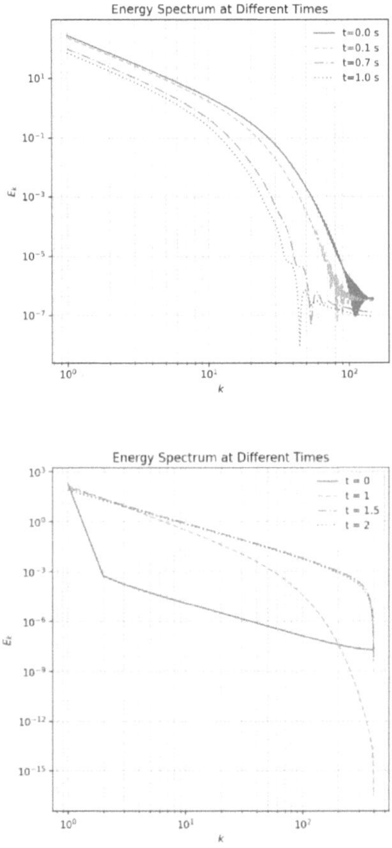

Figure 12.2: **Comparison of energy spectrum for Burgers' equation over time. [Left] computed using second-order FD-O2 (Euler-explicit), and [Right] fourth-order FD-O4 (RK-O4) finite difference methods.**

allowing for more accurate representation of both energy transfer and dissipation. Consequently, high-order methods are especially valuable in simulations involving turbulent flows, shock development, or energy spectra, where accurate resolution of fine-scale dynamics is critical.

Figure 12.4 shows the time evolution of the quantity $\pi \nu \frac{\partial u}{\partial x}$ at $x = 0$ on a logarithmic scale. Initially, the value is very small, but it rises steeply around $t = 0.3$, indicating rapid gradient growth before eventually reaching a plateau. This reflects shock formation and steepening typical of Burgers' equation. The grid had 801 mesh

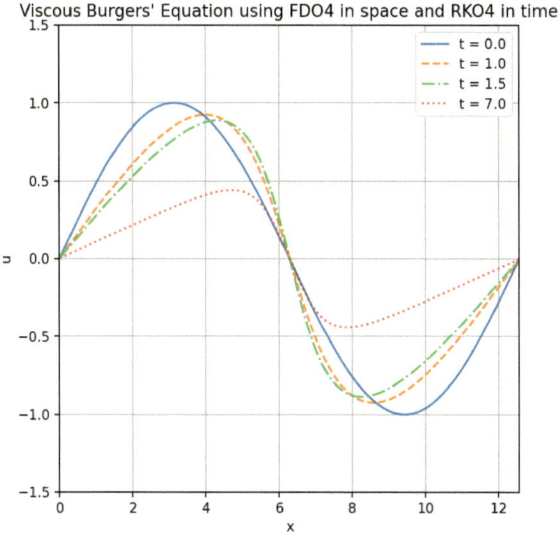

Figure 12.3: **Burgers' equation solved using FDO4 and RKO4.**

points and grid was non-uniform with stretching formula used:

$$y'(x) = \begin{cases} -1 + \dfrac{\tanh(2\alpha x)}{\tanh(\alpha)}, & x \leq 0.5 \\ 1 + \dfrac{\tanh(2\alpha(x-1))}{\tanh(\alpha)}, & x > 0.5 \end{cases}$$

Note that the solution of the Burgers' equation is highly sensitive to the quality of the computational grid because the equation often develops steep gradients or shock-like structures, particularly at low viscosity. Capturing these features accurately requires that the grid provides sufficient resolution in regions where the solution changes rapidly. If the grid is too coarse or lacks proper clustering near steep gradients, the shock region may be represented by only a few points, leading to large numerical errors, excessive artificial diffusion, or even nonphysical oscillations. Using grid stretching or adaptive mesh refinement enables more points to be concentrated where they are most needed, allowing for accurate representation of the shock thickness, peak gradients, and correct dissipation behavior. Without an adequately refined or well-designed grid, critical features of the solution are distorted, resulting in unreliable or unstable numerical results.

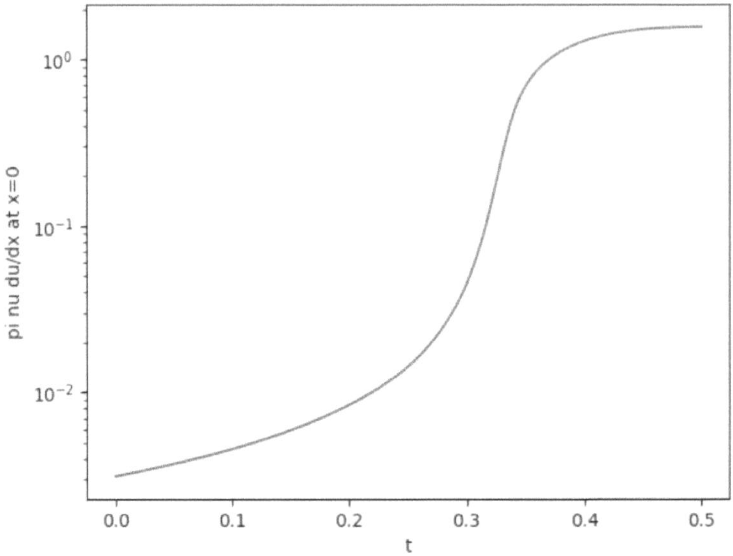

Figure 12.4: Time evolution of maximum velocity gradient with FD-O2 and RK-O3 method with timestep controlled by diffusion number.

12.2 SOLVING NONLINEAR BURGERS' EQUATION USING LIBRARIES

In this section, we demonstrate that solving Burger equation with useful Python library `SciPy` is a convenient option. `SciPy` library provides advanced scientific and numerical computing tools, including modules for optimization, integration, interpolation, linear algebra, and differential equation solving. We will solve the Burgers' equation with Dirichlet boundary conditions specify fixed values at the domain edges, which is physically meaningful for many problems such as prescribed inflow/outflow, wall, or initial profile anchoring.

See code `Ch12-Burger1` for solution of the Burger equation using `Python` Libraries.

The shock-like structure in solutions of the viscous Burgers' equation arises from a balance between nonlinear convection and viscous diffusion. This viscous shock sharpens as viscosity decreases, posing significant numerical challenges. Uniform grids often fail to resolve its steep gradients efficiently, requiring prohibitively fine resolution and high computational cost. Adaptive strategies such as non-uniform grids or coordinate transformations that cluster points near the shock can capture this feature accurately without excessive grid points, avoiding numerical artifacts like spurious oscillations or artificial smearing.

Notably, while the Burgers' equation models key nonlinear and diffusive phenomena of the compressible Navier-Stokes equations (e.g., shock formation and wave steepening), such viscous shocks do not occur in incompressible Navier-Stokes flows. This distinction stems from fundamental physical constraints:

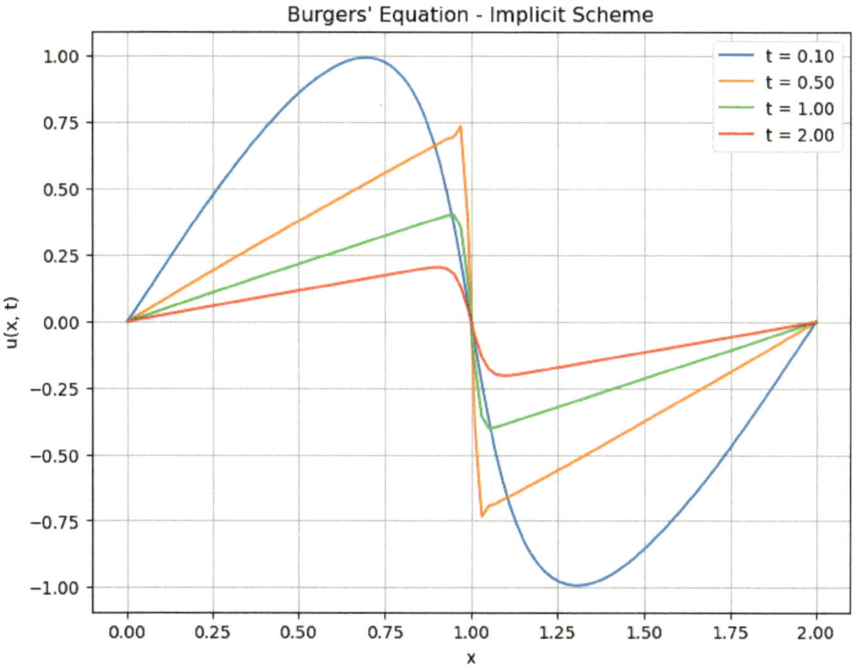

Figure 12.5: **Burgers' equations solved using the Sparse Matrix solver of python Scipy.**

☞ Incompressibility: The divergence-free condition ($\nabla \cdot \mathbf{u} = 0$) enforces global mass conservation.

☞ Pressure Role: Pressure acts as an instantaneous, non-local constraint, redistributing momentum to suppress localized steepening.

☞ Dimensionality and Complexity: Incompressible flows are inherently multidimensional and have flow structures like shear layers, vortices, and boundary layers.

Despite these differences, Burgers' equation remains a valuable prototype for studying nonlinear steepening, diffusive regularization, and turbulence mechanisms, offering insights into flow dynamics and computational methods for shock-dominated systems.

Lessons Learned

☞ Numerical schemes must satisfy both convective and diffusive CFL constraints. In explicit methods, the diffusive term typically imposes the stricter time step restriction due to the second-order spatial derivative.

☞ The energy spectrum provides a quantitative view of how energy is distributed and dissipated across spatial scales. High-order schemes better preserve the energy cascade and suppress artificial accumulation at high wavenumbers.

☞ High-order methods (e.g., FDO4 in space with RK4 in time) effectively minimize numerical dispersion and dissipation, making them especially suitable for resolving steep gradients, shock structures, or turbulence-like features.

☞ The choice of discretization—such as upwind versus central schemes, or explicit versus implicit time integration—directly affects both numerical stability and solution fidelity. Implicit schemes are advantageous for stiff systems but may introduce excessive numerical damping if not carefully implemented.

☞ Nonlinear aliasing leads to the generation of spurious high-frequency modes due to unresolved wavenumber interactions. While viscosity helps attenuate these components, complete mitigation requires higher spatial resolution or the application of spectral filtering.

PROBLEMS

12P-1 Convert the viscous Burgers' equation into the non-dimensional form:

$$\frac{\partial u'}{\partial t'} + u'\frac{\partial u'}{\partial x'} = \frac{1}{\text{Re}}\frac{\partial^2 u'}{\partial x'^2},$$

12P-2 Consider Burgers' equation given by

$$\frac{\partial u}{\partial t} + u\frac{\partial u}{\partial x} = \nu\frac{\partial^2 u}{\partial x^2}$$

where $u(x,t)$ represents the velocity field, ν is the viscosity coefficient, x is the spatial coordinate, and t denotes time.

Assume a periodic spatial domain defined as $x \in [-1,1]$, with periodic boundary conditions ensuring that the solution repeats itself outside the domain $u(-1,t) = u(1,t)$ for all $t \geq 0$. Analyze Burgers' equation with two distinct sets of initial conditions specified at time $t = 0$:

a. For the first initial condition, there is a discontinuous jump in the velocity field from $u = -1$ to $u = 1$ at $x = -\frac{1}{3}$, and another jump from $u = 1$ back to $u = -1$ at $x = \frac{1}{3}$. The initial velocity profile $u(x,0)$ is defined piecewise as

$$u(x,0) = \begin{cases} -1, & x < a_1 \\ 2 \cdot \frac{x-a_1}{a_2-a_1} - 1, & a_1 \leq x < a_2 \\ 1, & a_2 \leq x < a_3 \\ -1, & x \geq a_3 \end{cases}$$

where the parameters are defined as $t = 0.3, a_1 = -0.6333, a_2 = -0.0333, a_3 = 0.3333$.

b. For the second initial condition, there is a discontinuous jump in the velocity field from $u = 0$ to $u = 1$ at $x = -\frac{1}{3}$, and another jump from $u = 1$ back to $u = 0$ at $x = \frac{1}{3}$. The initial velocity profile $u(x,0)$ is defined piecewise as

$$u(x,0) = \begin{cases} 0, & x < a_1 \\ \frac{x-a_1}{a_2-a_1}, & a_1 \leq x < a_2 \\ 1, & a_2 \leq x < a_3 \\ 0, & x \geq a_3 \end{cases}$$

where the parameters are defined as $t = 0.6, a_1 = -0.3333, a_2 = 0.2667, a_3 = 0.6333$.

Solve Burgers' equation numerically using an appropriate numerical scheme. Investigate the evolution of the velocity field $u(x,t)$ over time, particularly the formation and propagation of shock waves resulting from the discontinuities in the initial conditions.

12P-3 The Burgers equation is given:

$$\frac{\partial u}{\partial t} + u\frac{\partial u}{\partial x} = \nu\frac{\partial^2 u}{\partial x^2}, \quad 0 \leq x \leq 2,$$

with the initial condition

$$u(x,0) = 1 + \cos(\pi x),$$

and the boundary conditions

$$\frac{\partial u}{\partial x}(0,t) = 0, \quad \frac{\partial u}{\partial x}(2,t) = 0.$$

Solve this equation numerically using explicit time stepping (Forward Euler), first-order upwind for the advection term, and second-order central difference for the diffusion term. Plot the result at $t = 0.2$ s and $t = 0.5$ s.

12P-4 A dye injected in flow stream, and the scalar concentration $u(x,t)$ is following the one-dimensional advection-diffusion equation given by

$$\frac{\partial u}{\partial t} + c\frac{\partial u}{\partial x} = \nu\frac{\partial^2 u}{\partial x^2}, \quad 0 \leq x \leq 1,$$

where, $u(x,t)$ represents the concentration of a passive scalar (such as ink), c is the constant advection velocity, and ν is the diffusion coefficient.

The initial condition is defined as a rectangular distribution:

$$u(x,0) = u_0(x) = \begin{cases} 0 & \text{for } 0 \leq x \leq x_1, \\ 1 & \text{for } x_1 < x \leq x_2, \\ 0 & \text{for } x_2 < x \leq 1, \end{cases}$$

where $0 < x_1 < x_2 < 1$.

The boundary conditions are of Neumann type (zero-gradient):

$$\frac{\partial u}{\partial x}(0,t) = \frac{\partial u}{\partial x}(1,t) = 0.$$

Discretize the governing equation using first-order upwind difference for the advection term $c\frac{\partial u}{\partial x}$, and second-order central difference for the diffusion term $\nu\frac{\partial^2 u}{\partial x^2}$. Divide the domain into 50 uniform segments (i.e., $\Delta x = 1/50$), and let u_j^n denote the numerical approximation of u at grid point x_j and time level t^n. Let $x_1 = 0.4\,\text{cm}$, and $x_2 = 0.6\,\text{cm}$. Analyze the time evolution of the scalar concentration $u(x,t)$. Investigate the influence of c on the result.

12P-5 Consider the one-dimensional PDE:

$$\frac{\partial u}{\partial t} + u\frac{\partial u}{\partial x} = \nu\frac{\partial^2 u}{\partial x^2},$$

where $u(x,t)$ is the scalar quantity being transported and diffused, and ν is the diffusion coefficient. Explicit finite-difference schemes are proposed to numerically solve this equation are

i Upwind Scheme

ii FTCS Scheme

iii Leapfrog Scheme

Write the FD formulation for this equation using first-order scheme for time and adection, and second-order scheme for diffusion.

REFERENCES

L. F. Richardson, The approximate arithmetical solution by finite differences of physical problems involving differential equations, with an application to the stresses in a masonry dam, Philosophical Transactions of the Royal Society of London, Series A, 210, 307-357, 1910.

L. F. Richardson, Weather Prediction by Numerical Process, Cambridge: Cambridge University Press, 1922.

J. M. Burgers, A mathematical model illustrating the theory of turbulence, Advances in Applied Mechanics, 1, 171-199, 1948.

E. Hopf, The partial differential equation $u_t + uu_x = \mu u_{xx}$, Communications on Pure and Applied Mathematics, 3, 201-230, 1950.

J. D. Cole, On a quasi-linear parabolic equation occurring in aerodynamics, Quarterly of Applied Mathematics, 9 (3), 225-236, 1951.

J. M. Burgers, The Nonlinear Diffusion Equation: Asymptotic Solutions and Statistical Problems, Dordrecht: Springer, 1974.

J. Bec, K. Khanin, Burgers turbulence, Physics Reports, 447 (1-2), 1-66, 2007.

13 Elliptic Partial Differential Equations and Grid Arrangement Strategies

In this chapter, we will explore the concept of elliptic partial differential equations (PDEs), focusing on their mathematical structure, physical interpretations, and computational treatment in fluid mechanics and related fields. Elliptic equations arise in steady-state problems where equilibrium is maintained, such as potential flow, electrostatics, and heat conduction. We study classic forms like Laplace's and Poisson's equations, as well as their role in incompressible Navier-Stokes equations. The global dependence of elliptic PDEs on boundary conditions is emphasized, and we discuss how staggered and collocated grids are used to discretize and solve these problems in CFD.

- Grasp the mathematical structure and physical significance of elliptic PDEs.
- Derive the pressure-Poisson equation from the incompressible Navier-Stokes system.
- Compare pressure-velocity coupling on staggered and collocated grids.
- Assess how boundary conditions globally influence solutions of elliptic systems.

Elliptic partial differential equations (PDEs) represent a fundamental class of equations central to steady-state phenomena across physics and engineering. These equations model systems that are in equilibrium, where changes in one part of the domain influence the entire region instantly. The most prominent examples include Laplace's equation, which governs potential flow and electrostatics in source-free regions, and Poisson's equation, which generalizes it to include distributed sources. The Helmholtz equation, another example, incorporates wave-like behavior with spatial attenuation and is widely used in acoustics and structural vibration. Unlike parabolic or hyperbolic PDEs that exhibit time dependence or wave propagation, elliptic PDEs are inherently spatial and global. Their solutions are continuous, smooth, and sensitive to boundary conditions, making the formulation and specification of appropriate boundary data a critical aspect of any problem. This global sensitivity means that a change at the boundary influences the solution throughout the entire domain, highlighting the importance of mathematical rigor and numerical consistency.

DOI: 10.1201/9781003663171-13

A key application of elliptic PDEs lies in the solution of the incompressible Navier-Stokes equations, particularly in computing the pressure field. In incompressible flows, the pressure does not evolve explicitly with time but is implicitly tied to the divergence-free condition of the velocity field through the Poisson equation. By applying the divergence operator to the momentum equation and using the incompressibility constraint, a Poisson equation for pressure is derived. This equation is elliptic in nature, meaning it requires boundary conditions across the domain and must be solved globally at every time step to ensure a divergence-free velocity field.

We also discuss the implications of grid arrangement in solving elliptic equations, particularly in CFD. Two common strategies are collocated and staggered grid arrangements. Collocated grids store all variables at the same spatial locations, offering simplicity but introducing the potential for pressure-velocity decoupling and checkerboard patterns. Staggered grids, on the other hand, store velocities and pressure at different spatial locations, enhancing numerical stability and preserving physical coupling. We illustrate these concepts with schematic diagrams and explain how they influence the formation of the pressure Poisson equation, its matrix structure, and its numerical solution.

Elliptic partial differential equations (PDEs) are a class of equations characterized by their behavior and the conditions under which they are valid. These equations typically arise in problems involving steady states or equilibrium states. Here are some classic examples of elliptic equations:

Laplace's Equation:

$$\nabla^2 u = 0$$

This equation appears in problems involving potential theory, electrostatics, and incompressible fluid flow, where u can represent a potential function in steady state.

Poisson's Equation:

$$\nabla^2 u = f$$

Here, f is a known function. Poisson's equation generalizes Laplace's equation and appears in contexts like gravitational fields and electrostatic potentials with charge distributions.

Helmholtz Equation:

$$\nabla^2 u + k^2 u = 0$$

This equation is used in problems involving vibration and acoustics, where $k^2 > 0$ is a parameter related to the properties of the medium and the nature of the wave-like phenomena.

Elliptic equations, such as the Laplace and Poisson equations, typically produce smooth solutions. Elliptic problems represent phenomena where information propagates instantly across the entire domain. As a result, the solution at each point in the domain depends globally on all boundary conditions and sources. Such global dependence ensures that any irregularities or discontinuities in boundary data or sources become rapidly smoothed out, resulting in inherently smooth and continuous solutions within the domain.

The Laplace operator, commonly represented as ∇^2, is fundamental in mathematical physics, appearing in numerous equations that describe physical phenomena. It involves second-order spatial derivatives, specifically, the sum of second derivatives in both spatial dimensions $u_{xx} + u_{yy}$. Equations utilizing the Laplace operator include the Laplace equation, which models potential flow in fluid dynamics, and the Poisson equation, which describes potential flows influenced by source terms or steady-state heat conduction with sources. Other important examples include the Helmholtz equation, pertinent to vibrational modes in structures such as plates.

For elliptic equations, the solution at any point within the domain is determined by the boundary conditions and source terms specified over the entire region, underscoring the intrinsically global nature of these phenomena.

Example 13.1

Solve the one-dimensional Laplace partial differential equation $\frac{d^2u}{dx^2} = 0$ numerically across the interval $x \in [0,1]$. The Laplace equation is indicative of systems without internal sources or sinks and is fundamental in understanding the behavior of scalar fields such as potential or concentration within a specified domain. The boundary conditions are $u(0) = 0$ at $x = 0$ and $u(1) = 1$ at $x = 1$. Apply the finite difference method to discretize the spatial domain and reformulate the continuous PDE into a solvable algebraic system. Investigate the resolution of the computational grid and its effect on the resulting linear system

Solution

$$\frac{d^2u}{dx^2} = 0$$

on the interval $[0,1]$ with boundary conditions

$$u(0) = 0, \quad u(1) = 1.$$

Exact Solution: The governing equation is

$$\frac{d^2u}{dx^2} = 0.$$

A zero second derivative means that the function $u(x)$ does not bend or curve; mathematically, it must be a linear function of x. Specifically, integrating $\frac{d^2u}{dx^2} = 0$ once with respect to x yields

$$\frac{du}{dx} = A,$$

for some constant A. Integrating again gives

$$u(x) = Ax + B,$$

where A and B are constants. The boundary conditions $u(0) = 0$ and $u(1) = 1$ fix these constants. Substituting $x = 0$ gives $B = 0$, and substituting $x = 1$

gives $A+B=1$. Thus $A=1$ and $B=0$, so the solution is $u(x)=x$.. This shows that the solution is a straight line between the two boundary values.

Numerical Solution

First, note that the exact analytical solution to

$$\frac{d^2u}{dx^2} = 0$$

with those boundary conditions is

$$u(x) = x.$$

To solve numerically, divide the domain $[0,1]$ into $N+1$ equal segments of size

$$h = \frac{1}{N+1},$$

so the grid points are

$$x_0 = 0,\ x_1 = h,\ x_2 = 2h,\ \ldots,\ x_N = Nh,\ x_{N+1} = 1.$$

We seek approximations $u_1, u_2, \ldots, u_N$ for the unknown solution at $x_1,\ldots,x_N$. The values at x_0 and x_{N+1} are the boundary conditions,

$$u_0 = 0, \quad u_{N+1} = 1.$$

The standard second-order finite difference discretization for $\frac{d^2u}{dx^2}$ at interior node x_i is

$$\frac{u_{i+1} - 2u_i + u_{i-1}}{h^2} = 0,$$

which rearranges to

$$-u_{i-1} + 2u_i - u_{i+1} = 0.$$

For $i=1$, the equation becomes

$$-u_0 + 2u_1 - u_2 = 0 \Rightarrow 2u_1 - u_2 = 0,$$

since $u_0 = 0$.

For $i = 2,\ldots,N-1$, the equations are similar:

$$-u_{i-1} + 2u_i - u_{i+1} = 0.$$

For the last interior node $i=N$,

$$-u_{N-1} + 2u_N - u_{N+1} = 0 \Rightarrow -u_{N-1} + 2u_N = u_{N+1} = 1.$$

Collecting all interior equations in matrix form:

$$\begin{pmatrix} 2 & -1 & 0 & \dots & 0 \\ -1 & 2 & -1 & \dots & 0 \\ 0 & -1 & 2 & \dots & 0 \\ \vdots & \vdots & \vdots & \ddots & -1 \\ 0 & 0 & 0 & -1 & 2 \end{pmatrix} \begin{pmatrix} u_1 \\ u_2 \\ u_3 \\ \vdots \\ u_N \end{pmatrix} = \begin{pmatrix} 0 \\ 0 \\ 0 \\ \vdots \\ 1 \end{pmatrix}.$$

When this tridiagonal linear system is solved, the resulting solution vector $(u_1, u_2, \dots, u_N)$ will approximate the straight line $u(x) = x$. As N increases (finer grid), the solution quickly converges toward the analytical solution.

```
import numpy as np

# Number of interior points
N = 5

# Step size
h = 1.0/(N+1)

# Construct the coefficient matrix A (N x N)
A = np.zeros((N, N))
for i in range(N):
    A[i, i] = 2.0
    if i > 0:
        A[i, i-1] = -1.0
    if i < N-1:
        A[i, i+1] = -1.0

# Construct right-hand side vector b (N x 1)
b = np.zeros(N)
# The last equation includes boundary condition u_{N+1} = 1
b[-1] = 1.0

# Solve the linear system
u_interior = np.linalg.solve(A, b)

# Include boundary conditions to get full solution
u = np.concatenate(([0.0], u_interior, [1.0]))

# Create array of x values
x = np.linspace(0.0, 1.0, N+2)

# Print numerical solution
print('Grid points x:', x)
print('Solution u:', u)
```

This gives

```
Grid points x:
[0.   0.16666667 0.33333333 0.5   0.66666667 0.83333333  1.]
Solution u:
[0.   0.16666667 0.33333333 0.5   0.66666667 0.83333333  1.]
```

The numerical analysis confirms the exact solution.
We have learned through this example the behavior of elliptic equation on numerical grid. The example shows that the solution of one-dimensional Laplace equation depends on the boundary conditions across the entire domain i.e., if we change the value at one boundary, the entire solution would readjust according to updated boundary condition. Also, we learned how grid resolution affects both accuracy and the numerical properties of the system, such as the condition number.

The Navier-Stokes equations describe the motion of fluids and are fundamental to fluid dynamics. Their mathematical classification—whether elliptic, parabolic, or hyperbolic—depends on the flow regime and assumptions such as incompressibility. For incompressible flows, the equations exhibit an elliptic character in the pressure field, which arises from the incompressibility condition. The governing equations for an incompressible fluid are:

$$\frac{\partial \vec{v}}{\partial t} + (\vec{v}\cdot\nabla)\vec{v} = -\frac{1}{\rho}\nabla p + \nu\nabla^2\vec{v} + \vec{f}, \quad \nabla\cdot\vec{v} = 0,$$

where $\vec{v}$ is the velocity, p is pressure, ρ is density, ν is kinematic viscosity, and $\vec{f}$ represents body forces like gravity or electromagnetic effects. The divergence-free condition $\nabla\cdot\vec{v} = 0$ enforces incompressibility, leading to an instantaneous adjustment of pressure to maintain this constraint.

To derive the elliptic equation governing the pressure field, the divergence operator is applied to the momentum equation. The nonlinear term simplifies using the identity:

$$\nabla\cdot\left((\vec{v}\cdot\nabla)\vec{v}\right) = (\partial_k v_j)(\partial_j v_k),$$

since $\nabla\cdot\vec{v} = 0$ eliminates higher-order derivatives. This yields the pressure Poisson equation:

$$\nabla^2 p = -\rho(\partial_k v_j)(\partial_j v_k) + \nabla\cdot\vec{f}.$$

Here, the Laplacian ∇^2 makes the equation elliptic, meaning pressure depends on global boundary conditions rather than local or time-dependent effects. If the body force $\vec{f}$ is irrotational (e.g., uniform gravity), its divergence vanishes, leaving the pressure determined solely by velocity gradients.

The elliptic character of the pressure formulation highlights its role in enforcing incompressibility by instantaneously communicating velocity adjustments throughout the fluid domain. This global coupling distinguishes it from hyperbolic or parabolic components, which govern wave propagation or diffusion. Thus, in incompressible flows, pressure acts as a constraint, adjusting dynamically to maintain continuity while linking velocity variations throughout the domain.

13.1 PRESSURE FIELD IN CFD

In computational fluid dynamics (CFD), when dealing with incompressible flow, the velocity is influenced by both inflow/outflow conditions and expansion effects. Pressure is determined by the conservation of mass and the force balance throughout the entire system, ensuring that these quantities are in equilibrium. Setting proper boundary conditions is crucial for CFD simulations. Boundary conditions, must be chosen to reflect the real-world physical constraints of the system, ensuring the solution's accuracy and reliability. The flow should be influenced by these conditions in a way that maintains physical realism and avoids non-physical outcomes.

In compressible flow, the behavior of velocity, pressure, and temperature becomes more complex due to the changes in fluid density, which can vary significantly within the flow field. The velocity is still influenced by the inflow/outflow conditions and expansion, but it is further affected by the compressibility of the fluid, meaning the density changes as the flow accelerates or decelerates. Pressure is governed by the conservation of mass and momentum equations, and temperature is determined through energy conservation, which accounts for the conversion between kinetic energy, internal energy, and work done by the fluid.

In CFD, the conservation equations (mass, momentum, and energy) are the primary tools for describing the flow. These equations involve pressure as a term, but the absolute pressure is often less important than the pressure difference between different regions of the flow. The pressure difference drives the flow and governs the dynamics of shock waves, expansion fans, and other compressibility effects. This is why the equation of state (ideal gas equation), which provides an absolute relationship between pressure, density, and temperature, is not always necessary for the CFD solver. Instead, solvers typically focus on the pressure gradient or pressure differential that drives the flow, rather than using the absolute pressure value provided by the equation of state. This approach is particularly useful because, in many cases, the thermodynamic properties like temperature and density can be derived indirectly from the conservation of energy and momentum without needing to solve an equation of state directly. Hence, the key information for simulating the flow field comes from how pressure changes spatially and temporally, not from its absolute values.

13.2 CHOICE OF GRID ARRANGEMENT

CFD solvers usually employ either collocated or staggered grid arrangements. A collocated grid arrangement refers to a discretization strategy in which all primary variables—most notably pressure and velocity components—are stored at the same set of points in the computational domain. In two-dimensional cases, for example, both p (pressure) and (u, v) (velocity components) are assigned to the centers of the same finite volume (or finite difference) cells. This contrasts with a staggered grid, where velocities and pressures occupy different points (e.g., velocities on cell faces and pressure at cell centers).

One advantage of a collocated approach is the relative simplicity of data management. Because all variables share the same index structure, code implementation can be more straightforward: each computational cell directly accesses its local velocity and pressure values without extra interpolation steps. This can reduce overhead in memory lookups and potentially lead to clearer, more concise programming logic. The collocated scheme is common in many finite volume codes, especially when combined with special techniques to handle incompressible flow.

Staggered grid arrangement for finite-difference and finite-volume solvers for incompressible flows.
In a staggered (staggered-grid) variable arrangement, the locations where u, v, and p are stored differ. The total number of variables also differs. Including the boundary nodes, we have:

u: $(nx+1) \times ny$

v: $nx \times (ny+1)$

p: $nx \times ny$

In a two-dimensional setting, nx and ny typically refer to the number of grid points (in FD) or control volumes (in FV) along the x- and y-directions. In a conventional collocated grid, all flow variables, such as pressure and velocity components, are stored at the same locations: these are grid nodes in finite difference methods and cell centers in finite volume methods. However, many CFD approaches adopt a staggered grid layout to better capture incompressible flow behavior and eliminate numerical artifacts such as pressure-velocity decoupling. In a staggered configuration, pressure is stored at grid points (FD) or cell centers (FV), while velocity components are defined at cell faces, offset from the scalar variable locations. In a staggered grid, the velocity components (u, v) and pressure (p) are stored at different spatial locations to prevent pressure-velocity decoupling (e.g., checkerboard oscillations in collocated grids).
For example, if there are nx grid points or cells in the x-direction and ny in the y-direction, the pressure field comprises $nx \times ny$ unknowns. The horizontal velocity component u is defined on vertical cell faces, requiring $nx+1$ columns and ny rows, while the vertical component v is defined on horizontal cell faces, needing nx columns and $ny+1$ rows. This results in data arrays of size $(nx+1) \times ny$ for u and $nx \times (ny+1)$ for v.

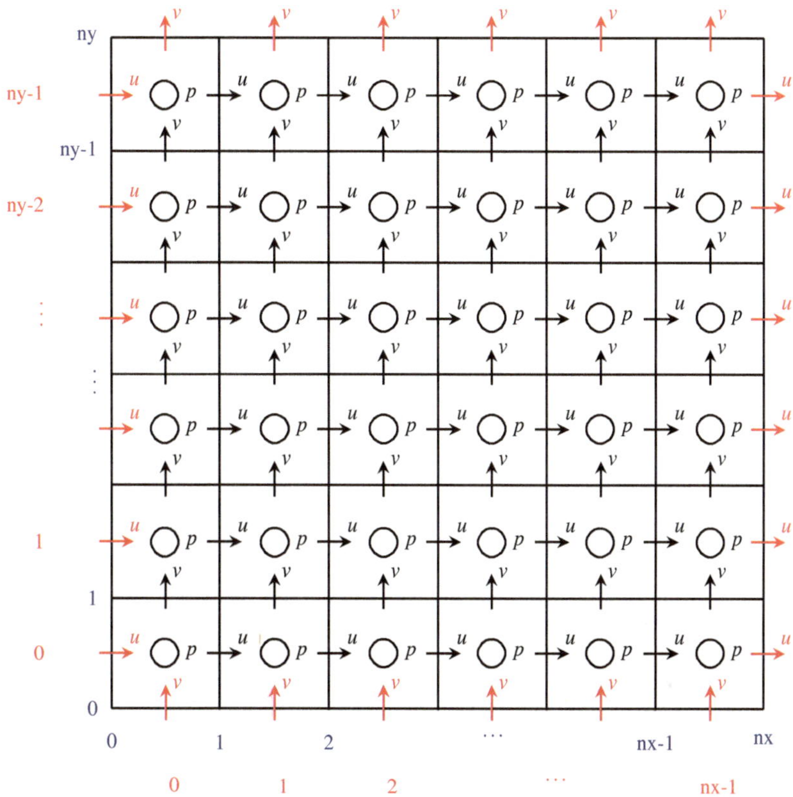

Figure 13.1: **Comparison of collocated and staggered grid indexing with zero-based convention including physical boundaries (without ghost cells). Pressure (p) is defined at cell centers (FV) or grid nodes (FD) using collocated indices nx and ny (shown in purple). In the staggered grid layout, velocity components u and v are defined on cell faces: u on vertical faces requiring $nx+1$ points in the x-direction, and v on horizontal faces requiring $ny+1$ points in the y-direction (indices shown in red). This arrangement avoids pressure-velocity decoupling. In the collocated FD or FV approach, all variables—u, v, and p—are stored at the same spatial locations.**

In programming languages such as C, C++, Java, Python (for lists/arrays), modern Fortran, and Julia, array indexing typically starts at zero. For example, an array declared as a[N] is indexed from a[0] to a[N-1], yielding N total elements. The grid layout in Figure 13.1 follows this zero-based indexing convention in programming languages (see Table 13.1). The grid layout for collocated grid is tabulated in Table 13.2.

> u (horizontal velocity): Defined at vertical cell faces (i.e., staggered in the x-direction); in FD, these are midpoints between adjacent grid nodes; in FV, they are the shared faces between adjacent control volumes.

Table 13.1
Array Dimensions in a Staggered Grid Layout (Applicable to FD and FV)

Variable	Physical Location	Index Range (0-based)	Total Elements
Pressure (p)	Grid nodes (FD); Cell centers (FV)	`p[0:nx-1][0:ny-1]`	$nx \times ny$
Horizontal velocity (u)	Vertical cell faces	`u[0:nx][0:ny-1]`	$(nx+1) \times ny$
Vertical velocity (v)	Horizontal cell faces	`v[0:nx-1][0:ny]`	$nx \times (ny+1)$

Table 13.2
Array Dimensions in a Collocated Grid Layout (Applicable to FD and FV)

Variable	Physical Location	Index Range (0-based)	Total Elements
Pressure (p)	Grid nodes (FD); Cell centers (FV)	`p[0:nx-1][0:ny-1]`	$nx \times ny$
Horizontal velocity (u)	Same as pressure	`u[0:nx-1][0:ny-1]`	$nx \times ny$
Vertical velocity (v)	Same as pressure	`v[0:nx-1][0:ny-1]`	$nx \times ny$

v (vertical velocity): Defined at horizontal cell faces (i.e., staggered in the y-direction); in FD, these are between adjacent grid nodes; in FV, they lie on the horizontal faces of control volumes.

p (pressure): In FD, defined at grid nodes; in FV, defined at cell centers (control volume centroids).

When forming the pressure Poisson equation or a pressure-correction equation in a finite-difference code, the discretized operator typically involves only the pressure values at cell centers. Each cell center receives a pressure update based on its immediate neighbors in the x- and y-directions. As a result, the coefficient matrix has dimensions $(nx \times ny) \times (nx \times ny)$, where nx and ny are the number of pressure control volumes in the respective directions. Each row of this matrix contains at most five nonzero entries: one on the main diagonal and up to four corresponding to the neighboring cells (left, right, top, bottom). The remaining entries are zero, resulting in a large, sparse linear system. While implementing a full dense matrix is straightforward for small problems, it becomes prohibitively memory-intensive as the grid size increases. Therefore, it is common practice to employ sparse matrix storage formats such as Compressed Sparse Row (CSR) or Compressed Sparse Column (CSC), together with iterative solvers optimized for sparse linear systems, to efficiently solve the elliptic equation for pressure. Even though some boundary cells or boundary faces might not require explicit storage, especially if their values are prescribed or remain fixed, it is common in early or instructional versions of finite difference codes to include them for coding simplicity. This means an array in memory might cover all indices from `0` up to `nx` (inclusive) in the x-direction, even if not all are truly needed. In languages like C or Python, arrays indeed start at index zero, which matches how the discrete nodes can be labeled. Over time, more advanced implementations or performance-focused CFD codes may optimize away unused boundary entries. Nevertheless, the key concept remains that a staggered finite difference mesh defines distinct storage patterns for pressure and velocity components, ensuring more stable and accurate incompressible flow simulations.

Figure 13.2 shows a typical 2D staggered-grid arrangement for incompressible flow computations. Circles (p) at cell centers represent pressure nodes. The arrows labeled u are the x direction velocity components, while arrows labeled v are the y direction velocity components. The hatched region marks the control volume for one representative cell, whose center is at (i, j). The dashed diamond region indicates the stencil or neighborhood of relevant points for updating variables in that cell. Storing pressure and velocities in this staggered manner helps avoid spurious decoupling and checkerboard patterns in incompressible flow simulations.

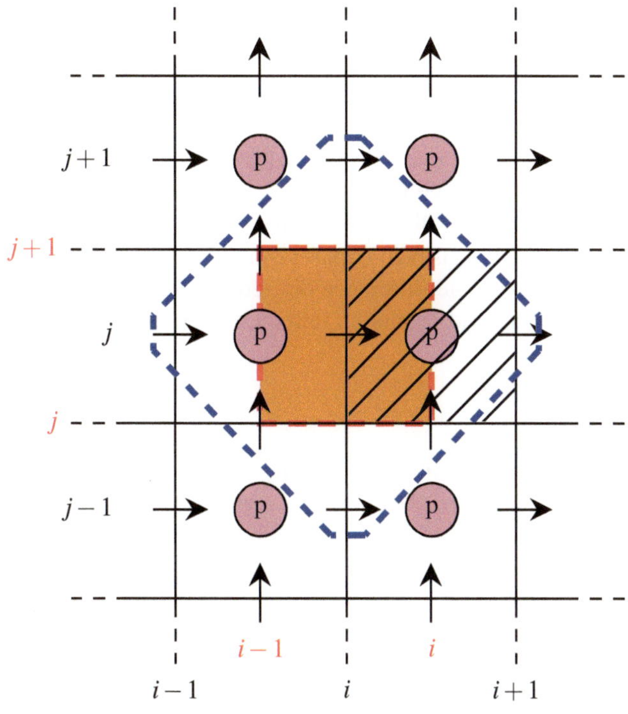

Figure 13.2: **Finite-difference stencil for the horizontal velocity component u on a staggered grid using zero-based indexing. The orange region outlined by the red dashed box marks the stencil center associated with $u[i, j]$, while the larger region enclosed by the blue dashed diamond includes the neighboring values required for discretization. The hatched region spans between two adjacent pressure nodes in the x-direction. In this staggered arrangement, $u[i, j]$ is defined at a point halfway between pressure nodes along the x-axis, $v[i, j]$ at a midpoint along y, and $p[i, j]$ at the grid node. This configuration is typical of finite-difference methods for incompressible flows. Black indices (i, j) refer to grid nodes where pressure is stored, while red indices are used visually to indicate the grid location for the horizontal velocity component u.**

PROBLEMS

13P-1 For the given system of partial differential equations

$$u_y - v_x = 0$$

$$au_x + \frac{b}{2}(u_y + v_x) + cv_y = 0$$

i Determine the condition when this set of equations becomes elliptic in type.

ii Transform this set of equations into a scalar partial differential equation of second order.

iii Discretize using finite-difference technique the scalar equation for an equidistant grid with second-order accuracy and determine the truncation error.

13P-2 Explain why the pressure Poisson equation derived from the incompressible Navier-Stokes equations is elliptic in nature. How does this elliptic character affect the formulation and numerical solution of incompressible flow problems?

13P-3 Given the one-dimensional Laplace equation $\frac{d^2u}{dx^2} = 0$ with Dirichlet boundary conditions, derive the finite difference discretization and form the resulting linear system. How does the grid resolution influence the condition number and accuracy of the solution?

REFERENCES

P. Bartholomew, F. Denner, M. H. S. A. Azis, A. Marquis, Unified formulation of the momentum-weighted interpolation for collocated variable arrangements, Journal of Computational Physics, Volume 375, Pages 1137-1159, 2018.

S. Majumdar, Role of Underrelaxation in Momentum Interpolation for Calculation of Flow with Non-Staggered Grids, Numerical Heat Transfer, 13(1), pp. 125-132, 1988.

K. C. Karki and S. V. Patankar, A Pressure-Based Calculation Procedure for Viscous Flows at All Speeds in Arbitrary Configurations, AIAA Journal, 27(9), pp. 1167-1174, 1989.

D. L. Brown, R. Cortez, M. L. Minion, Accurate projection methods for the incompressible Navier-Stokes equations. Journal of Computational Physics, 168(2), 464-499, 2001.

F. Sotiropoulos, X. Yang, Immersed boundary methods for simulating fluid-structure interaction. Progress in Aerospace Sciences, 65, 1-21, 2014.

14 Iterative Methods for Elliptic PDEs: Laplace and Poisson Solvers

In this chapter, we address the numerical solution of elliptic partial differential equations (PDEs), especially the two-dimensional Laplace and Poisson equations. These fundamental equations appear across a wide spectrum of steady-state physics and engineering problems, including heat conduction, electrostatics, and incompressible flow.

After working through this chapter, you should be able to:

- Formulate both Laplace and Poisson problems on a two-dimensional domain, enforcing Dirichlet or Neumann boundary conditions as needed.
- Use the finite difference method (FDM) to derive discrete algebraic equations from second-order elliptic PDEs, thereby constructing large sparse matrices suitable for numerical solution.
- Choose among direct solvers (such as LU factorization) and various iterative solvers (e.g., Jacobi, Gauss-Seidel, SOR, Conjugate Gradient, GMRES).
- Apply preconditioners (like ILU, Jacobi scaling, or Algebraic Multigrid) to reduce the condition number of large systems, thereby accelerating iterative convergence.
- Understand why pressure referencing (pinning a single cell or imposing a zero-mean constraint) is necessary in incompressible CFD to avoid singular systems and ensure a unique pressure field.
- Evaluate solver performance and scalability by analyzing residual norms, L^2 errors, execution times, and the impact of mesh refinement or ill-conditioning on convergence.

This chapter introduces numerical methods for solving the two-dimensional Laplace equation, which serves as a cornerstone in modeling physical phenomena such as steady-state heat conduction, electrostatics, and incompressible fluid flow. The Laplace equation

$$\frac{\partial^2 p}{\partial x^2} + \frac{\partial^2 p}{\partial y^2} = 0$$

is central to many engineering problems where a scalar field must satisfy equilibrium conditions under certain boundary constraints. By focusing on a representative example with Dirichlet conditions, one gains a strong foundation for handling more elaborate geometries and real-world applications. In practice, the core challenge lies

DOI: 10.1201/9781003663171-14

not only in discretizing this equation but also in choosing efficient algorithms that converge reliably to an accurate solution.

One natural choice for converting the continuous Laplace equation into a system of linear equations is the finite difference method (FDM). Through central differencing, second derivatives with respect to x and y become simple algebraic relationships linking each interior point to its neighbors. This yields a discrete version of the Laplace operator, which can be stored in a matrix form. The boundary conditions then appear as either explicit constraints or incorporated values at the edges of the computational mesh. While the discrete system is typically large and sparse for two-dimensional grids, methods such as Jacobi, Gauss-Seidel, and Successive Over-Relaxation (SOR) provide iterative ways to solve for the unknowns.

Each of these iterative solvers has distinct properties. Jacobi updates a grid node by relying on values from the previous iteration across the entire domain, making it conceptually simple but sometimes slow to converge. Gauss-Seidel improves convergence by employing newly updated node values as soon as they become available, hence distributing corrections more rapidly through the domain. SOR goes further by introducing a relaxation factor, ω, which can accelerate convergence if chosen properly (usually in the range $1 < \omega < 2$). Still, one must carefully tune ω to avoid poor convergence or numerical oscillations.

Another technique, the fractional-step method, attempts to split updates into horizontal and vertical sweeps. Although effective in certain contexts (notably spectral methods for incompressible Navier-Stokes flows), it can be less efficient when applied directly to a standard Laplace problem on a finite difference grid. Each sweep only partially adjusts the field, often requiring more sub-iterations or specialized relaxation factors to achieve the same accuracy as Gauss-Seidel or SOR in a comparable number of steps.

In addition to practical considerations about convergence speed and memory requirements, one must evaluate the final solution accuracy. A standard way to measure discrepancy is via the L^2 error, comparing the numerical solution to an exact or reference solution. This highlights whether the discretization and iterative method are adequately capturing the continuous physics. As the grid becomes finer, Jacobi, Gauss-Seidel, and SOR typically achieve low L^2 errors on the order of 10^{-4} or smaller, while fractional-step methods can plateau at higher residuals unless carefully tuned.

Throughout this chapter, we thus explore the 2D Laplace equation, from the core mathematical foundations and boundary conditions to the discretization via finite differences and subsequent iterative solution. By examining representative results and error metrics, readers gain insight into how each solver performs in practice, reinforcing key concepts that can be extended to more general elliptic problems in computational physics and engineering.

Finally, we address additional considerations like sparse matrix storage, preconditioning, and the importance of pressure referencing in incompressible computational fluid dynamics. Compressed Sparse Row (CSR) and Compressed Sparse Column (CSC) formats optimize memory usage for large, sparse systems, which is crucial when scaling up to big grids. Preconditioners, such as ILU or algebraic multigrid, help improve spectral properties of the linear system, reducing the overall iteration count and potentially preventing divergence. In incompressible flows, pressure referencing or pinning removes the nullspace inherent in pure Neumann problems, ensuring a unique, physically meaningful solution. Taken together, these sections provide a thorough overview of how to move from a simple PDE specification to robust numerical solutions and analyses across a broad range of engineering and scientific applications.

14.1 CHOICE OF SOLVERS

We will now solve a PDE—2D Laplace equation, which is widely used to model physical phenomena such as heat conduction, electrostatics, and steady-state fluid flow. The equation is:

$$\frac{\partial^2 p}{\partial x^2} + \frac{\partial^2 p}{\partial y^2} = 0$$

where:
$p(x,y)$: The dependent variable, e.g., temperature, potential, or fluid velocity.
x and y: Spatial coordinates.

This is a second-order, linear, elliptic partial differential equation.

Boundary Conditions (BCs): the boundary conditions for this problem are:

1. Dirichlet Boundary Condition (Fixed values):

$$p(x,y) = \sin\left(\frac{3\pi x}{2L_x}\right) \quad \text{on the top boundary } (y = L_y).$$

2. Zero Dirichlet Boundary Condition:

$$p(x,y) = 0 \quad \text{on the bottom boundary } (y = 0), \text{and the side boundaries } (x = 0, x = L_x).$$

These conditions define a problem with a fixed value at the top boundary and zero values on all other sides, creating a potential gradient across the domain.

Exact Solution: The analytical solution for this problem is given as:

$$p(x,y) = \sin\left(\frac{3\pi x}{2L_x}\right) \cdot \frac{\sinh\left(\frac{3\pi y}{2L_y}\right)}{\sinh\left(\frac{3\pi L_y}{2L_x}\right)},$$

Using finite difference methods (FDM), we discretize and solve the Laplace equation on a 2D grid. The computational domain is discretized into a uniform grid with $nx \times ny$ points. The Laplace operator is approximated using the central difference scheme (see Figure 14.1):

$$\frac{\partial^2 p}{\partial x^2} \approx \frac{p_{i-1,j} - 2p_{i,j} + p_{i+1,j}}{\Delta x^2},$$

$$\frac{\partial^2 p}{\partial y^2} \approx \frac{p_{i,j-1} - 2p_{i,j} + p_{i,j+1}}{\Delta y^2}.$$

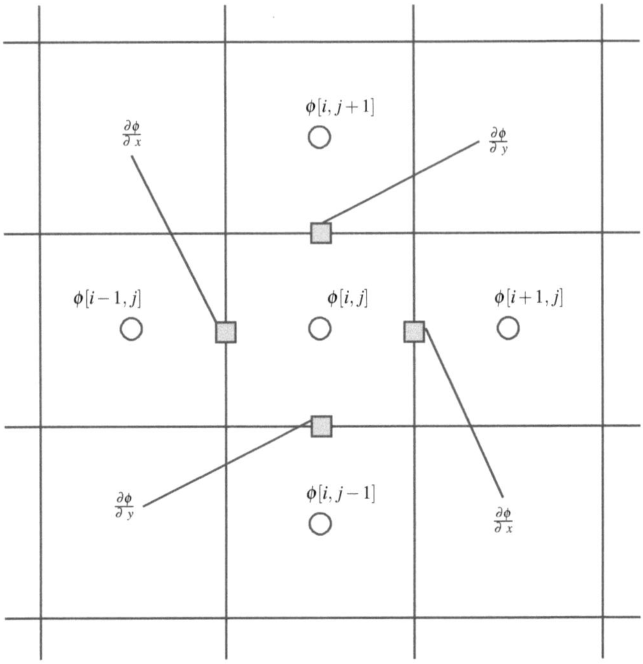

Figure 14.1: **The Poisson Stencil in terms of variable ϕ.**

Combining these, the discretized form of the Laplace equation becomes:

$$p_{i,j} = \frac{\Delta y^2 \left(p_{i-1,j} + p_{i+1,j}\right) \; + \; \Delta x^2 \left(p_{i,j-1} + p_{i,j+1}\right)}{2\left(\Delta x^2 + \Delta y^2\right)}.$$

We will solve the discretized Laplace equation iteratively using:

Jacobi Method: Updates $p_{i,j}$ using values from the previous iteration.

Gauss-Seidel Method: Updates $p_{i,j}$ sequentially using already updated values in the current iteration.

Successive Over-Relaxation (SOR): Accelerates convergence using a relaxation parameter ω (here $\omega = 1.8$).

Fractional Step Method: Splits updates into horizontal and vertical sweeps to improve convergence.

Each solver iterates until the relative error (L^2 norm of the residual) falls below a specified tolerance (10^{-8}). The code compares the numerical methods by calculating the L^2 error:

$$L^2\text{Error} = \frac{\|p_{\text{numerical}} - p_{\text{exact}}\|}{\|p_{\text{exact}}\|}.$$

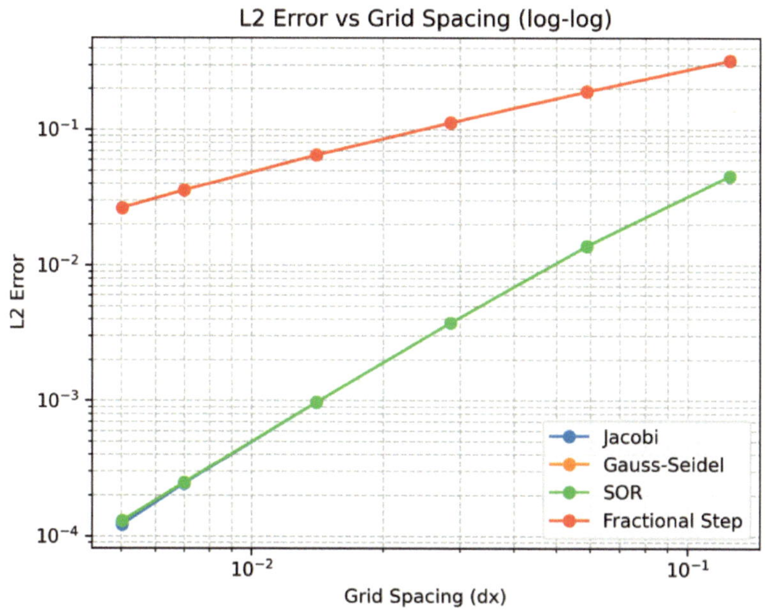

Figure 14.2: **Comparison of solvers.**

Figure 14.2 shows that SOR, Jacobi, and Gauss-Seidel converge fast. However, fractional step method, even with fine grids was failed to reach lower levels of L^2 norm. These results show a typical pattern for a 2D Laplace solver on a sufficiently refined mesh: Jacobi, Gauss-Seidel, and SOR all converge to a very low error (on the order of 10^{-4}), while the simple fractional-step approach lingers at a higher residual ≈ 0.026 (see Table 14.1). This is because the row-column partial updates in fractional-step are far less efficient for a purely elliptic problem: each pass only incorporates one-dimensional averaging instead of updating the full 2D stencil at once. Consequently, fractional-step often requires many more sub-sweeps and a carefully tuned relaxation factor to approach the accuracy that Gauss-Seidel or SOR achieve with fewer iterations.

Fractional-step methods are particularly effective in spectral or pseudo-spectral solvers for incompressible Navier-Stokes problems, because the transformation into Fourier or Chebyshev modes decouples most operators into simple diagonal systems. In each fractional step, one solves for an intermediate velocity field and then projects it onto a divergence-free space by solving a Poisson equation for the pressure. Due to the orthogonality of the basis functions, these subproblems become much simpler, often reducing to sets of independent equations for each mode. This diagonalization, coupled with fast transforms, allows fractional-step approaches in spectral codes to be both elegant and highly efficient, yielding accurate solutions with relatively few operations compared to real-space iterative schemes of equivalent resolution.

Table 14.1
Comparison of L^2 Errors for Different Iterative Methods

Method	L^2 Error
Jacobi	0.000121
Gauss-Seidel	0.000129
SOR	0.000129
Fractional Step	0.026303

Table 14.2
Convergence rates of iterative methods for the discrete Poisson equation with Dirichlet boundary conditions. Definitions used: $a = (i_{max} + 1)\Delta x$, $b = (j_{max} + 1)\Delta y$, $\delta^2 = \dfrac{\Delta x^2 \cdot \Delta y^2}{2(\Delta x^2 + \Delta y^2)}$

Method	r	r/r_J	ω
Jacobi	$r_J = \delta^2\pi^2\left(\frac{1}{a^2} + \frac{1}{b^2}\right)$	1	1
Gauss–Seidel	$r_{GS} = 2r_J$	2	1
Gauss–Seidel (optimized)	$r_{BGS} = 2\delta\pi\sqrt{2\left(\frac{1}{a^2} + \frac{1}{b^2}\right)}$	$\dfrac{2\sqrt{2}}{\sqrt{r_J}}$	$\omega_{\text{opt}} = 2 - r_{BGS}$

14.2 STEADY LAMINAR FLOW THROUGH DUCT

We would like to analyze the fully developed laminar flow in a rectangular duct. We start from the Navier-Stokes equations under the assumptions of steady, incompressible flow with no-slip boundary conditions. The axial velocity profile in the duct, $\bar{w}$, is governed by the balance of pressure gradient and viscous diffusion.

The Navier-Stokes equations for incompressible flow are:

$$\rho\left(\frac{\partial \mathbf{u}}{\partial t} + \mathbf{u}\cdot\nabla\mathbf{u}\right) = -\nabla P + \mu\nabla^2\mathbf{u},$$

where:
$\mathbf{u} = (u, v, w)$ is the velocity vector in the x-, y-, and z-directions.
ρ is the density of the fluid.
μ is the dynamic viscosity.
P is the pressure.
∇^2 is the Laplacian operator.

For incompressible flow, the continuity equation is:

$$\nabla\cdot\mathbf{u} = 0.$$

Assumptions

To simplify the equations, we make the following assumptions: Fully developed flow: The velocity profile does not vary in the axial (z) direction, so:

$$\frac{\partial w}{\partial z} = 0, \quad \frac{\partial u}{\partial z} = 0, \quad \frac{\partial v}{\partial z} = 0.$$

This also implies $u = 0$ and $v = 0$, as there is no velocity in the x- or y-directions in the fully developed state.

Steady flow: The flow is time-independent, so:

$$\frac{\partial w}{\partial t} = 0.$$

No pressure variation in x- or y-directions: The pressure gradient is only along the axial (z) direction:

$$\frac{\partial P}{\partial x} = 0, \quad \frac{\partial P}{\partial y} = 0.$$

Uniform pressure gradient along z: The pressure gradient in the z-direction is constant:

$$\frac{\partial P}{\partial z} = -G,$$

where G is a constant.

No-slip boundary condition: The velocity is zero at the duct walls:

$$w = 0 \quad \text{at } x = \pm a,\ y = \pm b.$$

Under these assumptions, the axial momentum equation becomes:

$$0 = -\frac{\partial P}{\partial z} + \mu \left(\frac{\partial^2 w}{\partial x^2} + \frac{\partial^2 w}{\partial y^2} \right).$$

Substituting $\frac{\partial P}{\partial z} = -G$, we get:

$$G = \mu \left(\frac{\partial^2 w}{\partial x^2} + \frac{\partial^2 w}{\partial y^2} \right).$$

Rewriting:

$$\frac{\partial^2 w}{\partial x^2} + \frac{\partial^2 w}{\partial y^2} = \frac{G}{\mu}.$$

To simplify the problem, the velocity w and spatial coordinates are normalized: Normalize the axial velocity:

$$\bar{w} = \frac{w}{w_{\max}},$$

where $w_{\max}$ is the maximum axial velocity.

Normalize the spatial coordinates:

$$\bar{x} = \frac{x}{a}, \quad \bar{y} = \frac{y}{b}.$$

In terms of normalized coordinates:

$$\frac{\partial^2 w}{\partial x^2} = \frac{1}{a^2}\frac{\partial^2 \bar{w}}{\partial \bar{x}^2}, \quad \frac{\partial^2 w}{\partial y^2} = \frac{1}{b^2}\frac{\partial^2 \bar{w}}{\partial \bar{y}^2}.$$

Substituting these into the previous equation:

$$\frac{1}{a^2}\frac{\partial^2 \bar{w}}{\partial \bar{x}^2} + \frac{1}{b^2}\frac{\partial^2 \bar{w}}{\partial \bar{y}^2} = \frac{G}{\mu w_{\max}}.$$

Let $C = \frac{G}{\mu w_{\max}}$. Then:

$$\frac{\partial^2 \bar{w}}{\partial \bar{x}^2} + \left(\frac{a}{b}\right)^2 \frac{\partial^2 \bar{w}}{\partial \bar{y}^2} = a^2 C.$$

To simplify further, let $a^2 C = -1$. This is a convenient choice to normalize the forcing term. Then, the equation becomes:

$$\frac{\partial^2 \bar{w}}{\partial \bar{x}^2} + \left(\frac{a}{b}\right)^2 \frac{\partial^2 \bar{w}}{\partial \bar{y}^2} + 1 = 0.$$

Using the aspect ratio $\beta = \frac{b}{a}$, this becomes:

$$\left(\frac{b}{a}\right)^2 \frac{\partial^2 \bar{w}}{\partial x^2} + \frac{\partial^2 \bar{w}}{\partial y^2} + 1 = 0.$$

This is a second-order, linear, elliptic partial differential equation.
The boundary conditions are:

$$\bar{w} = 0 \quad \text{at } x = \pm 1,\ y = \pm 1.$$

These represent the no-slip condition at the duct walls.
The governing Partial Differential Equation (PDE) for this problem is:

$$\left(\frac{b}{a}\right)^2 \frac{\partial^2 \bar{w}}{\partial x^2} + \frac{\partial^2 \bar{w}}{\partial y^2} + 1 = 0$$

where:

$\bar{w}$ represents the normalized axial velocity in the duct. $\frac{b}{a}$ is the aspect ratio of the rectangular duct (here a and b are the half-widths of the duct in the x- and y-directions, respectively). - The equation describes the fully developed, steady-state, laminar flow of a viscous fluid in a square or rectangular duct.

The equation results from the Navier-Stokes equations under steady-state, incompressible, and fully developed flow conditions. It models the balance between viscous diffusion and the pressure gradient driving the flow.

The boundary conditions for this problem are:

$$\bar{w} = 0 \quad \text{at } x = \pm 1,\ y = \pm 1$$

These Dirichlet boundary conditions state that the velocity is zero at all the walls of the duct (no-slip condition), as expected for viscous flows near solid surfaces.

The PDE is solved numerically using the Finite Difference Method (FDM), which discretizes the spatial domain into a uniform grid and replaces derivatives with finite difference approximations.

The PDE is approximated on a uniform grid using central difference approximations:

$$\left(\frac{b}{a}\right)^2 \frac{\bar{w}_{j,i-1} - 2\bar{w}_{j,i} + \bar{w}_{j,i+1}}{\Delta x^2} + \frac{\bar{w}_{j-1,i} - 2\bar{w}_{j,i} + \bar{w}_{j+1,i}}{\Delta y^2} + 1 = 0$$

Rearranging for $\bar{w}_{j,i}$:

$$\bar{w}_{j,i} = \frac{\left(\frac{b}{a}\right)^2 \frac{\bar{w}_{j,i-1}+\bar{w}_{j,i+1}}{\Delta x^2} + \frac{\bar{w}_{j-1,i}+\bar{w}_{j+1,i}}{\Delta y^2} - 1}{2\left(\frac{b}{a}\right)^2/\Delta x^2 + 2/\Delta y^2}$$

Here:
Δx: Grid spacing in the x-direction.
Δy: Grid spacing in the y-direction.
$\bar{w}_{j,i}$: Velocity at grid point (j,i).

The computational domain is discretized into a uniform grid with $n_x \times n_y$ grid points. Boundary conditions are directly enforced by setting $\bar{w} = 0$ at all boundary points.

The discretized equation forms a linear system that is solved iteratively using the following techniques:

Jacobi Method: The Jacobi method is an iterative solver where the solution at each grid point is updated based only on the values from the previous iteration.

$$\bar{w}_{j,i}^{(n+1)} = \frac{\left(\frac{b}{a}\right)^2 \frac{\bar{w}_{j,i-1}^{(n)}+\bar{w}_{j,i+1}^{(n)}}{\Delta x^2} + \frac{\bar{w}_{j-1,i}^{(n)}+\bar{w}_{j+1,i}^{(n)}}{\Delta y^2} - 1}{2\left(\frac{b}{a}\right)^2/\Delta x^2 + 2/\Delta y^2}$$

Gauss-Seidel method : improves upon the Jacobi method by using the most recently updated values during iterations. This typically leads to faster convergence.

$$\bar{w}_{j,i}^{(n+1)} = \frac{\left(\frac{b}{a}\right)^2 \frac{\bar{w}_{j,i-1}^{(n+1)}+\bar{w}_{j,i+1}^{(n)}}{\Delta x^2} + \frac{\bar{w}_{j-1,i}^{(n+1)}+\bar{w}_{j+1,i}^{(n)}}{\Delta y^2} - 1}{2\left(\frac{b}{a}\right)^2/\Delta x^2 + 2/\Delta y^2}$$

Successive Over-Relaxation (SOR) Method: The SOR method further accelerates convergence by applying a relaxation factor ω. It updates the solution as:

$$\bar{w}_{j,i}^{(n+1)} = (1-\omega)\bar{w}_{j,i}^{(n)} + \omega \times (\text{Updated Gauss-Seidel value})$$

Relaxation Factor (ω): $1 < \omega < 2$: Over-relaxation for faster convergence. $\omega = 1$: Reduces to Gauss-Seidel.

Convergence Criterion: The iterative methods stop when the root-mean-square (RMS) error between successive iterations falls below a specified tolerance:

$$\Delta w_{\text{rms}} = \sqrt{\frac{1}{N}\sum_{j,i}\left(w_{j,i}^{(n+1)} - w_{j,i}^{(n)}\right)^2} < \text{tol}$$

Here, tol $= 10^{-6}$ is the tolerance.

Example 14.1

Consider a rectangular duct through which fluid is flowing with a fully developed and constant streamwise pressure gradient in the z-direction. The flow aligned along the z-axis. The z-momentum equation simplifies to:

$$\mu\left(\frac{\partial^2 w}{\partial x^2} + \frac{\partial^2 w}{\partial y^2}\right) - \frac{\partial p}{\partial z} = 0$$

This is an elliptic partial differential equation and can be recognized as a form of the Poisson equation with respect to the velocity field w under certain flow conditions.

Solve this PDE equation using the SOR numerical scheme. The physical dimensions of the duct are $L = 1.5\,\text{m}$ in width and $h = 1.0\,\text{m}$ in height. The fluid's viscosity is $0.4\,\text{Ns/m}^2$, with different pressure gradients ($\frac{dp}{dz} = -10\,\text{N/m}^2$ is to be considered.

An analytical solution for comparison is

$$w_a(x,y) = \frac{16L^2}{\mu\pi^3}\left(-\frac{dp}{dz}\right)\sum_{m=1,3,5,\ldots}^{\infty}(-1)^{\frac{m-1}{2}}\frac{1 - \frac{\cosh(m\pi y/2L)}{\cosh(m\pi h/2L)}}{\cos(m\pi x/2L)m^3}$$

Solution

We start with the original momentum equation for the velocity component w (flow in the z-direction), assuming that u and v (velocity components in x and y directions) are negligible:

$$\mu\left(\frac{\partial^2 w}{\partial x^2} + \frac{\partial^2 w}{\partial y^2}\right) - \frac{\partial p}{\partial z} = 0$$

We solve this problem using SOR method and plot the normalized form of the analytical solution which can be expressed as:

$$w_{a,\text{normalized}}(y,z) = \frac{w_a(y,z) - w_{a,\min}}{w_{a,\max} - w_{a,\min}}$$

where $w_a(y,z)$ is given by:

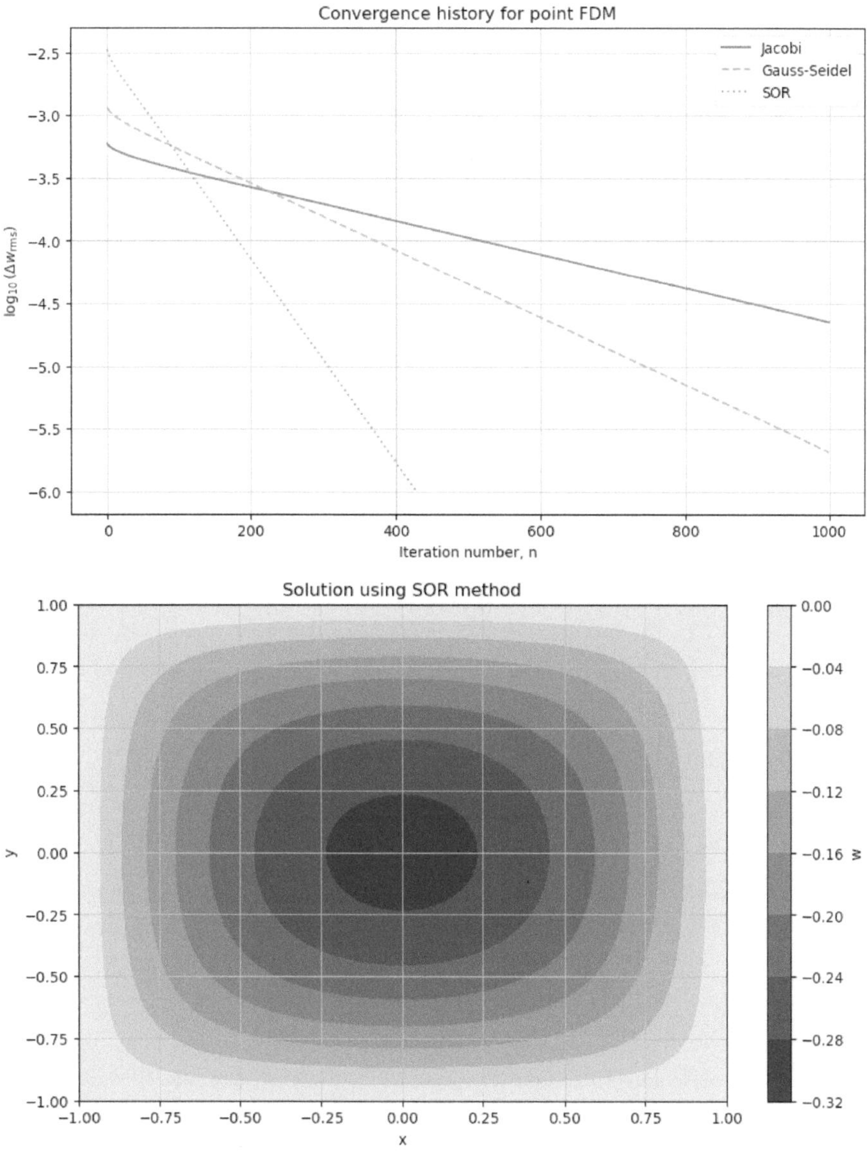

Figure 14.3: **Solver comparison for steady laminar flow through a duct and solution as obtained from SOR method.**

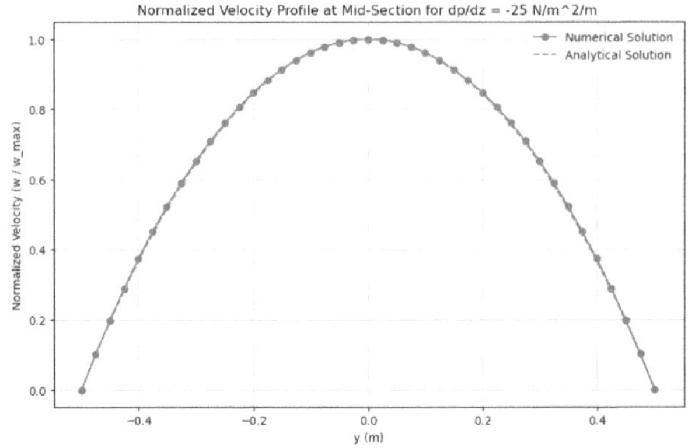

Figure 14.4: **Numerical solution of steady laminar flow through duct at the mid section and compared with analytical model.**

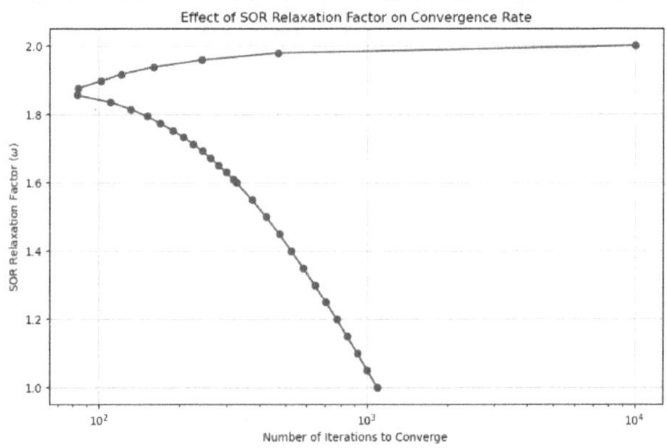

Figure 14.5: **The optimum ω value is around 1.876.**

$$w_a(y,z) =$$

$$\frac{16L^2}{\mu\pi^3}\left(-\frac{dp}{dz}\right)\sum_{m=1,3,5,\ldots}^{\infty}(-1)^{\frac{m-1}{2}}\left[\cos\left(\frac{m\pi y}{2L}\right)\cdot\frac{1-\exp\left(-\left|\frac{m\pi z}{2L}-\frac{m\pi h}{2L}\right|\right)}{m^3}\right]$$

Figure 14.4 shows the numerical solution of steady laminar flow through duct at the mid section and compared with analytical model. Figure 14.5 shows how the Successive Over-Relaxation (SOR) relaxation factor (ω) impacts convergence rates. Notably, convergence improves significantly as ω approaches an optimal value around 1.85, resulting in the fewest iterations. This behavior is consistent with the goal of optimizing iterative schemes to reduce computational costs. Beyond this optimal point, the number of iterations required for convergence escalates sharply, indicating that higher values

of ω can lead to instability or divergence. The graph highlights that careful tuning of ω is crucial for achieving efficient convergence, especially since a value that is too high can dramatically increase iterations, indicating slow or failed convergence. For numerical optimization, it is evident that selecting an appropriate ω is vital, as it can drastically impact both the computational cost and the feasibility of obtaining a solution.
See `Python` code `Ch14-Ex141` for solution of this problem.

Example 14.2

Consider the two-dimensional Poisson equation

$$\nabla^2 p(x,y) = f(x,y)$$

over the square domain $0 \le x \le 1$, $0 \le y \le 1$. Assume Dirichlet boundary conditions p=0 on all edges. Let the source term `f(x,y)` be nonzero only within a subregion (for example, $\frac{1}{4} \le x \le \frac{1}{2}$, $\frac{1}{4} \le y \le \frac{1}{2}$), and let `p(x,y)` be initialized to zero everywhere. (a) Apply a Gauss-Seidel iterative method to solve for `p(x,y)`. (b) Determine a suitable convergence criterion and verify that the numerical solution converges. (c) Plot the final pressure distribution `p(x,y)` and discuss how the source region affects its shape.

Solution
Step 1: Formulate the Problem.
Write the two-dimensional Poisson equation as

$$\nabla^2 p(x,y) = f(x,y),$$

on the domain $0 \le x \le 1, 0 \le y \le 1$, subject to Dirichlet boundary conditions "$p = 0$ on all boundaries." *Step 2: Discretize the Domain.*
Partition the domain into a mesh of `Nx` points in the x-direction and `Ny` points in the y-direction. Define $\Delta x = \frac{1}{\mathtt{Nx}-1}$ and $\Delta y = \frac{1}{\mathtt{Ny}-1}$. Arrange the unknowns $p_{i,j}$ at each grid node (x_i, y_j).
Step 3: Derive the Discrete Equations.
Approximate the Laplacian $\nabla^2 p$ at each interior node using central differences:

$$\nabla^2 p \approx \frac{p_{i+1,j} - 2p_{i,j} + p_{i-1,j}}{\Delta x^2} + \frac{p_{i,j+1} - 2p_{i,j} + p_{i,j-1}}{\Delta y^2}.$$

Enforce boundary values $p = 0$ at $i = 0$, $i = \mathtt{Nx} - 1$, $j = 0$, and $j = \mathtt{Ny} - 1$.
Step 4: Initialize.
Set the interior of the solution array p to zero. Specify a source term $f_{i,j}$ that is nonzero in some subregion (for example, $\frac{1}{4} \le x \le \frac{1}{2}, \frac{1}{4} \le y \le \frac{1}{2}$).

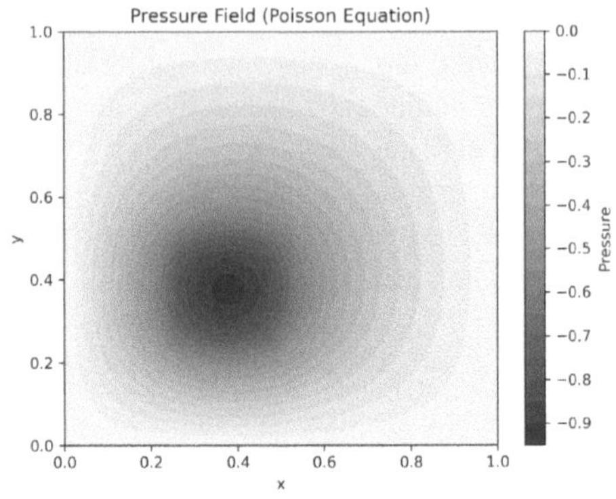

Figure 14.6: **Pressure distribution.**

Step 5: Apply Gauss–Seidel Iteration.
Iterate over all interior points, updating

$$p_{i,j} \leftarrow \frac{\frac{p_{i+1,j}+p_{i-1,j}}{\Delta x^2} + \frac{p_{i,j+1}+p_{i,j-1}}{\Delta y^2} - f_{i,j}}{2\left(\frac{1}{\Delta x^2} + \frac{1}{\Delta y^2}\right)}.$$

Keep the boundary points fixed at zero. Repeat until convergence.
Step 6: Check Convergence.
After each iteration, measure the relative change

$$\frac{\|p^{\text{new}} - p^{\text{old}}\|}{\|p^{\text{old}}\|}$$

and stop once this ratio falls below a chosen tolerance (for example, 10^{-6}).
See `Python` code `Ch14-Ex142` for solution of this problem.

14.3 SPARSE MATRIX STORAGE

CSR (Compressed Sparse Row) and CSC (Compressed Sparse Column) formats are two common sparse matrix storage formats designed to store matrices efficiently when most elements are zero. CSR (Compressed Sparse Row) and CSC (Compressed Sparse Column) formats are not specific to Python. They are general sparse matrix storage formats widely used across various programming languages and numerical computing libraries. In Python, libraries such as SciPy and PySparse provide built-in implementations of CSR and CSC.

Compressed Sparse Row (CSR) stores non-zero matrix elements in a row-wise manner, optimizing memory usage and computational speed for operations involving row access, such as matrix-vector multiplication. It is particularly favored for solving linear systems using iterative methods. The CSR format comprises three main arrays: data, which sequentially stores non-zero elements row-by-row; indices, containing the corresponding column indices for each element; and indptr, providing index pointers marking the starting positions of each row within the data array.

Consider matrix A:

$$A = \begin{bmatrix} 10 & 0 & 0 \\ 0 & 20 & 0 \\ 0 & 30 & 40 \end{bmatrix}$$

$$\text{data} = [10, 20, 30, 40], \quad \text{indices} = [0, 1, 1, 2], \quad \text{indptr} = [0, 1, 2, 4]$$

Compressed Sparse Column (CSC) efficiently stores non-zero matrix elements in a column-wise arrangement, making it particularly optimized for operations involving column access, such as matrix transpositions and matrix factorization techniques (e.g., LU decomposition). This storage scheme is preferred by solvers that perform column-oriented operations or direct factorization methods. The CSC format includes three main arrays: data, containing non-zero elements sequentially arranged column-by-column; indices, holding corresponding row indices for each stored value; and indptr, providing pointers that indicate the starting positions of each column within the data array.

For the previous matrix A:

$$\text{data} = [10, 20, 30, 40], \quad \text{indices} = [0, 1, 2, 2], \quad \text{indptr} = [0, 1, 3, 4]$$

The laplacian equation:

$$\Delta u = u_{xx} + u_{yy} = -2\pi^2 \cos^2(\pi x) \sin(\pi y)$$

is solved with boundary conditions:

$$u = 0, \quad \text{at} \quad x = 0,\, x = 1,\, y = 0,\, y = 1$$

using the python library for Sparse linear algebra solver `scipy.sparse.linalg`. We use a five-point discretization (also known as the five-point stencil) to approximate the LHS Laplacian (Δu) on a rectangular mesh:

$$\frac{\partial^2 u}{\partial x^2} + \frac{\partial^2 u}{\partial y^2} \approx \frac{u_{i+1,j} + u_{i-1,j} + u_{i,j+1} + u_{i,j-1} - 4u_{i,j}}{h^2}$$

Figure 14.8 shows the time taken by the solvers.

Table 14.3 shows that conjugate gradient method is fast compared to other methods. However, it is important to understand that the Conjugate Gradient (CG) method can indeed be sensitive to initial conditions, particularly when the problem is ill-conditioned, which can often be the case in large grids or poorly scaled problems. CG works well when the matrix is positive-definite, sparse, and well-conditioned. However, as the grid size grows to millions of points, the solver's performance might degrade due to the various reasons. CG's convergence can be slow or erratic if the initial guess is far from the true solution, especially on large grids with millions of

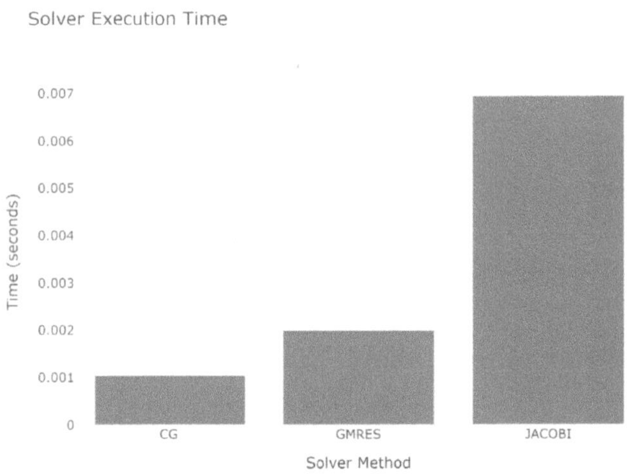

Figure 14.7: **Time taken to solve the Poisson equation by different solvers with CSC matrix.**

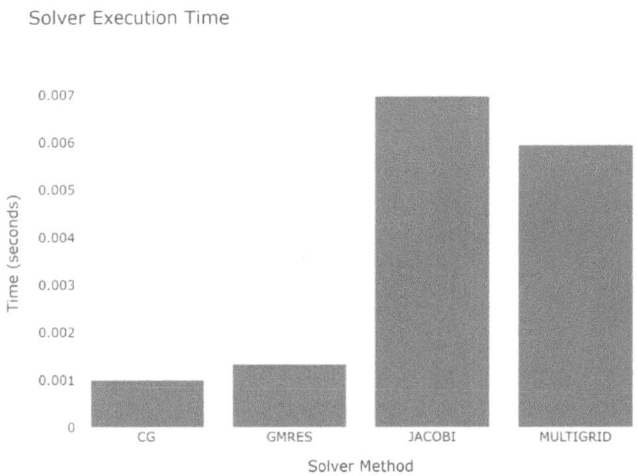

Figure 14.8: **Time taken to solve the Poisson equation by different solvers with CSR matrix.**

Table 14.3
Poisson Solver Execution Time for CSR and CSC Formats for 21 × 21 mesh

Method	Matrix Type	Time Taken (s)
CG	CSR	0.001664
GMRES	CSR	0.030489
JACOBI	CSR	0.030892
pyAMG	CSR	0.010613
CG	CSC	0.001109
GMRES	CSC	0.030381
JACOBI	CSC	0.030613
pyAMG	CSC→CSR	0.010529

unknowns. For large grids, the condition number of the system matrix (the ratio of the largest to smallest eigenvalues) tends to grow, making CG converge slower. If the matrix is poorly conditioned, CG might take many iterations to converge, or fail to converge within a reasonable number of iterations.

Multigrid preconditioning (as in this example) with pyAMG is effective for large-scale problems.

In a multigrid method, the computational domain is represented at multiple levels of resolution, as illustrated in this Figure 14.9. The highest level, labeled *Level 1*, features the finest mesh, containing the largest number of cells (and therefore nodes). Each subsequent level (Levels 2, 3, and so on) has a coarser grid, with fewer cells in each spatial dimension. Restriction and prolongation (sometimes called interpolation) operators allow data to be transferred between these levels: residuals, corrections, or approximate solutions are "restricted" from a finer grid to the next coarser grid, whereas solutions or corrections are "prolonged" back from a coarse grid to a finer grid. The rationale behind this approach is that smooth errors—those that do not vary rapidly in space—are hard to eliminate on a single fine grid using traditional relaxation methods like Gauss-Seidel or Jacobi. By moving to a coarser level, one can rapidly reduce low-frequency errors (which appear smooth on the fine mesh) because those same errors become higher-frequency on the coarser mesh. This complementary error reduction across multiple levels is the essence of multigrid's efficiency.

One can see, for example, that Level 1 with 24×16 cells might correspond to 25×17 pressure or scalar unknowns (if each cell center or node is counted). Then Level 2 could be half as fine in each dimension, meaning 12×8 cells and 13×9 nodes, and so forth down to Level 4, which has only 3×2 cells and 4×3 nodes. The figure suggests a classical setup where each coarser level has approximately half the resolution of the previous level in both x and y. In practical usage, these levels can be generated automatically, or the user can define custom coarsening ratios if the problem's geometry or domain complexity demands it. At the coarsest level, the grid is so small that one can often solve it directly, for instance using a simple LU decomposition or another direct method. By combining these coarse- and fine-level corrections in a systematic cycle, the multigrid approach typically achieves optimal–or near-optimal–complexity, meaning the computational cost grows linearly with the number of unknowns, in contrast with classic solvers whose costs might scale quadratically or worse.

Figure 14.9 also highlights that each level is still a fully valid mesh over the domain, just at a lower resolution. For example, the coarsest level (Level 4) may have only a handful of cells in each direction–here shown as 3×2–and is typically where the solver can capture global, low-frequency error modes most effectively. The synergy between fine and coarse levels arises because smoothing methods like Gauss-Seidel remove high-frequency errors well on fine grids but struggle with low-frequency components. Once a residual is restricted to a coarser grid, those "slow" variations in the fine-grid sense become relatively more oscillatory on the coarse mesh, allowing the same smoothing procedure to eliminate them rapidly. This cyclical interplay forms the basis of efficient multigrid algorithms.

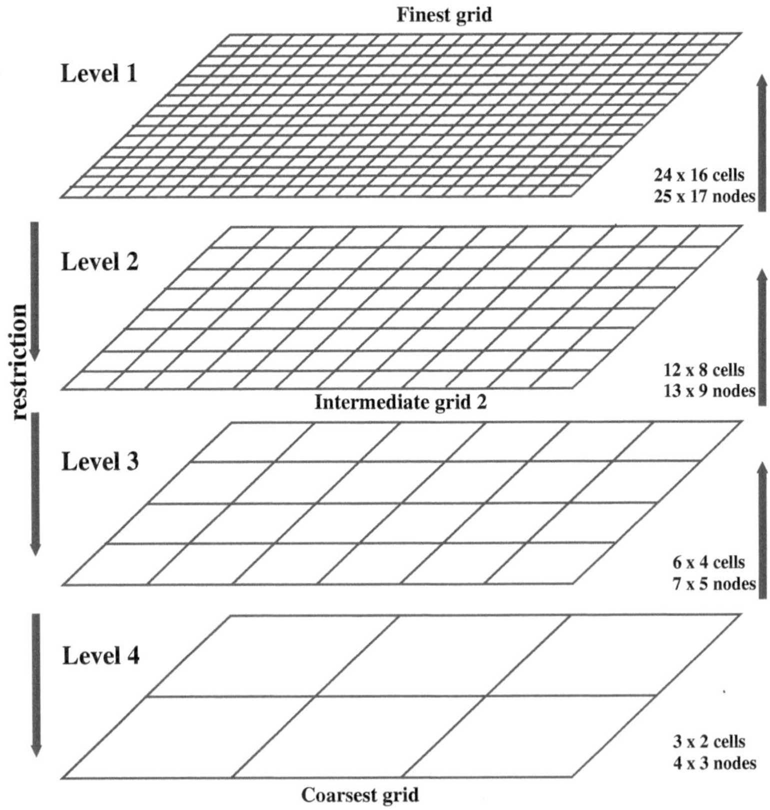

Figure 14.9: **Computational domain is seen as a multiple levels of resolution in multigrid methods.**

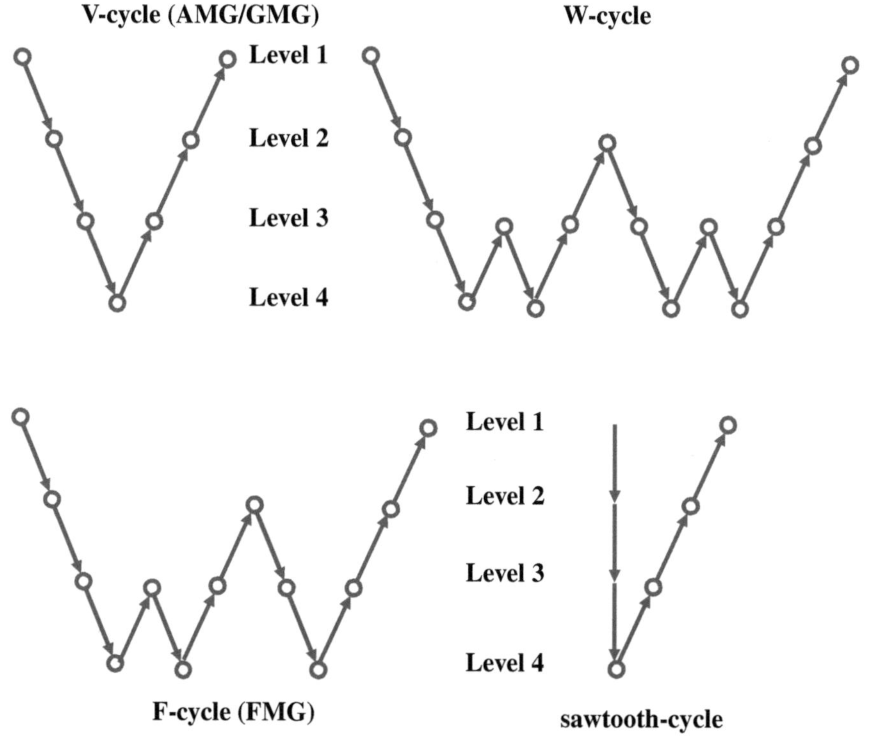

Figure 14.10: Four common cycling strategies used in multigrid methods: the V-cycle, W-cycle, F-cycle, and sawtooth-cycle .

Notably, the transition from one level to the next requires well-defined operators for both restriction and prolongation. Restriction aggregates information (such as residuals or updates) from a block of fine-grid cells into one coarse-grid cell, often via simple averaging or weighted schemes that respect mesh geometry. Prolongation (or interpolation) does the opposite: it maps coarse-grid corrections back onto a larger number of fine-grid points, usually applying linear or higher-order interpolation to fill in the additional degrees of freedom. By repeating cycles of restriction, smoothing on coarser levels, and prolongation, the overall algorithm systematically attacks error components at all wavelengths. The figure exemplifies how the mesh might be systematically halved in each dimension to create the typical $\log_2$ depth in a multigrid hierarchy. In practical CFD codes, multiple levels beyond four may be used if the domain is extremely large, or if certain error modes prove stubborn at intermediate resolutions. Ultimately, the objective is to achieve convergence rates that are nearly independent of the grid size, making multigrid one of the most powerful solvers for elliptic equations such as the Poisson or Helmholtz equations.

Figure 14.10 displays four common cycling strategies used in multigrid methods: the V-cycle, W-cycle, F-cycle, and sawtooth-cycle. A "cycle" denotes how the solver moves between levels (from fine to coarse and back) within one iteration of the overall multigrid algorithm. The vertical axis in each diagram corresponds to mesh levels, with Level 1 at the top (finest) and the last level–say Level 4 at the bottom (coarsest). In a **V-cycle**, the solver typically starts smoothing at the finest level, moves down the hierarchy applying smoothing or relaxation at each level, and upon reaching the coarsest level, solves or smooths there and then returns directly up, performing more smoothing passes at each intermediate level on the way back.

In a **W-cycle**, the solver visits the coarser levels more than once before returning to the finest level—hence the path zigzags downward and upward more frequently, forming a "W" shape when connected. This can be beneficial if the coarser levels need multiple sweeps to effectively remove certain error modes, but it increases computational cost per cycle compared to a V-cycle. Meanwhile, the **F-cycle** is somewhat of a hybrid that does an initial pass similar to a V-cycle but then adds additional passes through intermediate levels—like a partial W—before returning to the top. Finally, the **sawtooth-cycle** is simpler in that it only descends through all levels once before moving directly back to the finest level (skipping some intermediate "up" steps). Each of these approaches tries to balance the trade-off between computational expense and error reduction.

Selecting a cycle type often depends on how quickly the problem's errors are reduced at each level. The V-cycle is usually the default, as it offers a good compromise: each cycle transitions from fine to coarse, does a small number of sweeps on each level, and then returns directly back up, making it relatively inexpensive per iteration. The W-cycle, on the other hand, invests more work by descending and ascending the hierarchy multiple times within the same cycle. This additional effort can yield better convergence per cycle, but it also means each cycle costs more. If the problem is particularly challenging—where strong low-frequency errors linger after a standard V-cycle—some codes switch to a W-cycle for more aggressive coarse-grid correction.

The F-cycle sits between these extremes: it starts like a V-cycle, performing only a single descent to the coarse levels, but on the way back up it makes extra visits to intermediate levels. This can help in situations where partial re-smoothing on mid-level grids can effectively handle errors that arise from approximations on the coarsest grid. The sawtooth-cycle is at the opposite end of complexity: it descends through all levels but returns directly to the top, skipping re-smoothing on the in-between grids. This can be useful for certain specialized problems where only a single coarse-level update is needed to complement the fine-level smoothing, though it might not be as generally effective as a V- or W-cycle. In actual implementations, users or automatic heuristics can switch between these cycles based on observed convergence. Each cycle type has distinct advantages and potential drawbacks, and efficient multigrid codes often allow dynamic choices to adapt to evolving error characteristics as the solution converges.

Note that the ILU (Incomplete LU factorization) solver in `SciPy`, specifically `spilu()`, always requires a CSC (Compressed Sparse Column) formatted matrix. Even if we provide a CSR-formatted matrix, `SciPy` internally converts it to CSC format, resulting in a warning. Note that `PyAMG` explicitly requires CSR matrices. Iterative solvers like CG or GMRES are very sensitive to matrix conditioning; if the matrix is ill-conditioned, the residuals do not decay properly and convergence fails.

Table 14.4
Matrix Solvers and Corresponding Python Libraries

Direct Solvers		
Direct Sparse Solver (LU)	`scipy.sparse.linalg`	`spsolve`
UMFPACK (via SuiteSparse)	`scikit-umfpack`	`splu` or `factorized`
Iterative Solvers (CPU)		
Bi-Conjugate Gradient Stabilized (BiCGSTAB)	`scipy.sparse.linalg`	`bicgstab`
MINimum RESidual (MINRES)	`scipy.sparse.linalg`	`minres`
LSQR (Least Squares QR)	`scipy.sparse.linalg`	`lsqr`
Conjugate Gradient (CG)	`scipy.sparse.linalg`	`cg`
Generalized Minimal Residual (GMRES)	`scipy.sparse.linalg`	`gmres`
GPU-Accelerated Solvers		
Direct Sparse Solver (GPU)	`cupyx.scipy.sparse.linalg`	`spsolve`
Iterative Solvers (GPU)	`cupyx.scipy.sparse.linalg`	`cg`, `gmres`, `bicgstab`, etc.
Custom Solvers		
TriDiagonal Matrix Algorithm (TDMA)	`scipy.linalg`	(for banded matrices)
	Custom Implementation	Thomas Algorithm
Multigrid method	PyAMG	`pyamg.solve()`

Example 14.3

Implement the classical V-cycle multigrid approach to solve the 2D Poisson problem

$$\nabla^2 p = -2\pi^2 \sin(\pi x)\sin(\pi y),$$

subject to $p = 0$ on the boundaries.

Solution
See `Python` code `Ch14-Ex143` for solution of this problem.

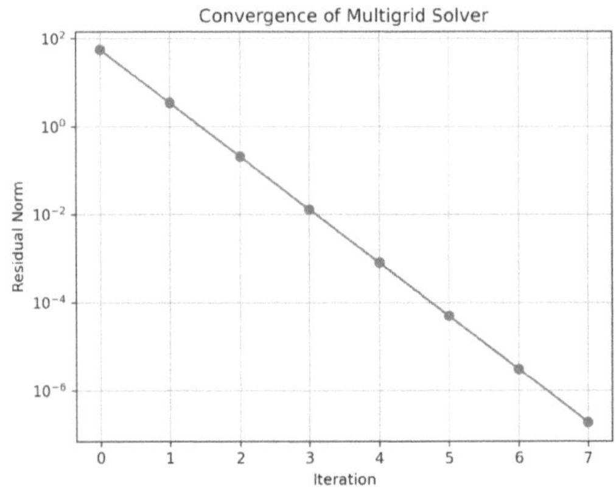

Figure 14.11: **Residual plot.**

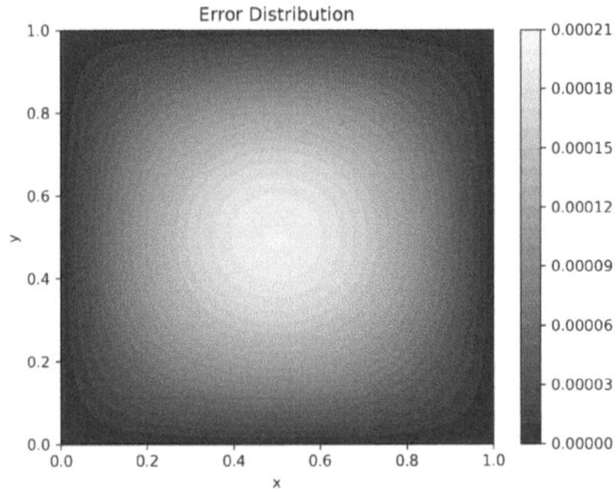

Figure 14.12: **Error between numerical and exact solution.**

Figures and show the residual plot and error between the numerical and exact solution.

14.4 PRECONDITIONER

In many large-scale numerical simulations, iterative solvers for systems such as

$$A\mathbf{x} = \mathbf{b}$$

often suffer from slow convergence or even divergence when the system is ill-conditioned. Preconditioners address this challenge by transforming the original system into one with improved spectral properties. Specifically, a preconditioner M is chosen such that the transformed system

$$M^{-1}A\mathbf{x} = M^{-1}\mathbf{b}$$

has a reduced condition number and its eigenvalues are more clustered. This clustering significantly accelerates the convergence of iterative methods like Conjugate Gradient, GMRES, or BiCGStab. Techniques such as Incomplete LU (ILU) factorization, Jacobi scaling, and Algebraic Multigrid (AMG) are commonly used in Python libraries to serve as preconditioners. They approximate the inverse of A either by partial factorization or by scaling the system, thus mitigating the effects of large disparities in matrix entries and spectral spread. As a result, these preconditioners enable iterative solvers to converge more reliably and in fewer iterations, making them essential components in the efficient numerical solution of sparse linear systems.

When we use an ILU preconditioner, it approximates the inverse of the matrix and transforms the original system into one with a lower condition number. In effect, the preconditioner "regularizes" the problem so that the iterative method—now solving the preconditioned system—can converge.

Solver	Python Library	Preconditioners
CG	`scipy.sparse.linalg.cg`	ILU, Jacobi, Incomplete Cholesky
GMRES	`scipy.sparse.linalg.gmres`	ILU, Jacobi, AMG (via PyAMG)
BiCGStab	`scipy.sparse.linalg.bicgstab`	ILU, Jacobi, AMG (via PyAMG)
MINRES	`scipy.sparse.linalg.minres`	Jacobi, Incomplete Cholesky (for SPD)
LGMRES	`scipy.sparse.linalg.lgmres`	ILU, Jacobi

14.5 PRESSURE REFERENCING IN INCOMPRESSIBLE CFD

In many CFD applications only the pressure gradient matters because the absolute pressure is defined up to an arbitrary constant. In incompressible or very low-Mach-number flows the governing pressure Poisson equation (derived from continuity and momentum equations) exhibits a nullspace; adding a constant to the pressure field everywhere still satisfies the equations. This leads to a singular linear system unless a reference is fixed. This requirement is valid for both finite-difference (FD) and finite-volume (FV) schemes.

Without a reference the pressure field can drift by an arbitrary constant, and iterative solvers may fail to converge or produce nonphysical pressure distributions. Even when only gradients are important, a stable and convergent pressure field is essential so that the computed pressure gradients remain reliable. Imposing a reference condition (or an additional constraint) ensures that the linear system is well conditioned.

Several methods are used to eliminate the nullspace and ensure a unique solution. The common strategies include:

☞ Single-Cell Reference (Point Fixing): A specific cell or node is chosen and its pressure is fixed (for example, $p = 0$). This approach directly removes the nullspace by anchoring the pressure at one point. It is widely used in FV schemes via a `pRefCell` or similar mechanism when no Dirichlet boundary (such as a pressure outlet) is available.

☞ Zero Mean Pressure (Global Average): The spatial average of the pressure is enforced to equal a prescribed value (often zero). For a uniform FD grid the reference pressure is computed as

$$p_{\mathtt{ref}} = \frac{1}{N_x N_y N_z} \sum_{i=1}^{N_x} \sum_{j=1}^{N_y} \sum_{k=1}^{N_z} p_{i,j,k}.$$

For nonuniform meshes (with different Δx_i, Δy_j, Δz_k) a volume-weighted average is more appropriate:

$$p_{\mathtt{ref}} = \frac{\sum_{i,j,k} \left(p_{i,j,k} \, \Delta x_i \, \Delta y_j \, \Delta z_k \right)}{\sum_{i,j,k} \left(\Delta x_i \, \Delta y_j \, \Delta z_k \right)}.$$

This technique is common in both FD and FV schemes as well as in finite-element formulations.

☞ A spatially varying reference function $p_{\mathtt{ref}}(x)$ is subtracted from the pressure field. This method is useful when the pressure field naturally follows a known baseline profile (for example, the hydrostatic variation with height). In buoyancy-driven flows one may define a modified pressure

$$p_{rgh} = p - \rho g h,$$

and then pin a reference cell for the modified field.

☞ Operating Pressure (External Reference): An external constant (often atmospheric pressure) is defined to set all input/output pressures as gauge values. Although this does not fix the gauge freedom internally, a closed system still requires an internal reference (by pinning one cell or enforcing a mean value) to obtain a unique solution.

Iterative solvers (such as GMRES, CG, or multigrid methods) can fail to converge if the system is singular because the residual cannot be minimized when the solution can slide in the nullspace. To ensure convergence, the nullspace must be removed by applying one of the reference constraints. Direct solvers (for example, sparse LU methods) may handle nearly singular matrices more robustly because they often detect the singularity and effectively set one degree of freedom automatically. However, even when using direct methods, it is best practice to include a reference pressure cell to guarantee the physical fidelity of the pressure solution.

In incompressible CFD both FD and FV schemes require a reference pressure to remove the nullspace inherent in the pressure Poisson equation under pure Neumann boundary conditions. Standard strategies include pinning a single cell, enforcing a zero mean pressure, subtracting a known pressure distribution, or using an external operating pressure. These methods ensure that the resulting linear system is well conditioned, thereby aiding the convergence of iterative solvers and ensuring reliable pressure gradient calculations. Direct solvers can sometimes tolerate near-singular systems, but robust CFD solvers typically incorporate a reference pressure to maintain stability and physical accuracy.

14.6 FFT SOLVER WITH PERIODIC DIRECTION

The Poisson equation can be solved efficiently using Fast Fourier Transform (FFT) by transforming the differential operator into an algebraic operator in Fourier space. This method is especially used when at least one direction in the problem domain is periodic like channel flow case, and has been employed in number of Direct Numerical Simulations and Large Eddy Simulation Studies.

Consider the Discrete Fourier Transform (DFT). For a vector

$$\mathbf{v} = [v_1, v_2, \ldots, v_N],$$

its DFT is given by

$$\hat{\mathbf{v}} = [\hat{v}_1, \hat{v}_2, \ldots, \hat{v}_N],$$

where each Fourier coefficient is defined as

$$\hat{v}_n = \sum_{j=1}^{N} v_j e^{-\frac{2\pi i(n-1)(j-1)}{N}}.$$

Similarly, the inverse DFT recovers the original vector via

$$v_j = \frac{1}{N}\sum_{n=1}^{N} \hat{v}_n e^{\frac{2\pi i(n-1)(j-1)}{N}}.$$

Because FFT algorithms compute these transforms in approximately $\mathcal{O}(N\log N)$ operations, they offer a significant computational advantage over direct summation methods.

1. *Linearity and Shifted Vectors*: The DFT is a linear operator. If one has two vectors such that

   $$\mathbf{v} = \mathbf{f} + \mathbf{g},$$

   then the Fourier transform satisfies

   $$\hat{\mathbf{v}} = \hat{\mathbf{f}} + \hat{\mathbf{g}}.$$

   This property greatly simplifies the treatment of linear differential operators.

Next, we define the shifted versions of a vector $\mathbf{v}$. For example, a left-shifted version is

$$\mathbf{v}^l = [v_2, \ldots, v_N, v_1],$$

and a right-shifted version is

$$\mathbf{v}^r = [v_N, v_1, \ldots, v_{N-1}].$$

The Fourier transforms of these shifted vectors become

$$\hat{v}_n^l = e^{\frac{2\pi i(n-1)}{N}} \hat{v}_n, \quad \hat{v}_n^r = e^{-\frac{2\pi i(n-1)}{N}} \hat{v}_n.$$

2. *1D Poisson Equation*: Consider the one-dimensional Poisson equation

$$\frac{d^2u}{dx^2} = f(x).$$

A standard discretization using central differences yields

$$\frac{u_{i-1} - 2u_i + u_{i+1}}{h^2} = f_i,$$

where h is the grid spacing. With Dirichlet boundary conditions, one typically sets

$$u_1 = a, \quad u_{N+1} = b.$$

3. *FFT-Compatible Extension*: To enable the use of FFTs for problems with Dirichlet boundary conditions (which are not naturally periodic), one extends the original vectors into a larger, symmetric domain. One common approach is to construct the extended solution vector as

$$U = [0, u_2, \ldots, u_N, 0, -u_N, \ldots, -u_2],$$

and similarly extend the right-hand side vector as

$$F = \left[0, f_2 - \frac{a}{h^2}, \ldots, f_N - \frac{b}{h^2}, 0, -\left(f_N - \frac{b}{h^2}\right), \ldots, -\left(f_2 - \frac{a}{h^2}\right)\right].$$

This extension, which involves mirroring and sign reversal, creates a periodic structure in the extended domain, ensuring that the FFT can be applied. In the extended vector, the discrete Laplacian is computed as

$$\frac{U^r - 2U + U^l}{h^2} = F,$$

where U^r and U^l denote the right-shifted and left-shifted versions of U, respectively.

4. *Fourier Space Solution*: With the extended vectors defined, take the DFT of both sides. In Fourier space, the second derivative operator transforms into an algebraic multiplier. In particular, the transformed Laplacian operator is given by
$$\frac{1}{h^2}\left(e^{-\frac{2\pi i(n-1)}{m}} - 2 + e^{\frac{2\pi i(n-1)}{m}}\right),$$
where m is the length of the extended vector. This expression simplifies to
$$-\frac{4}{h^2}\sin^2\left(\frac{\pi(n-1)}{m}\right).$$
Thus, one can write the Fourier-space equation as
$$-\frac{4}{h^2}\sin^2\left(\frac{\pi(n-1)}{m}\right)\hat{U}_n = \hat{F}_n,$$
which is rearranged to yield the solution in Fourier space:
$$\hat{U}_n = \frac{\hat{F}_n}{\mu_n}, \quad \text{with } \mu_n = -\frac{4}{h^2}\sin^2\left(\frac{\pi(n-1)}{m}\right).$$
If the partial differential equation includes a reaction term such as $-u$, the operator is modified accordingly:
$$\mu_n = -\frac{4}{h^2}\sin^2\left(\frac{\pi(n-1)}{m}\right) - 1.$$
5. *Recovery of the Interior Solution*: After computing $\hat{U}_n$ in Fourier space, the inverse FFT is used to transform back to the physical domain, yielding the extended solution vector U. The solution of interest, corresponding to the original interior points $u_2, \ldots, u_N$, is then extracted from the first half of U.
6. *Boundary and Symmetry Considerations*: A key requirement of the FFT-based method is that it assumes periodicity. In cases where the physical problem has non-periodic boundary conditions (such as Dirichlet conditions), the domain must be extended or mirrored appropriately. This mirroring introduces the symmetry necessary to convert the original non-periodic problem into one that is effectively periodic.

The FFT-based solver provides a rapid and efficient solution method. The approach transforms the differential equation into an algebraic one in Fourier space, solving for the Fourier coefficients before returning to the physical domain via the inverse FFT. This method is particularly well-suited as a benchmark or direct solver for structured problems where one or more dimensions are periodic.

PROBLEMS

14P-1 Solve the following Poisson equation:

$$u_{xx} + u_{yy} = -2\pi^2 \cos^2(\pi x) \sin(\pi y)$$

on a unit square domain defined by $0 < x < 1$ and $0 < y < 1$, with homogeneous Dirichlet boundary conditions ($u = 0$) on all four boundaries. Use a five-point finite difference discretization scheme on uniform rectangular grids.

Implement a python code and evaluate the performance of different iterative solvers, namely Jacobi, Conjugate Gradient (CG), GMRES, and Multigrid methods, by measuring the computational time required to achieve a user-defined convergence criterion of 10^{-4}. Conduct this analysis on two different mesh resolutions, 11×11, 21×21, and 200×200. Discuss how the performance and computational efficiency of these solvers vary with mesh size, and present findings graphically.

14P-2 Solve the following Poisson equation:

$$u_{xx} + u_{yy} = -2\pi^2 \cos^2(\pi x) \sin(\pi y)$$

on a unit square domain defined by $0 < x < 1$ and $0 < y < 1$, with homogeneous Dirichlet boundary conditions ($u = 0$) on all four boundaries. Use a five-point finite difference discretization scheme on uniform rectangular grids. Implement solution of this problem with convergence tolerance of 1E-4 and comment on the result.

14P-3 Condition number of a matrix is the ratio of the largest and smallest eigenvalues of the matrix. The condition number $\kappa(A)$ for a matrix A is defined as:

$$\kappa(A) = \|A\| \times \|A^{-1}\|$$

Where $\|A\|$ is the norm of the matrix (typically the largest singular value) and $\|A^{-1}\|$ is the norm of the inverse of the matrix (or the smallest singular value of A). For Problem 14P-2 estimate the condition number of matrix associated with CSR and CSC storage.

14P-4 Consider the Poisson equation given by:

$$\nabla^2 u(x,y) = -2\pi^2 \sin(\pi x) \sin(\pi y)$$

defined on a two-dimensional square domain $0 < x < 1, 0 < y < 1$. The boundary conditions are homogeneous Dirichlet conditions ($u = 0$) on the left, top, and bottom boundaries, and a homogeneous Neumann boundary condition ($\frac{\partial u}{\partial y} = 0$) on the right boundary.

Using a five-point finite difference discretization on a uniform rectangular grid with $n \times n$ interior points, solve the equation numerically. Implement and compare the convergence speed and solution accuracy for four iterative numerical methods: Conjugate Gradient (CG), Generalized Minimal Residual (GMRES), Jacobi, and Algebraic Multigrid (AMG).

Present:

Graphical plots of the numerical solution obtained using the Multigrid solver.

A comparison plot of execution times for each solver.

A tabulated summary of solver execution times.

14P-5 Consider the two-dimensional Laplace equation

$$\nabla^2 p(x,y) = 0, \quad 0 < x < 1, \quad 0 < y < 1.$$

The boundary conditions are given by

$$p(0,y) = 101325\,\text{Pa}, \quad p(1,y) = 101300\,\text{Pa}, \quad \frac{\partial p}{\partial y}(x,0) = 0, \quad \frac{\partial p}{\partial y}(x,1) = 0.$$

Discretize the domain using a uniform rectangular mesh with $n_x \times n_y$ physical grid points. To enforce the homogeneous Neumann boundary conditions at y=0, xt and y=1, add two ghost rows and use a standard five-point finite difference scheme for the Laplacian. Solve the resulting linear system using a direct solver (`spsolve`) e.g., LU factorization and compare the numerical pressure distribution with the exact solution

$$p_{\text{exact}}(x,y) = 101325 + (101300 - 101325)x$$

Produce a plot of the computed pressure field and report the L^2 error relative to p_{exact}.

14P-6 Consider a 2D incompressible flow through a rectangular domain with pressure distribution governed by the Laplace equation:

$$\nabla^2 p = 0$$

where p represents the pressure in the domain. The boundary conditions are:

Dirichlet boundary condition: $p = 101325\,\text{Pa}$ at $x = 0$ (left boundary).
Neumann boundary condition: $\frac{\partial p}{\partial x} = 0$ at $x = 1$ (right boundary).
Neumann boundary boundary conditions at the top and bottom: $\frac{\partial p}{\partial y} = 0$ at $y = 0$ (bottom) and $y = 1$ (top).

The problem is discretized on a grid of size 22×22 with uniform spacing. The finite difference method is used to solve for the pressure distribution. Write down the discretized form of the Laplace equation on a uniform grid with the specified boundary conditions. Solve it using using BiCGStab and ILU preconditioning in Python.

14P-7 Solve the two-dimensional Poisson equation

$$\nabla^2 p(x,y) = cos(\pi x)cos(\pi y)$$

on a rectangular domain of width L in x and height H in y. L is 5 and H is 1.

The boundary conditions are periodic in x (i.e., p(0,y) = p(L,y)) and homogeneous Neumann in y, enforcing

$$\frac{\partial p}{\partial y}(x,0) = 0, \quad \frac{\partial p}{\partial y}(x,H) = 0.$$

Since a pure Neumann problem for the Poisson equation determines p only up to an additive constant, impose the extra condition

$$\int_{x=0}^{L}\int_{y=0}^{H} p\,dx\,dy = 0$$

to fix a unique solution. Use Fourier spectral method via the Fast Fourier Transform (FFT) to solve this problem.

14P-8 Solve the nonlinear steady-state thermal conduction problem in one dimension with temperature-dependent thermal conductivity $\lambda(T)$ where the heat flux is given by:

$$q_x = -\lambda(T)\frac{\partial T}{\partial x}$$

The thermal conductivity is specified as: $\lambda(T) = 1 + 450T$. The governing equation for steady-state thermal conduction is:

$$\frac{d}{dx}[(1+450T)\frac{dT}{dx}] = 0 \quad (0 \le x \le 1)$$

The boundary conditions are: $T(0) = 0 \quad T(1) = 1$. The domain is discretized with $N = 50$ grid points over a length $L = 1.0$. The problem is solved using a finite difference method by approximating the nonlinear differential equation and iteratively solving for the temperature distribution.

14P-9 We aim to solve the two-dimensional Poisson equation on a unit square domain $[0,1] \times [0,1]$ using a second-order finite difference method (FD-O2) combined with pseudo-time relaxation via the explicit Euler method. The governing equation is:

$$\frac{\partial^2 u}{\partial x^2} + \frac{\partial^2 u}{\partial y^2} = -g,$$

where $g = 3000\,\mathrm{W/m^3}$ is a uniform source term. We use the Pseudo-Time Relaxation Approach, in which we iteratively converge to the steady-state solution, and we introduce a fictitious time derivative, transforming the problem into:

$$\frac{\partial u}{\partial t} = \frac{\partial^2 u}{\partial x^2} + \frac{\partial^2 u}{\partial y^2} + g.$$

This pseudo-time-dependent equation is evolved using the explicit Euler method until the solution stabilizes (steady state). Following are the Dirichlet boundary and initial Conditions:

Bottom boundary ($y = 0$): $u = 600$

Top boundary ($y = 1$): $u = 373$

Left and right boundaries ($x = 0, x = 1$): $u = 373$

Initial Guess: $u(x,y) = 0$ at all interior points.

The iteration continues until the maximum absolute change in the solution between successive pseudo-time steps falls below 10^{-8}, indicating that the solution has reached steady state. Once converged, the resulting temperature distribution can be compared with the solution obtained using the FFT-based Poisson solver with properly enforced Dirichlet boundary conditions. This comparison verifies the consistency and accuracy of the pseudo-time relaxation method against a spectral solution approach.

14P-10 We consider the incompressible two-dimensional viscous flow inside a square cavity of side length $L = 1.0$, driven by the motion of the top lid at constant velocity U. The governing equations are the incompressible Navier-Stokes equations written in terms of the vorticity-stream function formulation:

$$\frac{\partial \omega}{\partial t} + u\frac{\partial \omega}{\partial x} + v\frac{\partial \omega}{\partial y} = \nu\left(\frac{\partial^2 \omega}{\partial x^2} + \frac{\partial^2 \omega}{\partial y^2}\right),$$

and

$$\nabla^2 \psi = -\omega,$$

where $\omega(x,y,t)$ is the scalar vorticity, $\psi(x,y,t)$ is the stream function, and (u,v) are the velocity components recovered from ψ via

$$u = \frac{\partial \psi}{\partial y}, \qquad v = -\frac{\partial \psi}{\partial x}.$$

The fluid is confined within the square domain $[0,L] \times [0,L]$, and the flow is initialized from rest. The boundary conditions for the stream function are set to $\psi = 0$ on all boundaries, ensuring no flow penetration. For the vorticity, the boundary values are evaluated based on the stream function gradients near the walls using second-order finite differences. The top lid moves with horizontal velocity U, while the other three walls are stationary, which induces circulation in the domain. Solve this problem numerically using a finite difference scheme on a uniform Cartesian grid with $N \times N$ points, using a second-order central difference for spatial derivatives, and a two-stage Runge-Kutta method. At each time step, the Poisson equation for the stream function is solved using a sparse linear solver. The simulation runs until a final time $T = 5.0s$.

14P-11 Consider a square domain where the temperature distribution $T(x,y)$ satisfies Laplace's equation:

$$\frac{\partial^2 T}{\partial x^2} + \frac{\partial^2 T}{\partial y^2} = 0,$$

with boundary conditions given by:

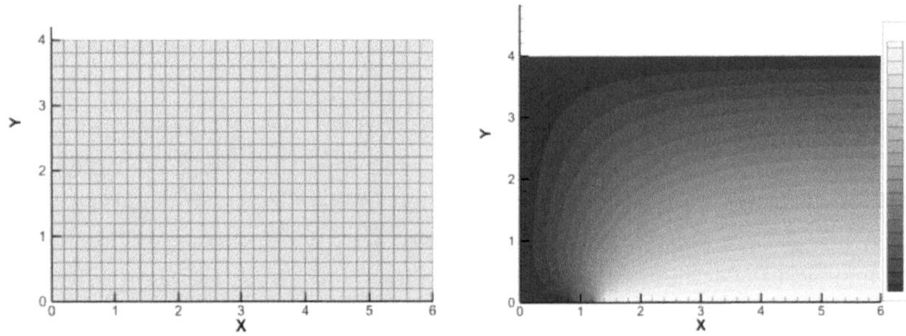

Figure 14.13: Rectangular grid for Laplace equation.

$$\begin{cases} \text{at } x=0, & \dfrac{\partial T}{\partial x}=0 \\ \text{at } x=1, & T=0 \\ \text{at } y=0, & \dfrac{\partial T}{\partial y}=0 \\ \text{at } y=1, & T=1 \end{cases}$$

Apply the Alternating Direction Implicit (ADI) method to numerically solve this boundary value problem. Outline the discretization steps involved, specify the finite difference equations for the ADI scheme, and clearly discuss how to implement the given boundary conditions numerically.

14P-12 A grid file showing the following domain and boundary conditions in is made in `Tecplot` (Figure 14.13). The governing equation for the streamfunction $\Psi(x,y)$ is the 2D Laplace equation:

$$\frac{\partial^2 \Psi}{\partial x^2}+\frac{\partial^2 \Psi}{\partial y^2}=0$$

Boundary conditions:

i Along $y=0$ (the bottom edge), we divide it into three segments:

$$\Psi(x=0..0.8, y=0)=0,\ \Psi(x=1.0..1.2, y=0)=50,\ \Psi(x=1.4..6, y=0)=200.$$

ii Along $x=0$ (the left edge), for $y=0.2..3.8$:

$$\Psi(x=0, y=0.2..3.8)=0.$$

iii Along $y=4$ (the top edge), for $x=0..6$:

$$\Psi(x=0..6, y=4)=0.$$

iv Along $x=6$ (the right edge), for $y=0.2..3.8$, we impose a zero-gradient (parallel outflow) boundary:

$$\frac{\partial \Psi}{\partial x}(x=6, y=0.2..3.8)=0.$$

Solve using Successive over-relaxation method (SOR) using $\omega = 1.75$.

14P-13 Consider a two-dimensional rectangular plate of length L (in the x-direction) and width W (in the y-direction). Let $\theta(x,y)$ be a dimensionless temperature field that satisfies the steady-state Laplace equation:

$$\frac{\partial^2\theta}{\partial x^2}+\frac{\partial^2\theta}{\partial y^2}=0$$

subject to the following boundary conditions:

$$\theta(0,y)=0,\quad \theta(L,y)=0,\quad \theta(x,0)=0,\quad \theta(x,W)=1.$$

1. Impose the above boundary conditions and using SOR method, determine the dimensionless temperature distribution $\theta(x,y)$.
2. Plot the solution for values of L=1 and W=2, and comment on the physical meaning of the resulting dimensionless temperature field.

compare solution with the exact solution:

$$\theta=2\sum_{n=1}^{\infty}\sin\left(\frac{n\pi}{L}x\right)\sinh\left(\frac{n\pi}{L}y\right)\left[\frac{2\left[\cos(n\pi)-1\right]}{\sinh\left(\frac{n\pi}{L}W\right)\left(-\cos(n\pi)\sin(n\pi)+n\pi\right)}\right]$$

14P-14 According to fractional-step or operator splitting methods the PDE

$$\frac{\partial u}{\partial t}=-a\frac{\partial u}{\partial x}+\nu\frac{\partial^2 u}{\partial x^2},$$

can be split into two sub-problems corresponding to the advection and diffusion components:

$$\frac{1}{2}\frac{\partial u}{\partial t}=-a\frac{\partial u}{\partial x}$$

$$\frac{1}{2}\frac{\partial u}{\partial t}=\nu\frac{\partial^2 u}{\partial x^2}$$

Comment on the physical and numerical interpretation of such a decomposition. Why is this approach particularly useful in modern computational methods such as particle-in-cell or Lagrangian-Eulerian schemes?

REFERENCES

E. Isaacson, H. B. Keller, Analyse numerischer Verfahren, Verlag Deutsch, Zürich, 1973

N. N. Yanenko, The Method of Fractional Steps. Springer-Verlag. Berlin, 1971.

C. K. Chu, Numerical Methods in Fluid Dynamics, Advances in Applied Mechanics, Volume 18, Academic Press, 1978.

15 Pressure-Velocity Coupling in Incompressible Flows: Finite Difference Method

This chapter explores the principle of pressure-velocity coupling in incompressible flow solvers. We examine the foundational algorithms that address the challenge of enforcing continuity in the absence of an explicit pressure evolution equation. Emphasis is placed on pressure correction via the pressure Poisson equation, the use of predictor-corrector sequences, and historical developments such as MAC, SMAC, and some new projection methods.

- Understand the physical origin of pressure-velocity coupling in incompressible Navier-Stokes solvers.
- Study classical schemes based on projection-type algorithms.
- Formulate the pressure Poisson equation and velocity correction step.
- Trace the sequence of predictor and corrector steps in fractional-step methods.

The Navier-Stokes equations govern the motion of incompressible fluids and embody the conservation of mass and momentum. Numerically, the central difficulty is the tight coupling between pressure and velocity: pressure lacks its own evolution equation yet must enforce the divergence-free condition on the velocity field. Various algorithms therefore compute pressure indirectly while preserving this constraint. This chapter surveys both historical and modern pressure-velocity coupling schemes, emphasising seminal developments at Los Alamos National Laboratory (MAC, SMAC, HSMAC) and Imperial College London (SIMPLE and its extensions).

We trace the progression from the Marker-and-Cell (MAC) method—which pioneered staggered finite-difference grids—to projection and fractional-step approaches, first formalised by Chorin and Temam, that advance momentum without pressure, solve a pressure Poisson equation, and then project velocity onto a divergence-free space. This procedure maintains mass conservation and removes any spurious divergence generated during the velocity prediction.

DOI: 10.1201/9781003663171-15

15.1 PROJECTION METHODS

The Navier-Stokes equations describe the motion of fluid such as liquids and gases, are a set of nonlinear partial differential equations fundamental to the field of fluid dynamics. These equations express conservation of momentum and mass, and solving them is crucial for predicting weather patterns, designing aircraft, understanding ocean currents, and simulating blood flow, among other applications. Due to the complexity of these equations, especially in three-dimensional and turbulent flow conditions, various numerical methods have been developed for their approximation and solution.

Among the most prominent techniques are those developed at Los Alamos National Laboratory and Imperial College, which have made significant contributions to the computational methods used in fluid dynamics. The Marker and Cell (MAC) method, introduced by Harlow and Welch in 1965, was one of the first approaches used to solve the Navier-Stokes equations for incompressible flows. This method discretizes the fluid domain into a fixed grid, and fluid properties such as velocity and pressure are calculated at discrete points, making it suitable for complex geometries. The method was further elaborated by Welch, Harlow, Shannon, and Daly in 1966.

Following the MAC method, the Simplified Marker and Cell (SMAC) method, developed by Amsden and Harlow in 1970, improved computational efficiency by simplifying the pressure-correction step, which is critical in ensuring that the computed flow field satisfies the continuity equation. The Highly Simplified MAC method (HSMAC), also known as SOLA, developed by Hirt et al. in 1975.

A foundational theoretical treatment was presented by N. N. Yanenko in his 1971 monograph The Method of Fractional Steps, which laid the mathematical groundwork for contemporary operator-splitting techniques, providing the rigorous convergence and stability proofs that later CFD schemes rely on. The Fractional Step method, or the Two Step Projection method by Kothe et al. in 1992, further refines these approaches by splitting the solution process into fractional steps, allowing for more stable and accurate solutions.

Each of these methods has its own set of advantages and limitations, making them suitable for specific types of fluid flow problems. The choice of method often depends on the specific requirements of the simulation, such as the need for accuracy in capturing sharp interfaces, the computational cost, and the geometric complexity of the problem domain. As research continues and computational resources expand, these methods are continuously refined and adapted, pushing the boundaries of what can be modeled and predicted with the Navier-Stokes equations.

Table 15.1
Grid type requirements for different pressure-velocity coupling approaches in FD methods

Method	Institution / Reference	Grid Arrangement
MAC	Harlow & Welch, 1965 Welch et al., 1966	Staggered
Artificial Compressibility	Chorin, 1968	Staggered or Collocated
SMAC	Amsden & Harlow, 1970	Staggered
FS (Fractional Step)	Chorin, 1967; Temam, 1969	Staggered or Collocated
HSMAC (SOLA)	Hirt et al., 1975	Staggered

Solution Methods for Pressure-Velocity Coupling (FD method)
Los Alamos National Laboratory (USA):

* **MAC method** (Marker and Cell): incompressible flow only (Harlow and Welch, 1965; Welch et al., 1966)
* **SMAC method** (Simplified Marker and Cell): incompressible flow only (Amsden and Harlow, 1970)
* **HSMAC method** (Highly Simplified MAC), also known as SOLA: incompressible flow only (Hirt et al., 1975)

Courant Institute of Mathematical Sciences, New York University (USA):

* **FS method** (Fractional-Step or Two-Step Projection): incompressible flow only (Chorin, 1967)
* **Artificial-Compressibility method**: incompressible flow, steady or dual-time (Chorin, 1968)

Table 15.1 summarizes the evolution of pressure-velocity coupling schemes and their relationship with grid arrangements. Early methods such as MAC, SMAC, and HSMAC, developed at Los Alamos, were designed for staggered finite-difference meshes, where face-centered velocities and cell-centered pressures naturally avoid checkerboarding. Later, Chorin's fractional-step and artificial-compressibility methods, introduced at the Courant Institute, generalized this concept and could operate on either staggered or collocated grids.

In this chapter, we focus on methods designed for finite-difference methods. pressure-velocity couplings in finite-voulme framework will be discussed in chapter.

15.1.1 MAC METHOD

The Marker and Cell (MAC) method represents a significant advancement in the numerical analysis of free surface problems involving fluid dynamics. Developed to address the complexities of time-dependent, viscous, incompressible flows in two dimensions with Cartesian coordinates, the MAC method utilizes the finite difference approach to solve the full Navier-Stokes equations. It specifically caters to scenarios where uniform body forces, such as gravity, impact the fluid within an irregular box-shaped domain. This technique stands out by discretizing the fluid domain into cells where variables like pressure and two-component velocity are represented as single average values for each cell, facilitating simpler and more accurate computations.

Historically, the Marker-and-Cell (MAC) method was introduced by Harlow and Welch in 1965 to simulate incompressible, viscous flows—particularly with free surfaces. The method's name derives from its use of "marker" particles to track the fluid interface and the "cell" grid arrangement for solving the governing equations. Modern terminology often refers to this same approach as a projection method for incompressible flow.

In the continuous form, the incompressible Navier-Stokes equations are

$$\frac{\partial \mathbf{u}}{\partial t} + (\mathbf{u}\cdot\nabla)\mathbf{u} = -\frac{1}{\rho}\nabla p + \nu\nabla^2\mathbf{u} + \mathbf{f},$$

$$\nabla\cdot\mathbf{u} = 0.$$

Here $\mathbf{u}$ is velocity, p is pressure, ρ is density, ν is kinematic viscosity, and $\mathbf{f}$ is any body force. Both the MAC scheme and the Harlow-Welch approach solve these same equations in a "projection" manner: velocity is updated in a two-step fashion each time step, and pressure appears as a Lagrange multiplier to enforce the divergence-free condition.

A key feature is the staggered grid (often called the MAC grid) in which the pressure (and other scalars) are stored at cell centres, while each velocity component is stored at the corresponding cell faces. This arrangement prevents decoupling of pressure and velocity that might otherwise arise on a collocated grid. In two dimensions, each cell centre holds a pressure $p_{i,j}$; the horizontal velocities u lie on the vertical faces, and the vertical velocities v lie on the horizontal faces. This layout ensures that the discrete divergence in each cell can be evaluated cleanly and that the pressure gradients directly drive the face-centred velocities.

The MAC (projection) method advances the flow in time with a fractional-step procedure:

1. Predictor (tentative velocity). Solve the momentum equations without the new pressure to obtain a provisional field $\mathbf{u}^*$:

$$\frac{\mathbf{u}^* - \mathbf{u}^n}{\Delta t} = -(\mathbf{u}\cdot\nabla)\mathbf{u}^n + \nu\nabla^2\mathbf{u}^n + \mathbf{f}^n,$$

where $\mathbf{f}^n$ may include explicit body-force terms such as gravity. The intermediate velocity $\mathbf{u}^*$ generally does not satisfy incompressibility.

2. Pressure-Poisson equation (PPE). Correct the velocity with the yet-unknown pressure:
$$\mathbf{u}^{n+1} = \mathbf{u}^* - \frac{\Delta t}{\rho}\nabla p^{n+1}.$$
Take the divergence and impose $\nabla\cdot\mathbf{u}^{n+1} = 0$; because ρ is known at time level n, this yields
$$\boxed{\nabla^2 p^{n+1} = \frac{\rho}{\Delta t}\nabla\cdot\mathbf{u}^*}$$
for constant density. (For variable density, keep the factor ρ^{-1} under the divergence: $\nabla\cdot\left[\rho^{-1}\nabla p^{n+1}\right] = (\Delta t)^{-1}\nabla\cdot\mathbf{u}^*$.)
3. Velocity correction. Once p^{n+1} is obtained, project the tentative field:
$$\mathbf{u}^{n+1} = \mathbf{u}^* - \frac{\Delta t}{\rho}\nabla p^{n+1},$$
which by construction satisfies $\nabla\cdot\mathbf{u}^{n+1} = 0$.
4. Boundary conditions and update. Apply appropriate velocity or pressure boundary conditions to $\mathbf{u}^{n+1}$ and p^{n+1}; advance to the next time step.

The MAC scheme was the first practical projection method, demonstrating that pressure acts as a Lagrange multiplier to enforce $\nabla\cdot\mathbf{u} = 0$.

Regarding time-stepping strategy, the method originated as explicit or semi-explicit method with stability restrictions tied to convective and diffusive time-step constraints. The original Harlow-Welch scheme performed these updates in an explicit manner. Further refinements and higher-order time integration approaches may alter details of stability and accuracy, but they do not change the fundamental equivalence of the projection procedure and the staggered layout.

MAC Algorithm

V: Velocity vector field.	**f**: External body force per unit mass.
ρ: Density of the fluid.	∇: Gradient operator.
p: Pressure field.	Δt: Time step.
τ: Stress tensor.	n: Time step index.
g: Gravitational force per unit mass.	$n+1$: Next time step index.

Step 1: Predict an intermediate velocity field (ignoring the new pressure):
$$\frac{\mathsf{V}^* - \mathsf{V}^n}{\Delta t} = -\nabla\left(\mathsf{V}^n\mathsf{V}^n\right) - \frac{1}{\rho^n}\nabla p^n + \mathbf{g}^n + \frac{\mathbf{f}^n}{\rho^n}$$

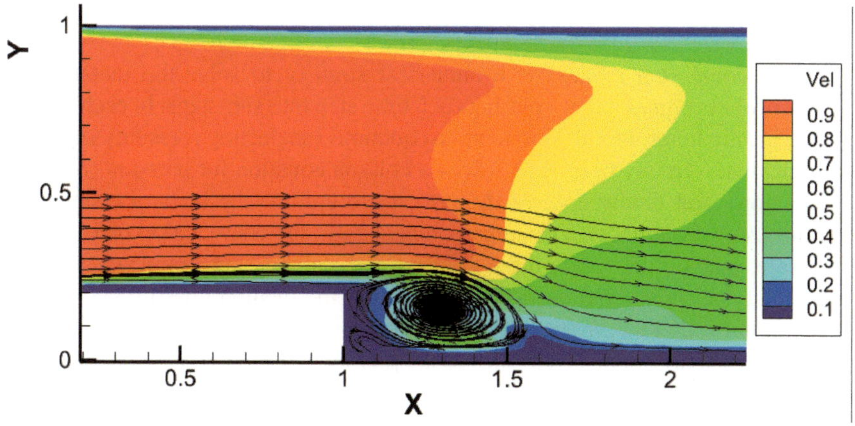

Figure 15.1: **Flow over a Backward-Facing Step using MAC method. The pathlines shows the formation of recirculation zones.**

Here, V^* is the predicted (or "star") velocity obtained without the contribution of the updated pressure p^{n+1}.

Step 2 Enforce incompressibility by solving the pressure-Poisson equation Take the divergence of the velocity-correction formula

$$V^{n+1} = V^* - \frac{\Delta t}{\rho^n}\nabla p^{n+1},$$

impose $\nabla \cdot V^{n+1} = 0$, and use the fact that ρ^n is known from the previous time level. We obtain the standard pressure-Poisson equation

$$\boxed{\nabla\cdot\left(\frac{1}{\rho^n}\nabla p^{n+1}\right) = \frac{1}{\Delta t}\nabla\cdot V^*} \qquad \text{(PPE)}$$

Solve (PPE) on staggered grid (cell-centred pressure, face-centred velocity) with the appropriate Neumann or Dirichlet boundary conditions to obtain p^{n+1}; then proceed to Step 3.

Step 3: Correct the velocity field with the newly computed pressure.

$$V^{n+1} = V^* - \Delta t \frac{1}{\rho^n}\nabla p^{n+1}$$

This correction ensures that $\nabla \cdot V^{n+1} = 0$ is satisfied.

Step 4: Apply boundary conditions and update solution variables. Impose the appropriate boundary conditions (velocity or pressure) on V^{n+1} and p^{n+1}. This completes one time step of the MAC algorithm.

15.1.2 FRACTIONAL-STEP METHOD

The fractional step method is a numerical strategy to solve the incompressible Navier-Stokes equations by splitting each time step into sub-steps. In each time step, one typically first solves the momentum equations (neglecting pressure) to predict an intermediate velocity field, then solves a Poisson equation for pressure (or pressure correction), and finally updates the velocity to enforce the incompressibility constraint (zero divergence). This approach decouples the computation of velocity and pressure, making the solver more efficient and stable for incompressible flows. In computational fluid dynamics (CFD) literature, the method goes by several names - it is often called the projection method (because the intermediate velocity is projected" onto a divergence-free field) or the pressure-correction method, among other similar terms. All these terms refer to the same basic idea of a fractional (split) time-step scheme where pressure is computed in a separate step to enforce continuity.

The origins of the fractional step method can be traced back to the mid-1960s. A significant early development was the Marker-and-Cell (MAC) method introduced by F. Harlow and J. Welch in 1965 for simulating incompressible flows. The MAC scheme pioneered the approach of enforcing the incompressibility constraint by solving a discrete Poisson equation for pressure, thereby decoupling the velocity and pressure computations. This ensured that the corrected velocity field satisfied the divergence-free condition. The method marked one of the earliest implementations in which the momentum equations were solved independently from a pressure correction step—a foundational concept that would later evolve into modern fractional step and projection methods.

Building on these ideas, a formal fractional step approach for Navier-Stokes was first proposed in the late 1960s. In 1967-1968, Alexandre J. Chorin introduced what came to be known as Chorin's projection method. Chorin's method treated the incompressibility condition via a two-step algorithm: advance the velocities without pressure, then solve a pressure Poisson equation and project the velocity onto a divergence-free space. This effectively splits each time step into a fractional step for velocity and one for pressure. Independently, Roger Temam developed a similar fractional-step scheme around the same time, publishing his results in 1969. Temam's work (in French) explicitly referred to the méthode de pas fractionnaires (method of fractional steps) for the Navier-Stokes equations, and is often cited alongside Chorin's as the co-invention of the approach. These pioneering works established the core idea that one could solve for velocity and pressure in separate substeps without sacrificing accuracy for incompressible flows.

The term fractional step itself was popularized by the Soviet mathematician Nikolai N. Yanenko, who, along with colleagues, developed a general operator-splitting technique in the 1960s. Yanenko 's work culminated in the book The Method of Fractional Steps (first published in Russian and later in English in 1971). This book presented the fractional step (or splitting) method as a powerful and general numerical strategy for solving multidimensional problems in mathematical physics, including fluid dynamics. As described in Yanenko's introduction, the fractional step method showed great promise for tackling incompressible flow problems over a wider range of conditions than earlier methods, and it was expected to supplant traditional relaxation methods as computers grew more powerful. In essence, Yanenko's work provided a theoretical foundation and broader context for the approach that Chorin and Temam had applied to Navier-Stokes. The fractional-step idea became well-established in the CFD community, with later researchers often citing Yanenko's book as a fundamental reference on splitting methods.

Chorin's algorithm is often referred to as a projection method because it relies on the mathematical idea of projecting an intermediate velocity onto a divergence-free subspace (via the pressure Poisson equation). In parallel, the similar ideas under the banner of pressure-correction methods become popular. The fractional step method quickly became a cornerstone of incompressible flow simulation, and numerous improvements and variations emerged in subsequent decades.

The Fractional Step Method is a numerical approach designed to efficiently solve the incompressible Navier-Stokes equations while enforcing the divergence-free condition ($\nabla\cdot\vec{v}=0$). A key aspect of this method is how it handles the nonlinear convection term ($-(\vec{v}\cdot\nabla)\vec{v}$) and its interaction with the pressure field. The nonlinear convection term contributes to the source term in the Pressure Poisson Equation, ensuring that it is implicitly accounted for during the pressure correction step rather than being explicitly solved in every time step. This approach improves computational efficiency and stability.

The Pressure Poisson Equation is central to the Fractional Step Method, as it ensures incompressibility by enforcing the condition $\nabla\cdot\vec{v}^{n+1}=0$. This leads to the elliptic equation:

$$\nabla^2 p = -\rho\nabla\cdot[(\vec{v}\cdot\nabla)\vec{v}] - \nabla\cdot\vec{f}.$$

Here, the nonlinear convection term ($(\vec{v}\cdot\nabla)\vec{v}$) and external forces ($\vec{f}$) contribute to the source term for pressure. Solving this equation adjusts the pressure field p to implicitly account for nonlinear advection and enforce divergence-free velocity in the fluid.

The Predictor Step of the Fractional Step Method computes an intermediate velocity field $\vec{v}^*$, neglecting the pressure gradient. This velocity is calculated as:

$$\vec{v}^* = \vec{v}^n + \Delta t\left[-(\vec{v}\cdot\nabla)\vec{v} + \nu\nabla^2\vec{v} + \vec{f}\right].$$

Since the nonlinear convection term has already been included in the source term of the Poisson equation, it does not need to be explicitly solved here. Instead, the pressure field p computed from the Poisson equation accounts for its effects, ensuring a more efficient and stable process.

The method proceeds with a Corrector Step, where the intermediate velocity $\vec{v}^*$ is adjusted using the pressure field:

$$\vec{v}^{n+1} = \vec{v}^* - \frac{\Delta t}{\rho}\nabla p.$$

This step ensures that the final velocity field $\vec{v}^{n+1}$ is divergence-free and incorporates the effects of pressure and nonlinear advection.

By solving the nonlinear convection term implicitly through the pressure Poisson equation, the Fractional Step Method avoids expensive direct computation of advection terms in every step. This ensures computational efficiency while maintaining stability and accuracy. The method's separation of velocity prediction, pressure correction, and incompressibility enforcement makes it a robust tool for solving incompressible Navier-Stokes equations.

In the Fractional Step Method, the CFL condition and diffusion number play crucial roles in ensuring numerical stability during the predictor step, where an intermediate velocity field is computed. The CFL condition governs the maximum allowable time step (Δt) based on the velocity field and grid resolution, ensuring stability for the advection term ($(\vec{v}\cdot\nabla)\vec{v}$). Similarly, the diffusion number constrains Δt based on viscosity (ν) and grid spacing, stabilizing the diffusion term ($\nu\nabla^2\vec{v}$). However, these constraints are not directly relevant to the Pressure Poisson Solver, as the Poisson equation is an elliptic equation solved to enforce incompressibility ($\nabla\cdot\vec{v}=0$). The computation of the pressure field is independent of the time step size, as it arises from enforcing the incompressibility constraint rather than advancing a physical quantity in time. This pressure field is subsequently used to correct the intermediate velocity field in the corrector step. Therefore, while the CFL and diffusion numbers are critical for the stability and accuracy of the time-stepping scheme, they do not influence the pressure field computation.

Fractional-Step (Two-Step Projection) Method

V : velocity vector field	$\mathbf{f}$: body-force per unit mass
ρ : density	Δt : time step
p : pressure	n : current time index
ν : kinematic viscosity	$n+1$: next time index

Step 1 - Predictor (intermediate velocity V^*)

$$\mathsf{V}^* = \mathsf{V}^n + \Delta t\left[-(\mathsf{V}^n\cdot\nabla)\mathsf{V}^n + \nu\nabla^2\mathsf{V}^n + \mathbf{f}^n\right]$$

(The pressure gradient is omitted, so $\nabla\cdot\mathsf{V}^* \neq 0$ in general.)
Step 2 - Pressure-Poisson equation (enforce incompressibility)

$$\boxed{\nabla^2 p^{n+1} = \frac{\rho^n}{\Delta t}\nabla\cdot\mathsf{V}^*} \tag{PPE}$$

(Solve on the staggered grid with suitable Neumann/Dirichlet boundary conditions.)
Step 3 - Corrector (projection)

$$\mathsf{V}^{n+1} = \mathsf{V}^* - \frac{\Delta t}{\rho^n}\nabla p^{n+1}$$

(After this update, $\nabla\cdot\mathbf{V}^{n+1} = 0$.)
Step 4 - Boundary update Apply the prescribed wall or outflow conditions to $\mathbf{V}^{n+1}$ and p^{n+1}; advance to the next time step.

15.1.3 SIMPLIFIED MARKER-AND-CELL (SMAC) METHOD

The Simplified Marker-and-Cell (SMAC) method, introduced by Amsden and Harlow in 1970, is a form of fractional-step or projection method for incompressible flow. In their report, Amsden and Harlow describe splitting each time-step of the incompressible Navier-Stokes solution into two stages: (1) a provisional velocity calculation (predictor step), and (2) a velocity correction using an auxiliary potential function (often interpreted as a pressure or pressure-correction) to enforce incompressibility. This two-step predictor-corrector strategy is characteristic of fractional-step methods. In essence, the SMAC algorithm computes an intermediate velocity field that does not necessarily satisfy the divergence-free condition, then projects this field onto a divergence-free space by solving a pressure-related equation -hence it falls under the class of projection methods.

Contemporary literature often uses the term *projection method* for schemes that enforce the divergence-free constraint by projecting the velocity onto a solenoidal field. Chorin's 1968 scheme and Temam's 1969 scheme were early examples of this approach, splitting the momentum update and pressure update into separate steps. The SMAC method specifically "splits the calculation cycle into two parts, namely: a provisional velocity field calculation followed by a velocity revision employing an auxiliary potential function to ensure incompressibility. That auxiliary potential function serves the role of a pressure correction. In modern terms, SMAC is a pressure-correction projection method, meaning it includes a pressure update to correct the provisional velocity field. This places SMAC firmly in the family of fractional-step methods, since the time-integration of the Navier-Stokes equations is performed in fractional steps (first an explicit advection-diffusion step, then an implicit pressure correction step).

In a pressure-correction scheme, one typically uses the pressure from the previous time step as an initial guess and computes a pressure increment that corrects the velocity field to enforce continuity. This is essentially equivalent to the projection method formulation -the difference is largely in whether the pressure term is partially retained in the predictor step. In the original Chorin projection method, the momentum equation is solved without any pressure term (resulting in a purely provisional velocity), and then a Poisson equation for pressure is solved to project the velocity onto divergence-free form. In SMAC (and similar pressure-correction methods), one may include the known old pressure in the momentum prediction and solve for a pressure correction (the auxiliary potential) to adjust the velocities. Despite this slight difference in formulation, both approaches accomplish the same fractional-step goal: the end-of-step velocity satisfies incompressibility.

Thus, SMAC falls under the category of fractional-step projection methods, using a predictor-corrector (velocity-pressure) update each time step to couple the momentum and continuity equations. The SMAC method was developed to solve the incompressible Navier-Stokes equations, which consist of the momentum balance and the incompressibility (continuity) condition. In continuous form, the governing equations are:

Continuity (incompressibility):

$$\nabla \cdot \mathbf{u} = 0$$

ensuring the velocity field $\mathbf{u}(x,y,z,t)$ has zero divergence (no volume change).

Momentum (Navier-Stokes):

$$\frac{\partial \mathbf{u}}{\partial t} + (\mathbf{u} \cdot \nabla)\mathbf{u} = -\frac{1}{\rho}\nabla p + \nu \nabla^2 \mathbf{u}$$

which represents conservation of momentum for a Newtonian fluid of density ρ and kinematic viscosity ν. Here $p(x,y,z,t)$ is the pressure field. (For clarity, body forces are neglected in this form, and $\mathbf{u} \cdot \nabla \mathbf{u}$ denotes the convective acceleration.)

The SMAC algorithm advances the solution of these equations over a time step Δt by splitting the update into two sub-steps using a predictor-corrector approach:

1. Provisional Velocity Step (Predictor): Compute an intermediate velocity $\mathbf{u}^*$ at the new time level without enforcing the divergence-free condition. This is done by integrating the momentum equation without the unknown new pressure. For example, one may use the pressure from the previous time step (or drop the pressure term entirely) to advance the velocity explicitly. In a simple explicit formulation, this predictor step can be written as:

$$\frac{\mathbf{u}^* - \mathbf{u}^n}{\Delta t} = -(\mathbf{u}^n \cdot \nabla)\mathbf{u}^n + \nu \nabla^2 \mathbf{u}^n - \frac{1}{\rho}\nabla p^n,$$

producing a tentative velocity $\mathbf{u}^*$ at time t^{n+1} that generally is not divergence-free. (If the pressure term $-\nabla p^n$ is omitted entirely, this corresponds to Chorin's original projection approach; including p^n corresponds to a pressure-correction variant. Either way, $\mathbf{u}^*$ will usually have nonzero divergence.)

2. Pressure-Poisson Equation: To enforce the continuity constraint at the new time, solve a Poisson equation for the *pressure correction* (or for the new pressure). This equation arises from taking the divergence of the momentum equation and requiring $\nabla \cdot \mathbf{u}^{n+1} = 0$. In SMAC, an auxiliary potential function ϕ is introduced such that $\phi = p^{n+1} - p^n$ (the pressure increment) or equivalently $p^{n+1} = p^n + \phi$. The Poisson equation for ϕ is obtained by applying $\nabla \cdot$ to the provisional momentum equation:

$$\nabla^2 \phi = \frac{\rho}{\Delta t} \nabla \cdot \mathbf{u}^*,$$

subject to appropriate boundary conditions (typically Neumann boundary conditions for ϕ on solid walls, derived from the normal momentum condition). This pressure Poisson equation (PPE) ensures that the correction ϕ will remove any divergence in the provisional field. (Notably, this step is the mathematical "projection"; ϕ acts as a Lagrange multiplier field enforcing $\nabla \cdot \mathbf{u}^{n+1} = 0$.)

3. Velocity Correction Step (Corrector): Adjust the provisional velocity by the gradient of the solved potential to obtain the divergence-free velocity at the new time. The corrected velocity is given by:

$$\mathbf{u}^{n+1} = \mathbf{u}^* - \frac{\Delta t}{\rho^n} \nabla \delta p$$

By construction, this updated $\mathbf{u}^{n+1}$ now satisfies $\nabla \cdot \mathbf{u}^{n+1} = 0$. The pressure field is simultaneously updated: $p^{n+1} = p^n + \phi$. In other words, the auxiliary potential ϕ from the Poisson solve effectively becomes the new pressure (or pressure correction) that enforces incompressibility. After this correction, the algorithm advances to the next time step.

The above set of equations constitutes the core of the original SMAC algorithm as formulated by Amsden and Harlow. In their implementation, the velocities are stored on a staggered grid (like the original MAC) and the pressure (or pseudo-pressure) is stored at cell centers. The provisional momentum equations are solved on that grid, and then the pressure-correction Poisson equation is solved for the cell-center potential ϕ, which is used to adjust face-centered velocities. This approach was novel in that era for decoupling the velocity and pressure calculation while ensuring mass conservation each step. By splitting the calculation cycle into a predictor and corrector, the SMAC method avoided the instabilities and complexities that arose in the fully coupled original MAC method, while still maintaining incompressibility at every step.

In summary, the governing equations employed in SMAC are the standard incompressible Navier-Stokes equations (continuity and momentum), supplemented by a derived Poisson equation for pressure (or pressure correction) and an update formula for velocity. These can be listed as follows, combining the above steps:

Continuity: $\nabla \cdot \mathbf{u}^{n+1} = 0$ (enforced by the correction step). Momentum predictor: $\mathbf{u}^*$ from $\frac{\mathbf{u}^* - \mathbf{u}^n}{\Delta t} = -(\mathbf{u}^n \cdot \nabla)\mathbf{u}^n + \nu \nabla^2 \mathbf{u}^n - \frac{1}{\rho} \nabla p^n$ (or without p^n term). Pressure Poisson: $\nabla^2 \phi = \frac{\rho}{\Delta t} \nabla \cdot \mathbf{u}^*$ (solves for ϕ, the pressure increment). Velocity and pressure correction: $\mathbf{u}^{n+1} = \mathbf{u}^* - \frac{\Delta t}{\rho} \nabla \phi$, and $p^{n+1} = p^n + \phi$.

These equations are essentially those given (either explicitly or implicitly) in the foundational SMAC documentation by Amsden and Harlow. The result is a stable algorithm that at each time step first predicts the flow and then projects it onto a divergence-free solution by solving a pressure (potential) equation. This ensured incompressibility "throughout" the computation, a key requirement for accurate incompressible flow simulation.

The SMAC method is one of the early projection schemes for incompressible flows, and its principles are shared with other fractional-step methods developed subsequently. In essence, any method that splits the Navier-Stokes solution into a sequence of substeps (one handling advection-diffusion of momentum, and one enforcing continuity via pressure) falls into the same family. The differences among these methods lie in details such as time integration order, treatment of the pressure term in the predictor, and how boundary conditions are applied. Below we clarify how SMAC relates to and differs from some other well-known approaches:

The original Marker-and-Cell (MAC) method of Harlow and Welch (1965) was a coupled algorithm using marker particles for free-surface tracking and a staggered grid for pressure-velocity computation. It required solving the pressure through a Poisson equation as well, but the procedure was somewhat entangled with the marker motion and had stability limitations on the time step. The SMAC method simplified the MAC approach by explicitly splitting the pressure solve into a separate step. By doing so, SMAC circumvented difficulties of the original MAC scheme and improved stability. Essentially, SMAC formalized the predictor-corrector nature: MAC became a true two-step process (prediction + pressure correction) rather than a single complex update. This is why SMAC (Simplified MAC) is considered a direct application of the fractional-step idea to the MAC method.

Chorin introduced the idea of treating the incompressibility via an artificial compressibility or projection approach around the same time. His method neglects pressure during the intermediate velocity update and then computes the pressure by enforcing $\nabla \cdot \mathbf{u}^{n+1} = 0$. Temam independently proposed a similar split. SMAC and Chorin's method are conceptually the same in that both solve a Poisson equation to enforce continuity at each step. One distinction is that SMAC was implemented on a staggered grid finite-difference framework, whereas Chorin's original work was more general and later adapted to finite-volume grids. Both are first-order accurate in time in their original forms (because the pressure at the new time is effectively treated explicitly).

Today's state-of-the-art projection methods still follow the template set by methods like SMAC. They may use higher-order Runge-Kutta for the predictor, treat viscous terms implicitly, or implement advanced boundary condition treatments, but they still have a step that solves $\nabla^2 p =$ (source) to enforce $\nabla \cdot \mathbf{u} = 0$. The SMAC method's legacy is evident -it demonstrated that one could compute incompressible flows by decoupling the update of velocity and pressure in a stable way.

SMAC Algorithm

$\mathbf{V}$: velocity vector field	$\mathbf{f}$: external body force
ρ : density	∇ : gradient operator
p : pressure	Δt : time step
τ : viscous stress tensor	n : current time index
$\mathbf{g}$: gravitational acceleration	$n+1$: next time index

Step 1 — Predictor (tentative velocity $\mathbf{V}^*$)

$$\frac{\mathbf{V}^* - \mathbf{V}^n}{\Delta t} = -(\mathbf{V}^n \cdot \nabla)\mathbf{V}^n - \frac{1}{\rho^n}\nabla p^n + \frac{1}{\rho^n}\nabla \cdot \tau^n + \mathbf{g}^n + \frac{\mathbf{f}^n}{\rho^n}$$

(The tentative field generally has $\nabla \cdot \mathbf{V}^* \neq 0$.)

Step 2 — Pressure-correction Poisson equation Write $p^{n+1} = p^n + \delta p$ and require $\nabla \cdot \mathbf{V}^{n+1} = 0$:

$$\nabla^2 \delta p = \frac{\rho^n}{\Delta t} \nabla \cdot \mathbf{V}^*$$

Solve this equation with suitable Neumann/Dirichlet boundary conditions for δp.

Step 3 — Update the pressure

$$p^{n+1} = p^n + \delta p.$$

Step 4 — Correct the velocity

$$\mathbf{V}^{n+1} = \mathbf{V}^* - \frac{\Delta t}{\rho^n} \nabla \delta p.$$

After this projection step, $\nabla \cdot \mathbf{V}^{n+1} = 0$.

15.1.4 DIRECTION-SPLITTING FRACTIONAL-STEP METHOD FOR COLLOCATED GRIDS

This section presents the formulation of a second-order-accurate, direction-splitting fractional-step algorithm developed by Chiarini et al. (2021) for advancing the incompressible Navier-Stokes equations on a collocated Cartesian mesh. Spatial derivatives are discretised using centred second-order finite differences, while time integration applies a Crank-Nicolson scheme to the viscous term combined with a leapfrog update for pressure. The symmetric tensor-product Helmholtz operator

$$A = (1 - \partial_{xx})(1 - \partial_{yy})(1 - \partial_{zz})$$

reduces each three-dimensional solve into three one-dimensional tridiagonal systems, which are efficiently inverted via the Thomas algorithm, thereby enabling straightforward parallelisation through domain decomposition and Schur-complement techniques.

$$\partial_t \mathbf{u} + (\mathbf{u} \cdot \nabla)\mathbf{u} - Re^{-1}\nabla^2 \mathbf{u} + \nabla p = \mathbf{f}, \qquad \nabla \cdot \mathbf{u} = 0$$

Here $\mathbf{u} = (u, v, w)$ is velocity, p is kinematic pressure, $\mathbf{f}$ is a prescribed forcing, and Re denotes the Reynolds number. Dirichlet, periodic, or symmetry conditions apply to $\mathbf{u}$ on $\partial\Omega$; pressure is fixed up to a constant level (zero spatial mean).

Introducing the time step $\varepsilon = \Delta t$, an auxiliary divergence ϕ, and a projection parameter $\chi \in [0, 1]$,

$$\partial_t \mathbf{u}^\varepsilon + (\mathbf{u}^\varepsilon \cdot \nabla)\mathbf{u}^\varepsilon - Re^{-1}\nabla^2 \mathbf{u}^\varepsilon + \nabla p^\varepsilon = \mathbf{f},$$

$$\varepsilon, A\phi^\varepsilon + \nabla \cdot \mathbf{u}^\varepsilon = 0,$$

$$\varepsilon, \partial_t p^\varepsilon = \phi^\varepsilon - \chi, Re^{-1} \nabla \cdot \mathbf{u}^\varepsilon.$$

The symmetric positive-definite operator used for pressure correction is

$$A = (1-\partial_{xx})(1-\partial_{yy})(1-\partial_{zz}),$$

whose tensor-product structure is key to the directional splitting.

Define the explicit pressure predictor

$$p^{*n+\frac{1}{2}} = p^{*n-\frac{1}{2}} + \phi^{,n-\frac{1}{2}},$$

and a second-order extrapolation of the velocity for the nonlinear term,

$$\mathbf{u}^{n+\frac{1}{2}} = \tfrac{3}{2}\mathbf{u}^n - \tfrac{1}{2}\mathbf{u}^{n-1}.$$

The time-stepping equations read

$$\frac{\mathsf{V}^{n+1}-\mathsf{V}^{n}}{\Delta t} - \frac{1}{2Re}\nabla^2\left(\mathsf{V}^{n+1}+\mathsf{V}^{n}\right) = f^{n+\frac{1}{2}} - \nabla p^{*n+\frac{1}{2}} - \mathcal{N}\left(\mathsf{V}^{*n+\frac{1}{2}}\right)$$

$$A\phi^{n+\frac{1}{2}} = -\frac{1}{\Delta t}\nabla\cdot\mathbf{u}^{n+1}$$

$$p^{n+\frac{1}{2}} = p^{n-\frac{1}{2}} + \phi^{n+\frac{1}{2}} - \frac{\chi}{2Re}\nabla\cdot\left(\mathbf{u}^{n+1}+\mathbf{u}^{n}\right)$$

The parameter χ controls how strongly the pressure-correction step enforces divergence-free flow, with values between 0 and 1 that allow tuning of stability and conservation properties. In the Chiarini et al. algorithm, it appears in the pressure-update equation as:

$$p^{n+1} = p^n + \phi - \chi\cdot\frac{\Delta t}{Re}(\nabla\cdot\mathbf{u}^*)$$

Where, ϕ is the pressure correction, Δt is the timestep, Re is the Reynolds number, and $\mathbf{u}^*$ is the intermediate velocity field

χ=0: Recovers the classic incremental pressure-correction method, where pressure evolves solely through ϕ:

$$p^{n+1} = p^n + \phi$$

χ=1: Creates a fully consistent projection by including the divergence term:

$$p^{n+1} = p^n + \phi - \frac{\Delta t}{Re}(\nabla\cdot\mathbf{u}^*)$$

This improves kinetic energy conservation but may reduce stability.

$0 < \chi < 1$: Provides a tunable compromise between stability (lower χ) and conservation (higher χ).

Start with χ = 1 for maximum accuracy in laminar flows. Reduce χ (e.g., 0.5-0.7) if encountering stability issues. For turbulent flows, lower χ often improves robustness.

Direction-Splitting Fractional-Step Method

V: Velocity vector field	Δt: Time step
p: Pressure field	n: Current time index
ϕ: Pressure correction	$n+1$: Next time index
ρ: Fluid density	$*$: Extrapolated value
Re: Reynolds number	$\diamond$: intermediate value
f: Body-force vector	Re: Reynolds number
A: Pressure-Poisson operator	$\mathcal{N}(\mathsf{V})$: $(\mathsf{V}\cdot\nabla)\mathsf{V}$
χ: Projection parameter	

Step 1: Implicit momentum update to obtain the provisional velocity $\mathsf{V}^{\diamond}$ (direction splitting).

$\frac{\mathsf{V}^{\diamond}-\mathsf{V}^{n}}{\Delta t}-\frac{1}{2Re}\nabla^{2}(\mathsf{V}^{\diamond}+\mathsf{V}^{n})=\mathbf{f}^{n+\frac{1}{2}}-\nabla p^{*n+\frac{1}{2}}-\mathcal{N}(\mathsf{V}^{*n+\frac{1}{2}})$

The Helmholtz operator is inverted by three successive one-dimensional tridiagonal solves along x, y, and z.

Step 2: Enforce incompressibility by solving the pressure-correction equation.

$A\phi^{n+\frac{1}{2}}=-\frac{1}{\Delta t}\nabla\cdot\mathsf{V}^{\diamond}$

Step 3: Update the pressure field.

$p^{n+\frac{1}{2}}=p^{n-\frac{1}{2}}+\phi^{n+\frac{1}{2}}-\frac{\chi}{2Re}\nabla\cdot(\mathsf{V}^{\diamond}+\mathsf{V}^{n})$

Step 4: Correct the velocity to obtain a divergence-free field.

$\mathsf{V}^{n+1}=\mathsf{V}^{\diamond}-\Delta t\nabla\phi^{n+\frac{1}{2}}$

Step 5: Apply boundary conditions and advance all variables to the next time level.

15.2 ARTIFICIAL COMPRESSIBILITY METHOD (ACM)

The artificial compressibility method (ACM) is a numerical technique for solving incompressible flow problems by introducing a *pseudo-time* derivative of pressure into the continuity equation. Originally proposed by A. J. Chorin in 1967, the method's key idea is to add an artificial compressibility term (i.e., a fictitious time-dependent term) to the incompressible continuity equation. This modifies the continuity equation to a hyperbolic form with a pseudo-time evolution of pressure (or density), allowing the combined Navier-Stokes system to be marched in time *as if* the fluid is compressible. Conceptually, the added term plays the role of an artificial acoustic compressibility. In this case, the pressure is no longer a purely instantaneous

Lagrange multiplier, but instead evolves in pseudo-time such that its steady-state satisfies the divergence-free (incompressibility) constraint. By marching the equations in pseudo-time until the pressure derivative vanishes, the method drives the velocity field toward a divergence-free solution, effectively decoupling the pressure-velocity coupling issue and obviating the need for solving a Poisson equation for pressure.

Historically, ACM provided a breakthrough in computing steady incompressible flows by converting the elliptic pressure constraint into a tractable time-marching process. It remains relevant in modern CFD, not only for steady flows but also for unsteady incompressible simulations (via dual time-stepping), as it ensures that when the pseudo-time iterations converge, the incompressible continuity equation is satisfied and the correct pressure field consistent with the velocity field is obtained.

The continuity equation asserts that the temporal variation of density, in conjunction with the divergence of mass flux, equates to zero. In index notation, this indicates that the density ρ and the velocity components u_j must fulfill the equation $\frac{\partial \rho}{\partial t} + \frac{\partial}{\partial x_j}(\rho u_j) = 0$.

This articulates the principle of mass conservation for liquids and gasses alike.

Momentum conservation in tensor form, representing Newton's second rule for a fluid, is expressed as

$$\frac{\partial(\rho \mathbf{u})}{\partial t} + \nabla \cdot \left(\rho \mathbf{u}\mathbf{u}^T + p\mathbf{I} - \tau\right) = \mathbf{0}$$

where, p denotes pressure, $\mathbf{I}$ represents the identity tensor, and τ signifies the viscous stress tensor.

In index notation, the i-th momentum component is expressed as

$$\frac{\partial(\rho u_i)}{\partial t} + \frac{\partial}{\partial x_j}(\rho u_i u_j + p\delta_{ij} - \tau_{ij}) = 0, \quad i = 1,2,3.$$

Here, δ_{ij} denotes the Kronecker delta, while τ_{ij} signifies the deviatoric (viscous) stress components. This yields three interrelated equations, one corresponding to each velocity component.

The comprehensive energy conservation equation for total energy in a compact representation is

$$\frac{\partial(\rho E)}{\partial t} + \nabla \cdot (\rho E\mathbf{u} + p\mathbf{u} - \tau \cdot \mathbf{u} + \mathbf{q}) = 0,$$

where E represents the total energy per unit mass (the sum of internal energy e and kinetic energy $\frac{|\mathbf{u}|^2}{2}$), and $\mathbf{q}$ denotes the heat flux vector. In index notation, the equation becomes

$$\frac{\partial(\rho E)}{\partial t} + \frac{\partial}{\partial x_j}(\rho E u_j + p u_j - \tau_{ij} u_i + q_j) = 0.$$

The expression $p u_j$ represents work done by pressure, $-\tau_{ij} u_i$ signifies viscous dissipation, and q_j pertains to heat conduction. For a Newtonian fluid, the viscous stress is expressed as

$$\tau_{ij} = \mu\left(\frac{\partial u_i}{\partial x_j} + \frac{\partial u_j}{\partial x_i} - \tfrac{2}{3}\delta_{ij}\frac{\partial u_k}{\partial x_k}\right), \quad q_j = -k\frac{\partial T}{\partial x_j}.$$

In these equations, ρ denotes density, u_i represents velocity components, p signifies pressure, E indicates total energy per unit mass, T refers to temperature, μ stands for dynamic viscosity, k symbolizes thermal conductivity, and δ_{ij} is the Kronecker delta. The Einstein summation convention, wherein repeated indices denote summing, is employed. The governing equations of mass, momentum, and energy conservation constitute the foundation of the majority of CFD codes for liquids and gases. An equation of state (linking p, ρ, and T) is typically necessary to complete the system.

To further understand the structure of Chorin's Artificial Compressibility Method (ACM), it is important to state that the classical incompressibility condition $\nabla \cdot \mathbf{u} = 0$ is modified by introducing a pseudo-time derivative of pressure. This leads to the artificial continuity equation:

$$\frac{\partial p}{\partial \tau} + \nabla \cdot \mathbf{u} = 0$$

where τ is a fictitious pseudo-time. This transforms the incompressible flow system into a form that can be advanced explicitly in pseudo-time, avoiding the need to solve a pressure Poisson equation. At each physical time level, the velocity field is updated from the momentum equations using a guessed or previous pressure field, and then the pressure is corrected iteratively in pseudo-time until the velocity becomes nearly divergence-free. The iteration is terminated when the condition

$$|\nabla \cdot \mathbf{u}| < \varepsilon$$

is satisfied for a chosen convergence threshold ε. This ensures that the velocity field satisfies the incompressibility constraint and the corresponding pressure field is consistent.

In Chorin's implementation, the momentum equation is first used to compute an intermediate velocity $\mathbf{u}^*$ without enforcing incompressibility:

$$\frac{\mathbf{u}^* - \mathbf{u}^n}{\Delta t} = -(\mathbf{u}^n \cdot \nabla)\mathbf{u}^n - \nabla p^n + \nu \nabla^2 \mathbf{u}^n$$

Then, pseudo-time iterations are performed using the artificial compressibility equation to evolve pressure and correct the velocity field. This iterative process resembles projection methods, although no global Poisson solve is required. Notably, ACM treats pressure as a dynamic variable that evolves in pseudo-time rather than being imposed instantly as a constraint.

In the steady limit, the ACM iterations converge to a divergence-free velocity field, and the resulting pressure approximates the Lagrange multiplier that enforces incompressibility. In this sense, ACM can be interpreted as a preconditioned iterative approximation of the pressure-Poisson solution, particularly effective for structured finite difference solvers.

15.2.1 THE PSEUDO-TRANSIENT PRESSURE GRADIENT METHOD

Consider a dimensionless incompressible Navier-Stokes equations expressed in primitive variables:

$$\nabla \cdot \vec{V} = 0, \tag{1}$$

$$\frac{\partial \vec{V}}{\partial t} + \vec{V} \cdot \nabla \vec{V} = -\nabla P + \frac{1}{Re} \nabla^2 \vec{V}$$

where, $\vec{V}$ denotes velocity, t represents time, P signifies pressure, and Re indicates the Reynolds number inside a spatial region $\Omega \subset \mathbb{R}^3$. A primary challenge in resolving these equations is the lack of a straight updating formula for the pressure variable.

Adhering to the idea of artificial compressibility and recognizing that incompressible flows are influenced by pressure gradients instead of absolute pressure, the continuity equation is revised to create a linkage between velocity and pressure gradient. A pseudo-transient pressure gradient term is included following the gradient of the continuity equation:

$$\frac{1}{\delta} \frac{\partial P}{\partial t} + \nabla \cdot \vec{V} = 0,$$

where $\mathscr{D} = \nabla \cdot \vec{V}$ represents the volumetric dilation (or strain rate), and $\delta > 1$ denotes a relaxation factor. The dilation $\mathscr{D}$ is employed to revise the pressure gradient derived from the velocity field obtained in the momentum equation. In a stable state, the pseudo-transient term $\frac{\partial P}{\partial t}$ becomes null, and $\nabla \mathscr{D} = 0$. Resolving $\nabla \mathscr{D} = 0$ results in $\mathscr{D} = \text{constant}$. The divergence-free condition is fulfilled by assigning that constant a value of zero at a reference point within the flow domain. This results in the revised system:

$$\frac{1}{\delta} \frac{\partial P}{\partial t} + \nabla \mathscr{D} = 0,$$

$$\frac{\partial \vec{V}}{\partial t} + \vec{V} \cdot \nabla \vec{V} = -\nabla P + \frac{1}{Re} \nabla^2 \vec{V}.$$

These equations, contingent upon suitable boundary conditions, are applicable to both two-dimensional and three-dimensional problems. The pressure gradient is computed exclusively at inside grid points, hence eliminating the necessity for pressure gradient boundary conditions.

A finite difference approximation is applied on a collocated $(N \times M)$ grid. An explicit time-marching approach is used for the momentum problem, employing Euler-explicit temporal discretization to reach a steady-state solution. A second-order central difference method is adopted to discretize the velocity-field divergence (dilation). However, using second-order central differences for the spatial terms in the pseudo-transient pressure gradient equation can lead to an odd-even decoupling, where the gradient-divergence operator acts only on odd or even grid points separately, causing fluctuations in the computed velocity field. To avoid this, first-order forward and backward difference methods are employed in the boundary-adjacent grid cells, ensuring communication between odd and even points. Second-order central differences are then maintained for the remaining interior grid cells.

For the momentum equation, a three-point central difference scheme handles the viscous term, and a two-point central difference scheme is used for the convective terms. The discretized momentum equation subsequently updates the velocity components by using the newly obtained pressure field.

When the second-order central difference scheme is applied to the spatial derivative terms, it introduces an odd-even decoupling. In other words, the combined gradient-divergence operator acting on the velocity components $\nabla(\nabla \cdot \vec{V})$ fails to ensure proper communication between odd and even grid points. For instance, if the x-gradient operator is discretized for $\mathscr{D}$ (the volumetric dilation), its discrete form

$$\frac{\partial \mathscr{D}}{\partial x} = \frac{u_{i+2} - 2u_{i+1} + u_i}{4\Delta x} + \dots$$

illustrates how odd grid points only interact with other odd points, while even grid points do the same among themselves. A similar effect occurs in the y-gradient operator, generating spurious oscillations in the velocity field.

To remedy this problem, first-order forward and backward difference schemes are employed in cells directly adjacent to the boundaries. For example, at $i = 2$ and $i = N - 1$,

$$i = 2: \quad \frac{1}{\delta} p_{n+1,x_j,x_i} - p_{n,x_j,x_i} + \frac{\mathscr{D}_{n,j,i}}{\Delta x} = 0,$$

$$i = N-1: \quad \frac{1}{\delta} p_{n+1,x_j,x_i} - p_{n,x_j,x_i} + \frac{\mathscr{D}_{n,j,N-2}}{\Delta x} = 0.$$

This ensures interaction between the odd and even points in the interior. For

$$\frac{1}{\delta}\frac{\partial P}{\partial t} + \nabla \mathscr{D} = 0, \tag{5}$$

a second-order central difference method is still employed on the remaining points ($i = 3$ to $i = N - 2$, $j = 3$ to $j = N - 2$).

In the momentum equation, the viscous term is discretized using a three-point central difference approach, and the convective terms use a two-point central difference approach. The resulting discretized form of the momentum equation is

$$u_{n+1,i,j} - u_{n,i,j} = \frac{u_n(i+1,j) - u_n(i-1,j)}{2\Delta x}$$

$$+ \frac{v_n(i,j+1) - v_n(i,j-1)}{2\Delta y} - \left(\frac{p_{n+1}(i+1,j) - p_{n+1}(i-1,j)}{2\Delta x}\right),$$

$$v_{n+1,i,j} - v_{n,i,j} = \frac{u_n(i+1,j) - u_n(i-1,j)}{2\Delta x}$$

$$+ \frac{v_n(i,j+1) - v_n(i,j-1)}{2\Delta y} - \left(\frac{p_{n+1}(i,j+1) - p_{n+1}(i,j-1)}{2\Delta y}\right).$$

where, u_n is the x-direction velocity at step n, and v_n is the y-direction velocity at step n.

In a fully consistent semi-discretization, we get

$$\begin{aligned} u_{n+1,i,j} \;=\;\; & u_{n,i,j} + \Delta t \Bigg[-\frac{p_{n+1}(i+1,j) - p_{n+1}(i-1,j)}{2\Delta x} \\ & + (\text{convective terms}) \\ & + \frac{1}{Re}(\text{diffusive terms}) \Bigg] \end{aligned}$$

15.3 VORTICITY TRANSPORT EQUATION

Although, we are discussing the pressure-velocity coupling, but in certain CFD problem it is easy to just use the stream funcation formulation. In computational fluid dynamics (CFD), particularly for incompressible flows, the traditional approach for solving the Navier-Stokes equations involves setting up of an equation relating velocity and pressure fields directly. However, for two-dimensional, incompressible flows, an alternative formulation using the stream function (ψ) and vorticity (ζ) offers several computational advantages.

The stream function-vorticity approach simplifies the Navier-Stokes equations by automatically satisfying the incompressibility condition ($\nabla \cdot \vec{u} = 0$). In this formulation, the stream function, ψ, relates to the velocity components as follows:

$$u = \frac{\partial \psi}{\partial y}, \quad v = -\frac{\partial \psi}{\partial x}$$

Here, u and v represent the velocity components in the x and y directions, respectively. The vorticity, ζ, is defined as the curl of the velocity field:

$$\zeta = \frac{\partial v}{\partial x} - \frac{\partial u}{\partial y}$$

which simplifies to $\zeta = -\Delta\psi$ in terms of the stream function.

With the definitions of ψ and ζ, the dynamics of the fluid can be described by two main equations: Vorticity Transport Equation:

$$\frac{\partial \zeta}{\partial t} + u\frac{\partial \zeta}{\partial x} + v\frac{\partial \zeta}{\partial y} = \nu \left(\frac{\partial^2 \zeta}{\partial x^2} + \frac{\partial^2 \zeta}{\partial y^2} \right)$$

This equation accounts for the transport and diffusion of vorticity in the fluid, where ν is the kinematic viscosity.

Poisson Equation for the Stream Function:

$$\Delta\psi = -\zeta$$

Solving this equation provides the stream function, which in turn gives the velocity field.

The numerical solution of these equations typically involves discretizing them using finite difference methods or other suitable numerical techniques. Boundary conditions are critical for the correct simulation of physical scenarios:

1. For solid boundaries, no-slip conditions ($u = v = 0$) typically translate into $\psi =$ constant and $\frac{\partial \psi}{\partial n} = 0$ where n is the direction normal to the boundary.

2. At inflows or outflows, more complex conditions might be applied depending on the specific problem setup.

The solution procedure involves an iterative approach where the vorticity equation is solved to update ζ at each time step, followed by solving the Poisson equation for ψ. The updated stream function is then used to compute the new velocity field, which feeds back into the vorticity equation in the next iteration.

This method is highly effective for 2D incompressible flows and is favored in many practical applications due to its robustness in enforcing the continuity equation and its relative simplicity in handling complex boundary geometries. Through this approach, the intricate dance of fluid elements is captured, allowing scientists and engineers to predict and analyze fluid behavior under a vast array of conditions.

Case Study 15A: Lid-driven Cavity Flow at Re=100 solved using FS algorithm

Problem Definition

This case considers the steady incompressible flow in a lid-driven square cavity of size 1×1 at Reynolds number $Re = 100$. The top lid moves rightward with a constant velocity of 1 m/s, while the other three walls are stationary. The fluid is assumed to be Newtonian and incompressible with constant viscosity. The primary vortex is expected to form in the center, and secondary eddies develop depending on grid resolution and Reynolds number. This is a celebrated benchmark study investigated by many researchers like Botella and Peyret (1998) and others.

Governing Equations

The simulation solves the 2D unsteady incompressible Navier-Stokes equations in primitive variables. The continuity equation enforces incompressibility:

$$\nabla \cdot \mathrm{V} = 0$$

The momentum equations in x and y directions are given by:

$$\frac{\partial u}{\partial t} + u\frac{\partial u}{\partial x} + v\frac{\partial u}{\partial y} = -\frac{1}{\rho}\frac{\partial p}{\partial x} + \nu\nabla^2 u$$

$$\frac{\partial v}{\partial t} + u\frac{\partial v}{\partial x} + v\frac{\partial v}{\partial y} = -\frac{1}{\rho}\frac{\partial p}{\partial y} + \nu\nabla^2 v$$

The flow is laminar and viscous, and the simulation evolves in time until a steady solution is reached.

Numerical Setup

A finite-difference formulation is used on a staggered Cartesian grid with 32×32 cells. The pressure field is defined at cell centers, while u and v velocities are stored at vertical and horizontal faces, respectively. The solver is based on the classical MAC (Marker-And-Cell) staggered grid arrangement. It uses classical fractional-step method for pressure-velocity coupling.
Time advancement is performed with an explicit time step $\Delta t = 10^{-4}$. The Poisson equation for pressure correction is solved via sparse matrix factorization using the 'scipy.sparse' linear algebra solver. The convergence tolerance is set to 10^{-8}.

Results and Discussion

The steady-state solution at $Re = 100$ is obtained using the criterion:

$$\frac{\|f^{n+1} - f^n\|}{\|f^{n+1}\|} < \varepsilon, \quad \varepsilon = 10^{-8}$$

Rather than targeting a fixed maximum physical time $t_{\max}$, the simulation was ran until the relative change in the velocity field per step dropped below 10^{-7}. If a fixed time step of $\Delta t = 10^{-4}$ is used, the user may need 30,000-60,000 time steps (i.e., 3-6 seconds of physical time) to satisfy the same convergence criterion. The number of required steps depends on grid resolution. The convergence is set to be achieved at

> The recommended setting is `tol = 1e-7` which is close to `max_iter = 20000`.
>
> Monitor convergence with residual history, stagnation of centerline velocity, and the stabilization of vorticity at $(0.5, 0.5)$.

Lessons Learned / Numerical Challenges

The flow at $Re = 100$ takes considerable simulated time to fully develop when using a low-resolution finite-difference solver. While convergence in divergence is fast, the physical development of the vortex requires longer time integration. One must distinguish between mathematical convergence of residuals and physical development of the flow. Grid resolution plays a key role, and 32×32 may under-resolve features near the corners. Numerical symmetry enforcement through correct boundary condition handling at the lid is also essential.

Problem Definition

This case involves the simulation of two-dimensional incompressible lid-driven cavity flow using the streamfunction-vorticity formulation. The geometry is a square cavity with a no-slip boundary on all sides and a top lid moving horizontally at unit velocity. The flow is expected to form a primary

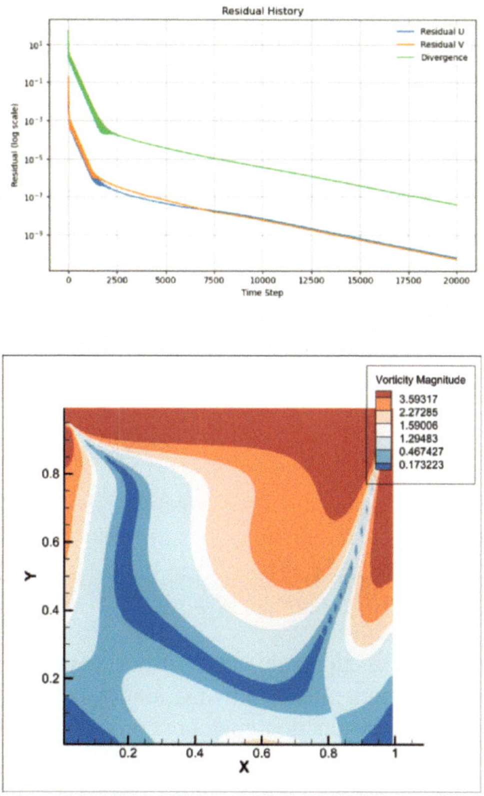

Figure 15.2: **Top: Residual history showing logarithmic convergence. Bottom: Vorticity contour of developing cavity vortex.**

clockwise vortex in the center, with possible secondary corner eddies depending on Reynolds number and grid resolution. The top boundary drives circulation, while the other walls impose zero velocity, creating sharp velocity gradients and vortex structures. The flow is unsteady at initialization but converges to a steady-state for Re = 100.

Case Study 15B: Lid-Driven Cavity Problem at Re=100 (Streamfunction-Vorticity Approach)

Governing Equations

The simulation solves the incompressible Navier-Stokes equations reformulated in terms of the streamfunction ψ and vorticity ω. The streamfunction satisfies the Poisson equation $\nabla^2\psi = -\omega$, ensuring incompressibility. The vorticity transport equation $\frac{\partial \omega}{\partial t} + u\frac{\partial \omega}{\partial x} + v\frac{\partial \omega}{\partial y} = \nu\nabla^2\omega$ governs the evolution of vorticity. Here, $u = \frac{\partial \psi}{\partial y}$ and $v = -\frac{\partial \psi}{\partial x}$. The assumptions include laminar, incompressible, unsteady flow with no turbulence model.

Numerical Setup

A finite difference method is used with second-order central differencing for spatial derivatives. The computational domain is a uniform Cartesian grid with $N = 16$ interior cells per side (total $N+2 = 18$ including ghost points for boundaries). The Poisson equation for the streamfunction is discretized using a five-point Laplacian and solved via sparse LU factorization. Time integration of the vorticity equation is performed explicitly using forward Euler with a small timestep $\Delta t = 0.001$. Vorticity boundary conditions are derived from wall velocities and streamfunction gradients using second-order approximations. Velocities are reconstructed from streamfunction gradients at each timestep.

Results and Discussion

The solution reveals a steady clockwise rotating primary vortex, whose core is identified by the global minimum of the streamfunction. The velocity field smoothly transitions from the driven lid to the stationary boundaries. Vorticity contours show sharp gradients near walls and decay toward the vortex center.

The minimum streamfunction value in lid-driven cavity flow represents the center of the primary vortex formed by the motion of the top lid. In incompressible 2D flow, streamfunction contours correspond to streamlines, and the vortex center occurs at the point where the streamfunction reaches its minimum. This point lies within a closed streamline loop and typically has near-zero velocity. The magnitude of the minimum streamfunction reflects the strength of circulation in the cavity. It is a critical diagnostic quantity used to assess solution accuracy, validate numerical solvers, and compare results against benchmark data like those reported by Ghia et al.

Comparison with Ghia et al. (1982) shows close agreement in the location of the vortex center. The streamfunction minimum from the present simulation approaches Ghia's benchmark with increasing grid resolution. Similarly, the vorticity at the geometric center matches closely with the values reported by Botella and Peyret (1998) using spectral methods.

Comparison with Ghia et al. work

Simulations	Grid	Primary vortex center x	Primary vortex center y	Stream function (ψ_{min})
Ghia et al. 1982	129	0.7344	0.6172	-0.103423
Run 1	137	0.7319	0.6159	-0.078001
Run 2	145	0.7329	0.6164	-0.077834
Run 3	153	0.7338	0.6169	-0.077681

Ghia et al. (1982) solved the incompressible Navier–Stokes equations using the streamfunction–vorticity formulation on a uniform Cartesian grid with second-order central differences. They employed a multigrid solver combined with a strongly implicit procedure (SIP) for steady-state convergence. A 129×129 grid was used for Re = 100, with vorticity boundary conditions derived from finite difference expressions. In contrast, Botella and Peyret (1998) used a Chebyshev collocation spectral method with high-order spatial accuracy. Singularities at cavity corners were subtracted using asymptotic expansions to preserve spectral convergence. Their formulation also solved a Poisson equation for the streamfunction with Dirichlet boundary conditions.

Comparison with Botella and Peyret work

Simulations	Grid x Grid	Vorticity at (0.5, 0.5)
Botella and Peyret 1998	64 x 64	1.174412
Run 1	137 x 137	1.172338
Run 2	145 x 145	1.176942
Run 3	153 x 153	1.181045

Lessons Learned / Numerical Challenges

The solver demonstrates stable and accurate behavior for low Reynolds numbers using explicit time-stepping. However, convergence slows for coarser grids and high Reynolds numbers. Vorticity boundary conditions require careful implementation to avoid nonphysical artifacts. The explicit scheme imposes a strict time step limit, and the solution is sensitive to grid resolution. A five-point Laplacian and LU factorization are efficient for the Poisson solve, but the vorticity evolution remains the limiting factor. Future improvements could include higher-order spatial discretization, implicit schemes, or adaptive meshing near wall layers to resolve gradients more efficiently.

PROBLEMS

15P-1 Describe the physical significance of the continuity equation

$$\nabla \cdot \mathrm{V} = 0$$

in incompressible flow and explain how pressure acts as a Lagrange multiplier to enforce this constraint in numerical methods.

15P-2 Explain how taking the divergence of the provisional momentum update in the projection method leads to the pressure Poisson equation

$$\nabla^2 p = \frac{\rho}{\Delta t} \nabla \cdot \mathrm{V}^*$$

and discuss the role of each term on the right-hand side.

15P-3 Why Vorticity transport equations are rarely used in modern CFD applications?

15P-4 Explain the principle of projection methods in incompressible flow solvers, detailing how the intermediate velocity is computed, how the pressure Poisson equation is derived.

15P-5 Discuss why Fourier-transform-based solvers achieve optimal performance for the discrete pressure Poisson equation on uniform grids, emphasizing their $\mathcal{O}(N \log N)$ complexity, spectral accuracy, and natural handling of periodic or homogeneous boundary conditions.

15P-6 Outline the predictor-corrector sequence in the fractional-step projection method, detailing how the intermediate velocity prediction, pressure Poisson solve, and final velocity correction combine to decouple and then re-couple pressure and velocity fields.

15P-7 Discuss the implementation of boundary conditions for the pressure Poisson equation

$$\nabla^2 p = \frac{\rho}{\Delta t} \nabla \cdot \mathbf{V}^*$$

comparing Dirichlet and Neumann conditions and their impact on global mass conservation and solver stability.

15P-8 Solve the two-dimensional lid-driven cavity flow problem using a fractional step scheme, with the following parameters: the top wall velocity $U_{\text{top}} = 1.5$, total simulation time T (input by the user), desired under-relaxation factor α (input by the user), and convergence tolerance tol (input by the user). Set the fluid properties with a density $\rho = 1.0$, kinematic viscosity $\nu = 0.1$, CFL number for stability CFL $=$ 0.5, and diffusion number for stability Diff $= 0.4$. In the predictor step, compute an intermediate velocity field based on diffusion terms while neglecting the pressure. Then, solve the pressure Poisson equation iteratively to enforce incompressibility. Finally, in the corrector step, update the velocity field using the computed pressure, ensuring the solution satisfies both the incompressibility condition and boundary conditions, and advance the solution in time using explicit time-stepping.

15P-9 Consider the two-dimensional, incompressible, laminar flow inside a square cavity of side length L=1 driven by a tangential lid velocity U. Let the flow be described in terms of the stream function ψ and the vorticity ω. The governing equations in vorticity-stream function form are

$$\frac{\partial \omega}{\partial t} + \left(u\frac{\partial \omega}{\partial x} + v\frac{\partial \omega}{\partial y}\right) = \nu \nabla^2 \omega,$$

$$\nabla^2 \psi = -\omega,$$

where $u = \frac{\partial \psi}{\partial y}$, $v = -\frac{\partial \psi}{\partial x}$, and ν is the kinematic viscosity. The boundary conditions on the stream function and vorticity are derived from the no-penetration, no-slip conditions on the walls, with the top boundary moving at U=1 m/s.

Use a uniform mesh of size N $\times$ N, with grid spacing $\Delta x = \Delta y = \frac{L}{N-1}$. Employ a simple time-stepping method to evolve the vorticity equation in time, and solve the Poisson equation for ψ via a five-point stencil for the Laplacian in conjunction with an appropriate sparse solver. Integrate up to T=5 using Δt=0.0005. Plot and discuss:

(a) The final contour plot of the stream function ψ(x,y).

(b) The final contour plot of the vorticity ω(x,y).

(c) The flow pattern inside the cavity and the influence of ν=0.01 on the recirculation zones.

Comment on the accuracy and stability of numerical results, especially near the moving lid and in the bottom corners of the cavity.

15P-10 Do Problem **15P-9** using v FD-O2 method for space and RK4 method for time-stepping.

15P-11 Given is a system of equations for a two-dimensional flow

$$\nabla \cdot \vec{v} = 0$$

$$\nabla p = \nu \nabla^2 \vec{v}$$

i Derive an equation for the pressure, assuming a known velocity field $\vec{v}(x,y)$, and comment on its type.

ii Discretize the resulting equation and suggest a solution scheme.

iii Check the convergence of the resulting finite difference equations.

15P-12 Consider the two-dimensional, unsteady, incompressible flow inside a square cavity domain $\Omega = [0,1] \times [0,1]$, commonly referred to as the lid-driven cavity problem. The top wall moves horizontally with a constant velocity U_{top}, while the bottom and side walls remain stationary. The governing equations are the incompressible Navier-Stokes equations:

$$\frac{\partial \mathbf{u}}{\partial t} + \mathbf{u} \cdot \nabla \mathbf{u} = -\frac{1}{\rho} \nabla p + \nu \nabla^2 \mathbf{u}, \quad \nabla \cdot \mathbf{u} = 0,$$

where $\mathbf{u} = (u, v)$ is the velocity field in the x- and y-directions, p is the pressure, ρ is the fluid density, and ν is the kinematic viscosity. The boundary conditions are:

$$u(x,1,t) = U_{\text{top}}, \quad u(x,0,t) = u(0,y,t) = u(1,y,t) = 0,$$

$$v(x,y,t) = 0 \quad \text{on all boundaries.}$$

To maintain incompressibility, a pressure-velocity coupling strategy is applied using a projection (fractional step) method. In this approach, an intermediate velocity field is first computed by advancing the momentum equations without the pressure term. The intermediate velocity is then used to compute the source term in a Poisson equation for pressure:

$$\nabla^2 p = \frac{\rho}{\Delta t} \left(\frac{\partial u^*}{\partial x} + \frac{\partial v^*}{\partial y} \right),$$

subject to homogeneous Neumann boundary conditions on all walls. The pressure field is used to correct the intermediate velocity, yielding a divergence-free velocity field:

$$\mathbf{u}^{n+1} = \mathbf{u}^* - \frac{\Delta t}{\rho}\nabla p.$$

Spatial discretization is based on a second-order central difference method on a uniform grid. The pressure Poisson equation is solved iteratively using Jacobi iteration. Plot the velocity contours.

15P-13 To evaluate the accuracy and stability of the Artificial Compressibility Method (ACM) Chorin (1967) applied his scheme to a canonical test case with a known analytical solution to the incompressible Navier–Stokes equations in two dimensions. The computational domain is the square $[0\pi] \times [0\pi]$ and the initial conditions at $t = 0$ are

$$u_1(x_1x_20) = -cosx_1 sinx_2 \quad u_2(x_1x_20) = sinx_1 cosx_2 \quad p(x_1x_20) = -\tfrac{1}{4}(cos2x_1 + cos2x_2)$$

Boundary conditions: The fields are periodic in both x_1 and x_2 consistent with the structure of the analytical solution.

These expressions also define the exact solution for all times $t > 0$:

$$u_1(x_1x_2t) = -cosx_1 sinx_2 e^{-2t}$$

$$u_2(x_1x_2t) = sinx_1 cosx_2 e^{-2t}$$

$$p(x_1x_2t) = -\tfrac{1}{4}(cos2x_1 + cos2x_2)\,e^{-4t}$$

Discuss the rationale for using this problem as a validation case. Explain why the convective term vanishes how the governing equations are modified in the artificial compressibility method and describe the observed numerical performance. Comment on the significance of accurate results on a collocated grid.

REFERENCES

H. H. Harlow, J. E. Welch, Numerical Calculation of Time-Dependent Viscous Incompressible Flow of Fluid with Free Surface, Physics of Fluids, 8(12): 2182-2189. 1965.

J. E. Welch, F. H. Harlow, J. P. Shannon, B. J. Daly, The MAC Method: A Computing Technique for Solving Viscous, Incompressible, Transient Fluid-Flow Problems Involving Free Surfaces, Los Alamos Scientific Lab Report LA-3425, 1966.

A. J. Chorin, A Numerical Method for Solving Incompressible Viscous Flow Problems, Journal of Computational Physics, 2: 12-26, 1967.

A. A Amsden, F. H. Harlow, The SMAC Method: A Numerical Technique for Calculating Incompressible Fluid Flows, Los Alamos Scientific Lab Report LA-4370, 1970.

C. W. Hirt, B. D. Nichols, N. C. Romero, SOLA: A Numerical Solution Algorithm for Transient Fluid Flows, Los Alamos Scientific Lab Report LA-5852, 1975.

A. A Amsden, F. H. Harlow, Fluid Dynamics, Los Alamos Scientific Lab Monograph LA-4700, 1971.

Hirt et al. An Arbitrary Lagrangian-Eulerian Computing Method for All Flow Speeds, J. Comput. Phys., 14(3): 227-253, 1974.

N. N. Yanenko, The Method of Fractional Step, Springer-Verlag, Berlin, 1971

Patankar and Spalding, A Calculation Procedure for Heat, Mass and Momentum Transfer in Three-Dimensional Parabolic Flows, Int. J. Heat Mass Transfer, 15(10): 1787-1806, 1972.

Rhie and Chow, A Numerical Study of the Turbulent Flow Past an Isolated Airfoil with Trailing Edge Separation, AIAA Journal, 21(11): 1525-1532, 1983.

Issa, Solution of the Implicitly Discretised Fluid Flow Equations by Operator-Splitting, J. Comput. Phys., 62(1): 40-65, 1986.

Weller et al., A Tensorial Approach to Computational Continuum Mechanics Using Object-Oriented Techniques, Computers in Physics, 12(6): 620-631, 1998.

F. Chiarini, M. Quadrio, P. Luchini, Semi-implicit direction-splitting method for the incompressible Navier-Stokes equations on co-located grids, Journal of Computational Physics, 436, 110297, 2021.

O. Botella, R. Peyret, Benchmark spectral results on the lid-driven cavity flow, Computers and Fluids, Vol. 27, pp. 421-433, 1998.

16 Classical Finite Difference Methods for Hyperbolic PDEs and Discontinuities

In this chapter, we explore classical finite difference methods for solving hyperbolic partial differential equations, with a focus on their ability to resolve discontinuities and steep gradients. These methods play a key role in gas dynamics and compressible flow simulations, where nonlinear conservation laws often produce shock waves and other abrupt solution features. The schemes covered in this chapter provide stability and improved accuracy for problems involving discontinuities. After completing this chapter, you should be able to:

- Explain the concept of characteristic lines and their role in determining solution behavior for conservation laws.
- Describe how shocks form from intersecting characteristics and how shock-capturing schemes resolve them.
- Understand the challenges of resolving discontinuities and the origin of the Gibbs phenomenon in numerical solutions.
- Learn about finite difference schemes such as Lax-Friedrichs, Lax-Wendroff, MacCormack, and Von Neumann-Richtmyer artificial viscosity methods.

The equation

$$u_t + uu_x = 0$$

is a fundamental equation in fluid dynamics, and we aim to solve this equation using a geometric approach. The characteristic curves of this equation are defined by the solutions to the ordinary differential equation (ODE)

$$\frac{dx}{dt} = u(x,t)$$

Since the equation is nonlinear, the characteristic equation depends directly on the function $u(x,t)$, which we have yet to determine. Each specific solution of $u(x,t)$ yields a different set of characteristic curves. According to the existence and uniqueness theorem for ODEs, through every point (x_0,t_0) passes a unique characteristic curve. If we knew the solution $u(x,t)$, we could populate the x-t plane with a network of non-intersecting curves.

DOI: 10.1201/9781003663171-16

At first glance, without knowing $u(x,t)$, it might seem impossible to ascertain any properties of a characteristic curve $x(t)$. However, we can observe that the value of u remains constant along each characteristic curve. This is shown by:

$$\frac{d}{dt}[u(x(t),t)] = u_t + \frac{dx}{dt}u_x = u_t + uu_x = 0,$$

where we used the chain rule and substituted $\frac{dx}{dt} = u$. The result $\frac{d}{dt}[u(x(t),t)] = 0$ indicates that $u(x(t),t)$ is constant along each characteristic curve. Consequently, $\frac{dx}{dt} = u(x(t),t)$ is also constant along the curve.

From this analysis, we deduce three key principles:

a Each characteristic curve is a straight line. This is because $\frac{dx}{dt}$ is constant, implying that every solution $u(x,t)$ corresponds to a family of straight lines with various slopes, which are the characteristics.

b The solution $u(x,t)$ is constant along each characteristic line.

c The slope of each characteristic line equals the value of $u(x,t)$ at any point on that line.

These points illustrate how we can analyze the characteristics of the partial differential equation using geometric methods. They show that in the *x*-*t* plane, each characteristic is straight while it remains within a region of constant u.

Let us now assume that we are seeking a solution to the PDE that satisfies the initial condition $u(x,0) = \phi(x)$. That is, the solution is delineated on the line $t = 0$. The characteristic line that passes through $(x_0,0)$ must have a slope of $\phi(x_0)$. In the same vein, the slope of the characteristic line through $(x_1,0)$ must be $\phi(x_1)$. We are in trouble if the two lines intersect. For $u = \phi(x_0)$ on one line and $u = \phi(x_1)$ on the other line, it is impossible for $\phi(x_0) = \phi(x_1)$ to be true due to the fact that they have distinct slopes.

$$u(x,0) = \phi(x)$$

Let us now consider solving the partial differential equation (PDE) $u_t + uu_x = 0$ with the initial condition $u(x,0) = \phi(x)$. This means we are specifying the value of u along the line $t = 0$ in the *x*-*t* plane. Our goal is to determine how this initial information propagates over time.

From our earlier analysis, we established three key principles:

☞ Each characteristic curve is a straight line-because $\frac{dx}{dt} = u$ is constant along each characteristic.

☞ The solution $u(x,t)$ is constant along each characteristic line-since $\frac{d}{dt}[u(x(t),t)] = 0$.

☞ The slope of each characteristic line equals the value of $u(x,t)$ along that line-that is, the slope is $u = \phi(x_0)$ for the characteristic passing through $(x_0,0)$.

Using these principles, let's delve deeper into how the initial condition $u(x,0) = \phi(x)$ influences the solution and the behavior of the characteristic curves. For each point x_0 on the line $t = 0$, the value of u is given by $\phi(x_0)$. According to principle (γ), the slope of the characteristic line passing through $(x_0,0)$ is $\phi(x_0)$. Therefore, the equation of this characteristic line is:

$$x(t) = x_0 + \phi(x_0)t.$$

Similarly, for another point x_1 on $t = 0$, the characteristic line passing through $(x_1,0)$ has the equation:

$$x(t) = x_1 + \phi(x_1)t.$$

Now, consider the possibility that these two characteristic lines intersect at some point (x,t) in the x-t plane. Setting the two equations equal to each other, we have:

$$x_0 + \phi(x_0)t = x_1 + \phi(x_1)t.$$

Solving for t, we find the time at which the intersection occurs:

$$t = \frac{x_1 - x_0}{\phi(x_0) - \phi(x_1)}.$$

This equation shows that if $\phi(x_0) \neq \phi(x_1)$, the characteristic lines will intersect at a finite time t. The intersection time t is positive if $\phi(x_0) > \phi(x_1)$ and $x_1 > x_0$, indicating that faster characteristics (with larger ϕ) are overtaking slower ones.

At the point of intersection (x,t), we face a significant problem: there are two characteristic lines passing through the same point, each carrying a different constant value of u:

Along the characteristic from $(x_0,0)$, $u = \phi(x_0)$.

Along the characteristic from $(x_1,0)$, $u = \phi(x_1)$.

This leads to a contradiction because the solution $u(x,t)$ cannot simultaneously take on two different values at the same point. The uniqueness of the solution is violated, and the PDE no longer has a classical (smooth) solution beyond this point.

The intersection of characteristic lines indicates the formation of a discontinuity or shock wave in the solution $u(x,t)$. Physically, this corresponds to the phenomenon where characteristics carrying different values of u converge, resulting in a sudden change (jump) in the value of u.

In fluid dynamics, such discontinuities represent shock waves or breaking waves, where quantities like velocity or pressure change abruptly. The mathematical description of the solution must then be extended to include these discontinuities, leading to the concept of weak solutions and the use of conservation laws.

To properly handle the solution beyond the point of characteristic intersection, we introduce additional conditions, such as the Rankine-Hugoniot condition, which governs the behavior of discontinuities. This condition ensures that the conservation laws are satisfied across the shock.

If the initial condition $\phi(x)$ is such that $\phi(x)$ is a constant function, then all characteristic lines are parallel (having the same slope), and they do not intersect. In this special case, the solution remains smooth for all time, and no discontinuities form.

However, for a general non-constant initial condition $\phi(x)$, characteristic lines will inevitably intersect at some finite time, leading to discontinuities in the solution. To predict when and where these intersections occur, we analyze the equation for the intersection time t derived earlier.

In general, specifying the initial condition $u(x,0) = \phi(x)$ determines the slopes of the characteristic lines emanating from the line $t = 0$. If $\phi(x)$ is not constant, characteristic lines with different slopes will eventually intersect, causing the solution $u(x,t)$ to become discontinuous at the points of intersection. This reflects the physical reality of phenomena like shock waves in fluid dynamics, where abrupt changes occur due to the nonlinear nature of the governing equations.

By recognizing this behavior, we understand the limitations of classical solutions for nonlinear PDEs like $u_t + uu_x = 0$ and the necessity of considering weak solutions and additional conditions to fully describe the system's evolution over time.

16.0.1 DETERMINATION OF CHARACTERISTICS

Characteristic lines are mathematically found by considering hyperbolic partial differential equations (PDEs), specifically first-order PDEs like the advection equation or conservation laws. They provide a path along which the solution to the PDE can be easily determined. For a PDE of the form:

$$u_t + a(u)u_x = 0,$$

where $u(x,t)$ is the unknown function of space x and time t, and $a(u)$ is a speed function that depends on u, the characteristic lines are the curves along which the PDE reduces to an ordinary differential equation (ODE). These curves describe the direction in the x-t plane along which information travels.

To determine the characteristic curves, we need to set up equations that describe the paths in the x-t plane along which the solution remains constant, and solve ODEs. The characteristic lines are found by solving these ODEs.

The method of characteristics transforms the PDE into an ODE along specific paths called characteristic lines. The steps are as follows:

Define Characteristics: Assume that $x(t)$ represents a path in the x-t plane, along which the solution u is constant. If u is constant along $x(t)$, then we have:

$$\frac{d}{dt}u(x(t),t) = u_t + u_x\frac{dx}{dt} = 0.$$

Find Characteristic Equation: Substitute $\frac{dx}{dt}$ by assuming it equals the propagation speed $a(u)$:

$$\frac{dx}{dt} = a(u).$$

This equation represents the characteristic lines along which u remains constant.

Given the initial condition $u(x,0) = \phi(x)$, solve the ODE to find the characteristic lines. Each initial point x_0 defines a unique characteristic curve starting from that point at $t = 0$:

$$x(t) = x_0 + a(\phi(x_0))t.$$

Thus, the characteristic lines are straight lines with slopes determined by the function $a(\phi(x_0))$. Along each characteristic line $x(t) = x_0 + a(u)t$, the value of u remains constant and is equal to the initial condition $\phi(x_0)$.

Using Jacobians in Characteristics (Systems of PDEs)

The concept of Jacobian matrices is used when dealing with systems of PDEs, especially in higher-order systems or coupled equations. For example:

$$\mathbf{U}_t + \mathbf{A}(\mathbf{U})\mathbf{U}_x = 0,$$

where $\mathbf{U}$ is a vector of conserved quantities, and $\mathbf{A}(\mathbf{U})$ is a Jacobian matrix representing the coupling between components.

Using Jacobians for Finding Characteristics:

Jacobian Matrix: In a system of hyperbolic conservation laws, the matrix $\mathbf{A}(\mathbf{U})$ is the Jacobian of the flux function:

$$\mathbf{A}(\mathbf{U}) = \frac{\partial \mathbf{f}}{\partial \mathbf{U}},$$

where $\mathbf{f}(\mathbf{U})$ represents the flux function.

To find the characteristic directions for a system of equations, calculate the eigenvalues and eigenvectors of the matrix $\mathbf{A}(\mathbf{U})$. The eigenvalues of $\mathbf{A}$ represent the characteristic speeds along which information propagates. The eigenvectors determine the characteristic directions in the space of the dependent variables.

Characteristic Equations for Systems: For each eigenvalue λ_i of $\mathbf{A}$, there is an associated characteristic equation:

$$\frac{dx}{dt} = \lambda_i,$$

which describes the characteristic curves for the system.

When the speed of advection is varying the shock or jump will happen. The key point in detecting a shock is that different characteristics need to have different propagation speeds, leading to faster characteristics overtaking slower ones. The convergence and intersection of characteristics indicate a shock.

Figure 16.1 shows the characteristics and shock formation in Burgers' equation. The Initial profile was $\phi(x) = 1.5 - \tanh(5x)$. Shock forms at $t = 0.2$ with speed $s = 1.5$ via Rankine-Hugoniot condition ($s = (u_L + u_R)/2$). Characteristics terminate upon intersecting the shock line, demonstrating nonlinear wave steepening.

Example 16.1

The shallow water equations are given by:

$$\begin{cases} h_t + (hu)_x = 0, \\ (hu)_t + \left(hu^2 + \frac{1}{2}gh^2\right)_x = 0, \end{cases}$$

where $h(x,t)$ is the water height, $u(x,t)$ is the velocity, and g is the

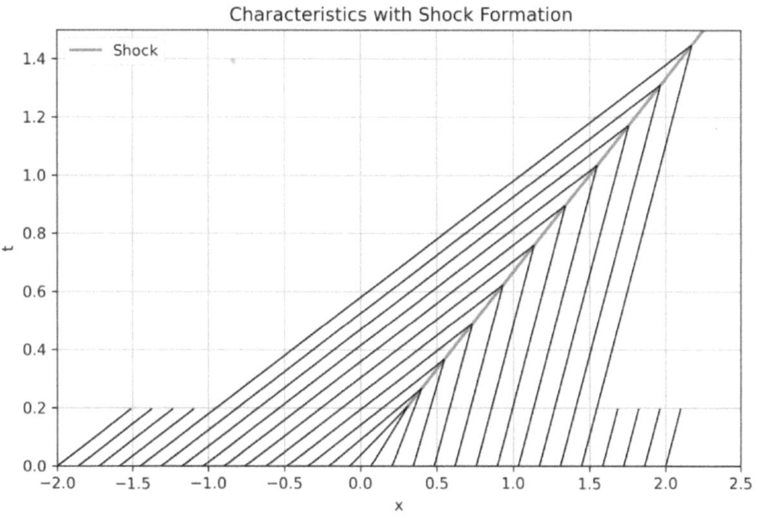

Figure 16.1: **Characteristics and Shock Formation in Burgers' Equation.**

gravitational constant.

i Derive the Jacobian matrix of the flux function and determine its eigenvalues.

ii Show that the characteristic speeds are $u \pm \sqrt{gh}$, and write the equations for the characteristic curves.

iii Briefly explain the physical significance of these characteristic speeds in terms of wave propagation.

Solution

To determine the characteristic speeds of the shallow water system, we linearize the equations and compute the Jacobian matrix **A** of the flux function with respect to the conserved variables. The eigenvalues of this matrix are:

$$\lambda_1 = u - \sqrt{gh}, \quad \lambda_2 = u + \sqrt{gh}$$

which represent the speeds of left- and right-traveling gravity waves. The corresponding characteristic equations are:

$$\frac{dx}{dt} = u \pm \sqrt{gh}$$

These curves in the x-t plane, known as characteristic lines, describe the paths along which information or disturbances propagate.
For a general first-order quasi-linear PDE of the form $u_t + a(u)\, u_x = 0$, characteristics are obtained by solving:

$$\frac{dx}{dt} = a(u)$$

In systems of conservation laws, the Jacobian matrix encapsulates how fluxes change with state variables. Its eigenvalues define the characteristic speeds, while the eigenvectors determine the directions of wave propagation.
Understanding characteristics is essential for analyzing wave motion, shock formation, and the overall structure of solutions to nonlinear hyperbolic PDEs—particularly in fields such as shallow water theory and fluid dynamics.

Example 16.2

Solve the scalar conservation law

$$\frac{\partial u}{\partial t} + \frac{\partial}{\partial x}(uR(u)) = 0, \qquad R(u) = b + \sqrt{u}, \qquad b = \text{const.} \qquad (1)$$

(a) Write the conservation law in quasilinear form $u_t + a(u)u_x = 0$. Determine $a(u)$ and state its physical meaning.

(b) Calculate, in the case of an unsteady shock, the shock speed s as a function of the right and left states u_R and u_L.

(c) Draw an $x-t$ diagram of the solution to the following initial value problem for this equation. Set $b = 1$ and use the initial condition

$$u_0(x) = \begin{cases} u_L, & x < 0, \\ u_R, & x > 0, \end{cases} \qquad (u_L, u_R) = (1, 0).$$

Solution
The given conservation law is

$$\frac{\partial u}{\partial t} + \frac{\partial}{\partial x}(uR(u)) = 0, \quad R(u) = b + \sqrt{u}, \quad b = \text{const.}$$

Since $R(u) = b + \sqrt{u}$, one can write the flux function as $f(u) = uR(u) = bu + u^{3/2}$. This leads to the differential equation $u_t + \partial_x f(u) = 0$, which can be re-expressed in quasilinear form as

$$u_t + f'(u)\, u_x = 0,$$

so identifying

$$a(u) = f'(u) = b + \tfrac{3}{2}\sqrt{u}.$$

Physically, this $a(u)$ is called the characteristic speed because small perturbations or discontinuities in u travel along curves $x = a(u)t + \text{constant}$.
Next, we find the propagation speed of a shock (an unsteady discontinuity) between two states u_L and u_R, one uses the Rankine-Hugoniot condition, which states

$$s = \frac{f(u_R) - f(u_L)}{u_R - u_L}.$$

For the present flux $f(u) = bu + u^{3/2}$, this gives

$$s = b + \frac{u_R^{3/2} - u_L^{3/2}}{u_R - u_L}.$$

That formula describes the shock speed in terms of the left and right states. For the specific initial-value problem with $b = 1$ and the piecewise-constant initial data

$$u_0(x) = \begin{cases} u_L = 1, & x < 0, \\ u_R = 0, & x > 0, \end{cases}$$

the flux evaluated at those states is $f(1) = 1 \cdot (1 + \sqrt{1}) = 2$ and $f(0) = 0$. Hence the Rankine-Hugoniot formula yields

$$s = \frac{2 - 0}{1 - 0} = 2.$$

Meanwhile, the characteristic speed on the left state is $a(u_L) = 1 + \tfrac{3}{2}\sqrt{1} = 2.5$ and on the right state $a(u_R) = 1 + \tfrac{3}{2}\sqrt{0} = 1$. Since the shock speed $s = 2$ lies between 1 and 2.5, the well-known Lax entropy condition (which requires characteristics to converge on the shock from one side and diverge from the other) is satisfied.
Thus the solution that satisfies both the conservation law and the entropy condition is a shock separating the state $u = 1$ on the left from the state $u = 0$ on the right, moving with speed 2. In other words, for all $t > 0$,

$$u(x,t) = \begin{cases} 1, & x < 2t, \\ 0, & x > 2t. \end{cases}$$

Geometrically, in the x-t plane, this corresponds to a single straight shock line starting at $(x,t) = (0,0)$ and moving to the right with slope 2. Characteristics from the left side (speed 2.5) catch up with the shock, while those from the right side (speed 1) trail behind it.

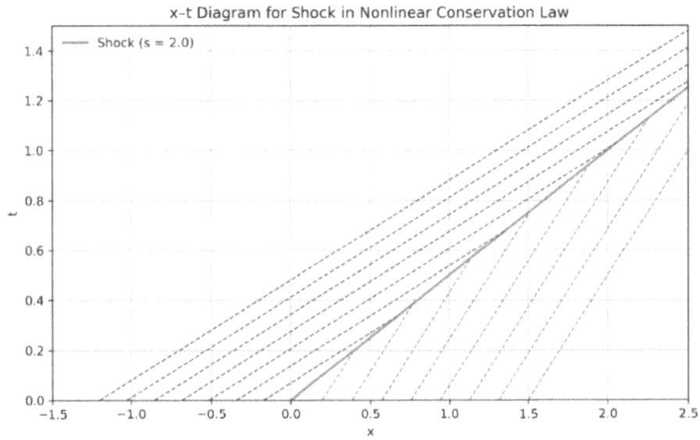

Figure 16.2: $x{-}t$ **diagram.**

i For $x < 0$, the initial condition is $u = 1$. The characteristic speed there is a_L = 2.5, so those characteristics are straight lines in the diagram with slope 1 / a_L. The quantity a_L is the constant characteristic speed of the solution on the left side of the Riemann problem, where $u = 1$.

ii For $x > 0$, the initial condition is $u = 0$. The characteristic speed is a_R = 1, so those lines have slope 1 / a_R.

iii The shock moves at speed $s = 2$, hence the line $x = 2t$. For $t > 0$, the solution is $u = 1$ on the left side of the shock and $u = 0$ on the right side, consistent with the Lax entropy condition.

Example 16.3

Consider the second-order wave equation:

$$u_{tt} - c^2 u_{xx} = 0$$

defined on the domain $-\infty < x < \infty$, with initial conditions: $u(x,0) = f(x), \quad u_t(x,0) = g(x)$.

i Use the change of variables $\xi = x + ct,\ \eta = x - ct$ to transform the wave equation into a simpler form in the new variables.

ii Using the initial conditions, derive the explicit form of the solution

iii Explain the domain of dependence of the solution $u(x_0,t_0)$. What region of the initial data (at $t = 0$) determines the value of the solution at the point (x_0,t_0)? Sketch or describe this domain geometrically.

Solution

Since

$$\xi = x + ct, \quad \eta = x - ct$$

and $u(x,t) = u(\xi,\eta)$, we apply the chain rule to transform derivatives. First, compute the partial derivatives in terms of ξ,η:

$$\frac{\partial u}{\partial t} = \frac{\partial u}{\partial \xi}\cdot\frac{\partial \xi}{\partial t} + \frac{\partial u}{\partial \eta}\cdot\frac{\partial \eta}{\partial t} = cu_\xi - cu_\eta$$

$$\frac{\partial^2 u}{\partial t^2} = c\frac{\partial}{\partial t}(u_\xi) - c\frac{\partial}{\partial t}(u_\eta) = c^2 u_{\xi\xi} - 2c^2 u_{\xi\eta} + c^2 u_{\eta\eta}$$

Similarly:

$$\frac{\partial u}{\partial x} = u_\xi + u_\eta$$

$$\frac{\partial^2 u}{\partial x^2} = u_{\xi\xi} + 2u_{\xi\eta} + u_{\eta\eta}$$

Now substitute into the PDE:

$$u_{tt} - c^2 u_{xx} = \left(c^2 u_{\xi\xi} - 2c^2 u_{\xi\eta} + c^2 u_{\eta\eta}\right) - c^2\left(u_{\xi\xi} + 2u_{\xi\eta} + u_{\eta\eta}\right)$$

Simplifying:

$$u_{tt} - c^2 u_{xx} = -4c^2 u_{\xi\eta}$$

So the transformed PDE becomes:

$$u_{\xi\eta} = 0$$

General solution and application of initial conditions: The equation

$$u_{\xi\eta} = 0$$

has the general solution:

$$u(\xi,\eta) = F_1(\xi) + F_2(\eta)$$

Returning to original variables:

$$u(x,t) = F_1(x+ct) + F_2(x-ct)$$

Use the initial conditions to determine F_1 and F_2:

At $t = 0$:

$$u(x,0) = F_1(x) + F_2(x) = f(x)$$

Also, compute:

$$u_t(x,t) = cF_1'(x+ct) - cF_2'(x-ct) \Rightarrow u_t(x,0) = cF_1'(x) - cF_2'(x) = g(x)$$

So we now solve the system:

$$F_1(x) + F_2(x) = f(x)$$
$$cF_1'(x) - cF_2'(x) = g(x)$$

Differentiate the first equation:

$$F_1'(x) + F_2'(x) = f'(x)$$

Now solve:

$$F_1'(x) + F_2'(x) = f'(x)$$
$$F_1'(x) - F_2'(x) = \frac{g(x)}{c}$$

Add and subtract:

$$2F_1'(x) = f'(x) + \frac{g(x)}{c} \quad \Rightarrow \quad F_1'(x) = \frac{1}{2}f'(x) + \frac{1}{2c}g(x)$$

$$2F_2'(x) = f'(x) - \frac{g(x)}{c} \quad \Rightarrow \quad F_2'(x) = \frac{1}{2}f'(x) - \frac{1}{2c}g(x)$$

Integrate:

$$F_1(x) = \frac{1}{2}f(x) + \frac{1}{2c}\int g(x)\,dx, \quad F_2(x) = \frac{1}{2}f(x) - \frac{1}{2c}\int g(x)\,dx$$

Thus, the complete solution is:

$$u(x,t) = \frac{1}{2}f(x+ct) + \frac{1}{2}f(x-ct) + \frac{1}{2c}\int_{x-ct}^{x+ct} g(s)\,ds$$

This is known as the D'Alembert solution.

Domain of dependence

$$u(x,t) = \frac{1}{2}f(x+ct) + \frac{1}{2}f(x-ct) + \frac{1}{2c}\int_{x-ct}^{x+ct} g(s)\,ds$$

we observe that the value of the solution $u(x,t)$ depends only on the values of f at $x-ct$ and $x+ct$, and - the values of g on the interval $[x-ct, x+ct]$. So the domain of dependence of the point (x_0, t_0) is the spatial interval:

$$[x_0 - ct_0, x_0 + ct_0]$$

This means the value of $u(x_0, t_0)$ depends only on the initial data within that interval.

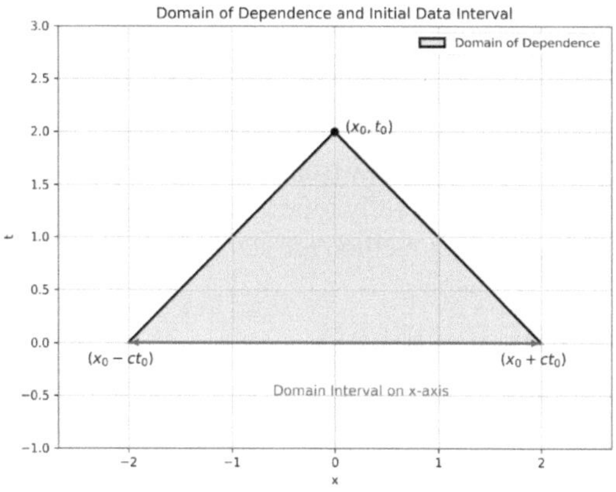

Figure 16.3: **The domain of dependence for location of observation set at $x_o = 0$, observation time as $t_o = 2$, and wave speed $c = 1$.**

Figure 16.3 show the domain of dependence the region in space and time that influences the value of the solution at a particular point. On a space-time diagram with x on the horizontal axis and t on the vertical axis, consider the point (x_o, t_o). If we draw two lines through this point with slopes $\pm c$, representing the characteristic lines of the wave equation, these lines intersect the initial time level $t = o$ at the points $x = x_o - ct_o$ and $x = x_o + ct_o$. These intersection points mark the limits of the interval on the x-axis from which information can propagate to influence the solution at (x_o, t_o). The triangular region enclosed between these two lines and the interval on the x-axis is called the domain of dependence. This region captures the principle of finite propagation speed: the value of $u(x_o, t_o)$ depends only on the initial conditions $f(x)$ and $g(x)$ within the interval $[x_o - ct_o, x_o + ct_o]$, and not on any data outside it.

According to CFL condition (Courant-Friedrichs-Lewy stability criterion)

$$\text{CFL} = c\frac{\Delta t}{\Delta x}$$

the scheme is stable only if $\text{CFL} \leq 1$. This requirement ensures that the numerical domain of dependence encloses the domain of dependence of the governing partial differential equation.

Case $\text{CFL} > 1$: The numerical domain of dependence is narrower than that of the governing PDE, omitting essential portions of the initial data. As a result, the discrete solution fails to converge, the scheme becomes unstable, and Lax's Equivalence Theorem is violated.

Case CFL = 1: The numerical and analytical domains of dependence coincide, so all information required by the PDE is incorporated. Under this condition the scheme can remain stable.

Case CFL < 1: The analytical domain of dependence is completely contained within the wider numerical domain of dependence. Every characteristic arriving at a grid point originates from nodes included in the stencil, ensuring that the numerical solution fully reflects the initial condition. Many schemes, including CTCS, leapfrog method are stable in this regime.

16.1 SHOCK-TUBE PROBLEM

The governing equations for 1D compressible flow are The given system of equations is:

Mass Conservation (Continuity Equation):

$$\frac{\partial \rho}{\partial t}+u\frac{\partial \rho}{\partial x}+\rho\frac{\partial u}{\partial x}=0$$

This represents the conservation of mass for a compressible flow. where, ρ is the density and u is the velocity.

Momentum Conservation:

$$\frac{\partial u}{\partial t}+u\frac{\partial u}{\partial x}+\frac{1}{\rho}\frac{\partial p}{\partial x}=0$$

This equation represents the conservation of momentum in the flow. The pressure term p introduces compressibility effects, which are essential for understanding shock dynamics.

Conservation of Energy :

$$\frac{\partial E}{\partial t}+\frac{\partial((E+p)u)}{\partial x}=0$$

where, ρ is the density, u is the velocity, p is the pressure, and E is the total energy per unit volume

To close the equations, we also need an equation of state for gas:

$$p=k\rho^{\gamma}$$

This is the isentropic equation of state relating pressure, density, and a constant k. The parameter γ is the ratio of specific heats (c_p/c_v). The speed of sound can be estimated as

$$a^2=\frac{\gamma p}{\rho}$$

Shock Tube Application: These equations are indeed related to the study of shock tubes:

A shock tube is an apparatus used to generate and study shock waves in a controlled environment. The governing equations for a shock tube involve the conservation of mass, momentum, and energy for a compressible medium. When a shock is initiated in the tube (by, for example, rupturing a diaphragm separating two sections with different pressures), a shock wave propagates into the low pressure region, and an expansion wave propagates into the high-pressure region. The behavior of these waves can be described using the hyperbolic system of equations for compressible, inviscid flow.

The given system is hyperbolic, meaning it has real eigenvalues and linearly independent eigenvectors, which allows us to determine characteristic lines. The characteristics for this system are given by:

$$\frac{dx}{dt} = u \pm a$$

These characteristic lines indicate the directions along which information propagates in the flow. In the context of a shock tube, these characteristics represent the propagation of shock waves and expansion waves. The fact that the characteristics are $u \pm a$ shows that information propagates at the local flow speed u, modified by the speed of sound a. This behavior is crucial for predicting shock locations, expansion fans, and the interactions between them.

16.2 EXACT RIEMANN SOLVER

The Euler equations for a one-dimensional, unsteady (time-dependent) flow of an inviscid (no viscosity) compressible fluid are given by:

$$\frac{\partial}{\partial t}\begin{bmatrix} \rho \\ \rho u \\ E \end{bmatrix} + \frac{\partial}{\partial x}\begin{bmatrix} \rho u \\ \rho u^2 + p \\ u(E+p) \end{bmatrix} = 0$$

Where:
ρ is the density,
u is the velocity,
p is the pressure,
E is the total energy per unit volume, $E = \rho e + \frac{1}{2}\rho u^2$,
e is the internal energy per unit mass.

These equations represent the conservation of mass, momentum, and energy.

The piecewise constant initial condition, typically involving two states separated by a discontinuity at $x = 0$. This setup is classic in Riemann problems: Left State ($x < 0$ at $t = 0$): Specified by a higher pressure, density, and a velocity.

Right State ($x > 0$ at $t = 0$): Specified by a lower pressure, density, and a velocity.

A one-dimensional tube initially divided into two stagnant fluid regions at different pressures experiences a sudden removal of the separating diaphragm. This instant change triggers an unsteady flow that develops three distinct wave features:

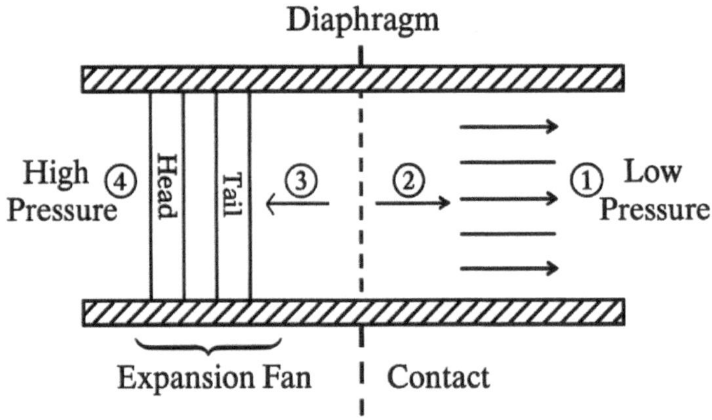

Figure 16.4: **Shock tube arrangement. The shock is between 1 and 2.**

a propagating shock wave, a simple centered expansion fan, and a contact discontinuity moving between them. Each wave front marks a boundary between uniform flow states. This configuration, known as a shock tube, serves as a fundamental example illustrating how pressure differences create and sustain traveling shock and rarefaction structures in a controlled environment.

If:

Region 1 refers to the uniform state on the low-pressure side to the far right, downstream of the shock wave.

Region 2 is the uniform state immediately to the right of the contact discontinuity, but to the left of the shock.

Region 3 is the uniform state immediately to the left of the contact discontinuity, but to the right of the expansion fan.

Region 4 is the uniform, high-pressure state to the far left, upstream of the expansion fan

For $k = c_p/c_v$, we can write:

$$\frac{a_2^2}{a_1^2} = \frac{\frac{p_2}{p_1}\frac{k+1}{k-1} + \frac{p_2}{p_1}}{1 + \frac{k+1}{k-1}\frac{p_2}{p_1}}$$

$$u_2 = u_1 + \frac{a_1}{k}\frac{\dfrac{p_2}{p_1} - 1}{\sqrt{\dfrac{k+1}{2k}\left(\dfrac{p_2}{p_1} - 1\right) + 1}}$$

$$S = u_1 + a_1\sqrt{\frac{k+1}{2k}\left(\frac{p_2}{p_1} - 1\right) + 1}$$

$$u_2 = u_3$$

$$p_2 = p_3$$

$$u(x,t) = \frac{2}{k+1}\left(\frac{x}{t} + \frac{k-1}{2}u_4 + a_4\right)$$

$$a(x,t) = u(x,t) - \frac{x}{t} = \frac{2}{k+1}\left(\frac{x}{t} + \frac{k-1}{2}u_4 + a_4\right) - \frac{x}{t}$$

$$p = p_4\left(\frac{a(x,t)}{a_4}\right)^{\frac{2k}{k-1}}$$

$$u_3 + \frac{2a_3}{k-1} = u_4 + \frac{2a_4}{k-1}$$

$$u_3 = u_4 + \frac{2a_4}{k-1}\left[1 - \left(\frac{p_3}{p_4}\right)^{\frac{k-1}{2k}}\right]$$

$$u_2 = u_4 + \frac{2a_4}{k-1}\left[1 - \left(\frac{p_2}{p_4}\right)^{\frac{k-1}{2k}}\right]$$

$$\frac{p_4}{p_1} = \frac{p_2}{p_1}\left[1 + \frac{k-1}{2a_4}(u_4 - u_2)\right]^{-\frac{2k}{k-1}}$$

$$\frac{p_4}{p_1} = \frac{p_2}{p_1}\left\{1 + \frac{k-1}{2a_4}\left[u_4 - u_1 - \frac{a_1}{k}\frac{\left(\frac{p_2}{p_1} - 1\right)}{\sqrt{\frac{k+1}{2k}\left(\frac{p_2}{p_1} - 1\right) + 1}}\right]\right\}^{-\frac{2k}{k-1}}$$

Toro proposed five Riemann test, which are widely used to benchmark numerical methods for solving hyperbolic conservation laws, especially the Euler equations of gas dynamics (see Table 16.1). Each test case is initialized with a discontinuity at the center of the domain and evolves distinct wave structures depending on the left and right states. Test 1, the Sod problem, is a classical and relatively mild case involving a left rarefaction wave, a contact discontinuity, and a right-moving shock. Test 2, known as the 123 problem, is symmetric and features two strong rarefactions separated by a stationary contact discontinuity; it is numerically challenging due to the near-vacuum pressure at the center. Test 3 represents the left half of the Woodward-Colella blast wave problem. It contains a strong pressure gradient, producing a left rarefaction, contact discontinuity, and right shock—making it a severe test for capturing wave interactions. Test 4 is the right half of the same blast wave problem and features a left shock, contact, and right rarefaction. Finally, Test 5 is a composite case formed from the shocks in Tests 3 and 4. It involves the collision of two strong shocks, leading to a configuration with a slow-moving left shock, a contact discontinuity, and a fast right shock.

Table 16.1
Initial data and physical description for five Riemann problem tests.

Test	ρ_L	u_L	p_L	ρ_R	u_R	p_R	Description
1	1.0	0.0	1.0	0.125	0.0	0.1	Sod problem: left rarefaction, contact discontinuity, right shock
2	1.0	-2.0	0.4	1.0	2.0	0.4	123 problem: two strong rarefactions and a stationary contact
3	1.0	0.0	1000.0	1.0	0.0	0.01	Left half of Woodward-Colella blast wave: left rarefaction, contact, right shock
4	1.0	0.0	0.01	1.0	0.0	100.0	Right half of Woodward-Colella blast wave: left shock, contact, right rarefaction
5	5.99924	19.5975	460.894	5.99242	-6.19633	46.0950	Shock collision: slow left shock, contact discontinuity, right shock

These five Riemann tests are useful for evaluating the accuracy, robustness, and resolution capabilities of numerical schemes used in computational gas dynamics (CFD). Each test imposes distinct challenges that probe different numerical behaviors.

Test 1 (Sod problem) is widely regarded as a baseline verification case, as it includes all three fundamental wave types: rarefaction, contact, and shock. A method that fails to resolve any of these accurately is unsuitable for general use. Test 2 (123 problem) challenges the solver's performance in low-pressure, near-vacuum conditions, which often lead to numerical instabilities or convergence issues. This test is particularly useful for assessing robustness and the behavior of the pressure iteration schemes.

Test 3 and Test 4 represent strong blast wave cases that expose a scheme's ability to handle high-gradient, high-Mach flows without generating spurious oscillations or smearing discontinuities. They are also useful for assessing how numerical diffusion affects the shock capturing ability.

Test 5, involving strong shock collisions, pushes solvers to their limits in terms of shock tracking and preservation of symmetry. Together, these tests help identify numerical diffusion, dispersion, stability, and the treatment of nonlinearity—crucial for validating high-resolution and shock-capturing schemes.

16.3 ATTEMPTS TO CAPTURE DISCONTINUITIES USING FDM

16.3.1 VON NEUMANN AND RICHTMYER SCHEME (VNR)

In 1949, Von Neumann and R. D. Richtmyer have proposed the scheme:

The one-dimensional Von Neumann-Richtmyer scheme employs a staggered predictor-corrector approach with artificial viscosity to model hydrodynamic shocks:

Momentum Equation (Interface-Centered Velocity)

$$\frac{U_i^{n+1/2} - U_i^{n-1/2}}{\Delta t} = -\frac{p_{i+1/2}^n + \mathsf{Q}_{i+1/2}^{n-1/2} - p_{i-1/2}^n - \mathsf{Q}_{i-1/2}^{n-1/2}}{\Delta x}$$

Updates velocity U at interfaces using pressure p (at time n) and artificial viscosity Q (lagged by a half-step, $n-1/2$).

Continuity Equation (Cell-Centered Specific Volume)

$$\frac{\mathsf{V}_i^{n+1/2} - \mathsf{V}_i^n}{\Delta t} = \frac{U_{i+1/2}^{n+1/2} - U_{i-1/2}^{n+1/2}}{\Delta x}$$

Updates specific volume V in cell centers using the new velocities $U^{n+1/2}$.

Artificial Viscosity (Interface)

$$\mathsf{Q}_{i+1/2}^{n+1/2} = \begin{cases} -\dfrac{2(c\Delta x)^2}{\mathsf{V}_{i+1/2}^{n+1/2} + \mathsf{V}_{i+1/2}^n} \cdot \dfrac{\left(U_{i+1}^{n+1/2} - U_i^{n+1/2}\right)^2}{(\Delta x)^2}, & \text{if } U_{i+1}^{n+1/2} - U_i^{n+1/2} < 0 \\ 0, & \text{otherwise} \end{cases}$$

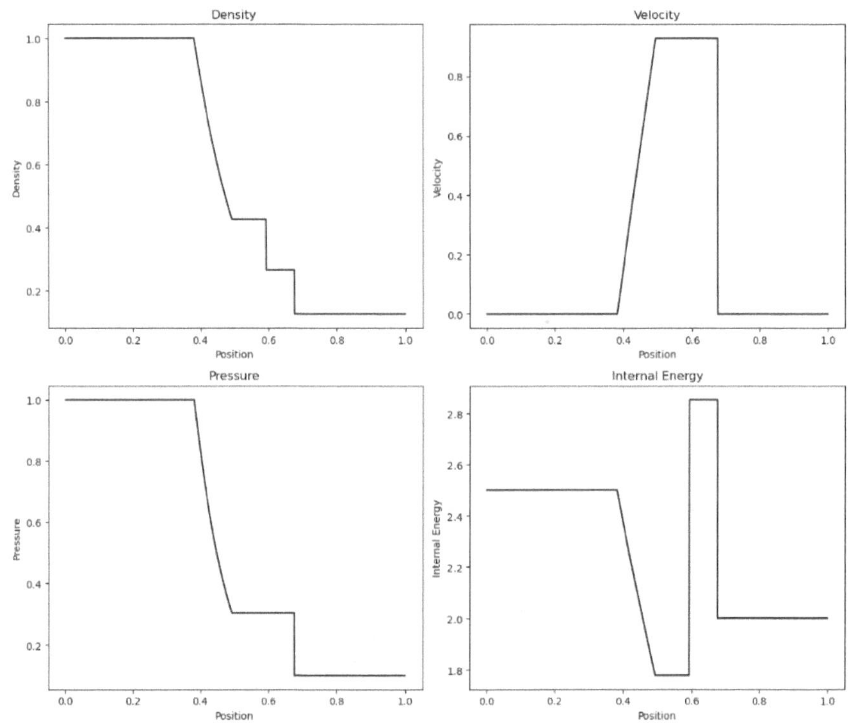

Figure 16.5: **Toro test 1.**

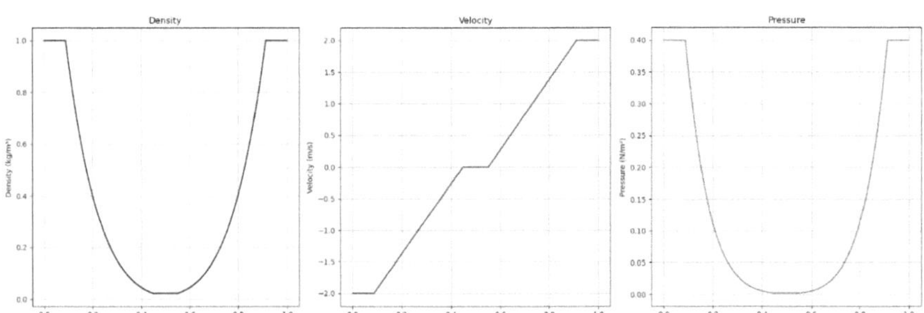

Figure 16.6: **Toro test 2.**

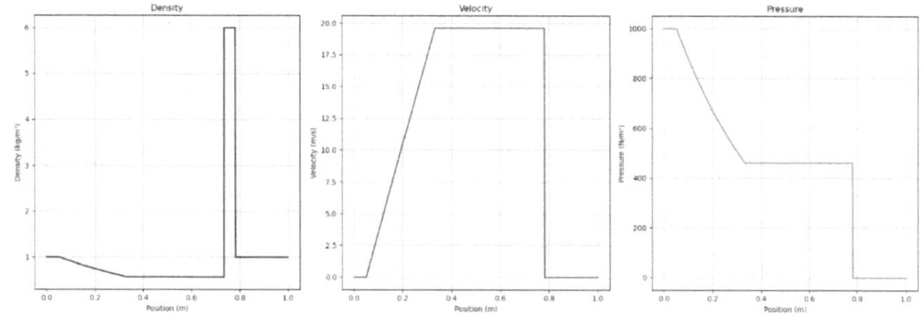

Figure 16.7: Toro test 3.

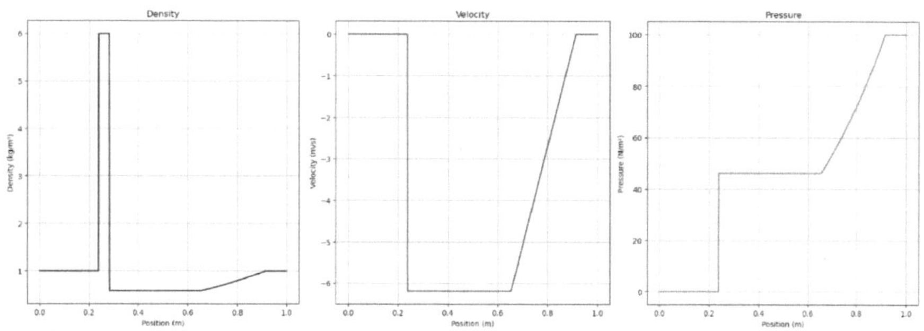

Figure 16.8: Toro test 4.

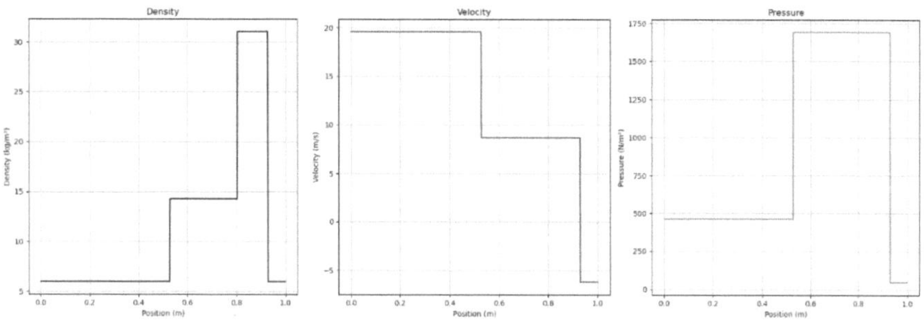

Figure 16.9: Toro test 5.

Activates only during compression (negative velocity gradient). The term $\mathsf{V}_{i+1/2}^{n+1/2} + \mathsf{V}_{i+1/2}^{n}$ averages specific volume at the interface over the current and half-step values. (Assumes corrected velocity time index $n+1/2$.)

Energy/Pressure Update (Cell Center)

$$\left[\frac{p_i^{n+1}+p_i^n}{2}+(\gamma-1)\mathsf{Q}_i^{n+1/2}\right]\frac{\mathsf{V}_i^{n+1/2}-\mathsf{V}_i^n}{\Delta t}+\frac{\mathsf{V}_i^{n+1/2}+\mathsf{V}_i^n}{2}\left(p_i^{n+1}-p_i^n\right)=0$$

Implicitly updates pressure p^{n+1} using the new specific volume $\mathsf{V}^{n+1/2}$ and artificial viscosity $\mathsf{Q}^{n+1/2}$.

Equation of State

$$p_i^n=(\gamma-1)\frac{e_i^n}{\mathsf{V}_i^n}$$

Relates pressure p, internal energy e, and specific volume V for an ideal gas.

$$p_{i+1/2}^n=\frac{1}{2}\left(p_i^n+p_{i+1}^n\right),\quad \mathsf{V}_{i+1/2}^n=\frac{1}{2}\left(\mathsf{V}_i^n+\mathsf{V}_{i+1}^n\right)$$

The scheme is based on first-order discretization of Euler's equation using finite-differences. Von Neumann and Richtmyer is a historical method that works well for relatively simple problems, but it is not as effective for shock-capturing or highly accurate simulations of compressible flows. The Von Neumann and Richtmyer (VNR) method exhibits limitations in shock capturing due to inherent numerical diffusion. It introduces artificial viscosity to stabilize shock calculations. This term, often quadratic in velocity gradients (nonlinear), is designed to mimic physical viscosity. However, its implementation may lack precise localization, applying diffusion broadly rather than exclusively at shock fronts. This results in a smoothed transition across the shock, diluting the sharp discontinuity.

As an early method, VNR likely employs first-order discretization in space and time, inherently diffusive and prone to smearing sharp gradients. The artificial viscosity coefficient may not dynamically adjust to local flow conditions, leading to excessive diffusion in smooth regions and insufficient sharpening at shocks. While VNR ensures monotonicity (no spurious oscillations), it sacrifices resolution for stability. Modern shock-capturing methods balance both by combining high-order accuracy with nonlinear limiters, achieving sharper shocks without oscillations.

16.3.2 UPWIND OR DONOR-CELL SCHEME

To discuss the gas dynamics scheme, it is helpful to first learn the upwind schemes that were developed to solve the 1D hyperbolic equation. Consider the 1D linear advection (or transport) equation:

$$\frac{\partial u}{\partial t}+a\frac{\partial u}{\partial x}=0,$$

where $u(x,t)$ is the unknown function, and a is a constant wave speed.

A first-order upwind method (FOU) approximates the spatial derivative using a backward difference:

$$\frac{\partial u}{\partial x} \approx \frac{u_i^n - u_{i-1}^n}{\Delta x}.$$

This method uses information from the upstream (previous) point, which is appropriate when $a > 0$, since the information travels in the positive x-direction. The upwind method can be written as:

$$u_i^{n+1} = u_i^n - \text{CFL} \cdot (u_i^n - u_{i-1}^n),$$

where $\text{CFL} = \frac{a\Delta t}{\Delta x}$ is the Courant number. The upwind method is first-order accurate in both space and time. For $a > 0$, the method is called a donor-cell scheme because it uses the value of u_{i-1} (the "donor" point) to compute the value of u_i at the next time step. When applying this upwind scheme to a general hyperbolic partial differential equation of the form:

$$\frac{\partial u}{\partial t} + a\frac{\partial u}{\partial x} = 0,$$

where $u = u(x,t)$ represents a quantity like density or velocity, the upwind method approximates the flux $f(u) = au$ in a simple manner using backward differences. The upwind scheme is stable under the condition that the CFL condition is satisfied:

$$|\text{CFL}| = \left|\frac{a\Delta t}{\Delta x}\right| \leq 1.$$

This CFL condition ensures stability for the numerical solution of the hyperbolic equation. When the CFL condition is violated (i.e., if CFL > 1), the scheme may become unstable, leading to growing oscillations and inaccurate results.

16.3.3 COURANT-ISAACSON-REES CIR SCHEME

The Courant-Isaacson-Rees (CIR) scheme is a foundational method for solving hyperbolic partial differential equations, particularly conservation laws. It is a first-order accurate finite difference scheme that employs the upwind principle, where numerical derivatives are aligned with the direction of wave propagation. This ensures both numerical stability and consistency with the physical characteristics of the problem. The CIR scheme highlights the importance of directional bias in spatial discretization for maintaining the correct domain of dependence. Consider the linear advection equation:

$$\frac{\partial u}{\partial t} + a\frac{\partial u}{\partial x} = 0,$$

where a is a constant advection speed.

For $a > 0$: The CIR scheme uses a backward difference for the spatial derivative:

$$\frac{u_i^{n+1} - u_i^n}{\Delta t} + a\frac{u_i^n - u_{i-1}^n}{\Delta x} = 0.$$

Rearranging yields the update formula:

$$u_i^{n+1} = u_i^n - \mathsf{CFL} \cdot (u_i^n - u_{i-1}^n),$$

where the Courant number is defined as:

$$\mathsf{CFL} = a\frac{\Delta t}{\Delta x}.$$

For $a < 0$: The scheme switches to a forward difference:

$$\frac{u_i^{n+1} - u_i^n}{\Delta t} + a\frac{u_{i+1}^n - u_i^n}{\Delta x} = 0,$$

leading to:

$$u_i^{n+1} = u_i^n - \mathsf{CFL} \cdot (u_{i+1}^n - u_i^n),$$

where $\mathsf{CFL} = a\frac{\Delta t}{\Delta x} < 0$.

Stability Analysis (Von Neumann Method) : Substitute a Fourier mode $u_i^n = \xi^n e^{jki\Delta x}$ into the difference equation, where ξ is the amplification factor and k is the wavenumber.

Case $a > 0$: The amplification factor satisfies:

$$\xi = 1 - \mathsf{CFL}\,(1 - \cos(k\Delta x) + j\sin(k\Delta x))\,.$$

Its magnitude is:

$$|\xi|^2 = 1 - 2\mathsf{CFL}(1 - \mathsf{CFL})(1 - \cos(k\Delta x)).$$

Stability requires:

$$0 \leq \mathsf{CFL} \leq 1.$$

Case $a < 0$: A similar analysis gives:

$$-1 \leq \mathsf{CFL} \leq 0.$$

Thus, the unified Courant-Friedrichs-Lewy (CFL) condition for stability is:

$$|\mathsf{CFL}| = |a|\frac{\Delta t}{\Delta x} \leq 1.$$

The CIR scheme, while only first-order accurate in both space and time, has several important numerical properties. Its low-order accuracy is offset by its robustness and simplicity. The scheme inherently introduces numerical diffusion, which suppresses spurious oscillations and ensures monotonicity, particularly useful in problems with discontinuities such as shocks. This makes the CIR method a reliable baseline scheme in the study of hyperbolic problems, and its influence persists in the development of more advanced upwind and flux-limiting methods.

16.3.4 LAX FRIEDRICHS SCHEME

The Lax-Friedrichs method is a simple finite difference method where the spatial derivatives are computed using central differences, and the time derivative is averaged using the average of the neighboring points. It is a more basic central-difference approach compared to the CIR scheme.

The Lax Friedrichs scheme for the convection equation is given as:

$$u_i^{n+1} = \frac{1}{2}(u_{i+1}^n + u_{i-1}^n) - \frac{CFL}{2}(u_{i+1}^n - u_{i-1}^n)$$

where $CFL = \frac{a\Delta t}{\Delta x}$ is the Courant number, and the convection equation we are solving is:

$$u_t + au_x = 0$$

This scheme consists of taking the average value of u_{i+1}^n and u_{i-1}^n to stabilize the scheme and adding a flux term that involves CFL.

We assume the solution u_i^n can be expressed as a Fourier mode (plane wave):

$$u_i^n = \xi^n e^{iki\Delta x}$$

where ξ^n is the amplification factor we seek to find, k is the wave number, and $i\Delta x$ represents the spatial location. Substituting this into the Lax Friedrichs scheme:

$$\xi^{n+1} e^{iki\Delta x} = \frac{1}{2}(\xi^n e^{ik(i+1)\Delta x} + \xi^n e^{ik(i-1)\Delta x}) - \frac{CFL}{2}(\xi^n e^{ik(i+1)\Delta x} - \xi^n e^{ik(i-1)\Delta x})$$

Using the exponential properties, we simplify the terms $e^{ik(i+1)\Delta x}$ and $e^{ik(i-1)\Delta x}$:

$$e^{ik(i+1)\Delta x} = e^{iki\Delta x} e^{ik\Delta x}$$

$$e^{ik(i-1)\Delta x} = e^{iki\Delta x} e^{-ik\Delta x}$$

Substituting these into the Lax Friedrichs scheme:

$$\xi^{n+1} e^{iki\Delta x} = \frac{1}{2}\left(\xi^n e^{iki\Delta x}(e^{ik\Delta x} + e^{-ik\Delta x})\right) - \frac{CFL}{2}\left(\xi^n e^{iki\Delta x}(e^{ik\Delta x} - e^{-ik\Delta x})\right)$$

Factor out $\xi^n e^{iki\Delta x}$:

$$\xi^{n+1} = \xi^n \left[\frac{1}{2}\left(e^{ik\Delta x} + e^{-ik\Delta x}\right) - \frac{CFL}{2}\left(e^{ik\Delta x} - e^{-ik\Delta x}\right)\right]$$

Recall the trigonometric identities:

$$e^{ik\Delta x} + e^{-ik\Delta x} = 2\cos(k\Delta x)$$

$$e^{ik\Delta x} - e^{-ik\Delta x} = 2i\sin(k\Delta x)$$

Substitute these into the equation:

$$\xi^{n+1} = \xi^n [\cos(k\Delta x) - iCFL\sin(k\Delta x)]$$

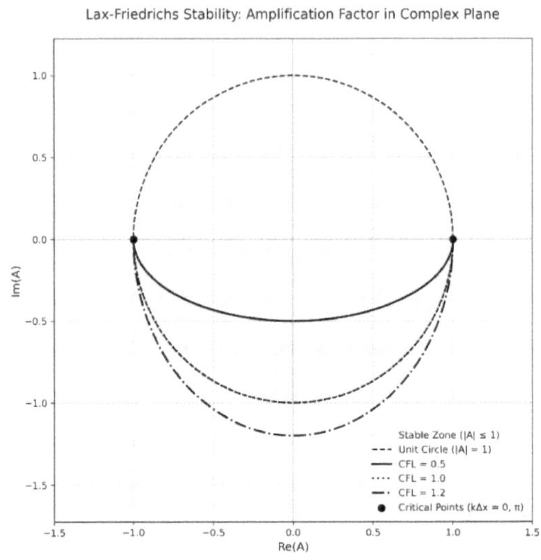

Figure 16.10: **Lax Friedrich Stability Zone.**

Thus, the amplification factor A is:

$$A = \cos(k\Delta x) - i, \mathrm{CFL}, \sin(k\Delta x)$$

Its modulus satisfies

$$|A|^2 = \cos^2(k\Delta x) + \mathrm{CFL}^2 \sin^2(k\Delta x)$$

$$|A|^2 = 1 - \left(1 - \mathrm{CFL}^2\right) \sin^2(k\Delta x)$$

Because $0 \leq \sin^2(k\Delta x) \leq 1$, hence

$$|A| \leq 1 \Longleftrightarrow |\mathrm{CFL}| \leq 1$$

The Lax Friedrichs scheme is conditionally stable if the Courant number CFL satisfies the stability condition derived from the magnitude of A. The scheme introduces dissipation, which stabilizes the solution but reduces the accuracy to first-order in space.

This scheme is only first-order accurate in space. While the LF scheme stabilizes the forward-in-time central scheme, it introduces numerical diffusion by replacing the value of the function u_i^n by the average of its neighbors. This introduces an additional error term proportional to Δx, leading to a loss in accuracy for smooth solutions.

Stability: The LF scheme is conditionally stable under the Courant-Friedrichs-Lewy (CFL) condition:

$$CFL = \frac{a\Delta t}{\Delta x} \leq 1$$

where a is the wave speed. Stability is ensured only when this condition is satisfied, making the LF scheme useful for a specific range of CFL. However, the artificial diffusion it introduces can make it overly dissipative, which limits its accuracy for finer simulations.

The Lax-Friedrichs scheme is known for being dissipative, and it is more stable than some other methods, particularly for problems involving shocks or discontinuities. However, this comes at the cost of reduced resolution in smooth regions. The Lax-Friedrichs scheme has inherent numerical viscosity (dissipation) that smooths oscillations around shocks. This dissipation makes it stable but also reduces the sharpness of the shock and smooth regions of the solution.

16.3.5 LAX WENDROFF SCHEME

We solve the linear advection equation:

$$\frac{\partial u}{\partial t} + a\frac{\partial u}{\partial x} = 0, \quad a = \text{constant}.$$

Express $u(x, t + \Delta t)$ as a Taylor series:

$$u_i^{n+1} = u_i^n + \Delta t \frac{\partial u}{\partial t}\bigg|_i^n + \frac{\Delta t^2}{2}\frac{\partial^2 u}{\partial t^2}\bigg|_i^n + \mathcal{O}(\Delta t^3).$$

From the PDE:

$$\frac{\partial u}{\partial t} = -a\frac{\partial u}{\partial x}, \quad \frac{\partial^2 u}{\partial t^2} = a^2\frac{\partial^2 u}{\partial x^2}.$$

Substituting these:

$$u_i^{n+1} = u_i^n - a\Delta t \frac{\partial u}{\partial x}\bigg|_i^n + \frac{a^2\Delta t^2}{2}\frac{\partial^2 u}{\partial x^2}\bigg|_i^n + \mathcal{O}(\Delta t^3).$$

Approximate $\frac{\partial u}{\partial x}$ and $\frac{\partial^2 u}{\partial x^2}$ using central differences:

$$\frac{\partial u}{\partial x}\bigg|_i^n \approx \frac{u_{i+1}^n - u_{i-1}^n}{2\Delta x}, \quad \frac{\partial^2 u}{\partial x^2}\bigg|_i^n \approx \frac{u_{i+1}^n - 2u_i^n + u_{i-1}^n}{\Delta x^2}.$$

Substituting into the Taylor expansion:

$$u_i^{n+1} = u_i^n - \frac{a\Delta t}{2\Delta x}(u_{i+1}^n - u_{i-1}^n) + \frac{a^2\Delta t^2}{2\Delta x^2}(u_{i+1}^n - 2u_i^n + u_{i-1}^n).$$

Let $\text{CFL} = \frac{a\Delta t}{\Delta x}$:

$$u_i^{n+1} = u_i^n - \frac{\text{CFL}}{2}(u_{i+1}^n - u_{i-1}^n) + \frac{\text{CFL}^2}{2}(u_{i+1}^n - 2u_i^n + u_{i-1}^n).$$

Assume a Fourier mode $u_i^n = A^n e^{iki\Delta x}$. Substitute into the scheme:

$$A^{n+1}e^{iki\Delta x} = A^n e^{iki\Delta x}\left[1 - \frac{\text{CFL}}{2}(e^{ik\Delta x} - e^{-ik\Delta x}) + \frac{\text{CFL}^2}{2}(e^{ik\Delta x} - 2 + e^{-ik\Delta x})\right].$$

Cancel $A^n e^{iki\Delta x}$:

$$A = 1 - \frac{\text{CFL}}{2}(2i\sin(k\Delta x)) + \frac{\text{CFL}^2}{2}(2\cos(k\Delta x) - 2).$$

Simplify using $\sin\theta = \frac{e^{i\theta} - e^{-i\theta}}{2i}$ and $\cos\theta = \frac{e^{i\theta} + e^{-i\theta}}{2}$:

$$A = 1 - i\cdot\text{CFL}\sin(k\Delta x) + \text{CFL}^2(\cos(k\Delta x) - 1).$$

Compute $|A|^2 = \text{Re}(A)^2 + \text{Im}(A)^2$:

$$|A|^2 = \left[1 + \text{CFL}^2(\cos\theta - 1)\right]^2 + [\text{CFL}\sin\theta]^2.$$

Expand and simplify using $\cos^2\theta + \sin^2\theta = 1$:

$$|A|^2 = 1 - \text{CFL}^2(1 - \text{CFL}^2)(1 - \cos\theta)^2.$$

For stability ($|A| \leq 1$):

$$\text{CFL}^2(1 - \text{CFL}^2) \geq 0 \implies \text{CFL}^2 \leq 1.$$

Thus, the CFL condition is:

$$|a|\frac{\Delta t}{\Delta x} \leq 1.$$

The Lax-Wendroff scheme is second-order accurate and stable if $|\text{CFL}| \leq 1$. Its amplification factor A includes a dispersion term ($-i\cdot\text{CFL}\sin(k\Delta x)$), causing oscillations, and a diffusion term ($\text{CFL}^2(\cos(k\Delta x) - 1)$), which stabilizes the scheme.

It is often favored for smooth solutions, but near sharp gradients, oscillations can appear due to numerical dispersion. The Lax-Wendroff method, introduced in 1960, was the pioneering scheme to attain second-order accuracy in both space and time, marking a significant milestone in the evolution of computational fluid dynamics.

Example 16.4

Consider the one-dimensional linear convection equation $u_t + au_x = 0$, where $u(x,t)$ represents a transported quantity and $a > 0$ is a constant wave speed. Numerically solve this equation using four explicit finite difference schemes: First-Order Upwind (FOU), Lax-Friedrichs, Leapfrog and Lax-Wendroff schemes. The initial condition is a triangular pulse centered near the middle of the domain:

$$u(x,0) = \begin{cases} 0 & \text{for } x \leq 0.9, \\ 10(x - 0.9) & \text{for } 0.9 < x \leq 1.0, \\ 10(1.1 - x) & \text{for } 1.0 < x \leq 1.1, \\ 0 & \text{for } x > 1.1. \end{cases}$$

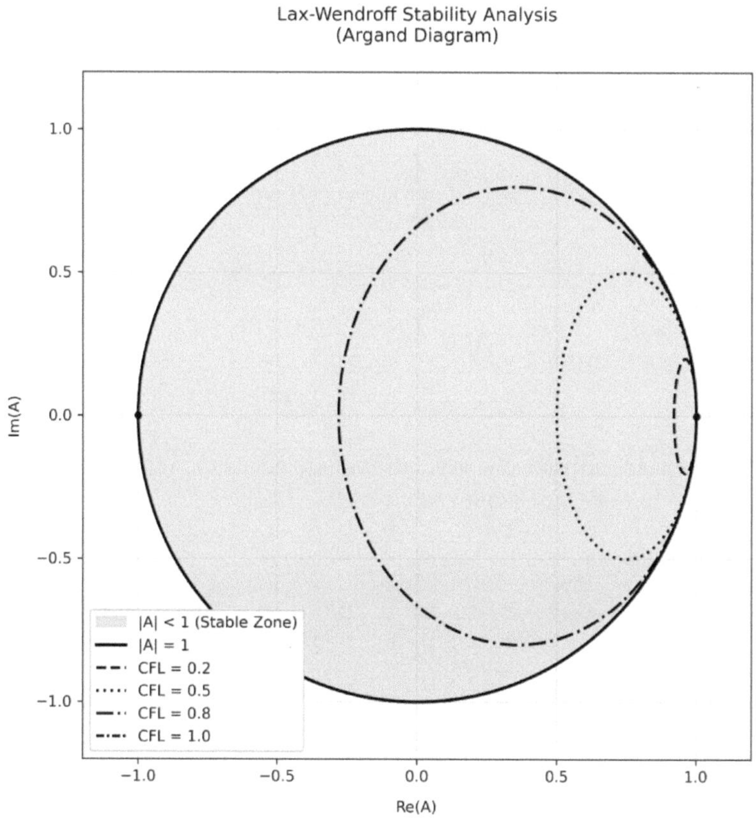

Figure 16.11: **Stability zone for Lax Wendroff scheme.**

The computational domain is $x \in [0, 2]$ meters, discretized into $nx = 100$ equally spaced grid points. Use a Courant number $CFL = 0.25$ to compute the time step Δt, and advance the solution for 20 time steps.

(a) Implement all four numerical schemes independently.

(b) Plot the numerical solution after 20 time steps and compare with the initial condition.

(c) Discuss the performance of each scheme in terms of Dissipative and dispersive behavior, Stability and accuracy, Phase and amplitude errors.

(d) Conclude which scheme offers the best balance between accuracy and stability for this problem.

Solution

See Python code `Ch16-Ex164` for solution of this problem.

The exact solution after twenty steps is the initial triangular profile convected a distance at with periodic wrapping. Using this reference, three discrete error norms are evaluated

$$L_1 = \frac{1}{N}\sum_{j=1}^{N}\left|u_j^{\text{num}} - u_j^{\text{exact}}\right|$$

$$L_2 = \sqrt{\frac{1}{N}\sum_{j=1}^{N}\left(u_j^{\text{num}} - u_j^{\text{exact}}\right)^2}$$

$$L_\infty = \max_{1\le j\le N}\left|u_j^{\text{num}} - u_j^{\text{exact}}\right|$$

which measure, respectively, average absolute deviation, root-mean-square deviation and worst-case pointwise error.

Scheme	L_1	L_2	L_∞
Lax-Wendroff	5.97×10^{-3}	1.76×10^{-2}	9.01×10^{-2}
FOU	1.35×10^{-2}	4.22×10^{-2}	2.28×10^{-1}
Lax-Friedrichs	3.75×10^{-2}	9.91×10^{-2}	4.95×10^{-1}
Leapfrog	4.33×10^{-2}	1.34×10^{-1}	5.49×10^{-1}

Lax-Wendroff emerges as the most accurate scheme across all norms, benefiting from its second-order accuracy and minimal dispersion. First-Order Upwind (FOU) follows with roughly two- to threefold larger errors than Lax-Wendroff, as its inherent numerical diffusion suppresses oscillations at the expense of peak attenuation.

The more dissipative Lax-Friedrichs scheme exhibits substantially larger errors, especially in the L_∞ norm, where it reaches 49.5% of the original peak magnitude. Surprisingly, Leapfrog ranks worst in all integral measures despite its apparent advantage in peak preservation (Figure 16.12). This contradiction arises because Leapfrog's dispersive nature, while maintaining excellent amplitude near the central peak, also generates high-frequency oscillations that inflate the global error metrics.

Figure 16.12 illustrates this core trade-off: Leapfrog's strong peak retention comes at the cost of oscillatory tails. For applications emphasizing local amplitude accuracy (such as shock detection or peak-sensitive measurements), Leapfrog's peak fidelity may be preferable despite its poor overall error statistics. Conversely, for solutions requiring global smoothness, Lax-Wendroff provides a better balance of accuracy with only moderate peak attenuation.

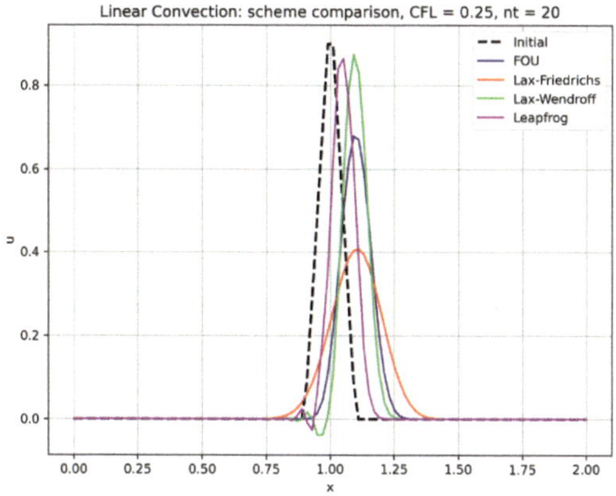

Figure 16.12: **Comparison of schemes with 100 grid points and CFL =0.25 and 20 time steps.**

Practical Implications: The choice of scheme depends critically on the application's priorities. For simulations requiring global smoothness, the Lax-Wendroff method is optimal, as it balances second-order accuracy with controlled peak attenuation, minimizing dispersive artifacts. In contrast, Leapfrog's superior peak preservation—despite its high-frequency oscillations—makes it suitable for peak-sensitive tasks like shock detection or waveform analysis, where local amplitude fidelity outweighs global error metrics. Meanwhile, the more dissipative First-Order Upwind (FOU) and Lax-Friedrichs schemes are advantageous when robustness to oscillations is paramount, such as in low-resolution simulations or cases where numerical stability trumps precision. Each method thus embodies a trade-off: Lax-Wendroff for accuracy, Leapfrog for peak resolution, and FOU/Lax-Friedrichs for stability at the cost of diffusion.

First-Order Schemes and Dissipation: First-order schemes like FOU (First Order Upwind) and Lax-Friedrichs introduce significant numerical dissipation, which smooths out the solution over time. This makes them unsuitable for problems where maintaining the sharpness of the solution is critical, such as shock waves or wave propagation. Dissipation dampens out high-frequency components, leading to a loss of sharp gradients and important features. In these contexts, higher-order schemes (like Lax-Wendroff) perform better because they introduce less dissipation. However, we still need to understand that dispersion error is not totally eliminated.

16.3.6 MACCORMACK SCHEME

The MacCormack scheme is applied to a linear advection equation:

$$\frac{\partial \rho}{\partial t} + c\frac{\partial \rho}{\partial x} = 0$$

where c is the wave speed. The scheme can be written as:

The MacCormack scheme involves two steps:

1. Predictor step:

$$\rho_i^* = \rho_i^n - CFL(\rho_{i+1}^n - \rho_i^n)$$

2. Corrector step:

$$\rho_i^{n+1} = \frac{1}{2}\left[\rho_i^n + \rho_i^* - CFL(\rho_i^* - \rho_{i-1}^*)\right]$$

where: $CFL = \frac{c\Delta t}{\Delta x}$ is the Courant number. $F_i = c\rho_i$.

We analyze stability by substituting a Fourier mode:

$$\rho_i^n = \hat{\rho}^n e^{ikx_i}, \quad \text{where } x_i = i\Delta x$$

Step 1: Predictor Step: Substitute $\rho_i^n = \hat{\rho}^n e^{ikx_i}$ into the predictor equation:

$$\rho_i^* = \rho_i^n - CFL(\rho_{i+1}^n - \rho_i^n)$$

For ρ_{i+1}^n, substitute $e^{ikx_{i+1}} = e^{ikx_i}e^{ik\Delta x}$, i.e., $z = e^{ik\Delta x}$:

$$\rho_i^* = \hat{\rho}^n e^{ikx_i}(1 - CFL(z-1))$$

The amplification for predictor level is

$$A_{\text{predictor}} = 1 - CFL(z-1)$$

Step 2: Corrector Step: Substitute ρ_i^* into the corrector equation:

$$\rho_i^{n+1} = \frac{1}{2}\left[\rho_i^n + \rho_i^* - CFL(\rho_i^* - \rho_{i-1}^*)\right]$$

For ρ_{i-1}^*, substitute $e^{ikx_{i-1}} = e^{ikx_i}z^{-1}$:

$$\rho_{i-1}^* = \hat{\rho}^n e^{ikx_i}z^{-1}(1 - CFL(z-1))$$

Substitute ρ_i^* and ρ_{i-1}^*:

$$\rho_i^{n+1} = \frac{1}{2}\left[\rho_i^n + \rho_i^* - CFL\left(\rho_i^* - \rho_{i-1}^*\right)\right]$$

Expanding:

$$\rho_i^{n+1} = \frac{1}{2}\Bigg[\rho_i^n + \hat{\rho}^n e^{ikx_i}(1 - CFL(z-1)) - CFL\Big(\hat{\rho}^n e^{ikx_i}(1 - CFL(z-1))$$

$$-\hat{\rho}^n e^{ikx_i} z^{-1} (1 - CFL(z-1)) \Big) \Big]$$

After simplifying:

$$\rho_i^{n+1} = \hat{\rho}^n e^{ikx_i} \left[1 - CFL(z-1) - \frac{CFL}{2} \Big((z-1) - z^{-1}(z-1) \Big) \right]$$

Simplify further using $z^{-1} = e^{-ik\Delta x}$:

$$A_{\text{corrector}} = \frac{1}{2} \left[2 - CFL(z-1) - CFL(1 - z^{-1})(z-1) \right]$$

Combine predictor and corrector:

$$A = \tfrac{1}{2} \Big[1 + \big(1 - \text{CFL}\,(z-1)\big)\big(1 - \text{CFL}\,(1 - z^{-1})\big) \Big], \qquad z = e^{ik\Delta x}$$

After algebra,

$$A = 1 - \, i\,\text{CFL}\,\sin(k\Delta x) \, - \, \text{CFL}^2 \left[1 - \cos(k\Delta x) \right].$$

Magnitude criterion for stability:

$$|A|^2 = \left[1 - \text{CFL}^2 (1 - \cos k\Delta x) \right]^2 + \text{CFL}^2 \sin^2(k\Delta x) = 1 - \left(1 - \text{CFL}^2 \right) \sin^2(k\Delta x) \le 1,$$

so the MacCormack scheme is stable whenever $|\text{CFL}| \le 1$.

16.3.7 BEAM-WARMING SCHEME

The Beam-Warming scheme, developed by R. M. Beam and R. F. Warming in the 1970s, is a second-order accurate implicit finite difference method for solving hyperbolic conservation laws. Introduced in their 1976 paper, the scheme employs a non-iterative, spatially-factored approach (similar to ADI methods) with a block tridiagonal solver for efficient time-stepping. It supports both central and upwind-biased differencing in a conservative framework. Originally tested on scalar models and the 2D Euler equations, the method was later extended to the Navier-Stokes equations. The scheme uses a delta-form update for linearization, ensuring unconditional stability in linear analyses, though artificial dissipation is needed for nonlinear stability with centered differences. Beam and Warming also incorporated hybrid differencing—mixing centered and upwind methods—to better capture shocks and discontinuities. This approach became influential in early computational fluid dynamics (CFD) for its balance of accuracy and efficiency.

Consider the **1D linear advection equation** $u_t + au_x = 0$, where a is a constant velocity. The Beam–Warming scheme is derived using a second-order Taylor expansion in time (similar to the Lax–Wendroff method) but with *upwind-biased finite*

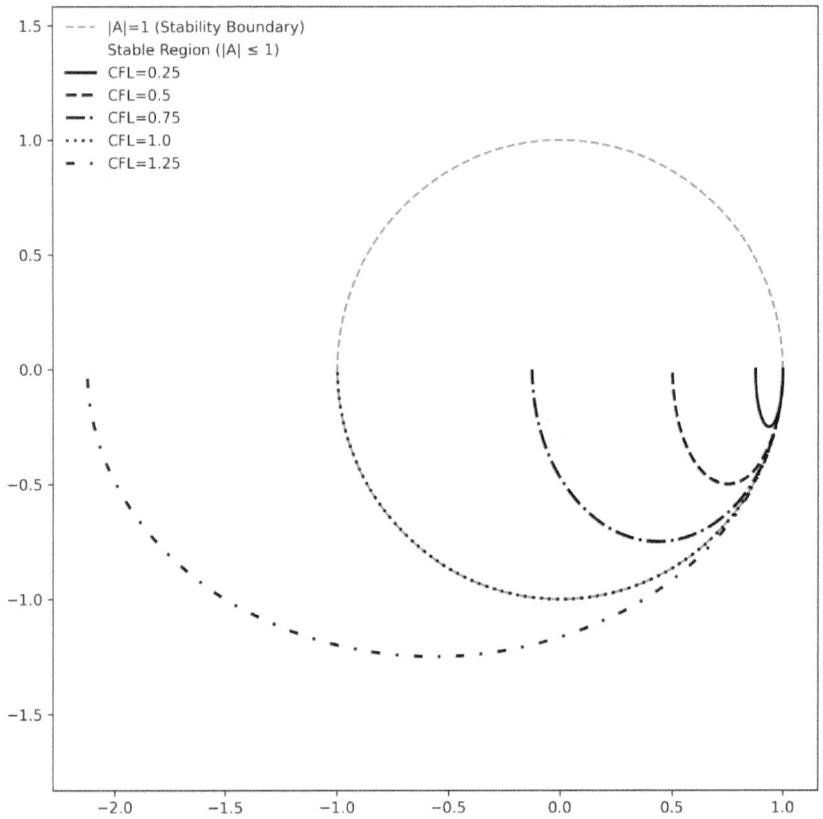

Figure 16.13: **Stability zone for MacCormack scheme.**

differences for spatial derivatives. For $a > 0$ (right-moving flow), the discretization uses a **three-point upwind stencil**:

$$u_j^{n+1} = u_j^n - \frac{a\Delta t}{2\Delta x}\left(3u_j^n - 4u_{j-1}^n + u_{j-2}^n\right) + \frac{(a\Delta t)^2}{2\Delta x^2}\left(u_j^n - 2u_{j-1}^n + u_{j-2}^n\right),$$

which is **second-order accurate** in both time and space. Here, Δx is the spatial step, Δt is the time step, and the superscripts n and $n+1$ denote the current and next time levels, respectively. If $a < 0$ (left-moving flow), the stencil shifts to the downwind side (using points $j, j+1, j+2$). Although derived from a three-level Taylor expansion, the Beam–Warming scheme is implemented as a **two-level method** through a factored delta-form approach, avoiding the need to store multiple time levels.

Von Neumann Stability Analysis of the Beam-Warming Scheme: The stability of the Beam-Warming scheme for the 1D linear advection equation $u_t + au_x = 0$ (where

$a > 0$) can be analyzed using the Von Neumann method. We begin by substituting a Fourier mode solution of the form $u_j^n = A^n e^{ikj\Delta x}$, where A is the amplification factor and k is the wavenumber. The discretized scheme, expressed in terms of the Courant number $C = \frac{a\Delta t}{\Delta x}$, becomes:

$$A = 1 - \frac{C}{2}\left(3 - 4e^{-i\xi} + e^{-i2\xi}\right) + \frac{C^2}{2}\left(1 - 2e^{-i\xi} + e^{-i2\xi}\right),$$

where $\xi = k\Delta x$. Simplifying using $z = e^{-i\xi}$, we obtain:

$$A = 1 - \frac{3C}{2} + \frac{C^2}{2} + (2C - C^2)\,z + \left(\frac{C^2}{2} - \frac{C}{2}\right)z^2.$$

For stability, we require $|A(\xi)| \leq 1$ for all ξ. Evaluating at the Nyquist frequency ($\xi = \pi$) gives the most restrictive condition:

$$A(\pi) = 1 - 4C + 2C^2.$$

Enforcing $|1 - 4C + 2C^2| \leq 1$ leads to the stability criterion:

$$0 \leq C \leq 2.$$

Thus, the Beam-Warming scheme is stable for Courant numbers C up to 2, a less restrictive condition than many explicit methods (e.g., Lax-Wendroff, which requires $C \leq 1$).

Figure 16.14 shows the stability region of the Beam-Warming scheme in the complex plane. For CFL ≤ 2.0, the amplification factor remains within the unit circle, indicating stability. At CFL = 2.1, the curve exits the stability zone, confirming the onset of instability.

Table 16.2 lists some useful Numerical Schemes made to address the hyperbolic character of advection equation $u_t + cu_x = 0$.

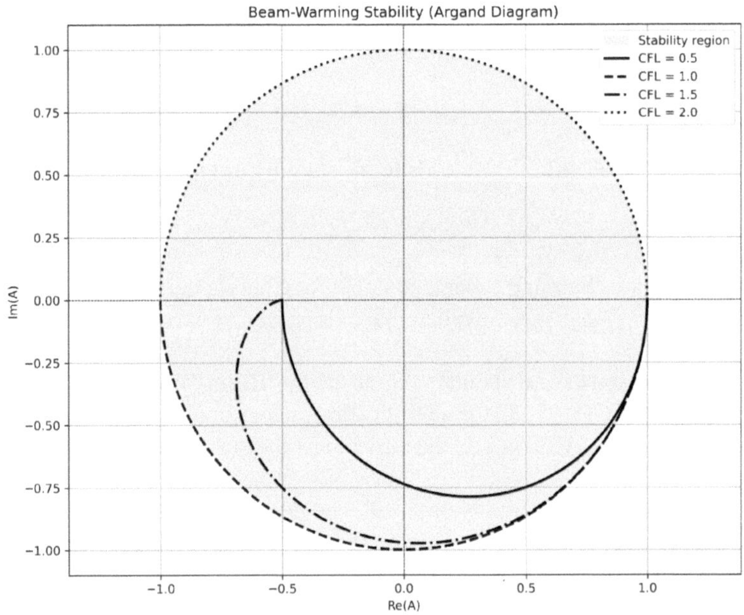

Figure 16.14: **Stability zone for Beam Warming scheme shaded in gray.**

Table 16.2
Comparison of Schemes for $u_t + cu_x = 0$ in Compressible Flow Applications

Numerical Scheme	Amplification Factor	Stability Condition	Order of Accuracy
Von Neumann-Richtmyer (VNR)	–	Stabilized by artificial viscosity	1st
First-Order Upwind (FOU)	$1 - \mathrm{CFL}(1 - e^{-ik\Delta x})$	$\mathrm{CFL} \leq 1$	1st
Courant-Isaacson-Rees (CIR)	$1 - \mathrm{CFL}(1 - \cos(k\Delta x)) - i\mathrm{CFL}\sin(k\Delta x)$	$\mathrm{CFL} \leq 1$	1st
Lax-Friedrichs (LF)	$\cos(k\Delta x) - i\,\mathrm{CFL}\,\sin(k\Delta x)$	$\mathrm{CFL} \leq 1$	1st
Lax-Wendroff (LW)	$1 - i\,\mathrm{CFL}\,\sin(k\Delta x) + \mathrm{CFL}^2(\cos(k\Delta x) - 1)$	$\mathrm{CFL} \leq 1$	2nd
Leapfrog (CTCS)	$1 - i\,\mathrm{CFL}\,\sin(k\Delta x)$	$\mathrm{CFL} \leq 1$	2nd
MacCormack	$1 - i\,\mathrm{CFL}\,\sin(k\Delta x) - \mathrm{CFL}^2(1 - \cos(k\Delta x))$	$\mathrm{CFL} \leq 1$	2nd
Beam-Warming (BW)	$1 - (\mathrm{CFL}/2)(3 - 4e^{-ik\Delta x} + e^{-i2k\Delta x})$ $+(\mathrm{CFL}^2/2)(1 - 2e^{-ik\Delta x} + e^{-i2k\Delta x})$	$\mathrm{CFL} \leq 2$	2nd

Case Study 16A: Inviscid Burgers' Equation Using Lax-Friedrichs Scheme

Problem Definition

This case studies the inviscid Burgers' equation, a nonlinear hyperbolic PDE that models wave propagation and shock formation in a 1D domain. The equation captures advection-dominated phenomena such as traffic flow, nonlinear acoustics, and hydrodynamics. The spatial domain is periodic, meaning the field values at the start and end of the domain are continuous, simulating an unbounded or cyclic flow. The initial condition is a smooth sine wave which evolves over time and leads to shock development due to nonlinear steepening. No viscosity is present in the governing equation, so discontinuities naturally emerge.

Governing Equations

The governing equation is the conservative form of the inviscid Burgers' equation:

$$\frac{\partial u}{\partial t} + \frac{\partial}{\partial x}\left(\frac{u^2}{2}\right) = 0$$

This is a first-order nonlinear PDE with no source or diffusive terms. The problem is unsteady, one-dimensional, and governed solely by nonlinear advection. The initial condition is $u(x,0) = \sin(x)$ and periodic boundary conditions are enforced such that $u(0,t) = u(2\pi,t)$.

Numerical Setup

The Lax-Friedrichs scheme, a first-order finite difference method with artificial viscosity, is used to discretize the governing PDE. The method is suitable for capturing shocks and discontinuities but introduces numerical diffusion. The spatial domain $[0,2\pi]$ is discretized using uniform grids with two configurations: 400 and 1000 points. Time integration uses an explicit scheme with a small enough time step to satisfy the CFL condition. The flux function is $F(u) = \frac{u^2}{2}$, and periodic BCs are implemented at boundaries by wrapping the values across the domain.

Results and Discussion

The initial sine wave deforms over time due to nonlinear advection. By $t = 1$, the profile steepens significantly. At later times (e.g., $t = 3, 5, 7$), a discontinuity forms, signaling the onset of a shock. This evolution is typical in inviscid nonlinear systems where the wave speed depends on the solution itself. The figure below illustrates this behavior over time.

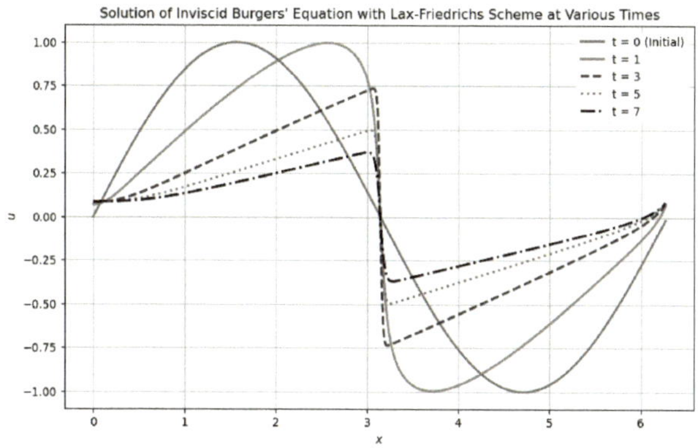

Figure 16.15: **Inviscid Burger equation solved using Lax-Friedrich method.**

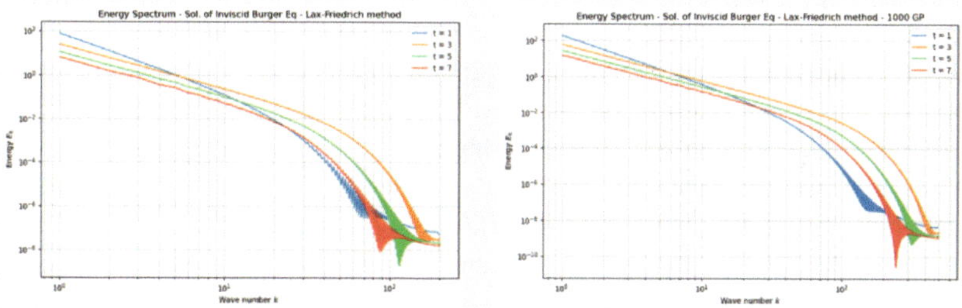

Figure 16.16: **Energy spectra at different resolutions showing improved fidelity with finer grid (right).**

The energy spectrum was analyzed for both grid resolutions. Energy migrates toward higher wavenumbers as nonlinear steepening proceeds. However, the Lax-Friedrichs scheme diffuses high-frequency components, evident from the broader spectrum. Increasing resolution from 400 to 1000 points improves the energy retention at higher wavenumbers, thereby capturing finer scales of the flow.

The higher-resolution simulation resolves more small-scale structures and delays the artificial dissipation seen in coarser grids. The difference between the spectra demonstrates the sensitivity of nonlinear problems to grid resolution.

Numerical Challenges

The Lax-Friedrichs scheme is easy to implement and handles shocks without oscillations, but its artificial diffusion smears out steep gradients. For shock-resolving problems, this method is useful for qualitative insights but lacks spectral accuracy. Grid refinement improves accuracy, but with higher computational cost. Energy spectra analysis provides a diagnostic tool for assessing resolution adequacy. Periodic BCs simplify the implementation, but in real problems with physical boundaries, more sophisticated techniques would be required. The tradeoff between stability and accuracy is evident, guiding the need for higher-order shock-capturing schemes in future investigations.

16.4 GIBBS ERROR (GIBBS PHENOMENON)

The Gibbs phenomenon is named after physicist Josiah Willard Gibbs, who provided the first comprehensive mathematical description in 1899, though it was initially observed by Henry Wilbraham in 1848. This oscillatory behavior occurs when approximating discontinuous functions with finite Fourier series or polynomial expansions, resulting in characteristic overshoots and undershoots near discontinuities.

In computational fluid dynamics (CFD), Gibbs error manifests as spurious oscillations near sharp gradients or discontinuities like shocks. These non-physical wiggles appear because high-order numerical schemes lack sufficient dissipation to properly represent the high-frequency components needed for abrupt transitions. The phenomenon is particularly pronounced in spectral methods and high-order finite difference/volume schemes, where the absence of numerical damping allows oscillations to persist. Notably, higher-order schemes tend to produce higher-frequency oscillations while maintaining the same overshoot magnitude.

These oscillations pose significant challenges in CFD simulations. Near shock waves, they can generate unphysical negative values for inherently positive quantities like density. In turbulent or aeroacoustic simulations, Gibbs-related noise contaminates the solution with high-frequency artifacts. The oscillations also violate monotonicity principles by introducing new extrema not present in the true solution, potentially leading to stability issues. This behavior fundamentally limits the reliability of numerical solutions near discontinuities without special treatment.

To address these challenges, shock-capturing schemes (such as TVD, ENO, WENO, and flux-limited methods) have been developed for solving hyperbolic conservation laws. We will discuss in details shock-capturing schemes for numerical compressible flows in Chapter 19.

Case Study 16B: Shallow Water Equations (SWE) and Dam Break Problem

Problem Definition

This case study addresses the classical one-dimensional dam-break problem governed by the shallow-water equations. A flat, impermeable bed supports an inviscid, incompressible fluid initially at rest with upstream depth h_L and downstream depth h_R. At time $t = 0$ the gate at $x = x_0$ is instantaneously removed, producing unsteady flow described by depth $h(x,t)$ and depth-averaged velocity $u(x,t)$. The resulting solution consists of a left-moving rarefaction fan, an intermediate plateau, and a right-moving hydraulic jump (shock). Transmissive boundary conditions at the domain ends allow waves to exit without reflection, and no symmetry assumptions are invoked. The dominant wave speeds are the nonlinear gravity-wave characteristics $u \pm \sqrt{gh}$.

Governing Equations

Mass conservation for a control volume of unit width:

$$\frac{\partial h}{\partial t} + \frac{\partial (hu)}{\partial x} = 0.$$

Depth-integrated momentum conservation (hydrostatic assumption, neglecting viscosity):

$$\frac{\partial (hu)}{\partial t} + \frac{\partial}{\partial x}\left(hu^2 + \frac{1}{2}gh^2\right) = 0.$$

Conservative system formulation:

The shallow-water system in conservative form and its corrected flux Jacobian are

$$\frac{\partial}{\partial t}\begin{bmatrix} h \\ hu \end{bmatrix} + \frac{\partial}{\partial x}\begin{bmatrix} hu \\ hu^2 + \frac{1}{2}gh^2 \end{bmatrix} = \mathbf{0}$$

with state vector

$$\mathbf{U} = \begin{bmatrix} h \\ hu \end{bmatrix}$$

and flux vector

$$\mathbf{F} = \begin{bmatrix} hu \\ hu^2 + \frac{1}{2}gh^2 \end{bmatrix}$$

The flux Jacobian is

$$\frac{\partial \mathbf{F}}{\partial \mathbf{U}} = \begin{bmatrix} 0 & 1 \\ gh - u^2 & 2u \end{bmatrix}$$

whose eigenvalues $u \pm \sqrt{gh}$ confirm the hyperbolic character of the system.

Numerical Setup

The numerical implementation employs a first-order Lax-Friedrichs scheme on a uniform computational grid comprising 1000 cells spanning the domain from x = 0 to x = 1. The solution update follows the standard finite-difference formulation, with time stepping controlled by a CFL condition targeting a stability value of 0.8. At the domain boundaries, ghost cells are implemented using simple copy conditions from the nearest interior points to provide non-reflecting outflow behavior.

$$\mathbf{U}_i^{n+1} = \frac{1}{2}\left(\mathbf{U}_{i+1}^n + \mathbf{U}_{i-1}^n\right) - \frac{\Delta t}{2\Delta x}\left(\mathbf{F}_{i+1}^n - \mathbf{F}_{i-1}^n\right)$$

$$\frac{\Delta t}{\Delta x} \max_i |u_i \pm \sqrt{gh_i}| \leq 1$$

Significance: The case study investigates the classical one-dimensional dam-break problem, a fundamental benchmark in shallow water flow analysis. This problem describes the unsteady motion of an inviscid, incompressible fluid initiated by the instantaneous removal of a barrier separating regions of differing water depth. Governed by the nonlinear shallow water equations, the flow evolution includes a leftward rarefaction fan, a central constant-depth region, and a rightward-propagating hydraulic jump (shock), depending on the initial depth configuration.

The foundational solution to this problem was first provided by A. Ritter in 1892. Ritter presented an analytical solution for the case where the dam separates a finite water depth from a dry downstream bed. His work introduced critical insights into rarefaction wave propagation and free-surface discontinuities under idealized conditions. This theoretical investigation was further explored by J. J. Stoker in 1957. Stoker extended Ritter's formulation to cases involving arbitrary initial depths on both sides of the dam, enabling the treatment of more complex wave patterns including both rarefaction waves and shock waves. His approach employed the method of characteristics and conservation principles to capture the nonlinear structure of the solution.

In the current study, the numerical results are compared against these classical solutions to validate the simulation approach. By reproducing the key features of the dam-break flow namely the rarefaction zone, intermediate plateau, and shock front—the case study demonstrates the capability of the numerical model to resolve nonlinear wave interactions and to reproduce historically significant shallow water flow behavior with high fidelity.

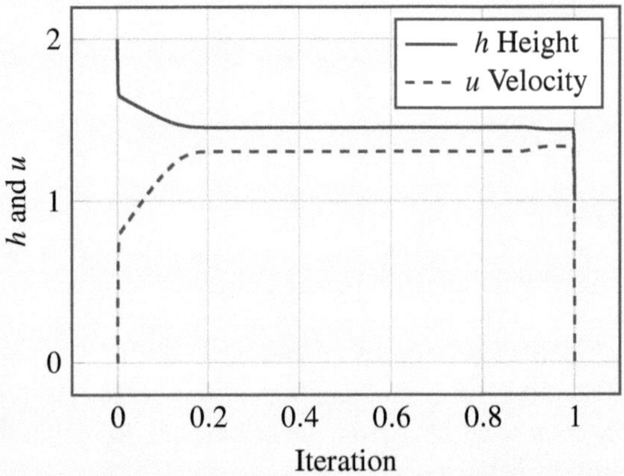

Figure 16.17: **Height and Velocity profiles with 1000 grid points.**

Numerical results at $t = 0.20$ s are tabulated. A rarefaction wave connects the initial upstream depth $h_L = 2$m to a plateau with depth $h_m = 1.454$ m (numerical: 1.45 m). The analytical values come from the classical shallow-water dam-break theory first presented by A. unequal upstream-downstream depths by J. J. Stoker in 1957. The plateau depth, plateau velocity and shock position values listed in table are those predicted by the Ritter-Stoker analytical solution for an initial depth ratio $h_L/h_R = 2$.

Quantity	Numerical	Analytical	Error
Plateau depth h_m (m)	1.45	1.454	0.28%
Plateau velocity u_m (m/s)	1.30	1.306	0.46%

Discussion

In open-channel hydraulics the steep front that surges into the quiescent downstream water after a dam collapses is traditionally called a hydraulic jump (or bore). Mathematically, however, the one-dimensional shallow-water equations form a nonlinear hyperbolic conservation law, and any moving discontinuity that satisfies the Rankine-Hugoniot jump conditions of that law is called a shock. The term Hydraulic jump is used to indicate the water-wave movement, while shock terminology comes from theory of hyperbolic PDEs used in gas and aerodynamics.

The Lax-Friedrichs scheme offers robust stability for this dam-break simulation but introduces some numerical diffusion in the solution. Using a 1000-cell spatial resolution proves sufficient to maintain accuracy, keeping depth errors below 1% compared to analytical solutions, while adaptive time stepping based on the evolving wave speeds ensures computational efficiency without unnecessary time step restrictions.

PROBLEMS

16P-1 The MacCormack method is a two-step predictor-corrector scheme commonly used for solving hyperbolic conservation laws such as the linear advection equation $\partial\rho/\partial t + a\partial\rho/\partial x = 0$. The predictor step uses a forward difference in space to compute an intermediate value, while the corrector step uses a backward difference on the predicted value, and the final result is obtained by averaging the predictor and corrector outcomes. Show that this two-step MacCormack scheme can be rewritten as a single update formula that includes both first- and second-order spatial derivatives. Then, by identifying these terms and performing a truncation error analysis, demonstrate that the scheme is second-order accurate in both space and time.

16P-2 The MacCormack method is a two-step predictor-corrector scheme commonly used for solving hyperbolic conservation laws such as the linear advection equation $\partial\rho/\partial t + a\partial\rho/\partial x = 0$. The predictor step uses a forward difference in space to compute an intermediate value, while the corrector step uses a backward difference on the predicted value, and the final result is obtained by averaging the predictor and corrector outcomes. Do the Von Neumann analysis and identify the stability limits.

16P-3 Simulate the MacCormack method for the 1D linear advection equation with a discontinuous initial condition and investigate the monotonicity-preserving for this scheme for

i a = 0.5 m/s, T = 0.1 s, CFL = 0.1

ii a = 1 m/s, T = 0.1 s, CFL = 0.8

16P-4 Using periodic boundary conditions on the interval $0 \leq x < 1$, investigate the numerical behaviour of two finite-difference schemes—the first-order upwind method, the second-order MacCormack predictor-corrector and second-order Beam-Warming scheme when applied to the one-dimensional linear advection equation $\partial\rho/\partial t + a\partial\rho/\partial x = 0$ with constant speed $a = 1$. Advect a smooth Gaussian pulse $\rho(x,0) = \exp[-100(x-0.5)^2]$ through one full period $t = 1$ while maintaining a fixed Courant number of CFL $= 0.5$. Perform the computation on uniform grids with $N = 50, 100, 200, 400$ cells; after each run compute the normalised L_2 error by comparing the numerical solution to the initial condition. Create a log-log convergence plot of error versus grid spacing to determine the observed order of accuracy for each scheme, then plot and compare the solution profiles obtained with $N = 50$ and $N = 400$ to highlight qualitative differences. Comment on the relative accuracy, dissipation, and dispersion characteristics demonstrated by the two methods.

16P-5 To address the instability issues that arise in advection problems, one can modify the centered difference scheme by replacing it with a one-sided gradient operator, taken in the upwind direction. Specifically, when the flow is from left to right ($v > 0$), the upwind scheme can be employed to stabilize the solution. How does this upwind scheme compare to the MacCormack method, which utilizes a two-step predictor-corrector approach that combines both forward and backward spatial differences, in terms of accuracy and numerical stability? Furthermore, how do different boundary conditions (e.g., Dirichlet or Neumann) and initial step function conditions affect the performance of both methods in terms of convergence, accuracy, and the mitigation of numerical instabilities such as oscillations or excessive diffusion?

16P-6 Starting from the steady Euler equations of mass and momentum, prove the relation:

$$\frac{\partial V}{\partial x} = \frac{V}{\sigma}\frac{\partial \sigma}{\partial x}\left(\frac{1}{M^2-1}\right)$$

where $M = v/c$ is the Mach number. Consider a nozzle with cross-sectional area ($\sigma = x^2 + 1$) in which fluid having $\gamma = 4/3$ is flowing. Take boundary conditions for the steady computation:

left boundary x = -1: ρ =1, v = 0.9, p = 0.75

right boundary x = +1: ρ =1, v = 1.1, p = 0.75.

Solve for the Mach number distribution inside nozzle till the solution tolerance reaches 8.5E-05. Use MacCormack predictor-corrector method.

16P-7 Given the equation for shock-smoothing methods:

$$\frac{\partial u}{\partial t} + u^2 \frac{\partial u}{\partial x} = \frac{\partial u}{\partial t} + \frac{\partial}{\partial x}\left(\frac{u^3}{3}\right),$$

with the initial conditions:

$$t = 0: \quad u = a \quad \text{for} \quad -\infty < x < 0,$$

$$u = b \quad \text{for} \quad 0 \leq x < \infty,$$

Solve for the solution u at any time t.

16P-8 The given partial differential equation (PDE) is

$$xu_t + u_x = 0, \quad x > 0$$

with the boundary conditions:

$$u(x, t = 0) = c(x)$$

$$u(x = 0, t) = b(t)$$

i Find the characteristic lines of the PDE, using the method of characteristics.

Rewrite the PDE:

$$x\frac{\partial u}{\partial t} + \frac{\partial u}{\partial x} = 0$$

The characteristic equations are: $\frac{dx}{dt} = \frac{1}{x}$, $\frac{du}{dt} = 0$. Solve these equations and determine the characteristic lines and the behavior of the solution along them.

ii Determine the solution along the characteristic line, note that $\frac{du}{dt} = 0$, which means that $u(x,t)$ remains constant along the characteristic lines.

iii Find the general solution $u(x,t)$ as a function of the given functions $c(x)$ and $b(t)$, we use the initial and boundary conditions. By solving the characteristic equations, we express $u(x,t)$ in terms of $c(x)$ and $b(t)$.

16P-9 For the given wave equation

$$u_{tt} - c^2 u_{xx} + au = 0$$

i Discretize the equation such that an explicit solution scheme with a truncation error $O(\Delta t^2, \Delta x^4)$ is obtained.

ii Transform the equation into a set of first-order equations and determine the slopes of the characteristic lines and the type of the equation.

iii Determine the maximum time step required based on the CFL condition.

iv Check the stability with the von Neumann analysis.

16P-10 For the given partial differential equation

$$\frac{\partial^2 u}{\partial t \partial x} + c\frac{\partial^2 u}{\partial x^2} = 0, \quad c > 0, x \geq 0, t \geq 0$$

The finite difference equation is setup for the solution on an equidistant grid $x_i = i\Delta x$, $t_n = n\Delta t$:

$$\frac{u_i^{n+1} - u_i^{n-1}}{2\Delta t} + c\frac{u_{i+1}^n - 2u_i^n + u_{i-1}^n}{\Delta x^2} = 0$$

i Determine the type and the characteristic lines of the PDE.

ii Determine the truncation error of the FDE.

iii Apply the CFL condition to determine a stability limit for the FDE.

iv Check the stability of the FDE with the von Neumann analysis.

16P-11 Given the following scalar conservation equation:

$$\frac{\partial u}{\partial t} + \frac{\partial}{\partial x} f(u) = 0 \quad \text{with} \quad f(u) = \frac{1}{4}\left(u^2 - \sin^2(u)\right)$$

i Write the conservation equation in quasilinear form $u_t + a(u)u_x = 0$. What is $a(u)$, and what is its physical meaning?

ii Compute the shock speed s as a function of the right and left states u_R and u_L for the equation (1) for an unsteady shock.

iii Sketch the solution of the following initial value problem for the conservation equation (1) in an x-t diagram.

The initial condition is given by:

$$u_0(x) = \begin{cases} u_L & \text{for} x < 0 \\ u_R & \text{for} x > 0 \end{cases}$$

with $(u_L, u_R) = (1, 0)$.

16P-12 Consider the linear advection equation

$$\frac{\partial u}{\partial t} + a\frac{\partial u}{\partial x} = 0$$

on the spatial domain $0 \leq x < 6$, with periodic boundary conditions and constant advection speed $a = 1$. The initial condition is a square pulse, defined as

$$u(x,0) = \begin{cases} 1, & 0.4 \leq x \leq 0.6 \\ 0, & \text{otherwise} \end{cases}$$

The domain is discretized into $N_x = 200$ uniform grid points, and the solution is advanced in time to $t = 1$. Use a CFL number of 2, and compute the time step based on CFL. Implement the Beam-Warming scheme (second-order accurate, upwind-biased).

16P-13 Investigate finite-difference numerical schemes for linear advection with spatially varying wave speed. The domain $0 \leq x \leq 2$ features a discontinuous velocity $a(x)$, changing from 1 to 2 at $x = 1$. Using a uniform grid with $\Delta x = 0.02$ and CFL $= 0.25$, apply Courant-Isaacson-Rees scheme over twenty time steps for a triangular pulse initially centered at $x = 1$. Compute L_1, L_2, and L_∞ error norms against the exact solution. Analyzes how the solution deforms and error evolves as the pulse crosses the discontinuity in $a(x)$. Compare the CIR solution with first-order upwind, Lax-Friedrichs, leapfrog, and Lax-Wendroff schemes, and highlighting their differences near discontinuities and any numerical artifacts observed.

16P-14 Consider a one-dimensional shallow water flow of an incompressible, inviscid fluid over a flat bottom. Starting from the integral forms of the mass and momentum conservation laws, derive the shallow water equations (SWE) in conservative form involving the fluid depth $h(x,t)$ and depth-averaged velocity $u(x,t)$. Identify the resulting system as a nonlinear hyperbolic system and derive expressions for the characteristic wave speeds, showing that they are given by $u \pm \sqrt{gh}$, where g is the gravitational acceleration. Then, linearize the SWE about a steady uniform flow and determine the linear wave speeds. Finally, apply the Lax-Friedrichs finite difference scheme to the conservative form of the SWE and derive the update formula for the numerical solution. Discuss the CFL condition required for stability of the scheme in terms of the maximum wave speed. Implement the method for a dam-break initial condition and interpret the physical propagation of the resulting wavefronts.

REFERENCES

J. Von Neumann, R. D. Richtmyer, A method for the numerical calculation of hydrodynamic shocks. Journal of Applied Physics, 21(3), 232-237, 1950.

R. Courant, E. Isaacson, M. Rees, On the solution of nonlinear hyperbolic differential equations by finite differences. Communications on Pure and Applied Mathematics, 5, 243-255, 1952.

E. F. Toro, Riemann Solvers and Numerical Methods for Fluid Dynamics, Springer, 2009.

G. A. Sod, Numerical Methods in Fluid Dynamics: Initial and Initial Boundary-Value Problems, Cambridge University Press, 1985.

P. D. Lax, Weak solutions of nonlinear hyperbolic equations and their numerical computation. Communications on Pure and Applied Mathematics, 7, 159-193, 1954.

R. J. LeVeque, Numerical Methods for Conservation Laws, Birkhäuser, 1992.

C. Hirsch, Numerical Computation of Internal and External Flows, Vol. 2: Computational Methods for Inviscid and Viscous Flows, Butterworth-Heinemann, 1990.

P. D. Lax, , B. Wendroff, Systems of conservation laws. Communications on Pure and Applied Mathematics, 13(2), 217-237, 1960.

K. S. Yee, Numerical solution of initial boundary value problems involving Maxwell's equations in isotropic media. IEEE Transactions on Antennas and Propagation, 14(3), 302-307, 1966.

R. W. MacCormack, The effect of viscosity in hypervelocity impact cratering, AIAA Paper 69-354 (1969)

R. W. MacCormack, Numerical solution of the interaction of a shock wave with a laminar boundary layer, Proceedings of the 2nd Int. Conf. on Numerical Methods in Fluid Dynamics, Lecture Notes in Physics, Vol. 8, Springer, Berlin, pp. 151-163, 1971.

R. M. Beam, R. F. Warming, An Implicit Finite-Difference Algorithm for Hyperbolic Systems in Conservation-Law Form, Journal of Computational Physics, 22(1): 87-110, 1976.

R. M. Beam, R. F. Warming, An Implicit Factored Scheme for the Compressible Navier-Stokes Equations, AIAA Journal, 16(4): 393-402, 1978.

R. F. Warming, R. M. Beam, An Extension of A-Stability to Alternating Direction Methods, BIT Numerical Mathematics, 19(4): 395-417, 1979.

J. P. Boris, D. L. Book, Flux-corrected transport I: SHASTA, a fluid transport algorithm that works. Journal of Computational Physics, 11(1), 38-69, 1973.

A. Ritter, Die Fortpflanzung der Wasserwellen, Sitzungsberichte der Königlich Preuischen Akademie der Wissenschaften, pp 839–852, 1892.

J. J. Stoker, Water Waves: The Mathematical Theory with Applications, Interscience Publishers, 1957, chap. 10, USA.

17 Finite Volume Method I: Introduction

This chapter will systematically introduce the finite volume formulation for different types of conservation laws. Emphasis will be placed on spatial discretization using control volumes, approximation of convective and diffusive fluxes, treatment of source terms, and the construction of discrete algebraic systems.

In this chapter, we will discuss the foundational concepts of finite volume in computational fluid dynamics (CFD).

After finishing this chapter, you should be able to:

- Understand and derive the finite volume method (FVM) formulation for conservation laws by discretizing the integral form over control volumes in computational domains.
- Apply spatial discretization techniques to transform governing partial differential equations (PDEs) into algebraic equations, handling convective, diffusive, and source fluxes accurately.
- Analyze the role of time discretization using methods like Crank-Nicolson and understand the stability criteria, including the significance of the Courant-Friedrichs-Lewy (CFL) condition.

The Finite Volume Method (FVM) is a widely used numerical technique in computational fluid dynamics (CFD) for solving partial differential equations representing conservation laws. Unlike the finite difference, FVM is based directly on the integral form of the conservation equations, ensuring the local conservation of mass, momentum, and energy across control volumes.

Sergei K. Godunov first proposed the finite-volume method in his 1959 paper on conservative difference schemes for fluid dynamics. This method divides the computational domain into a finite number of non-overlapping control volumes, and the conservation principles are applied over each volume individually. The solution involves approximating the fluxes of conserved quantities through the control volume faces and balancing these fluxes with the rate of change within the volume and any internal sources or sinks. By integrating the governing equations over discrete volumes and applying the Gauss divergence theorem, volume integrals of divergence terms are converted into surface integrals, significantly simplifying the numerical treatment of fluxes across control surfaces. This feature inherently guarantees conservation at the discrete level, which is crucial for the physical accuracy of CFD simulations.

The primary goal of Finite Volume discretization is to transform partial differential equations into an algebraic form. This is achieved by integrating the governing

DOI: 10.1201/9781003663171-17

transport equation, as given below:

$$\int_V \frac{\partial(\rho\psi)}{\partial t}dV + \int_V \frac{\partial(u_i\rho\psi)}{\partial x_i}dV = \int_V \frac{\partial}{\partial x_i}\left(\Gamma_\psi \frac{\partial\psi}{\partial x_i}\right)dV + \int_V S_\psi dV$$

Here, ψ represents an arbitrary conserved property, such as velocity or a scalar field.

Γ_ψ denotes the transport coefficient, such as laminar or turbulent viscosity, or diffusivity.

S_ψ encompasses all relevant source terms, for instance, the pressure gradient in the momentum equation.

Applying the Gauss divergence theorem allows the transformation of volume integrals into surface integrals, leading to:

$$\overbrace{\int_V \frac{\partial(\rho\psi)}{\partial t}dV}^{time} + \overbrace{\underbrace{\int_\sigma \left(\rho u_i\psi - \Gamma_\psi \frac{\partial\psi}{\partial x_i}\right)n_i d\sigma}_{Flux} = \underbrace{\int_V S_\psi dV}_{source}}^{Space} \tag{17.1}$$

where σ represents the closed surface surrounding the control volume V, and n_i is the component of the outward-pointing unit normal vector.

The computational domain is partitioned into a finite number of small control volumes. The discretized equation must be solved individually for each control volume. For simplicity the hexahedral control volumes are employed here. The computational grid defines the boundaries of these control volumes, and a computational node is assigned to each one. The flux term includes both convective and diffusive contributions.

17.1 SPATIAL DISCRETIZATION

In the FVM, the dependent variables are associated with the centroids of the control volumes. The fluxes at the control volume faces are typically interpolated from the neighboring nodal values using schemes such as central differencing, upwind differencing, or higher-order reconstructions. The choice of interpolation greatly affects the accuracy, stability, and numerical diffusion of the method. For instance, while first-order upwind schemes offer robustness, they suffer from excessive numerical diffusion, whereas second-order schemes like central differencing offer greater accuracy but require careful treatment to maintain stability, especially in convective-dominated flows. In practical CFD applications, hybrid schemes are often employed to balance stability and accuracy. The FVM also elegantly handles complex geometries and unstructured meshes, making it highly versatile across a wide range of engineering problems.

For sake of simplicity, we only consider a hexahedral control volume illustrated in Figure 17.1. The center of the control volume is labeled as **P**, surrounded by six neighboring control volumes: **E** (east), **W** (west), **N** (north), **S** (south), **T** (top), and **B** (bottom). Each neighbor shares a common face with **P**, identified as **ea** (east), **we** (west), **no** (north), **so** (south), **to** (top), and **bo** (bottom), respectively.

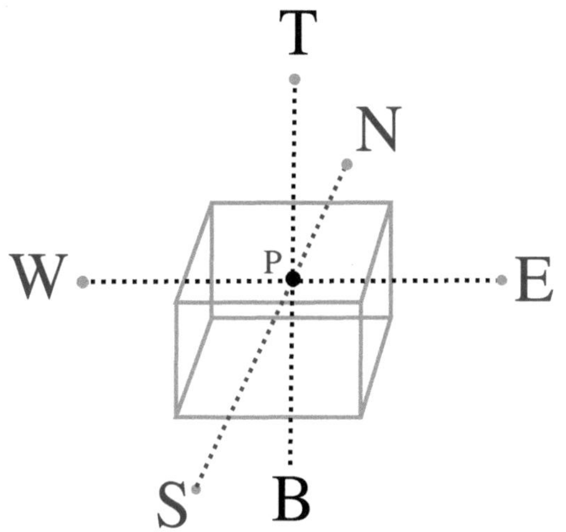

Figure 17.1: **The hexahedral control volume.**

The purpose of spatial and temporal discretization is to transform equation 17.1 into an algebraic form:

$$\underbrace{A_P\psi_P - \Sigma A_{N-CV}\psi_{N-CV}}_{\text{implicit part}} = \underbrace{S_P}_{\text{explicit part}} \tag{17.2}$$

In this formulation, the left-hand side represents the implicit terms, while the right-hand side accounts for the explicit contributions. Here, the subscript $N-CV$ refers to the six neighboring control volumes. The coefficients A_p and A_{N-CV} correspond to the dependent variable at **P** and its neighbors. Terms that cannot be incorporated into the implicit part are treated explicitly and grouped into the source term S_P.

Each hexahedral control volume is bounded by six faces, which may not necessarily lie on a plane since the four vertices defining a face might not be coplanar. However, it is generally assumed that each face is planar, with its orientation specified by a normal vector. The total flux through the control volume is the sum of the fluxes across all six faces. To evaluate the surface integrals, the values of the integrand must be determined at the surfaces. As the primary computation occurs at the control volume center, surface integrals are approximated using a second-order midpoint rule:

$$\int_\sigma F\,d\sigma = \sum_i \int_{\sigma_i} F_i\,d\sigma$$

where F_i represents the convective or diffusive flux across face σ_i.

Similarly, volume integrals are approximated using the midpoint rule:

$$\int_V F\,dV \approx F_p\,dV$$

where F_p denotes the value of F at the center **P**, and dV is the control volume.

17.1.1 CONVECTIVE FLUXES

Convective flux discretization is a central component of the finite volume method (FVM) for solving transport equations in fluid dynamics and heat transfer. In such problems, a variable ψ (for instance, temperature, species concentration, or velocity component) is advected across cell faces by the fluid motion. The net transport of ψ in each control volume depends on accurately approximating the fluxes through these faces. Because the advective term can strongly dominate over diffusion when fluid velocities are large, the choice of discretization scheme has major implications for stability and accuracy.

A typical control volume in the FVM is bounded by faces labeled "east" and "west" (in one dimension), or extended with "north," "south," "top," and "bottom" faces in two or three dimensions. Convective fluxes are computed via mass flow rates multiplied by the transported variable at each face. However, ψ is generally not known exactly at the faces, so interpolation must be used from the cell-center values. The method of interpolation defines the numerical scheme.

For relatively small cell Reynolds or Peclet numbers $\mathrm{Pe} = \frac{\rho u \Delta x}{\Gamma}$, a central difference approach—where ψ is interpolated linearly between neighboring cell centers—tends to be stable and accurate. However, for $\mathrm{Pe} > 2$, standard central differencing can produce negative coefficients in the finite volume discretization, violating the fundamental requirement that coefficients in the discrete equation all share the same sign for physically meaningful solutions. In such cases, the upwind approach can be used to ensure positivity by always taking ψ from the upstream cell. Though robust, upwind can introduce excessive numerical diffusion. QUICK (Quadratic Upstream Interpolation for Convective Kinetics) is a higher-order alternative combining elements of upwind and central differencing to achieve improved accuracy while controlling spurious oscillations. All these schemes reflect trade-offs between numerical diffusion, dispersion, and stability, making the choice of scheme context-dependent—especially in flows with high convective transport where the Peclet number can be large.

The convective flux of a transported variable ψ through the east face σ_{ea} can be expressed as

$$F_{ea}^{C} = \int_{\sigma_{ea}} (\rho\, u_i\, \psi\, n_i)_{ea}\, d\sigma \approx (\rho\, u_i n_i)_{ea}\, d\sigma\ \psi_{ea} = \dot{m}_{ea}\, \psi_{ea},$$

where $\dot{m}_{ea}$ is the mass flux. Numerically, ψ_{ea} is found via an interpolation scheme. Below are three primary approaches used in FVM for convective flux discretization:

Upwind Scheme: Upwind Scheme assigns the face value of the variable strictly based on the upwind cell, guaranteeing positive coefficients and numerical stability in convection-dominated flows, albeit at the cost of higher numerical diffusion.

$$\psi_{ea} = \begin{cases} \psi_P, & \text{if } u_e > 0, \\ \psi_E, & \text{if } u_e < 0, \end{cases}$$

selects the upstream cell value, thereby guaranteeing positive coefficients and stability but often introducing significant numerical diffusion if Pe is moderate or large.

For the approximation of ψ at the other control-volume boundaries in the Upwind approach is done by assuming a piecewise linear distribution for the diffusion terms. The coefficients calculated by the Upwind approach at faces e and w are given by

$$a_E = \left(\frac{\Gamma_e}{\Delta x_e} + \left[-(\rho\, u)_e, 0\right]\right) A_e$$

$$a_W = \left(\frac{\Gamma_w}{\Delta x_w} + \left[+(\rho\, u)_w, 0\right]\right) A_w$$

Here, the notation $\left[\ ,\ \right]$ indicates taking the maximum of the two enclosed values. With the Upwind approach, a_E and a_W are thus guaranteed to be greater than zero. Note that the use of Upwind can lead to serious discretization errors. Consequently, a variety of discretization schemes have been developed over the years, featuring smaller discretization errors than the Upwind approach and fewer oscillations than the central-difference approach when discretizing the convective terms.

Central Difference Scheme (CDS): Central Difference Scheme (CDS) is a second-order method that approximates the face value of a transported variable by averaging the neighboring cell-center values, making it accurate at relatively low Peclet numbers.

$$\psi_{ea} \approx \frac{\psi_P + \psi_E}{2}$$

in one dimension, or more generally a linear interpolation between surrounding cell centers. Accurate for small Pe but may yield unphysical oscillations or negative coefficients at larger Pe.

Hybrid Scheme: The Hybrid approach is a method that combines both the Upwind and central-difference schemes. To derive the Hybrid approach, we begin with the exact solution of the one-dimensional transport equation, which includes only convection and diffusion (without source terms). The idea is to define a simple interpolation rule inspired by the exact solution of the variable profile between computational points. When the Peclet number is large, indicating strong convection relative to diffusion, the exact solution of the one-dimensional transport equation can be captured by the Upwind scheme without any diffusive component. Conversely, when the Peclet number is below two, the central-difference scheme provides a better approximation. As a result, the Hybrid method employs a modified Upwind scheme for Peclet numbers greater than two and switches to the central-difference scheme for Peclet numbers less than two. For instance, the following formulas are used to compute the coefficients in the x-direction:

$$a_E = \left[-(\rho u)_e, \frac{\Gamma_e}{\Delta x_e} - \frac{1}{2}(\rho u)_e\right] \cdot A_e$$

$$a_W = \left[+(\rho u)_w, \frac{\Gamma_w}{\Delta x_w} + \frac{1}{2}(\rho u)_w\right] \cdot A_w$$

A critical review of the Hybrid approach reveals that it may also be subject to significant discretization errors, similar to the Upwind method. The Upwind scheme can introduce large discretization errors at high Peclet numbers. Indeed, at high Peclet numbers, the Hybrid approach switches to the Upwind scheme.

In practice, problems are rarely one-dimensional, and they often involve not just convection and diffusion, but also source terms. Therefore, the Hybrid discretization, which is tailored to the exact solution of a one-dimensional case, may not always provide a fully accurate solution in real-world scenarios. Another concern is the stability of iterative solution methods when applying the Hybrid approach.

Since this method can lead to coefficients with differing signs, the stability of an iterative solver is not guaranteed as it is with the central-difference scheme. Some efficient algorithms require a specific symmetry in the coefficient matrix. This symmetry is preserved in the Upwind method because all coefficients retain a diffusive component, ensuring that none of them becomes zero. However, under the Hybrid approach, there may be situations where a coefficient vanishes, breaking the required symmetry in the matrix.

QUICK Scheme: A higher-order interpolation using a three-point stencil (upstream, current, and downstream cells). QUICK aims to reduce numerical diffusion compared to pure upwind while limiting spurious oscillations common in straightforward central differencing at high Pe. It is especially popular in two- and three-dimensional flows where accuracy in boundary layers or shear layers can be critical.

The individual values of ψ at the control volume boundaries can be efficiently computed using a Lagrange interpolation polynomial. A polynomial of order n is constructed by determining the values of $f(x_i)$ at $n+1$ points x_i. The value of $p_n(x)$ at any arbitrary point x can then be obtained from the Lagrange interpolation polynomial:

$$p_n(x) = \sum_{i=0}^{n} L_i(x) f(x_i)$$

where

$$L_i(x) = \prod_{j=0, j\neq i}^{n} \frac{x - x_j}{x_i - x_j} \quad \text{for } i = 0, 1, \ldots, n$$

For interpolating the value ψ_e using a parabola as the interpolation polynomial, with $u_e > 0$ and ψ-values given at x_W, x_P, and x_E, the following relation is derived as

$$\psi_e = \frac{(x_e - x_P)(x_e - x_E)}{(x_W - x_P)(x_W - x_E)} \cdot \psi_W + \frac{(x_e - x_W)(x_e - x_E)}{(x_P - x_W)(x_P - x_E)} \cdot \psi_P + \frac{(x_E - x_W)(x_E - x_P)}{(x_E - x_W)(x_E - x_P)} \cdot \psi_E$$

For uniform grid spacings Δx and when $u_e > 0$, $u_w > 0$, the equation for a_E is given by:

$$a_E = \left(\frac{\Gamma_e}{\Delta x_e} - \frac{3}{8} \{\rho u\} \right) A_e$$

When $Pe > 8/3$, it follows that $a_E < 0$. This indicates that, similar to the central-difference method, the QUICK approach leads to negative coefficients, meaning coefficients with different signs. This can result in numerical issues and unphysical outcomes. Moreover, this method may also cause instability when applied to iterative solution techniques for solving the difference equations.

Table 17.1

Common entries for `gradSchemes` in OpenFOAM and their use in gradient computation.

Scheme	Description and Use
`Gauss linear`	Standard central differencing scheme. Uses face interpolation and the Gauss theorem to compute gradients. Second-order accurate on structured or mildly distorted meshes. Most commonly used.
`leastSquares`	Gradient computed via least-squares fit using neighboring cell values. More robust on highly skewed or unstructured grids. May be slightly more diffusive than Gauss-based schemes.
`Gauss cubic`	Uses higher-order interpolation for improved accuracy in smooth regions. Less robust on distorted meshes. Requires higher stencil connectivity.
`cellLimited Gauss linear`	Applies a limiter to the computed gradient magnitude to prevent overshoots or instabilities in regions with large gradients or discontinuities. Useful in multiphase and compressible flows.
`faceLimited leastSquares 1.0`	Limits the gradient reconstruction at face level to enforce boundedness. The number indicates the limiting coefficient (1.0 = full gradient).

Note that although pressure gradients (∇p) appear as source terms in the momentum equations, their numerical evaluation involves spatial derivatives. Therefore, `OpenFOAM` uses `gradSchemes` to control how gradients like ∇p are computed. This applies equally to pressure source terms, diffusive fluxes, and high-order convective reconstructions.

Local Peclet Number: A key parameter governing the behavior of these schemes is the local Peclet number, $\text{Pe} = \frac{\rho u \Delta x}{\Gamma}$, where Γ is the diffusion coefficient (for thermal or species diffusion). When $\text{Pe} \gg 1$, convection strongly dominates, rendering purely central differencing prone to physically incorrect solutions. Conversely, at small $\text{Pe} \ll 1$, diffusion is relatively strong, and central differencing can remain stable and accurate. Thus, the choice among upwind, central difference, QUICK, or other advanced schemes can hinge on balancing numerical dissipation against the risk of non-monotonic or unstable solutions in convection-dominated problems.

In finite volume simulations of fluid flows and related transport phenomena, the careful handling of convective fluxes is essential for both stability and accuracy. Unlike diffusion, which is naturally well-captured by central or second-order formulations, convection can transport errors rapidly downstream if the discretization scheme does not adequately respect the sign and magnitude of local flow velocities. The Peclet number $\text{Pe} = \frac{\rho u \Delta x}{\Gamma}$ is a convenient measure of the ratio between advective and diffusive effects: when Pe is small, diffusive processes dominate, and central difference approximations usually suffice; for larger Pe, more robust measures, such as upwind or QUICK, may be required to preserve physically reasonable results.

The upwind method assigns the face value of ψ based on the upstream cell, thereby guaranteeing that all coefficients in the discretized equations are non-negative and avoiding unphysical solutions. However, this approach can introduce an overly dissipative solution in regions where the true physics may exhibit sharp gradients or turbulent fluctuations. QUICK, by blending upwind and higher-order interpolation, aims to reduce this artificial diffusion, yet must be implemented with care to prevent numerical oscillations that can appear at high Pe.

In practice, low-Peclet-number regimes often do well with central differencing, while strongly convection-dominated flows may demand more specialized techniques, such as flux limiters or blended schemes. If the local cell Peclet number exceeds a threshold (commonly Pe > 2), central differencing will produce at least one negative coefficient, contravening the usual FVM requirement of coefficient positivity. Such sign inconsistencies can lead to non-physical oscillations, erroneous temperature or velocity fields, and, in extreme cases, outright divergence of the iterative solver. By selecting an upwind or QUICK scheme, the finite volume practitioner actively promotes numerical stability and overall fidelity to the governing equations while mitigating the potential pitfalls of excessive numerical diffusion or spurious oscillations.

17.1.2 DIFFUSIVE FLUXES

The integrated diffusive fluxes are approximated using the midpoint rule:

$$F_{ea}^{D} = \int_{\sigma_{ea}} \left(\Gamma_{\psi} \frac{\partial \psi}{\partial x_i} \right) n_i d\sigma \approx \left(\Gamma_{\psi} \frac{\partial \psi}{\partial x_i} \right)_{ea} \Delta\sigma$$

The spatial derivative $\frac{\partial \psi}{\partial x_i}$ is approximated using a central difference scheme for Cartesian grids:

$$\left(\frac{\partial \psi}{\partial x_i} \right) \approx \frac{\psi_E - \psi_P}{x_E - x_P}$$

This provides a second-order accurate approximation on a uniform grid. Table 17.3 lists common schemes for diffusive fluxes in `OpenFOAM`.

Table 17.2 lists common entries for `divSchemes` in `OpenFOAM` and their use in convective flux discretization. Although divergence schemes are required to discretize the continuity equation, they are also essential for all convective transport terms across momentum, energy, and scalar transport equations in CFD. Although the QUICK scheme is originally derived using a combination of convective and diffusive considerations, in `OpenFOAM` it is applied strictly to the discretization of the convective flux term $\nabla \cdot (\rho \psi \mathbf{u})$, and therefore classified under `divSchemes`. The actual physical diffusion $\nabla \cdot (\Gamma_{\psi} \nabla \psi)$ is handled separately using `laplacianSchemes`.

Table 17.2
Common entries for `divSchemes` in OpenFOAM and their use in convective flux discretization.

Scheme	Description and Use
`Gauss upwind`	First-order upwind differencing. Very stable and bounded, but introduces numerical diffusion. Useful for high Peclet number flows or when robustness is a priority.
`Gauss linear`	Second-order central differencing. Higher accuracy than upwind but can cause non-physical oscillations in convection-dominated flows. Suitable for low Peclet number regimes.
`Gauss QUICK`	Third-order upwind-biased scheme. Reduces numerical diffusion but may produce unbounded or oscillatory solutions unless combined with limiters.
`Gauss linearUpwind grad(`ϕ`)`	Higher-order scheme using gradients to reconstruct face values. Requires `gradSchemes` definition. More accurate but sensitive to mesh quality.
`bounded Gauss limitedLinear`	Total Variation Diminishing (TVD) scheme that blends central and upwind methods. Enforces boundedness and reduces spurious oscillations. A safe default for many problems.
`Gauss vanLeer`	Flux-limited TVD scheme based on van Leer limiter. Bounded and second-order accurate. Good for shock-capturing or steep gradients in compressible flows.
`Gauss SFCD`	Simplified Flux Corrected Differencing. A second-order bounded scheme blending central and upwind. Good compromise between accuracy and robustness.

Figure 17.2 illustrates the correspondence between physical terms in the finite volume formulation and the specific discretization scheme categories used in `OpenFOAM` via the `fvSchemes` dictionary. Each row connects a physical process such as unsteady accumulation, convection, diffusion, or pressure gradients to the appropriate scheme type responsible for its numerical treatment. For example, convective terms are handled through `divSchemes`, while diffusive fluxes and gradient calculations rely on `laplacianSchemes` and `gradSchemes`, respectively. Interpolation of face-centered values and treatment of non-orthogonality are managed by `interpolationSchemes` and `snGradSchemes`. This diagram illustrates how `OpenFOAM` links governing equation components to their corresponding discretization controls.

17.1.3 SOURCE TERM

Source terms are evaluated using the midpoint rule, with the integrand value at the control volume center:

$$S_\psi = \int_V F\,dV \approx F_P\,dV$$

Since all necessary variables are known at point **P**, no interpolation is required. In the momentum equations, the source term includes body forces and the pressure gradient. Body forces are directly approximated, whereas the pressure gradient is

Table 17.3
Common schemes for diffusive fluxes in OpenFOAM (`laplacianSchemes` and `snGradSchemes`).

Scheme Section	Scheme	Description and Use
`laplacianSchemes`	`Gauss linear corrected`	Standard scheme with non-orthogonal correction. Recommended for general unstructured grids.
	`Gauss linear limited`	Limits the non-orthogonal correction to improve stability while preserving accuracy.
	`Gauss linear uncorrected`	Assumes orthogonal mesh. Not recommended unless mesh is near-orthogonal.
`snGradSchemes`	`corrected`	Applies non-orthogonal correction to surface-normal gradient computation. Necessary for general meshes.
	`uncorrected`	Assumes orthogonality between cell centers and face normals. Fast but less accurate.
	`limited 0.33`	Partially applies correction with user-defined weight (e.g., 0.33). Balances stability and accuracy.

treated using the Gauss theorem:

$$\int_V \frac{\partial p}{\partial x_i} dV = \int_\sigma p n_i d\sigma \approx \sum_c \int_{\sigma_c} p n_i d\sigma_c$$

where the sum extends over all faces $c =$ ea, we, no, so, to, bo. Using the midpoint rule gives:

$$\sum_c \int_{\sigma_c} p n_i d\sigma_c \approx \sum_c (p n_i)_{m_c} d\sigma_c$$

The pressure at the face center is approximated via second-order linear interpolation between neighboring control volumes:

$$p_{ea} = \gamma_{e1} P_E + \gamma_{e2} P_P$$

where γ_{e1} and γ_{e2} are interpolation factors.

System of ODEs:

After approximating the time derivative in equation 17.1 using the midpoint rule, the system of ordinary differential equations can be written as:

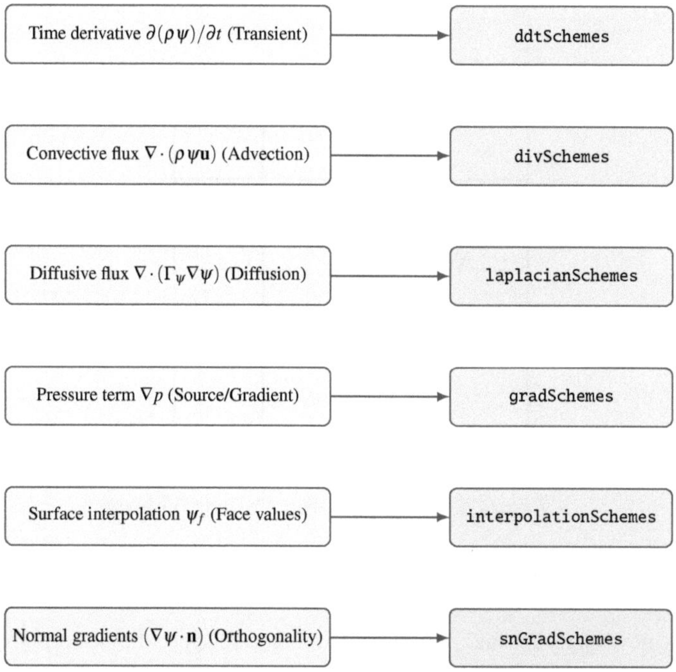

Figure 17.2: **Mapping of physical terms in the finite volume method to OpenFOAM `fvSchemes` configuration sections.**

$$\frac{\partial(\rho\psi_{i,P})}{\partial t}dV = \left(-a_{i,P}\psi_{i,P} + \sum_c a_{i,c}\psi_{i,c} + S_{\psi_i}\right) \tag{17.3}$$

where $i = 1,2,3,\ldots,N$, with N being the total number of control volumes. The right-hand side results from the rearrangement of the discretized convective and diffusive fluxes along with the source terms. Here, C denotes neighboring control volumes (E, W, N, S, T, B), and S_ψ represents the discretized source term.

Conservation principles impose the following relationship among the coefficients:

$$a_{i,P} = \sum_c a_{i,c}$$

Example 17.1

Starting from the continuity equation for two-dimensional steady flows in Cartesian coordinates

$$\frac{\partial(\rho v_j)}{\partial x_j} = 0,$$

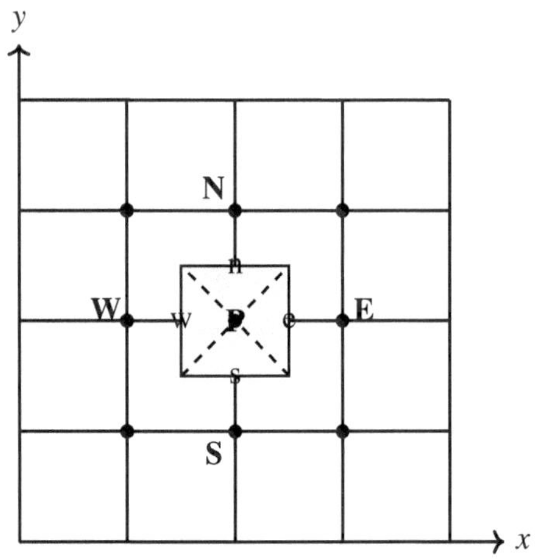

Figure 17.3: **Structured Control Volume.**

show how applying the finite volume method leads to the discrete mass balance

$$(\rho uA)_e - (\rho uA)_w + (\rho vA)_n - (\rho vA)_s = 0,$$

and explain how this further simplifies for one-dimensional flows.

Solution

FV-discretization of the continuity equation in Cartesian coordinates for two-dimensional steady flows (see Figure 17.3)

$$\frac{\partial(\rho v_j)}{\partial x_j} = 0 \quad \Rightarrow \quad \mathrm{div}(\rho \mathbf{v}) = 0$$

From Gauss Divergence Theorem, the net mass flux leaving a control volume (surface integral) is equal to the volume integral of the divergence of the mass flux inside the volume:

$$\int_V \mathrm{div}(\rho \mathbf{v})\, dV = \int_A \rho(\mathbf{v}\, d\mathbf{A})$$

$$\int_A \rho(\mathbf{v}\, d\mathbf{A}) = \int_{A_e} \rho u_e\, dA_e - \int_{A_w} \rho u_w\, dA_w + \int_{A_n} \rho v_n\, dA_n - \int_{A_s} \rho v_s\, dA_s$$

$$\Rightarrow (\rho uA)_e - (\rho uA)_w + (\rho vA)_n - (\rho vA)_s = 0$$

For one-dimensional flows, we can cast it into form

$$(\rho u)_e - (\rho u)_w = 0$$

Example 17.2

Transport Equation for One-Dimensional, Steady-State Heat Conduction inside a metallic rod is given by a governing equation

$$\frac{d}{dx}\left(\lambda\frac{dT}{dx}\right) = 0,$$

where, $\lambda = \lambda(T,x)$ denotes the thermal conductivity, which may vary with temperature and position. The thermal conductivity of the rod can generally depend on the local temperature, and may also vary spatially due to differences in material composition. Determine the temperature distribution along a long, adiabatically insulated rod, whose ends are maintained at different but constant and known temperatures (see Figure 17.4).

Solution
Assuming no internal heat sources, the heat flux must be continuous across the domain. The steady-state heat conduction equation to be solved is:

$$\frac{d\lambda}{dx}\frac{dT}{dx} + \lambda\frac{d^2T}{dx^2} = 0$$

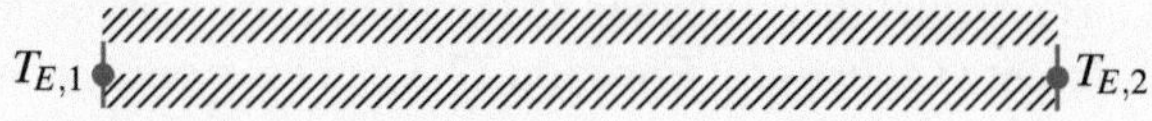

Figure 17.4: **The heat conduction through a rod.**

The solution is sought over a computational grid with uniform spacing Δx between adjacent nodes:
We consider two control volumes such that point 1 is located at the left boundary, and point 3 at the right boundary of the domain (see Figure 17.5). Assuming that the boundary temperatures T_1 and T_3 are specified, and only unknown is the temperature T_2 at the internal node. Although the discretization is intentionally coarse, it serves to highlight essential properties of the finite volume numerical approach.

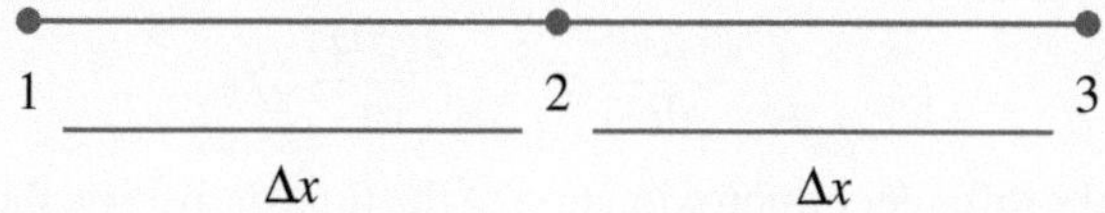

Figure 17.5: **Computational control volumes.**

Using a backward difference approximation for the first derivative and a central difference for the second derivative, the discretized equation at point 2 becomes:

$$\frac{\lambda_2-\lambda_1}{\Delta x}\cdot\frac{T_2-T_1}{\Delta x}+\lambda_2\cdot\frac{T_3-2T_2+T_1}{(\Delta x)^2}=0$$

which simplifies to:

$$\lambda_1\cdot\frac{T_2-T_1}{(\Delta x)^2}=\lambda_2\cdot\frac{T_3-T_2}{(\Delta x)^2}$$

The computed temperature T_2 can be determined as

$$T_2=\frac{\lambda_1 T_1+\lambda_2 T_3}{\lambda_1+\lambda_2}$$

It is notable that the thermal conductivity at point 3 does not influence the expression for T_2.

17.2 TIME DISCRETIZATION

In unsteady problems, the value of the flow variable $\psi(x,t)$ must be specified at the initial time t_o. The spatial discretization of the unsteady equations leads to a system of ordinary differential equations. Equation 17.3 can be expressed as:

$$\frac{\partial\Psi}{\partial t}=L(\Psi) \tag{17.4}$$

These equations can be integrated over a time step according to:

$$\psi^{n+1}-\psi^{n}=\int_{t_n}^{t_{n+1}}L(\psi)\,dt$$

The right-hand side is approximated using information from both the current and previous time levels:

$$\psi^{n+1}=g(\psi^{n+1},\psi^{n},\psi^{n-1},\ldots)$$

The temporal term arising from the transport equation is approximated as:

$$\int_V\frac{\partial(\rho\psi)}{\partial t}dV\approx\frac{\partial(\rho\psi)}{\partial t}\Delta V_p$$

For the finite difference approximation of the time derivative, the Crank-Nicolson method has been employed.

Time discretization plays a critical role when dealing with unsteady problems. Techniques such as explicit, implicit, and Crank-Nicolson methods are employed to discretize the temporal derivatives. The Crank-Nicolson method, for example, is a second-order accurate scheme that blends the benefits of explicit and implicit methods, offering unconditional stability while maintaining good accuracy for reasonably chosen time steps. However, caution must be taken with respect to the Courant-Friedrichs-Lewy (CFL) condition to avoid nonphysical oscillations, especially when large time steps are employed. Stability analysis and time step selection are therefore integral parts of a successful FVM simulation.

17.2.1 CRANK-NICOLSON METHOD

The Crank-Nicolson method is a two level approximation. The unknown variables at the current time step and the known values from the previous time step are required. It is unconditionally stable but gives oscillatory solutions if large time steps are used. However for accuracy the Courant-Friedrich-Lewy number must be kept below one.

$$CFL = \left(\frac{u_i}{\Delta x_i}\right)_{CV-max} \Delta t \tag{17.5}$$

where maximum is taken over all CVs.

$$\begin{aligned}\frac{(\rho\psi_P\delta V_P)^{n+1} - (\rho\psi_P\delta V_P)^n}{\Delta t} &= \beta\Big(\sum_{Nb} A_{Nb}^{n+1}\psi_{Nb}^{n+1} + S_P^{n+1} - A_P^{n+1}\psi_P^{n+1}\Big) \\ &\quad +(1-\beta)\Big(\sum_{Nb} A_{Nb}^n\psi_{Nb}^n + S_P^n - A_P^n\psi_P^n\Big)\end{aligned} \tag{17.6}$$

Here, Δt denotes the size of time step and superscript n+1 & n denotes the time step level i.e., current and previous time steps. The equation is a combination of fully implicit and fully explicit discretizations. If β is 1, the scheme is Euler explicit and if β is zero scheme is Euler implicit. The β of 0.5 is taken.

Example 17.3

Solve a 1-dimensional steady-state heat conduction problem and determine the temperature distribution across a material domain, assuming a cylindrical geometry. The domain is defined by inner and outer radii of $R_A = 0.5\,\text{m}$ and $R_B = 1.0\,\text{m}$, respectively. The material has a thermal conductivity of $\lambda = 25\,\text{W/m·K}$, a heat transfer coefficient at the boundary of $h_{\text{air}} = 100\,\text{W/m}^2 \cdot \text{K}$, and an ambient temperature of $T_{\text{air}} = 298.15\,\text{K}$. The initial temperature at the boundaries is $T_0 = 373.15\,\text{K}$. The domain is discretised into $N = 5$ control volumes, with the spatial step Δx calculated based on the radii. Using an appropriate solver, the program computes the temperature distribution at each node.

Solution
See `Python` code `Ch17-Ex173` for solution of this problem.
This gives the solution

Node 0 at r= 0.5 m: T= 373.15 K
Node 1 at r= 0.625 m: T= 355.41 K
Node 2 at r= 0.75 m: T= 340.91 K
Node 3 at r= 0.875 m: T= 328.65 K

Node 4 at r= 1.0 m: T= 318.03 K

Example 17.4

Solve the one-dimensional, steady advection-diffusion equation

$$\mathsf{V}\frac{\mathrm{d}\phi}{\mathrm{d}x} - D\frac{\mathrm{d}^2\phi}{\mathrm{d}x^2} = 0,$$

with a constant velocity $\mathsf{V} = 0.5\ \mathrm{m\,s^{-1}}$, a molecular diffusivity $D = 0.1\ \mathrm{m^2\,s^{-1}}$ and Dirichlet boundary conditions $\phi(0) = 1$, $\phi(L) = 0$ on a domain of length $L = 1$ m. Introducing $\mathrm{Pe} = \mathsf{V}L/D = 5$, the exact solution is

$$\phi_{\mathrm{exact}}(x) = 1 - \frac{\exp(\mathrm{Pe}\,x/L) - 1}{\exp(\mathrm{Pe}) - 1}.$$

Take the grid and march parameters: $\Delta x = 0.1$ m (11 nodes), an explicit time increment $\Delta t = 0.02$ s that satisfies both diffusive and advective stability limits, and 100 pseudo-time iterations—sufficient for the residual to fall below machine precision on this coarse mesh.
Discretisation of the convective flux through the west (W→P) and east (P→E) faces:

$$\text{Upwind: } \phi_w = \phi_{\mathtt{W}}, \quad \phi_e = \phi_{\mathtt{P}},$$

$$\text{Central: } \phi_w = \tfrac{1}{2}(\phi_{\mathtt{W}} + \phi_{\mathtt{P}}),\ \phi_e = \tfrac{1}{2}(\phi_{\mathtt{P}} + \phi_{\mathtt{E}}),$$

$$\text{QUICK: } \phi_w = \frac{3\phi_{\mathtt{W}} + 6\phi_{\mathtt{P}} - \phi_{\mathtt{E}}}{8},\ \phi_e = \frac{3\phi_{\mathtt{P}} + 6\phi_{\mathtt{E}} - \phi_{\mathtt{EE}}}{8},$$

where indices follow the usual finite-volume stencil.

Solution
See `Python` code `Ch17-Ex174` for solution of this problem.
Figure 17.6 shows the solution of 1-D steady advection-diffusion equation for various schemes at *Pe*=5.

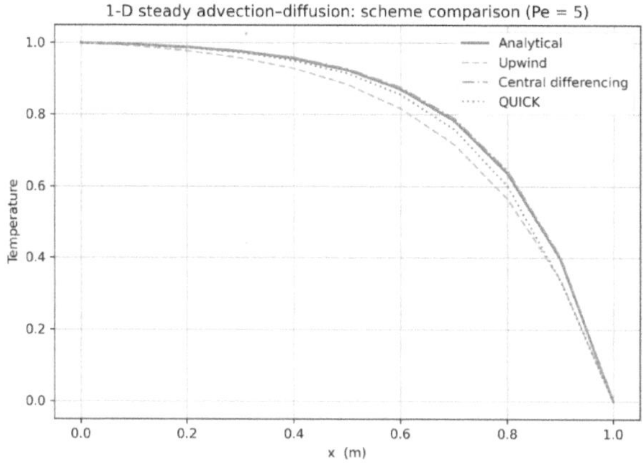

Figure 17.6: **Solution of 1-D steady advection-diffusion equation for various schemes at *Pe*=5.**

PROBLEMS

17P-1 An aluminum rod of length 2 meters is subjected to steady-state one-dimensional heat conduction with internal volumetric heat generation. The temperature at the left boundary is maintained at 600 K, and the right boundary is held at 300 K. The thermal conductivity of aluminum is constant at 237 W/m·K. To evaluate the internal source term, consider a control volume centered at $x = 1.0$m and assume the temperature values at three equally spaced nodes: 600 K at $x = 0.0$m, 450 K at $x = 1.0$m, and 325 K at $x = 2.0$m. Assuming a quadratic temperature distribution due to internal heating, compute the average volumetric heat generation rate $\overline{S}$ within the control volume centered at $x = 1.0$m. Apply the finite volume method by calculating the net heat flux across the faces of the control volume and use this to determine $\overline{S}$. Plot the quadratic temperature profile obtained from interpolation through the three known nodal values.

17P-2 A copper rod with thermal conductivity $k = 400\,\mathrm{W/mK}$ has a length of 1 m and a square cross-sectional area of $5 \times 10^{-3}\,\mathrm{m}^2$. The left boundary of the rod is maintained at $T_A = 150\,\mathrm{K}$ and the right boundary at $T_B = 450\,\mathrm{K}$. Assuming steady-state, one-dimensional heat conduction without internal heat generation, and using a uniform spatial discretization into 5 nodes, apply the finite volume method to compute the steady temperature distribution within the rod. Use central differencing for evaluating the temperature gradients at the faces and assume that fluxes at the ends are approximated using linear profiles between the nodes and the boundary temperatures. Plot the resulting discrete temperature distribution and compare it with the analytical linear temperature profile. Comapre with the general form of the analytical solution

$$T(x) = T_A + \left(\frac{T_B - T_A}{L}\right)x$$

where, $T(x)$ is the temperature at position x, T_A is the left boundary temperature, T_B is the right boundary temperature, and L is the total length of the rod.

17P-3 Consider one-dimensional steady-state heat conduction in a rod composed of two different adjacent materials. The thermal conductivity of the left material (Material 1) is $\lambda_1 = 200\,\text{W/mK}$, and the thermal conductivity of the right material (Material 2) is $\lambda_2 = 50\,\text{W/mK}$. The length of the entire rod is $L = 0.2\,\text{m}$, with the material interface located exactly at the midpoint, $x = 0.1\,\text{m}$. The left end of the rod at $x = 0$ is maintained at a temperature $T(0) = 400\,\text{K}$, and the right end at $x = 0.2\,\text{m}$ is maintained at $T(0.2) = 300\,\text{K}$. Assume the cross-sectional area $A = 1\,\text{m}^2$ for simplicity, and assume no internal heat generation.

i Using the harmonic mean for interpolating thermal conductivity at the material interface, calculate the effective conductivity λ_e at the interface.

ii Determine the heat flux q through the rod.

iii Calculate the temperature distribution in each material region assuming linear profiles within each region.

17P-4 consider a single interior node P with neighbor node W at a fixed temperature $T_W = 350$ K and neighbor node E at temperature T_E. The standard finite-volume equation for the energy balance at P is

$$a_P T_P = a_W T_W + a_E T_E + S_0.$$

Assuming that the conduction-diffusion coefficients are $a_W = 200$, $a_E = -30$, $a_P = 180$, and the constant source term is $S_0 = 5000$.

i Compute T_P for $T_E = 360$ K.

ii Investigate how T_P changes if T_E increases from 360 K to 370 K. Does T_P increase or decrease? Discuss if the result appears physically reasonable.

iii Assume a temperature-dependent (linearized) source term $S(\Phi) = S_0 + S'\Phi$ with $S' = -10$. The finite volume equation becomes $(a_P - S')T_P = a_W T_W + a_E T_E + S_0$. Investigate how $S' < 0$ influences the stability and physical realism of the solution, compared to the case without linearization.

REFERENCES

S. K. Godunov, A difference method for the numerical computation of discontinuous solutions of the equations of hydrodynamics, Matematicheskii Sbornik, 47, 271-306, 1959.

S. V. Patankar, D.B. Spalding, A calculation procedure for heat and momentum transfer in three-dimensional parabolic flows, Int. J. Heat Mass Transfer, 1972.

18 Finite Volume Method II: Pressure-Velocity Coupling in Incompressible Flows

In this chapter, we will discuss the foundational concepts of finite volume in computational fluid dynamics (CFD).

After finishing this chapter, you should be able to:

- Explain the key concepts of the Finite Volume Method (FVM) in computational fluid dynamics (CFD), focusing on its application to the solution of incompressible Navier-Stokes equations.
- Understand the implementation and differences between pressure-velocity coupling algorithms such as SIMPLE, SIMPLEC, PISO, and PIMPLE, and be able to apply them to steady and unsteady flow problems.
- Derive and solve the finite volume discretized equations for momentum, pressure correction, and velocity updates, and analyze the advantages and limitations of different algorithms in CFD simulations.

18.1 PRESSURE-VELOCITY COUPLING

The finite volume method (FVM) was originally formulated by Godunov (1959) for solving compressible flow problems. In these applications, the explicit thermodynamic relationship between pressure and density through an equation of state eliminated the need for specialized pressure-velocity coupling techniques. This made FVM particularly well-suited for shock-capturing and other compressible flow phenomena.

However, the application of FVM to incompressible flows presented significant theoretical and computational challenges. The absence of a direct pressure-density relationship requires implicit coupling between the pressure and velocity fields through the solution of an elliptic Poisson equation. This fundamental difference delayed the adoption of FVM for incompressible flow simulations until the 1970s, when key developments emerged:

☞ The introduction of staggered-grid formulations (Harlow and Welch, 1965) that naturally prevented pressure-velocity decoupling

DOI: 10.1201/9781003663171-18

☞ The development of the SIMPLE algorithm (Patankar and Spalding, 1972) for pressure-velocity coupling

☞ The mathematical framework of projection methods (Chorin, 1968) for enforcing incompressibility

These breakthroughs established FVM as a rigorous approach for incompressible flow simulations and laid the foundation for modern computational fluid dynamics.

The development of the SIMPLE algorithm (Semi-Implicit Method for Pressure-Linked Equations) by Patankar and Spalding in 1972 marked a pivotal advancement in computational fluid dynamics. This innovative approach introduced a robust iterative procedure for handling the fundamental pressure-velocity coupling problem in steady incompressible flows. The algorithm's elegant three-step process—pressure field estimation, correction to satisfy continuity, and iterative refinement—provided engineers with the first practical framework for solving the coupled Navier-Stokes equations.

While differing in implementation details from Chorin's contemporaneous projection method, SIMPLE shared the core philosophy of operator splitting, separating the solution into distinct momentum and pressure correction phases. This conceptual parallel demonstrated how both mathematical and engineering approaches converged on similar strategies during the 1970s to address the central challenge of enforcing mass conservation. The algorithm's success spawned numerous variants (SIMPLER, PISO, etc.), collectively establishing pressure-correction methods as a cornerstone of modern CFD practice.

In computational fluid dynamics literature, one sometimes distinguishes the pressure-correction methods (like SMAC, SIMPLE, PISO) from the pure projection methods. The difference is mostly semantic: pressure-correction schemes use the language of correcting an initial guessed pressure field, while projection schemes talk about projecting onto divergence-free space. In practice, they involve the same Poisson solve and velocity update. For example, the SIMPLE algorithm (Patankar and Spalding, 1972) is a steady-state pressure-correction method that guesses a pressure, corrects it to satisfy continuity, and iterates. PISO (Issa, 1986) is a multi-corrector extension for transient problems. SMAC can be viewed as a single-step pressure correction applied in time-marching: it uses the old pressure as a guess and applies one correction per time step.

In segregated pressure-velocity coupling methods, the momentum equations are first solved for a provisional velocity field, then a pressure correction equation is solved to enforce continuity, and finally velocities are updated. Below are the key discretized finite-volume equations for the SIMPLE, SIMPLEC, PISO, and PIMPLE algorithms. We use standard notation: a_P is the central coefficient in a cell P for the momentum equation, a_F are neighboring coefficients, d_P is a coefficient related to a_P (defined below), u_P is the cell-centered velocity (we write a single component for clarity), p_P is pressure at cell P, and A_f is face area (with subscripts f, N, etc. indicating faces/neighbors).

Table 18.1
Grid type requirements for different pressure-velocity coupling approaches in FV methods

Method	Institution / Reference	Grid Arrangement
SIMPLE (original)	Patankar & Spalding, 1972	Staggered or Collocated
PISO	Issa, 1986	Staggered or Collocated
PIMPLE	Weller et al., 1998	Staggered or Collocated

Solution Methods for Pressure-Velocity Coupling (FV method)
Imperial College London (UK):

* **SIMPLE method**: incompressible flows (Patankar and Spalding, 1972)
* **PISO method** (Pressure-Implicit with Splitting of Operators): incompressible flows (Issa, 1986)

OpenFOAM / broader CFD community:

* **PIMPLE method**: incompressible or weakly compressible flows (Weller et al., 1998)

The SIMPLE algorithm, originating from Imperial College, was initially formulated for staggered finite-volume grids, but the development of Rhie-Chow interpolation later enabled its widespread use on collocated meshes. PISO, also from Imperial College, was first implemented on a collocated finite-volume grid but can be adapted to staggered arrangements. The PIMPLE algorithm in OpenFOAM continues this preference for collocated grids.

Historically, staggered grids were dominant in early schemes (see Table 18.1), while projection methods like Chorin's provided flexibility in grid choice. Later pressure-correction methods, including SIMPLE, PISO, and PIMPLE, tend to favor collocated layouts, though some successful adaptations for staggered grids exist. This progression is reflected in the table's organization, highlighting how grid preferences have evolved alongside numerical methods.

18.1.1 SIMPLE ALGORITHM

1. Momentum (Predictor) Equation: The momentum equation (discretized Navier-Stokes without continuity enforcement) is solved using the pressure field from the previous iteration (or initial guess). For each cell P, for a given velocity component (e.g., u-velocity), we have:

$$a_P u_P^* = \sum_N a_N u_N^* + b_P - (p_P^{(old)} - p_N^{(old)}) A_{PN},$$

where u_P^* is the intermediate (predicted) velocity, b_P represents the contributions of explicit source terms, and the last term is the discretized pressure gradient (with $p^{(old)}$ as the current pressure field) acting across face PN. This yields an intermediate velocity field u^* that generally does not satisfy continuity.

2. Pressure-Correction Equation: A Poisson-like equation for the pressure correction p' is derived from the discrete continuity (mass conservation) requirement. The face-normal velocity is corrected as

$$u_f = u_f^* - \frac{1}{a_P}\left(p'_P - p'_N\right),$$

where f denotes a face between cells P and N, and $\frac{1}{a_P}$ (often denoted d_P) is a coefficient from the momentum equation:

$$a_P\, p'_P - \sum_F a_F\, p'_F = -\sum_{f\in P} \rho\, u_f^* A_f,$$

where the summation on the right is the net mass flux imbalance in cell P (with density ρ). This is the pressure correction equation. (In matrix form it has a singular matrix; a reference pressure or other stabilization is used to set the arbitrary pressure datum.)

3. Pressure and Velocity Update: The pressure field is updated as $p^{new} = p^{old} + p'$ (often with under-relaxation in practice). The cell velocities are then corrected using p' so that continuity is satisfied. For example, for the u-velocity in cell P (with west and east neighbors W and E):

$$u_P^{new} = u_P^* + d_P\left(p'_W - p'_E\right),$$

where $d_P = \frac{1}{a_P}$ for a collocated grid (or a similar coefficient on a staggered grid). This formula comes from the pressure-gradient term of the momentum equation; it projects the velocity field onto a divergence-free field using p'. After this correction, the mass fluxes through cell faces are conservative and the momentum and continuity equations are (approximately) satisfied.

SIMPLE Algorithm

1. Solve Momentum Equation using current pressure:
For each velocity component u^*, solve the discretized momentum equation:

$$a_P u_P^* = \sum_N a_N u_N^* + b_P - (p_P^{(old)} - p_N^{(old)})A_{PN}$$

This gives the intermediate velocity u^* which does not satisfy continuity.
2. Formulate Pressure-Correction Equation:
(i) Relate velocity corrections to pressure corrections (e.g., for face-normal velocity u_f):

$$u'_f = d_f(p'_W - p'_E), \quad v'_f = d_f(p'_S - p'_N)$$

where $d_f = \dfrac{\Delta t}{\rho \Delta x}$ is the correction coefficient.
(ii) Apply continuity (mass conservation) for each cell:

$$\sum_{\text{inflow}} (\mathbf{u}^* + \mathbf{u}') \cdot \mathbf{n}A = \sum_{\text{outflow}} (\mathbf{u}^* + \mathbf{u}') \cdot \mathbf{n}A$$

(iii) Substitute expressions for u' and v' into continuity to derive a linear system for p':

$$a_P p'_P - \sum_F a_F p'_F = -\sum_f \rho u^*_f A_f$$

3. Solve Pressure-Correction System:
Solve:

$$A \cdot \mathbf{p}' = \mathbf{b}$$

where $\mathbf{b}$ is the net mass imbalance from $\mathbf{u}^*$.
4. Correct Velocities and Pressures:
(i) Update velocity at face f:

$$u_f^{new} = u_f^* + d_f(p'_W - p'_E)$$

(ii) Update pressure:

$$p^{new} = p^{old} + \alpha_p p'$$

where α_p is a relaxation factor (usually ≤ 1).
The corrected velocity field approximately satisfies continuity, and pressure is updated for the next iteration.

Example 18.1

A one-dimensional incompressible flow is discretized on a staggered grid, where pressure p is stored at cell centers and velocity u at cell faces. At a certain iteration, the intermediate velocities are given as $u_B^* = 6$ and $u_C^* = 13$, with the boundary velocity specified as $u_A = 9$. The velocity correction is defined by

$$u = u^* + u', \quad \text{with} \quad u' = d(p'_{i-1} - p'_i)$$

where $d = 1.5$. Assuming the flow is incompressible—so that net inflow equals net outflow across each control volume and that the pressure correction datum is fixed at $p'_3 = 0$, calculate the values of pressure corrections p'_1, p'_2, p'_3 and the corrected velocities u_B and u_C.

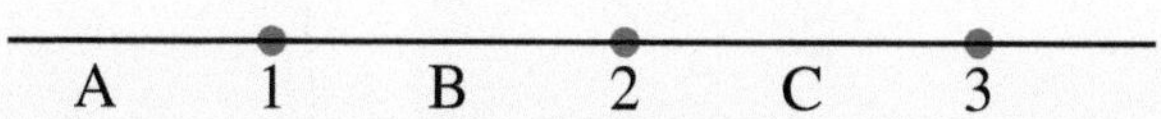

Solution

From the velocity correction formula at B and C:

$$u_B = u_B^* + 1.5(p_1' - p_2') = 6 + 1.5(p_1' - p_2')$$

$$u_C = u_C^* + 1.5(p_2' - p_3') = 13 + 1.5(p_2' - p_3')$$

Apply continuity at control volume between A and B:

$$u_B - u_A = 0 \Rightarrow u_B = u_A = 9$$

$$6 + 1.5(p_1' - p_2') = 9 \Rightarrow p_1' - p_2' = 2$$

Apply continuity at control volume between B and C:

$$u_C - u_B = 0 \Rightarrow u_C = u_B = 9$$

$$13 + 1.5(p_2' - p_3') = 9 \Rightarrow p_2' - p_3' = -\frac{8}{3}$$

With $p_3' = 0$, we get:

$$p_2' = -\frac{8}{3}, \quad p_1' = p_2' + 2 = -\frac{2}{3}$$

Now calculate corrected velocities:

$$u_B = 6 + 1.5\left(-\frac{2}{3} + \frac{8}{3}\right) = 6 + 1.5 \cdot 2 = 9$$

$$u_C = 13 + 1.5\left(-\frac{8}{3} - 0\right) = 13 - 4 = 9$$

hence,

Pressure corrections:

$$p_1' = -\frac{2}{3}, \quad p_2' = -\frac{8}{3}, \quad p_3' = 0$$

Corrected velocities:

$$u_B = u_C = 9$$

This confirms mass conservation is satisfied in both control volumes.

Example 18.2

Consider a 2D uniform staggered grid with the following arrangement:

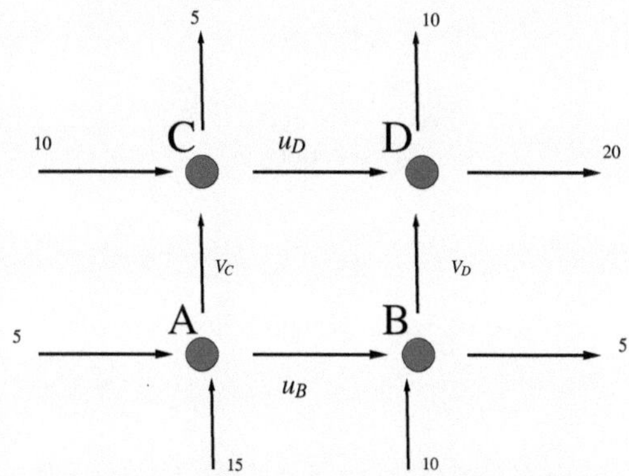

At an intermediate stage of a fluid flow simulation, the interior velocities are computed as:

$$u_B = 9, \quad u_D = 12, \quad v_C = 6, \quad v_D = 7.$$

To enforce continuity, assume pressure corrections p' are:

$$u' = 2(p'_w - p'_e), \quad v' = 3(p'_s - p'_n),$$

where subscripts w, e, s, n denote adjacent pressure nodes (west, east, south, north). If area of cell faces and fluid density are both unity:

i Derive the pressure correction equations by applying mass conservation to control volumes centered on nodes A-D.

ii Solve the resulting linear system for p'_A, p'_B, p'_C, p'_D.

iii Compute the corrected velocities to obtain a mass conserved flow field.

Solution

Step 1: Apply Continuity to Each Control Volume: The continuity condition (mass conservation) must hold for each cell:

Cell A (Bottom Left):

Inflows: u_B (right face), v_C (bottom face)

Outflows: 5 (left face), 15 (top face)

$$(u_B - 5) + (v_C - 15) = 0$$

Velocity corrections are defined by:

$$u_B = u_B^* + u'_B, \quad \text{where } u'_B = 2(p'_A - p'_B)$$

$$v_C = v_C^* + v_C', \quad \text{where } v_C' = 3(p_A' - p_C')$$

We plug in Intermediate Velocities $u_B^* = 9$, and $v_C^* = 6$: so,

$$u_B = 9 + 2(p_A' - p_B')$$

$$v_C = 6 + 3(p_A' - p_C')$$

Now substitute:

$$(9 + 2(p_A' - p_B') - 5) + (6 + 3(p_A' - p_C') - 15) = 0$$

We group constants and corrections, and rearrange:

$$(4 + 2(p_A' - p_B')) + (-9 + 3(p_A' - p_C')) = 0$$

$$(4 - 9) + 2(p_A' - p_B') + 3(p_A' - p_C') = 0$$

$$-5 + 2(p_A' - p_B') + 3(p_A' - p_C') = 0$$

Multiply through or move constant to RHS:

$$2(p_A' - p_B') + 3(p_A' - p_C') = 5$$

$$5p_A' - 2p_B' - 3p_C' = 5$$

$$5p_A' - 2p_B' - 3p_C' = 5 \quad \text{(Equation 1)}$$

Cell B (Bottom Right):
Inflows: 5 (left face), v_D (bottom face)
Outflows: u_B (right face), 10 (top face)

$$(5 - u_B) + (v_D - 10) = 0$$

Substitute corrections:

$$5 - 9 - 2(p_A' - p_B') + 7 + 3(p_B' - p_D') - 10 = 0$$

Simplify:

$$-2p_A' + 5p_B' - 3p_D' = 7 \quad \text{(Equation 2)}$$

Cell C (Top Left): Inflows: u_D (right face), v_C (bottom face) Outflows: 10 (left face), 5 (top face)

$$(u_D - 10) + (v_C - 5) = 0$$

Substitute corrections:

$$12 + 2(p_C' - p_D') - 10 + 6 + 3(p_A' - p_C') - 5 = 0$$

Simplify:

$$-3p'_A + 5p'_C - 2p'_D = -3 \quad \text{(Equation 3)}$$

Cell D (Top Right): Inflows: u_D (left face), v_D (bottom face) Outflows: 20 (right face), 10 (top face)

$$(u_D - 20) + (v_D - 10) = 0$$

Substitute corrections:

$$12 + 2(p'_C - p'_D) - 20 + 7 + 3(p'_B - p'_D) - 10 = 0$$

Simplify:

$$-3p'_B - 2p'_C + 5p'_D = 11 \quad \text{(Equation 4)}$$

Step 2: Assemble the Linear System : The four equations form a matrix system:

$$\begin{bmatrix} 5 & -2 & -3 & 0 \\ -2 & 5 & 0 & -3 \\ -3 & 0 & 5 & -2 \\ 0 & -3 & -2 & 5 \end{bmatrix} \begin{bmatrix} p'_A \\ p'_B \\ p'_C \\ p'_D \end{bmatrix} = \begin{bmatrix} 5 \\ 7 \\ -3 \\ 11 \end{bmatrix}$$

Step 3: Solve Numerically

```
import numpy as np

 Coefficient matrix
A = np.array([
    [5, -2, -3,  0],
    [-2, 5,  0, -3],
    [-3, 0,  5, -2],
    [0, -3, -2,  5]
], dtype=float)

 Right-hand side vector
b = np.array([5, 7, -3, 11], dtype=float)

 Solve for pressure corrections
p_corr = np.linalg.solve(A, b)
pA, pB, pC, pD = p_corr

 Compute corrected velocities
uB = 9 + 2 * (pA - pB)
uD = 12 + 2 * (pC - pD)
vC = 6 + 3 * (pA - pC)
vD = 7 + 3 * (pB - pD)
```

This gives pressure corrections:

$$p'_A = 2.0000, \quad p'_B = 1.0000, \quad p'_C = 1.0000, \quad p'_D = 0.5000$$

and corrected velocities:

$$\begin{aligned} u_B &= 9 + 2(2.0 - 1.0) = 11.0000 \\ u_D &= 12 + 2(1.0 - 0.5) = 13.0000 \\ v_C &= 6 + 3(2.0 - 1.0) = 9.0000 \\ v_D &= 7 + 3(1.0 - 0.5) = 8.5000 \end{aligned}$$

The solution satisfies continuity.

18.1.2 SIMPLEC ALGORITHM

The SIMPLE-Consistent (SIMPLEC) algorithm, developed by Van Doormaal and Raithby (1984), enhances the pressure-velocity coupling by retaining additional neighbor velocity correction terms that are neglected in the original SIMPLE formulation. While the overall algorithmic steps remain the same as SIMPLE, the pressure correction equation is reformulated to improve accuracy and convergence behavior.

Momentum equation: Identical to SIMPLE's predictor step (solve for u^* with old pressure).

Pressure-correction equation: In SIMPLEC, the relation between u' and p' is adjusted. Rather than assuming neighbor velocity corrections u'_N are negligible, SIMPLEC derives a correction factor. The velocity correction for cell P is approximated as:

$$u'_P \approx \frac{d_P}{1 - \sum_F (a_F/a_P)} \left(p'_{nbL} - p'_{nbR}\right),$$

where the denominator $1 - \sum_F (a_F/a_P)$ accounts for the influence of neighboring velocities (nbL and nbR are the cells on either side of P in the given velocity's direction). By contrast, SIMPLE assumed $u'_P \approx d_P(p'_{nbL} - p'_{nbR})$. This modification leads to the use of an adjusted central coefficient, denoted as a'_P, in the pressure correction formulation. In practical implementations, the SIMPLEC pressure correction equation can be expressed as:

$$a'_P \, p'_P - \sum_F a_F \, p'_F = -\sum_{f \in P} \rho \, u^*_f A_f,$$

with $a'_P = a_P - \sum_F a_F$ for each cell. (Note: In steady flows $\sum_F a_F \approx a_P$, so a small pseudo-time term or under-relaxation is added to avoid $a'_P = 0$ singularity. The resulting equation is then solved for the pressure correction p'.

Velocity update: The pressure and velocity are corrected as in SIMPLE, but using the refined coefficient. That is, $p^{new} = p^{old} + p'$ and

$$u_P^{new} = u_P^* + \frac{d_P}{1 - \sum_F (a_F/a_P)} \left(p'_{nbL} - p'_{nbR}\right),$$

so that continuity is satisfied. In SIMPLEC's equations have the same form as SIMPLE's, but with the effective inverse coefficient $d_P/(1-\sum_F a_F/a_P)$ replacing d_P in the pressure gradient terms. This yields a more consistent correction and often converges faster than SIMPLE.

18.1.3 PISO ALGORITHM

The Pressure-Implicit with Splitting of Operators (PISO) algorithm (Issa, 1986) extends SIMPLE for unsteady problems by performing multiple correction (pressure solve) steps *within* each time step, eliminating the need for iterations at each time level. It uses one momentum prediction and two (or more) pressure corrections in sequence to enforce mass conservation more fully.

1. Momentum Predictor: Same as SIMPLE—using the pressure from the previous time level (or iteration) to solve $a_P u_P^* = \sum_N a_N u_N^* + b_P - (p_P^* - p_N^*)A_{PN}$, yielding provisional velocity u^*.

2. First Pressure-Correction: Formulate and solve the pressure correction equation as in SIMPLE, using the divergence of u^* as the source term. This gives a first pressure correction p' (we'll call it p' for the first corrector) satisfying:

$$a_P\, p'_P \;-\; \sum_F a_F\, p'_F \;=\; -\sum_f \rho\, u_f^* A_f ,$$

which has the same structure as the pressure correction equation in the SIMPLE algorithm. Solve for p', then update pressure and velocity:

$$p = p^* + p' ,$$

$$u_P = u_P^* + d_P\,(p'_{nbL} - p'_{nbR}) ,$$

for each velocity component (this corresponds to $u^* + u'$). After this first correction, the velocity field u satisfies the discrete continuity equation (divergence-free) with respect to p.

3. Second Pressure-Correction: PISO introduces a second correction step to account for the fact that after the first correction, the momentum equation may not be fully consistent with the updated pressure (because we did not re-solve momentum after updating p). The momentum equations are therefore re-evaluated using the updated neighbor velocities from the first correction. In deriving the second pressure-correction equation, the term involving the neighbor velocity corrections (which SIMPLE/SIMPLEC would neglect) is now approximated using the known u values. This yields a second pressure correction p''. The form is analogous to the first, but with a different source term that accounts for any remaining mass imbalance due to previously ignored neighbor influences. Symbolically:

$$a_P\, p''_P \;-\; \sum_F a_F\, p''_F \;=\; -\sum_f \rho\, \left(u_f A_f - u_f^* A_f\right) ,$$

i.e., based on the residual flux after the first correction (some formulations derive this differently, but the idea is to correct for the momentum coupling error). This equation is solved to obtain p'', and then the fields are updated again:

$$p^* = p + p'',$$

$$u_P^* = u_P + d_P\left(p''_{nbL} - p''_{nbR}\right),$$

yielding the final corrected pressure p^* and velocity u^* at the new time step. After two corrections, the velocity and pressure satisfy continuity to a higher degree of consistency without additional outer iterations.

(In practice, PISO can include more than two pressure corrections—e.g., nCorrectors—especially for highly non-orthogonal grids, but two is the typical default.)

Each PISO cycle (steps 1-3) advances the solution in time by one time-step without any outer iteration, making it an efficient transient solver. The key equations used (momentum predictor, pressure Poissons, velocity updates) are the same linear systems as in SIMPLE, but PISO applies the pressure-correction twice (or more) per time-step to improve coupling.

PISO Algorithm (Pressure-Implicit with Splitting of Operators)

The PISO algorithm enhances SIMPLE by performing *multiple pressure corrections within each time step*, making it efficient for unsteady flows without requiring outer iterations.

1. **Momentum Predictor Step** (as in SIMPLE):
Solve the discretized momentum equation using pressure from the previous time level:

$$a_P u_P^* = \sum_N a_N u_N^* + b_P - (p_P^* - p_N^*)A_{PN}$$

This gives a provisional velocity field u^* that does not yet satisfy continuity.

2. **First Pressure-Correction Step:**
(i) Solve the pressure correction equation to enforce continuity:

$$a_P p'_P - \sum_F a_F p'_F = -\sum_f \rho\, u_f^* A_f$$

(ii) Update pressure and velocity using p':

$$p = p^* + p', \quad u_P = u_P^* + d_P(p'_W - p'_E)$$

After this step, the velocity field satisfies continuity with respect to the corrected pressure p, but the momentum equation is not yet consistent.

3. **Second Pressure-Correction Step:**
(i) Derive a second pressure correction equation to restore momentum-pressure consistency:

$$a_P p''_P - \sum_F a_F p''_F = -\sum_f \rho\left(u_f - u_f^*\right)A_f$$

(ii) Update pressure and velocity again:

$$p^* = p + p'', \quad u_P^* = u_P + d_P(p''_W - p''_E)$$

4. **Final Result:**
The final velocity u^* and pressure p^* satisfy both continuity and momentum to a higher degree of accuracy—without outer iterations.
Note: More than two pressure corrections can be applied (i.e., multiple PISO loops), especially for complex or non-orthogonal grids.

18.1.4 PIMPLE ALGORITHM

The PIMPLE algorithm is essentially a hybrid of SIMPLE and PISO—often described as SIMPLE + PISO or a *merged* PISO-SIMPLE approach. It is used for transient simulations where larger time-steps are desired. The idea is to perform PISO's multi-correction cycle *within an outer iteration loop* akin to SIMPLE's, until convergence at each time step.

PIMPLE uses the same discretized momentum, pressure-correction, and velocity update equations as PISO. In each outer iteration, it: (a) solves the momentum predictor, (b) solves the pressure-correction equation(s) (with possibly multiple PISO-like correctors), and (c) updates p and u. These steps may be repeated for a few outer iterations (with under-relaxation if necessary) before proceeding to the next time step. Thus, PIMPLE can attain a closer coupled solution each time step, improving stability for high Courant numbers.

Multiple Outer Iterations: In practice, one may perform the momentum solve followed by two or more pressure correction steps (as in the PISO algorithm). If the residuals remain above the convergence threshold, the system coefficients are updated and the cycle is repeated—similar to the iterative approach in SIMPLE. This iterative loop ensures that both the pressure Poisson equation and the momentum equations are adequately resolved at each time step. Although the pressure correction and velocity update formulas are identical to those in PISO, they are applied repeatedly until convergence is achieved. Each outer iteration progressively improves the pressure-velocity coupling, thereby rendering the method effectively implicit in time. This combined strategy permits the use of larger time steps while preserving both numerical stability and solution accuracy.

Thus, SIMPLE provides one pressure correction per iteration (suitable for steady problems), SIMPLEC modifies SIMPLE's correction for faster convergence, PISO applies multiple pressure corrections in a single time-step (no outer iterations for transient cases), and PIMPLE adds outer iterations around PISO's inner loop for extra stability. All methods share the same foundational equations: a discretized momentum equation, a derived pressure (or pressure correction) Poisson equation enforcing $\nabla \cdot \mathbf{u} = 0$, and a velocity update formula to correct the fluxes. Below is the summary of key formulas for each algorithm:

☞ SIMPLE: $a_P u_P^* = \sum_N a_N u_N^* + b_P - (p_P^{old} - p_N^{old}) A_{PN}$; $a_P p'_P - \sum_F a_F p'_F =$

$-\sum_f \rho u_f^* A_f$; $u_P^{new} = u_P^* + d_P(p_W' - p_E')$

☞ SIMPLEC: (Momentum same as SIMPLE); $a_P' p_P' - \sum_F a_F p_F' = -\sum_f \rho u_f^* A_f$, with $a_P' = a_P - \sum_F a_F$; $u_P^{new} = u_P^* + \frac{d_P}{1-\sum_F (a_F/a_P)}(p_W' - p_E')$

☞ PISO (1st corr): (Momentum same); $a_P p_P' - \sum_F a_F p_F' = -\sum_f \rho u_f^* A_f$; $u_P = u_P^* + d_P(p_W' - p_E')$

☞ PISO (2nd corr): $a_P p_P'' - \sum_F a_F p_F'' = -\sum_f \rho (u_f - u_f^*) A_f$; $u_P^* = u_P + d_P(p_W'' - p_E'')$

☞ PIMPLE: Uses PISO equations in an iterative loop (potentially under-relaxed), so equations are the same as PISO's .(repeated until convergence)

Each of these algorithms is a projection method at heart: they solve a pressure (or pressure correction) equation to project the intermediate velocity field onto the divergence-free space, ensuring incompressible continuity is satisfied. The differences lie in how the pressure correction is formulated and how many correction steps or iterations are performed for coupling at each time step.

PIMPLE Algorithm

Inputs: Initial fields $\mathbf{u}^0$, p^0; time step Δt; mesh and boundary data; tolerance ε.
PIMPLE: Outer-Inner Loop Structure for Transient Flows

1. **For each time step** $n \to n+1$:

   a. **Outer Loop (PIMPLE Iteration)**: repeat until convergence:

      i. **Momentum Predictor Step:** Solve

$$a_P u_P^* = \sum_N a_N u_N^* + b_P - (p_P - p_N) A_{PN}$$

      ii. **Pressure Correction Loop (PISO Inner Loop)**:
      Repeat *nCorrectors* times:

         A. **Solve Pressure Correction Equation:**

$$a_P p_P' - \sum_F a_F p_F' = -\sum_f \rho u_f^* A_f$$

         B. **Update Pressure and Velocity:**

$$p \leftarrow p + \alpha_p p', \quad u_f \leftarrow u_f^* + d_f(p_W' - p_E')$$

         C. **Update Residual:** $\mathscr{R} = \|\nabla \cdot \mathbf{u}\|$

      iii. **Check Residual and Under-Relaxed Update:**
      If $\mathscr{R} > \varepsilon$, recompute a_P, b_P, and repeat outer loop.

b. Advance time: $t \leftarrow t + \Delta t$

Notes:
The inner loop (PISO-style corrections) improves continuity satisfaction. The outer loop (SIMPLE-style relaxation) ensures consistent momentum-pressure coupling. PIMPLE allows larger Δt than pure PISO while remaining stable.

18.2 RHIE-CHOW INTERPOLATION: ELIMINATING CHECKERBOARD PRESSURE MODES

A well-known challenge with collocated grids in incompressible fluid dynamics is the potential for spurious decoupling (sometimes called *checkerboard pressure*). Because pressure and velocity are collocated at identical cell centers, certain discretization formulations can fail to provide a strong linkage between pressure gradients and velocity divergence constraints. This can lead to oscillatory, non-physical pressure fields that do not convincingly drive correct velocity distributions. To address such pitfalls, practitioners frequently employ additional interpolation or smoothing steps, for example, Rhie-Chow interpolation in finite volume codes—to better couple pressure and velocity.

Modern numerical techniques and carefully crafted interpolation schemes can largely mitigate the checkerboard problem, and many commercial CFD packages use collocated approaches by default. From a user perspective, it can be more intuitive to think of each computational cell having a single set of variables. Overall, the collocated grid arrangement balances coding simplicity with the need to implement specialized measures to prevent decoupling, making it an established choice for simulating both compressible and incompressible flows.

On collocated FV grids, where both pressure and velocity are stored at cell centers, simple interpolation of velocity to the faces can lead to decoupling between pressure and velocity. This results in nonphysical oscillations in the pressure field, often called the checkerboard problem. In such cases, alternating pressure patterns can exist without influencing the computed mass fluxes, creating a null space in the pressure solution (see Figure 18.1).

To overcome this, Rhie and Chow (1983) introduced a correction that directly incorporates the pressure gradient into the interpolation of face velocities. This technique, also known as momentum-weighted or pressure-weighted interpolation, is fundamental for pressure-based solvers using collocated variable storage.

Stage 1: Linear Interpolation (Baseline Estimate): Let P and E be two adjacent cell centers, and f the face between them. The naïve linear interpolation for face velocity is

$$\overline{U}_f = l_x U_P + (1 - l_x) U_E$$

where l_x is the fractional distance from P to f. This expression gives a provisional, purely interpolated face velocity, which fails to account for pressure-driven effects and can produce spurious solutions.

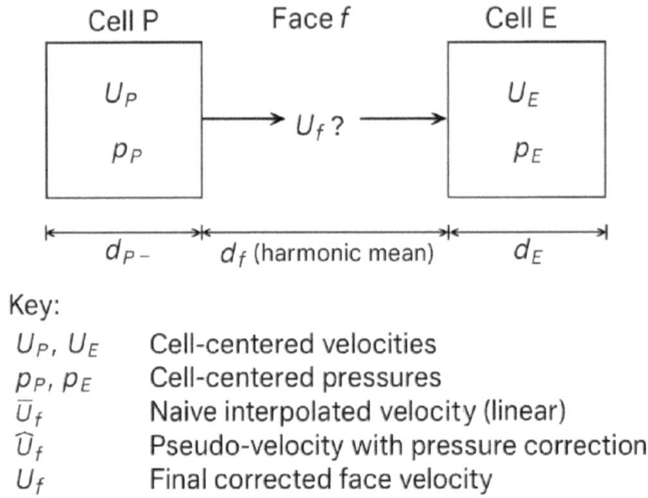

Figure 18.1: **Schematic illustrating pressure and velocity variables for Rhie-Chow interpolation on a collocated grid. Face velocity U_f is corrected using pressure gradients at cell centers and interpolated face values.**

Stage 2: Pressure-Gradient Correction: To correctly represent the influence of pressure, Rhie and Chow introduced a two-step correction. First, define a pseudo-velocity incorporating cell-centered pressure gradients:

$$\widehat{U}_f = l_x U_P + (1 - l_x) U_E + l_x d_P \left.\frac{\partial p}{\partial x_j}\right|_P + (1 - l_x) d_E \left.\frac{\partial p}{\partial x_j}\right|_E$$

where d_P and d_E are the distances from cell centers P and E to the face f.

Next, apply the correction to enforce a consistent pressure drop at the face:

$$U_f = \widehat{U}_f - d_f \left.\frac{\partial p}{\partial x_j}\right|_f$$

Here, d_f is the harmonic mean of d_P and d_E, and $\left.\frac{\partial p}{\partial x_j}\right|_f$ is the interpolated pressure gradient at the face.

Recognizing that $l_x d_P + (1 - l_x) d_E = d_f$, the final expression can be compactly written as:

$$U_f = \overline{U}_f + d_f \left(\left.\frac{\partial p}{\partial x_j}\right|_{\text{coll}} - \left.\frac{\partial p}{\partial x_j}\right|_{\text{interp}} \right)$$

where $\left.\frac{\partial p}{\partial x_j}\right|_{\text{coll}}$ is the central-difference pressure gradient based on P and E, and $\left.\frac{\partial p}{\partial x_j}\right|_{\text{interp}}$ is the interpolated gradient at the face.

If the pressure field is linear, the gradients cancel and $U_f = \overline{U}_f$; otherwise, the correction prevents decoupling and stabilizes the solution.

Rhie-Chow interpolation constructs a local one-dimensional momentum balance at each face. Instead of just averaging cell-centered velocities, the scheme corrects the face velocity based on how it should respond to local pressure variations. The corrected flux is then used in the pressure-correction (Poisson) equation, suppressing spurious numerical modes. This restores the coupling found in staggered grid arrangements but retains the simplicity of the collocated approach. Notably, the scheme does not introduce any artificial damping for smooth, well-resolved pressure fields; it is active only in the presence of irregular or oscillatory pressure patterns. This makes it a targeted and efficient solution for pressure-velocity coupling on collocated grids. Brown etal. (2001), and Sotiropoulos and Yang (2014) have applied this interpolation in finite-difference framework.

Case Study 18A: Lid-Driven Cavity Flow at Re = 100 using SIMPLE Algorithm

Problem Definition

The lid-driven cavity problem is a classical benchmark for validating incompressible flow solvers. The geometry is a square cavity with stationary walls on three sides and a uniformly moving lid on the top. The top wall drives the internal circulation, inducing vortices within the cavity. The flow is expected to be steady, laminar, and incompressible for Reynolds number $Re = 100$. The primary vortex forms near the center with secondary eddies near the corners.

Governing Equations

The flow is governed by the steady incompressible Navier–Stokes equations in two dimensions:

$$\frac{\partial u}{\partial x} + \frac{\partial v}{\partial y} = 0$$

$$\rho\left(u\frac{\partial u}{\partial x} + v\frac{\partial u}{\partial y}\right) = -\frac{\partial p}{\partial x} + \mu\left(\frac{\partial^2 u}{\partial x^2} + \frac{\partial^2 u}{\partial y^2}\right)$$

$$\rho\left(u\frac{\partial v}{\partial x} + v\frac{\partial v}{\partial y}\right) = -\frac{\partial p}{\partial y} + \mu\left(\frac{\partial^2 v}{\partial x^2} + \frac{\partial^2 v}{\partial y^2}\right)$$

The assumptions include steady-state behavior, laminar flow regime, incompressibility, and constant fluid properties.

Numerical Setup

The computational domain is a unit square discretized on a collocated grid with 129×129 control volumes. No-slip conditions are enforced on all walls except the top lid, which moves with unit horizontal velocity of 1 m/s. The SIMPLE algorithm is used for pressure–velocity coupling. The convection terms use upwind differencing, and diffusion is discretized using central differences. A sparse matrix solver is used to compute pressure corrections from the pressure Poisson equation. The time stepping is pseudo-transient with CFL and viscous time step criteria. Velocity residuals are monitored for convergence.

Results and Discussion

The residual history shows monotonic decay, confirming numerical convergence. A convergence threshold of 10^{-6} was used as the stopping criterion. Velocity magnitude contours also capture the expected circulation pattern (see Figure). Figure shows th computed horizontal velocity profile along the vertical centerline was compared against the benchmark data of Ghia et al. (1982). The simulation accurately reproduces the primary vortex and matches the benchmark values closely in both core and near-wall regions.

Lessons Learned

The SIMPLE algorithm proved robust for steady laminar flow, but grid resolution had a strong influence on vortex location and strength (see table). Convergence was sensitive to time step size and boundary condition treatment. Care was required in solving the pressure Poisson equation accurately, especially at high resolution. Collocated grid pressure–velocity decoupling was handled implicitly. The comparison with Ghia et al. validated the solution quality, confirming that the implementation is suitable for further extension to higher Reynolds numbers and more complex geometries.

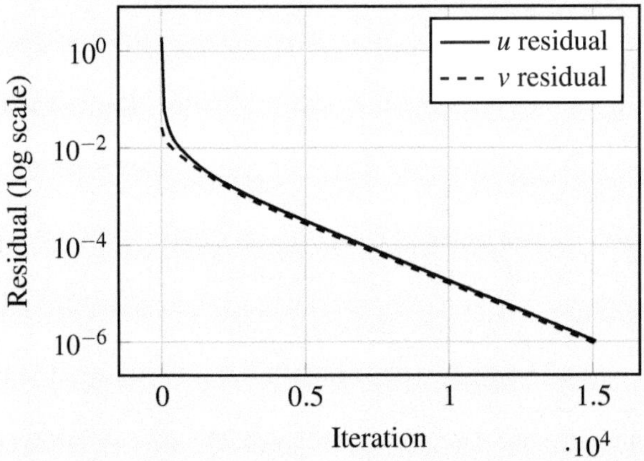

Figure 18.2: **Residual convergence history with 10^{-6} threshold.**

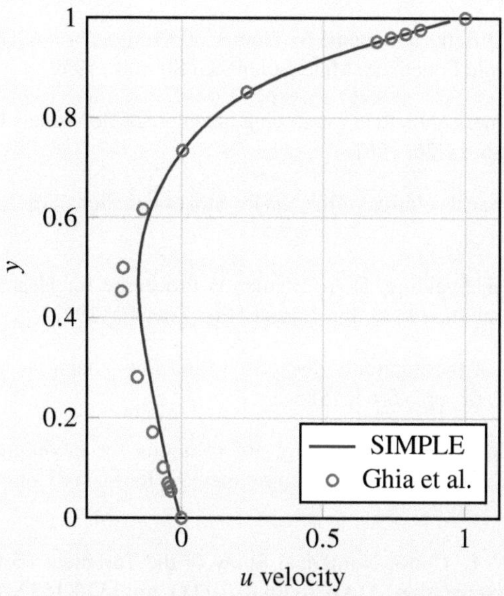

Figure 18.3: **Computed u-velocity along vertical centerline compared with Ghia et al. (1982).**

Runs	Grid x Grid	Primary vortex center x	Primary vortex center y
Ghia et al. 1982	129 × 129	0.6172	0.7344
SIMPLE	129 × 129	0.6302	0.7310

Figure 18.4: **Comparison of primary vortex center location between Ghia et al. (1982) and SIMPLE-based solution at** $Re = 100$.

PROBLEMS

18P-1 Compare the SIMPLE and PISO algorithms in terms of their pressure-velocity coupling loops, detailing how each handles pressure correction and under-relaxation, and analyze their relative strengths for steady versus unsteady simulations.

18P-2 Analyze the origin of checkerboard pressure modes on collocated grids, and describe how Rhie-Chow momentum interpolation modifies the discrete equations to prevent this decoupling.

18P-3 Discuss the impact of the under-relaxation factor on convergence behaviour in segregated solvers, and derive how a change in under-relaxation alters the pressure correction step in the SIMPLE procedure.

REFERENCES

S. K. Godunov, A Difference Scheme for Numerical Computation of Discontinuous Solutions of Hydrodynamic Equations. Matematicheskii Sbornik, 1959.

A. J. Chorin, A numerical method for solving incompressible viscous flow problems, J. Comput. Phys., 2, pp. 12-26, 1967.

A. J. Chorin, Numerical solution of the Navier-Stokes equations, Math. Comput., 22, pp. 745-762, 1968.

S. V. Patankar,D. B. Spalding, D A Calculation Procedure for Heat, Mass and Momentum Transfer in Parabolic Flows. Int. J. Heat Mass Transfer, 1972.

R.I. Issa, Solution of the implicitly discretised fluid flow equations by operator-splitting, J. Comput. Phys., 62, pp. 40-65, 1986.

U. Ghia, K. N. Ghia, C. T. Shin, High-Re solutions for incompressible flow using the Navier–Stokes equations and a multigrid method, Journal of Computational Physics, Vol. 48, pp. 387–411, 1982.

C. M. Rhie and W. L. Chow, Numerical Study of the Turbulent Flow Past an Airfoil with Trailing Edge Separation, AIAA Journal, 21(11), pp. 1525-1532, 1983.

19 Finite Volume Methods III: Flow and Discontinuities

Finite Volume (FV) schemes are widely used for solving compressible hyperbolic flows because of their inherent conservation properties and ability to accurately capture shock waves. These schemes integrate the governing equations over discrete control volumes, ensuring that fluxes across cell boundaries are balanced. In compressible flows, the governing equations, such as the Euler equations, are hyperbolic in nature. FV schemes efficiently handle wave propagation and shock interactions by employing Riemann solvers that compute inter-cell fluxes based on approximate solutions of local Riemann problems. After finishing this chapter, you should be able to:

- Identify and recall the fundamental concepts behind finite-volume schemes and their role in simulating discontinuities in nonlinear hyperbolic partial differential equations.
- Comprehend the formation of finite-volume methods and the underlying principles of discontinuity capturing, including the derivation and application of the Roe approximation with flux correction.
- Implement numerical techniques for compressible flow, utilising reconstruction procedures such as MUSCL and WENO alongside slope limiters to improve accuracy and reduce numerical diffusion.
- Examine the comparative performance of different flux-splitting methods, with particular focus on the advantages and limitations of the Roe scheme when augmented by flux-correction and reconstruction techniques.
- Critically assess alternative approaches, such as the HLL scheme, evaluating their effectiveness, stability, and robustness in handling discontinuities and complex flow features.

The finite volume method (FVM) is a powerful and widely used approach for solving partial differential equations, particularly in fluid dynamics. The method is built upon the integral form of the conservation law, which relates the local rate of change of a quantity within a control volume to the net flux crossing its boundaries. By discretizing the domain into small cells, FVM converts continuous problems into manageable algebraic equations that are easier to solve numerically. This approach naturally ensures conservation properties, making it ideal for problems involving mass, momentum, energy, or other conserved quantities. FVM is well-suited for complex geometries, as it only requires specifying volumes and their faces, allowing flexibility in handling irregular domains. The accuracy of the method depends heavily on the choice of numerical flux functions, which determine how the fluxes are approximated at the cell interfaces.

DOI: 10.1201/9781003663171-19

The method begins by expressing the conservation law in its general form, where the rate of change of a conserved quantity u(x,t) is related to the divergence of the flux function f(u(x,t)). The spatial domain is then divided into discrete cells, and the integral form of the conservation law is applied to each cell. This involves integrating both the time derivative of $u(x,t)$ and the flux term over the cell and the time interval. The flux term is then discretized using numerical approximations, often relying on the values of u at the cell boundaries. The update rule for the cell average is derived by considering contributions from all faces of the cell, leading to a discrete formulation that can be solved iteratively. For multidimensional problems, the method is extended by applying the conservation law over control volumes in higher dimensions, integrating the fluxes across the surface of each volume. The fully discrete update incorporates the fluxes at the cell interfaces, where different numerical schemes may be applied to approximate the flux function. Because the FVM is based on a cell-centered approach, it is particularly advantageous for unstructured grids, where the domain may not conform to a regular grid. The flexibility in defining the numerical flux function allows the method to adapt to a wide range of applications, from simple scalar problems to complex multi-equation systems. The finite volume method provides a robust framework for solving conservation laws in both steady and unsteady regimes, making it a critical tool in computational fluid dynamics and many other engineering fields.

19.1 LINEAR ADVECTION EQUATION

For the linear advection equation:

$$u_t + au_x = 0, \quad a \in \mathbb{R}$$

The solution can be characterized by the characteristics :

$$C : x = x(t) \, \text{with} \, \frac{dx(t)}{dt} = a$$

This implies that along the characteristics, the function $u = u(x(t),t)$ satisfies the equation:

$$\frac{d}{dt}u(x(t),t) = u_t + au_x$$

Therefore, the solution is constant along the characteristic curves :

$$u \, \text{Solution} \quad \Rightarrow \quad u = \text{constant along} \, C$$

A key property of the finite volume method is that it is inherently conservative if the numerical fluxes between adjacent cells are equal and opposite (the flux leaving one cell is the flux entering the neighboring cell). This ensures that the global integral of u over the entire domain (except for boundary inflow/outflow) remains constant, consistent with conservation laws. The finite volume scheme is built on an integral approach that ensures local conservation in each control volume. By dividing the domain into cells (control volumes), integrating the governing PDE over space and time, and applying the divergence theorem, the flux across each cell boundary can be computed and used to update cell averages. Through careful selection of numerical flux functions and reconstruction techniques, one can achieve both stability and accuracy. This step-by-step procedure generalizes naturally from one dimension to multiple dimensions and is widely used in engineering, physics, and other scientific fields where the solution of hyperbolic conservation laws is central.

For simplicity, we use the conservation law:

$$u_t(x,t) + f(u(x,t))_x = 0 \quad x \in \Omega \subset \mathbb{R}, \quad t \in [0,T].$$

Here, $u(x,t)$ is the quantity of interest (for example, density or temperature), $u_t = \frac{\partial u}{\partial t}$, and $f(u)$ is the flux function with respect to x.

We divide the spatial domain Ω into discrete intervals (cells) $x_{i-\frac{1}{2}}, x_{i+\frac{1}{2}}$. For each cell [i], we integrate the conservation law over both space and time:

$$\int_{t^n}^{t^{n+1}} \int_{x_{i-\frac{1}{2}}}^{x_{i+\frac{1}{2}}} u_t(x,t)\,dx\,dt + \int_{t^n}^{t^{n+1}} \int_{x_{i-\frac{1}{2}}}^{x_{i+\frac{1}{2}}} f(u(x,t))_x\,dx\,dt = 0.$$

Let the time interval $[0,T]$ be subdivided into discrete steps $t^n, t^{n+1}, \ldots$. The width of the time step is $\Delta t = t^{n+1} - t^n$. We assume $u(x,t)$ is approximated by a representative value u_i^n in cell [i] at time t^n.

For the flux term, integrate with respect to x:

$$\int_{x_{i-\frac{1}{2}}}^{x_{i+\frac{1}{2}}} f(u(x,t))_x\,dx = f(u(x_{i+\frac{1}{2}},t)) - f(u(x_{i-\frac{1}{2}},t)).$$

Next, sum contributions over time Δt for the term involving u_t. Observing that

$$\int_{x_{i-\frac{1}{2}}}^{x_{i+\frac{1}{2}}} u_t(x,t)\,dx = \frac{d}{dt} \int_{x_{i-\frac{1}{2}}}^{x_{i+\frac{1}{2}}} u(x,t)\,dx,$$

we have:

$$\int_{t^n}^{t^{n+1}} \int_{x_{i-\frac{1}{2}}}^{x_{i+\frac{1}{2}}} u_t(x,t)\,dx\,dt = \left[\int_{x_{i-\frac{1}{2}}}^{x_{i+\frac{1}{2}}} u(x,\tau)\,dx \right]_{\tau=t^n}^{t^{n+1}} \approx \Delta x_i \left(u_i^{n+1} - u_i^n\right),$$

where $\Delta x_i = x_{i+\frac{1}{2}} - x_{i-\frac{1}{2}}$.

Likewise, the flux integral becomes:

$$-\int_{t^n}^{t^{n+1}} \left[f(u(x_{i+\frac{1}{2}},t)) - f(u(x_{i-\frac{1}{2}},t)) \right] dt.$$

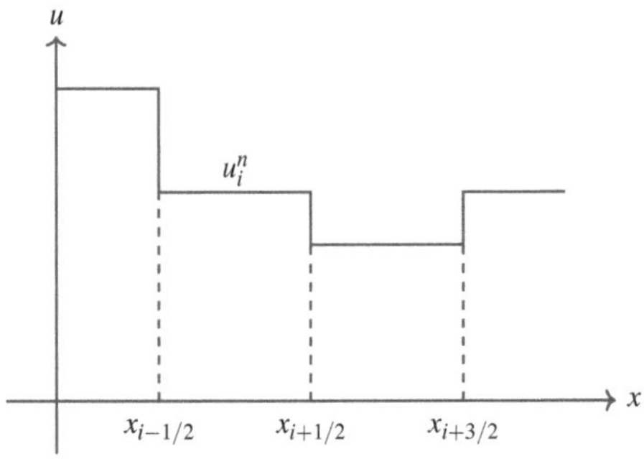

Figure 19.1: **Discretization of the domain with cell-averaged values u_i^n, separated by interfaces $x_{i-1/2}, x_{i+1/2}$.**

Putting these pieces together, the update for the cell average takes the form:

$$u_i^{n+1} = u_i^n - \frac{\Delta t}{\Delta x_i}\left(F^n_{i+\frac{1}{2}} - F^n_{i-\frac{1}{2}}\right)$$

with cell width $\Delta x_i = x_{i+\frac{1}{2}} - x_{i-\frac{1}{2}}$ and numerical fluxes $F^n_{i\pm\frac{1}{2}}$ taken as suitable time-averaged values over the interval $[t^n, t^{n+1}]$. The minus sign appears because the flux term is moved to the right-hand side when solving for u_i^{n+1}. It ensures that flux entering the cell increases the cell average, while flux leaving the cell decreases the cell average.

Figure 19.1 illustrates the finite volume discretization of a one-dimensional domain, where the solution is represented by piecewise-constant cell averages u_i^n at time level n. Each control volume, or cell, spans from $x_{i-1/2}$ to $x_{i+1/2}$, with the average value u_i^n assumed to be constant within the cell. The figure showd these cells and their interfaces, where numerical fluxes are computed. The fluxes capture how information enters or exits each cell, driving the time evolution of the cell averages.

Canonical finite-volume update

For any one-dimensional conservation law

$$\partial_t \mathsf{Q}(x,t) + \partial_x \mathscr{F}(\mathsf{Q}(x,t)) = 0,$$

the finite-volume discretisation updates the cell-average $\overline{\mathsf{Q}}_i^n = \frac{1}{\Delta x_i}\int_{x_{i-\frac{1}{2}}}^{x_{i+\frac{1}{2}}} \mathsf{Q}(x,t^n)\,dx$ according to

$$\boxed{\overline{\mathsf{Q}}_i^{n+1} = \overline{\mathsf{Q}}_i^n - \frac{\Delta t}{\Delta x_i}\left(F^n_{i+\frac{1}{2}} - F^n_{i-\frac{1}{2}}\right)} \tag{19.1}$$

☞ Q is a vector of conserved variables.

For compressible Euler flow in 1-D

$$\mathsf{Q} = \begin{bmatrix} \rho \\ \rho u \\ E \end{bmatrix}, \qquad \mathscr{F}(\mathsf{Q}) = \begin{bmatrix} \rho u \\ \rho u^2 + p \\ u(E+p) \end{bmatrix},$$

and equation 19.1 is equally valid for a single scalar (u) or any other hyperbolic system. ☞ $F^n_{i\pm\frac{1}{2}}$ are numerical fluxes that approximate $\mathscr{F}\left(\mathsf{Q}(x_{i\pm\frac{1}{2}},t)\right)$ during the time slab $[t^n, t^{n+1}]$.

☞ The minus sign ensures strict conservation: flux entering a cell ($F_{i-\frac{1}{2}}$) raises the average, flux leaving the cell ($F_{i+\frac{1}{2}}$) lowers it.

Equation 19.1 is the template into which every flux-based scheme in this chapter (Godunov, Roe, HLL, etc.) fits; what changes from method to method is how the interface fluxes $F_{i\pm\frac{1}{2}}$ are constructed and how left and right states are reconstructed before the flux is evaluated.

19.2 CIR SCHEME—THE FINITE-VOLUME INTERPRETATION

Originally developed as a finite difference method, the Courant-Isaacson-Rees (CIR) scheme also fits naturally into a finite volume framework. At its core, CIR relies on upwind differencing—a concept central to both formulations—making it particularly well-suited for hyperbolic equations like the linear advection equation, where conservation properties are crucial.

In its finite difference form, CIR discretizes spatial derivatives using forward or backward differences, dictated by the sign of the wave speed a. This directional bias ensures physical consistency by respecting wave propagation direction while maintaining stability under the Courant-Friedrichs-Lewy (CFL) condition.

When viewed as a finite volume method, CIR focuses on fluxes at cell interfaces. Here, the solution represents the average value of the conserved quantity within each control volume, and the numerical flux at each interface is determined by upwind values. For example, if $a > 0$, the flux at the left interface is computed as $F_{i-1/2} = au_{i-1}$. While this yields the same algebraic result as the finite difference version for linear problems, the finite volume perspective highlights conservation more explicitly.

In the finite volume approach, CIR updates cell averages via flux differences:

$$u_i^{n+1} = u_i^n - \frac{\Delta t}{\Delta x}\left(F_{i+1/2} - F_{i-1/2}\right),$$

The interface flux $F_{i+1/2}$ can be expressed using the numerical flux function $g(u^L, u^R)$ as follows:

$$F_{i+1/2} = \begin{cases} au_i^n, & \text{if } a > 0 \\ au_{i+1}^n, & \text{if } a < 0 \end{cases}$$

This form highlights that the CIR scheme applies upwinding by selecting the appropriate one-sided value based on the sign of the wave speed a. The stability is ensured via CFL condition $CFL = |a|\Delta t/\Delta x \leq 1$.

A key aspect of the finite volume formulation of the CIR method is its use of numerical fluxes at cell interfaces, computed using values on either side of the interface. The flux function $g(u^L, u^R)$ is determined from the system's eigenstructure, capturing wave propagation directions and speeds. As a first-order upwind method, CIR introduces numerical diffusion to stabilize the solution and suppress oscillations near discontinuities. This built-in dissipation makes it more robust than central schemes, though it limits accuracy to first order.

19.3 GODUNOV'S METHOD

Godunov's method (Godunov, 1959) is a conservative finite-volume scheme that was historically the next leap forward. In this method, the solution $u(x,t_n)$ at each time step is assumed to be piecewise constant per cell (equal to the cell average u_i^n on cell i). At a cell interface, the discontinuity between u_i^n and u_{i+1}^n defines a local Riemann problem—a Cauchy problem with two constant states. Godunov's key idea was to solve each of these Riemann problems exactly (for the given PDE) over one time step Δt and use the self-similar solution to compute fluxes into each cell (Godunov, 1959). Thus, rather than adding an ad hoc diffusive term, Godunov's scheme enforces conservation by using the exact physics of wave propagation at the interface. Godunov, in his construction of the *best* monotone scheme, has used the exact solutions of local Riemann problems to obtain the numerical fluxes (Godunov, 1959). At the end of each step, the solution is averaged back to piecewise constants (cell averages) and the process repeats (Godunov, 1959).

Godunov introduced an ingenious approach for approximating the numerical fluxes in Equation 19.1 as follows:

$$\overline{f}_{i+1/2}^{n} = \frac{1}{\Delta t}\int_{n}^{n+1} f(u(x_{i+1/2},t))\,dt$$

at each cell interface $x_{i+1/2}$. As the cell averages u_i^n are constant within each cell C_i at each time level, Godunov observed that they define a Riemann problem at each cell interface $x_{i+1/2}$:

$$\begin{aligned} &u_t + f(u)_x = 0 \\ &u(x,t_n) = \begin{cases} u_i & \text{if} x < x_{i+1/2} \\ u_{i+1} & \text{if} x > x_{i+1/2} \end{cases} \end{aligned} \tag{19.2}$$

Thus, at every time level, the cell averages define a superposition of Riemann problems of the form in Equation 19.2 at each interface. In Chapter 16, we solved the exact Riemann problem. The solution consists of shock waves, rarefactions, and compound waves. Therefore, the Riemann problem at each time level can be solved explicitly in terms of waves emanating from each interface. Moreover, the solution of each Riemann problem in Equation 19.2 is self-similar. That is, the solution $\overline{u_i}(x,t)$

can be expressed as a function $\overline{u_i}(\xi)$ of a single variable $\xi = \frac{x-x_{i+1/2}}{t-t_n}$:

$$u_i(x,t) = \overline{u_i}\left(\frac{x-x_{i+1/2}}{t-t_n}\right) \tag{19.3}$$

Waves originating from adjacent Riemann problems may intersect over time. However, each wave propagates at a finite speed, with the maximum speed of any wave in a Riemann problem being limited by

$$\max_i |f'(u_i)|$$

By imposing the CFL condition:

$$\max_i |f'(u_i)|\frac{\Delta t}{\Delta x} \leq \frac{1}{2}$$

we ensure that waves from neighboring problems do not interact before reaching the next time level. Assume that this condition holds. From Equation 19.3, the solution remains constant when ξ is constant. In particular, at the cell interface $\xi = 0$, the flux across the interface is given by the constant value:

$$f(u(x_{i+1/2},t)) = f(u_i(0^+)) = f(u_i(0^-))$$

At $\xi = 0$ (which corresponds to the curve $x = x_{i+1/2}$ for $t > t_n$), the function $\overline{u_i}(\xi)$ can either be continuous or discontinuous. If $\overline{u_i}$ is continuous at $\xi = 0$, we have

$$f(u_i(0^+)) = f(u_i(0^-))$$

On the other hand, if $\overline{u_i}$ is discontinuous at $\xi = 0$, a discontinuity arises along the line $x = x_{i+1/2}$ for all $t > t_n$, representing a stationary shock at the cell interface. The discontinuity must satisfy the Rankine-Hugoniot condition:

$$f(u_i(0^+)) - f(u_i(0^-)) = 0$$

For a stationary interface (i.e., wave speed $s = 0$), the Rankine-Hugoniot condition implies continuity of flux across the interface even if the state u is discontinuous. This condition shows that the numerical flux is identical on both sides of $\xi = 0$, so the interface behaves as a stationary discontinuity its propagation speed is zero, even though the conserved variable u itself may remain discontinuous.

This condition ensures that the term $f(u_i(0))$ is well-defined at the interface. We may thus define the edge-centered flux value as:

$$f^n_{i+1/2} := f(u_i(0^+)) = f(u_i(0^-))$$

$$f^n_{i+1/2} := f(u_i(0^+)) = f(u_i(0^-))$$

Consequently, the approximate flux remains constant over time and can be explicitly computed as:

$$\overline{f}^n_{i+1/2} = \frac{1}{\Delta t}\int_n^{n+1} f(u(x_{i+1/2},t))\,dt = f^n_{i+1/2}$$

Substituting this into Equation 19.1 yields the finite volume scheme:

$$u_i^{n+1} = u_i^n - \frac{\Delta t}{\Delta x}\left(F_{i+\frac{1}{2}}^n - F_{i-\frac{1}{2}}^n\right),$$

This represents the standard formulation of a finite volume scheme for conservation laws. The numerical flux f is expressed in terms of the Riemann solution and can be calculated explicitly for scalar conservation laws. The numerical flux at the interface $x_{i+\frac{1}{2}}$ is a function

$$F_{i+\frac{1}{2}}^n = g(u_i^n, u_{i+1}^n)$$

obtained from the exact Riemann solution.

For a scalar conservation law with a convex or concave flux $f(u)$ the Godunov flux is

$$g(u_L, u_R) = \begin{cases} \min\limits_{u \in [u_L, u_R]} f(u), & u_L \leq u_R, \\ \max\limits_{u \in [u_R, u_L]} f(u), & u_L > u_R. \end{cases}$$

If $f(u)$ is strictly monotone (increasing or decreasing) the extremum sits at an end-point, giving the upwind form

$$g(u_L, u_R) = \begin{cases} f(u_L), & u_L < u_R, \\ f(u_R), & u_L > u_R. \end{cases}$$

These expressions complete the first-order Godunov scheme for a scalar conservation law.

The method respects the directional propagation of information (characteristics), preventing unphysical oscillations. For shocks and rarefactions, exact or approximate Riemann solvers are used to compute $u(0)$. This approach inherently handles both linear and nonlinear hyperbolic systems, making it a cornerstone of computational fluid dynamics and shock-capturing schemes. Note that CIR uses a single frozen wave speed, whereas the Godunov flux is obtained from the exact nonlinear Riemann solution that distinguishes shocks, centred rarefactions and sonic points.

A major advantage of Godunov's method is its accurate handling of discontinuities (shocks and contact discontinuities). Because the fluxes are derived from exact Riemann solutions, the scheme captures the correct shock propagation speed and strength by construction. Each inter-cell discontinuity is resolved into waves (shocks, rarefactions, contacts) according to the true solution of the conservation law, satisfying the Rankine-Hugoniot jump conditions exactly over the time step. In contrast, the CIR (or any simpler diffusive scheme) smears out discontinuities over several cells due to numerical viscosity.

Godunov's method was in fact the first conservative scheme that could robustly capture shocks without additional artificial viscosity. It is often described as the first of the shock-capturing schemes in computational fluid dynamics. Harten et al. (1983)

remarked that Godunov's scheme is the best first-order monotone scheme—it introduces the minimum amount of numerical dissipation needed to satisfy entropy conditions and avoid oscillations (Harten et al., 1983).

We now perform a numerical experiment on the inviscid 1D Burgers' equation, using the Godunov finite volume method to compute the solution.

$$u_t + \left(\tfrac{1}{2}u^2\right)_x = 0,$$

Test Case 1 (Shock):

Initial data

$$u(x,0) = \begin{cases} 1, & x < 0, \\ 0, & x > 0. \end{cases}$$

The Rankine-Hugoniot condition yields a shock speed

$$s = \frac{u_L + u_R}{2} = 0.5,$$

so the exact solution is

$$u_{\text{exact}}(x,t) = \begin{cases} 1, & x < 0.5t, \\ 0, & x > 0.5t. \end{cases}$$

The Godunov scheme captures this discontinuity; the slight smearing across a few cells is the expected numerical dissipation (see Figure 19.3 (a)).

Test Case 2 (Rarefaction):

Initial data

$$u(x,0) = \begin{cases} -1, & x < 0, \\ 1, & x \geq 0. \end{cases}$$

Because the left state is smaller than the right, a centred rarefaction fan forms. Its self-similar exact solution reads

$$u_{\text{exact}}(x,t) = \begin{cases} -1, & x/t < -1, \\ x/t, & -1 \leq x/t \leq 1, \\ 1, & x/t > 1. \end{cases}$$

The numerical profile reproduces this linear interpolation smoothly and without spurious oscillations (see Figure 19.3 (b)).

The first-order Godunov method introduces excessive numerical dissipation, which leads to smearing of key flow features such as shock waves, contact discontinuities, and expansion fans. As a result, steep gradients are poorly resolved, and critical structures in compressible flows are artificially diffused. These limitations

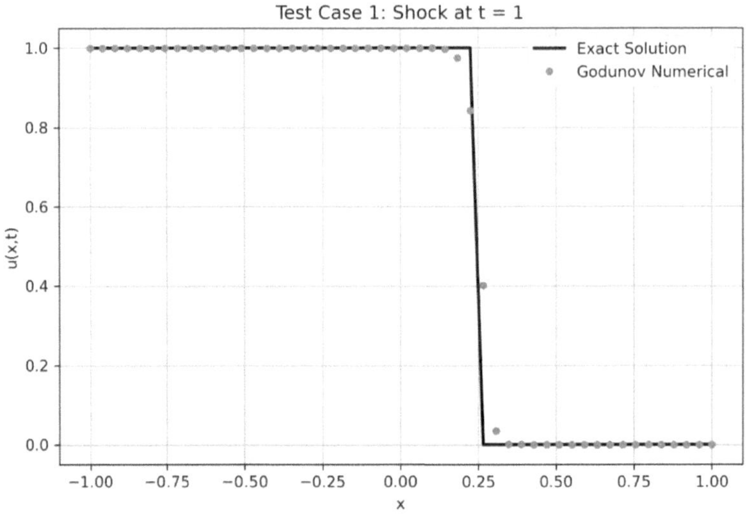

Figure 19.2: (a) Test case 1 - Shock (Godunov scheme using 50 cells).

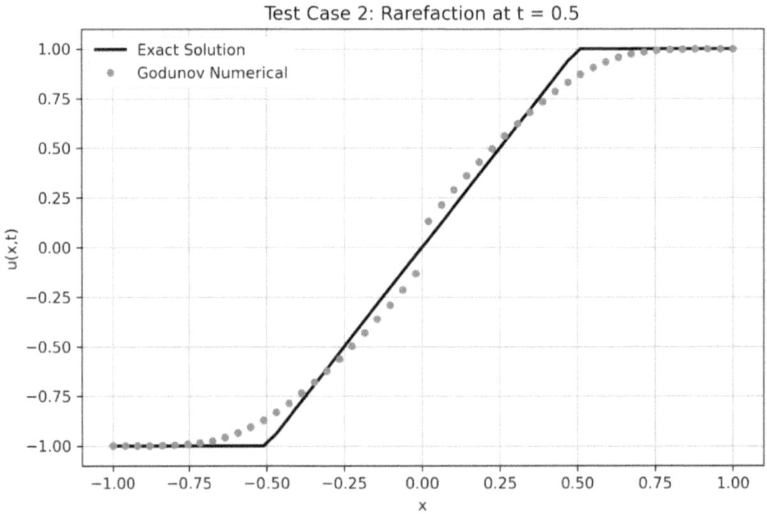

Figure 19.3: (b) Test case 2 - Rarefaction (Godunov scheme using 50 cells).

are especially pronounced in multidimensional problems. Symmetric flows can become asymmetric due to numerical artifacts, and the method may suffer from shock instabilities, especially when strong planar shocks align with the computational grid. A well-known manifestation of this issue is the carbuncle phenomenon (or odd-even decoupling), where an initially steady shock develops unphysical oscillations and loses its coherence. This typically happens in low-diffusion flux schemes when the flow is supersonic and the shock stands on a blunt body. The method also struggles to accurately capture moving or oblique shocks without excessive mesh refinement or additional stabilization techniques.

Another challenge arises in low-Mach-number flows. In such regimes, compressible solvers are expected to behave consistently with incompressible theory, introducing minimal numerical dissipation. However, Godunov-type schemes often overdamp the solution, leading to significant errors in pressure and velocity fields, making them unsuitable for nearly incompressible applications.

Godunov's theorem (1959) proves that any linear and monotonic numerical scheme cannot be more than first-order accurate. In other words, achieving higher-order accuracy requires relaxing monotonicity, which can introduce spurious oscillations near discontinuities such as the Gibbs phenomenon. Thus, the original Godunov method remains fundamentally limited by its design: its inherent numerical diffusion is a trade-off for monotonicity and stability. Improving accuracy, therefore, requires fundamentally new strategies, such as nonlinear reconstructions, slope limiters, or higher-order flux evaluations.

Example 19.1

Given is the linear hyperbolic system:

$$\rho_t + 4\rho_x - p_x = 0$$

$$p_t + 5\rho_x - 2p_x = 0$$

(a) Write it in the form:

$$u_t + Au_x = 0 \quad \text{with} \quad u = \begin{pmatrix} \rho \\ p \end{pmatrix}$$

(b) What is the flux $f(u)$ when the equation is written in the conservation form $u_t + f(u)_x = 0$?

(c) For the exact Riemann solver for linear hyperbolic systems using characteristic decomposition, calculate the numerical flux:

$$g(u^L, u^R) = A^+ u^L + A^- u^R$$

Solution

The linear hyperbolic system is:

$$\rho_t + 4\rho_x - p_x = 0$$

$$p_t + 5\rho_x - 2p_x = 0$$

Rewriting this in matrix form, define:

$$u = \begin{pmatrix} \rho \\ p \end{pmatrix}$$

So the system becomes:

$$u_t + Au_x = 0$$

Part (a) — Writing the system in the form $u_t + Au_x = 0$:

From the coefficients, the matrix A is:

$$A = \begin{pmatrix} 4 & -1 \\ 5 & -2 \end{pmatrix}$$

Part (b) — Finding the flux $f(u)$ when the equation is written in conservation form $u_t + f(u)_x = 0$:

Using the relationship:

$$A = \frac{\partial f}{\partial u}$$

we compute:

$$f(u) = Au = \begin{pmatrix} 4 & -1 \\ 5 & -2 \end{pmatrix} \begin{pmatrix} \rho \\ p \end{pmatrix} = \begin{pmatrix} 4\rho - p \\ 5\rho - 2p \end{pmatrix}$$

This is the flux vector.

Part (c) — Exact Riemann solver for linear hyperbolic systems using characteristic decomposition:

We compute the numerical flux as:

$$g(u^L, u^R) = A^+ u^L + A^- u^R$$

Step 1: Find eigenvalues of A:

$$A = \begin{pmatrix} 4 & -1 \\ 5 & -2 \end{pmatrix}$$

Compute:

$$\det(A - \lambda I) = \begin{vmatrix} 4-\lambda & -1 \\ 5 & -2-\lambda \end{vmatrix} = (4-\lambda)(-2-\lambda) + 5$$

$$= \lambda^2 - 2\lambda - 3 = 0 \Rightarrow \lambda = \frac{2 \pm \sqrt{16}}{2} = \frac{2 \pm 4}{2} \Rightarrow \lambda_1 = -1, \quad \lambda_2 = 3$$

Step 2: Find eigenvectors.
For $\lambda = -1$:

$$(A+I)r_1 = 0 \Rightarrow \begin{pmatrix} 5 & -1 \\ 5 & -1 \end{pmatrix} \begin{pmatrix} x_1 \\ x_2 \end{pmatrix} = 0 \Rightarrow 5x_1 - x_2 = 0 \Rightarrow r_1 = \begin{pmatrix} 1 \\ 5 \end{pmatrix}$$

For $\lambda = 3$:

$$(A-3I)r_2 = 0 \Rightarrow \begin{pmatrix} 1 & -1 \\ 5 & -5 \end{pmatrix} \begin{pmatrix} x_1 \\ x_2 \end{pmatrix} = 0 \Rightarrow x_1 = x_2 \Rightarrow r_2 = \begin{pmatrix} 1 \\ 1 \end{pmatrix}$$

Step 3: Construct matrix R of eigenvectors and its inverse:

$$R = \begin{pmatrix} 1 & 1 \\ 5 & 1 \end{pmatrix}, \qquad \det(R) = 1 \cdot 1 - 5 \cdot 1 = -4$$

$$R^{-1} = \frac{1}{\det(R)} \begin{pmatrix} 1 & -1 \\ -5 & 1 \end{pmatrix} = -\frac{1}{4} \begin{pmatrix} 1 & -1 \\ -5 & 1 \end{pmatrix} = \frac{1}{4} \begin{pmatrix} -1 & 1 \\ 5 & -1 \end{pmatrix}$$

Step 4: Compute A^+ and A^- using diagonalization:

$$\Lambda^+ = \begin{pmatrix} 0 & 0 \\ 0 & 3 \end{pmatrix}, \quad \Lambda^- = \begin{pmatrix} -1 & 0 \\ 0 & 0 \end{pmatrix}$$

$$A^+ = R\Lambda^+ R^{-1} = \begin{pmatrix} 1 & 1 \\ 5 & 1 \end{pmatrix} \begin{pmatrix} 0 & 0 \\ 0 & 3 \end{pmatrix} \frac{1}{4} \begin{pmatrix} -1 & 5 \\ 1 & -1 \end{pmatrix} = \frac{1}{4} \begin{pmatrix} 15 & -3 \\ 15 & -3 \end{pmatrix}$$

$$A^- = R\Lambda^- R^{-1} = \begin{pmatrix} 1 & 1 \\ 5 & 1 \end{pmatrix} \begin{pmatrix} -1 & 0 \\ 0 & 0 \end{pmatrix} \frac{1}{4} \begin{pmatrix} -1 & 5 \\ 1 & -1 \end{pmatrix} = \frac{1}{4} \begin{pmatrix} 1 & -1 \\ 5 & -5 \end{pmatrix}$$

Step 5: Compute the flux:

$$g(u^L, u^R) = A^+ u^L + A^- u^R$$

Substitute:

$$g(u^L, u^R) = \frac{1}{4} \begin{pmatrix} 15 & -3 \\ 15 & -3 \end{pmatrix} \begin{pmatrix} \rho^L \\ p^L \end{pmatrix} + \frac{1}{4} \begin{pmatrix} 1 & -1 \\ 5 & -5 \end{pmatrix} \begin{pmatrix} \rho^R \\ p^R \end{pmatrix}$$

$$= \frac{1}{4} \begin{pmatrix} 15\rho^L - 3p^L + \rho^R - p^R \\ 15\rho^L - 3p^L + 5\rho^R - 5p^R \end{pmatrix}$$

19.4 ROE APPROXIMATION FOR RIEMANN SOLVER

Researchers sought to improve on Godunov's original scheme. There was an explosion of research creating higher-order Godunov-type schemes that preserve the core idea (updating solutions via Riemann solvers) but with modifications to greatly increase accuracy and reduce numerical diffusion. One major advancement came from Roe who proposed approximate Riemann solver.

The Roe approximate Riemann solver represents a significant advancement over the original Godunov scheme, primarily through the use of linearization to simplify the solution of Riemann problems. While the original Godunov method computes the exact solution to the nonlinear Riemann problem at each cell interface—ensuring high fidelity at the cost of computational expense—Roe's solver introduces a locally linearized Jacobian matrix that satisfies the so-called Roe property, allowing the system to be solved approximately but efficiently. This approximation enables decomposition of the flux difference into characteristic waves using the eigen-matrix of the Jacobian, making the method computationally cheaper and more practical for complex systems such as the compressible Euler equations.

In Roe's approximate Riemann solver, we evaluate numerical fluxes at cell boundaries. Instead of explicitly solving the Riemann problem using exact physical states from each side of the boundary, this method employs simplified quantities computed through a procedure known as the Roe average. Roe's method provides accurate approximations with relatively simple computations, resulting in low CPU costs, and has therefore become widely adopted in recent years.

$$\frac{\partial \rho}{\partial t} + \frac{\partial}{\partial x_i}(\rho u_i) = 0,$$

$$\frac{\partial \rho u_j}{\partial t} + \frac{\partial}{\partial x_i}(\rho u_i u_j + p\delta_{ij}) = 0,$$

$$\frac{\partial e}{\partial t} + \frac{\partial}{\partial x_i}[(e+p)u_i] = 0$$

Here, the indices i, j, k run over x, y, z, and repeated indices imply summation (Einstein summation convention). The symbol δ_{ij} denotes the Kronecker delta. These equations can be compactly expressed as:

$$\frac{\partial \mathrm{Q}}{\partial t} + \frac{\partial \mathrm{F}_j}{\partial x_j} = 0 \tag{19.4}$$

For sake of simplification, we consider a one-dimensional case in which physical quantities vary only in the x-direction. In this case, equation 19.4 simplifies by retaining only derivatives in the x-direction, resulting in:

$$\frac{\partial \mathrm{Q}}{\partial t} + \frac{\partial \mathrm{F}}{\partial x} = 0$$

Expressing only the nonzero components explicitly, we have:

$$\mathrm{Q} = \begin{pmatrix} \rho \\ \rho u \\ e \end{pmatrix}, \quad \mathrm{F} = \begin{pmatrix} \rho u \\ \rho u^2 + p \\ (e+p)u \end{pmatrix}$$

where $u = u_x$. Additionally, equation 19.4 can be expressed using the Jacobian matrix $A = \partial \mathrm{F}/\partial \mathrm{Q}$ as:

$$\frac{\partial \mathrm{Q}}{\partial t} + A\frac{\partial \mathrm{Q}}{\partial x} = 0 \tag{19.5}$$

The matrix A can be expressed explicitly as a function of velocity u and enthalpy $H = (e+p)/\rho$:

$$A = A(u,H) = \begin{pmatrix} 0 & 1 & 0 \\ (\gamma-3)\frac{u^2}{2} & (3-\gamma)u & \gamma-1 \\ \{(\gamma-1)\frac{u^2}{2} - H\}u & H-(\gamma-1)u^2 & \gamma u \end{pmatrix} \tag{19.6}$$

where, γ is the specific heat ratio. Furthermore, the matrix A can be expressed in terms of the speed of sound c (where $c^2 = (\gamma-1)(H - u^2/2)$), using three eigenvectors r_1, r_2, r_3:

$$R = (r_1 \quad r_2 \quad r_3) = \begin{pmatrix} 1 & 1 & 1 \\ u-c & u & u+c \\ H-cu & u^2/2 & H+cu \end{pmatrix} \tag{19.7}$$

We diagonalize the matrix as follows:

$$\Lambda = \begin{pmatrix} \lambda_1 & 0 & 0 \\ 0 & \lambda_2 & 0 \\ 0 & 0 & \lambda_3 \end{pmatrix} = R^{-1}AR \tag{19.8}$$

where

$$\lambda_1 = u-c, \quad \lambda_2 = u, \quad \lambda_3 = u+c \tag{19.9}$$

are the eigenvalues of matrix A. Additionally, the inverse of matrix R is expressed as:

$$R^{-1} = \begin{pmatrix} (\eta u^2/2 + u/c)/2 & -(\eta u + 1/c)/2 & \eta/2 \\ -\eta u^2/2 + 1 & \eta u & -\eta \\ (\eta u^2/2 - u/c)/2 & -(\eta u - 1/c)/2 & \eta/2 \end{pmatrix} \tag{19.10}$$

where $\eta = (\gamma-1)/c^2$.

Flux Difference Splitting Method: Using flux differences, equation 19.4 can be written as

$$\mathrm{Q}_k^{n+1} = \mathrm{Q}_k^n - \frac{\Delta t}{\Delta x}\Delta \mathrm{F}_k \tag{19.11}$$

where, n denotes the time step, and k is the cell index. In what follows, we will omit the superscript indicating the n-th time step. Applying a first-order upwind scheme in space using the Flux Difference Splitting (FDS) method, equation 19.11 can be written as:

$$Q_k^{n+1} = Q_k - \frac{\Delta t}{\Delta x}(\Delta F^+_{k-1/2} + \Delta F^-_{k+1/2}) \tag{19.12}$$

where, subscript $k+1/2$, $k-1/2$ indicate that the quantities are evaluated at the cell boundaries. The superscripts $+$ and $-$ denote waves propagating to the right (positive) and left (negative), respectively. If the physical quantities on the left and right sides of the cell boundary $k+1/2$ are denoted as $Q^L_{k+1/2}$ and $Q^R_{k+1/2}$, respectively, then:

$$\Delta F^{\pm}_{k+1/2} = \left\{F(Q^R_{k+1/2}) - F(Q^L_{k+1/2})\right\}^{\pm}$$

Similarly, at the cell boundary $k-1/2$ following is valid:

$$\Delta F_{k+1/2} = F(Q^R_{k+1/2}) - F(Q^L_{k+1/2}) = \Delta F^+_{k+1/2} + \Delta F^-_{k+1/2}$$

Numerical Flux: To accurately satisfy the conservation law given by equation 19.4, the discretized form using a temporal expansion must be expressed as follows:

$$Q_k^{n+1} = Q_k - \frac{\Delta t}{\Delta x}(g_{k+1/2} - g_{k-1/2}) \tag{19.13}$$

where, $g_{k+1/2}$ is called the numerical flux. In the Godunov method, numerical fluxes are computed by exactly solving the shock tube problem (Riemann problem) with left and right states at the cell boundary, $Q^L_{k+1/2}$ and $Q^R_{k+1/2}$, and evaluating the resulting boundary state as $\tilde{Q}$. Thus, the numerical flux is obtained as $g_{k+1/2} = F(\tilde{Q})$. In Roe's method, numerical fluxes are computed based on the Flux Difference Splitting (FDS) method as follows:

$$\boxed{G_{k+1/2} = \frac{1}{2}\left[F(Q^L_{k+1/2}) + F(Q^R_{k+1/2})\right] - \frac{1}{2}|\tilde{A}|\left(Q^R_{k+1/2} - Q^L_{k+1/2}\right)} \tag{19.14}$$

By comparing equations 19.12 and 19.13 using equation 19.14, the flux difference terms become:

$$(G_{k+1/2} - G_{k-1/2}) - (\Delta F^+_{k-1/2} + \Delta F^-_{k+1/2}) = F(Q^L_{k+1/2}) - F(Q^R_{k-1/2})$$

This term arises solely from the use of higher-order spatial discretization. For a first-order spatial scheme, the states at the left and right sides of each cell boundary, $Q^L_{k+1/2}$ and $Q^R_{k+1/2}$, can simply be taken as Q_k and Q_{k+1}, respectively. Consequently, the flux differences become equal because $Q^L_{k+1/2} = Q^R_{k-1/2} = Q_k$.

Monotonicity, is the property of a numerical scheme that ensures the solution preserves the physical trends and does not introduce non-physical oscillations, particularly in the context of sharp gradients or discontinuities. A monotone scheme avoids oscillations that would violate physical principles like positivity or smooth behavior of certain quantities (e.g., pressure, density).

In cases where higher spatial accuracy is implemented using the MUSCL (Monotone Upstream-centered Schemes for Conservation Laws) method, physical quantities on both sides of the cell boundaries are computed by referencing cells other than the immediate neighboring ones $(k, k+1)$, resulting in corrected states for Q_k and Q_{k+1}.

We now focus on the cell boundary $k+1/2$ and, for brevity, omit the subscript $k+1/2$, using instead the simplified notation $\mathsf{Q}_L = \mathsf{Q}^L_{k+1/2}$, $\mathsf{Q}_R = \mathsf{Q}^R_{k+1/2}$.

Roe's method satisfies the following conditions, collectively known as Property U (Uniform validity across discontinuities):

(i) $\mathsf{Q}_L \to \mathsf{Q}, \quad \mathsf{Q}_R \to \mathsf{Q}$ implies $\tilde{A}(\mathsf{Q}_L, \mathsf{Q}_R) \to A(u, H)$,
(ii) For any states $\mathsf{Q}_L, \mathsf{Q}_R$, $F(\mathsf{Q}_R) - F(\mathsf{Q}_L) = \tilde{A}(\mathsf{Q}_L, \mathsf{Q}_R)(\mathsf{Q}_R - \mathsf{Q}_L)$,
(iii) The eigenvectors of $\tilde{A}$ are linearly independent.

Using the matrix $\tilde{A}$ that meets these conditions, the flux difference $\Delta F_{k+1/2}$ can be computed. Thus, the flux difference terms appearing in equation 19.14 are expressed as:

$$\Delta F^+_{k+1/2} - \Delta F^-_{k+1/2} = |\tilde{A}(\mathsf{Q}_L, \mathsf{Q}_R)|(\mathsf{Q}_R - \mathsf{Q}_L) \tag{19.15}$$

In this context, the absolute-value notation signifies that the matrix $\tilde{A}$ has been decomposed into eigenvectors and eigenvalues; only the eigenvalues have their absolute values taken, resulting in a proper upwind flux difference.

The matrix $\tilde{A}(\mathsf{Q}_L, \mathsf{Q}_R)$ can be determined through the following approach. Introduce a homogeneous parameter vector W, representing both the physical state vector Q and the flux vector F:

$$W = \begin{pmatrix} w_1 \\ w_2 \\ w_3 \end{pmatrix} = \begin{pmatrix} \sqrt{\rho} \\ \sqrt{\rho}\, u \\ \sqrt{\rho}\, H \end{pmatrix}$$

Then, using this parameter vector, the state vector Q is represented as:

$$\mathsf{Q} = \begin{pmatrix} w_1^2 \\ w_1 w_2 \\ \frac{1}{\gamma} w_1 w_3 + \frac{\gamma-1}{2\gamma} w_2^2 \end{pmatrix}$$

and the flux vector F is given by:

$$F = \begin{pmatrix} w_1 w_2 \\ \frac{1}{\gamma} w_1 w_3 + \frac{\gamma-1}{2\gamma} w_2^2 \\ \frac{\gamma-1}{\gamma} w_2 w_3 + \frac{\gamma+1}{2\gamma} \frac{w_2^3}{w_1} \end{pmatrix}$$

In fact, all the terms in vectors Q and F can be expressed as second-order functions of the parameter vector W. For arbitrary values p_L, p_R, q_L, q_R, the following identities hold:

$$(p_R+q_R)-(p_L+q_L)=(p_R-p_L)+(q_R-q_L),$$

$$(p_Rq_R)-(p_Lq_L)=\frac{p_L+p_R}{2}(q_R-q_L)+\frac{q_L+q_R}{2}(p_R-p_L),$$

$$\left(\frac{1}{p_R}\right)-\left(\frac{1}{p_L}\right)=-\frac{1}{p_Lp_R}(p_R-p_L).$$

Introducing the notation $\Delta(\cdot)=(\cdot)_R-(\cdot)_L$ for differences, $\overline{(\cdot)}$ for arithmetic averages defined by $\{(\cdot)_L+(\cdot)_R\}/2$, and the geometric average defined by $(\cdot)^*=\sqrt{(\cdot)_L(\cdot)_R}$, these identities can be concisely rewritten as:

$$\Delta(p+q)=\Delta p+\Delta q,$$

$$\Delta(pq)=\overline{p}\Delta q+\overline{q}\Delta p,$$

$$\Delta\left(\frac{1}{p}\right)=-\frac{\Delta p}{p^{*2}}.$$

Utilizing these relations, the difference $\Delta Q=Q_R-Q_L$ can be expressed explicitly as:

$$\Delta Q=\begin{pmatrix}\Delta(w_1^2)\\ \Delta(w_1w_2)\\ \Delta\left(\frac{1}{\gamma}w_1w_3+\frac{\gamma-1}{2\gamma}w_2^2\right)\end{pmatrix}=\begin{pmatrix}2\overline{w}_1\Delta w_1\\ \overline{w}_2\Delta w_1+\overline{w}_1\Delta w_2\\ \frac{1}{\gamma}\overline{w}_3\Delta w_1+\frac{\gamma-1}{\gamma}\overline{w}_2\Delta w_2+\frac{1}{\gamma}\overline{w}_1\Delta w_3\end{pmatrix} \quad (19.16)$$

This expression can be represented compactly by introducing the matrix $\bar{B}$:

$$\bar{B}=\begin{pmatrix}2\overline{w}_1 & 0 & 0\\ \overline{w}_2 & \overline{w}_1 & 0\\ \frac{1}{\gamma}\overline{w}_3 & \frac{\gamma-1}{\gamma}\overline{w}_2 & \frac{1}{\gamma}\overline{w}_1\end{pmatrix}$$

so that:

$$\Delta Q=\bar{B}\Delta W \quad (2.25)$$

Similarly, for the flux difference ΔF, a corresponding matrix $\bar{C}$ can be introduced as follows:

$$\bar{C}=\begin{pmatrix}\overline{w}_2 & \overline{w}_1 & 0\\ \frac{\gamma-1}{\gamma}\overline{w}_3 & \frac{\gamma+1}{\gamma}\overline{w}_2 & \frac{\gamma-1}{\gamma}\overline{w}_1\\ 0 & \overline{w}_3 & \overline{w}_2\end{pmatrix} \quad (2.26)$$

Using this definition, the flux difference $\Delta F = F_R - F_L$ is written as:

$$\Delta F = \bar{C}\Delta W \tag{2.27}$$

Next, using the inverse of matrix $\bar{B}$, denoted as $\bar{B}^{-1}$, and combining equations (2.25) and (2.27), we get:

$$\Delta F = \bar{C}\bar{B}^{-1}\Delta \mathsf{Q}$$

Comparing with the form $\Delta F = \tilde{A}\Delta Q$, we identify:

$$\tilde{A}(\mathsf{Q}_L, \mathsf{Q}_R) = \bar{C}\bar{B}^{-1} \tag{2.29}$$

Upon explicit calculation, this gives:

$$\bar{C}\bar{B}^{-1} = \begin{pmatrix} 0 & 1 & 0 \\ \frac{\gamma-3}{2}\frac{\overline{w}_2^2}{\overline{w}_1^2} & (3-\gamma)\frac{\overline{w}_2}{\overline{w}_1} & \gamma-1 \\ \left(\frac{\gamma-1}{2}\frac{\overline{w}_2^2}{\overline{w}_1^2} - \frac{\overline{w}_3}{\overline{w}_1}\right)\frac{\overline{w}_2}{\overline{w}_1} & \frac{\overline{w}_3}{\overline{w}_1} - (\gamma-1)\frac{\overline{w}_2^2}{\overline{w}_1^2} & \gamma\frac{\overline{w}_2}{\overline{w}_1} \end{pmatrix} \tag{2.30}$$

Now, substituting the arithmetic averages for $\overline{w}_i$ from the parameter vectors:

$$\begin{pmatrix} \overline{w}_1 \\ \overline{w}_2 \\ \overline{w}_3 \end{pmatrix} = \begin{pmatrix} (\sqrt{\rho_L} + \sqrt{\rho_R})/2 \\ (\sqrt{\rho_L}u_L + \sqrt{\rho_R}u_R)/2 \\ (\sqrt{\rho_L}H_L + \sqrt{\rho_R}H_R)/2 \end{pmatrix} \tag{2.31}$$

and defining the Roe-averaged quantities:

$$\tilde{u} = \frac{\overline{w}_2}{\overline{w}_1} = \frac{\sqrt{\rho_L}u_L + \sqrt{\rho_R}u_R}{\sqrt{\rho_L} + \sqrt{\rho_R}}$$
$$\tilde{H} = \frac{\overline{w}_3}{\overline{w}_1} = \frac{\sqrt{\rho_L}H_L + \sqrt{\rho_R}H_R}{\sqrt{\rho_L} + \sqrt{\rho_R}} \tag{2.32-2.33}$$

Then, the matrix becomes:

$$\bar{C}\bar{B}^{-1} = \begin{pmatrix} 0 & 1 & 0 \\ \frac{(\gamma-3)}{2}\tilde{u}^2 & (3-\gamma)\tilde{u} & \gamma-1 \\ \left((\gamma-1)\frac{\tilde{u}^2}{2} - \tilde{H}\right)\tilde{u} & \tilde{H} - (\gamma-1)\tilde{u}^2 & \gamma\tilde{u} \end{pmatrix} \tag{2.34}$$

This is structurally identical to the analytical Jacobian matrix:

$$\tilde{A}(\mathsf{Q}_L, \mathsf{Q}_R) = A(\tilde{u}, \tilde{H}) \tag{2.35}$$

Thus, Roe's method satisfies Property U:
* (i) As $\mathsf{Q}_L, \mathsf{Q}_R \to \mathsf{Q}$, we have $\tilde{u} \to u$, $\tilde{H} \to H$, hence $\tilde{A} \to A$ * (ii) The relation $\tilde{A} = \bar{C}\bar{B}^{-1}$ holds by construction * (iii) Eigenvectors of A are linearly independent

The numerical flux is computed as:

$$G_{k+1/2} = \frac{1}{2}\left\{F(Q^L_{k+1/2}) + F(Q^R_{k+1/2})\right\} - \frac{1}{2}\tilde{R}|\tilde{\Lambda}|\tilde{R}^{-1}(Q^R_{k+1/2} - Q^L_{k+1/2})$$

Where matrices $\tilde{R}, \tilde{\Lambda}, \tilde{R}^{-1}$ are evaluated at Roe-averaged values using equations (2.5), (2.6), and (2.8). The Roe-averaged sound speed is:

$$\tilde{c}^2 = (\gamma - 1)\left(\tilde{H} - \frac{\tilde{u}^2}{2}\right)$$

The absolute value of the eigenvalue matrix is:

$$|\tilde{\Lambda}| = \begin{pmatrix} |\tilde{\lambda}_1| & 0 & 0 \\ 0 & |\tilde{\lambda}_2| & 0 \\ 0 & 0 & |\tilde{\lambda}_3| \end{pmatrix} \tag{2.38}$$

with

$$\tilde{\lambda}_1 = \tilde{u} - \tilde{c}, \quad \tilde{\lambda}_2 = \tilde{u}, \quad \tilde{\lambda}_3 = \tilde{u} + \tilde{c}$$

The Roe-averaged density is determined from:

$$\Delta(\rho u) = \tilde{u}\Delta\rho + \tilde{\rho}\Delta u \quad \Rightarrow \quad \tilde{\rho} = \sqrt{\rho_L \rho_R}$$

Then:

$$\Delta(\rho u^2) = \tilde{u}\Delta(\rho u) + \tilde{\rho}\tilde{u}\Delta u \quad \Rightarrow \quad \widetilde{\rho u} = \tilde{\rho}\tilde{u}$$

and the energy variation becomes:

$$\Delta e = \frac{1}{2}\tilde{u}^2\Delta\rho + \tilde{\rho}\tilde{u}\Delta u + \frac{\Delta p}{\gamma - 1}$$

Note: $\tilde{\rho}$ is not needed inside matrices A, R, Λ, but is required to evaluate flux quantities.

Note that the matrices A, Λ, and those defined previously as R, R^{-1} do not explicitly contain density ρ. Hence, the Roe-averaged density is specifically required only to compute numerical fluxes.

The computed shock-tube problem solution using Roe's method is shown in Figure 19.4. The specific heat ratio is set as $\gamma = 7/5$, with the initial conditions:

$$\rho_L = 1.0, \quad p_L = 1.0, \quad u_L = 0$$

$$\rho_R = 0.1, \quad p_R = 0.125, \quad u_R = 0$$

Figures 19.4 and 19.5 depict numerical solutions obtained using the Roe approximate Riemann solver without MUSCL reconstruction, compared with exact Riemann solutions. In Figure 19.4, the initial conditions are set as $\rho_L = 1.0$, $u_L = 0.0$, $p_L = 1.0$, and $\rho_R = 0.125$, $u_R = 0.0$, $p_R = 0.1$, representing a classical Sod shock

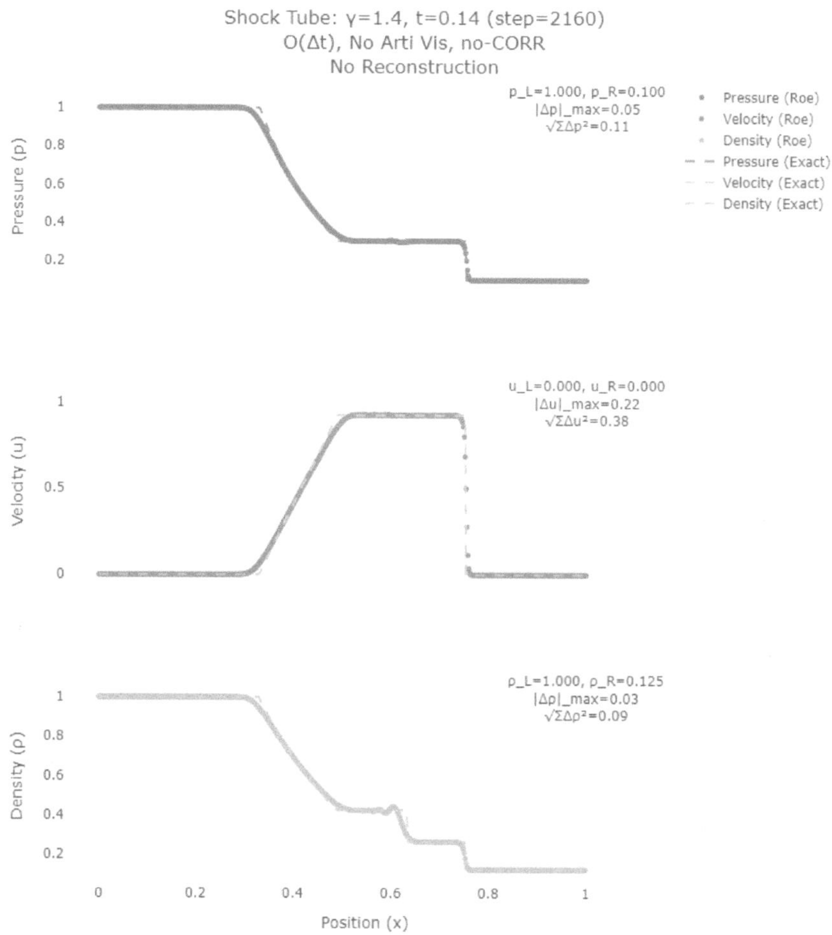

Figure 19.4: **Roe approximation without reconstruction compared with exact Riemann solver solution with 500 cells.**

tube test. With a relatively fine mesh of 500 cells, the Roe scheme closely matches the exact solution. Pressure and density jumps (shock and contact discontinuity) are captured sharply, although slight numerical dissipation is observed at the discontinuities. The velocity profile also shows excellent agreement, capturing accurately both magnitude and location of the velocity discontinuities. Thus, Roe's method demonstrates reliable performance with sufficient resolution, even without higher-order reconstruction.

Figure 19.5 examines a similar case but with modified initial velocity conditions: $\rho_L = 1.0$, $u_L = 0.6$, $p_L = 1.0$, and $\rho_R = 0.125$, $u_R = 0.6$, $p_R = 0.1$. Moreover,

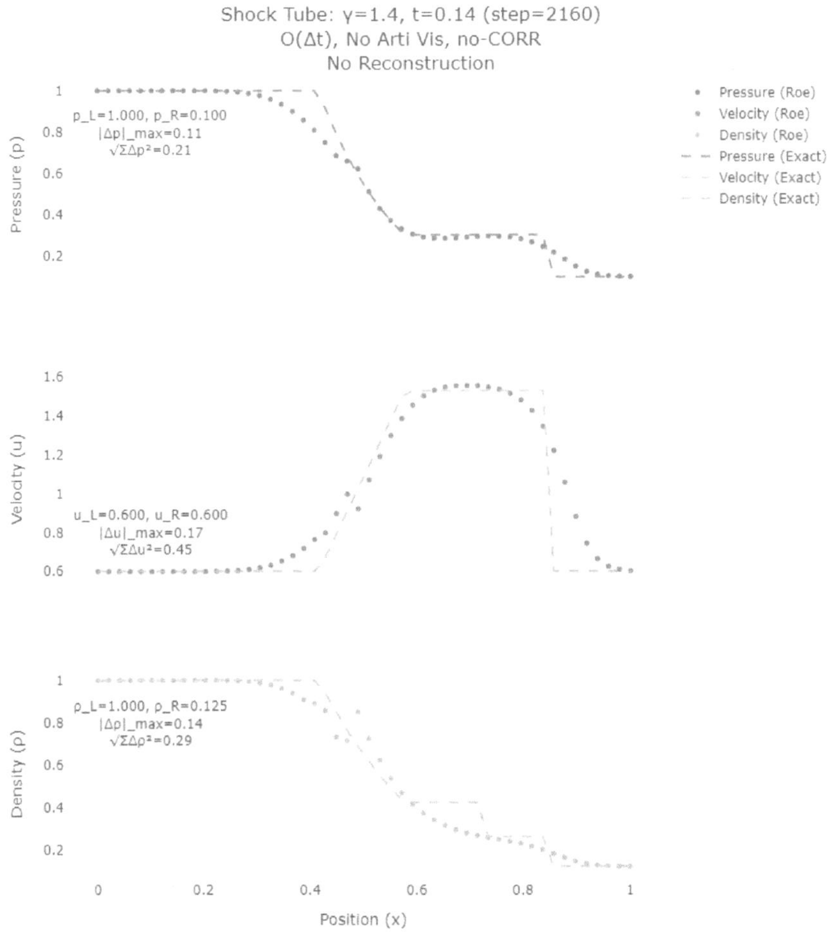

Figure 19.5: **Roe approximation without reconstruction compared with exact Riemann solver solution with 50 cells. No reconstruction applied.** $\rho_L = 1.0, \quad u_L = 0.6, \quad p_L = 1.0; \rho_R = 0.125, \quad u_R = 0.6, \quad p_R = 0.1.$

the grid resolution is significantly coarser (50 cells). Under these conditions, the Roe solver noticeably exhibits numerical diffusion, particularly evident in the density and pressure distributions, where sharp features appear significantly smeared and less accurately represented. The velocity distribution is also diffused but remains qualitatively correct. Such diffusion arises due to the reduced grid resolution and absence of higher-order reconstruction techniques, highlighting the necessity of mesh refinement or employing reconstruction methods such as MUSCL or WENO to improve accuracy.

Example 19.2

Consider system of equations:

$$\frac{\partial u_1}{\partial t}+\frac{\partial u_2}{\partial x}=0,\quad \frac{\partial u_2}{\partial t}+\frac{\partial u_3}{\partial x}=0,\quad \frac{\partial u_3}{\partial t}+4\frac{\partial u_1}{\partial x}-17\frac{\partial u_2}{\partial x}+8\frac{\partial u_3}{\partial x}=0$$

find if the system is hyperbolic.
Write in the form

$$\frac{\partial \mathbf{u}}{\partial t}+A\frac{\partial \mathbf{u}}{\partial x}=0$$

and find A^-, A, A^+.

Solution

Let $\mathbf{U}=\begin{pmatrix}u_1\\u_2\\u_3\end{pmatrix}$ and define the flux $\mathbf{F}(\mathbf{U})=\begin{pmatrix}u_2\\u_3\\4u_1-17u_2+8u_3\end{pmatrix}$.

In the form $\mathbf{U}_t+\mathbf{F}(\mathbf{U})_x=0$, the Jacobian $A=\frac{\partial \mathbf{F}}{\partial \mathbf{U}}$ is

$$A=\begin{pmatrix}0&1&0\\0&0&1\\4&-17&8\end{pmatrix}.$$

To check hyperbolicity, solve $\det(A-\lambda I)=0$. That determinant becomes

$$-\lambda^3+8\lambda^2-17\lambda+4=0.$$

The factorization shows $(\lambda-4)(\lambda^2-4\lambda+1)=0$, so the roots are

$$\lambda_1=4,\quad \lambda_2=2+\sqrt{3},\quad \lambda_3=2-\sqrt{3}.$$

All are real.
Since the system has three real eigenvalues and thus a complete set of real characteristics, it is hyperbolic.
Eigenvector Calculation for Hyperbolic System: We now cast it into the form

$$\frac{\partial \mathbf{u}}{\partial t}+A\frac{\partial \mathbf{u}}{\partial x}=0 \quad \text{where} \quad \mathbf{u}=\begin{pmatrix}u_1\\u_2\\u_3\end{pmatrix},\quad A=\begin{pmatrix}0&1&0\\0&0&1\\4&-17&8\end{pmatrix}.$$

We now find thr right eigenvectors. For each eigenvalue λ_j, solve

$$(A-\lambda_j I)\,\mathbf{r}_j=\mathbf{0}.$$

$\lambda_1 = 4$. Compute

$$A - 4I = \begin{pmatrix} -4 & 1 & 0 \\ 0 & -4 & 1 \\ 4 & -17 & 4 \end{pmatrix}.$$

Let $\mathbf{r}_1 = (r_1, r_2, r_3)^T$. From the first row, $-4r_1 + r_2 = 0$. Choose $r_1 = 1$ so $r_2 = 4$. From the second row, $-4r_2 + r_3 = 0$ implies $r_3 = 16$. The third row is then consistent. Thus

$$\mathbf{r}_1 = \begin{pmatrix} 1 \\ 4 \\ 16 \end{pmatrix}.$$

$\lambda_2 = 2 + \sqrt{3}$. Denote $\alpha = 2 + \sqrt{3}$. Let $\mathbf{r}_2 = (r_1, r_2, r_3)^T$. Choose $r_1 = 1$. From the first row, $-\alpha \cdot 1 + r_2 = 0$ so $r_2 = \alpha$. From the second row, $-\alpha r_2 + r_3 = 0$ so $r_3 = \alpha^2$. Hence

$$\mathbf{r}_2 = \begin{pmatrix} 1 \\ \alpha \\ \alpha^2 \end{pmatrix} = \begin{pmatrix} 1 \\ 2 + \sqrt{3} \\ (2 + \sqrt{3})^2 \end{pmatrix}.$$

$\lambda_3 = 2 - \sqrt{3}$. Denote $\beta = 2 - \sqrt{3}$. By the same steps,

$$\mathbf{r}_3 = \begin{pmatrix} 1 \\ \beta \\ \beta^2 \end{pmatrix} = \begin{pmatrix} 1 \\ 2 - \sqrt{3} \\ (2 - \sqrt{3})^2 \end{pmatrix}.$$

Form the right-eigenvector matrix and diagonal matrix.
Let

$$R = \begin{bmatrix} \mathbf{r}_1 \ \mathbf{r}_2 \ \mathbf{r}_3 \end{bmatrix} = \begin{pmatrix} 1 & 1 & 1 \\ 4 & \alpha & \beta \\ 16 & \alpha^2 & \beta^2 \end{pmatrix},$$

where $\alpha = 2 + \sqrt{3}$, $\beta = 2 - \sqrt{3}$. Then

$$\Lambda = \operatorname{diag}(\lambda_1, \lambda_2, \lambda_3) = \operatorname{diag}(4,\ 2 + \sqrt{3},\ 2 - \sqrt{3}),$$

and

$$A = R\Lambda R^{-1}.$$

Since $\lambda_1, \lambda_2, \lambda_3$ are all positive, define

$$\Lambda^+ = \operatorname{diag}\big(\max\{\lambda_1, 0\}, \max\{\lambda_2, 0\}, \max\{\lambda_3, 0\}\big)$$

$$\Lambda^+ = \operatorname{diag}(4,\ 2 + \sqrt{3},\ 2 - \sqrt{3})$$

$$\Lambda^- = 0.$$

Hence

$$A^+ = R\Lambda^+ R^{-1} = A$$

$$A^- = R\Lambda^- R^{-1} = 0$$

$$|A| = A^+ + (-A^-) = A.$$

19.4.1 FLUX-CORRECTION FOR EXPANSION WAVES

In Roe's approximate Riemann solver, unlike the Godunov method, the expansion waves that form at cell boundaries in the Riemann problem are approximated as simple waves, and the boundary states are computed based on this assumption. Due to this simplification, the solver fails to accurately handle situations where an expansion region remains stationary at the boundary for an extended period. This occurs because Roe's method does not properly account for the entropy increase inherent to expansion waves, resulting in persistent discontinuities at boundaries that should naturally dissipate over time. Such situations typically occur when $(\lambda_l)_L < 0 < (\lambda_l)_R$ for certain eigenvalues $\lambda_l (l = 1,2,3)$.

Roe's approximate Riemann solver is a popular shock-capturing method used in computational fluid dynamics (CFD) to compute fluxes between cells. The basic idea is to treat the discontinuity between two neighboring cells as a small Riemann problem (a wave problem) and solve it approximately. Roe's method does this by linearizing the equations about an averaged state (often called the Roe average) and decomposing the jump in variables into propagating waves. Each wave moves with some characteristic speed (eigenvalue) and represents how a certain disturbance (like a pressure or velocity change) travels across the interface. In simple terms, Roe's solver mimics the behavior of shock and expansion waves at a cell boundary by splitting the difference between left and right states into a set of waves that carry that difference into the neighboring cells. This approach is very good at capturing sharp features like shock waves and contact discontinuities in a stable, crisp way. However, its treatment of expansion waves (smooth rarefaction fans) is only approximate, and this is where some limitations arise.

At each cell interface, Roe's method computes flux by summing contributions of waves moving rightward and leftward. It calculates wave strengths and speeds such that if we were to combine all these linear waves, they would produce the jump from the left state to the right state. This linear combination of waves is an approximation to the true nonlinear solution. For shocks (which are true discontinuities), Roe's linear wave captures the jump well. For contact surfaces, one of Roe's waves corresponds to the contact and is handled exactly. For expansion fans, which in reality consist of a continuous spectrum of characteristic waves (a smooth gradient region), Roe's solver represents them as one or more discrete jumps moving outward. In effect, Roe's method replaces a smooth rarefaction fan with one equivalent jump (or a very sharp transition)—an approximation that can be problematic in certain cases.

Expansion waves (rarefactions) are smooth, spreading waves where flow variables change gradually across a region. A classic example is when a high-pressure region expands into a low-pressure region, creating a fan of characteristics along which information travels. Physically, rarefactions are the opposite of shocks: instead of a sudden jump that increases entropy, a rarefaction has a gradual change and satisfies the physical requirement that no shock forms when expanding. Roe's solver, being a linearized Riemann solver, doesn't inherently know how to make a smooth gradient—it produces discrete waves. In most cases, Roe's method works fine, but at an expansion wave that passes through a sonic point (where flow speed equals the wave speed), a trouble arises.

The specific problematic situation is a transonic expansion fan—for example, part of the flow is subsonic and part is supersonic, with the expansion fan straddling the sonic speed. In this case, one of the characteristic wave speeds changes sign across the interface (say, the left state has a wave speed that is negative, and the right state's corresponding wave speed is positive). Roe's averaging will find an intermediate wave speed (eigenvalue) that might be zero or very small. If Roe's solver assigns essentially zero speed to that wave, it means no spread for the fan—the method then collapses the rarefaction into a single stationary jump (like a shock at that point). In other words, the true expansion fan gets represented as an abrupt jump, which is not physical. As one source explains, Roe's scheme may inadvertently catch an expansion as if it were a shock—at a sonic expansion, the fan is replaced by a single discontinuity (expansion shock) because the scheme has no mechanism to spread out that wave. This happens because the numerical dissipation (artificial viscosity) in Roe's method can vanish at a sonic rarefaction, leaving nothing to smooth out the solution. The result is a non-physical expansion shock.

If this issue is not corrected, the solver produces an unphysical feature in the solution: a sharp jump where a smooth expansion should be. This spurious jump is often called an expansion shock. It violates the expected physics—according to the second law of thermodynamics (the entropy condition), an expansion should not form a shock that decreases entropy. Yet Roe's method, as originally formulated, can produce a stable shock in an expanding flow, essentially violating the entropy condition by allowing a shock where entropy would drop. This numerical artifact is not just a tiny glitch that goes away with a finer grid; it can persist even as we refine the mesh. For example, in a computed shock-tube problem involving a sonic expansion, Roe's basic method (without any fix) clearly showed an expansion shock that did not weaken or disappear with increased resolution. The analytic solution in that case has a smooth fan, so the presence of a shock in the numerical solution is purely an error. This kind of behavior can lead to incorrect predictions (e.g., a jump in pressure or density where there should be a smooth gradient), and it indicates that the solver is not strictly obeying the correct physical solution in those situations.

Users of Roe's solver noticed this problem early on. They would see a small kink or jump—sometimes called a *sonic glitch* when mild, or an outright expansion shock when severe—in regions of expansion. Such a feature is clearly non-physical, so something needed to be done to correct it.

To address the expansion shock issue, researchers introduced what is known as an entropy fix for Roe's Riemann solver. In simple terms, an entropy fix is a tweak to the numerical method to ensure it respects the entropy condition (so that it doesn't allow unphysical solutions like expansion shocks). The main idea is to modify how Roe's method handles waves at the sonic point (where the trouble occurs), usually by adding a bit of artificial dissipation or adjusting the wave speeds. By doing so, the solver is forced to produce a smooth expansion fan instead of a shock.

A simple entropy fix for Roe's method can be implemented in this way:

1 Detect the problematic condition: The code checks if an expansion wave is transonic at the interface—i.e., one state wants to send a wave to the right

while the other wants to send it to the left (a characteristic speed changes sign across the interface). This is the condition that could lead to an expansion shock.

2 Adjust the wave speed: Instead of using Roe's exact averaged eigenvalue (which might be zero or very small), the entropy fix slightly perturbs the wave speeds in that characteristic field. One common approach is to ensure a minimum amount of separation between the left-going and right-going portions of that wave. In practice, this might mean setting the wave speed to a small positive value for the right-going part and a small negative value for the left-going part (even if the Roe average was zero). This effectively introduces a tiny fan in place of a single jump.

3 Add a touch of numerical viscosity: By splitting the wave or giving it a minimum speed, the method adds a bit of smoothing (numerical viscosity) to that wave. This prevents the flux difference from dropping to zero diffusion at the sonic point. In essence, the fix smears out the would-be discontinuity just enough to mimic a physical rarefaction. It is similar to numerical viscosity which is added to the wave field at the sonic expansion to ensure the correct entropy behavior.

4 Result—a smooth expansion: With the adjusted speeds, the solver now produces a gradual change across what used to be an abrupt jump. The expansion shock disappears, replaced by a more correct smooth profile. The solution now satisfies the entropy condition (no unphysical entropy decrease), and matches the expected physical solution better. In the shock-tube example mentioned earlier, applying an entropy fix removed the spurious expansion shock (and the small glitch in another method), yielding a correct smooth expansion fan.

The entropy fix is a small modification that has a big impact on these edge cases. It ensures that Roe's method will not produce a non-physical expansion shock, by effectively respecting the rule that expansions should be smooth. Different entropy fixes have been proposed (by various researchers, often differing in the exact formula for adjusting the wave speeds), but they all share the same goal: cure Roe's expansion wave problem without spoiling its accuracy on shocks.

The Roe-averaged state quantities at each cell interface are computed first:

$$\rho_L, \rho_R, u_L, u_R, e_L, e_R, p_L, p_R, h_L, h_R \quad \text{(from reconstructed states)},$$

$$\sqrt{\rho_L},\ \sqrt{\rho_R}, \quad u_{\text{avg}} = \frac{\sqrt{\rho_L}\,u_L + \sqrt{\rho_R}\,u_R}{\sqrt{\rho_L} + \sqrt{\rho_R}},$$

$$h_{\text{avg}} = \frac{\sqrt{\rho_L}\,h_L + \sqrt{\rho_R}\,h_R}{\sqrt{\rho_L} + \sqrt{\rho_R}},$$

$$c_{\text{avg}} = \sqrt{(\gamma-1)\left(h_{\text{avg}} - \frac{u_{\text{avg}}^2}{2}\right)}.$$

The Roe eigenvalues at the interface are calculated as follows:

$$\lambda_1 = u_{\text{avg}} - c_{\text{avg}}, \quad \lambda_2 = u_{\text{avg}}, \quad \lambda_3 = u_{\text{avg}} + c_{\text{avg}}.$$

Next, entropy-fix methods are applied to modify these eigenvalues to prevent non-physical entropy violations (expansion shocks). The three common entropy fixes are:

(A) Harten Method (1983): introduces a nonlinear smoothing of eigenvalues near zero, using a parabolic blending function

$$|\lambda_l|_{\text{mod}} = \begin{cases} \varepsilon, & |\lambda_l| < \varepsilon \\ |\lambda_l|, & |\lambda_l| \geq \varepsilon \end{cases}$$

where, $\varepsilon = \max\left[0, \lambda_l - (\lambda_l)_L, (\lambda_l)_R - \lambda_l\right]$.

(B) Harten-Hyman Method (1983): is a smoothed version of the original fix. Instead of a piecewise formula, a smooth function is applied to blend between the corrected and uncorrected eigenvalue:

$$\varepsilon = \max\left[0, \lambda_l - (\lambda_l)_L, (\lambda_l)_R - \lambda_l\right],$$

$$|\lambda_l|_{\text{mod}} = \begin{cases} \dfrac{\lambda_l^2 + \varepsilon^2}{2\varepsilon}, & |\lambda_l| < \varepsilon \\ |\lambda_l|, & |\lambda_l| \geq \varepsilon \end{cases}$$

(C) Chakravarthy's Method (1985-1986): is a piecewise correction method that modifies Roe eigenvalues only when necessary (i.e., in transonic regions). It compares the Roe eigenvalue with the left and right reconstructed eigenvalues:

$$|\lambda_l|_{\text{mod}} = \begin{cases} |\lambda_l| + \frac{1}{2}\left[(\lambda_l)_R - (\lambda_l)_L\right], & (\lambda_l)_L < 0 < (\lambda_l)_R \\ |\lambda_l|, & \text{otherwise.} \end{cases}$$

Here, the subscripts L and R represent eigenvalues computed from the left and right reconstructed states, respectively, and the unsubscripted λ_l represents the Roe-averaged eigenvalue. These are eigenvalues of the same field evaluated on either side of the interface.

After correcting the eigenvalues, the wave decomposition coefficients (α-values) are computed using the differences at interfaces:

$$\alpha_2 = \frac{\gamma - 1}{c_{\text{avg}}^2}\left(\Delta\rho(h_{\text{avg}} - u_{\text{avg}}^2) + u_{\text{avg}}\Delta(\rho u) - \Delta e\right),$$

$$\alpha_1 = \frac{1}{2\,c_{\text{avg}}}\left(\Delta\rho(u_{\text{avg}} + c_{\text{avg}}) - \Delta(\rho u) - c_{\text{avg}}\alpha_2\right),$$

$$\alpha_3 = \Delta\rho - (\alpha_1 + \alpha_2),$$

with differences defined as:

$$\Delta\rho = \rho_R - \rho_L, \quad \Delta(\rho u) = (\rho u)_R - (\rho u)_L, \quad \Delta e = e_R - e_L.$$

The corresponding Roe eigenvectors are:

$$r_1 = \begin{pmatrix} 1 \\ u_{\text{avg}} - c_{\text{avg}} \\ h_{\text{avg}} - u_{\text{avg}} c_{\text{avg}} \end{pmatrix}, \quad r_2 = \begin{pmatrix} 1 \\ u_{\text{avg}} \\ \frac{u_{\text{avg}}^2}{2} \end{pmatrix}, \quad r_3 = \begin{pmatrix} 1 \\ u_{\text{avg}} + c_{\text{avg}} \\ h_{\text{avg}} + u_{\text{avg}} c_{\text{avg}} \end{pmatrix}.$$

Finally, the entropy-fix correction term is computed as:

$$\text{Correction} = \frac{1}{2}\left(|\lambda_1|_{\text{mod}}\alpha_1 r_1 + |\lambda_2|_{\text{mod}}\alpha_2 r_2 + |\lambda_3|_{\text{mod}}\alpha_3 r_3\right),$$

and the corrected Roe numerical flux is given by:

$$F_{\text{corr}} = F_{\text{Roe}} - \text{Correction}.$$

This provides robust numerical stability by ensuring entropy conditions are strictly satisfied, eliminating non-physical artifacts such as expansion shocks or stationary discontinuities. These methods, originally introduced by Harten and Hyman (1983) and further refined by Chakravarthy (1986), have been widely adopted for use with Roe's approximate Riemann solver in computational fluid dynamics. Stanley Osher and Syamroy Chakravarthy (1984) developed high-resolution upwind schemes that inherently satisfy the entropy condition. Osher and Chakravarthy avoid the expansion shock problem by constructing an entropy-satisfying flux, using Osher's solver path-integral approach from the outset. Although, not a direct fix to Roe's method, their work is often cited alongside Roe's entropy correction as it provided an alternative solver free of entropy violations.

Figures 19.6, 19.7, and 19.8 display the results obtained using the Roe approximation with flux-correction for the classical Sod shock tube problem ($\gamma = 1.4$). The numerical solutions for pressure, velocity, and density are indicated by discrete markers, whilst the dashed lines represent the exact solution. All three methods performed similarly, which indicates that in the absence of a high-order reconstruction procedure the solution exhibits pronounced numerical diffusion, causing the discontinuities to appear significantly smeared. Nonetheless, the applied correction is effective in preventing the occurrence of expansion shocks.

19.5 STEPS TOWARD HIGHER ACCURACY

By the 1970s, researchers recognized that classical second-order schemes (like Lax-Wendroff and MacCormack) produced non-physical oscillations near shock waves. Sergei Godunov's fundamental 1959 theorem had already proven that any linear scheme of order greater than one will inevitably generate oscillations at discontinuities. To overcome this "Godunov barrier," a new generation of nonlinear shock-capturing methods emerged in the 1970s.

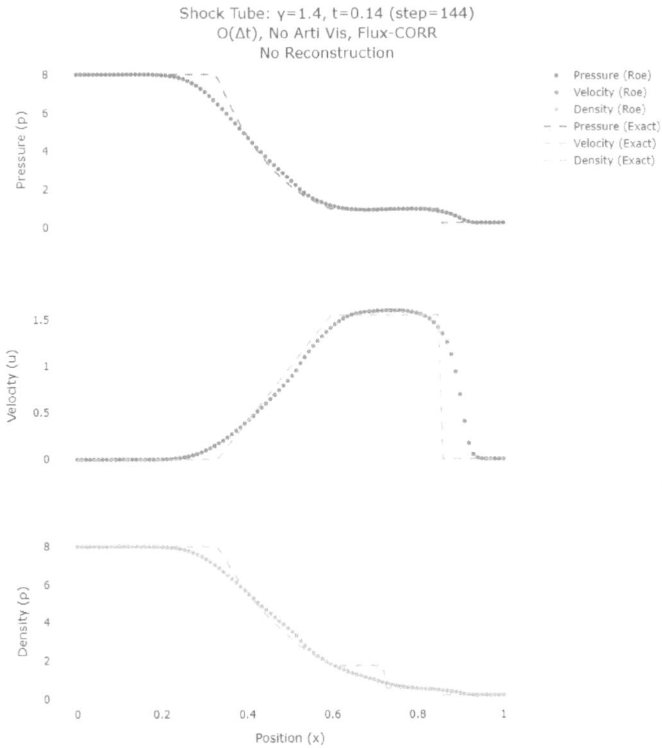

Figure 19.6: **Roe approximation with flux correction by Harten (A).**

MUSCL (Monotonic Upstream-Centered Scheme for Conservation Laws): The MUSCL scheme is an extension of the upwind method, providing higher-order accuracy by reconstructing the solution at cell faces. It uses limiters to prevent oscillations near discontinuities, which ensures that the solution remains monotonic. The MUSCL scheme is widely used for shock-capturing and offers improved accuracy over first-order upwind schemes while maintaining monotonicity.

Numerical methods aiming to achieve higher spatial accuracy in finite-volume methods typically focus on improving the computation of the numerical fluxes at cell boundaries $G_{k+1/2}$. Broadly speaking, two principal strategies exist for achieving higher-order spatial accuracy:

☞ Pre-processing (Interpolation-based) Method: This approach does not directly use values from cells immediately adjacent to a cell boundary (i.e., at k and $k+1$) to evaluate the numerical flux. Instead, additional neighboring cells (such as cells at $k-1, k+2$, etc.) are included to construct higher-order interpolations of the solution within each cell. Once the interpolated distribution within each cell is determined, the numerical fluxes at cell interfaces are

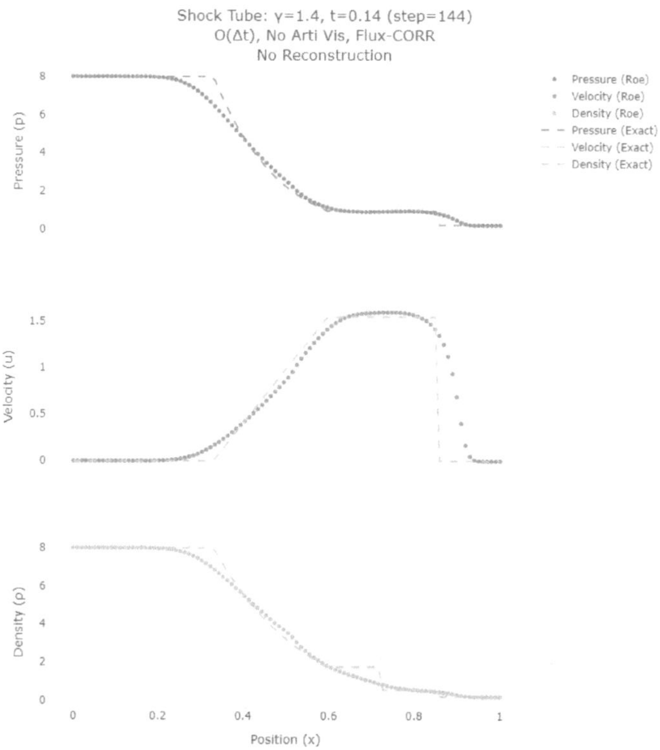

Figure 19.7: **Roe approximation with flux correction by Harten (B).**

computed directly from these interpolated values. A well-known representative of this approach is the MUSCL (Monotone Upstream-centered Schemes for Conservation Laws) method.

☞ Post-processing (Flux Reconstruction) Method: Initially, numerical fluxes at each cell boundary are computed using first-order accuracy, employing only the immediate neighboring cell values ($Q(k)$ and $Q(k+1)$). These first-order flux values, calculated at multiple neighboring cell interfaces (e.g., $k-1/2, k+1/2, k+3/2$), are then used to reconstruct a higher-order spatial distribution within each cell. Interpolations of these reconstructed flux distributions yield the desired higher-order numerical flux at each cell boundary.

The key distinction between these two methods lies in the stage at which high-order accuracy is implemented. Both approaches, however, tend to introduce numerical oscillations near discontinuities if simple interpolation procedures are used. To address this issue, flux-limiting techniques are applied to constrain interpolations, ensuring numerical stability and monotonicity. These limiters are typically governed

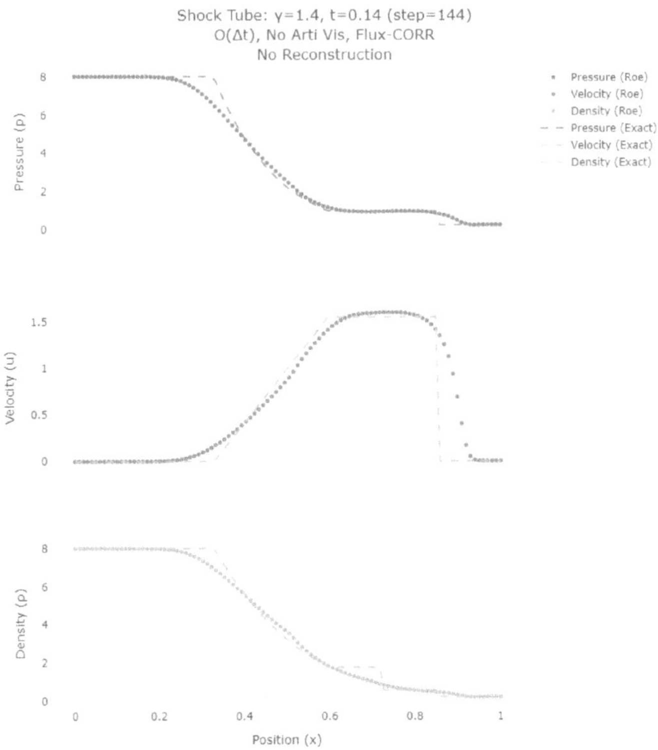

Figure 19.8: **Roe approximation with flux correction by Chakravarthy's method.**

by criteria such as the Total Variation Diminishing (TVD) condition, which effectively prevents unphysical oscillations while maintaining enhanced accuracy.

Reconstruction in the context of hyperbolic equations refers to the process of estimating the values of the conserved variables at cell interfaces from the values at the cell centers. This is crucial because numerical methods for solving hyperbolic equations, such as finite volume methods, require fluxes to be computed at the interfaces between computational cells. These fluxes are typically a function of the conserved variables, so to accurately compute these fluxes, the solution must be reconstructed at the interfaces.

Total Variation Diminishing (TVD) is a property of a numerical scheme or method which refers to a property where the total variation of the solution, which is a measure of the solution's oscillations or roughness, does not increase over time, ensuring that the solution remains smooth and does not develop spurious oscillations. It is a desirable property in numerical schemes for solving hyperbolic partial differential equations, particularly when dealing with shocks or discontinuities. While

TVD is not a limiter by itself, many limiters are designed to ensure that a numerical scheme is TVD. In other words, a scheme that incorporates a limiter is often designed to control oscillations and guarantee that the total variation of the solution does not increase, preserving monotonicity and avoiding unphysical solutions. For instance, the Minmod, Van Leer, and Superbee limiters are TVD limiters. These limiters ensure that the reconstruction of the solution does not result in non-physical oscillations, which could violate the TVD property.

Common types of reconstruction include:

i **MUSCL** (Monotonic Upstream-centered Scheme for Conservation Laws): A high-order scheme where the solution is reconstructed from neighboring cell averages using a slope-limited method to improve accuracy while avoiding oscillations near discontinuities.

ii **WENO** (Weighted Essentially Non-Oscillatory): A high-order reconstruction technique that is designed to handle sharp gradients and discontinuities without producing spurious oscillations.

iii **PPM** (Piecewise Parabolic Method): A higher-order reconstruction technique where the solution is approximated by parabolas in each cell, improving accuracy around discontinuities.

Van Leer introduced on monotonic upstream-centered schemes for conservation laws (MUSCL). In a series of papers (1974-1979), van Leer introduced slope limiters to produce the first high-order accurate scheme that preserved solution monotonicity. His 1979 paper (Toward the Ultimate Conservative Difference Scheme V) achieved second-order accuracy in space while preventing new extrema, effectively making it the first total variation diminishing scheme in spirit. These early innovations were developed heuristically, but they clearly demonstrated that nonlinear hybrid schemes could capture shock waves sharply without Gibbs oscillations, a breakthrough for computational fluid dynamics (CFD) simulations of compressible flows. By the late 1970s, test cases like Sod's shock tube showed that methods such as van Leer's MUSCL could propagate discontinuities over many grid cells with minimal smearing and no unphysical overshoots.

Despite the success of 1970s schemes, a rigorous theory was needed to solidify confidence in these methods. In the early 1980s, Ami Harten provided that theoretical foundation by formalizing the concept of Total Variation Diminishing (TVD). Harten's landmark paper High Resolution Schemes for Hyperbolic Conservation Laws (1983) introduced the TVD property as a criterion for nonlinear scheme stability. A scheme is TVD if the total variation of the solution (a measure of oscillatory content) does not increase in time. Harten proved two key theorems: (1) any monotone scheme (first-order upwind with positive coefficients) is TVD, and (2) any TVD scheme is monotonicity-preserving, meaning it will not create new extrema in the solution. These theorems connected the new high-resolution methods to established monotone schemes, giving a firm mathematical basis for why the nonlinear limiters prevent oscillations. In a follow-up paper (1984), Harten further developed a general class of high-resolution, TV-stable finite difference schemes, providing systematic design principles for achieving second-order accuracy without sacrificing stability. Harten's work brought clarity by showing that the excellent results of Boris, van Leer, and others were no fluke—they could be explained and reproduced within a unified TVD framework. This greatly increased trust in high-resolution shock-capturing schemes and accelerated their adoption in the CFD community.

The year 1984 saw further consolidation of TVD theory with practical algorithms. P. K. Sweby's influential paper High Resolution Schemes Using Flux Limiters for Hyperbolic Conservation Laws recast various existing schemes into a common form and introduced the now-standard terminology of flux limiters. Sweby showed that a second-order accurate scheme can be made TVD by adding a limited antidiffusive flux to a stable first-order scheme. He derived the necessary constraints on these nonlinear limiters to ensure the solution remains oscillation-free (TVD) while achieving high resolution. This analysis yielded the famous Sweby diagram, a graphical tool plotting limiter functions and the region in which they satisfy the TVD conditions. Sweby's compared a class of limiter functions—including van Leer's harmonic limiter, Roe's superbee limiter (an aggressively compressive limiter giving very sharp shocks), Chakravarthy-Osher's limiter, and others—all within the same framework. Notably, Sweby referenced Harten's introduction of the TVD concept and demonstrated that Harten's and others' schemes could be interpreted as specific choices of flux limiter. By presenting these limiters side by side, Sweby helped solidify best practices for designing high-resolution TVD schemes. Practitioners now had a menu of flux limiter options that could be plugged into a base scheme (typically a Godunov-type or upwind scheme) to obtain sharp, non-oscillatory solutions. This development was crucial in moving TVD schemes from theory into widespread use.

Figure 19.9 illustrates the Sweby diagram, which compares several limiter functions against the gradient ratio used in high-resolution finite volume schemes. The van Leer, MC (Monotonized Central), Superbee, and Minmod limiters represent different trade-offs between numerical diffusion and sharpness. Superbee is the most compressive, preserving steep gradients sharply but with potential for oscillations. Minmod is the most diffusive and stable, suitable for robust computations but at the cost of smeared solutions. MC and van Leer offer balanced alternatives that maintain accuracy in smooth regions while suppressing spurious oscillations. The diagram serves as a tool for understanding the design and selection of limiter functions in TVD-based schemes.

The advent of TVD schemes had a transformative impact on compressible flow simulations in CFD. Compared to earlier shock-fitting techniques (which explicitly tracked shock surfaces), TVD shock-capturing methods needed no special treatment of discontinuities the numerical diffusion added by the limiter is concentrated only where it's needed (near steep gradients), and smooth regions of the flow are left virtually unmodified. This nonlinear, solution-dependent dissipation is precisely what allows shocks to be resolved within a few mesh cells without spurious oscillations. For example, in a 1-D shock-tube problem, a TVD scheme can convect a shock or contact discontinuity with minimal numerical smearing, while avoiding the Gibbs phenomena that a classical high-order scheme would produce. As Sweby noted, the

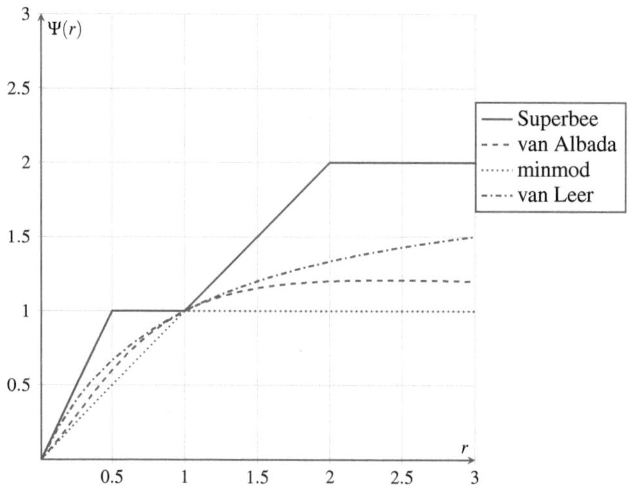

Figure 19.9: **Sweby diagram showing flux limiter functions. The van Leer, MC (monotonized central), van Albada, minmod, and Superbee limiters are plotted for comparison. The upper bound envelope is defined by $\Psi(r) \leq \min(2r, 2)$, and the condition $\Psi(1) = 1$, ensures second-order accuracy at smooth extrema.**

purpose of flux limiting is to achieve high resolution without the spurious oscillations associated with classical second-order schemes. TVD schemes like Harten's and Sweby's became known for capturing sharp shock profiles and preserving solution monotonicity even in extreme conditions (high Mach number shocks, blast waves, etc.).

By the mid-1980s, these methods were being integrated into aerospace and mechanical engineering CFD codes for transonic and supersonic flow simulations, where shock accuracy is paramount. Importantly, they also enforce the correct entropy conditions for shocks—for instance, Harten and others ensured that TVD schemes comply with the physical requirement that entropy increases across a shock. (Osher and Chakravarthy introduced an additional artificial compression step to guarantee entropy consistency in their high-resolution upwind scheme.)

It is worth noting that truly multi-dimensional TVD schemes faced some theoretical limitations. In 1985, Goodman and LeVeque proved that any genuinely two-dimensional scheme that is TVD in all directions cannot be higher than first order accurate on multi-dimensional problems. In practice, this meant that multi-D extensions of high-order TVD schemes had to be applied directionally (using dimensional splitting or limiters on each coordinate sweep). Despite such constraints, TVD methods remained effective and stable in multi-dimensional simulations, and researchers devised workarounds (like using TVD in one-dimension as a design guide for multi-D algorithms).

Following Harten and Sweby's foundational work, the late 1980s and 1990s brought a proliferation of algorithmic innovations building on the TVD concept.

Many researchers sought to attain even higher order accuracy without sacrificing the non-oscillatory property. Harten himself, along with colleagues (Osher, Engquist, Chakravarthy), developed the class of Essentially Non-Oscillatory (ENO) schemes in 1987. ENO schemes relaxed the strict TVD condition (which tends to limit accuracy to second order) in favor of a slightly weaker condition (essentially non-oscillatory), allowing third- or higher-order accuracy while still avoiding spurious oscillations. The ENO approach chooses the locally smoothest stencil dynamically to prevent oscillations, rather than limiting slopes in a fixed manner. This was followed in the 1990s by Weighted ENO (WENO) schemes (Liu, Osher, and Chan, 1994) that improved accuracy and robustness by adaptively blending multiple stencils. These high-order schemes were a direct outgrowth of the TVD revolution: they retained the same fundamental goal of high resolution and stability, but with more sophisticated logic to achieve up to 5th-order accuracy in smooth regions. By the end of the 1990s, WENO schemes and other high-order shock-capturing methods (e.g., the Piecewise Parabolic Method by Colella and Woodward, 1984) had become popular in research-oriented CFD for complex flows, although the simpler second-order TVD flux-limiter schemes remained the workhorse in many production codes due to their reliability.

Colella and Woodward (1984) developed the Piecewise Parabolic Method (PPM), extending the idea of limited reconstruction to a third-order piecewise-parabolic profile. PPM introduced nonlinear limiters on the quadratic interpolant to preserve monotonicity and achieved more precise resolution of contact discontinuities and shock fronts. Although more complex, PPM in 1D still adhered to the MUSCL spirit: using upstream-biased polynomial extrapolation constrained by a limiter. Toward the late 1990s, methods were proposed to push accuracy while still preserving monotonicity in a stricter sense. Notably, Suresh and Huynh (1997) introduced a class of Monotonicity-Preserving (MP) schemes that modify the limiter criteria to allow third-order accuracy in smooth regions yet still prevent any new extrema near steep gradients. The MP limiter was a refined algorithmic improvement that addressed the TVD order barrier, ensuring that even with higher-order polynomials the solution would remain free of wiggles. By the year 2000, MUSCL-type schemes and their progeny had become central to solving 1D hyperbolic conservation laws. These methods enabled reliable capture of shock waves, contact discontinuities, and rarefactions in one dimension with minimal numerical diffusion and no spurious oscillations. The algorithmic innovations—from basic slope limiters to advanced ENO/WENO—greatly enhanced the fidelity of simulations for gas dynamics, traffic flow, shallow water equations, and other 1D conservation law models. In summary, the period 1970-2000 saw a progression from the first rudimentary monotonic upwind differencing into a sophisticated family of high-resolution schemes that are second-order (or higher) accurate, use limiters or nonlinear reconstructions for stability, and preserve the true physics of 1D hyperbolic systems.

Reconstruction creates a higher-order approximation of the solution at cell interfaces, essential for accurately computing fluxes. Limiters control the reconstruction's behavior to avoid non-physical oscillations near discontinuities, ensuring stability and physical realism in the numerical solution.

Table 19.1
Spatial order of accuracy for various schemes

Method	Order of Accuracy	Notes
Godunov	1	Exact or approximate Riemann solver with piecewise constant reconstruction
Roe approximation (no reconstruction)	1	First-order, Roe flux without higher-order interface states
Roe approximation + MUSCL	2	Second-order, linear reconstruction with limiter
Roe approximation + PPM	3	Third order in smooth regions else 2nd order
Roe approximation + WENO (e.g., WENO5)	5	Fifth-order, nonlinear adaptive stencil reconstruction

19.5.1 RIEMANN SOLVER WITH MUSCL RECONSTRUCTION TECHNIQUE

The Monotonic Upstream-centered Schemes for Conservation Laws (MUSCL) method is a second-order accurate spatial reconstruction technique widely used for solving hyperbolic conservation laws. Originally introduced by van Leer, MUSCL enhances the first-order Godunov scheme by enabling piecewise linear reconstruction of solution variables within each control volume. This allows for higher-order accuracy in smooth regions of the flow, while preserving monotonicity and avoiding spurious oscillations near discontinuities. To explain the approach, we consider the scalar linear advection equation as a representative model. The objective is to approximate the solution at the cell interfaces with improved spatial accuracy, which is crucial for accurately computing numerical fluxes. MUSCL accomplishes this by constructing left and right interface values using Taylor series expansions around the cell center, followed by finite-difference approximations of the gradients. A key feature of the scheme is the introduction of a parameterized formulation that includes curvature effects and allows blending between different stencil contributions. However, higher-order interpolation may induce oscillations near steep gradients or shocks. To address this, MUSCL incorporates flux limiter functions, which adaptively reduce local gradients to preserve the total variation diminishing (TVD) property. These limiters ensure that the reconstructed solution remains physically realistic and numerically stable across discontinuities.

19.5.1.1 Taylor-Based Interface Reconstruction with Adjustable Curvature Weighting

For sake of simplicity, we consider a simple scalar wave equation:

$$\frac{\partial U}{\partial t} + a\frac{\partial U}{\partial x} = 0$$

First, consider the spatial distribution of physical quantity U within the range $x_{k-1/2} \leq x \leq x_{k+1/2}$:

$$U(x) = U(x_k) + (x - x_k)U'(x_k) + \frac{1}{2}(x - x_k)^2 U''(x_k) + O(\Delta x^3)$$

Integrating this over the cell and using the definition of the cell-average value, we have:

$$U_k = \frac{1}{\Delta x}\int_{x_{k-1/2}}^{x_{k+1/2}} U(x)\,dx$$

Carrying out the integration explicitly yields:

$$U_k = U(x_k) + \frac{\Delta x^2}{24}U''(x_k) + O(\Delta x^4)$$

Expanding the derivatives to first and second orders around cell x_k:

$$U'(x) = U'(x_k) + (x - x_k)U''(x_k) + O(\Delta x^2)$$

$$U''(x) = U''(x_k) + O(\Delta x^2)$$

We define cell averages for these derivatives as:

$$U'_k = U'(x_k) + O(\Delta x^2)$$

$$U''_k = U''(x_k) + O(\Delta x^2)$$

Substituting these results back into the distribution equation, we remove $U(x_k), U'(x_k), U''(x_k)$, obtaining:

$$U(x) = U_k + (x - x_k)U'_k + \frac{1}{2}\left\{(x - x_k)^2 - \frac{\Delta x^2}{12}\right\}U''_k + O(\Delta x^3)$$

Thus, defining the boundary values as:

$$U^L_{k+1/2} = U(x_{k+1/2}), \quad U^R_{k-1/2} = U(x_{k-1/2})$$

yields:

$$U^L_{k+1/2} = U_k + \frac{\Delta x}{2}U'_k + \frac{\Delta x^2}{12}U''_k + O(\Delta x^3)$$

$$U^R_{k-1/2} = U_k - \frac{\Delta x}{2}U'_k + \frac{\Delta x^2}{12}U''_k + O(\Delta x^3)$$

Generalizing with a parameter κ, we have:

$$U^L_{k+1/2} = U_k + \frac{\Delta x}{2} U'_k + \frac{\kappa \Delta x^2}{4} U''_k$$

$$U^R_{k-1/2} = U_k - \frac{\Delta x}{2} U'_k + \frac{\kappa \Delta x^2}{4} U''_k$$

Using finite difference approximations for the derivatives:

$$U'(x) = \frac{U(x+\Delta x) - U(x-\Delta x)}{2\Delta x} + O(\Delta x^2)$$

$$U''(x) = \frac{U(x+\Delta x) - 2U(x) + U(x-\Delta x)}{\Delta x^2} + O(\Delta x^2)$$

we obtain:

$$U'_k = \frac{1}{2\Delta x}(U_{k+1} - U_{k-1}) + O(\Delta x^2)$$

$$U''_k = \frac{1}{\Delta x^2}(U_{k+1} - 2U_k + U_{k-1}) + O(\Delta x^2)$$

Thus, the interface values become:

$$U^L_{k+1/2} = U_k + \frac{1-\kappa}{4} \Delta U_{k-1/2} + \frac{1+\kappa}{4} \Delta U_{k+1/2}$$

$$U^R_{k-1/2} = U_k - \frac{1-\kappa}{4} \Delta U_{k+1/2} - \frac{1+\kappa}{4} \Delta U_{k-1/2}$$

Here, $\Delta U_{k+1/2} = U_{k+1} - U_k$.

Flux Limiter : Direct interpolation can cause numerical oscillations. To prevent this, the following corrections using a limiter function $\Phi(r)$ are applied:

$$U^L_{k+1/2} = U_k + \frac{1-\kappa}{4} \Phi(r) \Delta U_{k-1/2} + \frac{1+\kappa}{4} \Phi(1/r) \Delta U_{k+1/2}$$

$$U^R_{k-1/2} = U_k - \frac{1-\kappa}{4} \Phi(1/r) \Delta U_{k+1/2} - \frac{1+\kappa}{4} \Phi(r) \Delta U_{k-1/2}$$

with:

$$r = \frac{\Delta U_{k+1/2}}{\Delta U_{k-1/2}}$$

Alternatively, this can be expressed as:

$$U^L_{k+1/2} = U_k + \frac{1}{2} \Psi(r) \Delta U_{k-1/2}$$

$$U^R_{k-1/2} = U_k - \frac{1}{2} \Psi(1/r) \Delta U_{k+1/2}$$

where $\Psi(r)$ is related to $\Phi(r)$:

$$\Psi(r) = \frac{1-\kappa}{2} \Phi(r) + \frac{1+\kappa}{2} r \Phi(1/r)$$

To satisfy monotonicity constraints, the limiter $\Psi(r)$ must fulfill the TVD condition.

19.5.1.2 Limiters

Limiter in the context of hyperbolic equations is used in the reconstruction process to control the slope of the reconstructed solution and prevent the development of non-physical oscillations, particularly near discontinuities (like shock waves). These oscillations can lead to numerical instability or unphysical results. The limiter modifies the slope of the reconstruction to ensure the solution remains physical and smooth.

Common types of limiters include:

Minmod Limiter: A nonlinear limiter that selects the smallest (in magnitude) of the input slopes, preserving monotonicity. It is defined as:

$$\text{minmod}(a,b) = \frac{1}{2}\left[\text{sgn}(a) + \text{sgn}(b)\right]\min(|a|,|b|)$$

The minmod function behaves as follows:

- ☞ If a and b have the same sign, it returns the value with smaller magnitude.
- ☞ If a and b have opposite signs, it returns 0, adding numerical dissipation.

For the given slope limiter, r is the ratio of consecutive gradients or slopes (a measure of local smoothness or steepness), and b is a tunable parameter controlling numerical dissipation:

- ☞ For $b \approx 0.5 - 1$: minimal dissipation, sharp resolution of shocks and discontinuities.
- ☞ For $b \approx 2 - 4$: balanced dissipation, effective in shock-capturing schemes (commonly used).
- ☞ For $b > 4$: excessive dissipation, leading to overly smooth solutions.

The constraint on b is:

$$1 \le b \le \frac{3-\kappa}{1-\kappa}$$

where κ influences numerical dissipation:

- ☞ $\kappa = \frac{1}{3}$ is standard, offering a balance between accuracy and dissipation.
- ☞ Smaller values ($\kappa < \frac{1}{3}$) increase dissipation.
- ☞ Larger values ($\kappa > \frac{1}{3}$) reduce dissipation, producing sharper solutions.

The slope limiter is designed to suppress spurious oscillations near discontinuities such as shocks or contact surfaces. Fine-tuning the parameters b and κ is necessary to balance stability and accuracy according to the characteristics of the specific flow.

Van Leer Limiter: A more sophisticated limiter that provides higher accuracy while maintaining monotonicity. It blends the slopes from neighboring cells in a way that reduces oscillations. Van Leer limiter :

$$\Phi(r) = \frac{r+|r|}{1+|r|}$$

Superbee Limiter: A limiter that attempts to use the maximum slope allowed while avoiding oscillations, often resulting in sharper shock fronts at the cost of potential overestimation of the gradients. Roe's Superbee Limiter:

$$\Phi(r) = \text{superbee}(r, b) = \max\{0, \min(xy, 1), \min(x, y)\}$$

with $1 \leq b \leq 2$.

Koren Limiter: A more advanced limiter designed for higher accuracy that adapts the reconstruction method to the flow's local behavior, particularly for smooth flows and sharp discontinuities. Koren limiter :

$$\Phi(r) = \max\left[0, \min\left(2r, \frac{2+r}{3}, 2\right)\right]$$

MC (Monotonized Central-difference) limiter :

$$\Phi(r) = \max\left[0, \min\left(\frac{1+r}{2}, 2, 2r\right)\right]$$

In practical numerical computations, to avoid division by zero or related numerical instability for $r \approx 0$, the limiter can be redefined as:

$$\text{limiter}(x, y, b) = \text{sgn}(y)\max[0, \min\{\text{sgn}(y)bx, |y|\}, \min\{\text{sgn}(y)x, b|y|\}]$$

Using the limiter functions, the physical values at the cell interfaces can be interpolated.

19.5.2 TVD-CONFORMING INTERPOLATION IN CHARACTERISTIC SPACE

It is important to understand that the variables to be interpolated should be the characteristic variables. This choice ensures that the limiter functions satisfy the TVD (Total Variation Diminishing) condition. The necessity of employing characteristic variables arises because enforcing the TVD condition requires the governing equations to be cast in a form similar to the simplified PDE given by:

$$\frac{\partial U}{\partial t} + a\frac{\partial U}{\partial x} = 0$$

Specifically, considering the Euler equations represented by equations 19.5 and 19.8:

$$R^{-1}\frac{\partial Q}{\partial t} + \Lambda R^{-1}\frac{\partial Q}{\partial x} = 0$$

If we keep R^{-1} fixed over a small Δt and Δx, then by defining the characteristic variables $W = R^{-1}\mathrm{Q}$, the equations simplify to:

$$\frac{\partial W}{\partial t} + \Lambda \frac{\partial W}{\partial x} = 0$$

When performing interpolation using these characteristic variables, it is important to note that the matrices R, R^{-1}, and Λ vary spatially. At cell k, the Euler equations 19.5 specifically become:

$$\frac{\partial \mathrm{Q}_k}{\partial t} + A_k \left.\frac{\partial \mathrm{Q}}{\partial x}\right|_k = 0$$

Hence, if interpolation is performed at cell k, the matrices R_k, R_k^{-1}, and Λ_k, obtained at cell k, must be used.

Thus, at the cell interfaces, the conserved variables $\mathrm{Q}^L_{k+1/2}, \mathrm{Q}^R_{k-1/2}$ can be expressed using the differences of characteristic variables as follows:

$$\begin{pmatrix} (\Delta w_1)_{j+1/2} \\ (\Delta w_2)_{j+1/2} \\ (\Delta w_3)_{j+1/2} \end{pmatrix} = \Delta W_{j+1/2} = R_k^{-1} \Delta \mathrm{Q}_{j+1/2}, \quad (j = k-1, k)$$

By interpolating these differences component-wise using limiter functions, the cell-interface variables become:

$$\mathrm{Q}^L_{k+1/2} = \mathrm{Q}_k + \frac{1-\kappa}{4} \sum_{i=1}^{3} \mathrm{limiter}\{(\Delta w_i)_{k+1/2}, (\Delta w_i)_{k-1/2}, b\} r_i$$

$$+ \frac{1+\kappa}{4} \sum_{i=1}^{3} \mathrm{limiter}\{(\Delta w_i)_{k-1/2}, (\Delta w_i)_{k+1/2}, b\} r_i$$

$$\mathrm{Q}^R_{k-1/2} = \mathrm{Q}_k - \frac{1-\kappa}{4} \sum_{i=1}^{3} \mathrm{limiter}\{(\Delta w_i)_{k-1/2}, (\Delta w_i)_{k+1/2}, b\} r_i$$

$$- \frac{1+\kappa}{4} \sum_{i=1}^{3} \mathrm{limiter}\{(\Delta w_i)_{k+1/2}, (\Delta w_i)_{k-1/2}, b\} r_i$$

Here, $r_i\,(i = 1, 2, 3)$ are the right eigenvectors evaluated at cell k:

$$(r_1 \quad r_2 \quad r_3) = R_k$$

Note: The inverse matrix R_k^{-1} must also be evaluated at cell k. The variables $\mathrm{Q}^L_{k+1/2}$ and $\mathrm{Q}^R_{k-1/2}$ are first computed for all k. Then, the numerical flux $G_{k+1/2}$ at each interface is computed using these interpolated values and equation:

$$G_{k+1/2} = \hat{R}|\hat{\Lambda}|\hat{R}^{-1}(\mathrm{Q}^L_{k+1/2}, \mathrm{Q}^R_{k+1/2})$$

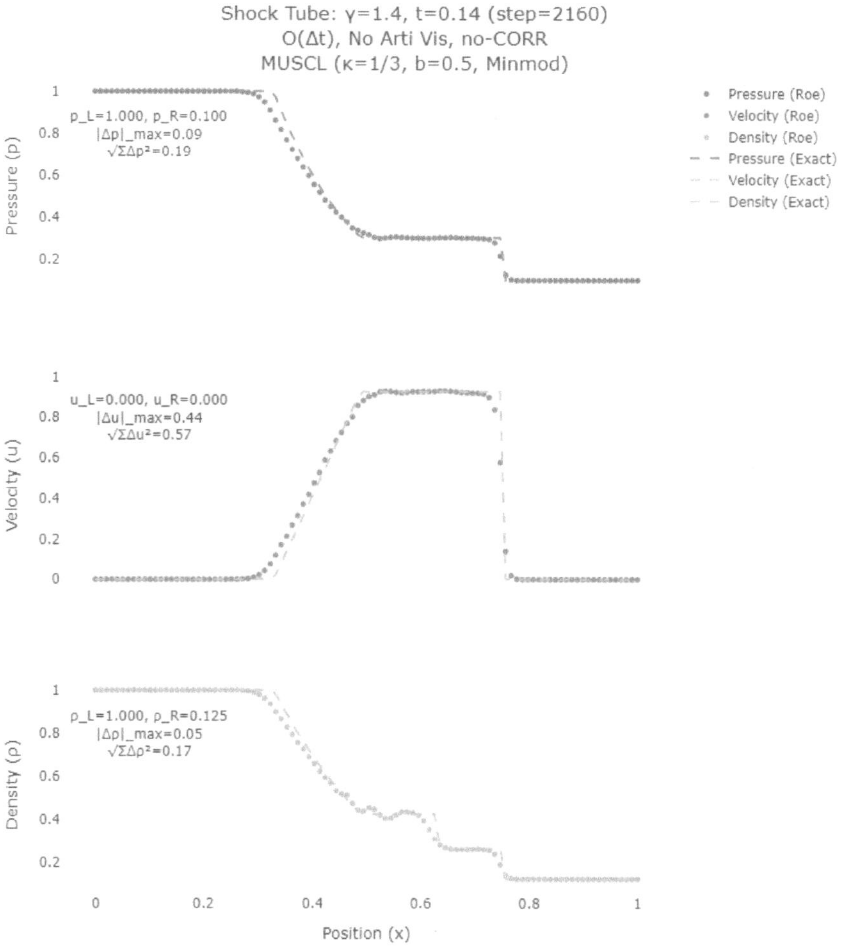

Figure 19.10: **Roe Approximation with MUSCL and limiter Minmod b=0.5. No flux correction applied. Grid cells=100 .**

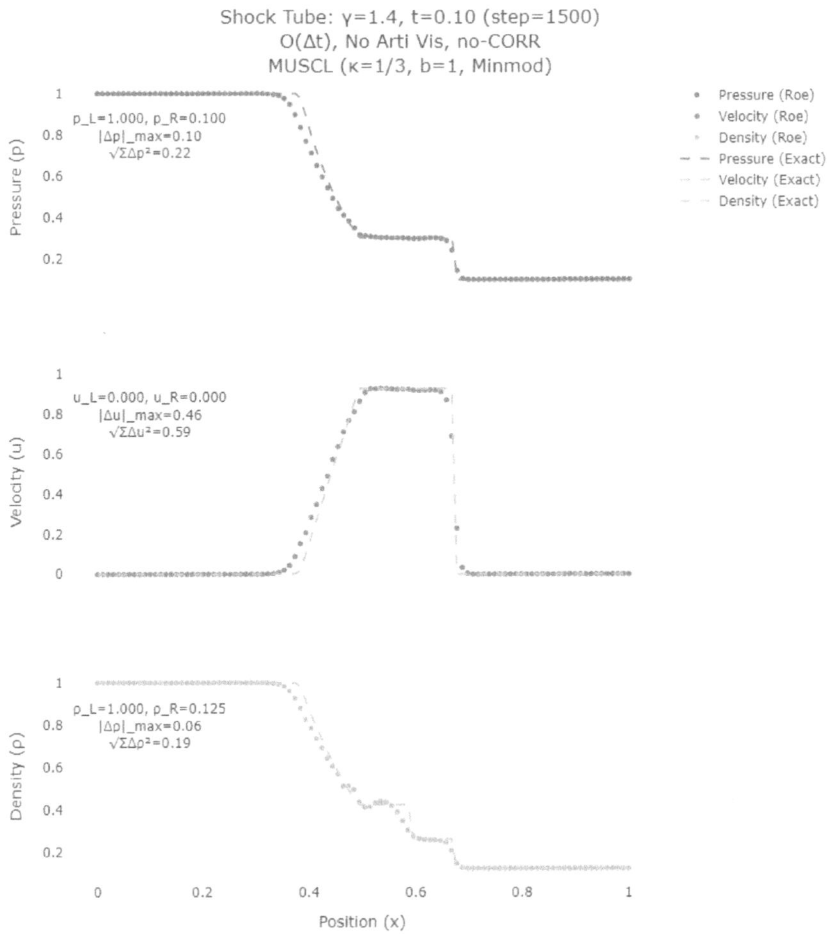

Figure 19.11: **Roe Approximation with MUSCL and limiter Minmod b=1. No flux correction applied. Grid cells=100.**

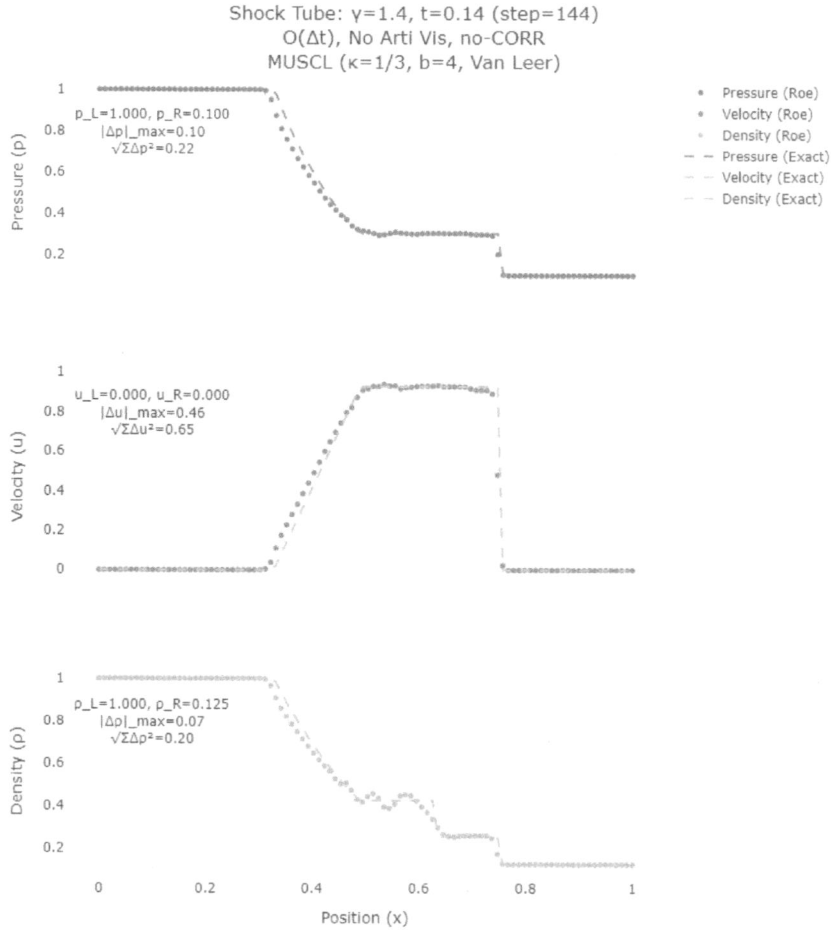

Figure 19.12: **Roe Approximation with limiter Van Leer and with no flux correction applied.**

Here, Roe averages are used. This numerical flux calculation, therefore, achieves second-order accuracy in space, and when using $\kappa = 1/3$, it attains third-order accuracy.

The MUSCL scheme thus provides a systematic approach to achieve higher-order accuracy while preserving the monotonicity of the solution, which is essential for resolving discontinuities without introducing non-physical oscillations. By reconstructing interface values with a tunable parameter κ and employing limiter functions such as $\Phi(r)$ or its hybrid form $\Psi(r)$, the scheme flexibly transitions between high-resolution and non-oscillatory behavior. The inclusion of curvature through second derivative terms and the control of interpolation via slope limiters make MUSCL particularly effective for problems involving shock waves, contact discontinuities, or steep gradients. This balance between accuracy and stability is what makes the MUSCL approach a robust and widely adopted method in the numerical solution of hyperbolic conservation laws.

Case Study 19A: Supersonic Flow Over a Wedge Using MUSCL-TVD with Roe Solver

Problem Definition

This case examines two-dimensional supersonic inviscid flow over a sharp wedge with an inclined lower wall — a canonical configuration in compressible aerodynamics. Two wedge geometries are analyzed: one with a relatively shallow deflection angle of 15° subjected to a freestream Mach number $M_\infty = 2.0$, and a second with a steeper 30.5° wedge under $M_s = 1.384$, corresponding to experimental dusty-gas shock studies. In the present case, however, the fluid is modeled as a single-phase, non-dusty ideal gas to isolate the shock structure without multiphase relaxation effects. The objective is to numerically resolve the formation of oblique shocks, expansion fans, and reflected wave systems using a high-resolution shock-capturing finite volume method. This configuration is selected for its analytical tractability, relevance to supersonic and hypersonic flow problems, and its suitability for evaluating the accuracy and robustness of MUSCL-TVD schemes with approximate Riemann solvers.

Governing Equations

The compressible Euler equations in conservative form are solved:

$$\frac{\partial \mathbf{U}}{\partial t} + \nabla \cdot \mathbf{F}(\mathbf{U}) = 0$$

where $\mathbf{U}$ contains the conserved variables, and fluxes $\mathbf{F}$ are calculated via Roe's approximate Riemann solver.

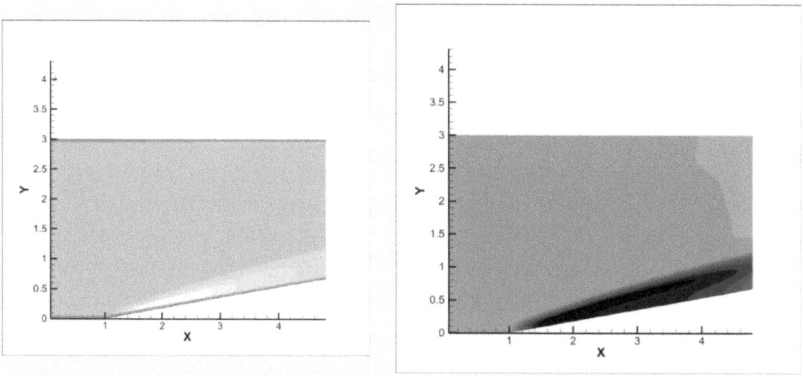

Figure 19.13: **Grayscale contours of density and velocity contours for a Mach-2 inviscid wedge flow with an inclined lower wall at 15 degrees, computed on a structured quadrilateral mesh. The second-order finite-volume scheme uses MUSCL reconstruction with a minmod limiter and Roe's approximate Riemann solver.**

Numerical Setup

A structured quadrilateral mesh discretizes the domain. MUSCL reconstruction with the minmod limiter enhances second-order accuracy while avoiding spurious oscillations. Time advancement is explicit, limited by a local CFL condition. Roe fluxes provide upwind-biased interface values. Slip boundary conditions are applied at walls; subsonic outflow conditions use characteristic-based extrapolation to minimize reflections.

Results and Discussion

Figure 19.13 shows the shock formation over a 15 degree wedge. The inlet Mach number is 2. The solver employs a classical shock-capturing strategy drawn from Godunov-type finite-volume methods. Cell-averaged conservative variables are advanced in time using fluxes evaluated at the faces of each control-volume, which guarantees global conservation. Inside every cell the solution is reconstructed with a piecewise-linear MUSCL approach that adopts the minmod limiter; this total-variation-diminishing limiter suppresses spurious oscillations at steep gradients yet retains second-order accuracy across smooth regions. The left and right reconstructed states on each face feed a Roe approximate Riemann solver, whose upwind dissipation depends on the largest local characteristic speed and therefore enforces the entropy condition while keeping the shock width small. An explicit time integrator respects a Courant restriction based on the smallest cell dimension and the maximum local signal speed, ensuring stability of nonlinear wave propagation.

At the boundaries, subsonic outflow surfaces apply a characteristic-based pressure condition that minimises reflections, whereas solid walls impose a slip reflection of the normal velocity component to preserve inviscid symmetry. Together, the finite-volume framework, MUSCL-minmod reconstruction, Roe flux, and carefully chosen boundary treatments yield a robust, second-order scheme capable of sharply resolving oblique shocks and expansion fans on the quadrilateral mesh without introducing unphysical oscillations.

Figure 19.14 shows the shadowgraph for supersonic flow over a 30.5° wedge, simulated using a second-order MUSCL-TVD scheme with Roe's approximate Riemann solver. A strong oblique shock forms at the wedge root, producing sharp density gradients and a thin compressed layer along the wall. A vertical reflected shock arises from the upper boundary. Compared to shallower wedges, the higher angle results in steeper shocks, stronger compression, and a more confined post-shock region. The numerical method captures these features with sharp resolution and minimal oscillations.

For the 30.5° wedge, Figure 19.14 shows a well-defined shock front and a compressed region near the surface. The reflected wave from the top boundary is captured without artificial noise. These results compare favorably with the dusty-gas wedge experiments and simulations of Igra et al. (2004), demonstrating that in the absence of dust-related relaxation effects, clean-gas simulations reproduce classical oblique shock reflection with clarity.

Lessons Learned

i The minmod limiter provides a stable yet non-oscillatory solution, but introduces slight clipping near steep gradients.

ii The Roe solver sharply resolves shocks but requires careful treatment of near-zero density or pressure.

iii Grid refinement with grid adaptation are needed for sharp shock structure.

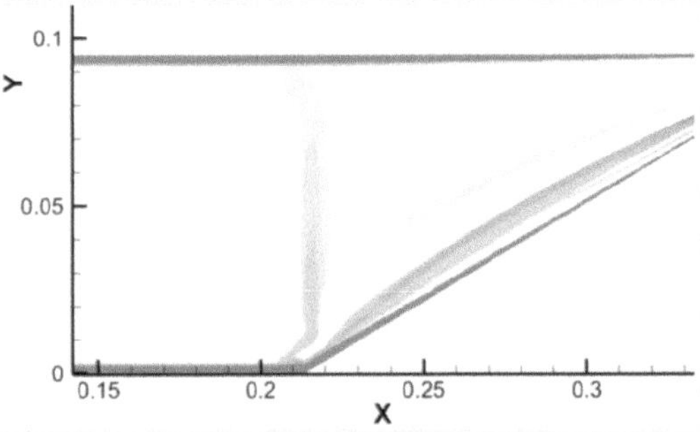

Figure 19.14: **Shadowgraph: The inflow is a supersonic stream with Mach number $M_s = 1.384$, shock strength = 2.069, and temperature $T_o = 294$ K. The wedge angle is 30.5°.**

19.5.3 RIEMANN SOLVER WITH WENO RECONSTRUCTION TECHNIQUE

Being a low-order method, the Roe scheme alone may introduce non-physical oscillations near discontinuities. To address this limitation, the Roe method is also often paired with another high-order scheme, called Weighted Essentially Non-Oscillatory (WENO) scheme. In the context of finite volume methods, accurate flux computation at cell interfaces requires reconstruction of variables from cell averages. Traditional linear reconstructions may lead to Gibbs-like oscillations near sharp gradients or discontinuities. To improve this, nonlinear reconstruction techniques are used, where the goal is to maintain high-order accuracy in smooth regions while suppressing spurious oscillations near discontinuities. WENO reconstruction achieves this by adaptively combining multiple candidate stencils using nonlinear weights derived from local smoothness indicators. These weights shift preference toward smoother stencils, effectively reducing contributions from stencils containing discontinuities. As a result, WENO schemes offer both high-order accuracy and excellent shock-capturing properties.

In the Roe-WENO procedure, the WENO scheme is used to reconstruct the conservative variables or physical fluxes at the left and right sides of each interface. The Roe solver then computes the interface flux using these reconstructed values. This combination merges the upwind precision of Roe's linearized Riemann solver with the robustness and accuracy of WENO, making it highly effective for solving compressible flow problems with strong shocks and complex wave interactions.

The Jiang and Shu (1996) WENO-JS scheme is a high-order finite difference method designed for solving hyperbolic conservation laws that exhibit sharp gradients or discontinuities, such as shock waves and contact surfaces. It improves upon the classical ENO schemes by providing better accuracy and robustness while maintaining non-oscillatory behavior near discontinuities. The scheme achieves fifth-order accuracy in smooth regions and automatically reduces its order near non-smooth features to suppress spurious oscillations.

The goal is to reconstruct the point value of a conserved variable at a cell interface, such as $q_{i+1/2}$, from cell-centered values q_i. To achieve this, WENO-JS constructs three third-order candidate polynomials using three-point stencils: Stencil 0:

$$q_{i+1/2}^{(0)} = \frac{1}{3}q_{i-2} - \frac{7}{6}q_{i-1} + \frac{11}{6}q_i$$

Stencil 1:

$$q_{i+1/2}^{(1)} = -\frac{1}{6}q_{i-1} + \frac{5}{6}q_i + \frac{1}{3}q_{i+1}$$

Stencil 2:

$$q_{i+1/2}^{(2)} = \frac{1}{3}q_i + \frac{5}{6}q_{i+1} - \frac{1}{6}q_{i+2}$$

Here, q represents a scalar conserved variable, like density, velocity, or pressure, that is being interpolated or reconstructed at the cell interface using WENO weights.

In a linear fifth-order scheme, the final value would be a weighted sum using fixed linear weights:

$$d_0 = \frac{1}{10}, \quad d_1 = \frac{6}{10}, \quad d_2 = \frac{3}{10}$$

However, WENO replaces these with nonlinear, adaptive weights based on local smoothness indicators. Each stencil is associated with a smoothness indicator β_k that quantifies local variation: For stencil 0:

$$\beta_0 = \frac{13}{12}(q_{i-2} - 2q_{i-1} + q_i)^2 + \frac{1}{4}(q_{i-2} - 4q_{i-1} + 3q_i)^2$$

For stencil 1:

$$\beta_1 = \frac{13}{12}(q_{i-1} - 2q_i + q_{i+1})^2 + \frac{1}{4}(q_{i-1} - q_{i+1})^2$$

For stencil 2:

$$\beta_2 = \frac{13}{12}(q_i - 2q_{i+1} + q_{i+2})^2 + \frac{1}{4}(3q_i - 4q_{i+1} + q_{i+2})^2$$

The nonlinear weights are then defined by:

$$\alpha_k = \frac{d_k}{(\varepsilon + \beta_k)^2}, \quad w_k = \frac{\alpha_k}{\sum_{j=0}^{2} \alpha_j}$$

where ε is a small constant (e.g., 10^{-6}) to avoid division by zero.

Finally, the left-biased reconstructed value at the interface is:

$$q^-_{i+1/2} = w_0 q^{(0)}_{i+1/2} + w_1 q^{(1)}_{i+1/2} + w_2 q^{(2)}_{i+1/2}$$

This formula adapts the weights w_k to favor stencils located in smooth regions and suppress those near discontinuities. As a result, the WENO-JS method maintains high-order accuracy in smooth regions while remaining stable and non-oscillatory near sharp gradients. This makes it an effective tool for capturing shock waves and complex flow structures in compressible fluid dynamics.

In the FV approach, WENO is applied to reconstruct the solution q at cell interfaces using cell averages.

The governing equation is written in integral form:

$$\frac{d}{dt}\bar{q}_i + \frac{1}{\Delta x}\left(F_{i+1/2} - F_{i-1/2}\right) = 0$$

WENO reconstructs left and right states $q^-_{i+1/2}, q^+_{i+1/2}$ from neighboring cell averages $\bar{q}_i$ These are then used in a Riemann solver to compute fluxes.

In Figure 19.15, the Roe flux approximation is combined with a fifth-order WENO reconstruction to solve the classical Sod shock tube problem on a one-dimensional domain discretized into 100 control volumes. The left state is $\rho_L = 1$, $u_L = 0$, $p_L = 1$, and the right state is $\rho_R = 0.125$, $u_R = 0$, $p_R = 0.1$. The circles (dots) represent the numerical results for pressure, velocity, and density, while the

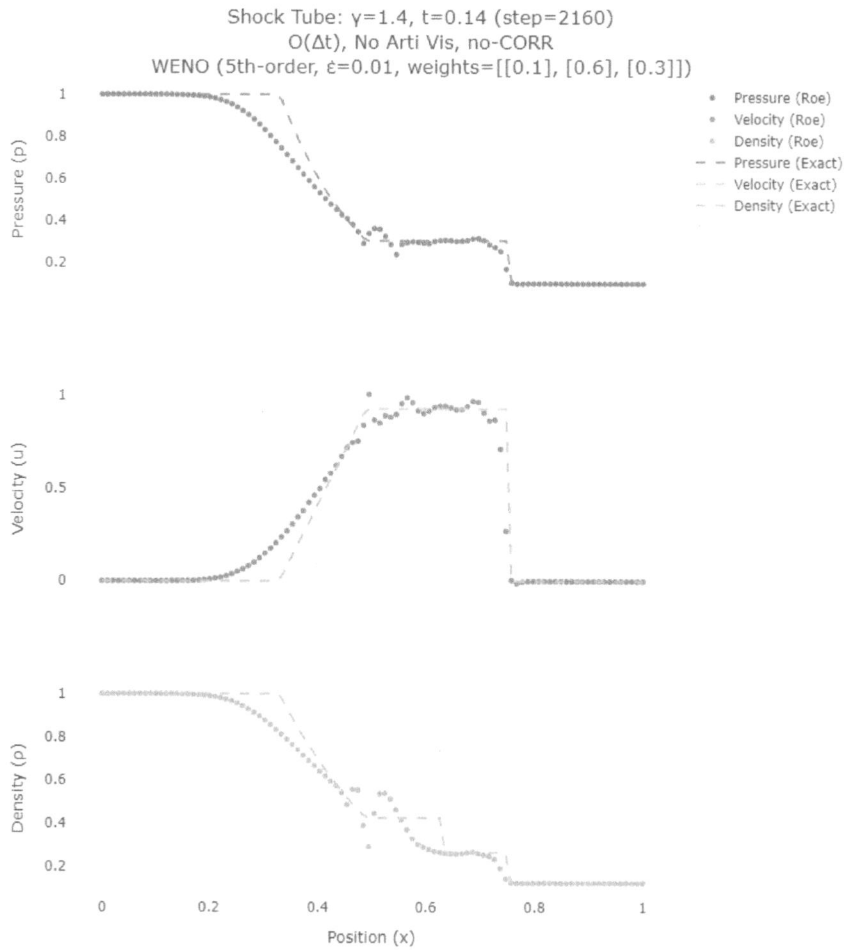

Figure 19.15: **Roe approximation with WENO with 100 CV and ε=10^{-2} for classical Sod shock tube initial conditions $\rho_L = 1, \quad u_L = 0, \quad p_L = 1$, and the right state is $\rho_R = 0.125, \quad u_R = 0, \quad p_R = 0.1$.**

dashed lines correspond to the exact solution. Overall, the numerical solution agrees well with the reference solution, capturing the rarefaction wave, contact discontinuity, and shock discontinuity. Small deviations near discontinuities can be attributed to numerical dissipation introduced by the WENO reconstruction and Roe flux.

The parameter $\varepsilon = 10^{-2}$ is introduced to prevent division by zero in the nonlinear weight formula, but such a large value can weaken the sensitivity of the weights to the smoothness indicators. This reduces the scheme's ability to distinguish between smooth and non-smooth regions, making it behave more like a linear scheme and degrading accuracy near discontinuities. Typically, a smaller value such as $\varepsilon = 10^{-6}$ is preferred to retain the non-oscillatory property and adaptive behavior of the WENO scheme.

Figure 19.16 illustrates the Roe-WENO solution for the Lax problem using 100 control volumes but with a smaller WENO parameter $\varepsilon = 10^{-6}$. The initial states are $\rho_L = 0.445$, $u_L = 0.698$, $p_L = 3.528$ and $\rho_R = 0.5$, $u_R = 0$, $p_R = 0.571$. The predicted pressure, velocity, and density fields match the exact solution closely, confirming that the fifth-order WENO scheme, coupled with Roe's flux, reasonably resolves the shock waves and contact discontinuities. Compared to Figure 19.15, the smaller ε value here yields slightly sharper discontinuities and reduces numerical viscosity. However, in practice, excessively small ε can lead to unwanted oscillations if not accompanied by other stabilizing mechanisms. In both test cases, the Roe flux combined with a high-order WENO reconstruction successfully provides non-oscillatory and stable solutions. The parameter ε controls the level of nonlinear dissipation, influencing the trade-off between sharp capturing of discontinuities and limiting spurious oscillations. These results illustrate the effectiveness of the fifth-order WENO approach for one-dimensional Riemann problems while showcasing how the choice of ε can be used to fine-tune the shock resolution and overall smoothness of the solution.

Classical WENO schemes in combination with the Roe flux do capture discontinuities in most standard test problems. However, there exhibit certain shortcomings that motivate the development of new methods. One issue is that in some cases excessive numerical dissipation in smooth regions, which can smear out flow structures such as contact discontinuities. In low-Mach regimes, WENO plus Roe may lose accuracy unless additional fixes are introduced. Furthermore, adapting WENO stencils on complex or unstructured meshes can be cumbersome, prompting simpler high-order polynomial approaches.

Researchers have proposed various WENO schemes over the years which are listet in Table 19.2.

The WENO-Z reconstruction scheme (the so-called third-order WENO-Z) was proposed by Borges et al. (2008) shows a close match for test case suggested by Lax. The initial condition involves a higher-pressure, right-moving gas on the left and lower-pressure gas at rest on the right. The solution shows a rightward-moving shock, a contact discontinuity, and a leftward rarefaction. The Comparison of some WENO variants is tabulated in Table 19.3.

Regardless of reconstruction procedure, when using Roe's approximate Riemann solver, several numerical difficulties can arise if the flow conditions are extreme or

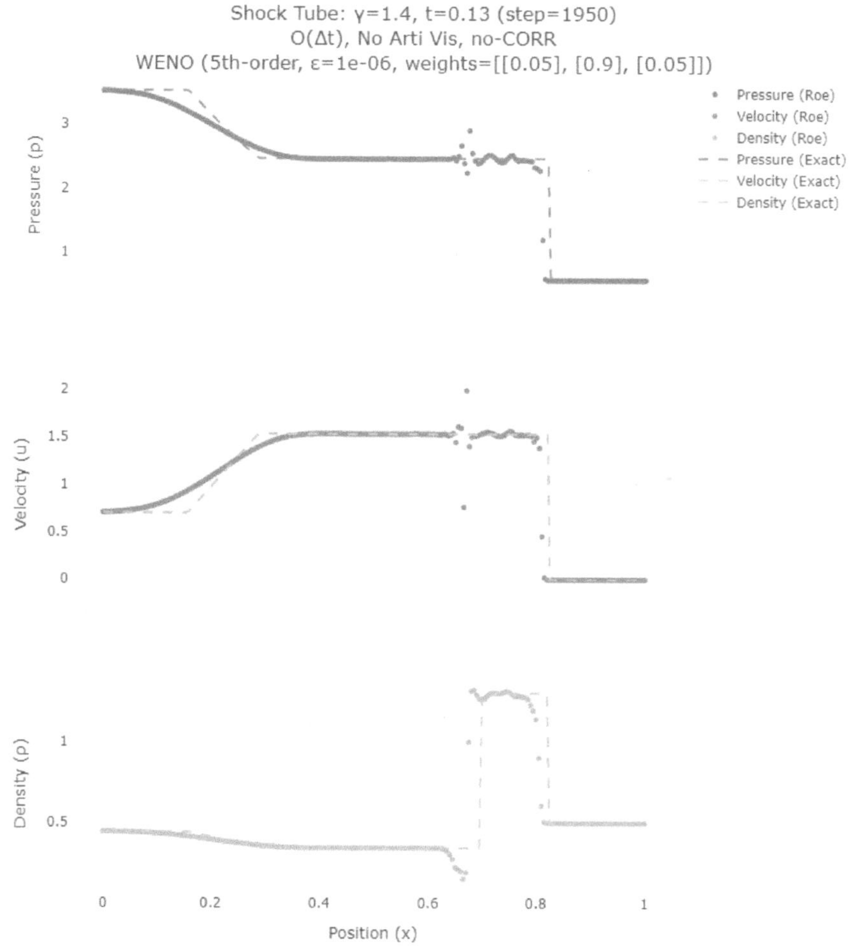

Figure 19.16: **Roe approximation with WENO with 100 CV and ε=10^{-6} for Lax problem** $(\rho_L, u_L, P_L) = (0.445, 0.698, 3.528)$, $(\rho_R, u_R, P_R) = (0.5, 0, 0.571)$.

Table 19.2
WENO Variants and their characteristics

Researchers (Year)	WENO Scheme	Comments
Jiang & Shu (1996)	WENO-JS	Classical 5th-order scheme; robust but suffers from excessive dissipation near smooth extrema.
Shu et al. (2000s)	WENO-CU	Combines central and upwind weights; better spectral resolution in smooth regions, but may underperform at discontinuities.
Henrick et al. (2005)	WENO-M	Introduced mapped weights to improve accuracy near critical points; reduces dissipation but may introduce oscillations.
Borges et al. (2008)	WENO-Z	Adds global smoothness indicator τ_5; more accurate and less dissipative than WENO-JS; strong general-purpose choice.
Hu et al. (2010s)	WENO-Z+	Further enhancement of WENO-Z for better smooth region resolution; good accuracy-stability balance.
Hu et al. (2013)	WENO-CU6-M2	Sixth-order central-upwind scheme; excellent low-dissipation properties; preferred for DNS/LES applications.
Balsara et al. (2016)	WENO-AO	Adaptive-order scheme switching between stencils; excellent shock-capturing with higher robustness near discontinuities.
Zhu & Qiu (2016)	WENO-ZQ	Variant of WENO-Z with better multi-dimensional behavior; improved accuracy in structured/unstructured grids.
Shu et al. (2017)	WENO-AOLW	Combines WENO with Lax-Wendroff-type time integration; improves temporal accuracy, but more complex implementation.
Li et al. (2020s)	WENO-R	Polynomial-based reconstruction approach; efficient and extendable to high-order variants.

if certain physical principles are not enforced. Below is an elaboration of the main points:

☞ *Entropy Violations*: Roe's method can produce non-physical expansion shocks, known as entropy violations, when encountering transonic rarefaction waves. An entropy fix (such as Harten's or Chakravarthy's method) modifies the wave speeds to ensure the physically correct wave structure. Without this fix, the solver may converge to an unphysical solution that violates the second law of thermodynamics.

☞ *High Mach Numbers and Large Density/Pressure Contrasts*: As Mach numbers increase or as density/pressure jumps become very large, numerical instabilities and inaccuracies become more likely. In particular, if one region of the domain approaches vacuum conditions, many standard approximate Riemann solvers (including Roe's) may fail or yield oscillatory, non-physical solutions. This is because extremely low densities amplify any numerical errors, potentially leading to negative densities or pressures.

Table 19.3
Chronological Comparison of WENO Variants

Feature	WENO-JS (1996)	WENO-Z (2005)	WENO-M (2016)
Weight Formula	$\alpha = \frac{d_k}{(\varepsilon+\beta_k)^2}$	$\alpha_k = d_k\left(1+\frac{\tau_5}{\varepsilon+\beta_k}\right)$	$\alpha_k = \frac{d_k \cdot g(d_k)}{(\varepsilon+\beta_k)^2}$
Accuracy	3rd-order (5th in smooth regions)	Improved at critical points	4th-order at extrema
Dissipation	Highest (conservative limiting)	Moderate	Lowest (optimized)
Key Innovation	Original WENO scheme	Global smoothness measurement	Nonlinear weight mapping

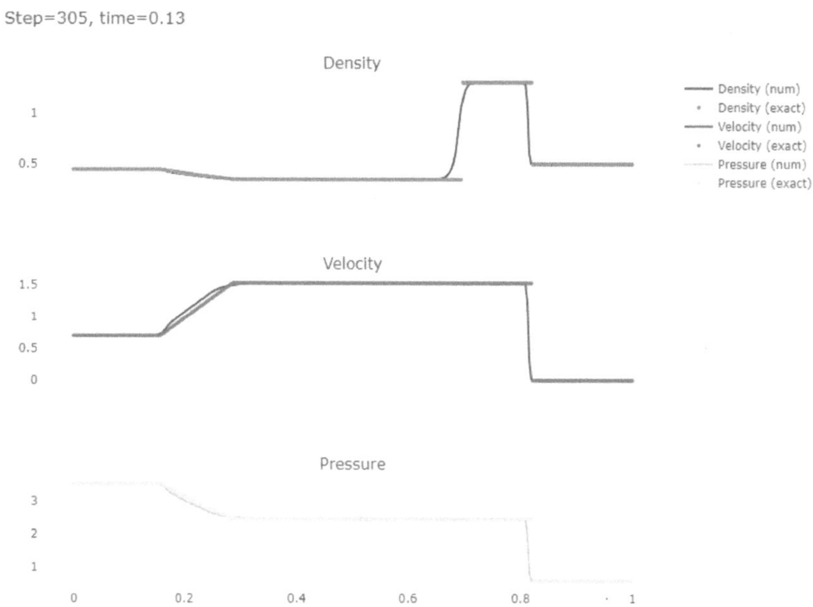

Figure 19.17: **WENO-Z reconstruction (third-order), Roe-averaged single-wave Rusanov flux scheme applied on test case from Lax (1954) with 201 cells.**

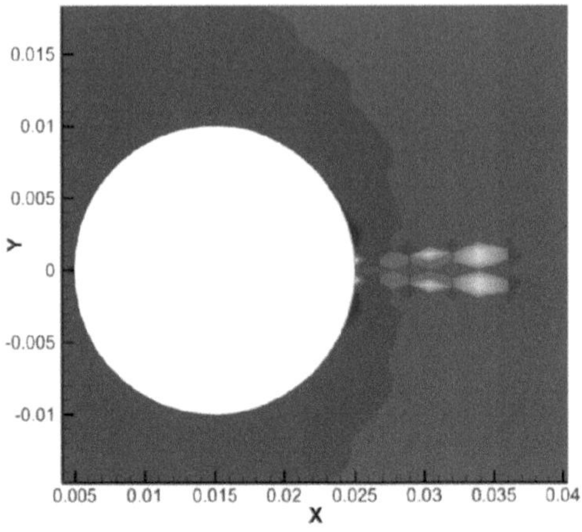

Figure 19.18: **Carbuncle formation near the stagnation zone of cylinder in a supersonic flow simulation.**

☞ *Spurious Artifacts*: Even when an entropy fix is applied, strong discontinuities or complex flow features (like shocks interacting with contact surfaces) can still generate small unphysical oscillations. These artifacts are often more pronounced on refined meshes (i.e., higher resolution) where the scheme's inherent numerical dissipation is reduced.

☞ *Remedies and Alternative Schemes*: In practice, it is best to compare results obtained with different approximate Riemann solvers or more robust methods. The dissipative schemes can help in stabilizing solutions in highly nonlinear or near-vacuum regimes. Another strategy is to ensure positivity-preserving algorithms for density and pressure, thereby avoiding negative values that lead to simulation failure.

☞ *Physical Considerations*: When setting up problems with extremely large density or pressure contrasts, one should assess whether such conditions are realistic or if they can be mitigated (for example, by cutting off vacuum regions). Understanding the actual physics is crucial, because no numerical solver can correct a fundamentally ill-posed setup.

☞ *carbuncle phenomenon*: The carbuncle phenomenon is a well-known numerical artifact in two-dimensional simulations that may appears when Roe's approximate Riemann solver is used to model a strong, steady bow shock aligned with the computational grid (see Figure 19.18).

In such case, a stagnation line forms ahead of a blunt body at high Mach numbers, and small perturbations can grow into a spurious *carbuncle* or bulge in the shock, distorting the solution. The carbuncle reflects insufficient transverse dissipation, leading to unphysical results in otherwise steady shock structures.

Table 19.4
Numerical Schemes Proposed Over the Years

Year	Scheme	Category
1940s (approx.)	Leapfrog Method	Central difference Scheme
1954	Lax-Friedrichs	Central difference Scheme
1959	Godunov	Flux-Difference Splitting (FDS)
1960	Lax-Wendroff	Central difference Scheme
1976	Beam-Warming	Higher-order upwind scheme
1981	Roe	FDS
1981	Steger-Warming	Flux-Vector Splitting (FVS)
1981	MUSCL (van Leer)	Higher-Order Reconstruction tool
1982	van Leer FVS	FVS
1983	HLL (Harten-Lax-van Leer)	FDS
1984	PPM (Colella-Woodward)	Higher-Order Reconstruction tool
1985	AUSM	FVS
1994	WENO	Higher-Order Reconstruction tool
1993	AUSM+	FVS
1997	HLLC (Toro)	FDS
2000	Kurganov-Tadmor	Central difference Scheme
2001	AUSM+-up	FVS

Various modifications, including entropy fixes, rotated or adaptive grids, and alternative flux functions, have been proposed to mitigate this instability in high-resolution simulations.

The current research trend is more focused on use of WENO with schemes other then Roe approximation.

19.6 OTHER SCHEMES

Over the years, researchers have proposed different schemes. Some of them are tabulated in Table 19.4.

It is interesting to know that the following schemes do not need a reconstruction procedure:

a. HLL (Harten, Lax, and van Leer)

b. HLLE (Harten, Lax, and van Leer Einfeldt)

c. CUSP (Convective Upstream Split Pressure)

d. AUSM (Advection Upstream Splitting Method)

Following schemes require a reconstruction procedure:

a. Kurganov-Tadmor scheme: uses piecewise linear reconstruction with slope limiters.

b. Multi-Stage Advection (or Approximate) Scheme (MUSTA): uses higher-order MUSTA schemes require reconstruction (e.g., MUSCL) for accuracy and stability.

c. Arbitrary DERivatives schemes (ADER): uses high-order polynomial reconstructions (e.g., WENO or discontinuous Galerkin-type interpolants) to build local space-time expansions.

19.6.1 STEGER-WARMING FLUX VECTOR SPLITTING

The Steger-Warming flux vector splitting scheme was introduced in 1981 by Joseph L. Steger and Richard F. Warming. Originally devised as a finite-difference upwind method for the compressible Euler equations, the Steger-Warming scheme was later reformulated in finite-volume form by splitting each flux into positive and negative characteristic contributions. In its finite-volume implementation, the numerical flux at each cell interface is assembled from upstream components of the left and right states, thereby ensuring strict conservation and robust stability—even in subsonic flow regimes.

Consider the one-dimensional Euler equations:

$$\frac{\partial \mathbf{U}}{\partial t} + \frac{\partial \mathbf{F}(\mathbf{U})}{\partial x} = 0, \tag{19.17}$$

where $\mathbf{U} = [\rho,\ \rho u,\ E]^T$ and $\mathbf{F}(\mathbf{U}) = [\rho u,\ \rho u^2 + p,\ (E+p)u]^T$. Here, pressure p is given by:

$$p = (\gamma - 1)\left(E - \frac{1}{2}\rho u^2\right). \tag{19.18}$$

The flux vector $\mathbf{F}(\mathbf{U})$ can be expressed in terms of the Jacobian matrix $A(\mathbf{U})$ as:

$$\mathbf{F}(\mathbf{U}) = A(\mathbf{U})\mathbf{U}, \tag{19.19}$$

where $A(\mathbf{U}) = \partial \mathbf{F}/\partial \mathbf{U}$.

The eigenvalues of the Jacobian $A(\mathbf{U})$ are the characteristic wave speeds:

$$\lambda_1 = u, \quad \lambda_2 = u + c, \quad \lambda_3 = u - c, \tag{19.20}$$

where $c = \sqrt{\gamma p/\rho}$ is the local speed of sound.

Steger-Warming flux splitting separates eigenvalues into positive and negative parts:

$$\lambda_i = \lambda_i^+ + \lambda_i^-, \quad \text{with} \quad \lambda_i^\pm = \frac{\lambda_i \pm |\lambda_i|}{2}, \quad i = 1,2,3. \tag{19.21}$$

The Jacobian is then split accordingly:

$$A = A^+ + A^-, \quad \text{where} \quad A^+ = R\Lambda^+ R^{-1}, \quad A^- = R\Lambda^- R^{-1}. \tag{19.22}$$

Consequently, the flux vectors are split:

$$\mathbf{F}^+(\mathbf{U}) = A^+\mathbf{U}, \quad \mathbf{F}^-(\mathbf{U}) = A^-\mathbf{U}. \tag{19.23}$$

Explicit eigenvalue splitting yields:

$$\begin{aligned}\lambda_1^+ &= \frac{u+|u|}{2}, & \lambda_1^- &= \frac{u-|u|}{2},\\ \lambda_2^+ &= \frac{u+c+|u+c|}{2}, & \lambda_2^- &= \frac{u+c-|u+c|}{2},\\ \lambda_3^+ &= \frac{u-c+|u-c|}{2}, & \lambda_3^- &= \frac{u-c-|u-c|}{2}.\end{aligned} \tag{19.24}$$

$$\mathbf{F}^+ = \frac{\rho}{2\gamma}\begin{bmatrix} 2\gamma u + (c-u) \\ 2(\gamma-1)u^2 + (u+c)^2 \\ \dfrac{(u+c)^3 - (u^3+c^3)}{2(\gamma-1)} + (\gamma-1)u^3 \end{bmatrix}, \quad \mathbf{F}^- = \frac{\rho}{2\gamma}\begin{bmatrix} u-c \\ (u-c)^2 \\ \dfrac{(u-c)^3 + (3-\gamma)c^2(u-c)}{2(\gamma-1)} \end{bmatrix}$$

The finite volume numerical flux at cell interfaces is constructed by combining positive flux from the left state $\mathbf{U}_L$ and negative flux from the right state $\mathbf{U}_R$:

$$\mathbf{F}_{i+\frac{1}{2}} = \mathbf{F}^+(\mathbf{U}_L) + \mathbf{F}^-(\mathbf{U}_R). \tag{19.25}$$

A first-order finite volume update is:

$$\frac{d\mathbf{U}_i}{dt} = -\frac{1}{\Delta x}(\mathbf{F}_{i+\frac{1}{2}} - \mathbf{F}_{i-\frac{1}{2}}), \tag{19.26}$$

ensuring conservation form and upwind stability.

The Steger-Warming flux-vector-splitting method does not solve a Riemann problem as the method never constructs intermediate states or wave strengths spanning U_L and U_R. It lacks the contact and shear-wave resolution that characterizes flux-difference-splitting Riemann solvers such as Roe, HLL or Osher. The smearing of contact discontinuities often observed due to the absence of Riemann-based coupling with Steger-Warming scheme.

19.6.2 HLL (HARTEN-LAX-VAN LEER) SCHEME

The 1D compressible Euler equations in conservative form are:

$$\frac{\partial \mathsf{Q}}{\partial t} + \frac{\partial \mathsf{F}(\mathsf{Q})}{\partial x} = 0$$

where the conservative variable vector is:

$$\mathsf{Q} = \begin{bmatrix} \rho \\ \rho u \\ e \end{bmatrix}$$

and the physical flux is:

$$\mathsf{F}(\mathsf{Q}) = \begin{bmatrix} \rho u \\ \rho u^2 + p \\ u(e+p) \end{bmatrix}$$

The HLL scheme approximates the solution of the Riemann problem by assuming that the solution consists of a single intermediate state between two waves with speeds S_L and S_R. It defines the numerical flux at an interface as:

$$\mathsf{F}_{\text{HLL}} = \begin{cases} \mathsf{F}_L, & \text{if } S_L > 0 \\ \frac{S_R \mathsf{F}_L - S_L \mathsf{F}_R + S_L S_R (\mathsf{Q}_R - \mathsf{Q}_L)}{S_R - S_L}, & \text{if } S_L \leq 0 \leq S_R \\ \mathsf{F}_R, & \text{if } S_R < 0 \end{cases}$$

Here:
$\mathsf{F}_L = \mathsf{F}(\mathsf{Q}_L)$, flux from the left state
$\mathsf{F}_R = \mathsf{F}(\mathsf{Q}_R)$, flux from the right state
S_L, S_R are estimates of the minimum and maximum signal velocities, given by:

$$S_L = \min(u_L - a_L, u_R - a_R), \quad S_R = \max(u_L + a_L, u_R + a_R)$$

with u being velocity, and $a = \sqrt{\gamma p/\rho}$ the local sound speed.

Figure 19.19 shows the HLL scheme predictions for Toro test 5 condition: $\rho_L = 5.99924$ kg/m^3, $u_L = 19.5975$ m/s, $p_L = 460.894$ N/m^2, $\rho_R = 5.99242$ kg/m^3, $u_R = -6.19633$ m/s, and $p_R = 46.0950$ N/m^2. Figure 19.19 illustrates the HLL scheme's ability to capture complex wave interactions in a shock tube simulation under Toro test 5 conditions. As a result of the initial discontinuities, the simulation produces a strong shock propagating toward the right, a contact discontinuity, and an expansion fan moving to the left. In the figure, the pressure profile shows a distinct high-pressure plateau behind the shock with a smooth transition to the lower pressure on the right, while the velocity profile reflects an abrupt transition across the contact discontinuity. The density profile reveals that the contact wave is somewhat smeared by the HLL scheme's numerical dissipation.

Although the HLL scheme performs well by capturing shocks and expansion fans robustly using averaged wave speeds, it inherently models only the fastest left- and right-going waves and perform not so perfect for the intermediate contact waves.

19.6.3 HARTEN-LAX-VAN LEER-EINFELDT (HLLE) SCHEME

Having already introduced the HLL (Harten-Lax-van Leer) scheme as a robust two-wave Riemann solver that approximates solutions using the fastest left- and right-propagating signals, we now consider a refinement introduced by Einfeldt. The HLL method performs well in capturing shocks and rarefactions but tends to degrade in accuracy near vacuum states or under extreme nonlinearities, particularly because it does not enforce strict positivity of key variables like density and pressure. To address these shortcomings, the HLLE (Harten-Lax-van Leer-Einfeldt) scheme was

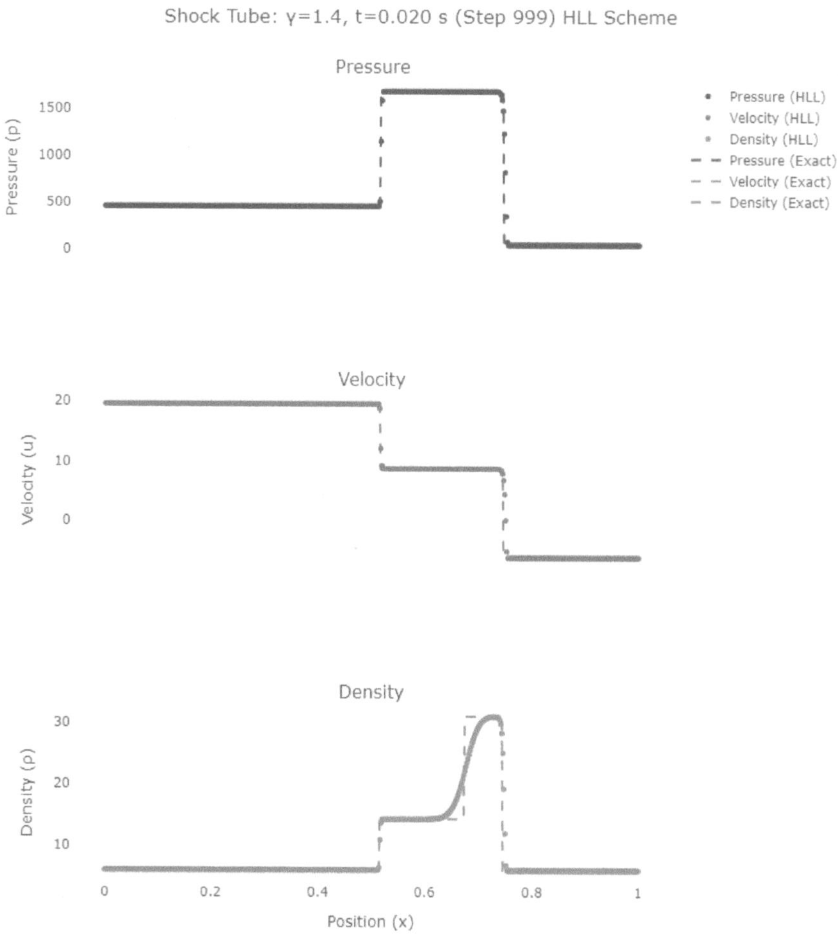

Figure 19.19: **HLL scheme applied on test case from Lax with 500 cells.**

proposed as a modification that guarantees positivity preservation and enhanced entropy compliance. Though HLLE shares the two-wave structure of HLL, it introduces more conservative estimates for signal speeds, making it especially suitable for highly nonlinear or transonic regimes where robustness is paramount.

The HLLE scheme is a variant of the HLL approximate Riemann solver introduced by Einfeldt (1988) to address shortcomings of HLL, particularly in preserving positivity of density and pressure in highly nonlinear flows. While the HLL method already provides a stable two-wave approximation, HLLE further modifies the wave speed estimates S_L and S_R to guarantee physically admissible solutions.

HLLE replaces the contact wave in the Riemann fan with a single intermediate state and computes the flux as:

$$\mathsf{F}_{\text{HLLE}} = \begin{cases} \mathsf{F}_L, & S_L > 0, \\ \dfrac{S_R \mathsf{F}_L - S_L \mathsf{F}_R + S_L S_R (\mathsf{Q}_R - \mathsf{Q}_L)}{S_R - S_L}, & S_L \le 0 \le S_R, \\ \mathsf{F}_R, & S_R < 0 \end{cases}$$

The wave-speed estimates are chosen conservatively to straddle the fan while ensuring one speed is non-positive and the other non-negative:

$$S_L = \min\big(0,\ u_L - a_L,\ u_R - a_R,\ \tilde{u} - \tilde{a}\big), \qquad S_R = \max\big(0,\ u_L + a_L,\ u_R + a_R,\ \tilde{u} + \tilde{a}\big).$$

$$S_L = \min(0, u_L - a_L, \tilde{u} - \tilde{a}), \quad S_R = \max(0, u_R + a_R, \tilde{u} + \tilde{a})$$

where $\tilde{u}$ and $\tilde{a}$ are Roe-averaged values. The HLLE scheme does not resolve the contact discontinuity but offers improved robustness under strong shocks and near-vacuum conditions. It is commonly used in applications where enforcing positivity and entropy consistency is more important than capturing contact wave accuracy.

Case Study 19B: Hypersonic Flow over a Cylinder Using HLLE Scheme

Problem Definition

We consider a two-dimensional Hypersonic inviscid flow past a circular cylinder of diameter 0.1 units. The freestream Mach number is set to $M_\infty = 10$, characterizing a strong supersonic regime. The objective of this case is to capture the bow shock formation and visualize the resulting shock structure around the blunt body. The domain is assumed to be symmetric about the horizontal midline. This setup mimics canonical blunt-body aerodynamics often encountered in high-speed aerospace applications.

Governing Equations

The flow is modeled using the compressible Euler equations in conservative form:

$$\frac{\partial \mathtt{W}}{\partial t} + \nabla \cdot \mathtt{Q}(\mathtt{W}) = 0$$

where $\mathtt{W}$ is the vector of conserved quantities and $\mathtt{Q}$ represents the flux vectors in Cartesian coordinates. The simulation neglects viscous and heat conduction effects, assuming an inviscid and adiabatic flow. The flow is steady-state, compressible, and treated as laminar in the absence of turbulence modeling.

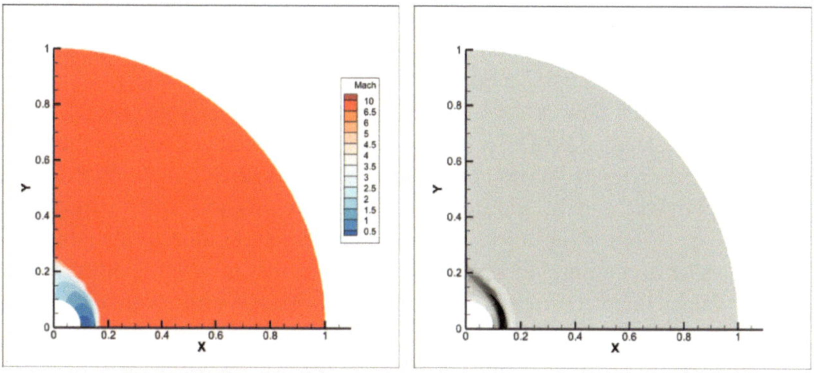

Figure 19.20: **Flow over cylinder at $M = 10$ computed using HLLE scheme. The shadowgraph reveals the sharp bow shock formation.**

Numerical Setup (Grid, BCs, Solver)

The computational domain is discretized using a body-fitted structured grid. A second-order MUSCL-type shock-capturing finite volume method is employed, with the HLLE (Harten-Lax-van Leer-Einfeldt) approximate Riemann solver for robust shock resolution. The boundary conditions include supersonic inflow at the left, extrapolated outflow at the right, and solid wall conditions on the cylinder surface. A constant $\gamma = 5/3$ is assumed for the ideal gas. The solution is advanced to steady state using a local time-stepping method.

Results and Discussion

Figure 19.20 depicts the Mach number contours and corresponding shadowgraph visualization around the cylinder. The HLLE solver effectively captures the detached bow shock directly upstream of the stagnation point. The shadowgraph clearly shows high gradient regions as alternating bands, typical of second-derivative visualization techniques. The symmetry of the shock and smoothness of the captured profile verify both numerical stability and spatial resolution. The stand-off distance of the shock is consistent with theoretical predictions for strong shocks in supersonic flows over blunt objects.

Lessons Learned

The HLLE scheme proved robust and stable for this high Mach number problem, with no oscillations near discontinuities. Key numerical challenges included grid resolution near the stagnation point to accurately capture the curvature and stand-off of the bow shock. While HLLE handled the shock robustly, the scheme is more diffusive compared to HLLC or Roe schemes and might under-resolve shear features in more complex configurations. Future extensions can explore adaptive mesh refinement or hybrid solvers to improve resolution near shocks while maintaining global stability.

Carbuncle instability, that manifests as an unphysical bulge or kink in shock capturing simulations, arises when discrete eigenmodes associated with pressure and shear waves decouple along grid aligned discontinuities. In this simulation this flaw is avoided through a confluence of geometric and algorithmic safeguards. The polar mesh continually rotates face normals relative to the shock front, breaking the alignment that normally amplifies odd even perturbations. In addition, the HLLE flux discards intermediate characteristic information, injecting transverse dissipation that suppresses pressure shear decoupling. First order spatial reconstruction, while diffusive, further damps incipient oscillations before they grow. A conservative CFL limit maintains small time steps, reducing the chance that spurious transverse velocities can coherently reinforce over successive iterations. Together these factors eliminated the feedback loop responsible for carbuncle, yielding a clean bow shock profile even at coarse resolution.

Case Study 19C: Transonic Flow in a Diverging Nozzle

Transonic diverging nozzle is a definitive benchmark for assessing rigorous high-resolution schemes, grid sensitivity, model validation, code verification and continual benchmarking in modern compressible-flow research. Yee et al. in 1981-83 used it to show that stable numerical boundary conditions are vital, because any mismatch quickly shifts the shock and amplifies error. We simulate quasi-one-dimensional compressible flow through a diverging nozzle section using the HLLE (Harten-Lax-van Leer-Einfeldt) approximate Riemann solver. The system is governed by the Euler equations with a geometric source term accounting for area variation:

$$\frac{\partial \mathbf{Q}}{\partial t} + \frac{\partial \mathbf{F}(\mathbf{Q})}{\partial x} = \mathbf{S}(\mathbf{Q}, A),$$

where the conserved variables, flux vector, and source term are respectively:

$$\mathbf{Q} = \begin{bmatrix} \rho \\ \rho u \\ e \end{bmatrix}, \quad \mathbf{F}(\mathbf{Q}) = \begin{bmatrix} \rho u \\ \rho u^2 + p \\ u(e+p) \end{bmatrix}, \quad \mathbf{S} = \begin{bmatrix} 0 \\ p\frac{dA}{dx} \\ 0 \end{bmatrix}.$$

The nozzle geometry is defined by a smooth area function modeling the post-throat diverging region:

$$A(x) = 1.398 + 0.347 \tanh(0.8x - 4.0),$$

which produces a continuously expanding duct without an explicit geometric throat.

The HLLE scheme provides robust shock-capturing capability for this nonlinear hyperbolic system. Unlike Roe's solver, HLLE resolves only the fastest

left- and right-moving waves, offering enhanced stability at the cost of slightly increased diffusivity at contact discontinuities.
The system is discretized on a uniform mesh of nx cells over domain $x \in [0, L]$. We solve for area-weighted conservative variables $\mathbf{W}_c = A \cdot \mathsf{Q}$ with the finite volume update:

$$\mathbf{W}_c^{n+1} = \mathbf{W}_c^n - \frac{\Delta t}{\Delta x} (\mathbf{F}_{i+1/2} - \mathbf{F}_{i-1/2}) + \Delta t \cdot \mathbf{S}_i,$$

where the HLLE numerical flux at interfaces is:

$$\mathbf{F}_{\text{HLLE}} = \begin{cases} \mathbf{F}_L, & S_L > 0, \\ \mathbf{F}_R, & S_R < 0, \\ \dfrac{S_R \mathbf{F}_L - S_L \mathbf{F}_R + S_L S_R (\mathsf{Q}_R - \mathsf{Q}_L)}{S_R - S_L}, & \text{otherwise.} \end{cases}$$

Wave speeds are estimated as:

$$S_L = \min(u_L - a_L, u_R - a_R), \quad S_R = \max(u_L + a_L, u_R + a_R),$$

with sound speed $a = \sqrt{\gamma p / \rho}$.
Boundary Conditions:
- **Supersonic Inflow:** All variables prescribed:

$$\mathbf{W}_c^{\text{inlet}} = A_{\text{inlet}} \cdot \begin{bmatrix} \rho_{\text{inlet}} \\ \rho_{\text{inlet}} u_{\text{inlet}} \\ e_{\text{inlet}} \end{bmatrix}$$

- **Subsonic Outflow:** Fixed back pressure p_{out} with Riemann invariant extrapolation:

$$\mathscr{R}^- = u - \frac{2a}{\gamma - 1}, \quad u_{\text{out}} = \mathscr{R}^- + \frac{2a_{\text{out}}}{\gamma - 1}$$

Solution Initialization:
Initial conditions respect the geometric source term:

$$\mathbf{W}_c^0(x) = A(x) \cdot \begin{bmatrix} \rho^0(x) \\ \rho^0(x) u^0(x) \\ e^0(x) \end{bmatrix},$$

typically initialized with uniform supersonic flow or isentropic profiles.

Table 19.5
Boundary Conditions

Location	Condition Type	Prescribed Quantities
Inlet	Dirichlet (fixed)	$\rho = 0.5, \quad u = 1.0, \quad E = 3.17059$
Outlet	Mixed (Dirichlet + Extrapolated)	$p = 1.7815$ (fixed), ρ, u from interior, $E = \frac{p}{\gamma-1} + \frac{1}{2}\rho u^2$

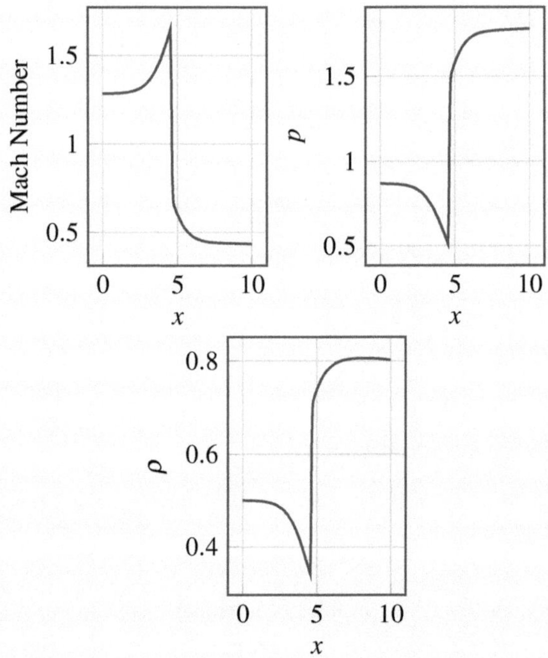

Figure 19.21: **Distribution of Mach number, pressure and density inside diverging section of a transonic nozzle investigated via HLLE scheme.**

Upon convergence, primitive variables are reconstructed:

$$\rho = \frac{\mathbf{W}_c[0]}{A}, \quad u = \frac{\mathbf{W}_c[1]}{\mathbf{W}_c[0]}, \quad p = (\gamma - 1)\left(\frac{\mathbf{W}_c[2]}{A} - \frac{1}{2}\rho u^2\right), \quad M = \frac{u}{a}.$$

Conservation diagnostics monitor solution quality:

$$\text{Mass Flux} = A\rho u, \quad \text{Energy Flux} = Au(e+p).$$

The HLLE scheme's inherent diffusivity provides entropy-satisfying solutions, making it particularly suitable for transonic flows with shocks in the diverging section, despite not modeling the upstream converging region explicitly.

Results and Discussion

Figure 19.21 shows the presence of a shock wave inside the diverging section of the nozzle. In the Mach number plot, there is a distinct discontinuity where the flow transitions abruptly from supersonic ($M > 1$) to subsonic ($M < 1$), which is an indication of a normal shock. Correspondingly, the pressure and density plots exhibit sharp jumps at the same axial location, with pressure

Table 19.6
Flux-Vector Splitting (FVS) vs. Flux-Difference Splitting (FDS) Schemes

Flux-Vector Splitting (FVS)	Flux-Difference Splitting (FDS)
Steger-Warming split	Godunov scheme (exact Riemann solver)
Van Leer split	Roe approximate Riemann solver
AUSM family	Osher approximate Riemann solver
Zha-Bilgen split	HLL family (HLL, HLLC, HLLE)
Harten-Yee TVD split	Kurganov-Tadmor central-upwind scheme

and density increasing suddenly across the shock. This behavior is consistent with the physics of a transonic nozzle flow where, under certain back pressure conditions, a shock forms in the diverging section to adjust the flow from supersonic to subsonic, ensuring compatibility with the imposed outlet boundary condition.

19.7 SUMMARY OF SCHEMES

The gas dynamics schemes can be classified as Flux-vector splitting (FVS) and flux-difference splitting (FDS) schemes, which provide two contrasting philosophies for building upwind finite-volume schemes for hyperbolic conservation laws (see Table 19.6).

Flux-Vector Splitting (FVS) decomposes the flux into parts moving with positive/negative wave speeds and upwinds them accordingly. It is first-order accurate (unless extended with reconstructions) and tends to smear contact discontinuities because it does not resolve the full Riemann fan.

Flux-Difference Splitting (FDS) (e.g., Roe, HLL, HLLC) computes the flux based on the solution to a Riemann problem (exact or approximate). It embeds wave structure information, allowing sharper resolution of discontinuities and better preservation of invariants. FDS can achieve higher-order accuracy with reconstructions (e.g., MUSCL, WENO).

Lessons Learned

☞ Finite volume methods ensure exact conservation of physical quantities by integrating balance laws over control volumes and approximating interface fluxes. This makes them ideal for discontinuous, compressible flows.

☞ The CIR scheme demonstrates how directional upwinding stabilizes first-order finite difference schemes and helps visualize the link between characteristics and finite volume updates.

☞ Godunov's scheme introduced shock-capturing without artificial viscosity by solving exact Riemann problems, but its first-order nature leads to excessive diffusion in smooth regions.

☞ Roe's method provides an efficient linearized alternative to exact solvers, maintaining wave decomposition and satisfying conservation, but it requires entropy fixes near sonic rarefactions.

☞ Expansion shocks are numerical artifacts in Roe-type solvers; applying entropy corrections like Harten or Chakravarthy methods restores physical consistency near sonic points.

☞ MUSCL reconstruction increases spatial accuracy to second order using slope limiters to prevent spurious oscillations, preserving monotonicity while sharply resolving discontinuities.

☞ WENO schemes achieve 5th-order accuracy in smooth regions while adaptively suppressing oscillations near discontinuities using smoothness-based nonlinear weights.

☞ Although easy to implement and stable, flux-vector-splitting methods like Steger-Warming are overly diffusive and unable to resolve contact discontinuities or expansion fans sharply.

☞ HLL and HLLE approximate Riemann solvers avoid complex eigen-decompositions and enforce entropy and positivity conditions but trade off contact accuracy for robustness.

☞ Interpolating in characteristic variables ensures that limiters satisfy the TVD condition, which is essential for stable, non-oscillatory reconstruction in nonlinear systems.

PROBLEMS

19P-1 Consider the following system of partial differential equations:

$$\frac{\partial u_1}{\partial t} + \frac{\partial u_2}{\partial x} = 0,$$

$$\frac{\partial u_2}{\partial t} + \frac{\partial u_3}{\partial x} = 0,$$

$$\frac{\partial u_3}{\partial t} + 3\frac{\partial u_1}{\partial x} - 19\frac{\partial u_2}{\partial x} + 6\frac{\partial u_3}{\partial x} = 0.$$

Is this system hyperbolic?

19P-2 Solve the Riemann problems for Burgers' equation with two different initial data sets.

$$u_t + uu_x = 0$$

The Riemann problem for Burgers' equation involves initial data that is piecewise constant with a discontinuity at some point, typically x=0 in this case.

The initial data for the first case is $u_{\text{left}} = 1$ and $u_{\text{right}} = -1$. The second case swaps these values: $u_{\text{left}} = -1$ and $u_{\text{right}} = 1$.

19P-3 Provide clear explanation of the concepts used in numerical methods for hyperbolic partial differential equations:

i Monotonicity preserving schemes,

ii Reconstruction techniques,

iii Slope limiters, and

iv Total variation diminishing (TVD) methods.

19P-4 For an isothermal flow, demonstrate that the Euler equations can be expressed as:

$$\frac{\partial}{\partial t}\rho + \frac{\partial}{\partial x}(\rho u) = 0,$$

$$\frac{\partial}{\partial t}(\rho u) + \frac{\partial}{\partial x}(\rho u^2 + a^2\rho) = 0,$$

where $a^2 = \left(\frac{dp}{d\rho}\right)_T = RT$.

Write the isothermal Euler equations in the forms

$$\frac{\partial}{\partial t}\mathbf{u} + \frac{\partial}{\partial x}(\mathbf{w}) = 0$$

and

$$\frac{\partial}{\partial t}\mathbf{u} + A\frac{\partial}{\partial x}\mathbf{u} = 0,$$

where $\mathbf{w}$ is the flux vector and A is the Jacobian matrix.

19P-5 Given the Jacobian matrix A derived in Problem **19P-4**:

i Determine the characteristic values, the left characteristic vectors, and the right characteristic vectors.

ii Based on the results from part (a), determine a matrix Q that diagonalizes A, and also compute Q^{-1}.

iii Using the findings from parts (a) and (b), rewrite the isothermal Euler equations in the following form:

$$\frac{\partial}{\partial t}\mathbf{q} + \Lambda\frac{\partial}{\partial x}\mathbf{q} = 0,$$

where Λ is a diagonal matrix.

iv Using the results from parts (a)-(c), express the isothermal Euler equations as $d\mathbf{v}_i = 0$ when $dx = \lambda_i dt$. If possible, solve the compatibility relations $d\mathbf{v}_i = 0$.

19P-6 Consider a Sod shock tube filled with a gaseous mixture composed of Ethylene (C_2H_4) at 7.04%, Oxygen (O_2) at 21.1%, and Nitrogen (N_2) at 71.86%, by molar fractions. The resulting mixture has an average molar mass $M = 28.86189\,\text{kg/mol}$ and a specific heat ratio $\gamma = 1.38003$. The initial conditions across the diaphragm separating the driver (left) and driven (right) chambers are specified as follows:

Left (Driver Chamber):

$$T_L = 5000\,\text{K}, \quad \rho_L = 100\,\text{kg/m}^3, \quad p_L = \rho_L R T_L = 100 \times R \times 5000, \quad u_L = 0.0\,\text{m/s}$$

Right (Driven Chamber):

$$T_R = 3000\,\text{K}, \quad \rho_R = 10\,\text{kg/m}^3, \quad p_R = \rho_R R T_R = 10 \times R \times 3000, \quad u_R = 0.0\,\text{m/s}$$

where $R = R_u/M$. $R_u = 8314.462618\,\text{J/(kmol·K)}$ is the universal gas constant and M is the molar mass of the gas mixture. Determine the pressure ratio across the normal shock wave generated after diaphragm rupture. Compare result with Saurel et al. (1994) work.

19P-7 Consider the one-dimensional Euler equations on the domain $x \in [0,1]$ with 200 uniformly control volumes. Let the ratio of specific heats be $\gamma = 1.4$. Use the following initial conditions for the Lax problem, advanced in time until $t = 0.15$s.

Left state: $\rho_L = 0.445$, $u_L = 0.698$, $p_L = 3.528$

Right state: $\rho_R = 0.5$, $u_R = 0.0$, $p_R = 0.571$.

Solve this Riemann problem using the Roe flux function. Investigate using

i Roe approximation with flux correction

ii Roe approximation with MUSCL and limiters

iii Roe approximation with WENO

Comment on how each step influences the solution accuracy, and outline the expected behavior of the shock and contact discontinuity in this Lax-type problem.

19P-8 Consider the one-dimensional Euler equations on the domain $x \in [0,1]$ with 200 uniformly spaced control volumes. Let the ratio of specific heats be $\gamma = 1.4$. Use the following initial conditions for a Mach 3 test problem (adapted from Arora and Roe, 1997), advanced in time until $t = 0.09$ s.

Left state: $\rho_L = 3.857, \quad u_L = 0.92, \quad p_L = 10.333$

Right state: $\rho_R = 1.0, \quad u_R = 3.55, \quad p_R = 1.0$

Solve this Riemann problem using the Roe flux function. Investigate the solution by comparing the following approaches:

i Roe approximation with flux correction

ii Roe approximation with MUSCL and limiters

iii Roe approximation with WENO

Comment on how each approach influences solution accuracy, and outline the expected behavior of the shock and contact discontinuity for this Mach 3 problem.

19P-9 Consider the one-dimensional Euler equations on the domain $x \in [0,1]$ with 200 uniformly spaced control volumes. Let the ratio of specific heats be $\gamma = 1.4$. Use the following initial conditions for a supersonic shock tube problem, advanced in time until $t = 0.1562$ s:

Left state: $\rho_L = 8.0, \quad u_L = 0.0, \quad p_L = 8.0$
Right state: $\rho_R = 0.2, \quad u_R = 0.0, \quad p_R = 0.2$

Solve this Riemann problem using the Roe flux function. Investigate the solution by comparing Roe approximation with MUSCL and use the following slope limiter (Kim and Kim 2005):

$$\psi(r) = \max\left[0, \min\left(2r, 2, \beta\right)\right],$$

$$\beta = \left(-\tfrac{1}{4} - \tfrac{1}{2}\kappa + \tfrac{1}{12}\kappa^2\right) r + \left(\tfrac{1}{4}\kappa + \tfrac{1}{2} - \tfrac{1}{12}\kappa^2\right),$$

$$\kappa = 1 - |u| \tfrac{\Delta t}{\Delta x}.$$

Comment on how each approach influences the solution accuracy, and outline the expected behavior of the resulting shock waves and contact discontinuities in this supersonic shock tube problem.

19P-10 Consider the one-dimensional Euler equations on the domain $x \in [0,1]$ using 50,100, 400 uniformly spaced cells. The ratio of specific heats is given by $\gamma = \frac{5}{3}$. The flow is initialized uniformly with $\rho = 1.0$, $u = 0.0$, $p = 0.6$. At the left boundary (x=0) a time-dependent velocity driver is imposed such that

$$u(t) = A \sin\left(2 \cdot \pi \cdot f \cdot t\right)$$

with $A = 0.02$ and 0.2. At the right boundary the conditions are transmissive. The numerical method employs the HLL (Harten-Lax-van Leer) approximate Riemann solver without reconstruction to compute the fluxes between control volumes. The solution is advanced in time until $t = 4.0$s. A CFL number of 0.3 is used to determine the time step for stability. Write a Python code that solves this linear sound wave problem using the HLL scheme. Investigate the influence of grid cells on the solution.

19P-11 Consider the 1-D hyperbolic equation

$$u_t + c u_x = 0$$

where u is a scalar and c is a constant wave speed ($c > 0$). The computational domain is $0 \leq x \leq 1$, and the initial conditions are specified as:

$$u(x,0) = \begin{cases} 1 & \text{if } x \leq 0.3 \\ 0 & \text{if } x > 0.3 \end{cases}$$

Use the Courant-Isaacson-Rees (CIR) method to advance the solutions by a distance equal to half the length of the computational domain. Consider the CFL values

CFL $= 0.2, 0.5, 0.7, 1.0$ for the analysis. Plot both the exact and computed solutions for each CFL value.

19P-12 Consider the model equation $u_t + u_x = 0$ for $-1 \le x \le 1$ with periodic boundary conditions and the initial data $u(x,0) = u_0(x)$. The exact solution is given by:

$$u(x,t) = \begin{cases} 1 & \text{if } x < t, \\ 0 & \text{if } x \ge t. \end{cases}$$

The problem is solved using a WENO (Weighted Essentially Non-Oscillatory) scheme, with the domain divided into different numbers of grid points l and a fixed step size $h = 2/l$. Various initial conditions are tested, including
$u_0(x) = \sin(\pi x)$,
$u_0(x) = \sin^4(\pi x)$,
and the errors are measured at $t = 1$.

i Using the WENO scheme, solve the 1D advection equation for the initial condition $u_0(x) = \sin(\pi x)$. Plot the numerical solution at $t = 1$ and compare it with the exact solution.

ii Compute the errors at $t = 1$ using the L_1 and L_∞ norms for the initial condition $u_0(x) = \sin^4(\pi x)$. What do we observe about the accuracy of the method as the grid is refined?

iii Discuss the accuracy and convergence rates observed in the errors for the given initial conditions, as presented in the tables. How do the results compare with the theoretical orders of convergence for the WENO scheme?

19P-13 Toro (2009) have suggested the numerical tests for checking the implementation of exact Riemann solver (see Table 19.7). Conduct these numerical test on uniform grid and find the time required to achieve the solution.

Table 19.7
Toro's numerical test dataset

Test	ρ_{Left}	u_{Left}	p_{Left}	ρ_{Right}	u_{Right}	p_{Right}
1	1.0	0.75	1.0	0.125	0.0	0.1
2	1.0	-2.0	0.4	1.0	2.0	0.4
3	1.0	0.0	1000.0	1.0	0.0	0.01
4	5.99924	19.5975	460.894	5.99242	-6.19633	46.0950
5	1.0	-19.59745	1000.0	1.0	-19.59745	0.01

REFERENCES

R. Courant, E. Isaacson, M. Rees, On the solution of nonlinear hyperbolic differential equations by finite differences, Comm. Pure Appl. Math. 5, 243-255. MR 53336, 1952

M. S. Liou, The Evolution of Upwind Schemes - A Historical Perspective. Def. Sci. J. 60.6, 606-619, 2010.

S. K. Godunov, V. S. Ryabenkii, Introduction to the Theory of Difference Schemes, Wiley (Interscience), New York, 1964.

J. L. Steger, R. F. Warming, Flux vector splitting of the inviscid gas dynamic equations with applications to finite difference methods. J. Comput. Phys. 40, 263-293 (1981)

B. Van Leer, Flux vector splitting for the Euler equations. In: Proceedings of the 8th International Conference on Numerical Methods in Fluid Dynamics, Springer-Verlag, Berlin (1982)

P. L. Roe, Approximate Riemann Solvers, Parameter Vectors, and Difference Schemes, Journal of Computational Physics, Vol. 43, 1981, pp. 357-372.

M. Arora and P. L. Roe, A well-behaved TVD limiter for high-resolution calculations of unsteady flow. Journal of Computational Physics, 132(1): 3-11, 1997.

K. H. Kim, C. Kim, Accurate, efficient and monotonic numerical methods for multi-dimensional compressible flows: Part II: Multi-dimensional limiting process, J. Comput. Phys. 208 (2), 570-615, 2005.

K. Sawada, E. Shima, and T. Matsuda, A TVD scheme using Roe's flux and the ambient boundary condition, Memoirs of the Faculty of Engineering, Kyoto University, 1989 (?).

C. Hirsch, Numerical Computation of Internal and External Flows, Vol. 2: Computational Methods for Inviscid and Viscous Flows, Wiley-Interscience, 1992.

A. Harten and J. M. Hyman, Self-Adjusting Grid Methods for One-Dimensional Hyperbolic Conservation Laws, J. Comput. Phys., 50(2): 235-269, 1983.

P. L. Roe, J. Pike, Efficient Construction and Utilization of Approximate Riemann Solutions. In Computing Methods in Applied Science and Engineering VI, eds. R. Glowinski and J. L. Lions, North-Holland, pp. 499-518, 1984.

S. Osher, S. Chakravarthy, High Resolution Schemes and the Entropy Condition. SIAM J. Numer. Anal., 21(5): 955-984, 1984.

A. Harten, High Resolution Schemes for Hyperbolic Conservation Laws. J. Comput. Phys., 49(3): 357-393, 1983.

A. Harten, B. Engquist, S. Osher, S. R. Chakravarthy, Uniformly high order accurate essentially non-oscillatory schemes, III. Journal of Computational Physics, 71(2), 231-303, 1983.

S. R. Chakravarthy and S. Osher, A New Class of High Accuracy TVD Schemes for Hyperbolic Conservation Laws, AIAA Paper, No. 85-0363, 1985.

S. R. Chakravarthy, The versatility and reliability of Euler solvers based on high-accuracy TVD formulations, AIAA Paper, No. 86-0243, 1986.

G. S. Jiang, C. W. Shu, Efficient implementation of weighted ENO schemes. Journal of Computational Physics, 126(1), 202-228, 1996.

B. van Leer, Towards the Ultimate Conservative Difference Scheme II: Monotonicity and Conservation Combined in a Second-Order Scheme, J. Comput. Phys., 361-370, 1974.

B. van Leer, Towards the Ultimate Conservative Difference Scheme V: A Second-Order Sequel to Godunov's Method, J. Comput. Phys., 101-136, 1979.

A. Harten, J. M. Hyman, P. D. Lax, On finite-difference approximations and entropy conditions for shocks, Comm. Pure Appl. Math., 297-322, 1976.

A. Harten, On a Class of High Resolution Total-Variation-Stable Finite Difference Schemes, SIAM J. Numer. Anal., 1-23, 1984.

P. K. Sweby, High Resolution Schemes Using Flux Limiters for Hyperbolic Conservation Laws, SIAM J. Numer. Anal., 995-1011, 1984.

P. Colella, P. Woodward, The Piecewise Parabolic Method (PPM) for Gasdynamical Simulations, J. Comput. Phys., 174-201, 1984.

A. Harten, S. Osher, Uniformly High-Order Accurate Nonoscillatory Schemes, I, SIAM J. Numer. Anal., 279-309, 1987.

X.-D. Liu, S. Osher, T. Chan, Weighted Essentially Non-Oscillatory Schemes, J. Comput. Phys., 200-212, 1994.

E. F. Toro, Riemann Solvers and Numerical Methods for Fluid Dynamics, Springer, 2009.

J. Qiu and C.-W. Shu, Finite Difference WENO Schemes with Lax-Wendroff-Type Time Discretizations, SIAM J. Sci. Comput. 24(6): 2185-2198, 2003.

A. K. Henrick, T. D. Aslam, and J. M. Powers, Mapped Weighted Essentially Non-Oscillatory Schemes: Achieving Optimal Order Near Critical Points, J. Comput. Phys. 207(2): 542-567, 2005.

R. Borges, M. Carmona, B. Costa, and W. S. Don, An Improved Weighted Essentially Non-Oscillatory Scheme for Hyperbolic Conservation Laws, J. Comput. Phys. 227(6): 3191-3211, 2008.

X. Y. Hu, Q. Wang, and N. A. Adams, An Adaptive Central-Upwind Weighted Essentially Non-Oscillatory Scheme, J. Comput. Phys. 229(23): 8952-8965, 2010.

X. Y. Hu and N. A. Adams, Scale Separation for Implicit Large Eddy Simulation, J. Comput. Phys. 230(19): 7240-7249, 2011.

F. Acker, R. Borges, and B. Costa, An Improved WENO-Z Scheme, J. Comput. Phys. 313: 726-753, 2016.

D. S. Balsara, S. Garain, and C.-W. Shu, An Efficient Class of WENO Schemes with Adaptive Order, J. Comput. Phys. 326: 780-804, 2016.

J. Zhu and J.-X. Qiu, A New Fifth-Order Finite Difference WENO Scheme for Solving Hyperbolic Conservation Laws, J. Comput. Phys. 318: 110-121, 2016.

Y. Wang, Y. Du, K. Zhao, and L. Yuan, A Low-Dissipation Third-Order WENO Scheme with a New Reference Smoothness Indicator, Int. J. Numer. Meth. Fluids 92(8): 1212-1234, 2020.

P. D. Lax, Weak solutions of nonlinear hyperbolic equations and their numerical computation. Comm. Pure Appl. Math. 7(1): 159-193, 1954.

R. Saurel, M. Larini, and J.-C. Loraud, Exact and Approximate Riemann Solvers for Real Gases, J. Comput. Phys., 112(1): 137-162, 1994.

G. Strang, On the Construction and Comparison of Difference Schemes. SIAM Journal on Numerical Analysis, 5(3), 506-517, 1968.

O. Igra, G. Hu, J. Falcovitz, B.Y. Wang, "Shock wave reflection from a wedge in a dusty gas," *International Journal of Multiphase Flow*, vol. 30, pp. 1139-1169, 2004.

H. C. Yee, Numerical Approximation of Boundary Conditions with Applications to Inviscid Equations of Gas Dynamics. NASA Technical Memorandum 81265, 1981.

H. C. Yee, R. F. Warming, A. Harten, Implicit Total Variation Diminishing Schemes for Steady-State Calculations. NASA Technical Memorandum 84342, 1983.

20 Turbulence Models

In this chapter, we will explore the concept of turbulence and its modeling using Reynolds-averaged Navier-Stokes (RANS) equations. After finishing this chapter, you should be able to:

- Understand the derivation and physical meaning of RANS equations.
- Learn about Reynolds stress and the closure problem in turbulence.
- Develop the ability to apply first-order turbulence models like the mixing-length model.
- Be familiar with the concept of turbulent scalar transport and eddy diffusivity.
- Analyze the limitations of simple turbulence models and understand the need for higher-order closures.

Turbulence is one of the most complex and widely encountered phenomena in fluid mechanics, characterized by chaotic, three-dimensional, and time-dependent motions across a wide range of scales. Turbulent flows are prevalent in natural systems such as atmospheric circulation, ocean currents, and blood flow as well as in engineered systems like aircraft, combustion engines, and pipelines. Despite its ubiquity, turbulence remains a challenging subject, both theoretically and computationally. Its inherently unsteady and nonlinear behavior makes direct numerical simulation (DNS) computationally prohibitive for most practical applications. Therefore, turbulence modeling becomes necessary to bridge the gap between theory and application, allowing engineers and scientists to capture essential features of turbulent flows without resolving all scales.

This chapter introduces the fundamental concepts underlying turbulence modeling, with particular focus on the Reynolds-Averaged Navier-Stokes (RANS) equations. Starting from the Navier-Stokes equations for unsteady compressible flow, we derive the RANS formulation through Reynolds decomposition and averaging. This leads to the appearance of new terms known as Reynolds stresses, which account for the momentum transport due to velocity fluctuations. These terms introduce closure problems, requiring turbulence models to relate the turbulent stresses to mean flow quantities.

We begin by examining the RANS form of the continuity and momentum equations, followed by the turbulent energy equation. Each term is carefully analyzed in the context of incompressible flow, and the influence of turbulent transport and heat transfer is explained through the introduction of Reynolds heat flux and turbulent scalar transport. The Boussinesq approximation is presented as the foundation for first-order closure models, wherein turbulent stresses are modeled analogously to molecular diffusion using the concept of eddy viscosity. Prandtl's mixing length theory is introduced as a classical approach for estimating eddy viscosity, offering a conceptual analogy to kinetic theory.

DOI: 10.1201/9781003663171-20

Further, the chapter outlines first-order turbulence models, including zero-, one-, and two-equation models. Particular emphasis is placed on the limitations of simple eddy viscosity models and the importance of quantities like the turbulent Prandtl number in modeling scalar transport. The formulation of turbulent fluxes for momentum and thermal energy is discussed, along with their dependence on local flow properties.

The chapter concludes with a discussion of scalar transport equations in turbulent flows and the challenges of modeling the effects of turbulent fluctuations. These form the basis for more advanced models such as k-ε and k-ω, which will be discussed in subsequent chapters. By the end of this chapter, readers will be equipped with the foundational understanding of turbulence modeling using RANS equations, the physical significance of Reynolds stresses, and the necessity of closure models to simulate real-world turbulent flows effectively.

20.1 REYNOLDS AVERAGING

An ensemble averaging is the most general type of Reynolds averaging which is suitable for flows that decay in time. It is defined as (Fériet 1949):

$$< \Psi(x_i,t) >= \lim_{N\to\infty} \frac{1}{N} \sum_{n=1}^{N} \Psi(x_i,t) \tag{20.1}$$

Once the flow becomes statistical stationary (statistics become independent of time) it can be time averaged which is defined as:

$$\overline{\Psi(x_i)} = \lim_{t\to\infty} \int_{-t/2}^{t/2} \Psi(x_i,t)dt \tag{20.2}$$

According to the Reynolds decomposition of turbulence quantities, we have:

$$\psi = \overline{\Psi} + \psi^{'} \tag{20.3}$$

Figure 20.1 shows a real turbulent velocity data probed at a single point in flow field. According to Reynolds Averaging, all the flow quantities like all velocity components and pressure etc. are fluctuating as:

$$u = \bar{u} + u', \; v = \bar{v} + v', \; w = \bar{w} + w' \text{ and } p = \bar{p} + p'.$$

In Figure 20.1 we can bifurcate the the velocity $u(t)$ signal into fluctuating and a mean quantity. if the mean is not changing with time then such flow can be locally steady. However, if the mean is changing with the time then such flow is considered as unsteady. in turbulent flow, the velocity of the fluid is a continuous random function of time and space. It is a random function cause the instantaneous value of the fluctuating quantity at any point cannot be predicted from the mean value of the flow quantities. The value of the velocity at any point is distributed according to the laws of probability. If the instantaneous velocity is considered the sum of a mean velocity

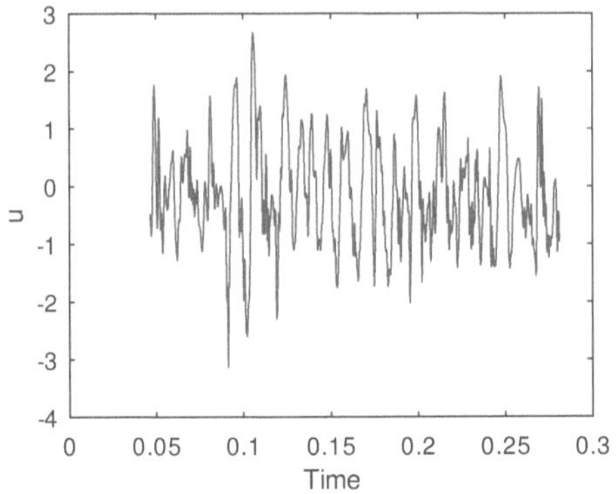

Figure 20.1: **A real turbulent velocity data probed at a single point in flow field.**

and fluctuating velocity, the distribution of fluctuating velocity follows the Gaussian or normal distribution.

Averaging rules: Fériet (1949) listed the following rules for averaging:

Rule i The time-average of all mean quantities is equal to the mean value $\overline{\overline{u}} = \overline{u}$.

Rule ii All fluctuating quantities are zero when time-averaged so $\overline{u'} = 0, \overline{v'} = 0, \overline{w'} = 0$, $\overline{\rho'} = 0$, $\overline{p'} = 0$

Rule iii The product of mean and fluctuating quantities will be zero as well.

Rule iv The fluctuating quantities will be zero when time averaged.

20.2 RANS EQUATIONS

Turbulent flow can be predicted through numerical calculations of above equations, provided we use non-dissipative numerical schemes and control time and progress of solution in consideration with Eddy Turn-over Time.

We can apply Reynolds Averaging on Conservation equations and in this section Reynolds Averaged Navier-Stokes Equations are developed.

History of RANS Turbulence Modeling

1941	Kolmogorov discovered the slope of turbulence spectrum
1942	Kolmogorov introduced ω as turbulence quantity
1945	Prandtl gave his k-ℓ model
1945	Chou formulated the Second moment Closure model or Reynolds Stress Transport (RSM) model

1951	Rotta modeled terms for RSM model and accomplished closure
1965	Van Driest devised a viscous damping correction for the mixing-length model
1967	Wolfstein gave one equation model for fluctuating turbulence kinetic energy
1968	Donaldson and Rosebaum advocated the invariant modeling for *realizability*
1968	Rotta proposed k-k$_\ell$ model
1970	Daly and Harlow modeled turbulent diffusion term
1974	Cebeci and Smith refined the eddy-viscosity/mixing-length model
1975	Launder, Reece and Rodi proposed the closure for Second Moment Closure model
1978	Lumley developed procedure to ensure *realizability* constraints for turbulence models
1978	Launder and Sharma proposed k-epsilon model

History of RANS Turbulence Modeling

1988	Wilcox offered his two equation k-ω model
1992	Yakhot etal. developed RNG model using Re-Normalisation Group (RNG) methods
1992	Spalart and Allmaras gave the most accurate one equation model for practical turbulent-flow applications
1992	Gatski and Speziale classified Algebraic Stress Models (ASM)
1994	Menter proposed Shear Stress Transport turbulence model (SST model)
1995	Robinson, Harris and Hassan proposed k-ζ (k-zeta), also called k-enstrophy model
2004	Menter and Egorov proposed Scale Adaptive model an improved version of Rotta's k-k$_\ell$ model
2009	Langtry and Menter proposed four equation SST model which uses additional γ (intermittency) and Re$_\theta$ equations, also called transition SST model

Turbulence Eddy Viscosity Modeling (EVM)

$$\tau_{ij} = -\rho \overline{u'_i u'_j} \equiv \rho \varepsilon_M \frac{\partial U}{\partial y}$$

Eddy Viscosity Models $\varepsilon_M = \nu_t = w_t \ell$
w_t is velocity scale and ℓ is length scale

0 Equation Models	1 Equation Models	2 Equations Models
Algebraic model	One more PDE	Two more PDE
$\nu_t = \frac{\partial w_t}{\partial x} \ell^2$	$\nu_t = c_\nu \ell \sqrt{k}$	$\nu_t = c_\nu c_z (k^2/\varepsilon)$
Dependence on local conditions	Dependence on specific energy of fluctuations	Dependence on length scale obtained from a transport equation

Turbulence models are classified as Low Reynolds number or High Reynolds number based on how they treat the near-wall region of turbulent flows, particularly the viscous sublayer. Low Reynolds number models are designed to resolve the entire boundary layer, including the viscous sublayer, buffer layer, and logarithmic region. These models incorporate damping functions or additional terms to accurately represent turbulence behavior close to the wall where viscous effects dominate. To function correctly, they require fine mesh resolution near walls, typically placing the first grid point at a non-dimensional wall distance of $y^+ \leq 1$. As a result, they do not rely on wall functions, allowing them to provide detailed predictions in flows with complex near-wall physics such as separation, reattachment, or strong adverse pressure gradients.

In contrast, High Reynolds number turbulence models do not attempt to resolve the viscous sublayer explicitly. Instead, they make use of empirical wall functions that assume a logarithmic velocity profile near the wall. These models are computationally more efficient in wall-bounded flows because they require coarser meshes, placing the first computational node at $y^+ > 30$, typically within the logarithmic layer. While this makes them suitable for industrial-scale simulations, they may not capture detailed wall behavior accurately. Examples include the standard $k\varepsilon$ model and classical algebraic models such as Baldwin-Lomax and Johnson-King. Hybrid or blended models like SST $k\omega$ bridge the two categories by switching formulations depending on wall proximity and mesh resolution.

The turbulence models are classified in Table 20.1. The table provides a chronological comparison of turbulence models categorized by Reynolds number applicability. It distinguishes models requiring near-wall resolution from those employing wall functions. The table shows that new turbulence models are mostly low Reynolds number models.

20.2.1 CONTINUITY EQUATION

In this section we perform Reynolds averaging on Continuity equation in tensor notation (Uddin 2023):

$$\frac{\partial \rho}{\partial t} + \frac{\partial}{\partial x_j}(\rho U_j) = 0 \tag{20.4}$$

Table 20.1
Low and High Reynolds number Turbulence models

Low Reynolds Number Models *Do not need wall function* ($y^+ \leq 1$)	**High Reynolds Number Models** *With wall function* (allowable $y^+ \geq 13$)
Low-Re k-ε (1972)	Standard k-ε (1974)
Launder-Sharma Low-Re k-ε (1974)	RSM (1975)
Chien Low-Re k-ε (1982)	LRR-RSM (1975)
Lam-Bremhorst Low-Re k-ε (1981)	Baldwin-Lomax (1978)
Chen-Kim Low-Re k-ε (1987)	Johnson-King (1980)
Standard k-ω (1988)	Speziale-Sarkar-Gatski RSM (1991)
v^2-f Durbin model (1991)	Algebraic Stress Model (ASM) (1976)
Spalart-Allmaras (1992)	RNG k-ε (1986)
Yang-Shih Low-Re k-ε (1993)	Realizable k-ε (1995)
Elliptic Blending RSM (1993)	Shih Quadratic k-ε (1993)
Abe-Kondoh-Nagano Low-Re k-ε (1994)	k-ε-$\overline{v^2}$ (Durbin, 1995)
SST k-ω (1994)	EARSM (1996)
Goldberg Transition Model (1998)	Wallin-Johansson EARSM (2000)
q-ζ (2004)	Hellsten EARSM k-ω (2005)
Walters-Cokljat Transition k-k_ℓ (2008)	
Transition SST (γ-Re$_\theta$) (2009)	
Abdol-Hamid etal. k-k_ℓ (2015)	

As per Reynolds decomposition we divide turbulent quantities into mean and fluctuating components like $\rho = \overline{\rho} + \rho^{'}$ and $u_j = U_j + u^{'}_j$. Here capital letter indicates mean component of turbulent velocity field signal and as per Reynolds averaging we can write $\overline{U_j} = U_j$.

$$\rho = \overline{\rho} + \rho^{'} \tag{20.5}$$

$$\overline{\frac{\partial \rho}{\partial t}} = \overline{\frac{\partial}{\partial t}[\overline{\rho} + \rho^{'}]} \tag{20.6}$$

$$\frac{\partial \overline{\rho}}{\partial t} = \frac{\partial \overline{\overline{\rho}}}{\partial t} + \frac{\partial \overline{\rho^{'}}}{\partial t} \tag{20.7}$$

As $\overline{\overline{\rho}} = \overline{\rho}$ we have

$$\frac{\partial \overline{\rho}}{\partial t} = \frac{\partial \overline{\rho}}{\partial t} + \frac{\partial \overline{\rho^{'}}}{\partial t} \tag{20.8}$$

Further we perform operation:

$$u_j \rho = (\overline{U_j} + u^{'}_j)(\overline{\rho} + \rho^{'})$$
$$u_j \rho = \overline{\rho} U_j + \overline{\rho} u^{'}_j + \rho^{'} U_j + \rho^{'} u^{'}_j$$

$$\overline{u_j\rho} = \overline{\overline{\rho}U_j} + \overline{\overline{\rho}u'_j} + \overline{\rho' U_j} + \overline{\rho' u'_j}$$

$$\overline{u_j\rho} = \overline{\rho}U_j + \overline{\rho' u'_j}$$

We substitute this result and eq. 20.8 into eq. 20.4:

$$\boxed{\frac{\partial\overline{\rho}}{\partial t} + \frac{\partial}{\partial x_j}[\overline{\rho}U_j + \overline{\rho'_j u'_j}] = 0} \tag{20.9}$$

This is RANS Continuity equation valid for unsteady and compressible flows.

In case of incompressible flows ρ is constant and there are no fluctuations in density field so $\rho' = 0$. For that case the applicable continuity equation is:

$$\begin{aligned} \frac{\partial}{\partial x_j}[\overline{U_j}] = 0 \\ \frac{\partial}{\partial x_j}[U_j] = 0 \end{aligned} \tag{20.10}$$

$$\boxed{\frac{\partial}{\partial x_j}[U_j] = 0} \tag{20.11}$$

As j=1,2,3 only we use tensor summation rule for repeated indices:

$$\frac{\partial U_j}{\partial x_j} = \frac{\partial U_1}{\partial x_1} + \frac{\partial U_2}{\partial x_2} + \frac{\partial U_3}{\partial x_3} \tag{20.12}$$

20.2.2 MOMENTUM EQUATION

Navier's momentum equation in tensor notation is:

$$\rho\frac{\partial u_i}{\partial t} + \rho u_j\frac{\partial u_i}{\partial x_j} = -\frac{\partial p}{\partial x_i} + \frac{\partial \tau_{ji}}{\partial x_j} \tag{20.13}$$

Convective term:

$$\frac{\partial u_i u_j}{\partial x_j} = u_j\frac{\partial}{\partial x_j}[u_i] + u_i\frac{\partial}{\partial x_j}[u_j] \tag{20.14}$$

for incompressible flows

$$\frac{\partial u_i u_j}{\partial x_j} = u_j\frac{\partial}{\partial x_j}[u_i] + u_i\cancel{\frac{\partial}{\partial x_j}[u_j]} \tag{20.15}$$

$$\frac{\partial u_i u_j}{\partial x_j} = u_j\frac{\partial u_i}{\partial x_j} \tag{20.16}$$

We substitute this result into eq. 20.13:

$$\rho\frac{\partial u_i}{\partial t}+\rho\frac{\partial u_i u_j}{\partial x_j}=-\frac{\partial p}{\partial x_i}+\frac{\partial \tau_{ji}}{\partial x_j} \tag{20.17}$$

The shear stress is described as for incompressible flows $\tau_{ij}=2\mu S_{ij}$, where S_{ij} is turbulent mean shear strain tensor.

Note that there are symmetric tensors so we have 6 unkowns instead of 9 unknows.

Doing Reynolds time averaging on whole equation yields:

$$\rho\frac{\partial U_i}{\partial t}+\rho\frac{\partial}{\partial x_j}(\overline{U_i U_j}+\overline{u_i^{'} u_j^{'}})=-\frac{\partial \overline{p}}{\partial x_i}+\frac{\partial}{\partial x_j}(2\mu S_{ij}) \tag{20.18}$$

Convective term is brought on RHS and treated as increase in shear stress:

$$\boxed{\rho\frac{\partial U_i}{\partial t}+\rho\frac{\partial}{\partial x_j}(\overline{U_i U_j})=-\frac{\partial \overline{p}}{\partial x_i}+\frac{\partial}{\partial x_j}(2\mu S_{ij}-\rho\overline{u_i^{'} u_j^{'}})} \tag{20.19}$$

$$\rho\frac{\partial U_i}{\partial t}+\rho\frac{\partial}{\partial x_j}(\overline{U_i U_j})=-\frac{\partial \overline{p}}{\partial x_i}+\frac{\partial}{\partial x_j}(\tau_{ij}+\tau_{ij}^{'}) \tag{20.20}$$

The convective fluctuations now treated as turbulent stresses arises due to velocity fluctuations and defined as:

$$\tau_{ij}^{'}=-\rho\begin{bmatrix}\overline{u^{'}u^{'}} & \overline{u^{'}v^{'}} & \overline{u^{'}w^{'}}\\ \overline{v^{'}u^{'}} & \overline{v^{'}v^{'}} & \overline{v^{'}w^{'}}\\ \overline{w^{'}u^{'}} & \overline{w^{'}v^{'}} & \overline{w^{'}w^{'}}\end{bmatrix}$$

Along with u,v,w,p and 6 strain tensors we have total 10 unknowns. However one mass and three momentum conservation equations are not enough to completely solve the set of equations. To close the problem we need model to related fluctuating velocity strain tensor with mean velocity components.

20.2.3 TURBULENT ENERGY EQUATION

We start from balance of enthalpy:

$$\frac{\partial(\rho H)}{\partial t}+\frac{\partial(\rho u_j H)}{\partial x_j}=\frac{\partial p}{\partial t}$$
$$\frac{\partial}{\partial x_j}(u_i\tau_{ij}-q_j)$$

where, $q_j=-k(\partial T/\partial x_j)$.

We substitute:

$$H = \overline{H} + H^{'}$$

and

$$u_j = U_j + u_j^{'}$$

also

$$\tau_{ij} = \mu\left(\frac{\partial U_i}{\partial x_j} + \frac{\partial U_j}{\partial x_i}\right) + \frac{2}{3}\delta_{ij}\frac{\partial U_k}{\partial x_k}$$

Taking time averaging and $H = mc_pT$ with substituting NS eq. into above equation

$$\frac{\partial T}{\partial t} + \frac{\partial \overline{(TU_j)}}{\partial x_j} = \frac{1}{\rho c_p}\left[\frac{\partial \overline{p}}{\partial t} + U_j\frac{\partial \overline{p}}{\partial x_j} + \overline{u_j^{'}\frac{\partial p^{'}}{\partial x_j}}\right] + \frac{\Phi}{\rho c_p} + \frac{\partial}{\partial x_j}\left[\alpha\frac{\partial T}{\partial x_j} - \overline{T^{'}u_j^{'}}\right]$$

We define Reynolds heat flux as:

$$q^R = \rho c_p \overline{T^{'}u_j^{'}}$$

In case of steady, incompressible flows (ignoring pressure fluctuations and viscous dissipation):

$$\boxed{\overline{U_j}\frac{\partial \overline{T}}{\partial x_j} = \frac{\partial}{\partial x_j}\left[\alpha\frac{\partial T}{\partial x_j} - \overline{T^{'}u_j^{'}}\right]} \tag{20.21}$$

RANS conservation equations needed to solve are:

Continuity Equation:

$$\frac{\partial \rho}{\partial t} + \frac{\partial}{\partial x_j}[\rho U_j] = 0 \tag{20.22}$$

Navier-Stokes Equation:

$$\rho\frac{\partial U_i}{\partial t} + \rho\frac{\partial}{\partial x_j}(\overline{U_iU_j}) = -\frac{\partial \overline{p}}{\partial x_i} + \frac{\partial}{\partial x_j}(2\mu S_{ij} - \rho\overline{u_i^{'}u_j^{'}}) \tag{20.23}$$

Energy Equation:

$$\frac{\partial \overline{T}}{\partial t} + \overline{U_j}\frac{\partial \overline{T}}{\partial x_j} = \frac{\partial}{\partial x_j}\left[\alpha\frac{\partial \overline{T}}{\partial x_j} - \overline{T'u'_j}\right] \tag{20.24}$$

20.3 EDDY VISCOSITY BASED MODELS

The shear stresses can be modeled as

$$-\rho\overline{u'v'} = \mu_t\left[\frac{\partial U}{\partial y} + \frac{\partial V}{\partial x}\right]$$

in tensor notation

$$-\rho\overline{u'_i u'_j} = \mu_t\left[\frac{\partial U_i}{\partial x_j} + \frac{\partial U_j}{\partial x_i}\right]$$

The normal stress component in similar way can be

$$-(\rho\overline{u'u'} + \rho\overline{v'v'} + \rho\overline{w'w'}) \neq 2\mu_t\left[\underbrace{\frac{\partial U}{\partial x} + \frac{\partial V}{\partial y} + \frac{\partial W}{\partial z}}_{Zero\ \ see\ continuity\ eq.}\right]$$

Here we can notice that the RHS is zero according to continuity equation, so the correction normally done in model is to subtract the third of error from each of stress component

$$-\rho\overline{u'u'} = 2\mu_t\left\{\frac{\partial U}{\partial x} - \frac{1}{3}\left[\frac{\partial U}{\partial x} + \frac{\partial V}{\partial y} + \frac{\partial W}{\partial z}\right]\right\} - \frac{2}{3}(\rho k)$$

$$-\rho\overline{v'v'} = 2\mu_t\left\{\frac{\partial V}{\partial y} - \frac{1}{3}\left[\frac{\partial U}{\partial x} + \frac{\partial V}{\partial y} + \frac{\partial W}{\partial z}\right]\right\} - \frac{2}{3}(\rho k)$$

$$-\rho\overline{w'w'} = 2\mu_t\left\{\frac{\partial W}{\partial z} - \frac{1}{3}\left[\frac{\partial U}{\partial x} + \frac{\partial V}{\partial y} + \frac{\partial W}{\partial z}\right]\right\} - \frac{2}{3}(\rho k)$$

and in tensorial notation

$$-\rho\overline{u'_i u'_j} = \mu_t\left\{\frac{\partial U_i}{\partial x_j} + \frac{\partial U_j}{\partial x_i} - \frac{2\delta_{ij}}{3}\left[\frac{\partial U_k}{\partial x_k}\right]\right\} - \frac{2}{3}(\rho k)\,\delta_{ij}$$

Normally this equation is written in terms of shear strain tensor as

$$S_{ij} = \frac{1}{2}\left(\frac{\partial U_i}{\partial x_j} + \frac{\partial U_j}{\partial x_i}\right)$$

$$-\rho\overline{u'_i u'_j} = 2\mu_t\left\{S_{ij} - \frac{1}{3}\left[\frac{\partial U_k}{\partial x_k}\right]\delta_{ij}\right\} - \frac{2}{3}(\rho k)\,\delta_{ij}$$

In Eddy Viscosity Models the eddy viscosity is calculated using some other quantities like turbulent kinetic energy, dissipation etc.

By the early 1950's, turbulence models had evolved into four categories like:

Algebraic (Zero-Equation) Models

One-Equation Models

Two-Equation Models

Stress-Transport Models

20.3.1 FIRST ORDER MODELS

Transport phenomena in turbulent flow are much more complicated than those in laminar flow and we cannot obtain any analytical solution of the time-averaged governing equations, unless we describe turbulent shear stress. In most simple terms turbulent shear stress can be modeled as

$$\tau'_{turb} = \rho\overline{u'v'} \equiv \rho\varepsilon_M\left(\frac{du}{dy}\right) \equiv \rho\ell^2\left(du/dy\right)^2$$

Boussinesq Approximation: According to Boussinesq approximation (1877) stress tensors can be related with gradient as:

$$\tau \propto \varepsilon\frac{du}{dy}$$

In context of turbulent flows, L. Prandtl (1925) proposed the concept of mixing length in turbulent transport phenomena, in analogy to the concept of mean free path in the kinetic theory of gases. Prandtl proposed that ε can be related with a mixing length i.e., the turbulent eddy act like a molecule and exchange its momentum.

Let us consider a two-dimensional flow and a change of momentum of a small element of fluid at the plane B, as shown in Fig. We assume that small eddies at planes A and C, which are ℓ [m] from plane B at time t, will reach plane B after t [s] by the motion of the transverse velocity fluctuation v without a loss of momentum in the x-direction.

The difference in the velocity at plane A is given by:

$$\Delta u_1 = \ell\left(u/y\right)$$

The difference in the velocity at plane C is given by:

$$\Delta u_1 = -\ell\left(u/y\right)$$

The net change in momentum in the x-direction at plane B due to the eddy transport mechanism is then given by the following equation:

$$\Delta u_1 + \Delta u_1 = \ell\left(u/y\right)$$

The x-component of the velocity fluctuation is related to the y-component thereof by the continuity equation, and we may assume the following equation:

$$u' \equiv \ell(u/y)$$

and

$$v' \equiv \ell(u/y)$$

$$\tau_{turb} = \rho\overline{u'v'} \equiv \rho\varepsilon_M\left(\frac{du}{dy}\right) \equiv \rho\ell^2(du/dy)^2$$

The ℓ needs to be determined from experimental data which makes this model less effective. However as it is a basic and simple model in this book we focus on this model mainly.

The turbulent stress and scalar fluxes can be modeled as:

$$\boxed{\tau_{turb} = -\rho(\nu+\varepsilon_M)\frac{\partial u}{\partial y}} \tag{20.25}$$

$$\boxed{q_{turb} = -\rho c(\alpha+\varepsilon_H)\frac{\partial T}{\partial y}} \tag{20.26}$$

Prandtl mixing length model belong to class of first order closure models. First order closure models are described as:

$$\begin{aligned}\varepsilon_M &= \ell^2(du/dy)\ (0\ \text{Eq model})\\ &= \ell k^{1/2}\ (1\ \text{Eq model})\\ &= k^2/\varepsilon\ (2\ \text{Eq model})\end{aligned} \tag{20.27}$$

The first option uses Boussinesq approximation and ℓ is referred to as Prandtl's mixing length. Analogus to Boussinesq model of the turbulent transport of momentum, the turbulent scalar fluxes are assumed to be proportional to the gradient of the transport quantity. We introduce here a parameter called turbulent Prandtl number defined as

$$Pr_t = \frac{\varepsilon_M}{\varepsilon_H}$$

where ε_M is turbulent eddy diffusivity and ε_H is turbuent thermal diffusivity. Hence, we can write $\varepsilon_H = \frac{\varepsilon_M}{\mathrm{Pr}_t}$ and the turbulent heat transfer shall be

$$q_{turb} = -\rho c_p(\alpha + _H)\frac{\partial T}{\partial y}$$

$$q_{turb} = -\rho c_p(\frac{\nu}{\mathrm{Pr}} + \frac{\varepsilon_M}{\mathrm{Pr}_t})\frac{\partial T}{\partial y}$$

here can write it in simplified manner as

$$q_{turb} = -\overline{u'_iT'} = \Gamma_t\frac{\partial T}{\partial y}$$

and Γ_t is defined as

$$\Gamma_t = -\rho c_p(\frac{\nu}{\text{Pr}} + \frac{\varepsilon_M}{\text{Pr}_t})$$

Γ_t is a property of the flow and depends on the local state of the turbulence.

Both the temperature (or enthalpy) and the concentration (or the mass fraction of a chemical species) are scalar quantities and can be represented by a general variable Θ. The time averaged form of the equation governing the transport of Θ can be written as:

$$U_i \frac{\partial \Theta}{\partial x_i} = \frac{\partial}{\partial x_i}\left(\Gamma_\Theta \frac{\partial \Theta}{\partial x_i} - \overline{u_i^{'}\theta^{'}}\right) \tag{20.28}$$

The equation represents a balance between the rates at which the scalar quantity Θ is transported by the mean flow field (advection) and it diffusion by the combined action of molecular and turbulent effects. We will discuss more more complex modeling approaches used chapter later.

20.3.2 SPALART-ALLMARAS MODEL

A one-equation model designed for aerodynamic flows, solving a transport equation for modified eddy viscosity $\tilde{\nu}$:

$$\frac{D\tilde{\nu}}{Dt} = c_{b1}\tilde{S}\tilde{\nu} - c_{w1} f_w \left(\frac{\tilde{\nu}}{d}\right)^2 + \frac{1}{\sigma}\left[\nabla\cdot((\nu+\tilde{\nu})\nabla\tilde{\nu}) + c_{b2}(\nabla\tilde{\nu})^2\right].$$

Ideal for boundary layers and mildly separated flows, but less accurate for complex shear.

20.3.3 K-ε MODEL

Turbulent kinetic energy is defined as kinetic energy of turbulent fluctuations per unit mass and mathematically:

$$k = \frac{1}{2}\overline{u_i^{'}u_i^{'}} = \frac{1}{2}\overline{(u_i^{'})^2}$$

We approximate the shear stress tensor as:

$$\tau_{ij} \approx -\rho\overline{(u_i^{'})^2} \approx -2\rho k$$

We can differentiate this equation in time:

$$\frac{\partial \tau_{ij}}{\partial t} = -2\rho\frac{\partial k}{\partial t} \tag{20.29}$$

and in space:

$$\frac{\partial \tau_{ij}}{\partial x_k} = -2\rho\frac{\partial k}{\partial x_k} \tag{20.30}$$

Multiply both sides by U_k:

$$U_k \frac{\partial \tau_{ij}}{\partial x_k} = -2U_k \rho \frac{\partial k}{\partial x_k} \tag{20.31}$$

We utilized RSM Eq. and take $i = j$ so

$$\begin{aligned}\tau_{ii,t} + U_k \tau_{ii,k} = &-\tau_{ik} U_{i,k} - \tau_{jk} U_{i,k} \\ &+ 2\mu \overline{\left(\frac{\partial u_i'}{\partial x_k}\frac{\partial u_i'}{\partial x_k}\right)} \\ &+ \frac{\partial}{\partial x_k}\left[\nu \frac{\partial \tau_{ii}}{\partial x_k} + \rho(\overline{u_i' u_i' u_k'})\right] + 2\frac{\partial(\overline{u'_i p'})}{\partial x_i} - \Pi_{ii}\end{aligned}$$

and

$$\begin{aligned}-2\rho \frac{\partial k}{\partial t} - 2\rho U_k \frac{\partial k}{\partial x_k} = &-(-2\rho k)U_{i,k} - (-2\rho k)U_{i,k} \\ &+ 2\mu \overline{\left(\frac{\partial u_i'}{\partial x_k}\frac{\partial u_i'}{\partial x_k}\right)} \\ &+ \frac{\partial}{\partial x_k}\left[\nu \frac{\partial \tau_{ii}}{\partial x_k} + \rho(\overline{u_i' u_i' u_k'})\right] + 2\frac{\partial(\overline{u'_i p'})}{\partial x_i}\end{aligned}$$

Divide whole eq by -2ρ:

$$\begin{aligned}\frac{\partial k}{\partial t} + U_k \frac{\partial k}{\partial x_k} = &-kU_{i,k} - kU_{i,k} \\ &- \nu \overline{\left(\frac{\partial u_i'}{\partial x_k}\frac{\partial u_i'}{\partial x_k}\right)} \\ &- \frac{\partial}{\partial x_k}\left[-\nu \frac{\partial k}{\partial x_k} + \frac{1}{2}(\overline{u_i' u_i' u_k'})\right] - \frac{1}{\rho}\frac{\partial(\overline{u'_i p'})}{\partial x_i}\end{aligned}$$

As k is used to represent the turbulent kinetic energy, the dummy index k is replaced often by j and the equation is

$$\begin{aligned}\frac{\partial k}{\partial t} + U_j \frac{\partial k}{\partial x_j} = &-2k \frac{\partial U_i}{\partial x_j} \\ &- \nu \overline{\left(\frac{\partial u_i'}{\partial x_j}\frac{\partial u_i'}{\partial x_j}\right)} \\ &- \frac{\partial}{\partial x_j}\left[-\nu \frac{\partial k}{\partial x_j} + \frac{1}{2}(\overline{u_i' u_i' u_j'})\right] - \frac{1}{\rho}\frac{\partial(\overline{u'_i p'})}{\partial x_i}\end{aligned}$$

Kinetic Energy Budget of Turbulent Flow

$$\frac{\partial k}{\partial t}+U_j\frac{\partial k}{\partial x_j} \quad (Advection)$$

$$= -2k\frac{\partial U_i}{\partial x_j} \quad (ShearProduction)$$

$$-\nu\overline{\left(\frac{\partial u_i^{'}}{\partial x_j}\frac{\partial u_i^{'}}{\partial x_j}\right)} \quad (ViscousDissipation)$$

$$+\frac{\partial}{\partial x_j}\left[\nu\frac{\partial k}{\partial x_j}\right] \quad (MolecularViscousTransport)$$

$$-\frac{\partial}{\partial x_j}\left[\frac{1}{2}(\overline{u_i^{'}u_i^{'}u_j^{'}})\right] \quad (TurbulentTransport)$$

$$-\frac{1}{\rho}\frac{\partial(\overline{u'_ip'})}{\partial x_i} \quad (PressureDiffusion)$$

It is customary to use the following notation for turbulence quantities

$$s_{ij}=\frac{1}{2}\left(\frac{\partial u'_i}{\partial x_j}+\frac{\partial u'_j}{\partial x_i}\right)=\frac{\partial u'_i}{\partial x_j}$$

$$\varepsilon_{ij}=\nu\overline{s_{ij}s_{ij}}$$

$$\mathscr{S}_{ij}=\frac{1}{2}\left(\frac{\partial U_i}{\partial x_j}+\frac{\partial U_j}{\partial x_i}\right)=\frac{\partial U_i}{\partial x_j}$$

$$\mathscr{P}_{ij}=-\overline{u'_iu'_j}S_{ij}$$

$$\mathscr{D}_{ij}=\frac{-1}{\rho}\overline{p'u'_j}-\frac{1}{2}\overline{u'_iu'_iu'_j}+\nu\frac{\partial k}{\partial x_j}$$

$$\frac{\partial k}{\partial t}+U_j\frac{\partial k}{\partial x_j}=\mathscr{P}_{ij}+\frac{\partial \mathscr{D}_{ij}}{\partial x_i}-\varepsilon_{ij}$$

For steady conditions we can cast a balance between the quantities as

$$Production + Diffusion - Dissipation - Convection \approx 0$$

$$\mathscr{P}_{ij}+\frac{\partial \mathscr{D}_{ij}}{\partial x_j}-\varepsilon_{ij}-\mathscr{C}_{ij}=0$$

Note that in terms of order of magnitude $S_{ij}\approx\frac{U}{\ell}$, where, V is representative velocity scale and ℓ is representative length scale. Also $\overline{u_iu_j}\approx u^2$. If we do the balance between Production and dissipation we have

$$\mathscr{P}_{ij}=\varepsilon_{ij}$$

$$-\overline{u'_iu'_j}S_{ij}=\nu\overline{s_{ij}s_{ij}}$$

It can be extracted that

$$\mathrm{Re} \propto \frac{\overline{s_{ij}s_{ij}}}{S_{ij}S_{ij}}$$

and

$$\overline{s_{ij}s_{ij}} \gg S_{ij}S_{ij}$$

This shows that viscous dissipation fluctuations are much larger than the mean flow shear or viscous dissipation. We associate Λ to large scale eddies and η to smallest scale eddies and in terms of order of magnitude or scales

$$\mathscr{P} = V^2(V/\Lambda)$$

where, V is velocity of large scales and for sake of comparison we also consider as Large scale vortices size $\Lambda \rightarrow \eta$ as V$\rightarrow$u and we can do balance as

$$\mathscr{P} = V^2(V/\Lambda) = u^3/\eta$$

For balance between production and dissipation

$$\mathscr{P} = \varepsilon = u^3/\eta$$

This leads to $u = (\varepsilon\eta)^{1/3}$

From small scales we have

$$\varepsilon = \nu(u/\eta)^2$$

$$u \approx \left(\frac{\varepsilon}{\nu}\right)^{1/2} \eta$$

Comparing the large scale and smallest scale velocities we have

$$(\varepsilon\eta)^{1/3} \propto \left(\frac{\varepsilon}{\nu}\right)^{1/2} \eta$$

$$(\eta) \propto \left(\frac{\varepsilon}{\nu}\right)^{3/2} \eta^3$$

$$\eta \propto \left(\frac{\nu^3}{\varepsilon}\right)^{1/4}$$

The is the order of magnitude of smallest turbulent scale also called the Kolmogorov scale.

20.3.3.1 Dissipation Rate (ε) Equation

The exact equation for turbulence dissipation rate ε can be arrived at by doing the following operation:

$$\overline{2\nu \frac{\partial u'_i}{\partial x_j} \frac{\partial_i}{\partial x_j} [N(u_i)]} = 0$$

However, the resulting equation is far more complex than the turbulent kinetic energy equations.

Launder and Sharma Standard k-ε model

Turbulence-modelers have devised simplified version of k-ε equation on heuristic approach to represent transport equations.

$$\frac{\partial \varepsilon}{\partial t}+U_j\frac{\partial \varepsilon}{\partial x_j}=C_{\varepsilon 1}\left(\frac{\varepsilon}{k}\right)\tau_{ij}\frac{\partial U_i}{\partial x_j}-C_{\varepsilon 2}\left(\frac{\varepsilon^2}{k}\right)+\frac{\partial}{\partial x_j}\left[\left(\nu+\frac{\varepsilon_M}{\sigma_\varepsilon}\right)\frac{\partial \varepsilon}{\partial x_j}\right]$$

Also, the k equation is used in its reduced model form as

$$\frac{\partial k}{\partial t}+U_j\frac{\partial k}{\partial x_j}=\tau_{ij}\frac{\partial U_i}{\partial x_j}-\varepsilon+\frac{\partial}{\partial x_j}\left[\left(\nu+\frac{\varepsilon_M}{\sigma_k}\right)\frac{\partial k}{\partial x_j}\right]$$

where, eddy turbulent diffusivity is

$$\varepsilon_M=C_\mu\frac{k^2}{\varepsilon}$$

with the coefficient as

$$C_\mu=0.09 \quad C_{\varepsilon 1}=1.44 \quad C_{\varepsilon 2}=1.92 \quad \sigma_\varepsilon=1.3 \quad \sigma_k=1$$

Note that in development of this model the local isotropy is assumed and these transport equations are not applicable all the way to the physical boundaries or walls. Model thus should be used with appropriate wall function. Further, it is assumed that the deviatoric Reynolds stresses are linearly related to the local mean strain rate. Due to this the model is well known to perform low for complex flows. Once k and ε are calculated, they be used for calculation of eddy viscosity.

$$\mu_t=\rho C_\mu\left(\frac{k^2}{\varepsilon}\right)$$

20.3.3.2 k-ω SST Model

Menter's Shear Stress Transport model blends k-ε (far-field) and k-ω (near-wall) to predict separation:

$$\frac{Dk}{Dt}=P_k-\beta^* k\omega+\nabla\cdot\left[(\nu+\sigma_k\nu_t)\nabla k\right],$$

$$\frac{D\omega}{Dt}=\alpha\frac{\omega}{k}P_k-\beta\omega^2+\nabla\cdot\left[(\nu+\sigma_\omega\nu_t)\nabla\omega\right]+\text{Cross-Diffusion Term.}$$

Limits eddy viscosity in regions of adverse pressure gradients, improving separation predictions.

20.4 COMPLEX FLOWS AND REYNOLDS STRESS EQUATIONS

Most engineering applications involve shear flows—such as boundary layers, jets, and wakes—which are the most frequently encountered flow types. In idealized cases like two-dimensional thin shear flows, mean deformation rates are relatively mild, allowing turbulence to reach equilibrium with the mean flow. Under these conditions, the principal axes of the Reynolds stress tensor typically align with those of the mean strain, and an eddy-viscosity modeling approach is sufficient.

However, real engineering flows are inherently three-dimensional and are influenced by adjacent turbulent fields, body forces, shock waves, and various disturbances. Such complexities characterize practical shear flows, where the principal axes of the Reynolds stress tensor are no longer aligned with those of the mean deformation tensor. This misalignment leads to significant deviations from the universal logarithmic law near walls, as turbulence departs from local equilibrium. Furthermore, the statistical properties of the flow can depend on multiple spatial coordinates and, in many cases, also vary in time. Examples of complex shear flows include

Shear layers on rotating bodies, such as those found on turbomachinery rotors, helicopter blades, and meteorological flows, where Coriolis forces directly impact turbulence structures.

Shear layers developing on curved surfaces, for example, boundary layers on turbine blades, wall jets used in cooling applications, and curved flows in turbomachinery passages, where centrifugal forces introduce additional strain.

Three-dimensional boundary layers and wakes, such as flow over finite wings, fuselages, marine vehicles, and hub regions in turbomachinery.

Shear layers with lateral convergence or divergence, as observed in annular diffusers or nozzles.

Separated flows, including those resulting from shock/boundary-layer interactions.

Swirling flows and vortices, such as tornadoes, hurricanes, and annular flows in turbomachinery.

Flows exhibiting strong compressibility effects.

Flows influenced by buoyancy, chemical reactions, and related phenomena.

These complexities directly influence the turbulence stresses and scales. In some scenarios, large-scale turbulent structures are particularly sensitive to these additional strains, necessitating more advanced turbulence modeling. As a result, prediction methods for such flows continue to evolve and improve. A critical requirement for turbulence models is that they accurately represent both the flow and turbulence physics across a broad spectrum of engineering scenarios. This need is especially acute in industrial applications, where achieving higher accuracy and greater reliability in modeling complex flows is of paramount importance.

Due to the limitations of simpler turbulence models in capturing the complex physics of three-dimensional and non-equilibrium flows, researchers have developed Reynolds Stress Models (RSM). The RSM approach addresses the shortcomings of eddy-viscosity models by directly solving transport equations for each component of the Reynolds stress tensor. This enables a more accurate representation of turbulence anisotropy, the effects of streamline curvature, rotation, and strong inhomogeneity. As a result, RSM models provide a more physically consistent framework for predicting the behavior of complex flows frequently encountered in engineering applications.

In this section we will first develop the Stress-Transport Model equation and then we will see the simplified versions of these models. In quest for more equation we take moment of Navier-Stokes equation with fluctuating component and time average whole equation. Symbolically.

$$\overline{u' \times NS}$$

We define here NS operator to represent the whole Navier-Stokes equation:

$$N(u_i) = \rho\frac{\partial u_i}{\partial t} + \rho U_k \frac{\partial u_i}{\partial x_k} + \frac{\partial p}{\partial x_k} - \mu \frac{\partial^2 u_i}{\partial x_k \partial x_k}$$

so moment operation is:

$$\overline{u_i' N(u_j) + u_j' N(u_i)} = 0$$

This approach is also known as **Second Moment Closure** modeling and it result into an equation which is the basis for the **Reynolds Stress Model** (RSM). The primary advantage of a stress-transport model is the inclusion of non-local and history effects into turbulence modeling.

Obviously this procedure would lead to more unknowns. However the approach is worth taking.

Unsteady term:

$$\begin{aligned}
-\tau_{ij,t} =& \overline{u_i'(\rho u_j)_{,t} + u_j'(\rho u_i)_{,t}} \\
=& \overline{\rho u_i'(U_j + u_j')_{,t} + \rho u_j'(U_i + u_i')_{,t}} \\
=& \overline{\rho u_i'(U_j)_{,t} + \rho u_i'(u_j')_{,t} + \rho u_j'(U_i)_{,t} + \rho u_j'(u_i')_{,t}} \\
=& \overline{\rho u_i' U_{j,t}} + \overline{\rho u_i' u_{j,t}'} + \overline{\rho u_j' U_{i,t}} + \overline{\rho u_j' u_{i,t}'} \\
=& \cancel{\overline{\rho u_i' U}}_{j,t} + \overline{\rho u_i' u_{j,t}'} + \cancel{\overline{\rho u_j' U}}_{i,t} + \overline{\rho u_j' u_{i,t}'} \\
=& \overline{\rho u_i' u_{j,t}'} + \overline{\rho u_j' u_{i,t}'} \\
=& \overline{\rho (u_i' u_j')_{,t}} = \rho (\overline{u_i' u_j'})_{,t}
\end{aligned}$$

Convective Term:

$$\overline{u_i'[\rho u_k u_{j,k}] + u_j'[\rho u_k u_{i,k}]}$$

Note that we changed the index of convective term to ensure the correct operation of moment. After some cumbersome operation we get:

$$\begin{aligned}
\overline{u_i'[\rho u_k u_{j,k}] + u_j'[\rho u_k u_{i,k}]} =& \overline{\rho u_i'[(U_k + u_k')(U_j + u_j')_k] + u_j'[(U_k + u_k')(U_i + u_i')_k]} \\
=& -U_k \tau_{ij,k} - \tau_{ik} U_{j,k} - \tau_{jk} U_{i,k} + \rho(\overline{u_i' u_j' u_k'})_{,k}
\end{aligned}$$

Pressure gradient term We can simplify pressure term as

$$\overline{u'_i \frac{\partial p}{\partial x_j} + u'_j \frac{\partial p}{\partial x_i}} = \left[\overline{(u'_i p')_{,j} - (p' u'_{i,j})}\right] + \left[\overline{(u'_j p')_{,i} - (p' u'_{j,i})}\right]$$

We introduce the differentiation product role and fluctuating components

$$= \overline{\left[(u'_i p')_{,j} + (u'_j p')_{,i}\right]} - \underbrace{\overline{[(p' u'_{j,i}) + (p' u'_{i,j})]}}_{\Pi_{ij}}$$

$$\begin{aligned}
&= \overline{\left[(u'_i p')_{,j} + (u'_j p')_{,i}\right]} - \Pi_{ij} \\
&= \frac{\partial (u'_i p')}{\partial x_j} + \frac{\partial (u'_i p')}{\partial x_i}
\end{aligned}$$

Introducing Kronecker delta we have

$$= \frac{\partial}{\partial x_k} \left(\overline{p' u'_j} \delta_{ik} + \overline{p' u'_i} \delta_{jk}\right) - \Pi_{ij}$$

Note that Π_{ij} is zero if i=j and

$$\overline{u'_i \frac{\partial p}{\partial x_j} + u'_j \frac{\partial p}{\partial x_i}} = 2 \frac{\partial (u'_i p')}{\partial x_i} \qquad for \quad i = j$$

Viscous Term:

$$\begin{aligned}
\overline{\mu[u_i' u_{j,kk} + u_j' u_{i,kk}]} =& \mu[\overline{u_i' u'_{j,kk}} + \overline{u_j' u'_{i,kk}}] \\
=& \nu \frac{\partial^2 \tau_{ij}}{\partial x_k \partial x_k} + 2\mu \overline{\left(\frac{\partial u_i'}{\partial x_k} \frac{\partial u_j'}{\partial x_k}\right)}
\end{aligned}$$

Finally the moment operation on Navier-Stokes equation takes form:

$$\begin{aligned}
& -\tau_{ij,t} - U_k \tau_{ij,k} - \tau_{ik} U_{j,k} - \tau_{jk} U_{i,k} + \rho(\overline{u_i' u_j' u_k'})_{,k} \\
+& \frac{\partial \overline{(u_i' p')}_{,j} + \overline{(u_j' p')}_{,i}}{\partial x_k} - \Pi_{ij} + \nu \frac{\partial^2 \tau_{ij}}{\partial x_k \partial x_k} + 2\mu \overline{\left(\frac{\partial u_i'}{\partial x_k} \frac{\partial u_j'}{\partial x_k}\right)} = 0
\end{aligned} \tag{20.32}$$

This is known as Reynolds Stress Equation.

Note that $\frac{\partial u_j'}{\partial x_k}$ give rise to 6 unknowns. Term $\overline{u_i' p_{,j}' + u_j' p_{,j}'}$ give rise to 6 unknowns and triple velocity correlations would give rise to 10 unknowns. So total 22 unknowns are generated through this procedure and all require modeling.

It is customary to arrange Reynolds Stress Model (RSM) equation in following nomenclature:

$$
\begin{aligned}
\tau_{ij,t} + U_k \tau_{ij,k} = & -\tau_{ik} U_{j,k} - \tau_{jk} U_{i,k} \\
& + 2\mu \overline{\left(\frac{\partial u_i'}{\partial x_k} \frac{\partial u_j'}{\partial x_k} \right)} \\
& + \frac{\partial((\overline{u_i' p'})\delta_{jk} + (\overline{u_j' p'})\delta_{ik})}{\partial x_k} \\
& + \frac{\partial}{\partial x_k} \left[\nu \frac{\partial \tau_{ij}}{\partial x_k} + \rho (\overline{u_i' u_j' u_k'}) \right] - \Pi_{ij}
\end{aligned}
$$

$$
\begin{aligned}
\tau_{ij,t} + U_k \tau_{ij,k} = & -\tau_{ik} U_{j,k} - \tau_{jk} U_{i,k} \\
& + 2\mu \overline{\left(\frac{\partial u_i'}{\partial x_k} \frac{\partial u_j'}{\partial x_k} \right)} \\
& + \frac{\partial}{\partial x_k} \left[\nu \frac{\partial \tau_{ij}}{\partial x_k} + \rho (\overline{u_i' u_j' u_k'}) + ((\overline{u_i' p'})\delta_{jk} + (\overline{u_j' p'})\delta_{ik}) \right] - \Pi_{ij}
\end{aligned}
$$

where, ε_{ij} is called turbulence dissipation term

$$
\varepsilon_{ij} = 2\nu \overline{\left(\frac{\partial u_i'}{\partial x_k} \frac{\partial u_j'}{\partial x_k} \right)}
$$

C_{ijk} is diffusion due to pressure and velocity fluctuations
and Π_{ij} is $\overline{(p' u_{i,j}' + p' u_{j,i}')}$ term.

20.4.1 EXPLICIT ALGEBRAIC REYNOLDS STRESS MODEL (EARSM)

Turbulence modelers have long sought to improve upon the simple linear eddy-viscosity (Boussinesq) hypothesis, which cannot capture stress anisotropy in turbulent flows. Early insights in the 1970s laid the groundwork for explicit algebraic stress models. In particular, Pope (1975) introduced a nonlinear anisotropic eddy-viscosity hypothesis by formulating a constitutive relation for the Reynolds stresses as a finite polynomial in the strain-rate and rotation tensors. Pope's analysis (building on Lumley's integrity basis concepts) showed that, under certain restrictions (e.g., nearly homogeneous or equilibrium flows), one can relate the Reynolds stress tensor algebraically to the local mean velocity gradients. Around the same time, Launder, Reece and Rodi (1975) developed the first Reynolds-stress transport model (RSM), and Rodi (1976) demonstrated that by assuming weak convective and diffusive effects (local equilibrium), the transport equations could be simplified to yield an algebraic stress relation. These pioneering efforts established that an algebraic (zero-equation) closure for the stress tensor was feasible, offering a middle ground between simple two-equation models and full second-moment closures.

Through the 1980s, researchers refined these ideas. Notably, models began incorporating material frame indifference and quadratic nonlinear stress-strain terms. For example, Speziale (1987) formulated a nonlinear k-ε model (including rotation-awareness) that produced much more realistic normal stress ratios in channel flow than the standard k-ε model. A key outcome was that nonlinear algebraic models could predict phenomena that linear models miss: for instance, the emergence of secondary flows in non-circular ducts (driven by normal stress anisotropy in corners) was captured by Speziale's nonlinear model, whereas a standard isotropic eddy-viscosity model predicted no secondary flow. Such successes underscored the potential of algebraic Reynolds stress closures to improve accuracy for complex wall-bounded flows.

The early 1990s saw a significant breakthrough with general formulations of EARSM. Gatski and Speziale (1993) presented a systematic derivation of an explicit algebraic stress model valid for fully three-dimensional flows (including rotating frames). Their work generalized Pope's 1975 solution (which had been limited to 2D mean flows with a specific pressure-strain model) to a broader class of flows, using a hierarchical tensor basis and invariants approach. Around the same time, Taulbee (1992) developed an improved algebraic stress model in Phys. Fluids, and Speziale, Sarkar and Gatski (1991) introduced a new invariant pressure-strain correlation (often used as a basis for algebraic models). These efforts demonstrated that explicit algebraic Reynolds stress models could reproduce the results of full second-moment closures in many cases. The new explicit model of Gatski-Speziale yielded predictions virtually indistinguishable from the full differential RSM for certain benchmark flows. However, it was also recognized that early explicit models had some limitations. For example, when applied to strongly non-equilibrium flows, the models of Taulbee (1992) and Gatski-Speziale (1993) showed internal inconsistencies due to assumptions of strict equilibrium; Girimaji (1996) addressed this by formulating a fully explicit and self-consistent algebraic stress model that remained more robust away from equilibrium conditions. The 1990s thus established the viability of EARSMs and prompted research into consistency, realizability, and numerical stability of these models.

By the turn of the millennium, EARSM technology had matured into practical turbulence closures. A landmark was the Wallin and Johansson (2000) model, which presented an explicit algebraic Reynolds stress formulation (with an accompanying scalar-flux model) applicable to both incompressible and compressible flows. The Wallin-Johansson (WJ) EARSM built upon Rodi's earlier closure and enforced a condition of self-consistency, meaning the explicit stress solution satisfies the original Reynolds-stress transport equations exactly for 2D mean flows. This model became widely cited and has been used as a foundation for further extensions. In the years following, researchers integrated EARSMs into two-equation frameworks for use in commercial CFD codes. For example, Hellsten (2005) embedded an EARSM into a k-ω model for high-lift aerodynamic flows, and Rumsey and Gatski (2003) documented validations of EARSMs in the NASA CFL3D code. The WJ-EARSM and similar models have since been applied successfully to a range of engineering problems (as discussed below), and they continue to be improved. In the 2010s, Johansson and co-workers extended the EARSM approach to handle effects of variable density and compressibility (e.g., a realizable EARSM for compressible flow with dilatation was developed by Grigoriev et al. 2013). Very recent work has even leveraged data-driven techniques: for instance, Chen et al. (2023) used machine-learning (field inversion) to tune an algebraic stress model for separated flows, further improving its predictive accuracy. Today, EARSMs represent a well-established class of turbulence models, bridging the gap between simple eddy-viscosity models and full Reynolds-stress transport closures, with a rich literature documenting their development from the 1970s to the present.

EARSMs originate from the exact Reynolds stress transport equations, which for constant-density flow can be written (in index notation) as:

$$\frac{\partial \langle u_i u_j \rangle}{\partial t} + U_k \frac{\partial \langle u_i u_j \rangle}{\partial x_k} = P_{ij} + \Pi_{ij} - \varepsilon_{ij} + D_{ij}$$

where $P_{ij} = -\langle u_i u_k \rangle \frac{\partial U_j}{\partial x_k} - \langle u_j u_k \rangle \frac{\partial U_i}{\partial x_k}$ is production, Π_{ij} the pressure-strain correlation, ε_{ij} the dissipation, and D_{ij} transport by diffusion. In an algebraic model, one makes the equilibrium assumption that the slow processes of convection and diffusion can be neglected (i.e., in source-term dominated regions of the flow). This leads to a balance between production and dissipation/pressure terms, yielding an approximate algebraic equation for the Reynolds stress tensor (or equivalently the anisotropy $b_{ij} = \tau_{ij}/(2k) - \frac{1}{3}\delta_{ij}$). The guiding principle is that in flows where turbulence adjusts rapidly to local conditions, the Reynolds stresses at a point can be expressed as a function of the local mean velocity gradients and turbulence scales. Under these assumptions, one can derive a nonlinear constitutive relation of the form $\tau_{ij} = f_{ij}(S_{mn}, \Omega_{mn}, k, \varepsilon)$, where S_{ij} and Ω_{ij} are the mean strain-rate and rotation tensors. This approach was first demonstrated by Pope (1975) and Rodi (1976) for simple 2D shear flows, and later generalized to 3D flows by solving the algebraic stress equations symbolically (often involving invariant theory to ensure coordinate-independent form).

The algebraic transformations used in EARSMs rely on expressing the anisotropy tensor as a polynomial function of the non-dimensional strain and rotation tensors. A convenient approach is to introduce a turbulence time scale $\tau = \frac{k}{\varepsilon}$ (or similarly $\frac{1}{\omega}$ in k–ω formulations) and define non-dimensional tensors $S^{ij} = \tau S_{ij}$ and $W^{ij} = \tau \Omega_{ij}$. The Reynolds-stress anisotropy is then expanded in a basis of invariant tensors built from S and W. Pope (1975) identified a complete integrity basis of ten independent tensors up to quadratic order (e.g., S_{ij}, $S_{ik}S_{kj}$, $S_{ik}W_{kj}$, $W_{ik}W_{kj}$, etc.), such that a general second-moment closure can be represented as a linear combination of those basis tensors. In practice, later researchers found that a reduced basis (five to six tensors) can suffice for most flows, especially if one focuses on 2D mean flows or includes only the most significant nonlinear terms. The transformation from the implicit algebraic equations to an explicit formula for b_{ij} involves solving a system of nonlinear equations for the scalar coefficients (often exploiting tensor invariants like $II_S = S_{ij}S_{ji}$ and $II_\Omega = W_{ij}W_{ji}$). For example, one common derivation introduces the production-to-dissipation ratio as an unknown, leading to a cubic or quartic equation that must be solved to determine the coefficients in the stress-strain relation. Despite the mathematical complexity, the end result is a closed-form expression for the anisotropy. In words, an EARSM provides an explicit algebraic relation between the Reynolds stress tensor and the normalized strain-rate and rotation-rate tensors.

In its final form, an EARSM augments the linear Boussinesq model with additional nonlinear terms to capture anisotropy. For instance, in the k-ω based EARSM of Wallin and Johansson (2000), the stress tensor is given by:

$$\tau_{ij} = 2\,\mu_t\left(S_{ij} - \tfrac{1}{3}S_{kk}\delta_{ij}\right) - \tfrac{2}{3}\rho k\,\delta_{ij} \;-\; \rho k\, a_{ij}^{(ex)},$$

where the extra anisotropy $a_{ij}^{(ex)}$ is a sum of higher-order tensor terms. Specifically, $a_{ij}^{(ex)}$ contains quadratic and cubic combinations of S and W. For example, a term proportional to $\left(W_{ik}W^{kj} - \tfrac{1}{3}II_\Omega\,\delta_{ij}\right)$ represents pure rotation-induced anisotropy, a term proportional to $\left(S_{ik}W^{kj} - W_{ik}S^{kj}\right)$ represents the strain-vorticity interaction (important for swirling flows), and further terms account for higher-order effects. The coefficients (often denoted $\beta_1, \beta_2, \ldots$) are functions of invariants like II_S and II_Ω; these are determined by enforcing that the modeled b_{ij} satisfies the contracted form of the stress-equilibrium equations (which yields a polynomial equation for the coefficients). As an example, for 2D mean flows, one finds an exact solution with only one nonlinear term ($a_{ij}^{(ex,2D)} \propto S_{ik}W^{kj} - W_{ik}S^{kj}$), whereas for 3D flows additional terms are needed. The use of a tensor invariant basis ensures that the model respects symmetry and frame-independence, and additional constraints are usually applied to guarantee realizability (such as limiting the magnitude of anisotropy to keep Reynolds stresses physical). Researchers have noted the need for regularization of the nonlinear solution in certain extreme strain/rotation regimes. For instance, Gatski and Speziale (1993) introduced a clipping procedure to avoid non-unique roots, explaining why earlier algebraic models could fail in complex flows.

It is important to note that an EARSM is not a standalone turbulence model. It is typically coupled with two transport equations for turbulence scales. In other words, EARSMs are often implemented as a modified two-equation model: one still solves, say, the standard k-ε or k-ω equations for the turbulent kinetic energy and dissipation (or specific dissipation) to obtain the turbulence time scale. The difference lies in the constitutive relation for the Reynolds stress: instead of $\tau_{ij} = 2\mu_t S_{ij} - \frac{2}{3}k\delta_{ij}$, the EARSM provides τ_{ij} as described above. Thus, the form of the k and ε equations remains standard, but the eddy viscosity μ_t is no longer the sole factor determining stresses. The additional nonlinear terms contribute to each normal and shear stress component individually. This approach retains much of the numerical robustness and efficiency of two-equation models (EARSMs add only a minor computational overhead for evaluating the algebraic expressions), while significantly enriching the model's ability to represent anisotropic turbulence. All modern explicit algebraic stress models require the local turbulence time scale (from k-ε, k-ω, etc.), but in return they yield a stress tensor that mirrors what a full second-moment closure would predict in an equivalent situation, by accounting for history effects like rotation and rapid strain in an algebraic manner.

Explicit algebraic Reynolds stress models have been applied and validated in a wide range of turbulent flow scenarios.

Boundary Layer and Wall-Bounded Flows: One of the primary motivations for EARSM development was to improve predictions in wall-bounded flows (boundary layers, ducts, channels) where turbulence is highly anisotropic. Linear eddy-viscosity models assume isotropy of turbulence, which is known to be false especially near walls. For example, experiments show normal stress components in a shear flow are unequal ($u'^2 \neq v'^2 \neq w'^2$), something a standard k-ε model cannot reproduce. EARSMs address this by producing anisotropic stress ratios in response to mean strain. In a flat-plate boundary layer, an EARSM will naturally reduce the turbulent shear stress in the viscous sublayer (since the model can yield $u'v'$ that drops to zero at the wall without ad-hoc damping functions) and can adjust the normal stress distribution across the layer to better match DNS data or experiments. Lumley (1978) and others noted that a nonlinear stress model can capture the mechanisms (such as rapid pressure-strain redistributing energy among components) that lead to, say, $\langle u'^2 \rangle > \langle w'^2 \rangle$ in near-wall turbulence. This leads to improved predictions of mean velocity profiles, skin friction, and heat transfer in boundary layers under various pressure gradients.

Rotating and 3D wall-bounded flows: provide a stringent test for turbulence models. The linear models often fail dramatically, while EARSMs have shown clear improvements. Gatski and Speziale's (1993) model, for instance, was applied to a rotating channel flow and a spanwise-rotating boundary layer. The explicit algebraic model was able to reproduce the asymmetric turbulence suppression on the stabilizing (rotating) wall and enhancement on the destabilizing wall, consistent with experiments, whereas standard models largely miss this behavior. In secondary flows that develop in non-circular ducts (e.g., flow in a rectangular channel), EARSMs can generate the secondary circulation driven by anisotropic Reynolds normal stresses. Speziale's nonlinear k-ε model (which is a form of EARSM embedded in a two-equation model) famously demonstrated this: it predicted the formation of secondary flow cells in the corners of a square duct, in good agreement with observations, whereas a linear k-ε predicted a physically incorrect symmetry (no secondary flow). Such successes have been reported in the literature as evidence that EARSMs capture important 3D effects (like stress-induced secondary motions or the influence of system rotation) that are missed by isotropic models. In general, for wall-bounded flows with strong curvature, rotation or 3D boundary conditions (e.g., swirling pipe flow, centrifugal bends), EARSMs provide more accurate turbulence anisotropy, which translates into better predictions of velocity profiles and wall friction. Many studies have confirmed these improvements. For example, Wallin and Johansson (2000) showed that their EARSM yields correct asymptotic near-wall behavior and reflects the damping of turbulence in a rotating reference frame consistent with DNS data.

Separated and Adverse-Pressure-Gradient Flows: Flows with separation (such as flow over a backward-facing step, an airfoil at high angle of attack, or a diffuser) are notoriously challenging for turbulence models. In separated flows, the distribution of Reynolds stresses can significantly influence the size and reattachment point of the separation bubble. Linear two-equation models often struggle in this case and often they tend to over-predict eddy viscosity in the separated shear layer, which can delay separation and hasten reattachment artificially. EARSMs, by contrast, allow for anisotropy in the shear layer and typically predict lower eddy viscosity in regions of strong normal stress anisotropy (since production and pressure-strain can redistribute energy into transverse fluctuations, reducing shear stress). This generally leads to more accurate predictions of separation extent. For instance, Franke et al. (2005) applied the Wallin-Johansson EARSM to flow over an airfoil and reported better agreement in the stalled wake region compared to k-ω SST results (the EARSM captured the more pronounced turbulence anisotropy in the separated flow). In high-lift aerodynamics, Hellsten's EARSM-based k-ω model was shown to improve the prediction of flow separation on multi-element airfoils. Likewise, EARSMs have been used to simulate flows with vortex generators or control devices on walls (which induce local separations and streamwise vortices); studies (e.g., Jirásek 2005) found that the EARSM could reproduce the altered turbulence structure around the device more accurately than linear models.

A key finding across these studies is that EARSMs tend to yield more realistic shear stress and turbulent kinetic energy profiles in separated zones, which in turn give better pressure recovery and wall pressure predictions. That said, separated flows also test the limits of the equilibrium assumption inherent in algebraic models, if the flow has strong unsteadiness or non-local effects, even EARSMs may need further calibration. Recent research has targeted these challenges: for example, Chen et al. (2022) used a field-inversion machine learning approach to adjust the functional form of an algebraic stress model in separated flow regions. This led to notable improvements (reducing model-form errors in predicting the size of separation bubbles), highlighting that while EARSMs are a step up from linear models, continued refinement (even using data-driven methods) is beneficial for complex separation phenomena. Overall, in flows with adverse pressure gradients and separation, EARSMs have proven to predict the onset and extent of separation more reliably than linear eddy-viscosity models, making them valuable for aerodynamic stall simulations, diffuser flows, and other engineering applications where flow separation is present.

Free Shear Flows: Jets and Mixing Layers: Free shear flows such as jets and mixing layers present another area where EARSMs have shown advantages. These flows often exhibit anisotropy between the streamwise and cross-stream turbulence intensities and can involve rapid distortion (e.g., mixing layer growth, jet spreading). Standard two-equation models calibrated to wall-bounded flows frequently under-predict the spreading rate and turbulence levels in free shear flows. This is partly because such models assume a fixed relationship between shear stress and kinetic energy (through a constant C_μ in the linear eddy viscosity), which does not hold universally in free shear layers. An EARSM can overcome this by adjusting the stress-strain relationship based on the state of the turbulence.

Studies have found that applying an EARSM to jet flows yields better agreement with measurements of Reynolds stresses and mean flow. For instance, a NASA investigation (Yoder et al., 2019) evaluated a k-ε EARSM on a high-speed jet: by recalibrating pressure-strain coefficients in the algebraic stress model, they achieved closer agreement with experimental peak shear stress profiles in the jet's mixing layer. The EARSM was able to produce higher anisotropy (i.e., a different ratio of normal stresses to shear stress) in the shear layer, which in turn gave a more accurate jet centerline velocity decay and width of the jet plume than the baseline k-ε model. Similarly, explicit algebraic models have been tested in planar and round jets, showing that they can distinguish the different spreading rates of these flows (something a single constant C_μ cannot do). For example, Papp and Dash (2000s, as cited in Yoder 2019) showed that an algebraic Reynolds stress model could be tuned to capture the faster decay of turbulence in axisymmetric jets versus planar jets by altering the nonlinear stress terms.

Another example is turbulent wakes (which are essentially jet-like shear flows developing behind a body). Recent work in wind energy used the Wallin-Johansson EARSM to simulate wind turbine wakes and highly anisotropic, swirling free-shear flows. The EARSM was found to significantly improve the prediction of turbulence intensity and anisotropy in the far-wake, compared to standard two-equation models. The model captured the more realistic non-Gaussian velocity deficit profile and the enhanced mixing due to wake rotation. Researchers noted that the EARSM-based RANS was only slightly more expensive than a normal k-ε model, yet it alleviated many deficiencies of the linear model in wake prediction. In summary, for free shear flows such as jets, mixing layers, and wakes, EARSMs provide a better representation of how turbulence evolves (especially in capturing the proper level of shear stress and cross-stream turbulent transport), leading to improved predictions of flow spreading and centerline velocity decay. These improvements are crucial for engineering problems like jet noise prediction (where accurate turbulence stats are needed for acoustic analogies) and atmospheric dispersion modeling.

Compressible and Stratified Flows: The versatility of EARSMs has also been demonstrated in flows with additional physics, such as compressibility and buoyancy. Compressible turbulent flows (e.g., shock-boundary layer interactions, supersonic mixing layers) can exhibit phenomena like dilatation-dependent pressure-strain and variable density effects, which were not present in the incompressible formulations. The explicit algebraic framework has been extended to include these. Wallin and Johansson (2000) originally formulated their EARSM for compressible flows by including dependence on density fluctuations (Favre averaging). Subsequent researchers introduced modifications for flows with significant mean dilatation. For instance, Grigoriev et al. (2013) developed a realizable EARSM that incorporates terms for shock-induced anisotropy and variable density, showing good agreement with DNS of rapid compression and expansion flows. In practical terms, using an EARSM in a shock-laden flow helps capture the rise in normal Reynolds stresses across a shock and the accompanying suppression of shear stress (an effect linear models often mis-predict). Recent studies (e.g., Chen et al. 2021 in Physics of Fluids) have applied EARSMs to shock/boundary-layer interaction cases and found that the model's ability to adjust the turbulence structure (such as increasing the streamwise normal stress in the post-shock region) yields better shock foot prediction and separation bubble size, compared to eddy-viscosity models which lack such flexibility.

Buoyancy-affected flows (such as atmospheric boundary layers with stratification) are another arena where EARSMs have been applied. Stable stratification tends to damp vertical turbulent fluctuations, whereas convection (unstable stratification) enhances them; thus the turbulence anisotropy in a stratified flow can be quite pronounced. Algebraic stress models have been extended to include buoyancy production terms and anisotropic buoyancy damping. For example, Lazeroms et al. (2013) combined the WJ-EARSM with a density stratification term to model atmospheric boundary layers, and later Lazeroms et al. (2016) showed that the EARSM could handle both stable and convective regimes by appropriately coupling turbulence fluxes with the density gradient. The EARSM-based approach in these studies reproduced key features like the reduced turbulence intensities in strongly stable nighttime boundary layers and the rapid growth of the mixed layer in daytime convection, matching field observations more closely than simpler models. Another work by Želi et al. (2019) addressed consistent wall-boundary conditions for EARSM in an atmospheric boundary layer, facilitating realistic surface flux predictions. These efforts indicate that EARSMs, when extended with physics-informed terms, are capable of capturing complex turbulence physics in geophysical flows, such as the transition between different stability regimes in the atmosphere.

EARSMs offer improved predictions of turbulence anisotropy and related mean-flow effects, compared to conventional linear two-equation models. This comes at a moderate increase in model complexity, but without the heavy cost of full Reynolds-stress transport models.

20.5 WALL FUNCTIONS AND NEAR-WALL TREATMENT

Turbulent flows near solid boundaries exhibit steep gradients, requiring careful treatment in RANS simulations. Direct resolution of the viscous sublayer is computationally expensive. Wall functions bridge the gap between the wall and the fully turbulent region by providing empirical approximations for velocity and scalar profiles.

Log-Law of the Wall:

$$u^+ = \frac{1}{\kappa}\ln(y^+) + B, \tag{20.33}$$

where $u^+ = u/u_\tau$, $y^+ = yu_\tau/\nu$, κ is the von Kármán constant (0.41), and $B \approx 5.0$.

Enhanced Wall Treatment: Hybrid approaches combining low-Reynolds-number models with wall functions improve accuracy for separated flows. The non-dimensional wall distance y^+ dictates the choice:

$y^+ < 5$: Resolve viscous sublayer.

$30 < y^+ < 300$: Use log-law wall functions.

Heat Transfer Near Walls: Turbulent Prandtl number Pr_t models relate thermal and momentum boundary layers:

$$q_w = \rho c_p \frac{u_\tau}{\kappa} \left(\frac{T_w - T_\infty}{\ln(y^+)/Pr_t + C} \right).$$

Case Study 20A: Flow in a Square Duct Investigated using RANS

Problem Definition

This case investigates turbulent flow through a square duct at a friction Reynolds number $Re_\tau = 342$. The primary objective is to evaluate the predictive capabilities of various Reynolds-Averaged Navier-Stokes (RANS) turbulence models in capturing secondary motions induced by turbulence anisotropy. The flow geometry is a straight square duct with all four walls treated as no-slip boundaries. The simulations aim to replicate conditions from existing high-fidelity LES data at $Re_\tau = 360$, conducted by Madabhushi (1991), which serve as a benchmark for comparison. The focus is on examining the distribution of mean streamwise velocity and secondary cross-plane flows. The secondary flows, being a direct manifestation of turbulent stress anisotropy, serve as a critical indicator of model fidelity. All numerical simulations were performed using the finite-volume based solver `OpenFOAM-V2406`.

Governing Equations

The governing equations are the incompressible Reynolds-Averaged Navier-Stokes (RANS) equations. The continuity equation enforces the incompressibility constraint, given by

$$\frac{\partial \bar{u}_i}{\partial x_i} = 0.$$

The momentum equation for the mean flow, accounting for turbulent stresses, is given by

$$\frac{\partial \bar{u}_i}{\partial t} + \bar{u}_j \frac{\partial \bar{u}_i}{\partial x_j} = -\frac{1}{\rho} \frac{\partial \bar{p}}{\partial x_i} + \nu \frac{\partial^2 \bar{u}_i}{\partial x_j^2} - \frac{\partial \overline{u_i' u_j'}}{\partial x_j}.$$

The Reynolds stress term $\overline{u_i' u_j'}$ is modeled using different turbulence closure models. Linear eddy-viscosity models assume isotropy in turbulent stresses. In contrast, second-moment closure models such as RSM-LRR and RSM-SSG solve transport equations for each component of the Reynolds stress tensor, allowing for anisotropic turbulence representation.

Numerical Setup

The simulations are carried out using a finite volume method (FVM) based code `OpenFOAM-V2406`. A structured mesh is generated for the square duct, with finer grid clustering near the walls to resolve steep gradients in the viscous sublayer using `BlockMesh`. All walls are assigned no-slip boundary conditions, and the inlet and outlet periodic boundary conditions are applied.

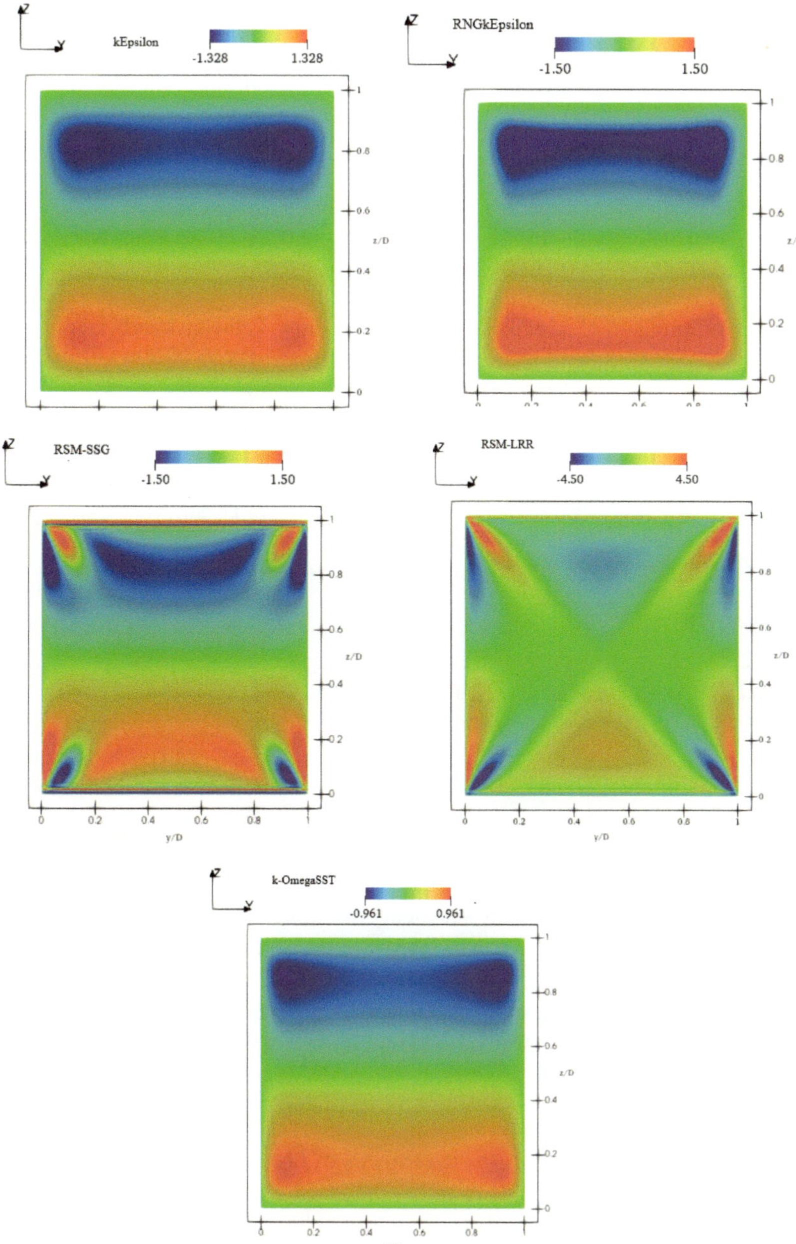

Figure 20.2: **Flow visualization for various RANS models showing secondary motion behavior at $Re_\tau = 342$. Contours of w/u_tau distribution.**

The solver used is `simpleFoam`, which is a solver employing the SIMPLE algorithm for pressure-velocity coupling. All simulations are run until residuals for velocity and pressure drop below 10^{-6}, ensuring numerical convergence. Turbulence models under evaluation include the standard k-ε, RNG k-ε, k-ω, SST, RSM-LRR, and RSM-SSG.

Results and Discussion

Table summarizes simulation results, presenting the centerline velocity and peak secondary velocity normalized by the bulk velocity, along with a comment on whether secondary flows were captured.

Comparison of Mean Flow and Secondary Velocity Predictions for RANS Models at $Re_\tau = 342$ Against LES at $Re_\tau = 360$

Model	Re_τ	U_{CL}/U_b	w_{max}/U_b (%)	Secondary Flow
LES (Madabhushi, 1991)	360	1.278	2.58	Yes
k-Epsilon	342	1.22	3.16	No
RNG k-Epsilon	342	1.24	2.59	No
k-Omega SST	342	1.16	7.61	No
RSM-LRR	342	1.23	3.91	Yes
RSM-SSG	342	1.25	3.79	Yes

Figure 21.7(a) shows the mean streamwise velocity contours, highlighting the low Reynolds number effect of wall-bisector velocity peaks, as observed in DNS by Gavrilakis (1992). The secondary flow structure, displayed in Figure 21.7(b), provides a qualitative benchmark for cross-plane velocity prediction.

The results show a stark difference in the ability of RANS models to predict secondary flow structures. The linear eddy-viscosity models largely fail to reproduce the characteristic secondary vortices. This failure arises from the inherent assumption of turbulence isotropy, which does not hold in duct flows dominated by cross-stream secondary motions. In contrast, the Reynolds Stress Models (RSM-LRR and RSM-SSG) successfully capture the corner vortex systems. Both models localize the peak secondary velocities with magnitudes consistent with benchmark LES data. This underscores the importance of stress anisotropy modeling in flows with strong secondary circulation.

Lessons Learned / Numerical Challenges

This case study highlights the inadequacy of linear eddy-viscosity models in predicting secondary motions in internal duct flows. These models are limited by their assumption of isotropic Reynolds stress distribution, which is unsuitable for capturing corner-driven vortices. The RSM models, though more computationally intensive, demonstrate improved accuracy due to their ability to resolve anisotropic effects explicitly. Grid resolution and wall treatment are found to be critical, especially for RSM simulations, which showed sensitivity to mesh refinement and solver under-relaxation parameters. The simulation results emphasize the need for careful turbulence model selection in flows where secondary motion and turbulence anisotropy play a dominant role.

20.6 TURBULENT HEAT TRANSFER

20.6.1 TURBULENT SCALAR-FLUXES TRANSPORT EQUATION

The turbulence correlations $\overline{u'_iT'}$ are called *turbulent scalar fluxes* in convective heat transfer literature. The terms are generated in Reynolds time-averaging procedure and basically they represent the convection terms in the energy equation for the instantaneous scalar. However, like turbulent shear stresses the scalar fluxes are often modeled as gradient of temperature fields. Turbulent scalar fluxes can be modeled either *algebraic* or *differential* models. The approach of Prandtl number and ε_H discussed in previous chapters is basically a simplistic algebraic modeling approach.

We can arrive at a scalar-flux transport equation by substituting $T = \bar{T} + T'$ and $u = \bar{u} + u'$ into the following procedure:

$$\overline{T'N(u_i) + u'_iE(T)} = 0$$

$$\begin{aligned}
\underbrace{\frac{\partial\left(U_k\overline{u'_iT'}\right)}{\partial x_k}}_{Conv_{Ti}} = & \underbrace{-\overline{u'_ku'_i}\frac{\partial T}{\partial x_k}}_{P_{Ti,1}} \underbrace{-\overline{u'_kT'}\frac{\partial U_i}{\partial x_k}}_{P_{Ti,2}} \underbrace{-\beta g_i\overline{T'^2}}_{G_{i,T}} \\
& \underbrace{-(\nu+\alpha)\overline{\frac{\partial T'}{\partial x_k}\frac{\partial u'_i}{\partial x_k}}}_{\varepsilon_{Ti}} \\
& \underbrace{-\frac{\partial}{\partial x_k}\left(\overline{u'_ku'_iT'} + \frac{\overline{p'T'}}{\rho}\delta_{ik} - \gamma\overline{u'_i\frac{\partial T'}{\partial x_k}} - \nu\overline{T'\frac{\partial u'_i}{\partial x_k}}\right)}_{Diff_{Ti}} \\
& \underbrace{-\overline{\frac{p'}{\rho}\frac{\partial T'}{\partial x_i}}}_{\Pi_{Ti}}
\end{aligned} \tag{20.34}$$

where β is the volumetric expansion coefficient, ν is the molecular diffusivity, ρ is the fluid density, and p' is the fluctuating pressure. The quantity $\overline{T'^2}$ is the mean square of the scalar fluctuations.

The term Conv_{Ti} represents the rate of transport (advection) of $\overline{u'_iT'}$. This term is exact.

The three terms:

$$P_{iT,1} \equiv -\overline{u'_ku'_i}\frac{\partial T}{\partial x_k} \tag{20.35}$$

$$P_{iT,2} \equiv -\overline{u_k' T'} \frac{\partial U_i}{\partial x_k} \tag{20.36}$$

$$G_{i\theta} \equiv -\beta g_i \overline{T'^2} \tag{20.37}$$

represent, respectively, the rates at which $\overline{u_i' T'}$ is generated by the turbulent interaction with the mean (scalar and velocity) fields and by the body forces (buoyancy in this case). These terms are exact and in no need for any approximation. They are also large, and hence the relative success of scalar-flux transport models.

The rate of dissipation of $\overline{u_i' T'}$ by viscous action (ε_{Ti}) vanishes at high turbulence Reynolds numbers. It is usually dropped from all models for $\overline{u_i' T'}$.

Of the three terms that constitute the 'diffusion' of $\overline{u_i' T'}$, on that which represents transport by the turbulence fluctuations is usually retained and modeled via. a gradient-transport relation, thus:

$$-\overline{u_k' u_i' T'} = C_\theta \frac{k}{\varepsilon} \overline{u_k' u_l'} \frac{\partial \overline{u_i \theta}}{\partial x_l} \tag{20.38}$$

where C_θ is a coefficient taken equal to 0.15 (Malin & Younis, 1990).

The fluctuating pressure-scalar-gradient correlation term ($\pi_{i\theta}$), may be viewed as the counterpart of the pressure-strain term in the $\overline{u_i u_j}$ transport equation. It is clearly an important agency which requires careful modelling and it is perhaps here that current scalar-fluxtransport models are least well developed.

An exact expression for this quantity (derived from the instantaneous equations for the scalar and the ith -component of momentum) is given as:

$$\pi_{i\theta} \equiv \overline{\frac{p' \partial \theta}{\rho \partial x_i}} = \frac{1}{4\pi} \int \left(\overline{\frac{\partial^2 u_l' u_m'}{\partial x_l \partial x_m} \frac{\partial \theta}{\partial x_i}} + 2 \frac{\partial U_l'}{\partial x_m} \overline{\frac{\partial u_m'}{\partial x_l} \frac{\partial \theta}{\partial x_i}} + \frac{\beta g_i}{\Theta'} \overline{\frac{\partial \theta}{\partial x_l} \frac{\partial \theta}{\partial x_l}} \right) \frac{dVol}{\mathbf{r}} \tag{20.39}$$

where the primed quantities are evaluated at distance $\mathbf{x} + \mathbf{r}$.

There are three distinct groupings of terms in Eq. 138:

a term which is associated with purely turbulence interactions (the 'slow' part, $\pi_{i\theta,1}$),

another which involves interactions between the mean velocity gradients and fluctuating quantities (the 'rapid' part, $\pi_{i\theta,2}$),

a third which involves the body forces $(\pi_{i\theta,3})$.

The conventional approach to modeling the integral expression in Eq. 138 has been to model each of its components separately:

1. Monin (1965), by analogy with Rotta's return-to-isotropy proposal, suggested the model:

$$\pi_{Ti,1} = -C_{1T} \frac{\varepsilon}{k} \overline{u_i' T'} \tag{20.40}$$

2. Launder (1975) models the rapid part in exact analogy to the pressure-strain models, thus:

$$\pi_{Ti,2} = -C_{2T} P_{Ti,2} \tag{20.41}$$

3. Launder (1975) also proposed the extension for buoyant flows:

$$\pi_{Ti,3} = -C_{3T} G_{Ti} \tag{20.42}$$

When a solid wall is present, the turbulent pressure fluctuations are modified as explained in the previous Chapter and this is accounted for in the model for $\pi_{i\theta}$ by a. correction of the form (Gibson and Launder, 1978):

$$\pi_{Ti,w} = -\left(C_{T1,w}\frac{\varepsilon}{k}\overline{u_i'T'}n_i n_k + C_{2T',w}\pi_{kT',2}n_i n_k + C_{3T',w}\pi_{k\theta,3}n_i n_k\right) f\left(\frac{L}{n_i r_i}\right) \tag{20.43}$$

where n_i is the unit vector from the surface and f is the wall-damping function.

The constants in the above equations can be determined by reference to homogeneous shear flow data on the streamwise and cross-stream heat fluxes. Their commonly-used values are:

C_{1T}	C_{2T}	C_{3T}	$C_{1T,w}$	$C_{2T,w}$	$C_{3T,w}$
2.85	0.55	0.55	1.2	0	0

Note that this piecewise approach to modeling the integral expression results in the absence from the model of any explicit dependence on terms containing $\overline{u_i u_j}$ and $\partial\Theta/\partial x_k$. This is a serious omission as the role of the fluctuating pressure-scalar-gradient correlation is to counter-balance the rate of production of $\overline{u_i\theta}$, including that due to the interaction of the Reynolds stresses with the scalar gradients.

Jones and Musonge (1987) restored such a dependence to the model by arguing that $\pi_{i\theta,1}$ and $\pi_{i\theta,2}$ should be modeled together since both ultimately depend on the mean field. Their proposal reads:

$$\pi_{Ti,1} + \pi_{iT,2} = -C_{1T}\frac{\overline{u_i'T'}}{\tau} + C_{2T}a_{ij}\frac{\partial T}{\partial x_j} + d_{ijk}\frac{\partial U_j}{\partial x_k} \tag{20.44}$$

where

$$a_{ij} = \left(\frac{\overline{u_i u_j}}{2k} - \frac{1}{3}\delta_{ij}\right)$$

$$\tau = \left(1 + C'_{1T}\sqrt{A_2}\right)\frac{q^2}{\varepsilon}$$

$$A_2 = a_{ij}a_{ij}$$

$$d_{ijk} = C_{3\theta}\delta_{ij}\overline{u_k'T'} + C_{4T}\delta_{ik}\overline{u_i'T'} + C_{5T}\delta_{jk}\overline{u_i'T'}$$

The model is of some complexity and involves six coefficients—it has not yet been subjected to extensive testing.

20.6.2 ALGEBRAIC SCALAR-FLUX MODELS

Note that the concept of turbulent Prandtl number fails to give the correct heat transfer distributions in strongly-buoyant flows and flows with significant transport effects. Also, the turbulent diffusivity is assumed to be isotropic, which is not realist in most of three-dimensional flows. Eddy-diffusivity based models fail badly in complex shear flows because relating eddy diffusivity to turbulent Prandtl number is a very simplistic approach. They lack the scalar fluxes explicit dependence on the Reynolds stresses or on the mean shear. In simple model it is assumed that the turbulent scalar fluxes and the mean scalar gradients are aligned. However, that too is not true in complex flows.

Batchelor Model: Batchelor (1949) proposed a generalization to the gradient-transport hypothesis involving the definition of a turbulent eddy-diffusivity tensor (D_{ij}), whereby

$$-\overline{u_i'T'} \equiv D_{ij}\frac{\partial T}{\partial x_j}. \tag{20.45}$$

Several attempts have been made at obtaining a practical representation for D_{ij} (e.g., Yoshizawa (1985, 1988), Rogers et al. (1989) and Rubinstein & Barton (1991)).

Daly and Harlow (1970) Model: proposed that. D_{ij} be made directly proportional to the Reynolds stresses as:

$$-\overline{u_i'T'} = C_\theta\frac{k}{\varepsilon}\overline{u_i'u_k'}\frac{\partial T}{\partial x_k}. \tag{20.46}$$

where C_θ is an empirical coefficient taken to be around 0.3 (Launder, 1988).

Daly and Harlow's simple explicit scalar flux model is able to account for the anisotropy of turbulent diffusivity. This model under-predicts the streamwise heat flux component and assumes that scalar fluxes are aligned with temperature gradient.

WET Model Launder (1988) made proposals for a so-called "WET" model wherein the scalar fluxes are obtained from:

$$-\overline{u_i'T'} = C_\theta\frac{k}{\varepsilon}\left(\overline{u_i'u_k'}\frac{\partial T}{\partial x_k} + \overline{u_k'T'}\frac{\partial U_i}{\partial x_k} + g_i\overline{T'^2}\right). \tag{20.47}$$

Launder's model is unique amongst the existing algebraic closures as it an implicit model in the scalar fluxes and thus pose a computational or stability problem in rapidly-evolving flows.

Abe and Suga model (2001)

$$-\overline{u'_iT'} = C_{\theta 2}\left(\frac{k}{k.\varepsilon}\right)\overline{u'_iu'_k}\cdot\overline{u'_ju'_k}\frac{\partial T}{\partial x_k}$$

where $C_{\theta 2}$ is 0.6.

Younis, Speziale and Clarke (1995) proposed an algebraic, explicit and non-linear model for the scalar fluxes based on the use of Representation Theory. According to their model the scalar-fluxes can be modeled via following terms:

$$
\begin{aligned}
-\overline{u_i'T'} = & C_1 \frac{k^2}{\varepsilon} \frac{\partial T}{\partial x_i} \\
& + C_2 \frac{k}{\varepsilon} \overline{u_i'u_j'} \frac{\partial T}{\partial x_j} \\
& + C_3 \frac{k^3}{\varepsilon^2} \frac{\partial U_i}{\partial x_j} \frac{\partial T}{\partial x_j} \\
& + C_4 \frac{k^2}{\varepsilon^2} \left(\overline{u_i'u_k'} \frac{\partial U_j}{\partial x_k} + \overline{u_j'u_k'} \frac{\partial U_i}{\partial x_k} \right) \frac{\partial T}{\partial x_j}
\end{aligned}
\tag{20.48}
$$

where the C's are dimensionless constants whose values were determined by curve-fitting from Direct Numerical Simulations data for channel or bondary layer flows.

C_1	C_2	C_3	C_4
−0.0455	0.373	−0.00373	−0.0235

It has been found that the coefficient C_1 has a smaller value than earlier reported. The recent recommendation for C_1 is given in the form of a damping function:

$$C_1^* = C_1[1 - exp(-A\beta Pe_t^{\alpha})] \tag{20.49}$$

where,

$$Pe_t = Re_t Pr$$

$$Re_t = \frac{k^2}{\nu\varepsilon}$$

$$
\begin{aligned}
A_{ij} &= \frac{\overline{\acute{u}_i\acute{u}_j}}{k} - \frac{2}{3}\delta_{ij} \\
A_2 &= A_{ij}A_{ij} \\
A_3 &= A_{ij}A_{jk}A_{ki} \\
A &= 1 - \frac{9(A_2 - A_3)}{8}
\end{aligned}
$$

The new coefficients α and β, recommended for the C_1 modification, are 0.1 and 1.5 respectively. The values for α and β have been determined via DNS data of wall-bounded flows. However, there is still a need to assess the coefficients for more complex flows.

The RANS equation based turbulence models are the most widely used CFD tools. They are now a standard industrial simulation tool for the design and analysis of engineering systems. Around 100 years ago (1895), O. Reynolds gave the idea of statistical time averaging. Through this technique, the Navier-Stokes equations are averaged in time, which introduces new terms in the equations which needs to be modeled. Over the years numerous investigators and researcher have proposed the *turbulence- models*. These models predict the flow field accurately in many cases, however, in many cases they also fail to predict the true flow physics. Therefore, the need to refine the models is always present.

The time averaged Navier-Stokes equations are given as:

$$\rho U_j \frac{\partial U_i}{\partial x_j} = -\frac{\partial P}{\partial x_i} + \mathbf{F}_i + \frac{\partial}{\partial x_j}\left(\mu\left(\frac{\partial U_i}{\partial x_j} + \frac{\partial U_j}{\partial x_i}\right) - \rho\overline{u_i' u_j'}\right) \tag{20.50}$$

The $\overline{u_i' u_j'}$ term is an unknown quantity, which originates due to averaging a non-linear term. It is called Reynolds stresses.

Similarly the Reynolds averaged equations for a mean scalar field (neglecting source term) is:

$$\frac{\partial(\rho T)}{\partial t} + \frac{\partial(\rho U_j T)}{\partial x_j} = \frac{\partial}{\partial x_j}\left(\Gamma\frac{\partial T}{\partial x_j} - \rho\overline{u_i' T'}\right) \tag{20.51}$$

where Γ is the thermal diffusivity of the fluid. The $\overline{u_j' T'}$ term is called passive scalar-flux and is an unknown quantity, which originates due to averaging a nonlinear term in transport equation of total scalar field. This is analogous to Reynolds stresses.

20.6.3 TEMPERATURE-VARIANCE TRANSPORT EQUATION

Chou (1945) derived the transport equation for the variance of a turbulent scalar (such as temperature). In his pioneering work on turbulence, Chou extended Reynolds' averaging method to derive equations for second-order correlations, effectively formulating the budget equation for $\overline{T'T'}$ (the mean-square temperature fluctuation) analogously to the Reynolds stress transport equation. This temperature-variance transport equation is obtained by multiplying the instantaneous scalar (energy) equation by the temperature fluctuation T' and then averaging. It balances the production of temperature variance by mean temperature gradients, the turbulent transport (redistribution) of T' fluctuations, and the dissipation of those fluctuations by molecular diffusion (thermal dissipation). Chou's 1945 analysis provided the first rigorous foundation for such second-moment turbulent heat transfer equations, which has since become a cornerstone in turbulence modeling and later works (e.g., Harlow and Nakayama 1967) on scalar variance transport in turbulent flows.

Like Reynolds stress equation one can develop Turbulent Scalar-fluxes equation by taking moment of Energy equation with temperature fluctuation:

$$\overline{T'E(T)} = 0$$

where

$$E(T) := \frac{\partial T}{\partial t} + \overline{U_j}\frac{\partial \overline{T}}{\partial x_j} - \alpha\frac{\partial^2 T}{\partial x_j^2}$$

This operation gives:

$$\frac{U_j}{2}\frac{\partial}{\partial x_j}(\overline{T'T'})+\frac{1}{2}\frac{\partial\overline{T'T'}}{\partial t}=-\frac{1}{2}\frac{\partial}{\partial x_j}\left[\overline{T'T'u_j'}-\alpha\frac{\partial\overline{T'T'}}{\partial x_j}\right]$$
$$-\overline{u_j'T'}\frac{\partial T}{\partial x_j}$$
$$-\alpha\overline{\frac{\partial T'}{\partial x_j}\frac{\partial T'}{\partial x_j}}$$

We define,

$\mathscr{E}_\theta=\frac{1}{2}\overline{T'T'}$ as turbulence scalar fluxes term

$\mathscr{P}_\theta=-\overline{u_j'T'}\frac{\partial T}{\partial x_j}$ as turbulence scalar-fluxes production

$\mathscr{C}_\theta=\frac{U_j}{2}\frac{\partial}{\partial x_j}(\overline{T'T'})=\frac{U_j\partial\mathscr{E}_\theta}{\partial x_j}$ as scalar-fluxes convection term

$\mathscr{D}_\theta=\overline{T'T'u_j'}-\alpha\frac{\partial\overline{T'T'}}{\partial x_j}$ as scalar-fluxes diffusion.

$\varepsilon_\theta=\alpha\overline{\frac{\partial T'}{\partial x_j}\frac{\partial T'}{\partial x_j}}$ as destruction term.

The equation can be cast into following form for understanding the balance of terms

$$\mathscr{P}_\theta-\varepsilon_\theta-\frac{\partial\mathscr{D}_\theta}{\partial x_j}-\mathscr{C}_\theta=0$$

Nagano and Kim (1988) pioneered this approach in wall-bounded flows. Subsequent refinements by Nagano's group (1991-1995) improved near-wall predictions for non-uniform thermal boundary conditions. So and co-workers developed similar two-equation scalar variance models with near-wall and compressibility corrections in the early 1990s. More recently, Dash and colleagues (Chidambaram et al. 2001; Brinckman et al. 2007) extended scalar variance transport to jet and high-speed boundary-layer flows.

PROBLEMS

20P-1 Explain the Reynolds averaging procedure and derive the Reynolds-averaged Navier-Stokes (RANS) momentum equation for incompressible turbulent flows. What is the physical meaning of the Reynolds stress tensor, and why does it introduce a closure problem?

20P-2 Compare and contrast zero-equation, one-equation, and two-equation eddy viscosity models. How do they estimate turbulent viscosity ν_t, and in what flow conditions might each be most appropriate?

20P-3 Discuss the modeling of turbulent scalar fluxes $\overline{u_i'T'}$. What are the limitations of the turbulent Prandtl number approach, and how do algebraic scalar-flux models such as Daly-Harlow or Younis-Speziale-Clarke address these limitations?

REFERENCES

J. Kampé de Fériet, Sur un probléme d'algébre abstraite posé par la définition de la moyenne dans la théorie de la turbulence, Annales de la Société Scientifique de Bruxelles, Série I (Sciences Mathématiques, Astronomiques et Physiques), 63, 165–180 (1949).

N. Uddin, Fluid Mechanics, A Problem Solving Approach, CRC Press, 2023.

J. C. Rotta, Turbulente Strömungen, Teubner, Stuttgart, 1972.

B. J. Daly, F. H. Harlow, Transport equations in turbulence, Phy. Fluids, Vol.13, No. 11, 2634-2649, 1970.

K. Abe, K. Suga, Towards the development of a Reynolds-averaged algebraic turbulent scalar-flux model, Int. J. Heat and Fluid Flow, Vol. 22, 19 - 29, 2001.

B. A. Younis, C. G. Speziale, T. T. Clark, 2005, A rationale model for turbulent scalar fluxes, Proceedings of the Royal Society A, Vol. 461, 575 - 594.

B. E. Launder, G. J. Reece and W. Rodi, Progress in the Development of a Reynolds-Stress Turbulence Closure, J. Fluid Mech, vol. 68, no. 03, p. 537, 2006.

F. Lien and M. Leschziner, Assessment of Turbulent Transport Models Including Non-Linear RNG Eddy-Viscosity Formulation and Second-Moment Closure, Comput. Fluids, vol. 23, no. 8, pp. 983-1004, 1994.

F. R. Menter, Zonal Two Equation k-ω Turbulence Models for Aerodynamic Flows, AIAA Paper 93-2906, 1993.

F. R. Menter, Two-Equation Eddy-Viscosity Turbulence Models for Engineering Applications, AIAA Journal, vol. 32, no 8. pp. 1598-1605, 1994.

S. B. Pope, A more general effective-viscosity hypothesis, J. Fluid Mech., 72(2), 331-340, 1975.

B. E. Launder, G. J. Reece, W. Rodi, Progress in the development of a Reynolds-stress turbulence closure, J. Fluid Mech., 68(3), 537-566, 1975.

J. L. Lumley, Computational modelling of turbulent flows, Adv. Appl. Mech., 18, 123-176, 1978.

C. G. Speziale, Analytical methods for the development of Reynolds-stress closures, Annu. Rev. Fluid Mech., 23, 107-157, 1991.

T. B. Gatski, C. G. Speziale, On explicit algebraic stress models for complex turbulent flows, J. Fluid Mech., 254, 59-78, 1993.

D. B. Taulbee, An improved algebraic Reynolds stress model and corresponding nonlinear stress model, Phys. Fluids A, 4(11), 2555-2561, 1992.

S. S. Girimaji, Fully explicit and self-consistent algebraic Reynolds stress model, Theor. Comput. Fluid Dyn., 8, 387-402, 1996.

S. Wallin, A. V. Johansson, An explicit algebraic Reynolds stress model for incompressible and compressible turbulent flows, J. Fluid Mech., 403, 89-132, 2000.

A. Hellsten, New advanced k-ω turbulence model for high-lift aerodynamics, AIAA Journal, 43(9), 1857-1869, 2005.

W. M. J. Lazeroms, G. Brethouwer, A. V. Johansson, An explicit algebraic Reynolds stress and scalar flux model for stably stratified flows, J. Fluid Mech., 723, 91-125, 2013.

I. A. Grigoriev, S. S. Girimaji, T. B. Gatski, A realizable explicit algebraic Reynolds stress model for compressible turbulent flow with significant mean dilatation, Physics of Fluids, 25(10), 105110, 2013.

D. A. Yoder, D. T. Papamoschou, T. B. Gatski, Improvements to an explicit algebraic stress model for turbulent jet flows, NASA Technical Memorandum, NASA TM-2019-219978, 2019.

Y. Chen, D. Tang, Improvement of the algebraic stress model for separated flows based on field inversion and machine learning, Phys. Fluids, 35(11), 115101, 2023.

K. Abe, T. Kondoh, Y. Nagano, A new turbulence model for predicting fluid flow and heat transfer in separating and reattaching flows—I. Flow field calculations, Int. J. Heat Mass Transfer, 37(1), 139-151, 1994.

P. A. Durbin, Separated flow computations with the $k-\varepsilon-v^2$ model, AIAA Journal, 33(4), 659-664, 1995.

F. R. Menter, Two-equation eddy-viscosity turbulence models for engineering applications, AIAA Journal, 32(8), 1598-1605, 1994.

S. Wallin, A. V. Johansson, An explicit algebraic Reynolds stress model for incompressible and compressible turbulent flows, J. Fluid Mech., 403, 89-132, 2000.

U. C. Goldberg, O. Peroomian, S. Chakravarthy, A transition model suitable for unstructured CFD codes, AIAA Paper 98-2648, 1998.

A. Hellsten, New advanced $k-\omega$ turbulence model for high-lift aerodynamics, AIAA Journal, 43(9), 1857-1869, 2005.

R. B. Langtry, F. R. Menter, Correlation-based transition modeling for unstructured parallelized computational fluid dynamics codes, AIAA Journal, 47(12), 2894-2906, 2009.

D. K. Walters, D. Cokljat, A three-equation eddy-viscosity model for Reynolds-averaged Navier-Stokes simulations of transitional flow, J. Fluids Eng., 130(12), 121401, 2008.

K. S. Abdol-Hamid, J.-R. Carlson, C. L. Rumsey, Verification and validation of the k-kL turbulence model in FUN3D and CFL3D codes, NASA Technical Memorandum 2015-218930, 2015.

P. Y. Chou, On velocity correlations and the solutions of the equations of turbulent fluctuation, Quarterly of Applied Mathematics, 3(1): 38-54, 1945.

M. Nagano, C. Kim, Turbulent heat transfer in channel flow, International Journal of Heat and Mass Transfer, 31(5), 979-990, 1988.

M. Nagano, Y. Hattori, Development of a two-equation heat transfer model based on the k-ε model and its application to separated and impinging flows, Numerical Heat Transfer, Part A, 19(2), 233-246, 1991.

B. J. So, H. Yoo, Near-wall second-moment modeling of turbulent scalar transport, International Journal of Heat and Mass Transfer, 36(16), 3973-3982, 1993.

A. Chidambaram, S. Dash, Extension of scalar variance models to compressible turbulent flows, AIAA Journal, 39(5), 859-867, 2001.

K. Brinckman, S. Dash, Scalar transport with variable Prandtl numbers in compressible boundary layers, Journal of Thermophysics and Heat Transfer, 21(2), 314-322, 2007.

21 Direct and Large Eddy Simulations

In this chapter, we explore the foundations and implementation of turbulence modeling using Direct and Large Eddy Simulation (DNS and LES). After completing this chapter, you should be able to:

- Understand the motivation behind DNS and LES for turbulent flow simulation, including their advantages and limitations.
- Explain the concept of energy cascade, inertial subrange, and Kolmogorov length scale in turbulent flows.
- Estimate DNS resolution requirements and time-step constraints based on Reynolds number and pipe flow parameters.
- Describe LES filtering operations, including top-hat, Gaussian, and spectral filters, and how they separate resolved and subgrid scales.
- Evaluate and compare subgrid-scale models (e.g., Smagorinsky, WALE, dynamic model, one-equation model) and their applications to LES.

Turbulence is a very complex flow, fortunately, it can be computed using the purely deterministic equations (Navier-Stokes equations). However, the solution is very sensitive to initial or boundary conditions. The perturbations or disturbances which come from initial conditions, boundary conditions and external noise can significantly alter the outcome. To study the turbulent flow field, the field variable are treated as random variable and useful information about the behavior can be obtained by the standard statistical analysis.

In practice usually a one-dimensional energy density spectrum is utilized for the study of turbulent flows. It is nearly impossible to directly measure the three-dimensional energy spectra E($\mathscr{k}$), it is relatively simple to measure the one dimensional energy spectrum.

A Fourier transformation of the autocorrelation function is the energy spectrum, defined as:

$$E(f) = \int_{-\infty}^{\infty} e^{-i(2\pi f)\Delta t}[R^t(t, t+\Delta t)]d(\Delta t) \tag{21.1}$$

Compared to laminar flow, the turbulent flow comprises eddies (fluid particles moving together) of different sizes and scales. These eddies significantly affect the overall flow field. The large eddies have high energy and from them this energy is transfered to small eddies. The energy of small scales can be dissipated due to viscous action. In Figure 21.1 a schematic diagram of a typical energy spectrum

DOI: 10.1201/9781003663171-21

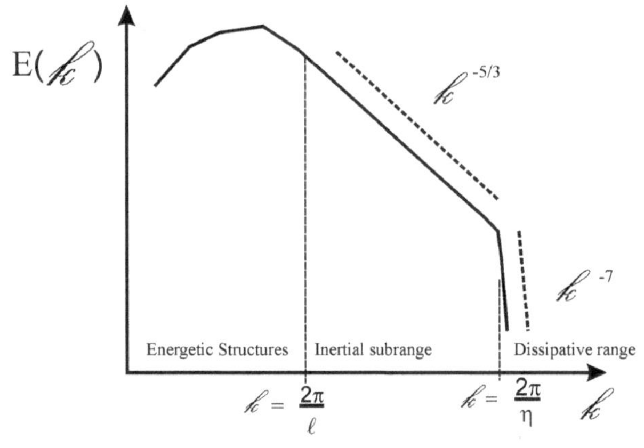

Figure 21.1: **The turbulence energy spectrum.**

is shown. The range between energetic eddies and small scale dissipative eddies is called inertial subrange. The wave number k, which is reciprocal of the length scale, shows the size of the eddies. A representative length scale for large eddies is called integral length scale (ℓ) and for small scales it is called Kolmogorov length scale (η).

21.1 DIRECT NUMERICAL SIMULATION

Turbulent flow problems at low Reynolds numbers (Re $<10^3$) can be computed using the Navier-Stokes equations directly without invoking any models. This approach is called the Direct Numerical Simulation (DNS) of turbulent flow. Various DNS datasets are also made available for researchers and students on the internet and can be used to understand turbulence dynamics more. As the Reynolds number increases, the numerical diffusion in codes is too large, and eventually, it is no longer possible to compute turbulence scale correctly. For high Reynolds number problems (Re $>10^6$) the available approach at the moment is to use Reynolds Averaged Navier-Stokes (RAN) based turbulence models. For industrial applications, the Reynolds Averaged Navier-Stokes (RAN) based turbulence models are still used, and in this approach, the flow is modeled using various assumptions. Since 1970 there has been significant development in RANS based modeling approach, and these models are available in Computational Fluid Dynamics (CFD) codes or applications like FLUENT, CFX, and OpenFOAM. A third option is the Large Eddy Simulation (LES) between these two extremes. In this approach, large turbulent scales are computed, and small scales are modeled. LES technique has been proved to be a fruitful option for a wide range of Reynolds numbers. Figure 21.2 shows the difference between the three approaches.

The direct measurement of turbulence by experimental techniques has tremendously helped in our understanding of the phenomenon. However, it is not a simple

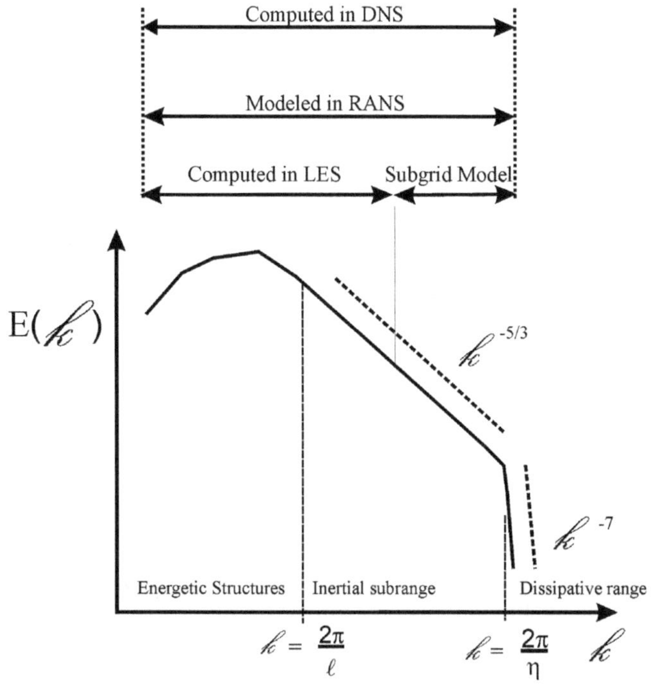

Figure 21.2: **Energy spectrum vs wave number. Various approaches used in CFD for turbulent flow computations.**

matter to perform fully satisfying measurements in turbulent flow. The measurement sensors are beset with difficulties like resolution, reliability, precision and cost. Moreover the variables which can be measured are very limited.

In spite of the difficulties, in the last 100 years, data has been accumulated and a large body of knowledge is available for turbulent flow. The continuous advances in data acquisition techniques is also helping in reducing the above mentioned problems.

In the early 1980s the advances in computational techniques and resources permitted the numerical solution of Navier-Stokes equations. The full field simulation called Direct Numerical Simulation (DNS) has become an important source of information and insight for turbulence. DNS helps in understanding turbulence physics and its modeling. This helps in understanding the time evolution of the turbulence field, life histories of eddies and multi point correlations. DNS is now applied for flows over a flat plate, pipe flow, jet and wakes.

The computational requirement for DNS increases with Reynolds number. This makes the applicability of DNS for industrial problems limited. The restrictions come from the fact that we want to resolve the scales having order of the Kolmogorov scale (η).

$$\eta = \left(\frac{\nu^3}{\varepsilon}\right)^{1/4} \tag{21.2}$$

In order to resolve the turbulent flow field for DNS computations, different resolution requirements are suggested in literature. According to Leonard, to simulate reliably the dissipation of turbulence energy, the grid spacing must be somewhat smaller than the length scale corresponding to the peak in dissipation spectrum and mean grid spacing corresponds to 3η (Leonard 1996). According to Ferziger, the grid must not be larger than the viscously determined η (Ferziger 1996). According to Moin and Mahesh, the smallest resolved scale is required to be of $O(\eta)$ not equal to η (Moin and Mahesh 1998).

The volume-averaged dissipation rate ε given by

$$\varepsilon = -\frac{4\overline{U}\nu}{D}\left.\frac{\partial U}{\partial r}\right|_{\text{wall}}$$

This represents the energy dissipation rate per unit mass in a pipe, where, $\overline{U}$ is the mean velocity, D is the pipe diameter, ν is the kinematic viscosity, and $\frac{\partial U}{\partial r}|_{\text{wall}}$ is the wall shear gradient.

We can estimate the Kolmogorov length scale η, which is given by

$$\eta = \left(\frac{\nu^3}{\varepsilon}\right)^{1/4}$$

Substituting ε we have

$$\eta = \left(\frac{\nu^3}{\frac{4\overline{U}\nu}{D}\frac{\partial U}{\partial r}\Big|_{\text{wall}}}\right)^{1/4} = \eta = \left(\frac{\nu^2 D}{4\overline{U}\,\frac{\partial U}{\partial r}\Big|_{\text{wall}}}\right)^{1/4}$$

From pipe flow theory, wall shear stress is related to friction factor f by:

$$\tau_w = \frac{f}{8}\rho\overline{U}^2 \quad \text{and} \quad \tau_w = \mu\left.\frac{\partial U}{\partial r}\right|_{\text{wall}}$$

So

$$\mu\left.\frac{\partial U}{\partial r}\right|_{\text{wall}} = \frac{f}{8}\rho\overline{U}^2 \Rightarrow \left.\frac{\partial U}{\partial r}\right|_{\text{wall}} = \frac{f}{8}\frac{\rho\overline{U}^2}{\mu}$$

Use $\nu = \mu/\rho$, thus:

$$\left.\frac{\partial U}{\partial r}\right|_{\text{wall}} = \frac{f}{8}\frac{\overline{U}^2}{\nu}$$

Substitute back into expression for η:

$$\eta = \left(\frac{\nu^2 D}{4\overline{U} \cdot \frac{f}{8} \cdot \frac{\overline{U}^2}{\nu}} \right)^{1/4} = \left(\frac{8\nu^3 D}{4f\overline{U}^3} \right)^{1/4} = \left(\frac{2\nu^3 D}{f\overline{U}^3} \right)^{1/4}$$

Use Reynolds number $\mathrm{Re} = \frac{\overline{U}D}{\nu} \Rightarrow \overline{U} = \frac{\mathrm{Re}\nu}{D}$, so:

$$\eta = \left(\frac{2\nu^3 D}{f\left(\frac{\mathrm{Re}\nu}{D}\right)^3} \right)^{1/4} = \left(\frac{2\nu^3 D^4}{f\mathrm{Re}^3 \nu^3} \right)^{1/4} = \left(\frac{2D^4}{f\mathrm{Re}^3} \right)^{1/4} = \frac{(2/f)^{1/4} D}{\mathrm{Re}^{3/4}}$$

We get

$$\eta = \frac{(2/f)^{1/4} D}{\mathrm{Re}^{3/4}}$$

We can use an implicit equation for f, commonly used in turbulent pipe flow to estimate the Darcy-Weisbach friction factor.

$$\sqrt{\frac{8}{f}} = 2.44 \ln\left(\sqrt{\frac{f}{32}} \mathrm{Re} \right) + 2.0$$

Once η is computed, you can calculate, η/D, and

$$\eta^+ = \eta\, u_\tau / \nu$$

where, $u_\tau = \sqrt{\tau_w/\rho}$. The number of grid points needed are

$$N = \frac{10\pi}{4} \left(\frac{D}{\Delta} \right)^3 \quad \text{where} \quad \Delta = 3\eta$$

The spatial step Δ is chosen to be on the order of the Kolmogorov scale, typically:

$$\Delta = 3\eta \Rightarrow \Delta \approx \frac{9D}{\mathrm{Re}^{3/4}}$$

The temporal resolution must ensure that fluid parcels do not traverse more than one grid cell per time-step, to maintain numerical stability and accuracy. This is similar to a CFL-type condition:

$$\Delta t \leq \frac{\Delta}{\overline{U}} \Rightarrow \Delta t \leq \frac{9D}{\overline{U}\mathrm{Re}^{3/4}}$$

We can estimates the minimum number of time steps N_s required to run a Direct Numerical Simulation (DNS) for a sufficiently long physical time T to gather meaningful statistics. The physical time considered is:

$$T = \frac{100D}{\overline{U}}$$

Table 21.1
DNS Grid and Time-Step Parameters for Turbulent Pipe Flow

Re	f	η/D	η^+	$\Delta^+ = 3\eta^+$	N_s (min)
1.0×10^3	0.06314	0.013341	1.19	3.56	1956
5.0×10^3	0.03776	0.004537	1.56	4.68	6541
1.0×10^4	0.03119	0.002830	1.77	5.30	11000
2.0×10^4	0.02615	0.001758	2.01	6.03	18500
5.0×10^4	0.02111	0.000933	2.40	7.19	36781
1.0×10^5	0.01818	0.000576	2.75	8.24	61858

Table 21.2
Grid Requirements for DNS and LES for channel flows (y is wall normal coordinate)

DNS	LES
$\Delta x^+ \approx 15$	$\Delta x^+ \approx 50$
$\Delta y^+_{min} \leq 1$	$\Delta y^+_{min} \leq 1$
$\Delta z^+ \approx 6$	$\Delta z^+ \approx 20$

which corresponds to following the flow for a time equal to the duration it takes to traverse 100 pipe diameters at the mean velocity $\overline{U}$.

From earlier, the maximum allowable time-step to resolve the smallest (Kolmogorov) scales is:

$$\Delta t \leq \frac{9D}{\overline{U}\,\mathrm{Re}^{3/4}}$$

So the minimum number of steps required for a simulation of duration T is:

$$N_s = \frac{T}{\Delta t} \geq \frac{100D/\overline{U}}{9D/(\overline{U}\,\mathrm{Re}^{3/4})} = \frac{100}{9}\mathrm{Re}^{3/4} \approx 11\,\mathrm{Re}^{3/4}$$

This yields

$$N_s \geq 11\,\mathrm{Re}^{3/4}$$

For second order accurate schemes it is recommended that:

For incompressible Direct Numerical Simulation (DNS), explicit Runge-Kutta methods (especially high-order versions like RK4) have become the standard for time integration. These methods excel at capturing the fine details of turbulent flows while remaining computationally efficient. Their explicit formulation provides two key advantages: they eliminate the need to solve complex nonlinear systems at each time step, and they work seamlessly with high-precision spatial techniques like spectral methods. Additionally, their structure allows for excellent parallel computing performance, a critical requirement for large-scale DNS calculations.

In some codes, Adams-Bashforth method is employed which offers reduced memory requirements at the cost of lower accuracy compared to RK4. For memory-intensive simulations, specialized low-storage RK variants such as Williamson's (2N) RK have gained popularity in large-scale DNS implementations.

While implicit methods provide unconditional stability independent of time step size, their substantial computational overhead generally makes them impractical for conventional DNS applications. However, they become indispensable for specific challenging scenarios featuring stiff systems—particularly reactive flows, and geophysical simulations. In these cases, hybrid IMEX (Implicit-Explicit) or semi-implicit formulations offer an effective solution.

An important implementation detail is that even predominantly explicit RK-based DNS solvers must incorporate an implicit pressure projection step, although this is treated as a decoupled component from the primary time integration scheme. For the majority of turbulent flow simulations, the optimal balance between accuracy and efficiency is achieved through either pure explicit Runge-Kutta methods or carefully designed IMEX hybrids (such as combining Crank-Nicolson for diffusion terms with RK for convective terms).

The DNS community continues to prefer explicit RK methods for most turbulence simulations, with the choice of specific method depending on how each application balances the competing demands of precision, stability, and computational expense. This practical approach reflects the field's emphasis on achieving accurate results while managing the enormous computational costs of fully resolved turbulence simulations.

21.2 LARGE EDDY SIMULATION

The rationale of LES is that the large scale structures in turbulent flow produced directly by the instability of the mean flow (shear or buoyancy effects) should be simulated directly as they are problem dependent, anisotropic and play a major role in transport of mass, momentum and energy. On the other hand, small scale structures can be modeled with less effect on the prediction quality.

The technique was originally proposed by the meteorologists Lilly (1962) and Smagorinsky (1963). They introduced the idea of relating the turbulent quantity, eddy viscosity, to the large scale dynamics of the atmospheric flows. Deardorff tested the technique for channel flow and planetary flows in the early 70s. The procedure is further improved by Schumann who used the idea of volume averaging over the computational cell for the results refinement (Schumann 1975).

As mentioned above, the philosophy of LES is to explicitly resolve the large-scale motion and to model the small scales. To distinguish the small scales eddies from resolved large scales, a spatial filtering process is applied to the Navier-Stokes equations.

21.2.1 FILTERING

There are two types of filtering approaches used in LES namely, implicit approach and Explicit approach. In explicit filtering one can use a computational grid finer than the width of filter and only the largest scales by some (explicit) filtering operation are retained. This approach reduces the numerical discretization errors as the retained motion is always well resolved. Further, in explicit filtering the filter width and the grid size can be changed independently in order to separate both error contributions and a grid independent solution can be achieved. Because of extra filtering operations the explicit filtering is computationally expensive but can deliver better solutions than the implicit filtering approach.

In case of the implicit LES approach, the only reference to the filter is through the filter width Δ, which is based on grid.

In LES each quantity is decomposed as:

$$\Psi = \hat{\Psi} + \tilde{\Psi} \tag{21.3}$$

This is analogous to the Reynolds decomposition of field quantities into mean ($\overline{\Psi}$) and fluctuating quantities ($\Psi^{'}$). However, here a caret ($\hat{\Psi}$) denotes the Large scale (resolved) component which is time dependent and a tilde ($\tilde{\Psi}$) denotes small-scale (unresolved) or subgrid components. This spatial separation is made by low-pass filter:

$$\hat{\Psi}(x,t) = \int\int\int G(x,x')\Psi(x',t)d^3x' \tag{21.4}$$

where Ψ can be u, p and T etc., G is a normalized weighting function or filter. Some commonly used filters are Top-hat filter, Gaussian filter and Sharp Fourier cut-off filter.

The Gaussian filter is smooth both in physical and Fourier space and is defined as:

$$\mathbf{G}(x,x') = \sqrt{\frac{6}{\pi\Delta^2}} exp(\frac{-6(x-x^{'})^2}{\Delta^2}) \tag{21.5}$$

The Cut-off filter is defined in Fourier space and eliminates all fourier coefficients belonging to wave numbers above a particular cut-off. It is natural to use this filter in spectral methods as it leaves more energy in the large scales field than other filters. It is difficult to apply it in homogeneous flows. The cut-off filter is defined as:

$$\mathbf{G}(x,x^{'}) = \frac{1}{\pi}\left[\frac{sin(\frac{\pi(x-x^{'})}{\Delta})}{(x-x^{'})}\right] \tag{21.6}$$

The Top-hat filter or Box filter is simply an average over a rectangular region. It is the most natural choice when finite volume or finite difference method is used. It is defined as:

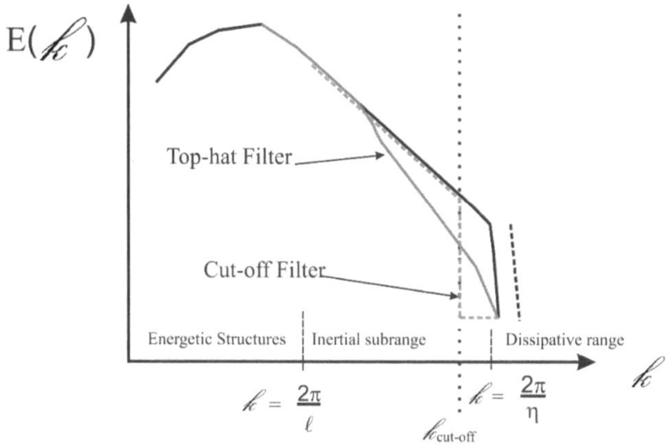

Figure 21.3: **Energy spectrum showing the resolved and subfilter portion in LES computation. The sharp Fourier cut off filter is shown by dashed line. The red line shows the top hat filter. Part of the real spectrum is lost due to numerical errors.**

$$\mathbf{G}(x,x') = \begin{Bmatrix} 1/\Delta & \forall & |x-x'| & \leq \frac{1}{2}\Delta \\ 0 & \forall & |x-x'| & > \frac{1}{2}\Delta \end{Bmatrix} \tag{21.7}$$

here, Δ is the characteristic filter width.

The box filter does not remove filter energy at the wave numbers beyond the highest resolvable in a satisfactory manner. This energy is aliased into lower wave number components and gives falsified amplitude to the resolvable scale waves.

The schematic representation of top hat filter in energy spectrum is shown as red line in Figure 21.3. It is interesting to see the application of the top-hat filter in the wave-number space. When the top-hat filter is used it will not sharply divide the wave number space into large and small scales, as in case of sharp Fourier cut off filter (dashed line).

Filtering Operations By low pass filtering the flow is divided into resolved and subfilter-scale motions. The subfilter-scale motion is modeled by the subgrid model. The following filter operations are used:

Linearity

$$\widehat{f+g} = \widehat{f} + \widehat{g} \tag{21.8}$$

Conservation of constant

$$\widehat{C} = C \tag{21.9}$$

Commutation

$$\widehat{\frac{\partial f}{\partial x}} = \frac{\partial \widehat{f}}{\partial x} \tag{21.10}$$

Researchers have propsed different filters for LES simulations, for example:
Schumann's Filter:

$$\Delta = \sqrt[3]{\Delta x_1 \Delta x_2 \Delta x_3} \tag{21.11}$$

where the $\Delta x_1, \Delta x_2$ and Δx_3 are mesh sizes along three directions.

Bardina Filter: Bardina et al. (1980) showed that in homogeneous turbulent flows the better filter definition is:

$$\Delta = \sqrt{\frac{\Delta x_1^2 + \Delta x_2^2 + \Delta x_3^2}{3}} \tag{21.12}$$

In LES with non-aeroacoustic applications, it is good enough to select the filter as large as the grid spacing. Therefore, the smallest filter width would be proportional to grid spacing.

Bogey and Bailly Explicit Filtering Approach: In Computational Aeroacoustics, resolving waves with minimal phase/group speed error is critical. Dispersion-Relation-Preserving (DRP) schemes were pioneered by Tam and Webb (1993) to ensure the finite-difference schemes dispersion relation matches the PDEs as closely as possible. Extending the DRP philosophy, Bogey and Bailly (2004) introduced explicit selective filters optimized in Fourier space to effectively remove numerical grid-to-grid oscillations typically produced by finite-difference discretizations. These selective filters utilize central finite-difference stencils whose coefficients are carefully determined by minimizing dissipation errors within a targeted wavenumber range. Specifically, the filters achieve high dissipation at short wavelengths near the grid-scale, significantly damping numerical noise, while ensuring negligible dissipation at longer wavelengths to preserve essential physical solution structures. This selective approach avoids unnecessary smoothing of critical features and prevents the introduction of spurious numerical dispersion. In advection-diffusion problems, such filtering is particularly relevant, as it prevents oscillations arising from sharp gradients and maintains fidelity in the simulation of physical transport processes. Among the filters proposed, the optimized 13-point selective filter stands out due to its superior balance between computational efficiency and spectral selectivity. The DDR-optimized finite difference approximation of the first derivative is given as:

$$\left.\frac{\partial f}{\partial x}\right|_j \approx \frac{1}{\Delta x} \sum_{m=-N}^{N} a_m f_{j+m}$$

where,

f_{j+m}: value at spatial node $j+m$.

a_m: optimized stencil coefficients.

N: stencil half-width (e.g., $N = 6$ for 13-point stencil).

Δx: grid spacing.

Selective Filtering (Bogey-Bailly approach): The selective filter procedure to damp out high-frequency numerical oscillations is applied as follows:

$$f_j^{\text{filtered}} = f_j - r_d \sum_{m=-N}^{N} d_m f_{j+m}$$

where,
f_j: solution before filtering at node j.
f_j^{filtered}: solution after filtering.
d_m: optimized filtering coefficients.
r_d: filtering strength parameter, typically between 0.1 and 0.2 for stability.
N: stencil half-width.

In LES, explicit filters remove high-frequency numerical oscillations, preventing contamination of resolved scales. The optimized selective filters (like Bogey-Bailly) ensure minimal damping of physically significant scales while effectively eliminating unwanted grid-scale numerical noise.

Example 21.1

Consider a turbulent flow field represented by the one-dimensional velocity component $u(x) = \sin(2x) + 0.5\sin(8x) + 0.2\sin(32x)$ with $x \in [0, 2\pi]$, discretized over 256 grid points. This signal contains multiple scales of motion, with the higher wavenumber components representing smaller eddies.

(a) Explain the purpose of filtering in Large Eddy Simulation (LES) and how it differs from Reynolds-Averaged Navier-Stokes (RANS) approaches. Discuss the concept of resolved scales and subgrid scales in this context.

(b) Implement and compare the following LES filters applied to $u(x)$:

i Box filter (top-hat filter) with a width of 8 grid points

ii Gaussian filter with standard deviation $\sigma = 2.0$

iii Spectral cutoff filter at wavenumber $k_c = 10$

iv Spectral Gaussian filter with spectral width $\sigma_k = 5.0$

Plot the original signal and filtered results, and comment on how each filter affects the different scales of motion. Which filter best preserves the large-scale features while effectively removing small-scale fluctuations?

(c) The transfer function of a filter describes its wavenumber-dependent attenuation. Derive the transfer functions for the box and Gaussian filters in physical space, and compare them to the spectral filters. How does the choice of filter influence the energy spectrum of the resolved field?

(d) In practice, the filter width Δ is often tied to the grid spacing. For the box filter, justify why $\Delta = 2\pi/k_c$ might be a reasonable choice when k_c is the cutoff wavenumber. How does this relate to the Nyquist-Shannon sampling theorem?

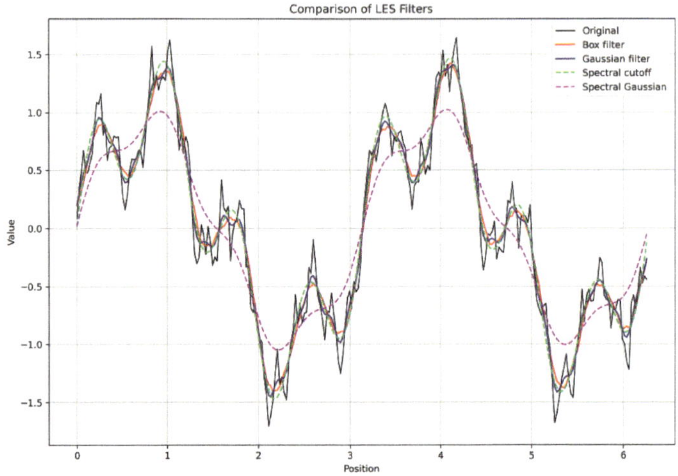

Figure 21.4: **Filtered representations of a turbulent signal using box, Gaussian, spectral cutoff, and spectral Gaussian filters. Spectral filters produce smoother results, while real-space filters retain more detail.**

Solution

Box filter: This filter reduces high-frequency content by averaging values over a fixed window. It smooths the signal but can still preserve some relatively sharp variations.

Gaussian filter: Offers smoother filtering than the box filter by applying a weighted average, giving more emphasis to central values. It effectively smooths transitions while maintaining the overall shape of the signal.

Spectral cutoff: Eliminates all frequency components beyond a defined threshold. This produces a clean separation between resolved and unresolved scales, though it may introduce mild Gibbs oscillations near discontinuities.

Spectral Gaussian: Yields the most smoothed profile by gradually damping high-frequency modes. While it avoids abrupt truncation artifacts, it also suppresses finer-scale features more strongly than the other filters.

Figure 21.4 compares the effects of different LES filters on a turbulent signal. The original signal exhibits sharp fluctuations, which are progressively smoothed by the filters. The box and Gaussian filters, applied in physical space, reduce noise while preserving some small-scale features. In contrast, the spectral cutoff and spectral Gaussian filters, applied in Fourier space, remove high-frequency content more aggressively. The spectral Gaussian filter yields the smoothest result due to its gradual attenuation of high wavenum-

bers. This comparison highlights the trade-off between resolution retention and smoothing strength among various LES filtering approaches.
In Large Eddy Simulation (LES), no single filter is optimal for all cases. The best choice depends on the simulation goals, the numerical method, and the flow characteristics. The box filter is well suited for structured grids with finite-volume methods. For finite-volume or finite-difference LES, the Gaussian filter applied in physical space provides a good compromise between accuracy and stability. In spectral LES codes, spectral filters such as the spectral cutoff or spectral Gaussian are commonly used, with the spectral Gaussian offering greater robustness against aliasing and high-frequency noise.

21.2.1.1 LES: Governing Equations

By filtering the continuity, Navier-Stokes, and energy equations and neglecting body forces and subgrid density fluctuations, we obtain:

$$\frac{\partial(\rho\hat{u}_i)}{\partial t}+\frac{\partial(\rho\hat{u}_i\hat{u}_j)}{\partial x_j}=-\frac{\partial\hat{p}}{\partial x_i}+\frac{\partial}{\partial x_j}\left\{\mu\left(\frac{\partial\hat{u}_i}{\partial x_j}+\frac{\partial\hat{u}_j}{\partial x_i}\right)\right\}-\frac{\partial\tau_{ij}^{sgs}}{\partial x_j} \tag{21.13}$$

where the additional term τ_{ij}^{sgs} is the subgrid-scale (SGS) stress that arises from filtering the nonlinear convective term. The decomposition of velocity into resolved and unresolved components is

$$u_i=\hat{u}_i+\widetilde{u}_i \tag{21.14}$$

so that

$$\widehat{u_iu_j}=\hat{u}_i\hat{u}_j+\hat{u}_i\widetilde{u}_j+\widetilde{u}_i\hat{u}_j+\widetilde{u}_i\widetilde{u}_j \tag{21.15}$$

The SGS stress tensor is defined as

$$\tau_{ij}^{sgs}=\rho\left(\widehat{u_iu_j}-\hat{u}_i\hat{u}_j\right)=\rho\left(L_{ij}+Cr_{ij}+R_{ij}\right) \tag{21.16}$$

with the decomposition

$$L_{ij}=\widehat{\hat{u}_i\hat{u}_j}-\hat{u}_i\hat{u}_j \tag{21.17}$$

is the Leonard term (resolved-resolved interactions),

$$Cr_{ij}=\widehat{\hat{u}_i\widetilde{u}_j}+\widehat{\widetilde{u}_i\hat{u}_j} \tag{21.18}$$

is the cross term (resolved-unresolved interactions),

$$R_{ij}=\widehat{\widetilde{u}_i\widetilde{u}_j} \tag{21.19}$$

is the subgrid (unresolved) Reynolds stress.

The above mentioned decomposition is now only mentioned for historical reasons. Previously, many researchers modeled the subgrid stresses separately. However in the recent past it has become common to model the entire subgrid scale stress as a single unit. In the context of LES the subgrid scale stress is in fact the large scale momentum flux caused by the action of small or unresolved scales. The subgrid is not related to grid size as name might incorrectly suggest, but related to the average of the small scale field.

One can write the mass and momentum equations as:

$$\frac{\partial \rho}{\partial t} + \frac{\partial (\rho \widehat{u}_i)}{\partial x_i} = 0 \tag{21.20}$$

$$\frac{\partial (\rho \widehat{u}_i)}{\partial t} + \frac{\partial (\rho \widehat{u}_i \widehat{u}_j)}{\partial x_j} = -\frac{\partial \widehat{p}}{\partial x_i} + \frac{\partial}{\partial x_j}\left\{ \mu \left[\frac{\partial \widehat{u}_i}{\partial x_j} + \frac{\partial \widehat{u}_j}{\partial x_i} \right] \right\} - \frac{\partial \tau_{ij}^{sgs}}{\partial x_j} \tag{21.21}$$

Similarly, one can write the energy equation as:

$$\frac{\partial (\rho c_p \widehat{T})}{\partial t} + \frac{\partial (\rho c_p \widehat{u}_j \widehat{T})}{\partial x_j} = \frac{\partial}{\partial x_j}\left[\Gamma c_p \frac{\partial \widehat{T}}{\partial x_j} \right] - \frac{\partial q_j^{sgs}}{\partial x_j} \tag{21.22}$$

For the LES computations, there are broadly following approaches:

i A LES computation with no subgrid model (quasi-DNS).

ii A LES computation with subgrid model.

iii A monotonically integrated LES (MILES), in which, numerical diffusion is used in place of the subgrid-scale model.

21.2.2 SUBGRID MODELS

The model used to approximate the subgrid-scale term is called subgrid-scale model. The large scales in the flow depend on the geometry of the problem and are produced by the mean flow, so they are inhomogeneous, anisotropic, diffusive and not universal. In contrast, the small scales are homogeneous, isotropic, dissipative, and universal. It is easier to model only small scales. If the trace of the tensor τ_{ij} is subtracted the subgrid scale Reynolds stress to be modeled is:

$$\tau_{ij}^{sgs} = \underbrace{\tau_{ij}^{a}}_{anisotropic} + \underbrace{\frac{1}{3}\delta_{ij}\tau_{kk}^{sgs}}_{isotropic} \tag{21.23}$$

This is the subgrid scale stress term which represents the effects of unresolved small scales on resolved large scales.

We can categorize the models used in LES as:

A Gradient—diffusion models

B Higher-order models

B Multilevel models / Scale similarity or Dynamic models

D Non-conventional models

21.2.2.1 Gradient Diffusion Model

A large class of subgrid closure models are based on the gradient-diffusion hypothesis. It can be expressed as a relation between anisotropic stress and the (large-scale)strain rate tensor.

$$\tau_{ij}^{sgs} - \frac{1}{3}\delta_{ij}\tau_{kk}^{sgs} = -2\mu_s\left[\widehat{S_{ij}} - \frac{1}{3}\delta_{ij}\widehat{S_{kk}}\right] \tag{21.24}$$

Where μ_s is the subgrid viscosity, which needs to be described by appropriate models. If the subgrid scale energy $\frac{1}{2}\widehat{\widetilde{u}_k\widetilde{u}_k}$ is indicated by E_s^{sgs} the second term in the above equation can be expressed as:

$$\frac{1}{3}\delta_{ij}\tau_{kk}^{sgs} = \frac{2}{3}\delta_{ij}\rho E_s^{sgs} \tag{21.25}$$

Substitute this in equation 21.24 and then in equation 21.21; by adding and subtracting $\frac{1}{3}\delta_{ij}\tau_{kk}^{sgs}$ we find a modified pressure with kinetic energy as:

$$\widehat{P} = \widehat{p} + \frac{2}{3}\rho E_s^{sgs} \tag{21.26}$$

Also,

$$\widehat{P} = \widehat{p} + \frac{1}{3}\tau_{kk}^{sgs} \tag{21.27}$$

Now the filtered Navier-Stokes equations can be written as:

$$\frac{\partial(\rho\widehat{u}_i)}{\partial t} + \frac{\partial(\rho\widehat{u}_i\widehat{u}_i)}{\partial x_j} = -\frac{\partial\widehat{P}}{\partial x_i} + \frac{\partial}{\partial x_j}\left\{(\mu+\mu_s)\left[\frac{\partial\widehat{u}_i}{\partial x_j} + \frac{\partial\widehat{u}_j}{\partial x_i}\right]\right\} \tag{21.28}$$

21.2.2.2 Smagorinsky-Lilly Model

This model is based on the work of Smagorinsky (1963) and Lilly (1966), and is therefore called the Smagorinsky-Lilly or simply Smagorinsky model. It is a non-linear eddy-viscosity model, where the subgrid viscosity is related to the magnitude of the resolved strain-rate tensor $\hat{S}_{ij}$. The model can be expressed as

$$\mu_s = \rho(c_s\Delta)^2\hat{S}, \tag{2.29}$$

where the characteristic strain-rate magnitude is defined as

$$\hat{S} = \sqrt{2\hat{S}_{ij}\hat{S}_{ij}}, \qquad \hat{S}_{ij} = \tfrac{1}{2}\left(\frac{\partial\hat{u}_i}{\partial x_j} + \frac{\partial\hat{u}_j}{\partial x_i}\right), \tag{2.30}$$

so that the subgrid viscosity becomes

Table 21.3
Optimum value of Smagorinsky constant

Researchers	Year	Value	Application
Lilly	1966	0.23	Homogeneous-isotropic turbulence
Deardorff	1970	0.17	Atmospheric boundary layer
Dardorff	1970	0.1	Channel flow
Piomelli et al.	1989	0.1	Channel flow
Moin and Kim	1981	0.065	Channel flow
Bardina at al.	1983	0.2 and 0.09	Homogeneous-isotropic turbulence
Mason and Callen	1986	0.2	Channel flow

$$\mu_s = \rho(c_s\Delta)^2\sqrt{2\hat{S}_{ij}\hat{S}_{ij}}. \tag{2.31}$$

Table 21.3 summarizes the disagreement in the reported values of the Smagorinsky constant. Most studies suggest that the optimal value of c_s is of the order 0.1, although its precise value depends on the flow configuration. The limitations of the Smagorinsky model are (Moin, 1991):

1 The optimal model constant must be changed in different flows.

2 The model does not have correct limiting behavior near the wall.

3 The model does not vanish in a laminar flow and it is dissipative in laminar-turbulent transition zone.

4 The model does not account for the backscatter of energy from small scales to large scales, which is important in the transition regime. In turbulence spectrum, on average, the energy is transferred from large scales to small scales, however, at some locations the reverse energy transfer is possible, as shown in Figure 21.5. Piomelli et al. (1990) have investigated large databases of channel flow DNS and shown that backscatter is happening at large fraction of grid points in transitional flow especially in near wall region. The precise amount of it depends on the flow Reynolds number. An accurate modeling of this phenomenon is important for the transitional and non-equilibrium flows. However, for fully developed channel flows it is not a crucial issue.

5 The model does not account for compressibility effects.

21.2.2.3 Dynamic Model

The dynamic model proposed by Germano et al. (1991) is regarded as one of the most effective subgrid-scale models in LES. Germano et al. proposed a procedure which allows the determination of the Smagorinsky constant dynamically. By this procedure the Smagorinsky constant is no longer a constant, but—being more realistic—a

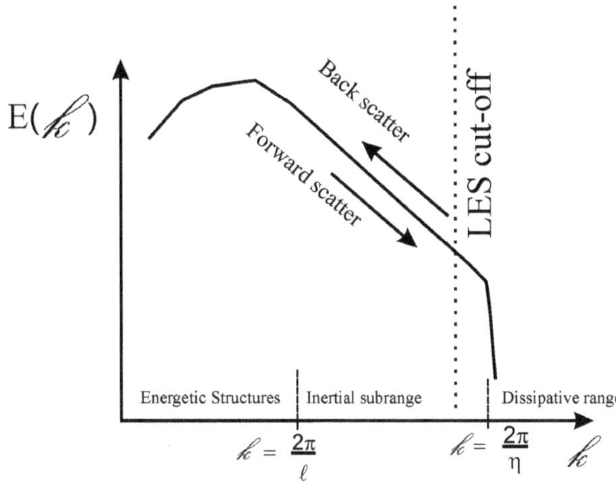

Figure 21.5: **Schematic presentation of the energy cascade and backscatter process.**

function of space and time: $C_s = C_s(\mathbf{x}, t)$. The dynamical computation of C_s is accomplished by defining a test filter with width $\widehat{\Delta}$ which is larger than the grid filter width Δ. The procedure is shown schematically in Figure 21.6.

Applying the test filter to the filtered momentum equation, the subtest-scale stresses Ω_{ij}^{test} can be written as:

$$\sigma_{ij}^{sgs} = \widehat{u}_i\widehat{u}_j - \widehat{u_i u_j} \tag{21.29}$$

$$\Omega_{ij}^{test} = \overline{\widehat{u}_i\widehat{u}_j} - \overline{\widehat{u}_j\widehat{u}_j} \tag{21.30}$$

$$Z_{ij} = \Omega_{ij}^{test} - \overline{\sigma_{ij}^{sgs}} \tag{21.31}$$

The last equation shows the relationship between resolved stresses Z_{ij} and the subtest-scale stress and Subgrid-scale stress. We can approximate the above stresses with the Smagorinsky model.

$$\sigma_{ij}^{sgs} - \frac{1}{3}\delta_{ij}\sigma_{kk}^{sgs} = 2\left[C_g\Delta^2\widehat{S}\right]\widehat{S_{ij}} = 2C_g\alpha_{ij}^{sgs} \tag{21.32}$$

$$\Omega_{ij}^{test} - \frac{1}{3}\delta_{ij}\Omega_{kk}^{test} = 2\left[C_g\widehat{\Delta}^2\overline{\widehat{S}}\right]\overline{\widehat{S_{ij}}} = 2C_g\alpha_{ij}^{sgs} \tag{21.33}$$

where C_g is square the value of C_s. $\widehat{S}$ and $\widehat{S_{ij}}$ are subgrid level and test level strain rate tensors. For incompressible flows S_{kk} is zero. Inserting the above equations in 21.31 relation we have:

$$Z_{ij} = 2C_g\alpha_{ij}^{test} - \overline{2C_g\alpha_{ij}^{sgs}} \tag{21.34}$$

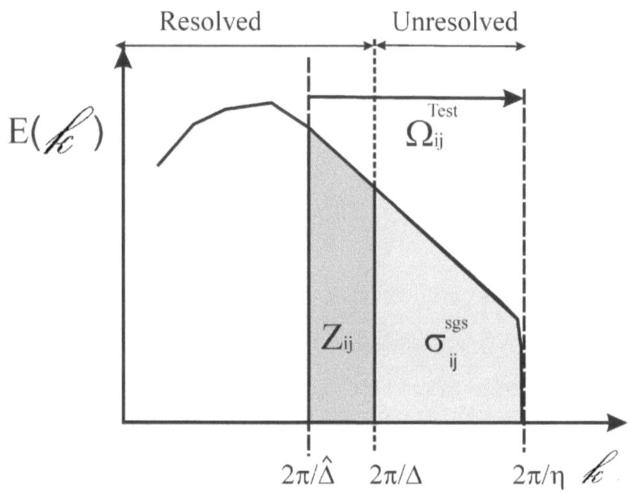

Figure 21.6: **Schematic representation of the Dynamic model.**

Approximating $\overline{C_g \alpha_{ij}^{sgs}} \approx C_g \overline{\alpha_{ij}^{sgs}}$ we can write:

$$Z_{ij} = 2C_g(\alpha_{ij}^{test} - \overline{\alpha_{ij}^{sgs}}) = 2C_g M_{ij} \tag{21.35}$$

$$M_{ij} = (\alpha_{ij}^{test} - \overline{\alpha_{ij}^{sgs}}) \tag{21.36}$$

$$Z_{ij}\widehat{S_{ij}} = 2C_g M_{ij}\widehat{S_{ij}} \tag{21.37}$$

The term M_{ij} in the above equation 21.37 can become zero. This solution is provided by Lilly that square of error can be minimized through:

$$Er = Z_{ij} - 2C_g M_{ij} \tag{21.38}$$

$$\frac{\partial Er^2}{\partial C_g} = 4(Z_{ij} - 2C_g M_{ij})M_{ij} = 0 \tag{21.39}$$

The model parameter to be computed is:

$$C_g(x,t) = \frac{Z_{ij}M_{ij}}{2M_{ij}M_{ij}} \tag{21.40}$$

The model parameter can take negative values. A negative value indicates a negative eddy viscosity and implies an energy transfer from small (subgrid) scales to the resolved scales or backscatter. However, the negative C_g can also lead to numerical instability. In order to avoid the instability, the present investigation uses the relation:

$$C_g(x,t) = max\left\{\frac{Z_{ij}M_{ij}}{2M_{ij}M_{ij}}, 0\right\} \tag{21.41}$$

21.2.2.4 Wall Adapting Local Eddy-viscosity (WALE) Model

The WALE model is known to perform well for flows with laminar to turbulence transition (Nicoud 1999). It is based on the square of the velocity gradient tensor to account for the influence of strain and $\nu_{sgs} = C_k \Delta \sqrt{k_{sgs}}$ where C_k denotes a model constant and k_{sgs} the subgrid-scale kinetic energy. The subgrid-scale kinetic energy is obtained from the expression:

$$k_{sgs} = \left(\frac{\Delta C_w^2}{C_k}\right)^2 \left(\frac{(S_{ij}^d S_{ij}^d)^3}{[(\widetilde{S}_{ij}\widetilde{S}_{ij})^{5/2} + (\widetilde{S}_{ij}\widetilde{S}_{ij})^{5/4}]^2}\right) \tag{21.42}$$

Here $\widetilde{S}_{ij}$ represents the resolved strain rate tensor.

$$\widetilde{S}_{ij} = \widetilde{S} = \frac{1}{2}\left(\frac{\partial \widetilde{u}_i}{\partial x_j} + \frac{\partial \widetilde{u}_j}{\partial x_i}\right)$$

The subgrid-scale viscosity is written as:

$$\nu_{sgs} = (C_w \Delta)^2 \left(\frac{(S_{ij}^d S_{ij}^d)^3}{[(\widetilde{S}_{ij}\widetilde{S}_{ij})^{5/2} + (\widetilde{S}_{ij}\widetilde{S}_{ij})^{5/4}]^2}\right)$$

21.2.2.5 One-equation Eddy Viscosity Model

In one equation, the subgrid-scale model a transport equation for resolved turbulence kinetic energy is solved to consider the effects of history due to production, dissipation, and diffusion that can improve their performance when applied to complex flow situations with non-equilibrium turbulence (Yoshizawa 1986):

$$\frac{D}{Dt}(\rho k) = \nabla.(\rho D_k \nabla k) + \rho G - \frac{2}{3}\rho k \nabla.\mathbf{u} - \frac{C_e \rho k^{1.5}}{\Delta} + S_k \tag{21.43}$$

The subgrid-scale kinematic viscosity (ν_{sgs}) can be defined as $\nu_{sgs} = C_k \Delta \sqrt{k_{sgs}}$ where $C_k = 0.094$. The equation represents (i) the production of turbulence by resolved scales, (ii) turbulent dissipation, (iii) turbulent diffusion, and (iv) viscous dissipation.

21.3 SUBGRID-SCALE HEAT TRANSFER MODELING

The gradient diffusion model is used for the subgrid-scale heat transfer modeling. Using the gradient diffusion assumption for the Energy equation, we can write:

$$q_{ij}^{sgs} = -\Gamma_s c_p \frac{\partial \widehat{T}}{\partial x_i} \tag{21.44}$$

Table 21.4
Subgrid Modeling Approaches

Modeling Approach	Representative Models
Eddy-Diffusivity Models (Functional Models)	Smagorinsky–Lilly (1963/1967)
	Deardorff One-Equation TKE (1974/1980)
	Dynamic Smagorinsky (Germano et al., 1991)
	Dynamic SGS Prandtl Model (Moin et al., 1991; Wong and Lilly, 1994)
Gradient (Nonlinear) Models (Structural Models)	Clark et al. Gradient Model (1979)
	Approximate Deconvolution Model (ADM, Stolz and Adams, 1999)
Scale-Similarity Models (Structural Models)	Bardina Similarity Model (1980/1983)
Mixed Models (Hybrid Structural–Functional Models)	Mixed Eddy-Diffusivity + Bardina Similarity (1980s–90s)
	Dynamic Mixed Model (Zang et al., 1993)
	Dynamic Nonlinear Models (Wang et al., 2008)
Anisotropic Diffusivity Models	Generalized Gradient Diffusion Hypothesis (GGDH) Models (2000s)
	Dynamic Anisotropic Model (Peng and Davidson, 2001–2002)
Data-Driven and Machine Learning Models	Neural Network-Based Models (Gamahara and Hattori, 2017)
	Deep Learning and Convolutional Neural Networks (Lapeyre et al., 2019)

Here, $\widehat{T}$ is resolved temperature and Γ_s is the subgrid thermal diffusivity. By substituting equation 21.44 into 21.22 one has the following form of energy equation:

$$\frac{\partial(\rho c_p \widehat{T})}{\partial t} + \frac{\partial(\rho c_p \widehat{u}_j \widehat{T})}{\partial x_j} = \frac{\partial}{\partial x_j}\left[(\Gamma_e + \Gamma_s) c_p \frac{\partial \widehat{T}}{\partial x_j}\right] \tag{21.45}$$

The subgrid thermal diffusivity Γ_s is expressed as:

$$\Gamma_s = \mu_s / Pr_s \tag{21.46}$$

where Pr_s is the subgrid Prandtl number.

There is no fixed value for the subgrid Prandtl number. Different researchers have used different values of subgrid Prandtl number depending on the grid quality, numerical methodology and problem investigated. There is no unanimous value for the subgrid Prandtl number. Small values of Pr_s means that at small scales the transfer of heat is much more intensive than that of momentum. The value of the subgrid Prandtl number can also be dynamically adjusted. For the consideration of the backscatter effect, the subgrid Prandtl number must be changed.

The subgrid-scale Prandtl number $\mathrm{Pr}_{\mathrm{sgs}}$ in the literature span a range, typically between 0.25 and 1.0, with most estimates clustering around 0.3–0.9. In recent investigations, values closer to 0.9–1.0 is suggested for simulation of complex flows or improved modeling fidelity.

21.3.0.1 Near-Wall Treatment for LES

An important issue in LES computations is the near-wall treatment. There are two approaches used in LES computations. One can increase the near wall resolution or can use a wall-model. In many cases good quality results can be obtained with out wall models. However, the issue becomes more important as the Reynolds number increases, which naturally demand higher near wall resolution. Several wall models had been proposed for the large eddy simulations, details of which can be found in Ciofalo (Ciofalo 1994). However most of these models are based on the ideas taken from channel or duct flow simulations and there performance for impinging jet flows is not investigated.

Case Study 21A: LES of Turbulent Flow in a Square Duct

Problem Definition

We investigate fully developed turbulent flow through a square duct using the finite-volume code FASTEST. The geometry is a straight square duct with all walls set to no-slip and impermeable conditions. The flow is assumed statistically steady and isothermal. The key objective is to capture the transverse secondary motions that arise purely due to turbulence,Äîan effect absent in laminar flows. These secondary flows, classified as "flows of the second kind" by Prandtl, form streamwise vortices at corners and influence shear stress and heat transfer. Their strength is weak (2-3

Governing Equations

We solve the filtered incompressible Navier-Stokes equations in LES form:

$$\frac{\partial \tilde{u}_i}{\partial t}+\frac{\partial \tilde{u}_i\tilde{u}_j}{\partial x_j}=-\frac{1}{\rho}\frac{\partial \tilde{p}}{\partial x_i}+\nu\frac{\partial^2\tilde{u}_i}{\partial x_j^2}-\frac{\partial \tau_{ij}}{\partial x_j},\quad \frac{\partial \tilde{u}_i}{\partial x_i}=0$$

i Filtering: LES (Large Eddy Simulation)

ii Closure model: Dynamic Smagorinsky model

iii Flow assumptions: incompressible, isothermal, statistically stationary

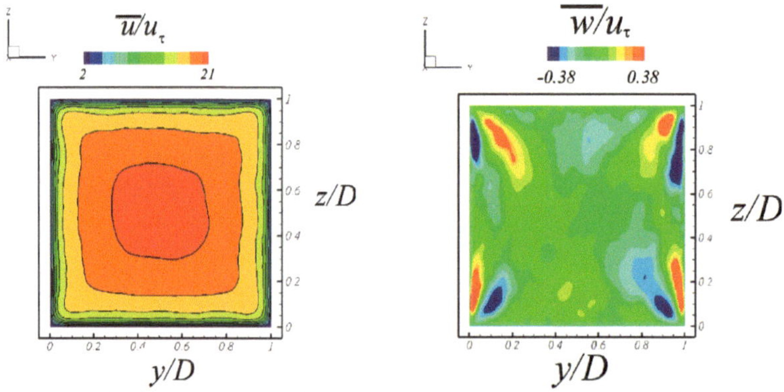

Figure 21.7: **(a) Streamwise velocity contours (u/u_τ) bending near corners. (b) Secondary velocity contours (w/u_τ) forming counter-rotating vortices.**

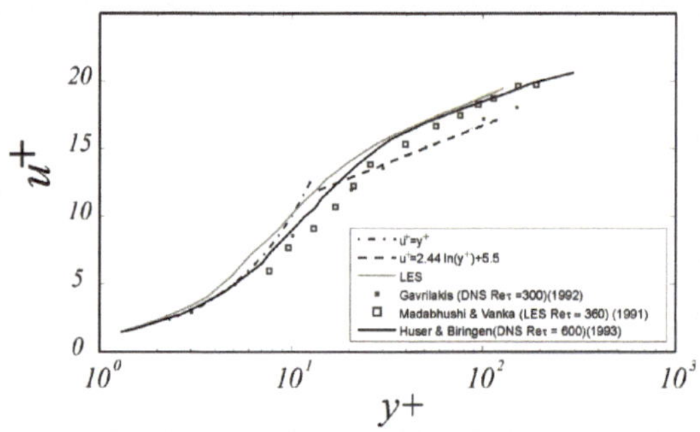

Figure 21.8: **Comparison with logarithmic velocity law at wall bisector.**

Numerical Setup (Grid, BCs, Solver)

i Discretization: Finite Volume Method (FVM), second-order central differencing

ii Grid: Structured hexadedral, wall-refined mesh made via `ICEM-CFD`

iii BCs: Periodic at inlet/outlet, no-slip walls

iv Time-stepping: Crank-Nicolson method

v Pressure-velocity coupling: SIMPLE algorithm

vi Simulation type: LES

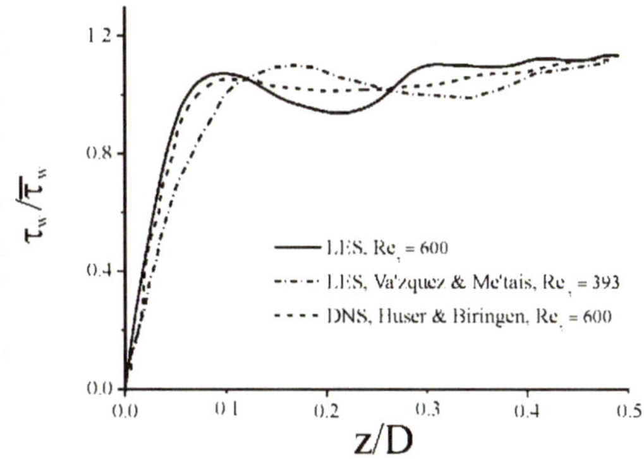

Figure 21.9: **Wall shear stress $\tau_w/\overline{\tau_w}$ profile, showing non-uniformity due to secondary flows.**

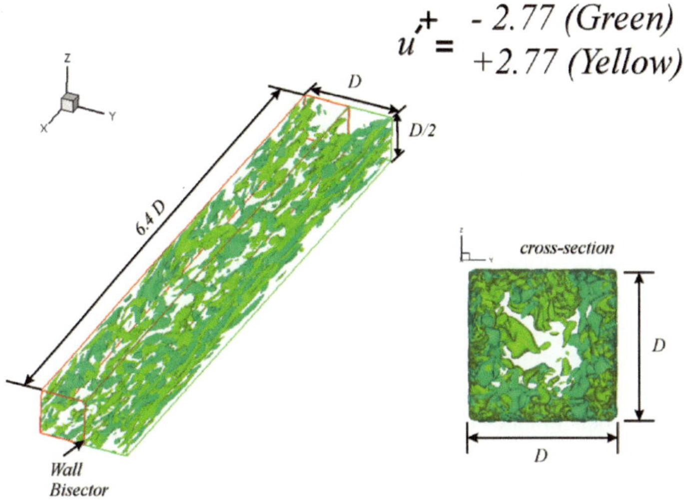

Figure 21.10: **Near-wall streak structures from LES. Absence of streaks in corner zones visible.**

Results and Discussion

Fully developed duct flow shows secondary vortices forming near each corner. These structures are shown in the figures below.

Comparison with the law of the wall at the wall bisector is shown below:

Wall shear stress distribution highlights peak at the wall bisector:

Near-wall streaks extracted via iso-velocity surfaces are shown below. These streaks lift away from the wall and eject fluid into the outer core, except at corners where streaks are absent.

Simulation Comparison with DNS and LES Benchmarks

Parameters	DNS (Gavrilakis, 1992)	DNS (Huser and Biringen, 1993)	DNS (Vazquez and Metais, 2002)	LES (Madabushi, 1991)	Uddin (2011)
Re_τ	300	600	-	360	600
U_{CL}/U_b	1.33	-	1.29	1.278	1.24
w_{max}/U_b (%)	1.9	-	1.92	2.58	2.12
$(\tau_w/\overline{\tau_w})_{max}$	1.18	1.14	1.11	-	1.14

The simulation results presented in the comparison table demonstrate strong agreement between the present LES and previously published DNS and LES benchmarks. The centerline to bulk velocity ratio U_{CL}/U_b, which is a key indicator of mean flow development, closely follows the DNS trends. The present LES predicts a value of 1.24, which is consistent with 1.33 reported by Gavrilakis (1992) and 1.29 by Vzquez and Metais (2002). The maximum magnitude of secondary velocity w_{max}/U_b is slightly higher in the LES by Madabushi (2.58 percent), but the present LES estimate of 2.12 percent remains well within the expected range, indicating successful resolution of the weak secondary motions. In addition, the normalized peak wall shear stress $(\tau_w/\overline{\tau_w})_{max}$, which reflects stress intensification due to secondary vortices, is captured accurately with a value of 1.14, in line with the DNS of Huser and Biringen (1993). The friction Reynolds number Re_τ used in the present study matches the turbulent regime explored in previous work, supporting the validity of the comparison. Overall, the table confirms that the current LES setup effectively reproduces both the mean and secondary flow features observed in benchmark DNS results, demonstrating its reliability in modeling turbulent square duct flows.

Lessons Learned / Numerical Challenges

The LES simulation accurately captured the subtle secondary motions that are typically missed by many RANS models. The secondary vortices near duct corners demonstrate isotach distortion that are consistent with benchmark DNS results, validating the fidelity of the LES. The secondary flow significantly influences the wall shear stress distribution and alters the structure of streamwise streaks, which are seen to lift and eject fluid from the wall region while vanishing near corners. Accurate resolution of these effects requires high grid sensitivity, particularly near the walls, underscoring the importance of wall-resolved LES for capturing turbulence anisotropy. Overall, the LES results exhibit excellent agreement with the benchmark DNS study by Gavrilakis (1992), confirming both qualitative and quantitative consistency.

21.4 ARAKAWA SCHEME

In the 1960s, a numerical scheme was developed for the nonlinear convective term with the specific goal of preserving enstrophy (mean square vorticity) across the computational grid. This property is particularly important for turbulence simulations, which fundamentally depend on the accurate representation of three-dimensional vorticity dynamics. While this scheme could have been introduced earlier during our discussion of central difference (CD) or QUICK schemes, we present it here instead due to its unique role in enstrophy conservation, which merits its importance for turbulence simulations using FD methods on staggered grids.

The Arakawa scheme refers to a family of finite-difference discretizations pioneered by Akio Arakawa for fluid flow equations, notable for preserving key physical invariants. In particular, Arakawas 9-point differencing of the Jacobian in 2D flow (developed in the 1960s) is a second-order centered scheme that exactly conserves domain-integrated kinetic energy and enstrophy (vorticity squared). These conservation properties prevent the spurious buildup of energy at small scales, ensuring nonlinear stability in long simulations. Over the last 20 years, research has focused on extending and integrating Arakawas approach with compact finite difference methods high-order schemes with minimal stencils to achieve greater accuracy while retaining physical fidelity. This report summarizes key advancements, methodologies, and applications of Arakawa-type schemes in the context of compact finite differences and computational fluid dynamics (CFD).

Arakawas original scheme was widely adopted in atmospheric and ocean modeling for its remarkable stability properties. By cleverly averaging several difference approximations, it yields a discrete Jacobian that is skew-symmetric and conserves quadratic quantities (energy, enstrophy, and mean vorticity). The discretized Jacobian operator $J(\zeta,\psi)$ for vorticity ζ and stream function ψ in 2D flow is given by:

$$J(\zeta,\psi)=\frac{1}{3}\left(J_1(\zeta,\psi)+J_2(\zeta,\psi)+J_3(\zeta,\psi)\right),$$

where J_1, J_2, and J_3 are three distinct second-order centered difference approximations of the Jacobian:

$$\begin{aligned}
J_1 &= \frac{1}{4\Delta x\Delta y}\left[(\zeta_{i+1,j}-\zeta_{i-1,j})(\psi_{i,j+1}-\psi_{i,j-1})-(\zeta_{i,j+1}-\zeta_{i,j-1})(\psi_{i+1,j}-\psi_{i-1,j})\right],\\
J_2 &= \frac{1}{4\Delta x\Delta y}\left[\zeta_{i+1,j}(\psi_{i+1,j+1}-\psi_{i+1,j-1})-\zeta_{i-1,j}(\psi_{i-1,j+1}-\psi_{i-1,j-1})\right.\\
&\quad\left.-\zeta_{i,j+1}(\psi_{i+1,j+1}-\psi_{i-1,j+1})+\zeta_{i,j-1}(\psi_{i+1,j-1}-\psi_{i-1,j-1})\right],\\
J_3 &= \frac{1}{4\Delta x\Delta y}\left[(\zeta_{i+1,j+1}-\zeta_{i-1,j-1})(\psi_{i-1,j+1}-\psi_{i+1,j-1})\right.\\
&\quad\left.-(\zeta_{i-1,j+1}-\zeta_{i+1,j-1})(\psi_{i+1,j+1}-\psi_{i-1,j-1})\right].
\end{aligned}$$

This averaging ensures discrete skew-symmetry, which enforces conservation. However, the scheme is formally second-order accurate in space. In contrast, compact finite differences (including Pade-type schemes) achieve high-order accuracy with a compact stencil, offering improved spectral resolution and efficiency. The challenge has been marrying these two approaches to preserve Arakawas invariant conservation in higher-order, compact schemes.

Researchers in the last two decades developed higher-order Arakawa-like schemes that retain conservation properties. Janji et al. (2011) introduced a fourth-order conservative advection scheme for atmospheric models that reduces to the Arakawa Jacobian in the non-divergent limit. Their fourth-order scheme is quadratic conservative (conserves second-order moments) and controls the nonlinear energy cascade similar to Arakawas method. Beyond fourth order, Sorgentone et al. (2015) formulated arbitrary high-order Arakawa-like discretizations using summation-by-parts (SBP) finite difference operators. For a grid function ζ, their high-order Jacobian $J^{(p)}(\zeta,\psi)$ generalizes Arakawas approach by combining skew-symmetric SBP operators $D_x^{(p)}$ and $D_y^{(p)}$ of order p:

$$J^{(p)}(\zeta,\psi)=\frac{1}{2}\left(D_x^{(p)}(\zeta)\cdot D_y^{(p)}(\psi)-D_y^{(p)}(\zeta)\cdot D_x^{(p)}(\psi)\right).$$

This formulation maintains discrete skew-symmetry and conservation of energy and enstrophy. Similarly, mimetic finite-difference approaches systematically construct conservative nonlinear operators, showing Arakawas scheme as a special case.

Recent studies highlight that combining high-order Arakawa-type schemes with explicit time-stepping imposes a restrictive Courant-Friedrichs-Lewy (CFL) condition. For a grid spacing Δx and flow velocity u, the CFL limit for stability is:

$$\Delta t\leq C\frac{\Delta x}{|u|},$$

where C depends on the scheme spectral properties. Rth and Hallatschek (2024) showed that high-order Arakawa operators excite high-frequency modes, reducing C significantly compared to low-order schemes.

Arakawa staggered-grid schemes and Jacobian remain foundational in climate/weather models. For example, the Arakawa-Lamb (1981) shallow-water scheme conserves potential enstrophy and energy via:

$$\frac{\partial\zeta}{\partial t}=-J(\zeta,\psi)-\beta\frac{\partial\psi}{\partial x},$$

where β is the planetary vorticity gradient. Modern extensions integrate compact schemes for improved wave dispersion in ocean models.

Recent advancements generalize Arakawas approach to arbitrary high-order schemes while retaining conservation. Hybrid methods combining compact finite differences and Arakawa-type operators now enable simulations that balance accuracy, stability, and physical fidelity.

PROBLEMS

21P-1 Derive the expression for the Kolmogorov length scale η in turbulent pipe flow in terms of Reynolds number Re, pipe diameter D, and friction factor f. Discuss how this scale governs grid spacing and time step requirements for DNS simulations.
21P-2 Compare and contrast DNS, LES, and RANS approaches in terms of the scales they resolve, computational requirements, and applicability across Reynolds numbers. Use the energy spectrum diagram (Figure 21.2) to support the answer.
21P-3 Explain the concept of spatial filtering in LES. What are the roles of the Leonard term, cross term, and subgrid-scale Reynolds stress in the decomposition of subgrid stress τ_{ij}^{sgs}? Why is the Smagorinsky model often used, and what are its key limitations?
21P-4 Consider the numerical solution of the one-dimensional linear advection equation using a dispersion and dissipation relation (DDR)-optimized finite-difference scheme (Bogey and Bailly 2004) coupled with explicit selective filtering. Initially, a Gaussian pulse with a peak located at $x = 0.3$ is given by $f(x,0) = e^{-(x-0.3)^2/0.005}$ within a domain length $L_x = 1.0$, discretized using 200 spatial grid points. The equation is solved numerically with periodic boundary conditions, zero physical diffusion ($D = 0$), and a constant advection velocity $U = 1.0$. For time advancement, the CFL number is initially set at 0.5, but numerical oscillations and amplitude distortions are observed. To achieve stability and minimize numerical dispersion errors, the CFL number is gradually reduced, and explicit selective filtering is applied with varying filter strength parameters r_d. Discuss the influence of reducing the CFL number, adjusting the filter strength r_d, and increasing the spatial resolution from 200 to 400 grid points on the stability, accuracy, and spectral fidelity of the numerical solution. Specifically, describe under what conditions amplitude distortion occurs, how explicit selective filtering helps control numerical oscillations, and why DDR-optimized schemes still require careful tuning of numerical parameters despite their spectral-like accuracy.
21P-5 How can the problem **21P-4** be useful in the development and validation of numerical methods intended for Large Eddy Simulation (LES), and what are the limitations of such a test case?

REFERENCES

N. Uddin, Fluid Mechanics, A Problem Solving Approach, CRC Press, 2023.

T. B. Gatski, M. Y. Hussaini, J. L. Lumley, (Eds.) Simulation and Modeling of Turbulent Flows. Oxford University Press, 1996.

D. K. Lilly, On the application of eddy viscosity concept in the inertial subrange of turbulence, Rep NCAR-123, National Center for Atmospheric Research, Boulder, USA, 1966.

J. S. Smagorinsky, General circulation experiments with the primitive equations: 1. The basic experiment, Mon. Weather. Rev. 91 (3), 99 - 164, 1963.

M. Ciofalo, Large-Eddy simulation: A critical survey of models and applications, Advances in Heat Transfer, Volume 25, 1994.

U. Schumann, Subgrid scale model for finite difference simulations of turbulent flows in a plane channels and annuli, J. Comp. Phys 18, 376- 404, 1975.

FASTEST User Manual, Department of Numerical Methods in Mechanical Engineering, Technische Universität Darmstadt, Germany, 2005.

J. W. Deardorff, A numerical study of three dimensional turbulent channel flow at large Reynolds number, J. Fluid Mech., Vol. 41, 453, 1970.

J. W. Deardorff, Three dimensional numerical modeling of planetary boundary layer, Workshop on Micrometerology, Chapter 7, 271-311, Am. Meteorol. Soc. Boston, 1973.

J. W. Deardorff, A three dimensional numerical investigation of the idealized planetary boundary layer, Geophy. Fluid Dynamics, 1, 377 -410, 1970.

J. W. Deardorff, Stratcumulus-capped mixed layers derived from a three-dimensional model, Boundary layer Meterorol., 32, 205 - 236, 1980.

U. Piomelli, J.H. Ferziger, P. Moin, J. Kim, New approximate boundary conditions for large eddy simulation of wall bounded flows, Phy. Fluids A 1 (6), 1061 - 1068, 1989.

U. Piomelli, H. Cabot, P. Moin, S. Lee, 1991, Subgrid-scale backscatter in turbulent and transitional flows, Phy. Fluids A, 3(7), 1766 - 1771, 1991.

P. Moin, J. Kim, 1981, Numerical investigation of turbulent channel flow, NASA Tech. Memo. NASA TM-81309, 1981.

J. Bardina, J. H. Ferziger, W. C. Reynolds, Improved turbulence modelsbased on large eddy simulation o homogeneous compressible turbulent flows, Rep no. TF-19, Thermoscience division, Standford University, USA, 1983.

P. J. Mason, N. S. Callon, On the magnitude of subgrid scale eddy coefficient in large edy simulations of turbulent channel flow, J. Fluid Mech., 162, 439 - 462, 1986.

M. Germano, U. Piomelli, P. Moin, W. H. Cabot, A dynamic subgrid-scale eddy viscosity model, Phy. Fluid A 3 (7), 1760 - 1765, 1991.

A. Yoshizawa. Statistical theory for compressible turbulent shear flows, with the application to subgrid modeling. Physics of Fluids, 29(7):2152-2164, 1986.

P. Sagaut, Large eddy simulation for incompressible flows: an introduction. Springer-Verlag, Berlin Heidelberg, 2006.

S. B. Pope, Turbulent Flows. Cambridge University Press, 2000.

J. H. Ferziger, M. Peric, Computational methods for fluid dynamics. Springer-Verlag, Berlin Heidelberg, 2002.

F. Nicoud, F. Ducros, Subgrid-scale stress modelling based on the square of the velocity gradient tensor. Flow Turbul Combust 62: 183-200, 1999.

S. V. Patankar, Numerical Heat Transfer and Fluid Flow, Washington: Hemisphere, 1980.

P. Moin, K. Squires, W. Cabot, S. Lee, 1991, A dynamic subgrid scale model for compressible turbulence and scalar transport, Phys. Fluids, 1991.

A. Leonard, Direct numerical Simulation of Turbulent flows, Simulation and modeling of Turbulent flows, Edited by Thomas B. Gatski, M. Yousuff Hussaini, John L. Lumley, Oxford University Press, 1996.

J. Ferziger, Large Eddy simulation, Simulation and modeling of Turbulent flows, Edited by Thomas B. Gatski, M. Yousuff Hussaini, John L. Lumley, Oxford University Press, 1996.

P. Moin, K. Mahesh, Direct Numerical Simulation: A tool in turbulence research, Ann. Rev. Fluid Mech., Vol. 30, 539 - 78, 1998.

R. A. Clark, J. H. Ferziger, W. C. Reynolds, Evaluation of Subgrid-Scale Models Using an Accurately Simulated Turbulent Flow. J. Fluid Mech. 91:1-16, 1979.

J. Bardina, J. H. Ferziger, W. C. Reynolds, Improved Turbulence Models Based on Large Eddy Simulation of Homogeneous Incompressible Turbulent Flows. Stanford Univ. Tech. Report TF-19, 1980.

T. M. Eidson, Numerical Simulation of the Turbulent Rayleigh-Bnard Problem Using Subgrid Modeling. J. Fluid Mech. 158:245-268, 1985.

C. G. Speziale, Galilean Invariance of Subgrid-Scale Stress Models in the Large-Eddy Simulation of Turbulence. J. Fluid Mech. 156:55-62, 1985.

M. Germano, U. Piomelli, P. Moin, W. H. Cabot, A Dynamic Subgrid-Scale Eddy Viscosity Model. Phys. Fluids A 3(7):1760-1765, 1991.

P. Moin, K. D. Squires, W. H. Cabot, S. Lee, A Dynamic Subgrid-Scale Model for Compressible Turbulence and Scalar Transport. Phys. Fluids A 3(11):2746-2757, 1991.

D. K. Lilly, A Proposed Modification of the Germano Subgrid-Scale Closure Method. Phys. Fluids A 4(3):633-635, 1992.

V. Wong, D. K. Lilly, A Comparison of Two Dynamic Subgrid Closure Methods for Turbulent Thermal Convection. Phys. Fluids 6(3):1016-1023, 1994.

Y. Zang, R. L. Street, J. R. Koseff, A Dynamic Mixed Subgrid-Scale Model and Its Application to Turbulent Recirculating Flows. Phys. Fluids A 5(12):3186-3196, 1993.

B. J. Geurts, D. D. Holm, Regulation of Burgers Equation Shocks by a Deterministic Subgrid Scale Model. Physica D 165:19-44, 2002.

J. Porte-Agel, M. B. Parlange, C. Meneveau, Scale-Dependent Dynamic Modeling for Large-Eddy Simulation of Turbulent Flows. J. Fluid Mech. 415:261-284, 2000.

Y. Mtais, M. Lesieur, Spectral Large-Eddy Simulation of Isotropic and Stably Stratified Turbulence. J. Fluid Mech. 239:157-194, 1992.

X. Shen, G. E. Karniadakis, A Hybrid Multidomain Method for Large-Eddy Simulations of Turbulent Flows and its Application to Heat Transfer. J. Comput. Phys. 207:568-590, 2005.

B. C. Wang, P. G. Huang, H. Xu, T. J. Lu, New Dynamic Subgrid-Scale Heat Flux Models for LES of Thermal Convection Based on the General Gradient Diffusion Hypothesis. J. Fluid Mech. 604:361-388, 2008.

A. M. Rasam, M. Breuer, F. Durst, A Dynamic Two-Equation Subgrid-Scale Model for Large Eddy Simulation of Turbulent Flow with Buoyancy. Phys. Fluids 25:015114, 2013.

K. Horiuti, The Role of the Bardina Model in Large Eddy Simulation of Turbulent Channel Flow. Phys. Fluids A 1(2):426-428, 1989.

H. Kobayashi, The Subgrid-Scale Models Based on Coherent Structures for Rotating Homogeneous Turbulence and Turbulent Channel Flow. Phys. Fluids 17:045104, 2005.

J. Meldi, P. Sagaut, O. Billard, DNS, LES and RANS of Microwave Heating in a Waveguide Cavity. J. Comput. Phys. 275:65-90, 2014.

A. Stroe, G. Iaccarino, P. Durbin, Large Eddy Simulation of Thermal Mixing in a T-Junction: Comparison of Eddy Diffusivity Models. Nucl. Eng. Des. 337:271-283, 2018.

A. Gamahara, Y. Hattori, Searching for Turbulence Models by Artificial Neural Network. Phys. Rev. Fluids 2:054604, 2017.

S. Lapeyre, G. Balarac, J. Le Sommer, R. Fablet, R. Lguensat, Training Deep Neural Networks for the Prediction of Turbulent Flows. J. Fluid Mech. 862:528-567, 2019.

F. Mesinger, A. Arakawa, Numerical Methods Used in Atmospheric Models (Volume I), Global Atmospheric Research Programme (GARP) Publications Series No. 17, WMO-ICSU Joint Organizing Committee, August 1976.

A. Arakawa, V. R. Lamb, A potential enstrophy and energy conserving scheme for the shallow water equations. Monthly Weather Review, 109 (1), 18-36, 1981.

A. Arakawa, Computational design for long-term numerical integration of the equations of fluid motion: Two-dimensional incompressible flow. Part 1 . J. Comput. Phys. 1, 119-143, 1966.

Z. Janjic, A. Gassmann, R. Zabkar, A class of conservative fourth-order advection schemes and impact of enhanced formal accuracy. Monthly Weather Review, 139 (5), 1556-1568, 2011.

A. Kacimi, M. H. Abidi, The Arakawa Jacobian method and a fourth-order ENO scheme for beta-plane barotropic equations. International Journal of Numerical Analysis and Modeling, 10(3), 571-587, 2013.

C. Bogey, C. Bailly, A family of low dispersive and low dissipative explicit schemes for flow and noise computations. Journal of Computational Physics, 194(1), 194-214, 2004.

V. Wong, D. K. Lilly, A comparison of two dynamic subgrid closure methods for turbulent thermal convection, Phys. Fluids, 6(3), 1016-1023, 1994.

S. Stolz, N. A. Adams, An approximate deconvolution procedure for large-eddy simulation, Phys. Fluids, 11(7), 1699-1701, 1999.

B. C. Wang, P. G. Huang, H. Xu, T. J. Lu, New dynamic subgrid-scale heat flux models for LES of thermal convection based on the general gradient diffusion hypothesis, J. Fluid Mech., 604, 361-388, 2008.

H. Kobayashi, The subgrid-scale models based on coherent structures for rotating homogeneous turbulence and turbulent channel flow, Phys. Fluids, 17, 045104, 2005.

S. Peng, L. Davidson, A dynamic anisotropic subgrid-scale model for large-eddy simulation, Phys. Fluids, 14(3), 1135-1145, 2002.

Appendix I: Grid Sensitivity Analysis and Grid Convergence Index (GCI)

A. GRID SENSITIVITY AND SOLUTION VERIFICATION

In computational fluid dynamics (CFD), grid sensitivity analysis is a crucial step in verifying that the numerical solution is independent of the mesh resolution. This ensures that observed flow features are due to physics, not artifacts of discretization. Grid convergence studies also help quantify the discretization error.

When numerical results are obtained on a series of systematically refined grids, one may use the method of Richardson extrapolation to estimate the order of accuracy and discretization error. This is commonly used in the GCI method.

A.1 RICHARDSON EXTRAPOLATION AND ERROR ESTIMATION

Let f_1, f_2, f_3 be the solutions for a quantity of interest (e.g., velocity, pressure, drag coefficient) on progressively refined grids with spacing $h_1 < h_2 < h_3$. Assuming the solution is smooth and in the asymptotic range, the Taylor expansion gives:

$$f_k = f_{\text{exact}} + g_p h_k^p + \mathcal{O}(h_k^{p+1})$$

where,
f_k: numerical solution on grid k
f_{exact}: exact (unknown) solution
g_p: constant
p: observed order of accuracy.

From three such results, the observed order of accuracy p is:

$$p = \frac{\log\left(\frac{f_3 - f_2}{f_2 - f_1}\right)}{\log(r)}$$

where $r = h_2/h_1 = h_3/h_2$ is the uniform grid refinement ratio.

Then, the Richardson-extrapolated estimate of the exact solution is:

$$f_{\text{ex}} \approx f_1 + \frac{f_1 - f_2}{r^p - 1}$$

The absolute discretization errors on each grid are:

$$E_1 = |f_{\text{ex}} - f_1|, \quad E_2 = |f_{\text{ex}} - f_2|, \quad E_3 = |f_{\text{ex}} - f_3|$$

And the relative error on the fine grid is:

DOI: 10.1201/9781003663171-A

$$e_{\text{rel}} = \left| \frac{f_2 - f_1}{f_1} \right|$$

A.2 GRID CONVERGENCE INDEX (GCI)

The Grid Convergence Index (GCI) provides a formal estimate of the uncertainty due to grid discretization. It defines a band around the numerical result on a given mesh:

For the fine grid (1):

$$\text{GCI}_{12} = F_s \cdot \left| \frac{f_1 - f_2}{f_1} \right| \cdot \frac{1}{r^p - 1}$$

For the coarse grid (2):

$$\text{GCI}_{21} = F_s \cdot \left| \frac{f_2 - f_1}{f_2} \right| \cdot \frac{1}{r^p - 1}$$

where F_s is a safety factor, usually 1.25 for three grids or 3.0 for two grids. The extrapolated solution range is:

$$f_1 \pm \text{GCI}_{12}$$

A.3 ASYMPTOTIC RANGE CHECK

To verify that the computed solutions lie in the asymptotic range (i.e., the convergence region), confirm:

$$\frac{f_3 - f_2}{f_2 - f_1} \approx r^p$$

This confirms that the solution differences are consistent with the expected rate of convergence.

A.4 GRID REFINEMENT AND EFFECTIVE RESOLUTION

When mesh spacings h_1 and h_2 are not known directly, the refinement ratio can be computed using cell counts:

$$r = \left(\frac{N_2}{N_1} \right)^{1/D}$$

where N_1, N_2 are the number of cells on the fine and coarse grids respectively, and D is the spatial dimension.

For LES (Large-Eddy Simulation), one cannot expect classical grid convergence. Instead, assess the resolved kinetic energy fraction:

$$\eta_{\text{resolved}} = \frac{k_{\text{resolved}}}{k_{\text{resolved}} + k_{\text{SGS}}}$$

where k_{SGS} is the modeled subgrid-scale kinetic energy.

A.5 ORDER OF ACCURACY FROM NORMS

If the exact solution is known or approximated using extrapolation, the observed order p can also be calculated via global error norms:

$$p = \frac{\log(E_2/E_1)}{\log(r)}$$

where:

$$L_2 = \left(\frac{1}{N}\sum_{i=1}^{N}(f_i - f_{\text{ex},i})^2\right)^{1/2}, \quad L_\infty = \max_i |f_i - f_{\text{ex},i}|$$

are the global L_2 and L_∞ norms of error.

A.6 PRACTICAL IMPLEMENTATION GUIDELINES

To ensure credible and reproducible CFD results:

☞ Use at least three systematically refined grids (preferably with $r = 2$)

☞ Apply refinement in all spatial directions

☞ Maintain grid quality by keeping cell skewness below 0.3 and the expansion ratio between neighboring cells under 1.2 to ensure numerical accuracy and stability.

☞ Perform both local and global convergence checks

☞ Avoid excessive artificial diffusion to suppress divergence

☞ For RANS, use schemes with at least second-order accuracy

☞ For LES, ensure $\eta_{\text{resolved}} \gtrsim 0.75$

☞ Ensure near-wall treatment complies with $y^+ \approx 1$ (low-Re) or $30 \leq y^+ \leq 300$ (wall-function)

☞ Monitor physical quantities (e.g., drag, pressure drop) in addition to residuals

☞ Average over cycles in unsteady simulations

☞ Keep solver settings constant during grid refinement

☞ Record grid statistics for reproducibility

Grid sensitivity analysis and GCI assessment are indispensable whenever CFD results are reported in peer?reviewed publications, underpin engineering design and certification efforts, or are compared against experimental data for validation. They are equally critical when evaluating or developing turbulence models and in any critical applications. By rigorously quantifying discretization error and demonstrating numerical convergence, these practices ensure that simulation outputs are reliable and that any remaining discrepancies arise from physical modeling limitations rather than from numerical artifacts.

REFERENCES

P.J. Roache, Perspective: A Method for Uniform Reporting of Grid Refinement Studies, Journal of Fluids Engineering, 116, 405–413, 1994

I.B. Celik, U. Ghia, P.J. Roache, C.J. Freitas, H. Coleman, and P.E. Raad, Procedure for Estimation and Reporting of Uncertainty Due to Discretization in CFD Applications, Journal of Fluids Engineering, 130, 078001, 2008

Appendix II: Grid Independence Analysis in LES Investigation

In large-eddy simulation (LES), achieving grid-independence is fundamentally different from Reynolds-averaged approaches: there is no true mesh-independent solution, because as the grid is refined more scales of the turbulence become resolved. In practice one applies quantitative criteria to judge whether a chosen grid resolves a sufficiently large fraction of the turbulent kinetic energy (TKE) so that further refinement yields diminishing returns. A seminal contribution in this regard is the Index of Resolution Quality (LES_{IQ}) introduced by Celik et al. (2005). LES_{IQ} is defined as the ratio of resolved TKE to total TKE (including subgrid and numerical contributions), so that:

$$LES_{IQ} = \frac{k_{\text{res}}}{k_{\text{tot}}}$$

where

$$k_{\text{tot}} = k_{\text{res}} + k_{\text{sgs}} + k_{\text{num}}$$

Celik et al. recommended that an LES grid be judged adequate when LES_{IQ} exceeds about 0.8 across the domain, indicating that at least 80 percent of the kinetic energy lies in the resolved scales. To estimate the unresolved contributions, they proposed a systematic grid-and-model variation procedure: run pairs of LES on fine and coarse grids related by a refinement factor α (typically $\alpha \approx 2$), assume that the combined subgrid plus numerical dissipation scales as Δ^m (with m≈2 for second-order schemes), and use Richardson extrapolation to solve for the constant in

$$k_{\text{tot}} - k_{\text{res}} = c_k \Delta^m$$

thus recovering an estimate of k_{tot} and computing LES_{IQ} according to

$$LES_{IQ} = \frac{k_{\text{res}}}{k_{\text{res}} + c_k \Delta^m}$$

Additional approaches for assessing grid independence have also been introduced. For example, Davidson (2009) demonstrated that analyzing two-point spatial correlations offers a reliable estimate of how many grid cells are required to resolve the principal eddies in wall-bounded flows; by examining the rate at which the correlation decays, one can determine the grid spacing needed to capture turbulent structures down to a chosen fraction of the integral length scale. Other researchers

DOI: 10.1201/9781003663171-B

recommend tracking the ratio of modeled to resolved turbulent kinetic energy, ensuring that the modeled portion remains minimal in critical regions, and verifying that mean velocity profiles and Reynolds stresses converge as the mesh is refined.

Another common practice is to express grid spacing in wall units for channel and boundary-layer simulations, with recommendations such as $\Delta x^+ = 50\text{-}100$, $\Delta z^+ = 20\text{-}50$, and first-cell $\Delta y^+ \leq 1$ for wall-resolving LES. These criteria are often combined with non-dimensional filter-width ratios (Δ/η) in homogeneous directions. A robust LES grid-independence study typically proceeds by:

- ☞ computing the LES_{IQ} index and verifying it exceeds approximately 0.8 everywhere
- ☞ performing a paired fine/coarse simulation set and applying Richardson extrapolation to estimate unresolved energy
- ☞ evaluating two-point correlations to ensure physical scales are adequately captured
- ☞ checking that modeled TKE fractions remain small and that key mean and fluctuating statistics converge with grid refinement.

These combined approaches originating with Celik et al. and extended by subsequent researchers form the backbone of grid-independence protocols in modern LES, and are essential material for any CFD textbook appendix on verification and validation of turbulence simulations.

REFERENCES

I. B. Celik, U. Ghia, P. J. Roache, C. J. Freitas, H. Coleman, P. E. Raad, Procedure for estimation and reporting of uncertainty due to discretization in CFD applications. Journal of Fluids Engineering 124(1):4-10, 2005.

P. P. Sullivan, E. G. Patton, The Effect of Mesh Resolution on Convective Boundary Layer Statistics and Structures Generated by Large-Eddy Simulation. Journal of the Atmospheric Sciences 68(10):2395-2415, 2011.

Index